Introductory

Statistics

Table of Contents

PREFACE

About *Introductory Statistics*

Introductory Statistics is designed for the one-semester, introduction to statistics course and is geared toward students majoring in fields other than math or engineering. This text assumes students have been exposed to intermediate algebra, and it focuses on the applications of statistical knowledge rather than the theory behind it.

The foundation of this textbook is *Collaborative Statistics*, by Barbara Illowsky and Susan Dean. Additional topics, examples, and ample opportunities for practice have been added to each chapter. The development choices for this textbook were made with the guidance of many faculty members who are deeply involved in teaching this course. These choices led to innovations in art, terminology, and practical applications, all with a goal of increasing relevance and accessibility for students. We strove to make the discipline meaningful, so that students can draw from it a working knowledge that will enrich their future studies and help them make sense of the world around them.

Coverage and Scope

Chapter 1 Sampling and Data
Chapter 2 Descriptive Statistics
Chapter 3 Probability Topics
Chapter 4 Discrete Random Variables
Chapter 5 Continuous Random Variables
Chapter 6 The Normal Distribution
Chapter 7 The Central Limit Theorem
Chapter 8 Confidence Intervals
Chapter 9 Hypothesis Testing with One Sample
Chapter 10 Hypothesis Testing with Two Samples
Chapter 11 The Chi-Square Distribution
Chapter 12 Linear Regression and Correlation
Chapter 13 F Distribution and One-Way ANOVA

Alternate Sequencing

Introductory Statistics was conceived and written to fit a particular topical sequence, but it can be used flexibly to accommodate other course structures. One such potential structure, which will fit reasonably well with the textbook content, is provided. Please consider, however, that the chapters were not written to be completely independent, and that the proposed alternate sequence should be carefully considered for student preparation and textual consistency.

Chapter 1 Sampling and Data
Chapter 2 Descriptive Statistics
Chapter 12 Linear Regression and Correlation
Chapter 3 Probability Topics
Chapter 4 Discrete Random Variables
Chapter 5 Continuous Random Variables
Chapter 6 The Normal Distribution
Chapter 7 The Central Limit Theorem
Chapter 8 Confidence Intervals
Chapter 9 Hypothesis Testing with One Sample
Chapter 10 Hypothesis Testing with Two Samples
Chapter 11 The Chi-Square Distribution
Chapter 13 F Distribution and One-Way ANOVA

Pedagogical Foundation and Features

- **Examples** are placed strategically throughout the text to show students the step-by-step process of interpreting and solving statistical problems. To keep the text relevant for students, the examples are drawn from a broad spectrum of practical topics; these include examples about college life and learning, health and medicine, retail and business, and sports and entertainment.

- **Try It** practice problems immediately follow many examples and give students the opportunity to practice as they read the text. **They are usually based on practical and familiar topics, like the Examples themselves**.

- **Collaborative Exercises** provide an in-class scenario for students to work together to explore presented concepts.

- **Using the TI-83, 83+, 84, 84+ Calculator** shows students step-by-step instructions to input problems into their calculator.

- **The Technology Icon** indicates where the use of a TI calculator or computer software is recommended.

- **Practice, Homework, and Bringing It Together** problems give the students problems at various degrees of difficulty while also including real-world scenarios to engage students.

Statistics Labs

These innovative activities were developed by Barbara Illowsky and Susan Dean in order to offer students the experience of designing, implementing, and interpreting statistical analyses. They are drawn from actual experiments and data-gathering processes, and offer a unique hands-on and collaborative experience. The labs provide a foundation for further learning and classroom interaction that will produce a meaningful application of statistics.

Statistics Labs appear at the end of each chapter, and begin with student learning outcomes, general estimates for time on task, and any global implementation notes. Students are then provided step-by-step guidance, including sample data tables and calculation prompts. The detailed assistance will help the students successfully apply the concepts in the text and lay the groundwork for future collaborative or individual work.

Ancillaries

- **Instructor's Solutions Manual**

- **Webassign Online Homework System**

- **Video Lectures (http://cnx.org/content/m18746/latest/?collection=col10522/latest)** delivered by Barbara Illowsky are provided for each chapter.

About Our Team
Senior Contributing Authors

Barbara Illowsky	De Anza College
Susan Dean	De Anza College

Contributing Authors

Abdulhamid Sukar	Cameron University
Abraham Biggs	Broward Community College
Adam Pennell	Greensboro College
Alexander Kolovos	
Andrew Wiesner	Pennsylvania State University
Ann Flanigan	Kapiolani Community College
Benjamin Ngwudike	Jackson State University
Birgit Aquilonius	West Valley College
Bryan Blount	Kentucky Wesleyan College
Carol Olmstead	De Anza College
Carol Weideman	St. Petersburg College
Charles Ashbacher	Upper Iowa University, Cedar Rapids
Charles Klein	De Anza College
Cheryl Wartman	University of Prince Edward Island
Cindy Moss	Skyline College
Daniel Birmajer	Nazareth College
David Bosworth	Hutchinson Community College
David French	Tidewater Community College

Dennis Walsh	Middle Tennessee State University
Diane Mathios	De Anza College
Ernest Bonat	Portland Community College
Frank Snow	De Anza College
George Bratton	University of Central Arkansas
Inna Grushko	De Anza College
Janice Hector	De Anza College
Javier Rueda	De Anza College
Jeffery Taub	Maine Maritime Academy
Jim Helmreich	Marist College
Jim Lucas	De Anza College
Jing Chang	College of Saint Mary
John Thomas	College of Lake County
Jonathan Oaks	Macomb Community College
Kathy Plum	De Anza College
Larry Green	Lake Tahoe Community College
Laurel Chiappetta	University of Pittsburgh
Lenore Desilets	De Anza College
Lisa Markus	De Anza College
Lisa Rosenberg	Elon University
Lynette Kenyon	Collin County Community College
Mark Mills	Central College
Mary Jo Kane	De Anza College
Mary Teegarden	San Diego Mesa College
Matthew Einsohn	Prescott College
Mel Jacobsen	Snow College
Michael Greenwich	College of Southern Nevada
Miriam Masullo	SUNY Purchase
Mo Geraghty	De Anza College
Nydia Nelson	St. Petersburg College
Philip J. Verrecchia	York College of Pennsylvania
Robert Henderson	Stephen F. Austin State University
Robert McDevitt	Germanna Community College
Roberta Bloom	De Anza College
Rupinder Sekhon	De Anza College
Sara Lenhart	Christopher Newport University
Sarah Boslaugh	Kennesaw State University
Sheldon Lee	Viterbo University
Sheri Boyd	Rollins College
Sudipta Roy	Kankakee Community College
Travis Short	St. Petersburg College

Valier Hauber	De Anza College
Vladimir Logvenenko	De Anza College
Wendy Lightheart	Lane Community College
Yvonne Sandoval	Pima Community College

Sample TI Technology

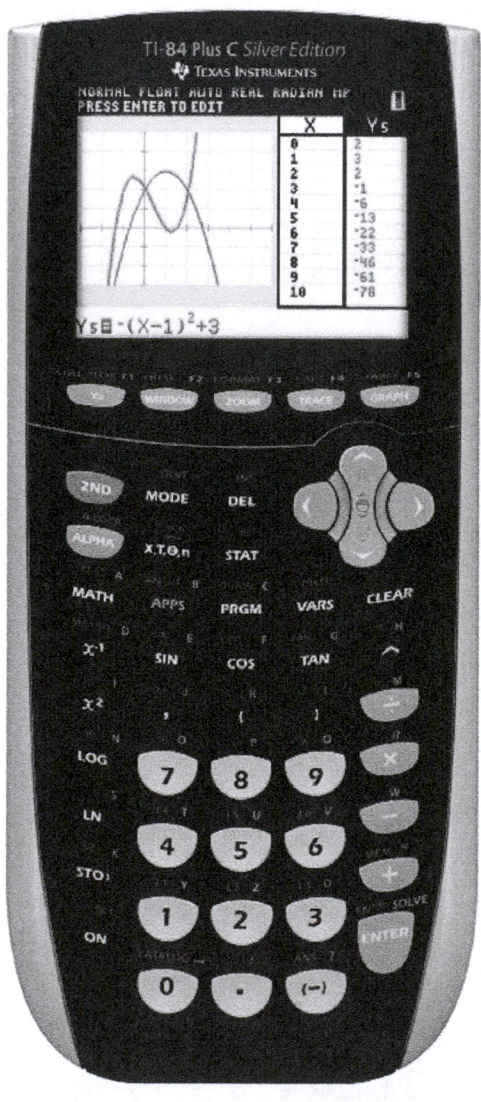

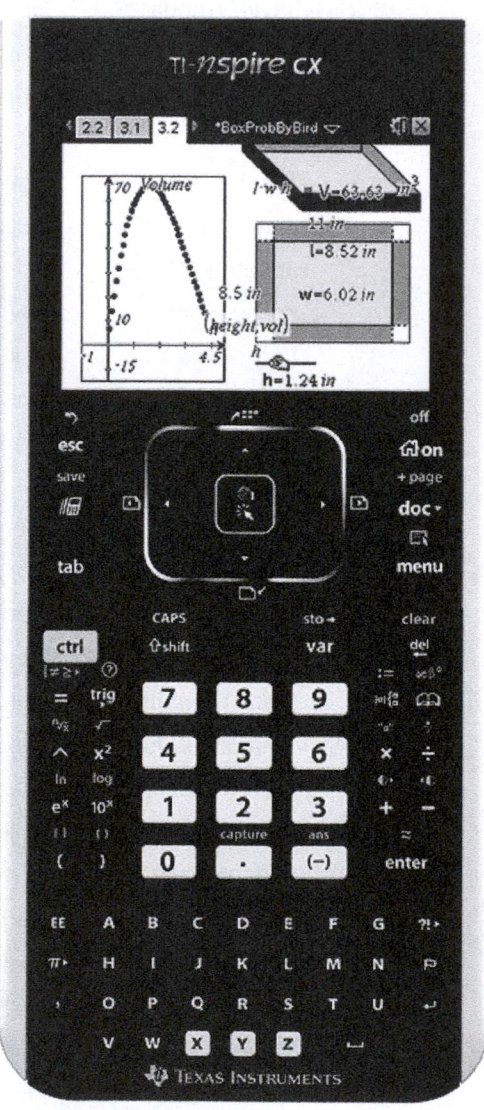

Disclaimer: The original calculator image(s) by Texas Instruments, Inc. are provided under CC-BY. Any subsequent modifications to the image(s) should be noted by the person making the modification. (Credit: ETmarcom TexasInstruments)

1 | SAMPLING AND DATA

Figure 1.1 We encounter statistics in our daily lives more often than we probably realize and from many different sources, like the news. (credit: David Sim)

Introduction

Chapter Objectives
By the end of this chapter, the student should be able to: • Recognize and differentiate between key terms. • Apply various types of sampling methods to data collection. • Create and interpret frequency tables.

You are probably asking yourself the question, "When and where will I use statistics?" If you read any newspaper, watch television, or use the Internet, you will see statistical information. There are statistics about crime, sports, education, politics, and real estate. Typically, when you read a newspaper article or watch a television news program, you are given sample information. With this information, you may make a decision about the correctness of a statement, claim, or "fact." Statistical methods can help you make the "best educated guess."

Since you will undoubtedly be given statistical information at some point in your life, you need to know some techniques for analyzing the information thoughtfully. Think about buying a house or managing a budget. Think about your chosen profession. The fields of economics, business, psychology, education, biology, law, computer science, police science, and early childhood development require at least one course in statistics.

Included in this chapter are the basic ideas and words of probability and statistics. You will soon understand that statistics and probability work together. You will also learn how data are gathered and what "good" data can be distinguished from "bad."

1.1 | Definitions of Statistics, Probability, and Key Terms

The science of **statistics** deals with the collection, analysis, interpretation, and presentation of **data**. We see and use data in our everyday lives.

Collaborative Exercise

In your classroom, try this exercise. Have class members write down the average time (in hours, to the nearest half-hour) they sleep per night. Your instructor will record the data. Then create a simple graph (called a **dot plot**) of the data. A dot plot consists of a number line and dots (or points) positioned above the number line. For example, consider the following data:

5; 5.5; 6; 6; 6; 6.5; 6.5; 6.5; 6.5; 7; 7; 8; 8; 9

The dot plot for this data would be as follows:

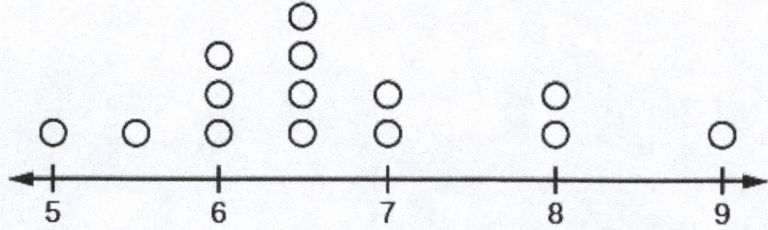

Figure 1.2

Does your dot plot look the same as or different from the example? Why? If you did the same example in an English class with the same number of students, do you think the results would be the same? Why or why not?

Where do your data appear to cluster? How might you interpret the clustering?

The questions above ask you to analyze and interpret your data. With this example, you have begun your study of statistics.

In this course, you will learn how to organize and summarize data. Organizing and summarizing data is called **descriptive statistics**. Two ways to summarize data are by graphing and by using numbers (for example, finding an average). After you have studied probability and probability distributions, you will use formal methods for drawing conclusions from "good" data. The formal methods are called **inferential statistics**. Statistical inference uses probability to determine how confident we can be that our conclusions are correct.

Effective interpretation of data (inference) is based on good procedures for producing data and thoughtful examination of the data. You will encounter what will seem to be too many mathematical formulas for interpreting data. The goal of statistics is not to perform numerous calculations using the formulas, but to gain an understanding of your data. The calculations can be done using a calculator or a computer. The understanding must come from you. If you can thoroughly grasp the basics of statistics, you can be more confident in the decisions you make in life.

Probability

Probability is a mathematical tool used to study randomness. It deals with the chance (the likelihood) of an event occurring. For example, if you toss a **fair** coin four times, the outcomes may not be two heads and two tails. However, if you toss the same coin 4,000 times, the outcomes will be close to half heads and half tails. The expected theoretical probability of heads in any one toss is $\frac{1}{2}$ or 0.5. Even though the outcomes of a few repetitions are uncertain, there is a regular pattern of outcomes when there are many repetitions. After reading about the English statistician Karl **Pearson** who tossed a coin 24,000 times with a result of 12,012 heads, one of the authors tossed a coin 2,000 times. The results were 996 heads. The fraction $\frac{996}{2000}$ is equal to 0.498 which is very close to 0.5, the expected probability.

The theory of probability began with the study of games of chance such as poker. Predictions take the form of probabilities. To predict the likelihood of an earthquake, of rain, or whether you will get an A in this course, we use probabilities. Doctors use probability to determine the chance of a vaccination causing the disease the vaccination is supposed to prevent. A stockbroker uses probability to determine the rate of return on a client's investments. You might use probability to decide to buy a lottery ticket or not. In your study of statistics, you will use the power of mathematics through probability calculations to analyze and interpret your data.

Key Terms

In statistics, we generally want to study a **population**. You can think of a population as a collection of persons, things, or objects under study. To study the population, we select a **sample**. The idea of **sampling** is to select a portion (or subset) of the larger population and study that portion (the sample) to gain information about the population. Data are the result of sampling from a population.

Because it takes a lot of time and money to examine an entire population, sampling is a very practical technique. If you wished to compute the overall grade point average at your school, it would make sense to select a sample of students who attend the school. The data collected from the sample would be the students' grade point averages. In presidential elections, opinion poll samples of 1,000–2,000 people are taken. The opinion poll is supposed to represent the views of the people in the entire country. Manufacturers of canned carbonated drinks take samples to determine if a 16 ounce can contains 16 ounces of carbonated drink.

From the sample data, we can calculate a statistic. A **statistic** is a number that represents a property of the sample. For example, if we consider one math class to be a sample of the population of all math classes, then the average number of points earned by students in that one math class at the end of the term is an example of a statistic. The statistic is an estimate of a population parameter. A **parameter** is a number that is a property of the population. Since we considered all math classes to be the population, then the average number of points earned per student over all the math classes is an example of a parameter.

One of the main concerns in the field of statistics is how accurately a statistic estimates a parameter. The accuracy really depends on how well the sample represents the population. The sample must contain the characteristics of the population in order to be a **representative sample**. We are interested in both the sample statistic and the population parameter in inferential statistics. In a later chapter, we will use the sample statistic to test the validity of the established population parameter.

A **variable**, notated by capital letters such as X and Y, is a characteristic of interest for each person or thing in a population. Variables may be **numerical** or **categorical**. **Numerical variables** take on values with equal units such as weight in pounds and time in hours. **Categorical variables** place the person or thing into a category. If we let X equal the number of points earned by one math student at the end of a term, then X is a numerical variable. If we let Y be a person's party affiliation, then some examples of Y include Republican, Democrat, and Independent. Y is a categorical variable. We could do some math with values of X (calculate the average number of points earned, for example), but it makes no sense to do math with values of Y (calculating an average party affiliation makes no sense).

Data are the actual values of the variable. They may be numbers or they may be words. **Datum** is a single value.

Two words that come up often in statistics are **mean** and **proportion**. If you were to take three exams in your math classes and obtain scores of 86, 75, and 92, you would calculate your mean score by adding the three exam scores and dividing by three (your mean score would be 84.3 to one decimal place). If, in your math class, there are 40 students and 22 are men and 18 are women, then the proportion of men students is $\frac{22}{40}$ and the proportion of women students is $\frac{18}{40}$. Mean and proportion are discussed in more detail in later chapters.

NOTE

The words " **mean**" and " **average**" are often used interchangeably. The substitution of one word for the other is common practice. The technical term is "arithmetic mean," and "average" is technically a center location. However, in practice among non-statisticians, "average" is commonly accepted for "arithmetic mean."

Example 1.1

Determine what the key terms refer to in the following study. We want to know the average (mean) amount of money first year college students spend at ABC College on school supplies that do not include books. We randomly survey 100 first year students at the college. Three of those students spent $150, $200, and $225, respectively.

Solution 1.1

The **population** is all first year students attending ABC College this term.

The **sample** could be all students enrolled in one section of a beginning statistics course at ABC College (although this sample may not represent the entire population).

The **parameter** is the average (mean) amount of money spent (excluding books) by first year college students at ABC College this term.

The **statistic** is the average (mean) amount of money spent (excluding books) by first year college students in the sample.

The **variable** could be the amount of money spent (excluding books) by one first year student. Let X = the amount of money spent (excluding books) by one first year student attending ABC College.

The **data** are the dollar amounts spent by the first year students. Examples of the data are $150, $200, and $225.

Try It Σ

1.1 Determine what the key terms refer to in the following study. We want to know the average (mean) amount of money spent on school uniforms each year by families with children at Knoll Academy. We randomly survey 100 families with children in the school. Three of the families spent $65, $75, and $95, respectively.

Example 1.2

Determine what the key terms refer to in the following study.

A study was conducted at a local college to analyze the average cumulative GPA's of students who graduated last year. Fill in the letter of the phrase that best describes each of the items below.

1._____ Population 2._____ Statistic 3._____ Parameter 4._____ Sample 5._____ Variable 6._____ Data

a) all students who attended the college last year
b) the cumulative GPA of one student who graduated from the college last year
c) 3.65, 2.80, 1.50, 3.90
d) a group of students who graduated from the college last year, randomly selected
e) the average cumulative GPA of students who graduated from the college last year
f) all students who graduated from the college last year
g) the average cumulative GPA of students in the study who graduated from the college last year

Solution 1.2
1. f; 2. g; 3. e; 4. d; 5. b; 6. c

Example 1.3

Determine what the key terms refer to in the following study.

As part of a study designed to test the safety of automobiles, the National Transportation Safety Board collected and reviewed data about the effects of an automobile crash on test dummies. Here is the criterion they used:

Speed at which Cars Crashed	Location of "drive" (i.e. dummies)
35 miles/hour	Front Seat

Table 1.1

Cars with dummies in the front seats were crashed into a wall at a speed of 35 miles per hour. We want to know the proportion of dummies in the driver's seat that would have had head injuries, if they had been actual drivers. We start with a simple random sample of 75 cars.

Solution 1.3

The **population** is all cars containing dummies in the front seat.

The **sample** is the 75 cars, selected by a simple random sample.

The **parameter** is the proportion of driver dummies (if they had been real people) who would have suffered head injuries in the population.

The **statistic** is proportion of driver dummies (if they had been real people) who would have suffered head injuries in the sample.

The **variable** X = the number of driver dummies (if they had been real people) who would have suffered head injuries.

The **data** are either: yes, had head injury, or no, did not.

Example 1.4

Determine what the key terms refer to in the following study.

An insurance company would like to determine the proportion of all medical doctors who have been involved in one or more malpractice lawsuits. The company selects 500 doctors at random from a professional directory and determines the number in the sample who have been involved in a malpractice lawsuit.

Solution 1.4

The **population** is all medical doctors listed in the professional directory.

The **parameter** is the proportion of medical doctors who have been involved in one or more malpractice suits in the population.

The **sample** is the 500 doctors selected at random from the professional directory.

The **statistic** is the proportion of medical doctors who have been involved in one or more malpractice suits in the sample.

The **variable** X = the number of medical doctors who have been involved in one or more malpractice suits.

The **data** are either: yes, was involved in one or more malpractice lawsuits, or no, was not.

Collaborative Exercise

Do the following exercise collaboratively with up to four people per group. Find a population, a sample, the parameter, the statistic, a variable, and data for the following study: You want to determine the average (mean) number of glasses of milk college students drink per day. Suppose yesterday, in your English class, you asked five students how many glasses of milk they drank the day before. The answers were 1, 0, 1, 3, and 4 glasses of milk.

1.2 | Data, Sampling, and Variation in Data and Sampling

Data may come from a population or from a sample. Small letters like x or y generally are used to represent data values. Most data can be put into the following categories:

- Qualitative
- Quantitative

Qualitative data are the result of categorizing or describing attributes of a population. Hair color, blood type, ethnic group, the car a person drives, and the street a person lives on are examples of qualitative data. Qualitative data are generally described by words or letters. For instance, hair color might be black, dark brown, light brown, blonde, gray, or red. Blood type might be AB+, O-, or B+. Researchers often prefer to use quantitative data over qualitative data because it lends itself more easily to mathematical analysis. For example, it does not make sense to find an average hair color or blood type.

Quantitative data are always numbers. Quantitative data are the result of **counting** or **measuring** attributes of a population. Amount of money, pulse rate, weight, number of people living in your town, and number of students who take statistics are examples of quantitative data. Quantitative data may be either **discrete** or **continuous**.

All data that are the result of counting are called **quantitative discrete data**. These data take on only certain numerical values. If you count the number of phone calls you receive for each day of the week, you might get values such as zero, one, two, or three.

All data that are the result of measuring are **quantitative continuous data** assuming that we can measure accurately. Measuring angles in radians might result in such numbers as $\frac{\pi}{6}$, $\frac{\pi}{3}$, $\frac{\pi}{2}$, π, $\frac{3\pi}{4}$, and so on. If you and your friends carry backpacks with books in them to school, the numbers of books in the backpacks are discrete data and the weights of the backpacks are continuous data.

Example 1.5 Data Sample of Quantitative Discrete Data

The data are the number of books students carry in their backpacks. You sample five students. Two students carry three books, one student carries four books, one student carries two books, and one student carries one book. The numbers of books (three, four, two, and one) are the quantitative discrete data.

1.5 The data are the number of machines in a gym. You sample five gyms. One gym has 12 machines, one gym has 15 machines, one gym has ten machines, one gym has 22 machines, and the other gym has 20 machines. What type of data is this?

Example 1.6 Data Sample of Quantitative Continuous Data

The data are the weights of backpacks with books in them. You sample the same five students. The weights (in pounds) of their backpacks are 6.2, 7, 6.8, 9.1, 4.3. Notice that backpacks carrying three books can have different weights. Weights are quantitative continuous data because weights are measured.

1.6 The data are the areas of lawns in square feet. You sample five houses. The areas of the lawns are 144 sq. feet, 160 sq. feet, 190 sq. feet, 180 sq. feet, and 210 sq. feet. What type of data is this?

Example 1.7

You go to the supermarket and purchase three cans of soup (19 ounces) tomato bisque, 14.1 ounces lentil, and 19 ounces Italian wedding), two packages of nuts (walnuts and peanuts), four different kinds of vegetable (broccoli, cauliflower, spinach, and carrots), and two desserts (16 ounces Cherry Garcia ice cream and two pounds (32 ounces chocolate chip cookies).

Name data sets that are quantitative discrete, quantitative continuous, and qualitative.

Solution 1.7

One Possible Solution:

- The three cans of soup, two packages of nuts, four kinds of vegetables and two desserts are quantitative discrete data because you count them.
- The weights of the soups (19 ounces, 14.1 ounces, 19 ounces) are quantitative continuous data because you measure weights as precisely as possible.
- Types of soups, nuts, vegetables and desserts are qualitative data because they are categorical.

Try to identify additional data sets in this example.

Example 1.8

The data are the colors of backpacks. Again, you sample the same five students. One student has a red backpack, two students have black backpacks, one student has a green backpack, and one student has a gray backpack. The colors red, black, black, green, and gray are qualitative data.

Try It Σ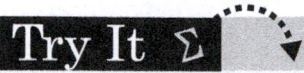

1.8 The data are the colors of houses. You sample five houses. The colors of the houses are white, yellow, white, red, and white. What type of data is this?

NOTE

You may collect data as numbers and report it categorically. For example, the quiz scores for each student are recorded throughout the term. At the end of the term, the quiz scores are reported as A, B, C, D, or F.

Download for free at http://cnx.org/content/col11562/latest/

Example 1.9

Work collaboratively to determine the correct data type (quantitative or qualitative). Indicate whether quantitative data are continuous or discrete. Hint: Data that are discrete often start with the words "the number of."

  a. the number of pairs of shoes you own

  b. the type of car you drive

  c. where you go on vacation

  d. the distance it is from your home to the nearest grocery store

  e. the number of classes you take per school year.

  f. the tuition for your classes

  g. the type of calculator you use

  h. movie ratings

  i. political party preferences

  j. weights of sumo wrestlers

  k. amount of money (in dollars) won playing poker

  l. number of correct answers on a quiz

  m. peoples' attitudes toward the government

  n. IQ scores (This may cause some discussion.)

Solution 1.9

Items a, e, f, k, and l are quantitative discrete; items d, j, and n are quantitative continuous; items b, c, g, h, i, and m are qualitative.

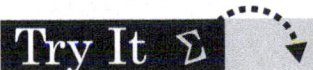
Try It Σ

1.9 Determine the correct data type (quantitative or qualitative) for the number of cars in a parking lot. Indicate whether quantitative data are continuous or discrete.

Example 1.10

A statistics professor collects information about the classification of her students as freshmen, sophomores, juniors, or seniors. The data she collects are summarized in the pie chart **Figure 1.2**. What type of data does this graph show?

Classification of Statistics Students

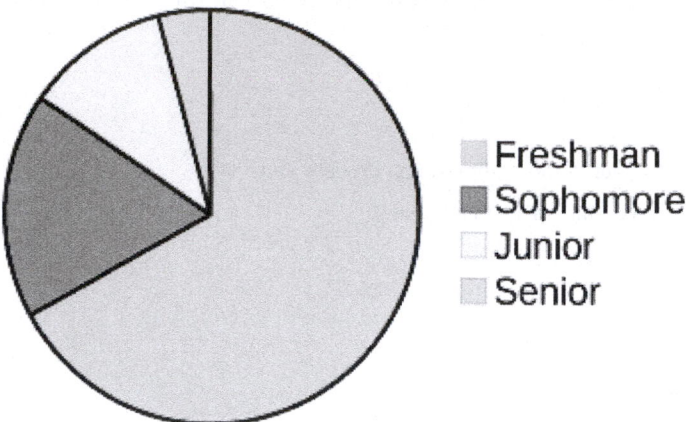

Figure 1.3

Solution 1.10
This pie chart shows the students in each year, which is **qualitative data**.

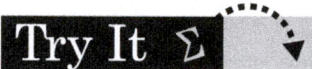

1.10 The registrar at State University keeps records of the number of credit hours students complete each semester. The data he collects are summarized in the histogram. The class boundaries are 10 to less than 13, 13 to less than 16, 16 to less than 19, 19 to less than 22, and 22 to less than 25.

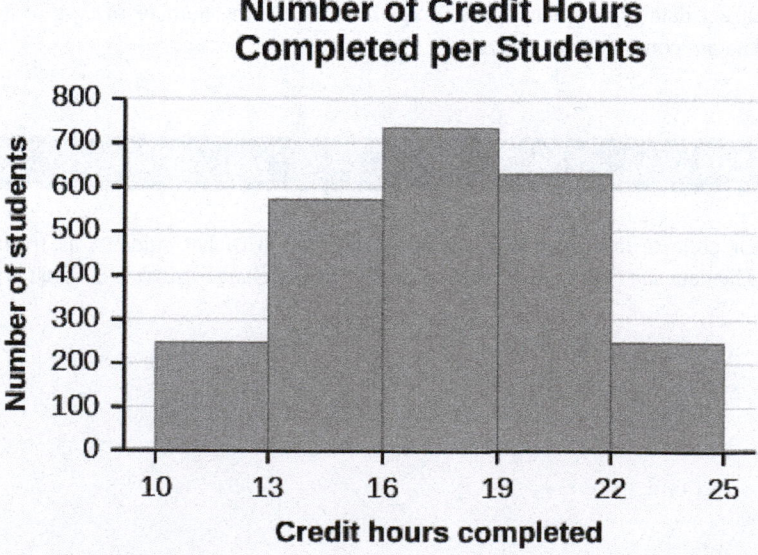

Figure 1.4

What type of data does this graph show?

Qualitative Data Discussion

Below are tables comparing the number of part-time and full-time students at De Anza College and Foothill College enrolled for the spring 2010 quarter. The tables display counts (frequencies) and percentages or proportions (relative frequencies). The percent columns make comparing the same categories in the colleges easier. Displaying percentages along with the numbers is often helpful, but it is particularly important when comparing sets of data that do not have the same totals, such as the total enrollments for both colleges in this example. Notice how much larger the percentage for part-time students at Foothill College is compared to De Anza College.

De Anza College				Foothill College		
	Number	Percent			Number	Percent
Full-time	9,200	40.9%		Full-time	4,059	28.6%
Part-time	13,296	59.1%		Part-time	10,124	71.4%
Total	22,496	100%		Total	14,183	100%

Table 1.2 Fall Term 2007 (Census day)

Tables are a good way of organizing and displaying data. But graphs can be even more helpful in understanding the data. There are no strict rules concerning which graphs to use. Two graphs that are used to display qualitative data are pie charts and bar graphs.

In a **pie chart**, categories of data are represented by wedges in a circle and are proportional in size to the percent of individuals in each category.

In a **bar graph**, the length of the bar for each category is proportional to the number or percent of individuals in each category. Bars may be vertical or horizontal.

A **Pareto chart** consists of bars that are sorted into order by category size (largest to smallest).

Look at **Figure 1.5** and **Figure 1.6** and determine which graph (pie or bar) you think displays the comparisons better.

It is a good idea to look at a variety of graphs to see which is the most helpful in displaying the data. We might make different choices of what we think is the "best" graph depending on the data and the context. Our choice also depends on what we are using the data for.

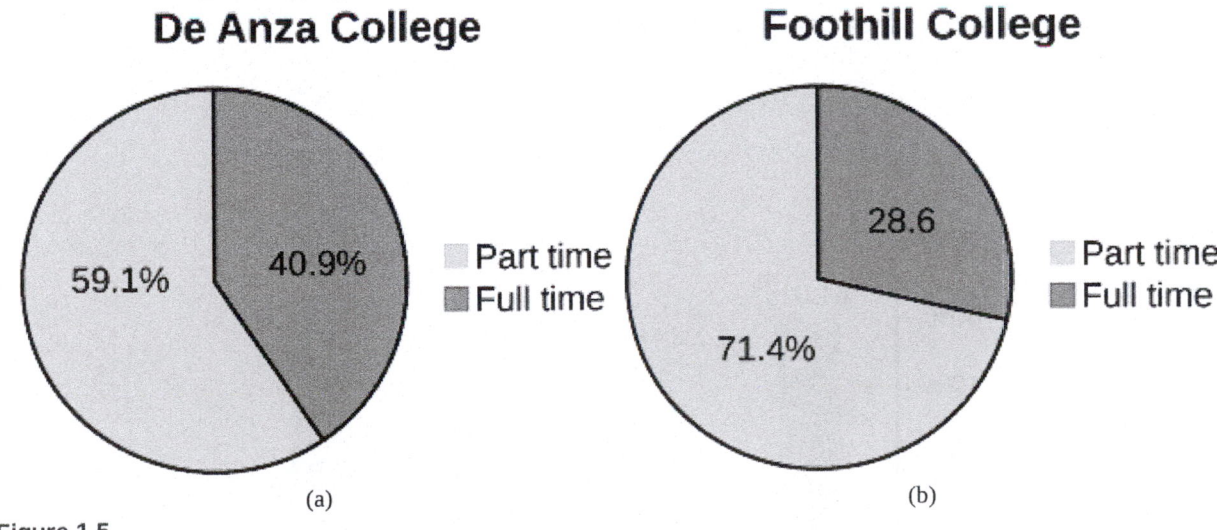

(a)   (b)

Figure 1.5

Student Status

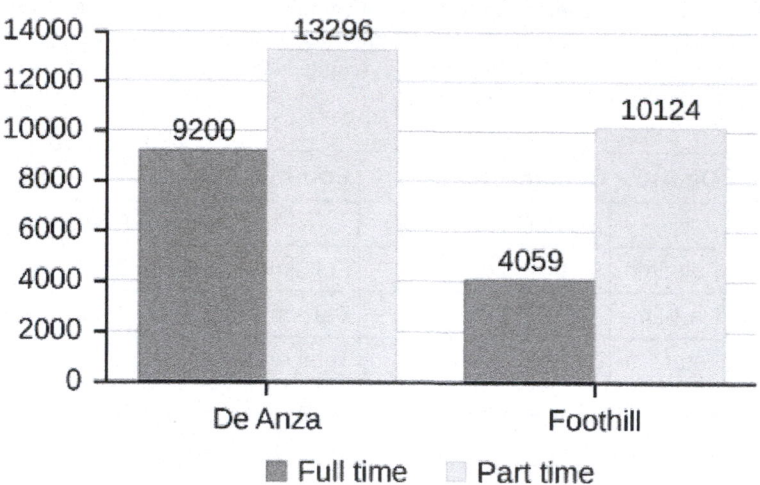

Figure 1.6

Percentages That Add to More (or Less) Than 100%

Sometimes percentages add up to be more than 100% (or less than 100%). In the graph, the percentages add to more than 100% because students can be in more than one category. A bar graph is appropriate to compare the relative size of the categories. A pie chart cannot be used. It also could not be used if the percentages added to less than 100%.

Characteristic/Category	Percent
Full-Time Students	40.9%
Students who intend to transfer to a 4-year educational institution	48.6%
Students under age 25	61.0%
TOTAL	150.5%

Table 1.3 De Anza College Spring 2010

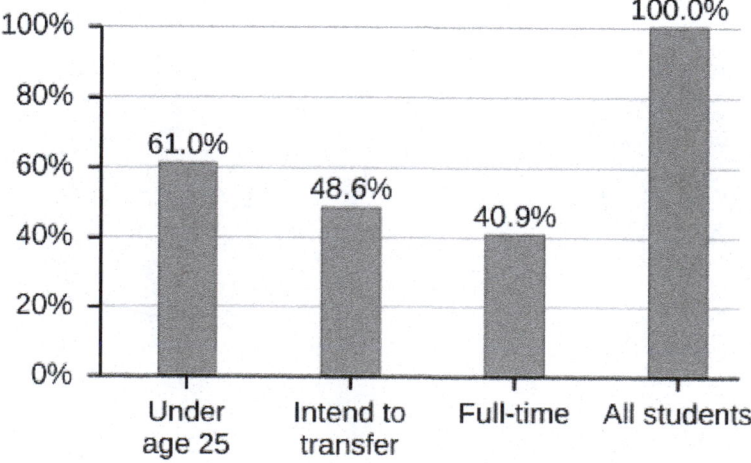

Figure 1.7

Omitting Categories/Missing Data

The table displays Ethnicity of Students but is missing the "Other/Unknown" category. This category contains people who did not feel they fit into any of the ethnicity categories or declined to respond. Notice that the frequencies do not add up to the total number of students. In this situation, create a bar graph and not a pie chart.

	Frequency	Percent
Asian	8,794	36.1%
Black	1,412	5.8%
Filipino	1,298	5.3%
Hispanic	4,180	17.1%
Native American	146	0.6%
Pacific Islander	236	1.0%
White	5,978	24.5%
TOTAL	22,044 out of 24,382	90.4% out of 100%

Table 1.4 Ethnicity of Students at De Anza College Fall Term 2007 (Census Day)

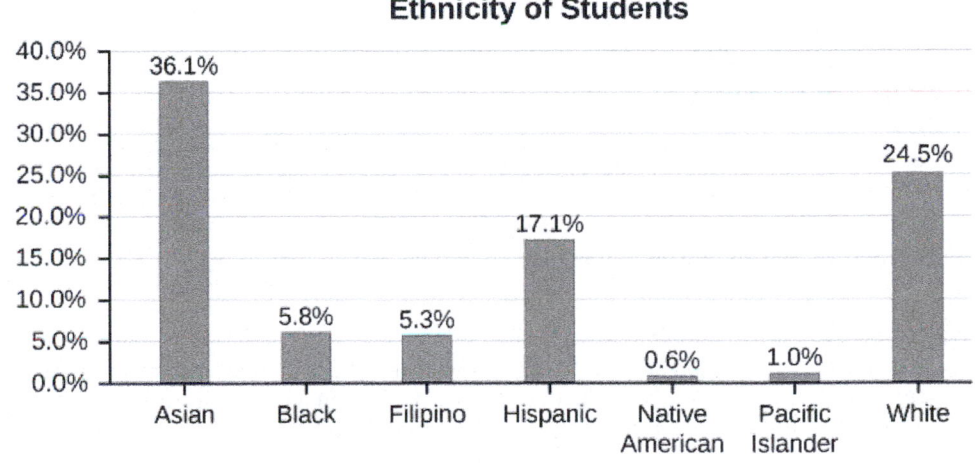

Figure 1.8

The following graph is the same as the previous graph but the "Other/Unknown" percent (9.6%) has been included. The "Other/Unknown" category is large compared to some of the other categories (Native American, 0.6%, Pacific Islander 1.0%). This is important to know when we think about what the data are telling us.

This particular bar graph in **Figure 1.9** can be difficult to understand visually. The graph in **Figure 1.10** is a Pareto chart. The Pareto chart has the bars sorted from largest to smallest and is easier to read and interpret.

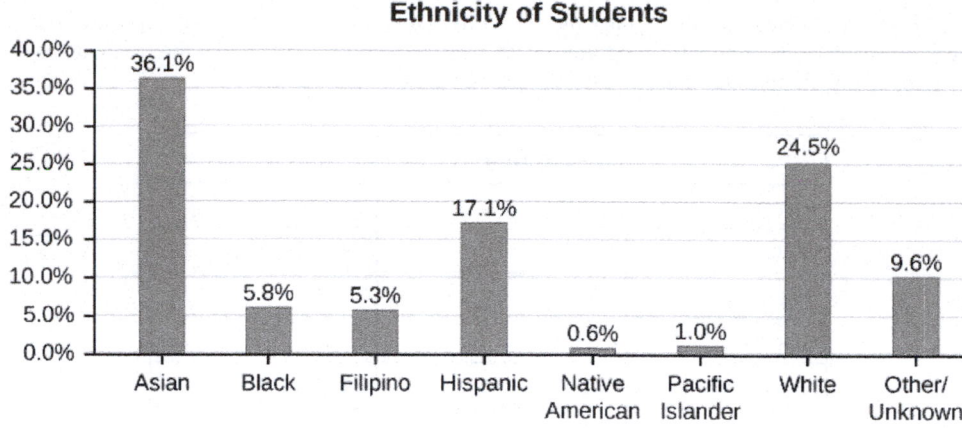

Figure 1.9 Bar Graph with Other/Unknown Category

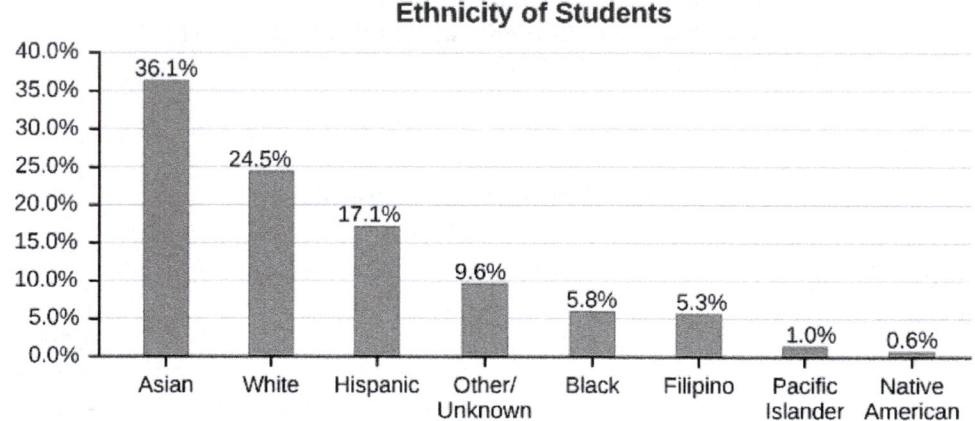

Figure 1.10 Pareto Chart With Bars Sorted by Size

Pie Charts: No Missing Data

The following pie charts have the "Other/Unknown" category included (since the percentages must add to 100%). The chart in **Figure 1.11b** is organized by the size of each wedge, which makes it a more visually informative graph than the unsorted, alphabetical graph in **Figure 1.11a**.

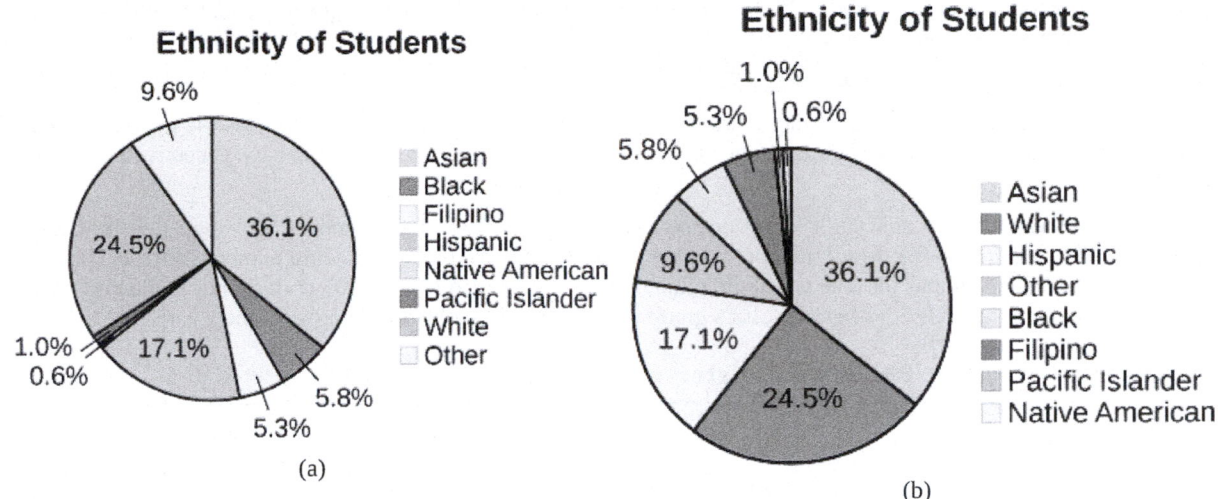

Ethnicity of Students

Ethnicity of Students

Figure 1.11

Sampling

Gathering information about an entire population often costs too much or is virtually impossible. Instead, we use a sample of the population. **A sample should have the same characteristics as the population it is representing.** Most statisticians use various methods of random sampling in an attempt to achieve this goal. This section will describe a few of the most common methods. There are several different methods of **random sampling**. In each form of random sampling, each member of a population initially has an equal chance of being selected for the sample. Each method has pros and cons. The easiest method to describe is called a **simple random sample**. Any group of n individuals is equally likely to be chosen by any other group of n individuals if the simple random sampling technique is used. In other words, each sample of the same size has an equal chance of being selected. For example, suppose Lisa wants to form a four-person study group (herself and three other people) from her pre-calculus class, which has 31 members not including Lisa. To choose a simple random sample of size three from the other members of her class, Lisa could put all 31 names in a hat, shake the hat, close her eyes, and pick out three names. A more technological way is for Lisa to first list the last names of the members of her class together with a two-digit number, as in **Table 1.5**:

ID	Name	ID	Name	ID	Name
00	Anselmo	11	King	21	Roquero
01	Bautista	12	Legeny	22	Roth
02	Bayani	13	Lundquist	23	Rowell
03	Cheng	14	Macierz	24	Salangsang
04	Cuarismo	15	Motogawa	25	Slade
05	Cuningham	16	Okimoto	26	Stratcher
06	Fontecha	17	Patel	27	Tallai
07	Hong	18	Price	28	Tran
08	Hoobler	19	Quizon	29	Wai
09	Jiao	20	Reyes	30	Wood
10	Khan				

Table 1.5 Class Roster

Lisa can use a table of random numbers (found in many statistics books and mathematical handbooks), a calculator, or a computer to generate random numbers. For this example, suppose Lisa chooses to generate random numbers from a calculator. The numbers generated are as follows:

0.94360; 0.99832; 0.14669; 0.51470; 0.40581; 0.73381; 0.04399

Lisa reads two-digit groups until she has chosen three class members (that is, she reads 0.94360 as the groups 94, 43, 36, 60). Each random number may only contribute one class member. If she needed to, Lisa could have generated more random numbers.

The random numbers 0.94360 and 0.99832 do not contain appropriate two digit numbers. However the third random number, 0.14669, contains 14 (the fourth random number also contains 14), the fifth random number contains 05, and the seventh random number contains 04. The two-digit number 14 corresponds to Macierz, 05 corresponds to Cuningham, and 04 corresponds to Cuarismo. Besides herself, Lisa's group will consist of Marcierz, Cuningham, and Cuarismo.

 Using the TI-83, 83+, 84, 84+ Calculator

To generate random numbers:

- Press MATH.
- Arrow over to PRB.
- Press 5:randInt(. Enter 0, 30).
- Press ENTER for the first random number.
- Press ENTER two more times for the other 2 random numbers. If there is a repeat press ENTER again.

Note: randInt(0, 30, 3) will generate 3 random numbers.

```
randInt(0,30)
                29
randInt(0,30)
                28
randInt(0,30)
                 4
```

Figure 1.12

Besides simple random sampling, there are other forms of sampling that involve a chance process for getting the sample. **Other well-known random sampling methods are the stratified sample, the cluster sample, and the systematic sample.**

To choose a **stratified sample**, divide the population into groups called strata and then take a **proportionate** number from each stratum. For example, you could stratify (group) your college population by department and then choose a proportionate simple random sample from each stratum (each department) to get a stratified random sample. To choose a simple random sample from each department, number each member of the first department, number each member of the second department, and do the same for the remaining departments. Then use simple random sampling to choose proportionate numbers from the first department and do the same for each of the remaining departments. Those numbers picked from the first department, picked from the second department, and so on represent the members who make up the stratified sample.

To choose a **cluster sample**, divide the population into clusters (groups) and then randomly select some of the clusters. All the members from these clusters are in the cluster sample. For example, if you randomly sample four departments from your college population, the four departments make up the cluster sample. Divide your college faculty by department. The departments are the clusters. Number each department, and then choose four different numbers using simple random sampling. All members of the four departments with those numbers are the cluster sample.

To choose a **systematic sample**, randomly select a starting point and take every n^{th} piece of data from a listing of the population. For example, suppose you have to do a phone survey. Your phone book contains 20,000 residence listings. You must choose 400 names for the sample. Number the population 1–20,000 and then use a simple random sample to pick a number that represents the first name in the sample. Then choose every fiftieth name thereafter until you have a total of 400 names (you might have to go back to the beginning of your phone list). Systematic sampling is frequently chosen because it is a simple method.

A type of sampling that is non-random is convenience sampling. **Convenience sampling** involves using results that are readily available. For example, a computer software store conducts a marketing study by interviewing potential customers who happen to be in the store browsing through the available software. The results of convenience sampling may be very good in some cases and highly biased (favor certain outcomes) in others.

Sampling data should be done very carefully. Collecting data carelessly can have devastating results. Surveys mailed to households and then returned may be very biased (they may favor a certain group). It is better for the person conducting the survey to select the sample respondents.

True random sampling is done **with replacement**. That is, once a member is picked, that member goes back into the population and thus may be chosen more than once. However for practical reasons, in most populations, simple random sampling is done **without replacement**. Surveys are typically done without replacement. That is, a member of the population may be chosen only once. Most samples are taken from large populations and the sample tends to be small in comparison to the population. Since this is the case, sampling without replacement is approximately the same as sampling with replacement because the chance of picking the same individual more than once with replacement is very low.

In a college population of 10,000 people, suppose you want to pick a sample of 1,000 randomly for a survey. **For any particular sample of 1,000**, if you are sampling **with replacement**,

- the chance of picking the first person is 1,000 out of 10,000 (0.1000);
- the chance of picking a different second person for this sample is 999 out of 10,000 (0.0999);
- the chance of picking the same person again is 1 out of 10,000 (very low).

If you are sampling **without replacement**,

- the chance of picking the first person for any particular sample is 1000 out of 10,000 (0.1000);
- the chance of picking a different second person is 999 out of 9,999 (0.0999);
- you do not replace the first person before picking the next person.

Compare the fractions 999/10,000 and 999/9,999. For accuracy, carry the decimal answers to four decimal places. To four decimal places, these numbers are equivalent (0.0999).

Sampling without replacement instead of sampling with replacement becomes a mathematical issue only when the population is small. For example, if the population is 25 people, the sample is ten, and you are sampling **with replacement for any particular sample**, then the chance of picking the first person is ten out of 25, and the chance of picking a different second person is nine out of 25 (you replace the first person).

If you sample **without replacement**, then the chance of picking the first person is ten out of 25, and then the chance of picking the second person (who is different) is nine out of 24 (you do not replace the first person).

Compare the fractions 9/25 and 9/24. To four decimal places, 9/25 = 0.3600 and 9/24 = 0.3750. To four decimal places, these numbers are not equivalent.

When you analyze data, it is important to be aware of **sampling errors** and nonsampling errors. The actual process of sampling causes sampling errors. For example, the sample may not be large enough. Factors not related to the sampling process cause **nonsampling errors**. A defective counting device can cause a nonsampling error.

In reality, a sample will never be exactly representative of the population so there will always be some sampling error. As a rule, the larger the sample, the smaller the sampling error.

In statistics, **a sampling bias** is created when a sample is collected from a population and some members of the population are not as likely to be chosen as others (remember, each member of the population should have an equally likely chance of being chosen). When a sampling bias happens, there can be incorrect conclusions drawn about the population that is being studied.

Example 1.11

A study is done to determine the average tuition that San Jose State undergraduate students pay per semester. Each student in the following samples is asked how much tuition he or she paid for the Fall semester. What is the type of sampling in each case?

 a. A sample of 100 undergraduate San Jose State students is taken by organizing the students' names by classification (freshman, sophomore, junior, or senior), and then selecting 25 students from each.

 b. A random number generator is used to select a student from the alphabetical listing of all undergraduate students in the Fall semester. Starting with that student, every 50th student is chosen until 75 students are included in the sample.

 c. A completely random method is used to select 75 students. Each undergraduate student in the fall semester has the same probability of being chosen at any stage of the sampling process.

 d. The freshman, sophomore, junior, and senior years are numbered one, two, three, and four, respectively. A random number generator is used to pick two of those years. All students in those two years are in the sample.

 e. An administrative assistant is asked to stand in front of the library one Wednesday and to ask the first 100 undergraduate students he encounters what they paid for tuition the Fall semester. Those 100 students are the sample.

Solution 1.11
a. stratified; b. systematic; c. simple random; d. cluster; e. convenience

1.11 You are going to use the random number generator to generate different types of samples from the data.

This table displays six sets of quiz scores (each quiz counts 10 points) for an elementary statistics class.

#1	#2	#3	#4	#5	#6
5	7	10	9	8	3
10	5	9	8	7	6
9	10	8	6	7	9
9	10	10	9	8	9
7	8	9	5	7	4
9	9	9	10	8	7
7	7	10	9	8	8
8	8	9	10	8	8
9	7	8	7	7	8
8	8	10	9	8	7

Table 1.6

Instructions: Use the Random Number Generator to pick samples.

 1. Create a stratified sample by column. Pick three quiz scores randomly from each column.

    ◦ Number each row one through ten.

- ◦ On your calculator, press Math and arrow over to PRB.
- ◦ For column 1, Press 5:randInt(and enter 1,10). Press ENTER. Record the number. Press ENTER 2 more times (even the repeats). Record these numbers. Record the three quiz scores in column one that correspond to these three numbers.
- ◦ Repeat for columns two through six.
- ◦ These 18 quiz scores are a stratified sample.

2. Create a cluster sample by picking two of the columns. Use the column numbers: one through six.
   - ◦ Press MATH and arrow over to PRB.
   - ◦ Press 5:randInt(and enter 1,6). Press ENTER. Record the number. Press ENTER and record that number.
   - ◦ The two numbers are for two of the columns.
   - ◦ The quiz scores (20 of them) in these 2 columns are the cluster sample.

3. Create a simple random sample of 15 quiz scores.
   - ◦ Use the numbering one through 60.
   - ◦ Press MATH. Arrow over to PRB. Press 5:randInt(and enter 1, 60).
   - ◦ Press ENTER 15 times and record the numbers.
   - ◦ Record the quiz scores that correspond to these numbers.
   - ◦ These 15 quiz scores are the systematic sample.

4. Create a systematic sample of 12 quiz scores.
   - ◦ Use the numbering one through 60.
   - ◦ Press MATH. Arrow over to PRB. Press 5:randInt(and enter 1, 60).
   - ◦ Press ENTER. Record the number and the first quiz score. From that number, count ten quiz scores and record that quiz score. Keep counting ten quiz scores and recording the quiz score until you have a sample of 12 quiz scores. You may wrap around (go back to the beginning).

Example 1.12

Determine the type of sampling used (simple random, stratified, systematic, cluster, or convenience).

   a. A soccer coach selects six players from a group of boys aged eight to ten, seven players from a group of boys aged 11 to 12, and three players from a group of boys aged 13 to 14 to form a recreational soccer team.
   b. A pollster interviews all human resource personnel in five different high tech companies.
   c. A high school educational researcher interviews 50 high school female teachers and 50 high school male teachers.
   d. A medical researcher interviews every third cancer patient from a list of cancer patients at a local hospital.
   e. A high school counselor uses a computer to generate 50 random numbers and then picks students whose names correspond to the numbers.
   f. A student interviews classmates in his algebra class to determine how many pairs of jeans a student owns, on the average.

Solution 1.12
a. stratified; b. cluster; c. stratified; d. systematic; e. simple random; f.convenience

1.12 Determine the type of sampling used (simple random, stratified, systematic, cluster, or convenience).

A high school principal polls 50 freshmen, 50 sophomores, 50 juniors, and 50 seniors regarding policy changes for after school activities.

If we were to examine two samples representing the same population, even if we used random sampling methods for the samples, they would not be exactly the same. Just as there is variation in data, there is variation in samples. As you become accustomed to sampling, the variability will begin to seem natural.

Example 1.13

Suppose ABC College has 10,000 part-time students (the population). We are interested in the average amount of money a part-time student spends on books in the fall term. Asking all 10,000 students is an almost impossible task.

Suppose we take two different samples.

First, we use convenience sampling and survey ten students from a first term organic chemistry class. Many of these students are taking first term calculus in addition to the organic chemistry class. The amount of money they spend on books is as follows:

$128; $87; $173; $116; $130; $204; $147; $189; $93; $153

The second sample is taken using a list of senior citizens who take P.E. classes and taking every fifth senior citizen on the list, for a total of ten senior citizens. They spend:

$50; $40; $36; $15; $50; $100; $40; $53; $22; $22

It is unlikely that any student is in both samples.

a. Do you think that either of these samples is representative of (or is characteristic of) the entire 10,000 part-time student population?

Solution 1.13
a. No. The first sample probably consists of science-oriented students. Besides the chemistry course, some of them are also taking first-term calculus. Books for these classes tend to be expensive. Most of these students are, more than likely, paying more than the average part-time student for their books. The second sample is a group of senior citizens who are, more than likely, taking courses for health and interest. The amount of money they spend on books is probably much less than the average parttime student. Both samples are biased. Also, in both cases, not all students have a chance to be in either sample.

b. Since these samples are not representative of the entire population, is it wise to use the results to describe the entire population?

Solution 1.13
b. No. For these samples, each member of the population did not have an equally likely chance of being chosen.

Now, suppose we take a third sample. We choose ten different part-time students from the disciplines of chemistry, math, English, psychology, sociology, history, nursing, physical education, art, and early childhood development. (We assume that these are the only disciplines in which part-time students at ABC College are enrolled and that an equal number of part-time students are enrolled in each of the disciplines.) Each student is chosen using simple random sampling. Using a calculator, random numbers are generated and a student from a particular discipline is selected if he or she has a corresponding number. The students spend the following amounts:

$180; $50; $150; $85; $260; $75; $180; $200; $200; $150

c. Is the sample biased?

Solution 1.13

c. The sample is unbiased, but a larger sample would be recommended to increase the likelihood that the sample will be close to representative of the population. However, for a biased sampling technique, even a large sample runs the risk of not being representative of the population.

Students often ask if it is "good enough" to take a sample, instead of surveying the entire population. If the survey is done well, the answer is yes.

Try It Σ

1.13 A local radio station has a fan base of 20,000 listeners. The station wants to know if its audience would prefer more music or more talk shows. Asking all 20,000 listeners is an almost impossible task.

The station uses convenience sampling and surveys the first 200 people they meet at one of the station's music concert events. 24 people said they'd prefer more talk shows, and 176 people said they'd prefer more music.

Do you think that this sample is representative of (or is characteristic of) the entire 20,000 listener population?

Collaborative Exercise

As a class, determine whether or not the following samples are representative. If they are not, discuss the reasons.

1. To find the average GPA of all students in a university, use all honor students at the university as the sample.

2. To find out the most popular cereal among young people under the age of ten, stand outside a large supermarket for three hours and speak to every twentieth child under age ten who enters the supermarket.

3. To find the average annual income of all adults in the United States, sample U.S. congressmen. Create a cluster sample by considering each state as a stratum (group). By using simple random sampling, select states to be part of the cluster. Then survey every U.S. congressman in the cluster.

4. To determine the proportion of people taking public transportation to work, survey 20 people in New York City. Conduct the survey by sitting in Central Park on a bench and interviewing every person who sits next to you.

5. To determine the average cost of a two-day stay in a hospital in Massachusetts, survey 100 hospitals across the state using simple random sampling.

Variation in Data

Variation is present in any set of data. For example, 16-ounce cans of beverage may contain more or less than 16 ounces of liquid. In one study, eight 16 ounce cans were measured and produced the following amount (in ounces) of beverage:

15.8; 16.1; 15.2; 14.8; 15.8; 15.9; 16.0; 15.5

Measurements of the amount of beverage in a 16-ounce can may vary because different people make the measurements or because the exact amount, 16 ounces of liquid, was not put into the cans. Manufacturers regularly run tests to determine if the amount of beverage in a 16-ounce can falls within the desired range.

Be aware that as you take data, your data may vary somewhat from the data someone else is taking for the same purpose. This is completely natural. However, if two or more of you are taking the same data and get very different results, it is time for you and the others to reevaluate your data-taking methods and your accuracy.

Variation in Samples

It was mentioned previously that two or more **samples** from the same **population**, taken randomly, and having close to the same characteristics of the population will likely be different from each other. Suppose Doreen and Jung both decide to study the average amount of time students at their college sleep each night. Doreen and Jung each take samples of 500 students. Doreen uses systematic sampling and Jung uses cluster sampling. Doreen's sample will be different from Jung's sample. Even if Doreen and Jung used the same sampling method, in all likelihood their samples would be different. Neither would be wrong, however.

Think about what contributes to making Doreen's and Jung's samples different.

If Doreen and Jung took larger samples (i.e. the number of data values is increased), their sample results (the average amount of time a student sleeps) might be closer to the actual population average. But still, their samples would be, in all likelihood, different from each other. This **variability in samples** cannot be stressed enough.

Size of a Sample

The size of a sample (often called the number of observations) is important. The examples you have seen in this book so far have been small. Samples of only a few hundred observations, or even smaller, are sufficient for many purposes. In polling, samples that are from 1,200 to 1,500 observations are considered large enough and good enough if the survey is random and is well done. You will learn why when you study confidence intervals.

Be aware that many large samples are biased. For example, call-in surveys are invariably biased, because people choose to respond or not.

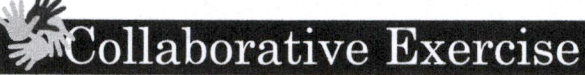

Collaborative Exercise

Divide into groups of two, three, or four. Your instructor will give each group one six-sided die. Try this experiment twice. Roll one fair die (six-sided) 20 times. Record the number of ones, twos, threes, fours, fives, and sixes you get in **Table 1.7** and **Table 1.8** ("frequency" is the number of times a particular face of the die occurs):

Face on Die	Frequency
1	
2	
3	
4	
5	
6	

Table 1.7 First Experiment (20 rolls)

Face on Die	Frequency
1	
2	
3	
4	
5	
6	

Table 1.8 Second Experiment (20 rolls)

Did the two experiments have the same results? Probably not. If you did the experiment a third time, do you expect the results to be identical to the first or second experiment? Why or why not?

Which experiment had the correct results? They both did. The job of the statistician is to see through the variability and draw appropriate conclusions.

Critical Evaluation

We need to evaluate the statistical studies we read about critically and analyze them before accepting the results of the studies. Common problems to be aware of include

- Problems with samples: A sample must be representative of the population. A sample that is not representative of the population is biased. Biased samples that are not representative of the population give results that are inaccurate and not valid.

- Self-selected samples: Responses only by people who choose to respond, such as call-in surveys, are often unreliable.

- Sample size issues: Samples that are too small may be unreliable. Larger samples are better, if possible. In some situations, having small samples is unavoidable and can still be used to draw conclusions. Examples: crash testing cars or medical testing for rare conditions

- Undue influence: collecting data or asking questions in a way that influences the response

- Non-response or refusal of subject to participate: The collected responses may no longer be representative of the population. Often, people with strong positive or negative opinions may answer surveys, which can affect the results.

- Causality: A relationship between two variables does not mean that one causes the other to occur. They may be related (correlated) because of their relationship through a different variable.

- Self-funded or self-interest studies: A study performed by a person or organization in order to support their claim. Is the study impartial? Read the study carefully to evaluate the work. Do not automatically assume that the study is good, but do not automatically assume the study is bad either. Evaluate it on its merits and the work done.

- Misleading use of data: improperly displayed graphs, incomplete data, or lack of context

- Confounding: When the effects of multiple factors on a response cannot be separated. Confounding makes it difficult or impossible to draw valid conclusions about the effect of each factor.

1.3 | Frequency, Frequency Tables, and Levels of Measurement

Once you have a set of data, you will need to organize it so that you can analyze how frequently each datum occurs in the set. However, when calculating the frequency, you may need to round your answers so that they are as precise as possible.

Answers and Rounding Off

A simple way to round off answers is to carry your final answer one more decimal place than was present in the original data. Round off only the final answer. Do not round off any intermediate results, if possible. If it becomes necessary to round off intermediate results, carry them to at least twice as many decimal places as the final answer. For example, the average of the three quiz scores four, six, and nine is 6.3, rounded off to the nearest tenth, because the data are whole numbers. Most answers will be rounded off in this manner.

It is not necessary to reduce most fractions in this course. Especially in **Probability Topics**, the chapter on probability, it is more helpful to leave an answer as an unreduced fraction.

Levels of Measurement

The way a set of data is measured is called its **level of measurement**. Correct statistical procedures depend on a researcher being familiar with levels of measurement. Not every statistical operation can be used with every set of data. Data can be classified into four levels of measurement. They are (from lowest to highest level):

- **Nominal scale level**
- **Ordinal scale level**
- **Interval scale level**
- **Ratio scale level**

Data that is measured using a **nominal scale** is **qualitative**. Categories, colors, names, labels and favorite foods along with yes or no responses are examples of nominal level data. Nominal scale data are not ordered. For example, trying to classify people according to their favorite food does not make any sense. Putting pizza first and sushi second is not meaningful.

Smartphone companies are another example of nominal scale data. Some examples are Sony, Motorola, Nokia, Samsung and Apple. This is just a list and there is no agreed upon order. Some people may favor Apple but that is a matter of opinion. Nominal scale data cannot be used in calculations.

Data that is measured using an **ordinal scale** is similar to nominal scale data but there is a big difference. The ordinal scale data can be ordered. An example of ordinal scale data is a list of the top five national parks in the United States. The top five national parks in the United States can be ranked from one to five but we cannot measure differences between the data.

Another example of using the ordinal scale is a cruise survey where the responses to questions about the cruise are "excellent," "good," "satisfactory," and "unsatisfactory." These responses are ordered from the most desired response to the least desired. But the differences between two pieces of data cannot be measured. Like the nominal scale data, ordinal scale data cannot be used in calculations.

Data that is measured using the **interval scale** is similar to ordinal level data because it has a definite ordering but there is a difference between data. The differences between interval scale data can be measured though the data does not have a starting point.

Temperature scales like Celsius (C) and Fahrenheit (F) are measured by using the interval scale. In both temperature measurements, 40° is equal to 100° minus 60°. Differences make sense. But 0 degrees does not because, in both scales, 0 is not the absolute lowest temperature. Temperatures like -10° F and -15° C exist and are colder than 0.

Interval level data can be used in calculations, but one type of comparison cannot be done. 80° C is not four times as hot as 20° C (nor is 80° F four times as hot as 20° F). There is no meaning to the ratio of 80 to 20 (or four to one).

Data that is measured using the **ratio scale** takes care of the ratio problem and gives you the most information. Ratio scale data is like interval scale data, but it has a 0 point and ratios can be calculated. For example, four multiple choice statistics final exam scores are 80, 68, 20 and 92 (out of a possible 100 points). The exams are machine-graded.

The data can be put in order from lowest to highest: 20, 68, 80, 92.

The differences between the data have meaning. The score 92 is more than the score 68 by 24 points. Ratios can be calculated. The smallest score is 0. So 80 is four times 20. The score of 80 is four times better than the score of 20.

Frequency

Twenty students were asked how many hours they worked per day. Their responses, in hours, are as follows: 5; 6; 3; 3; 2; 4; 7; 5; 2; 3; 5; 6; 5; 4; 4; 3; 5; 2; 5; 3.

Table 1.9 lists the different data values in ascending order and their frequencies.

DATA VALUE	FREQUENCY
2	3
3	5
4	3
5	6
6	2
7	1

Table 1.9 Frequency Table of Student Work Hours

A **frequency** is the number of times a value of the data occurs. According to **Table 1.9**, there are three students who work two hours, five students who work three hours, and so on. The sum of the values in the frequency column, 20, represents the total number of students included in the sample.

A **relative frequency** is the ratio (fraction or proportion) of the number of times a value of the data occurs in the set of all outcomes to the total number of outcomes. To find the relative frequencies, divide each frequency by the total number of students in the sample–in this case, 20. Relative frequencies can be written as fractions, percents, or decimals.

DATA VALUE	FREQUENCY	RELATIVE FREQUENCY
2	3	$\frac{3}{20}$ or 0.15
3	5	$\frac{5}{20}$ or 0.25
4	3	$\frac{3}{20}$ or 0.15
5	6	$\frac{6}{20}$ or 0.30
6	2	$\frac{2}{20}$ or 0.10
7	1	$\frac{1}{20}$ or 0.05

Table 1.10 Frequency Table of Student Work Hours with Relative Frequencies

The sum of the values in the relative frequency column of **Table 1.10** is $\frac{20}{20}$, or 1.

Cumulative relative frequency is the accumulation of the previous relative frequencies. To find the cumulative relative frequencies, add all the previous relative frequencies to the relative frequency for the current row, as shown in **Table 1.11**.

DATA VALUE	FREQUENCY	RELATIVE FREQUENCY	CUMULATIVE RELATIVE FREQUENCY
2	3	$\frac{3}{20}$ or 0.15	0.15
3	5	$\frac{5}{20}$ or 0.25	0.15 + 0.25 = 0.40
4	3	$\frac{3}{20}$ or 0.15	0.40 + 0.15 = 0.55
5	6	$\frac{6}{20}$ or 0.30	0.55 + 0.30 = 0.85
6	2	$\frac{2}{20}$ or 0.10	0.85 + 0.10 = 0.95
7	1	$\frac{1}{20}$ or 0.05	0.95 + 0.05 = 1.00

Table 1.11 Frequency Table of Student Work Hours with Relative and Cumulative Relative Frequencies

The last entry of the cumulative relative frequency column is one, indicating that one hundred percent of the data has been accumulated.

NOTE

Because of rounding, the relative frequency column may not always sum to one, and the last entry in the cumulative relative frequency column may not be one. However, they each should be close to one.

Table 1.12 represents the heights, in inches, of a sample of 100 male semiprofessional soccer players.

HEIGHTS (INCHES)	FREQUENCY	RELATIVE FREQUENCY	CUMULATIVE RELATIVE FREQUENCY
59.95–61.95	5	$\frac{5}{100} = 0.05$	0.05
61.95–63.95	3	$\frac{3}{100} = 0.03$	0.05 + 0.03 = 0.08
63.95–65.95	15	$\frac{15}{100} = 0.15$	0.08 + 0.15 = 0.23
65.95–67.95	40	$\frac{40}{100} = 0.40$	0.23 + 0.40 = 0.63
67.95–69.95	17	$\frac{17}{100} = 0.17$	0.63 + 0.17 = 0.80
69.95–71.95	12	$\frac{12}{100} = 0.12$	0.80 + 0.12 = 0.92
71.95–73.95	7	$\frac{7}{100} = 0.07$	0.92 + 0.07 = 0.99
73.95–75.95	1	$\frac{1}{100} = 0.01$	0.99 + 0.01 = 1.00
	Total = 100	Total = 1.00	

Table 1.12 Frequency Table of Soccer Player Height

The data in this table have been **grouped** into the following intervals:

- 59.95 to 61.95 inches
- 61.95 to 63.95 inches
- 63.95 to 65.95 inches
- 65.95 to 67.95 inches
- 67.95 to 69.95 inches
- 69.95 to 71.95 inches
- 71.95 to 73.95 inches
- 73.95 to 75.95 inches

NOTE

This example is used again in **Descriptive Statistics**, where the method used to compute the intervals will be explained.

In this sample, there are **five** players whose heights fall within the interval 59.95–61.95 inches, **three** players whose heights fall within the interval 61.95–63.95 inches, **15** players whose heights fall within the interval 63.95–65.95 inches, **40** players whose heights fall within the interval 65.95–67.95 inches, **17** players whose heights fall within the interval 67.95–69.95 inches, **12** players whose heights fall within the interval 69.95–71.95, **seven** players whose heights fall within the interval 71.95–73.95, and **one** player whose heights fall within the interval 73.95–75.95. All heights fall between the endpoints of an interval and not at the endpoints.

Example 1.14

From **Table 1.12**, find the percentage of heights that are less than 65.95 inches.

Solution 1.14
If you look at the first, second, and third rows, the heights are all less than 65.95 inches. There are 5 + 3 + 15 = 23 players whose heights are less than 65.95 inches. The percentage of heights less than 65.95 inches is then $\frac{23}{100}$ or 23%. This percentage is the cumulative relative frequency entry in the third row.

1.14 Table **1.13** shows the amount, in inches, of annual rainfall in a sample of towns.

Rainfall (Inches)	Frequency	Relative Frequency	Cumulative Relative Frequency
2.95–4.97	6	$\frac{6}{50} = 0.12$	0.12
4.97–6.99	7	$\frac{7}{50} = 0.14$	0.12 + 0.14 = 0.26
6.99–9.01	15	$\frac{15}{50} = 0.30$	0.26 + 0.30 = 0.56
9.01–11.03	8	$\frac{8}{50} = 0.16$	0.56 + 0.16 = 0.72
11.03–13.05	9	$\frac{9}{50} = 0.18$	0.72 + 0.18 = 0.90
13.05–15.07	5	$\frac{5}{50} = 0.10$	0.90 + 0.10 = 1.00
	Total = 50	Total = 1.00	

Table 1.13

From **Table 1.13**, find the percentage of rainfall that is less than 9.01 inches.

Example 1.15

From **Table 1.12**, find the percentage of heights that fall between 61.95 and 65.95 inches.

Solution 1.15
Add the relative frequencies in the second and third rows: 0.03 + 0.15 = 0.18 or 18%.

1.15 From **Table 1.13**, find the percentage of rainfall that is between 6.99 and 13.05 inches.

Example 1.16

Use the heights of the 100 male semiprofessional soccer players in **Table 1.12**. Fill in the blanks and check your answers.

a. The percentage of heights that are from 67.95 to 71.95 inches is: _____.

b. The percentage of heights that are from 67.95 to 73.95 inches is: _____.

c. The percentage of heights that are more than 65.95 inches is: _____.

d. The number of players in the sample who are between 61.95 and 71.95 inches tall is: _____.

e. What kind of data are the heights?

f. Describe how you could gather this data (the heights) so that the data are characteristic of all male semiprofessional soccer players.

Remember, you **count frequencies**. To find the relative frequency, divide the frequency by the total number of data values. To find the cumulative relative frequency, add all of the previous relative frequencies to the relative frequency for the current row.

Solution 1.16

a. 29%

b. 36%

c. 77%

d. 87

e. quantitative continuous

f. get rosters from each team and choose a simple random sample from each

Try It Σ

1.16 From **Table 1.13**, find the number of towns that have rainfall between 2.95 and 9.01 inches.

Collaborative Exercise

In your class, have someone conduct a survey of the number of siblings (brothers and sisters) each student has. Create a frequency table. Add to it a relative frequency column and a cumulative relative frequency column. Answer the following questions:

1. What percentage of the students in your class have no siblings?

2. What percentage of the students have from one to three siblings?

3. What percentage of the students have fewer than three siblings?

Example 1.17

Nineteen people were asked how many miles, to the nearest mile, they commute to work each day. The data are as follows: 2; 5; 7; 3; 2; 10; 18; 15; 20; 7; 10; 18; 5; 12; 13; 12; 4; 5; 10. **Table 1.14** was produced:

DATA	FREQUENCY	RELATIVE FREQUENCY	CUMULATIVE RELATIVE FREQUENCY
3	3	$\frac{3}{19}$	0.1579
4	1	$\frac{1}{19}$	0.2105
5	3	$\frac{3}{19}$	0.1579
7	2	$\frac{2}{19}$	0.2632
10	3	$\frac{4}{19}$	0.4737
12	2	$\frac{2}{19}$	0.7895
13	1	$\frac{1}{19}$	0.8421
15	1	$\frac{1}{19}$	0.8948
18	1	$\frac{1}{19}$	0.9474
20	1	$\frac{1}{19}$	1.0000

Table 1.14 Frequency of Commuting Distances

a. Is the table correct? If it is not correct, what is wrong?

b. True or False: Three percent of the people surveyed commute three miles. If the statement is not correct, what should it be? If the table is incorrect, make the corrections.

c. What fraction of the people surveyed commute five or seven miles?

d. What fraction of the people surveyed commute 12 miles or more? Less than 12 miles? Between five and 13 miles (not including five and 13 miles)?

Solution 1.17

a. No. The frequency column sums to 18, not 19. Not all cumulative relative frequencies are correct.

b. False. The frequency for three miles should be one; for two miles (left out), two. The cumulative relative frequency column should read: 0.1052, 0.1579, 0.2105, 0.3684, 0.4737, 0.6316, 0.7368, 0.7895, 0.8421, 0.9474, 1.0000.

c. $\frac{5}{19}$

d. $\frac{7}{19}$, $\frac{12}{19}$, $\frac{7}{19}$

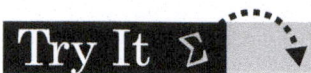

1.17 **Table 1.13** represents the amount, in inches, of annual rainfall in a sample of towns. What fraction of towns surveyed get between 11.03 and 13.05 inches of rainfall each year?

Example 1.18

Table 1.15 contains the total number of deaths worldwide as a result of earthquakes for the period from 2000 to 2012.

Year	Total Number of Deaths
2000	231
2001	21,357
2002	11,685
2003	33,819
2004	228,802
2005	88,003
2006	6,605
2007	712
2008	88,011
2009	1,790
2010	320,120
2011	21,953
2012	768
Total	823,356

Table 1.15

Answer the following questions.

   a. What is the frequency of deaths measured from 2006 through 2009?

   b. What percentage of deaths occurred after 2009?

   c. What is the relative frequency of deaths that occurred in 2003 or earlier?

   d. What is the percentage of deaths that occurred in 2004?

   e. What kind of data are the numbers of deaths?

   f. The Richter scale is used to quantify the energy produced by an earthquake. Examples of Richter scale numbers are 2.3, 4.0, 6.1, and 7.0. What kind of data are these numbers?

Solution 1.18
   a. 97,118 (11.8%)

   b. 41.6%

   c. 67,092/823,356 or 0.081 or 8.1 %

   d. 27.8%

  e. Quantitative discrete

  f. Quantitative continuous

1.18 **Table 1.16** contains the total number of fatal motor vehicle traffic crashes in the United States for the period from 1994 to 2011.

Year	Total Number of Crashes	Year	Total Number of Crashes
1994	36,254	2004	38,444
1995	37,241	2005	39,252
1996	37,494	2006	38,648
1997	37,324	2007	37,435
1998	37,107	2008	34,172
1999	37,140	2009	30,862
2000	37,526	2010	30,296
2001	37,862	2011	29,757
2002	38,491	Total	653,782
2003	38,477		

Table 1.16

Answer the following questions.

  a. What is the frequency of deaths measured from 2000 through 2004?

  b. What percentage of deaths occurred after 2006?

  c. What is the relative frequency of deaths that occurred in 2000 or before?

  d. What is the percentage of deaths that occurred in 2011?

  e. What is the cumulative relative frequency for 2006? Explain what this number tells you about the data.

1.4 | Experimental Design and Ethics

Does aspirin reduce the risk of heart attacks? Is one brand of fertilizer more effective at growing roses than another? Is fatigue as dangerous to a driver as the influence of alcohol? Questions like these are answered using randomized experiments. In this module, you will learn important aspects of experimental design. Proper study design ensures the production of reliable, accurate data.

The purpose of an experiment is to investigate the relationship between two variables. When one variable causes change in another, we call the first variable the **explanatory variable**. The affected variable is called the **response variable**. In a randomized experiment, the researcher manipulates values of the explanatory variable and measures the resulting changes in the response variable. The different values of the explanatory variable are called **treatments**. An **experimental unit** is a single object or individual to be measured.

You want to investigate the effectiveness of vitamin E in preventing disease. You recruit a group of subjects and ask them if they regularly take vitamin E. You notice that the subjects who take vitamin E exhibit better health on average than those who do not. Does this prove that vitamin E is effective in disease prevention? It does not. There are many differences between the two groups compared in addition to vitamin E consumption. People who take vitamin E regularly often take other steps to improve their health: exercise, diet, other vitamin supplements, choosing not to smoke. Any one of these factors could be influencing health. As described, this study does not prove that vitamin E is the key to disease prevention.

Additional variables that can cloud a study are called **lurking variables**. In order to prove that the explanatory variable is causing a change in the response variable, it is necessary to isolate the explanatory variable. The researcher must design her experiment in such a way that there is only one difference between groups being compared: the planned treatments. This is accomplished by the **random assignment** of experimental units to treatment groups. When subjects are assigned treatments randomly, all of the potential lurking variables are spread equally among the groups. At this point the only difference between groups is the one imposed by the researcher. Different outcomes measured in the response variable, therefore, must be a direct result of the different treatments. In this way, an experiment can prove a cause-and-effect connection between the explanatory and response variables.

The power of suggestion can have an important influence on the outcome of an experiment. Studies have shown that the expectation of the study participant can be as important as the actual medication. In one study of performance-enhancing drugs, researchers noted:

Results showed that believing one had taken the substance resulted in [performance] times almost as fast as those associated with consuming the drug itself. In contrast, taking the drug without knowledge yielded no significant performance increment.[1]

When participation in a study prompts a physical response from a participant, it is difficult to isolate the effects of the explanatory variable. To counter the power of suggestion, researchers set aside one treatment group as a **control group**. This group is given a **placebo** treatment–a treatment that cannot influence the response variable. The control group helps researchers balance the effects of being in an experiment with the effects of the active treatments. Of course, if you are participating in a study and you know that you are receiving a pill which contains no actual medication, then the power of suggestion is no longer a factor. **Blinding** in a randomized experiment preserves the power of suggestion. When a person involved in a research study is blinded, he does not know who is receiving the active treatment(s) and who is receiving the placebo treatment. A **double-blind experiment** is one in which both the subjects and the researchers involved with the subjects are blinded.

Example 1.19

Researchers want to investigate whether taking aspirin regularly reduces the risk of heart attack. Four hundred men between the ages of 50 and 84 are recruited as participants. The men are divided randomly into two groups: one group will take aspirin, and the other group will take a placebo. Each man takes one pill each day for three years, but he does not know whether he is taking aspirin or the placebo. At the end of the study, researchers count the number of men in each group who have had heart attacks.

Identify the following values for this study: population, sample, experimental units, explanatory variable, response variable, treatments.

Solution 1.19

The *population* is men aged 50 to 84.
The *sample* is the 400 men who participated.
The *experimental units* are the individual men in the study.
The *explanatory variable* is oral medication.
The *treatments* are aspirin and a placebo.
The *response variable* is whether a subject had a heart attack.

Example 1.20

The Smell & Taste Treatment and Research Foundation conducted a study to investigate whether smell can affect learning. Subjects completed mazes multiple times while wearing masks. They completed the pencil and paper mazes three times wearing floral-scented masks, and three times with unscented masks. Participants were

1. McClung, M. Collins, D. "Because I know it will!": placebo effects of an ergogenic aid on athletic performance. Journal of Sport & Exercise Psychology. 2007 Jun. 29(3):382-94. Web. April 30, 2013.

assigned at random to wear the floral mask during the first three trials or during the last three trials. For each trial, researchers recorded the time it took to complete the maze and the subject's impression of the mask's scent: positive, negative, or neutral.

    a. Describe the explanatory and response variables in this study.

    b. What are the treatments?

    c. Identify any lurking variables that could interfere with this study.

    d. Is it possible to use blinding in this study?

Solution 1.20

    a. The explanatory variable is scent, and the response variable is the time it takes to complete the maze.

    b. There are two treatments: a floral-scented mask and an unscented mask.

    c. All subjects experienced both treatments. The order of treatments was randomly assigned so there were no differences between the treatment groups. Random assignment eliminates the problem of lurking variables.

    d. Subjects will clearly know whether they can smell flowers or not, so subjects cannot be blinded in this study. Researchers timing the mazes can be blinded, though. The researcher who is observing a subject will not know which mask is being worn.

Example 1.21

A researcher wants to study the effects of birth order on personality. Explain why this study could not be conducted as a randomized experiment. What is the main problem in a study that cannot be designed as a randomized experiment?

Solution 1.21

The explanatory variable is birth order. You cannot randomly assign a person's birth order. Random assignment eliminates the impact of lurking variables. When you cannot assign subjects to treatment groups at random, there will be differences between the groups other than the explanatory variable.

Try It Σ

1.21 You are concerned about the effects of texting on driving performance. Design a study to test the response time of drivers while texting and while driving only. How many seconds does it take for a driver to respond when a leading car hits the brakes?

    a. Describe the explanatory and response variables in the study.

    b. What are the treatments?

    c. What should you consider when selecting participants?

    d. Your research partner wants to divide participants randomly into two groups: one to drive without distraction and one to text and drive simultaneously. Is this a good idea? Why or why not?

    e. Identify any lurking variables that could interfere with this study.

    f. How can blinding be used in this study?

Ethics

The widespread misuse and misrepresentation of statistical information often gives the field a bad name. Some say that "numbers don't lie," but the people who use numbers to support their claims often do.

A recent investigation of famous social psychologist, Diederik Stapel, has led to the retraction of his articles from some of the world's top journals including *Journal of Experimental Social Psychology, Social Psychology, Basic and Applied Social Psychology, British Journal of Social Psychology*, and the magazine *Science*. Diederik Stapel is a former professor at Tilburg University in the Netherlands. Over the past two years, an extensive investigation involving three universities where Stapel has worked concluded that the psychologist is guilty of fraud on a colossal scale. Falsified data taints over 55 papers he authored and 10 Ph.D. dissertations that he supervised.

Stapel did not deny that his deceit was driven by ambition. But it was more complicated than that, he told me. He insisted that he loved social psychology but had been frustrated by the messiness of experimental data, which rarely led to clear conclusions. His lifelong obsession with elegance and order, he said, led him to concoct sexy results that journals found attractive. "It was a quest for aesthetics, for beauty—instead of the truth," he said. He described his behavior as an addiction that drove him to carry out acts of increasingly daring fraud, like a junkie seeking a bigger and better high.[2]

The committee investigating Stapel concluded that he is guilty of several practices including:

- creating datasets, which largely confirmed the prior expectations,
- altering data in existing datasets,
- changing measuring instruments without reporting the change, and
- misrepresenting the number of experimental subjects.

Clearly, it is never acceptable to falsify data the way this researcher did. Sometimes, however, violations of ethics are not as easy to spot.

Researchers have a responsibility to verify that proper methods are being followed. The report describing the investigation of Stapel's fraud states that, "statistical flaws frequently revealed a lack of familiarity with elementary statistics."[3] Many of Stapel's co-authors should have spotted irregularities in his data. Unfortunately, they did not know very much about statistical analysis, and they simply trusted that he was collecting and reporting data properly.

Many types of statistical fraud are difficult to spot. Some researchers simply stop collecting data once they have just enough to prove what they had hoped to prove. They don't want to take the chance that a more extensive study would complicate their lives by producing data contradicting their hypothesis.

Professional organizations, like the American Statistical Association, clearly define expectations for researchers. There are even laws in the federal code about the use of research data.

When a statistical study uses human participants, as in medical studies, both ethics and the law dictate that researchers should be mindful of the safety of their research subjects. The U.S. Department of Health and Human Services oversees federal regulations of research studies with the aim of protecting participants. When a university or other research institution engages in research, it must ensure the safety of all human subjects. For this reason, research institutions establish oversight committees known as **Institutional Review Boards (IRB)**. All planned studies must be approved in advance by the IRB. Key protections that are mandated by law include the following:

- Risks to participants must be minimized and reasonable with respect to projected benefits.
- Participants must give **informed consent**. This means that the risks of participation must be clearly explained to the subjects of the study. Subjects must consent in writing, and researchers are required to keep documentation of their consent.
- Data collected from individuals must be guarded carefully to protect their privacy.

These ideas may seem fundamental, but they can be very difficult to verify in practice. Is removing a participant's name from the data record sufficient to protect privacy? Perhaps the person's identity could be discovered from the data that remains. What happens if the study does not proceed as planned and risks arise that were not anticipated? When is informed consent really necessary? Suppose your doctor wants a blood sample to check your cholesterol level. Once the sample has been tested, you expect the lab to dispose of the remaining blood. At that point the blood becomes biological waste. Does a researcher have the right to take it for use in a study?

It is important that students of statistics take time to consider the ethical questions that arise in statistical studies. How prevalent is fraud in statistical studies? You might be surprised—and disappointed. There is a website

2. *Yudhijit Bhattacharjee, "The Mind of a Con Man," Magazine, New York Times, April 26, 2013. Available online at: http://www.nytimes.com/2013/04/28/magazine/diederik-stapels-audacious-academic-fraud.html?src=dayp&_r=2& (accessed May 1, 2013).*
3. "Flawed Science: The Fraudulent Research Practices of Social Psychologist Diederik Stapel," Tillburg University, November 28, 2012, http://www.tilburguniversity.edu/upload/064a10cd-bce5-4385-b9ff-05b840caeae6_120695_Rapp_nov_2012_UK_web.pdf (accessed May 1, 2013).

(www.retractionwatch.com) (http://www.retractionwatch.com) dedicated to cataloging retractions of study articles that have been proven fraudulent. A quick glance will show that the misuse of statistics is a bigger problem than most people realize.

Vigilance against fraud requires knowledge. Learning the basic theory of statistics will empower you to analyze statistical studies critically.

Example 1.22

Describe the unethical behavior in each example and describe how it could impact the reliability of the resulting data. Explain how the problem should be corrected.

A researcher is collecting data in a community.

   a. She selects a block where she is comfortable walking because she knows many of the people living on the street.

   b. No one seems to be home at four houses on her route. She does not record the addresses and does not return at a later time to try to find residents at home.

   c. She skips four houses on her route because she is running late for an appointment. When she gets home, she fills in the forms by selecting random answers from other residents in the neighborhood.

Solution 1.22

   a. By selecting a convenient sample, the researcher is intentionally selecting a sample that could be biased. Claiming that this sample represents the community is misleading. The researcher needs to select areas in the community at random.

   b. Intentionally omitting relevant data will create bias in the sample. Suppose the researcher is gathering information about jobs and child care. By ignoring people who are not home, she may be missing data from working families that are relevant to her study. She needs to make every effort to interview all members of the target sample.

   c. It is never acceptable to fake data. Even though the responses she uses are "real" responses provided by other participants, the duplication is fraudulent and can create bias in the data. She needs to work diligently to interview everyone on her route.

Try It Σ

1.22 Describe the unethical behavior, if any, in each example and describe how it could impact the reliability of the resulting data. Explain how the problem should be corrected.

A study is commissioned to determine the favorite brand of fruit juice among teens in California.

  a. The survey is commissioned by the seller of a popular brand of apple juice.

  b. There are only two types of juice included in the study: apple juice and cranberry juice.

  c. Researchers allow participants to see the brand of juice as samples are poured for a taste test.

  d. Twenty-five percent of participants prefer Brand X, 33% prefer Brand Y and 42% have no preference between the two brands. Brand X references the study in a commercial saying "Most teens like Brand X as much as or more than Brand Y."

1.5 | Data Collection Experiment

Stats Lab

1.1 Data Collection Experiment

Class Time:

Names:

Student Learning Outcomes

- The student will demonstrate the systematic sampling technique.
- The student will construct relative frequency tables.
- The student will interpret results and their differences from different data groupings.

Movie Survey

Ask five classmates from a different class how many movies they saw at the theater last month. Do not include rented movies.

1. Record the data.

2. In class, randomly pick one person. On the class list, mark that person's name. Move down four names on the class list. Mark that person's name. Continue doing this until you have marked 12 names. You may need to go back to the start of the list. For each marked name record the five data values. You now have a total of 60 data values.

3. For each name marked, record the data.

Table 1.17

Order the Data

Complete the two relative frequency tables below using your class data.

Number of Movies	Frequency	Relative Frequency	Cumulative Relative Frequency
0			
1			
2			
3			
4			
5			
6			

Number of Movies	Frequency	Relative Frequency	Cumulative Relative Frequency
7+			

Table 1.18 Frequency of Number of Movies Viewed

Number of Movies	Frequency	Relative Frequency	Cumulative Relative Frequency
0–1			
2–3			
4–5			
6–7+			

Table 1.19 Frequency of Number of Movies Viewed

1. Using the tables, find the percent of data that is at most two. Which table did you use and why?
2. Using the tables, find the percent of data that is at most three. Which table did you use and why?
3. Using the tables, find the percent of data that is more than two. Which table did you use and why?
4. Using the tables, find the percent of data that is more than three. Which table did you use and why?

Discussion Questions

1. Is one of the tables "more correct" than the other? Why or why not?
2. In general, how could you group the data differently? Are there any advantages to either way of grouping the data?
3. Why did you switch between tables, if you did, when answering the question above?

1.6 | Sampling Experiment

Stats Lab

1.2 Sampling Experiment

Class Time:

Names:

Student Learning Outcomes

- The student will demonstrate the simple random, systematic, stratified, and cluster sampling techniques.
- The student will explain the details of each procedure used.

In this lab, you will be asked to pick several random samples of restaurants. In each case, describe your procedure briefly, including how you might have used the random number generator, and then list the restaurants in the sample you obtained.

NOTE

The following section contains restaurants stratified by city into columns and grouped horizontally by entree cost (clusters).

Restaurants Stratified by City and Entree Cost

Entree Cost	Under $10	$10 to under $15	$15 to under $20	Over $20
San Jose	El Abuelo Taq, Pasta Mia, Emma's Express, Bamboo Hut	Emperor's Guard, Creekside Inn	Agenda, Gervais, Miro's	Blake's, Eulipia, Hayes Mansion, Germania
Palo Alto	Senor Taco, Olive Garden, Taxi's	Ming's, P.A. Joe's, Stickney's	Scott's Seafood, Poolside Grill, Fish Market	Sundance Mine, Maddalena's, Spago's
Los Gatos	Mary's Patio, Mount Everest, Sweet Pea's, Andele Taqueria	Lindsey's, Willow Street	Toll House	Charter House, La Maison Du Cafe
Mountain View	Maharaja, New Ma's, Thai-Rific, Garden Fresh	Amber Indian, La Fiesta, Fiesta del Mar, Dawit	Austin's, Shiva's, Mazeh	Le Petit Bistro
Cupertino	Hobees, Hung Fu, Samrat, Panda Express	Santa Barb. Grill, Mand. Gourmet, Bombay Oven, Kathmandu West	Fontana's, Blue Pheasant	Hamasushi, Helios
Sunnyvale	Chekijababi, Taj India, Full Throttle, Tia Juana, Lemon Grass	Pacific Fresh, Charley Brown's, Cafe Cameroon, Faz, Aruba's	Lion & Compass, The Palace, Beau Sejour	
Santa Clara	Rangoli, Armadillo Willy's, Thai Pepper, Pasand	Arthur's, Katie's Cafe, Pedro's, La Galleria	Birk's, Truya Sushi, Valley Plaza	Lakeside, Mariani's

Table 1.20 Restaurants Used in Sample

A Simple Random Sample

Pick a **simple random sample** of 15 restaurants.

1. Describe your procedure.
2. Complete the table with your sample.

1. _____	6. _____	11. _____
2. _____	7. _____	12. _____
3. _____	8. _____	13. _____
4. _____	9. _____	14. _____
5. _____	10. _____	15. _____

Table 1.21

A Systematic Sample

Pick a **systematic sample** of 15 restaurants.

1. Describe your procedure.
2. Complete the table with your sample.

1. _____	6. _____	11. _____
2. _____	7. _____	12. _____
3. _____	8. _____	13. _____
4. _____	9. _____	14. _____
5. _____	10. _____	15. _____

Table 1.22

A Stratified Sample

Pick a **stratified sample**, by city, of 20 restaurants. Use 25% of the restaurants from each stratum. Round to the nearest whole number.

1. Describe your procedure.
2. Complete the table with your sample.

1. _____	6. _____	11. _____	16. _____
2. _____	7. _____	12. _____	17. _____
3. _____	8. _____	13. _____	18. _____
4. _____	9. _____	14. _____	19. _____
5. _____	10. _____	15. _____	20. _____

Table 1.23

A Stratified Sample

Pick a **stratified sample**, by entree cost, of 21 restaurants. Use 25% of the restaurants from each stratum. Round to the nearest whole number.

1. Describe your procedure.
2. Complete the table with your sample.

1. _____	6. _____	11. _____	16. _____
2. _____	7. _____	12. _____	17. _____
3. _____	8. _____	13. _____	18. _____
4. _____	9. _____	14. _____	19. _____
5. _____	10. _____	15. _____	20. _____
			21. _____

Table 1.24

A Cluster Sample

Pick a **cluster sample** of restaurants from two cities. The number of restaurants will vary.

1. Describe your procedure.
2. Complete the table with your sample.

1. ____	6. ____	11. ____	16. ____	21. ____
2. ____	7. ____	12. ____	17. ____	22. ____
3. ____	8. ____	13. ____	18. ____	23. ____
4. ____	9. ____	14. ____	19. ____	24. ____
5. ____	10. ____	15. ____	20. ____	25. ____

Table 1.25

KEY TERMS

Average also called mean; a number that describes the central tendency of the data

Blinding not telling participants which treatment a subject is receiving

Categorical Variable variables that take on values that are names or labels

Cluster Sampling a method for selecting a random sample and dividing the population into groups (clusters); use simple random sampling to select a set of clusters. Every individual in the chosen clusters is included in the sample.

Continuous Random Variable a random variable (RV) whose outcomes are measured; the height of trees in the forest is a continuous RV.

Control Group a group in a randomized experiment that receives an inactive treatment but is otherwise managed exactly as the other groups

Convenience Sampling a nonrandom method of selecting a sample; this method selects individuals that are easily accessible and may result in biased data.

Cumulative Relative Frequency The term applies to an ordered set of observations from smallest to largest. The cumulative relative frequency is the sum of the relative frequencies for all values that are less than or equal to the given value.

Data a set of observations (a set of possible outcomes); most data can be put into two groups: **qualitative** (an attribute whose value is indicated by a label) or **quantitative** (an attribute whose value is indicated by a number). Quantitative data can be separated into two subgroups: **discrete** and **continuous**. Data is discrete if it is the result of counting (such as the number of students of a given ethnic group in a class or the number of books on a shelf). Data is continuous if it is the result of measuring (such as distance traveled or weight of luggage)

Discrete Random Variable a random variable (RV) whose outcomes are counted

Double-blinding the act of blinding both the subjects of an experiment and the researchers who work with the subjects

Experimental Unit any individual or object to be measured

Explanatory Variable the independent variable in an experiment; the value controlled by researchers

Frequency the number of times a value of the data occurs

Informed Consent Any human subject in a research study must be cognizant of any risks or costs associated with the study. The subject has the right to know the nature of the treatments included in the study, their potential risks, and their potential benefits. Consent must be given freely by an informed, fit participant.

Institutional Review Board a committee tasked with oversight of research programs that involve human subjects

Lurking Variable a variable that has an effect on a study even though it is neither an explanatory variable nor a response variable

Nonsampling Error an issue that affects the reliability of sampling data other than natural variation; it includes a variety of human errors including poor study design, biased sampling methods, inaccurate information provided by study participants, data entry errors, and poor analysis.

Numerical Variable variables that take on values that are indicated by numbers

Parameter a number that is used to represent a population characteristic and that generally cannot be determined easily

Placebo an inactive treatment that has no real effect on the explanatory variable

Population all individuals, objects, or measurements whose properties are being studied

Probability a number between zero and one, inclusive, that gives the likelihood that a specific event will occur

Proportion the number of successes divided by the total number in the sample

Qualitative Data See **Data**.

Quantitative Data See **Data**.

Random Assignment the act of organizing experimental units into treatment groups using random methods

Random Sampling a method of selecting a sample that gives every member of the population an equal chance of being selected.

Relative Frequency the ratio of the number of times a value of the data occurs in the set of all outcomes to the number of all outcomes to the total number of outcomes

Representative Sample a subset of the population that has the same characteristics as the population

Response Variable the dependent variable in an experiment; the value that is measured for change at the end of an experiment

Sample a subset of the population studied

Sampling Bias not all members of the population are equally likely to be selected

Sampling Error the natural variation that results from selecting a sample to represent a larger population; this variation decreases as the sample size increases, so selecting larger samples reduces sampling error.

Sampling with Replacement Once a member of the population is selected for inclusion in a sample, that member is returned to the population for the selection of the next individual.

Sampling without Replacement A member of the population may be chosen for inclusion in a sample only once. If chosen, the member is not returned to the population before the next selection.

Simple Random Sampling a straightforward method for selecting a random sample; give each member of the population a number. Use a random number generator to select a set of labels. These randomly selected labels identify the members of your sample.

Statistic a numerical characteristic of the sample; a statistic estimates the corresponding population parameter.

Stratified Sampling a method for selecting a random sample used to ensure that subgroups of the population are represented adequately; divide the population into groups (strata). Use simple random sampling to identify a proportionate number of individuals from each stratum.

Systematic Sampling a method for selecting a random sample; list the members of the population. Use simple random sampling to select a starting point in the population. Let k = (number of individuals in the population)/(number of individuals needed in the sample). Choose every kth individual in the list starting with the one that was randomly selected. If necessary, return to the beginning of the population list to complete your sample.

Treatments different values or components of the explanatory variable applied in an experiment

Variable a characteristic of interest for each person or object in a population

CHAPTER REVIEW

1.1 Definitions of Statistics, Probability, and Key Terms
The mathematical theory of statistics is easier to learn when you know the language. This module presents important terms that will be used throughout the text.

1.2 Data, Sampling, and Variation in Data and Sampling

Data are individual items of information that come from a population or sample. Data may be classified as qualitative, quantitative continuous, or quantitative discrete.

Because it is not practical to measure the entire population in a study, researchers use samples to represent the population. A random sample is a representative group from the population chosen by using a method that gives each individual in the population an equal chance of being included in the sample. Random sampling methods include simple random sampling, stratified sampling, cluster sampling, and systematic sampling. Convenience sampling is a nonrandom method of choosing a sample that often produces biased data.

Samples that contain different individuals result in different data. This is true even when the samples are well-chosen and representative of the population. When properly selected, larger samples model the population more closely than smaller samples. There are many different potential problems that can affect the reliability of a sample. Statistical data needs to be critically analyzed, not simply accepted.

1.3 Frequency, Frequency Tables, and Levels of Measurement

Some calculations generate numbers that are artificially precise. It is not necessary to report a value to eight decimal places when the measures that generated that value were only accurate to the nearest tenth. Round off your final answer to one more decimal place than was present in the original data. This means that if you have data measured to the nearest tenth of a unit, report the final statistic to the nearest hundredth.

In addition to rounding your answers, you can measure your data using the following four levels of measurement.

- **Nominal scale level:** data that cannot be ordered nor can it be used in calculations

- **Ordinal scale level:** data that can be ordered; the differences cannot be measured

- **Interval scale level:** data with a definite ordering but no starting point; the differences can be measured, but there is no such thing as a ratio.

- **Ratio scale level:** data with a starting point that can be ordered; the differences have meaning and ratios can be calculated.

When organizing data, it is important to know how many times a value appears. How many statistics students study five hours or more for an exam? What percent of families on our block own two pets? Frequency, relative frequency, and cumulative relative frequency are measures that answer questions like these.

1.4 Experimental Design and Ethics

A poorly designed study will not produce reliable data. There are certain key components that must be included in every experiment. To eliminate lurking variables, subjects must be assigned randomly to different treatment groups. One of the groups must act as a control group, demonstrating what happens when the active treatment is not applied. Participants in the control group receive a placebo treatment that looks exactly like the active treatments but cannot influence the response variable. To preserve the integrity of the placebo, both researchers and subjects may be blinded. When a study is designed properly, the only difference between treatment groups is the one imposed by the researcher. Therefore, when groups respond differently to different treatments, the difference must be due to the influence of the explanatory variable.

"An ethics problem arises when you are considering an action that benefits you or some cause you support, hurts or reduces benefits to others, and violates some rule."[4] Ethical violations in statistics are not always easy to spot. Professional associations and federal agencies post guidelines for proper conduct. It is important that you learn basic statistical procedures so that you can recognize proper data analysis.

PRACTICE

1.1 Definitions of Statistics, Probability, and Key Terms

Use the following information to answer the next five exercises. Studies are often done by pharmaceutical companies to determine the effectiveness of a treatment program. Suppose that a new AIDS antibody drug is currently under study. It is given to patients once the AIDS symptoms have revealed themselves. Of interest is the average (mean) length of time in

---

4. Andrew Gelman, "Open Data and Open Methods," Ethics and Statistics, http://www.stat.columbia.edu/~gelman/research/published/ChanceEthics1.pdf (accessed May 1, 2013).

months patients live once they start the treatment. Two researchers each follow a different set of 40 patients with AIDS from the start of treatment until their deaths. The following data (in months) are collected.

Researcher A:

3; 4; 11; 15; 16; 17; 22; 44; 37; 16; 14; 24; 25; 15; 26; 27; 33; 29; 35; 44; 13; 21; 22; 10; 12; 8; 40; 32; 26; 27; 31; 34; 29; 17; 8; 24; 18; 47; 33; 34

Researcher B:

3; 14; 11; 5; 16; 17; 28; 41; 31; 18; 14; 14; 26; 25; 21; 22; 31; 2; 35; 44; 23; 21; 21; 16; 12; 18; 41; 22; 16; 25; 33; 34; 29; 13; 18; 24; 23; 42; 33; 29

Determine what the key terms refer to in the example for Researcher A.

1. population

2. sample

3. parameter

4. statistic

5. variable

1.2 Data, Sampling, and Variation in Data and Sampling

6. "Number of times per week" is what type of data?

a. qualitative; b. quantitative discrete; c. quantitative continuous

Use the following information to answer the next four exercises: A study was done to determine the age, number of times per week, and the duration (amount of time) of residents using a local park in San Antonio, Texas. The first house in the neighborhood around the park was selected randomly, and then the resident of every eighth house in the neighborhood around the park was interviewed.

7. The sampling method was

a. simple random; b. systematic; c. stratified; d. cluster

8. "Duration (amount of time)" is what type of data?

a. qualitative; b. quantitative discrete; c. quantitative continuous

9. The colors of the houses around the park are what kind of data?

a. qualitative; b. quantitative discrete; c. quantitative continuous

10. The population is _____

11. Table 1.26 contains the total number of deaths worldwide as a result of earthquakes from 2000 to 2012.

Year	Total Number of Deaths
2000	231
2001	21,357
2002	11,685
2003	33,819
2004	228,802
2005	88,003
2006	6,605
2007	712

Year	Total Number of Deaths
2008	88,011
2009	1,790
2010	320,120
2011	21,953
2012	768
Total	823,856

Table 1.26

Use **Table 1.26** to answer the following questions.

    a. What is the proportion of deaths between 2007 and 2012?
    b. What percent of deaths occurred before 2001?
    c. What is the percent of deaths that occurred in 2003 or after 2010?
    d. What is the fraction of deaths that happened before 2012?
    e. What kind of data is the number of deaths?
    f. Earthquakes are quantified according to the amount of energy they produce (examples are 2.1, 5.0, 6.7). What type of data is that?
    g. What contributed to the large number of deaths in 2010? In 2004? Explain.

For the following four exercises, determine the type of sampling used (simple random, stratified, systematic, cluster, or convenience).

12. A group of test subjects is divided into twelve groups; then four of the groups are chosen at random.

13. A market researcher polls every tenth person who walks into a store.

14. The first 50 people who walk into a sporting event are polled on their television preferences.

15. A computer generates 100 random numbers, and 100 people whose names correspond with the numbers on the list are chosen.

Use the following information to answer the next seven exercises: Studies are often done by pharmaceutical companies to determine the effectiveness of a treatment program. Suppose that a new AIDS antibody drug is currently under study. It is given to patients once the AIDS symptoms have revealed themselves. Of interest is the average (mean) length of time in months patients live once starting the treatment. Two researchers each follow a different set of 40 AIDS patients from the start of treatment until their deaths. The following data (in months) are collected.

Researcher A: 3; 4; 11; 15; 16; 17; 22; 44; 37; 16; 14; 24; 25; 15; 26; 27; 33; 29; 35; 44; 13; 21; 22; 10; 12; 8; 40; 32; 26; 27; 31; 34; 29; 17; 8; 24; 18; 47; 33; 34

Researcher B: 3; 14; 11; 5; 16; 17; 28; 41; 31; 18; 14; 14; 26; 25; 21; 22; 31; 2; 35; 44; 23; 21; 21; 16; 12; 18; 41; 22; 16; 25; 33; 34; 29; 13; 18; 24; 23; 42; 33; 29

16. Complete the tables using the data provided:

Survival Length (in months)	Frequency	Relative Frequency	Cumulative Relative Frequency
0.5–6.5			
6.5–12.5			
12.5–18.5			

Survival Length (in months)	Frequency	Relative Frequency	Cumulative Relative Frequency
18.5–24.5			
24.5–30.5			
30.5–36.5			
36.5–42.5			
42.5–48.5			

Table 1.27 Researcher A

Survival Length (in months)	Frequency	Relative Frequency	Cumulative Relative Frequency
0.5–6.5			
6.5–12.5			
12.5–18.5			
18.5–24.5			
24.5–30.5			
30.5–36.5			
36.5-45.5			

Table 1.28 Researcher B

17. Determine what the key term data refers to in the above example for Researcher A.

18. List two reasons why the data may differ.

19. Can you tell if one researcher is correct and the other one is incorrect? Why?

20. Would you expect the data to be identical? Why or why not?

21. How might the researchers gather random data?

22. Suppose that the first researcher conducted his survey by randomly choosing one state in the nation and then randomly picking 40 patients from that state. What sampling method would that researcher have used?

23. Suppose that the second researcher conducted his survey by choosing 40 patients he knew. What sampling method would that researcher have used? What concerns would you have about this data set, based upon the data collection method?

Use the following data to answer the next five exercises: Two researchers are gathering data on hours of video games played by school-aged children and young adults. They each randomly sample different groups of 150 students from the same school. They collect the following data.

Hours Played per Week	Frequency	Relative Frequency	Cumulative Relative Frequency
0–2	26	0.17	0.17
2–4	30	0.20	0.37
4–6	49	0.33	0.70
6–8	25	0.17	0.87
8–10	12	0.08	0.95

Hours Played per Week	Frequency	Relative Frequency	Cumulative Relative Frequency
10–12	8	0.05	1

Table 1.29 Researcher A

Hours Played per Week	Frequency	Relative Frequency	Cumulative Relative Frequency
0–2	48	0.32	0.32
2–4	51	0.34	0.66
4–6	24	0.16	0.82
6–8	12	0.08	0.90
8–10	11	0.07	0.97
10–12	4	0.03	1

Table 1.30 Researcher B

24. Give a reason why the data may differ.

25. Would the sample size be large enough if the population is the students in the school?

26. Would the sample size be large enough if the population is school-aged children and young adults in the United States?

27. Researcher A concludes that most students play video games between four and six hours each week. Researcher B concludes that most students play video games between two and four hours each week. Who is correct?

28. As part of a way to reward students for participating in the survey, the researchers gave each student a gift card to a video game store. Would this affect the data if students knew about the award before the study?

Use the following data to answer the next five exercises: A pair of studies was performed to measure the effectiveness of a new software program designed to help stroke patients regain their problem-solving skills. Patients were asked to use the software program twice a day, once in the morning and once in the evening. The studies observed 200 stroke patients recovering over a period of several weeks. The first study collected the data in **Table 1.31**. The second study collected the data in **Table 1.32**.

Group	Showed improvement	No improvement	Deterioration
Used program	142	43	15
Did not use program	72	110	18

Table 1.31

Group	Showed improvement	No improvement	Deterioration
Used program	105	74	19
Did not use program	89	99	12

Table 1.32

29. Given what you know, which study is correct?

30. The first study was performed by the company that designed the software program. The second study was performed by the American Medical Association. Which study is more reliable?

31. Both groups that performed the study concluded that the software works. Is this accurate?

32. The company takes the two studies as proof that their software causes mental improvement in stroke patients. Is this a fair statement?

33. Patients who used the software were also a part of an exercise program whereas patients who did not use the software were not. Does this change the validity of the conclusions from **Exercise 1.31**?

34. Is a sample size of 1,000 a reliable measure for a population of 5,000?

35. Is a sample of 500 volunteers a reliable measure for a population of 2,500?

36. A question on a survey reads: "Do you prefer the delicious taste of Brand X or the taste of Brand Y?" Is this a fair question?

37. Is a sample size of two representative of a population of five?

38. Is it possible for two experiments to be well run with similar sample sizes to get different data?

1.3 Frequency, Frequency Tables, and Levels of Measurement

39. What type of measure scale is being used? Nominal, ordinal, interval or ratio.
   a. High school soccer players classified by their athletic ability: Superior, Average, Above average
   b. Baking temperatures for various main dishes: 350, 400, 325, 250, 300
   c. The colors of crayons in a 24-crayon box
   d. Social security numbers
   e. Incomes measured in dollars
   f. A satisfaction survey of a social website by number: 1 = very satisfied, 2 = somewhat satisfied, 3 = not satisfied
   g. Political outlook: extreme left, left-of-center, right-of-center, extreme right
   h. Time of day on an analog watch
   i. The distance in miles to the closest grocery store
   j. The dates 1066, 1492, 1644, 1947, and 1944
   k. The heights of 21–65 year-old women
   l. Common letter grades: A, B, C, D, and F

1.4 Experimental Design and Ethics

40. Design an experiment. Identify the explanatory and response variables. Describe the population being studied and the experimental units. Explain the treatments that will be used and how they will be assigned to the experimental units. Describe how blinding and placebos may be used to counter the power of suggestion.

41. Discuss potential violations of the rule requiring informed consent.
   a. Inmates in a correctional facility are offered good behavior credit in return for participation in a study.
   b. A research study is designed to investigate a new children's allergy medication.
   c. Participants in a study are told that the new medication being tested is highly promising, but they are not told that only a small portion of participants will receive the new medication. Others will receive placebo treatments and traditional treatments.

HOMEWORK

1.1 Definitions of Statistics, Probability, and Key Terms

For each of the following eight exercises, identify: a. the population, b. the sample, c. the parameter, d. the statistic, e. the variable, and f. the data. Give examples where appropriate.

42. A fitness center is interested in the mean amount of time a client exercises in the center each week.

43. Ski resorts are interested in the mean age that children take their first ski and snowboard lessons. They need this information to plan their ski classes optimally.

44. A cardiologist is interested in the mean recovery period of her patients who have had heart attacks.

45. Insurance companies are interested in the mean health costs each year of their clients, so that they can determine the costs of health insurance.

46. A politician is interested in the proportion of voters in his district who think he is doing a good job.

47. A marriage counselor is interested in the proportion of clients she counsels who stay married.

48. Political pollsters may be interested in the proportion of people who will vote for a particular cause.

49. A marketing company is interested in the proportion of people who will buy a particular product.

Use the following information to answer the next three exercises: A Lake Tahoe Community College instructor is interested in the mean number of days Lake Tahoe Community College math students are absent from class during a quarter.

50. What is the population she is interested in?
   a. all Lake Tahoe Community College students
   b. all Lake Tahoe Community College English students
   c. all Lake Tahoe Community College students in her classes
   d. all Lake Tahoe Community College math students

51. Consider the following:

X = number of days a Lake Tahoe Community College math student is absent

In this case, X is an example of a:

   a. variable.
   b. population.
   c. statistic.
   d. data.

52. The instructor's sample produces a mean number of days absent of 3.5 days. This value is an example of a:
   a. parameter.
   b. data.
   c. statistic.
   d. variable.

1.2 Data, Sampling, and Variation in Data and Sampling

For the following exercises, identify the type of data that would be used to describe a response (quantitative discrete, quantitative continuous, or qualitative), and give an example of the data.

53. number of tickets sold to a concert

54. percent of body fat

55. favorite baseball team

56. time in line to buy groceries

57. number of students enrolled at Evergreen Valley College

58. most-watched television show

59. brand of toothpaste

60. distance to the closest movie theatre

61. age of executives in Fortune 500 companies

62. number of competing computer spreadsheet software packages

Use the following information to answer the next two exercises: A study was done to determine the age, number of times per week, and the duration (amount of time) of resident use of a local park in San Jose. The first house in the neighborhood around the park was selected randomly and then every 8th house in the neighborhood around the park was interviewed.

63. "Number of times per week" is what type of data?
   a. qualitative
   b. quantitative discrete
   c. quantitative continuous

64. "Duration (amount of time)" is what type of data?

a. qualitative
b. quantitative discrete
c. quantitative continuous

65. Airline companies are interested in the consistency of the number of babies on each flight, so that they have adequate safety equipment. Suppose an airline conducts a survey. Over Thanksgiving weekend, it surveys six flights from Boston to Salt Lake City to determine the number of babies on the flights. It determines the amount of safety equipment needed by the result of that study.
    a. Using complete sentences, list three things wrong with the way the survey was conducted.
    b. Using complete sentences, list three ways that you would improve the survey if it were to be repeated.

66. Suppose you want to determine the mean number of students per statistics class in your state. Describe a possible sampling method in three to five complete sentences. Make the description detailed.

67. Suppose you want to determine the mean number of cans of soda drunk each month by students in their twenties at your school. Describe a possible sampling method in three to five complete sentences. Make the description detailed.

68. List some practical difficulties involved in getting accurate results from a telephone survey.

69. List some practical difficulties involved in getting accurate results from a mailed survey.

70. With your classmates, brainstorm some ways you could overcome these problems if you needed to conduct a phone or mail survey.

71. The instructor takes her sample by gathering data on five randomly selected students from each Lake Tahoe Community College math class. The type of sampling she used is
    a. cluster sampling
    b. stratified sampling
    c. simple random sampling
    d. convenience sampling

72. A study was done to determine the age, number of times per week, and the duration (amount of time) of residents using a local park in San Jose. The first house in the neighborhood around the park was selected randomly and then every eighth house in the neighborhood around the park was interviewed. The sampling method was:
    a. simple random
    b. systematic
    c. stratified
    d. cluster

73. Name the sampling method used in each of the following situations:
    a. A woman in the airport is handing out questionnaires to travelers asking them to evaluate the airport's service. She does not ask travelers who are hurrying through the airport with their hands full of luggage, but instead asks all travelers who are sitting near gates and not taking naps while they wait.
    b. A teacher wants to know if her students are doing homework, so she randomly selects rows two and five and then calls on all students in row two and all students in row five to present the solutions to homework problems to the class.
    c. The marketing manager for an electronics chain store wants information about the ages of its customers. Over the next two weeks, at each store location, 100 randomly selected customers are given questionnaires to fill out asking for information about age, as well as about other variables of interest.
    d. The librarian at a public library wants to determine what proportion of the library users are children. The librarian has a tally sheet on which she marks whether books are checked out by an adult or a child. She records this data for every fourth patron who checks out books.
    e. A political party wants to know the reaction of voters to a debate between the candidates. The day after the debate, the party's polling staff calls 1,200 randomly selected phone numbers. If a registered voter answers the phone or is available to come to the phone, that registered voter is asked whom he or she intends to vote for and whether the debate changed his or her opinion of the candidates.

74. A "random survey" was conducted of 3,274 people of the "microprocessor generation" (people born since 1971, the year the microprocessor was invented). It was reported that 48% of those individuals surveyed stated that if they had $2,000 to spend, they would use it for computer equipment. Also, 66% of those surveyed considered themselves relatively savvy computer users.
    a. Do you consider the sample size large enough for a study of this type? Why or why not?

  b. Based on your "gut feeling," do you believe the percents accurately reflect the U.S. population for those individuals born since 1971? If not, do you think the percents of the population are actually higher or lower than the sample statistics? Why?

    Additional information: The survey, reported by Intel Corporation, was filled out by individuals who visited the Los Angeles Convention Center to see the Smithsonian Institute's road show called "America's Smithsonian."

  c. With this additional information, do you feel that all demographic and ethnic groups were equally represented at the event? Why or why not?

  d. With the additional information, comment on how accurately you think the sample statistics reflect the population parameters.

75. The Gallup-Healthways Well-Being Index is a survey that follows trends of U.S. residents on a regular basis. There are six areas of health and wellness covered in the survey: Life Evaluation, Emotional Health, Physical Health, Healthy Behavior, Work Environment, and Basic Access. Some of the questions used to measure the Index are listed below.

Identify the type of data obtained from each question used in this survey: qualitative, quantitative discrete, or quantitative continuous.

  a. Do you have any health problems that prevent you from doing any of the things people your age can normally do?

  b. During the past 30 days, for about how many days did poor health keep you from doing your usual activities?

  c. In the last seven days, on how many days did you exercise for 30 minutes or more?

  d. Do you have health insurance coverage?

76. In advance of the 1936 Presidential Election, a magazine titled Literary Digest released the results of an opinion poll predicting that the republican candidate Alf Landon would win by a large margin. The magazine sent post cards to approximately 10,000,000 prospective voters. These prospective voters were selected from the subscription list of the magazine, from automobile registration lists, from phone lists, and from club membership lists. Approximately 2,300,000 people returned the postcards.

  a. Think about the state of the United States in 1936. Explain why a sample chosen from magazine subscription lists, automobile registration lists, phone books, and club membership lists was not representative of the population of the United States at that time.

  b. What effect does the low response rate have on the reliability of the sample?

  c. Are these problems examples of sampling error or nonsampling error?

  d. During the same year, George Gallup conducted his own poll of 30,000 prospective voters. His researchers used a method they called "quota sampling" to obtain survey answers from specific subsets of the population. Quota sampling is an example of which sampling method described in this module?

77. Crime-related and demographic statistics for 47 US states in 1960 were collected from government agencies, including the FBI's *Uniform Crime Report*. One analysis of this data found a strong connection between education and crime indicating that higher levels of education in a community correspond to higher crime rates.

Which of the potential problems with samples discussed in **Section 1.2** could explain this connection?

78. YouPolls is a website that allows anyone to create and respond to polls. One question posted April 15 asks:

"Do you feel happy paying your taxes when members of the Obama administration are allowed to ignore their tax liabilities?"[5]

As of April 25, 11 people responded to this question. Each participant answered "NO!"

Which of the potential problems with samples discussed in this module could explain this connection?

79. A scholarly article about response rates begins with the following quote:

"Declining contact and cooperation rates in random digit dial (RDD) national telephone surveys raise serious concerns about the validity of estimates drawn from such research."[6]

The Pew Research Center for People and the Press admits:

---

5. lastbaldeagle. 2013. On Tax Day, House to Call for Firing Federal Workers Who Owe Back Taxes. Opinion poll posted online at: http://www.youpolls.com/details.aspx?id=12328 (accessed May 1, 2013).

6. Scott Keeter et al., "Gauging the Impact of Growing Nonresponse on Estimates from a National RDD Telephone Survey," Public Opinion Quarterly 70 no. 5 (2006), http://poq.oxfordjournals.org/content/70/5/759.full (http://poq.oxfordjournals.org/content/70/5/759.full) (accessed May 1, 2013).

"The percentage of people we interview – out of all we try to interview – has been declining over the past decade or more."[7]

a. What are some reasons for the decline in response rate over the past decade?
b. Explain why researchers are concerned with the impact of the declining response rate on public opinion polls.

1.3 Frequency, Frequency Tables, and Levels of Measurement

80. Fifty part-time students were asked how many courses they were taking this term. The (incomplete) results are shown below:

# of Courses	Frequency	Relative Frequency	Cumulative Relative Frequency
1	30	0.6	
2	15		
3			

Table 1.33 Part-time Student Course Loads

a. Fill in the blanks in **Table 1.33**.
b. What percent of students take exactly two courses?
c. What percent of students take one or two courses?

81. Sixty adults with gum disease were asked the number of times per week they used to floss before their diagnosis. The (incomplete) results are shown in **Table 1.34**.

# Flossing per Week	Frequency	Relative Frequency	Cumulative Relative Freq.
0	27	0.4500	
1	18		
3			0.9333
6	3	0.0500	
7	1	0.0167	

Table 1.34 Flossing Frequency for Adults with Gum Disease

a. Fill in the blanks in **Table 1.34**.
b. What percent of adults flossed six times per week?
c. What percent flossed at most three times per week?

82. Nineteen immigrants to the U.S were asked how many years, to the nearest year, they have lived in the U.S. The data are as follows: 2; 5; 7; 2; 2; 10; 20; 15; 0; 7; 0; 20; 5; 12; 15; 12; 4; 5; 10.

Table 1.35 was produced.

7. Frequently Asked Questions, Pew Research Center for the People & the Press, http://www.people-press.org/methodology/frequently-asked-questions/#dont-you-have-trouble-getting-people-to-answer-your-polls (accessed May 1, 2013).

This OpenStax book is available for free at http://cnx.org/content/col11562/1.17   Download for free at http://cnx.org/content/col11562/latest/

Data	Frequency	Relative Frequency	Cumulative Relative Frequency
0	2	$\frac{2}{19}$	0.1053
2	3	$\frac{3}{19}$	0.2632
4	1	$\frac{1}{19}$	0.3158
5	3	$\frac{3}{19}$	0.4737
7	2	$\frac{2}{19}$	0.5789
10	2	$\frac{2}{19}$	0.6842
12	2	$\frac{2}{19}$	0.7895
15	1	$\frac{1}{19}$	0.8421
20	1	$\frac{1}{19}$	1.0000

Table 1.35 Frequency of Immigrant Survey Responses

a. Fix the errors in **Table 1.35**. Also, explain how someone might have arrived at the incorrect number(s).
b. Explain what is wrong with this statement: "47 percent of the people surveyed have lived in the U.S. for 5 years."
c. Fix the statement in **b** to make it correct.
d. What fraction of the people surveyed have lived in the U.S. five or seven years?
e. What fraction of the people surveyed have lived in the U.S. at most 12 years?
f. What fraction of the people surveyed have lived in the U.S. fewer than 12 years?
g. What fraction of the people surveyed have lived in the U.S. from five to 20 years, inclusive?

83. How much time does it take to travel to work? **Table 1.36** shows the mean commute time by state for workers at least 16 years old who are not working at home. Find the mean travel time, and round off the answer properly.

24.0	24.3	25.9	18.9	27.5	17.9	21.8	20.9	16.7	27.3
18.2	24.7	20.0	22.6	23.9	18.0	31.4	22.3	24.0	25.5
24.7	24.6	28.1	24.9	22.6	23.6	23.4	25.7	24.8	25.5
21.2	25.7	23.1	23.0	23.9	26.0	16.3	23.1	21.4	21.5
27.0	27.0	18.6	31.7	23.3	30.1	22.9	23.3	21.7	18.6

Table 1.36

84. *Forbes* magazine published data on the best small firms in 2012. These were firms which had been publicly traded for at least a year, have a stock price of at least $5 per share, and have reported annual revenue between $5 million and $1 billion. **Table 1.37** shows the ages of the chief executive officers for the first 60 ranked firms.

Age	Frequency	Relative Frequency	Cumulative Relative Frequency
40–44	3		
45–49	11		
50–54	13		
55–59	16		
60–64	10		
65–69	6		
70–74	1		

Table 1.37

a.   What is the frequency for CEO ages between 54 and 65?
b.   What percentage of CEOs are 65 years or older?
c.   What is the relative frequency of ages under 50?
d.   What is the cumulative relative frequency for CEOs younger than 55?
e.   Which graph shows the relative frequency and which shows the cumulative relative frequency?

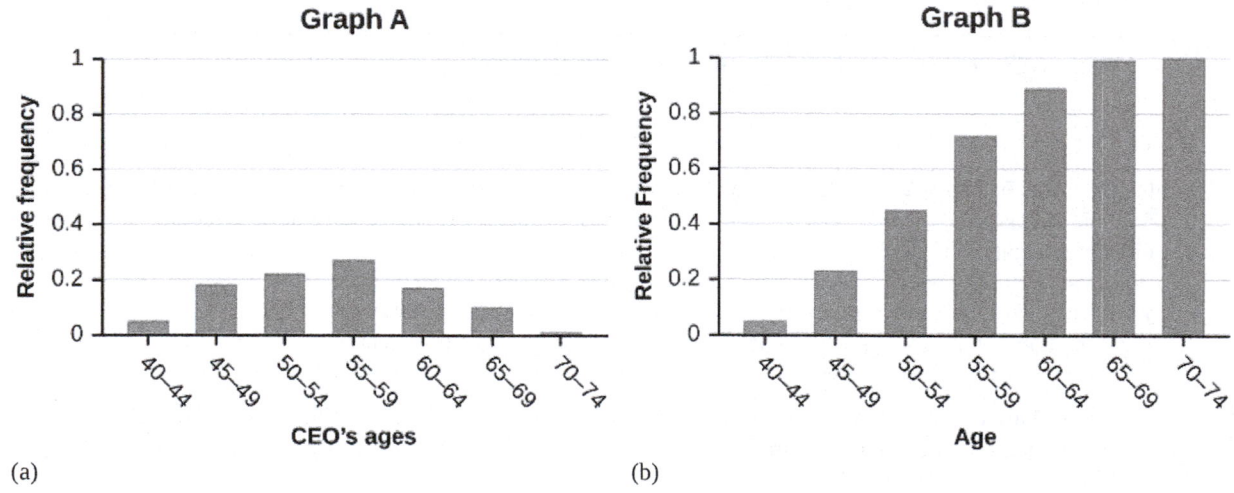

(a)   (b)

Figure 1.13

Use the following information to answer the next two exercises: **Table 1.38** contains data on hurricanes that have made direct hits on the U.S. Between 1851 and 2004. A hurricane is given a strength category rating based on the minimum wind speed generated by the storm.

Category	Number of Direct Hits	Relative Frequency	Cumulative Frequency
1	109	0.3993	0.3993
2	72	0.2637	0.6630
3	71	0.2601	
4	18		0.9890
	Total = 273		

Category	Number of Direct Hits	Relative Frequency	Cumulative Frequency
5	3	0.0110	1.0000
	Total = 273		

Table 1.38 Frequency of Hurricane Direct Hits

85. What is the relative frequency of direct hits that were category 4 hurricanes?
    a. 0.0768
    b. 0.0659
    c. 0.2601
    d. Not enough information to calculate

86. What is the relative frequency of direct hits that were AT MOST a category 3 storm?
    a. 0.3480
    b. 0.9231
    c. 0.2601
    d. 0.3370

1.4 Experimental Design and Ethics

87. How does sleep deprivation affect your ability to drive? A recent study measured the effects on 19 professional drivers. Each driver participated in two experimental sessions: one after normal sleep and one after 27 hours of total sleep deprivation. The treatments were assigned in random order. In each session, performance was measured on a variety of tasks including a driving simulation.

Use key terms from this module to describe the design of this experiment.

88. An advertisement for Acme Investments displays the two graphs in **Figure 1.14** to show the value of Acme's product in comparison with the Other Guy's product. Describe the potentially misleading visual effect of these comparison graphs. How can this be corrected?

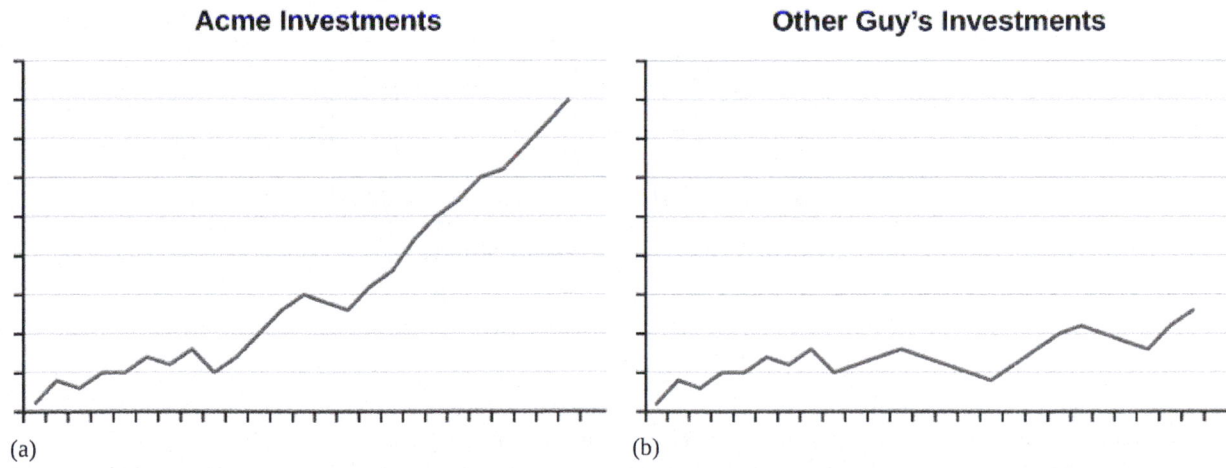

Figure 1.14 As the graphs show, Acme consistently outperforms the Other Guys!

89. The graph in **Figure 1.15** shows the number of complaints for six different airlines as reported to the US Department of Transportation in February 2013. Alaska, Pinnacle, and Airtran Airlines have far fewer complaints reported than American, Delta, and United. Can we conclude that American, Delta, and United are the worst airline carriers since they have the most complaints?

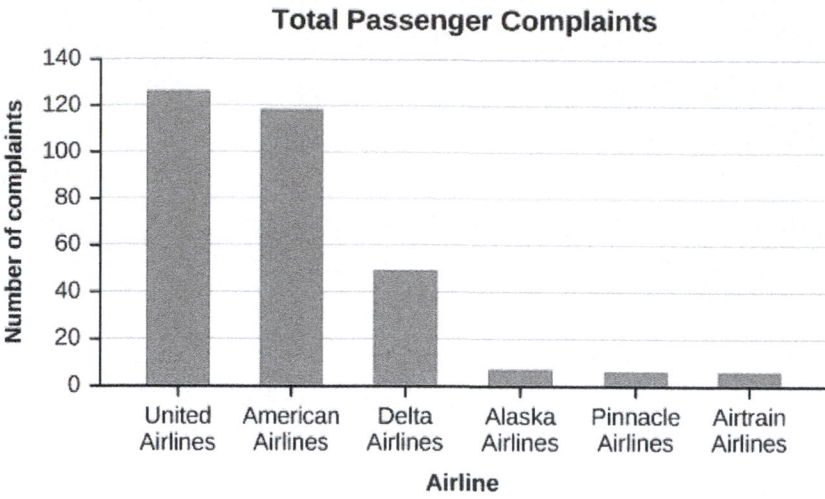

Figure 1.15

BRINGING IT TOGETHER: HOMEWORK

90. Seven hundred and seventy-one distance learning students at Long Beach City College responded to surveys in the 2010-11 academic year. Highlights of the summary report are listed in **Table 1.39**.

Have computer at home	96%
Unable to come to campus for classes	65%
Age 41 or over	24%
Would like LBCC to offer more DL courses	95%
Took DL classes due to a disability	17%
Live at least 16 miles from campus	13%
Took DL courses to fulfill transfer requirements	71%

Table 1.39 LBCC Distance Learning Survey Results

    a. What percent of the students surveyed do not have a computer at home?
    b. About how many students in the survey live at least 16 miles from campus?
    c. If the same survey were done at Great Basin College in Elko, Nevada, do you think the percentages would be the same? Why?

91. Several online textbook retailers advertise that they have lower prices than on-campus bookstores. However, an important factor is whether the Internet retailers actually have the textbooks that students need in stock. Students need to be able to get textbooks promptly at the beginning of the college term. If the book is not available, then a student would not be able to get the textbook at all, or might get a delayed delivery if the book is back ordered.

A college newspaper reporter is investigating textbook availability at online retailers. He decides to investigate one textbook for each of the following seven subjects: calculus, biology, chemistry, physics, statistics, geology, and general engineering. He consults textbook industry sales data and selects the most popular nationally used textbook in each of these subjects.

He visits websites for a random sample of major online textbook sellers and looks up each of these seven textbooks to see if they are available in stock for quick delivery through these retailers. Based on his investigation, he writes an article in which he draws conclusions about the overall availability of all college textbooks through online textbook retailers.

Write an analysis of his study that addresses the following issues: Is his sample representative of the population of all college textbooks? Explain why or why not. Describe some possible sources of bias in this study, and how it might affect the results of the study. Give some suggestions about what could be done to improve the study.

REFERENCES

1.1 Definitions of Statistics, Probability, and Key Terms

The Data and Story Library, http://lib.stat.cmu.edu/DASL/Stories/CrashTestDummies.html (accessed May 1, 2013).

1.2 Data, Sampling, and Variation in Data and Sampling

Gallup-Healthways Well-Being Index. http://www.well-beingindex.com/default.asp (accessed May 1, 2013).

Gallup-Healthways Well-Being Index. http://www.well-beingindex.com/methodology.asp (accessed May 1, 2013).

Gallup-Healthways Well-Being Index. http://www.gallup.com/poll/146822/gallup-healthways-index-questions.aspx (accessed May 1, 2013).

Data from http://www.bookofodds.com/Relationships-Society/Articles/A0374-How-George-Gallup-Picked-the-President

Dominic Lusinchi, "'President' Landon and the 1936 *Literary Digest* Poll: Were Automobile and Telephone Owners to Blame?" Social Science History 36, no. 1: 23-54 (2012), http://ssh.dukejournals.org/content/36/1/23.abstract (accessed May 1, 2013).

"The Literary Digest Poll," Virtual Laboratories in Probability and Statistics http://www.math.uah.edu/stat/data/LiteraryDigest.html (accessed May 1, 2013).

"Gallup Presidential Election Trial-Heat Trends, 1936–2008," Gallup Politics http://www.gallup.com/poll/110548/gallup-presidential-election-trialheat-trends-19362004.aspx#4 (accessed May 1, 2013).

The Data and Story Library, http://lib.stat.cmu.edu/DASL/Datafiles/USCrime.html (accessed May 1, 2013).

LBCC Distance Learning (DL) program data in 2010-2011, http://de.lbcc.edu/reports/2010-11/future/highlights.html#focus (accessed May 1, 2013).

Data from San Jose Mercury News

1.3 Frequency, Frequency Tables, and Levels of Measurement

"State & County QuickFacts," U.S. Census Bureau. http://quickfacts.census.gov/qfd/download_data.html (accessed May 1, 2013).

"State & County QuickFacts: Quick, easy access to facts about people, business, and geography," U.S. Census Bureau. http://quickfacts.census.gov/qfd/index.html (accessed May 1, 2013).

"Table 5: Direct hits by mainland United States Hurricanes (1851-2004)," National Hurricane Center, http://www.nhc.noaa.gov/gifs/table5.gif (accessed May 1, 2013).

"Levels of Measurement," http://infinity.cos.edu/faculty/woodbury/stats/tutorial/Data_Levels.htm (accessed May 1, 2013).

Courtney Taylor, "Levels of Measurement," about.com, http://statistics.about.com/od/HelpandTutorials/a/Levels-Of-Measurement.htm (accessed May 1, 2013).

David Lane. "Levels of Measurement," Connexions, http://cnx.org/content/m10809/latest/ (accessed May 1, 2013).

1.4 Experimental Design and Ethics

"Vitamin E and Health," Nutrition Source, Harvard School of Public Health, http://www.hsph.harvard.edu/nutritionsource/vitamin-e/ (accessed May 1, 2013).

Stan Reents. "Don't Underestimate the Power of Suggestion," athleteinme.com, http://www.athleteinme.com/ArticleView.aspx?id=1053 (accessed May 1, 2013).

Ankita Mehta. "Daily Dose of Aspiring Helps Reduce Heart Attacks: Study," International Business Times, July 21, 2011. Also available online at http://www.ibtimes.com/daily-dose-aspirin-helps-reduce-heart-attacks-study-300443 (accessed May 1, 2013).

The Data and Story Library, http://lib.stat.cmu.edu/DASL/Stories/ScentsandLearning.html (accessed May 1, 2013).

M.L. Jacskon et al., "Cognitive Components of Simulated Driving Performance: Sleep Loss effect and Predictors," Accident Analysis and Prevention Journal, Jan no. 50 (2013), http://www.ncbi.nlm.nih.gov/pubmed/22721550 (accessed May 1, 2013).

"Earthquake Information by Year," U.S. Geological Survey. http://earthquake.usgs.gov/earthquakes/eqarchives/year/ (accessed May 1, 2013).

"Fatality Analysis Report Systems (FARS) Encyclopedia," National Highway Traffic and Safety Administration. http://www-fars.nhtsa.dot.gov/Main/index.aspx (accessed May 1, 2013).

Data from www.businessweek.com (accessed May 1, 2013).

Data from www.forbes.com (accessed May 1, 2013).

"America's Best Small Companies," http://www.forbes.com/best-small-companies/list/ (accessed May 1, 2013).

U.S. Department of Health and Human Services, Code of Federal Regulations Title 45 Public Welfare Department of Health and Human Services Part 46 Protection of Human Subjects revised January 15, 2009. Section 46.111:Criteria for IRB Approval of Research.

"April 2013 Air Travel Consumer Report," U.S. Department of Transportation, April 11 (2013), http://www.dot.gov/airconsumer/april-2013-air-travel-consumer-report (accessed May 1, 2013).

Lori Alden, "Statistics can be Misleading," econoclass.com, http://www.econoclass.com/misleadingstats.html (accessed May 1, 2013).

Maria de los A. Medina, "Ethics in Statistics," Based on "Building an Ethics Module for Business, Science, and Engineering Students" by Jose A. Cruz-Cruz and William Frey, Connexions, http://cnx.org/content/m15555/latest/ (accessed May 1, 2013).

SOLUTIONS

1 AIDS patients.

3 The average length of time (in months) AIDS patients live after treatment.

5 X = the length of time (in months) AIDS patients live after treatment

7 b

9 a

11

a. 0.5242

b. 0.03%

c. 6.86%

d. $\frac{823,088}{823,856}$

e. quantitative discrete

f. quantitative continuous

g. In both years, underwater earthquakes produced massive tsunamis.

13 systematic

15 simple random

17 values for *X*, such as 3, 4, 11, and so on

19 No, we do not have enough information to make such a claim.

21 Take a simple random sample from each group. One way is by assigning a number to each patient and using a random number generator to randomly select patients.

23 This would be convenience sampling and is not random.

25 Yes, the sample size of 150 would be large enough to reflect a population of one school.

27 Even though the specific data support each researcher's conclusions, the different results suggest that more data need to be collected before the researchers can reach a conclusion.

29 There is not enough information given to judge if either one is correct or incorrect.

31 The software program seems to work because the second study shows that more patients improve while using the software than not. Even though the difference is not as large as that in the first study, the results from the second study are likely more reliable and still show improvement.

33 Yes, because we cannot tell if the improvement was due to the software or the exercise; the data is confounded, and a reliable conclusion cannot be drawn. New studies should be performed.

35 No, even though the sample is large enough, the fact that the sample consists of volunteers makes it a self-selected sample, which is not reliable.

37 No, even though the sample is a large portion of the population, two responses are not enough to justify any conclusions. Because the population is so small, it would be better to include everyone in the population to get the most accurate data.

39
   a. ordinal
   b. interval
   c. nominal
   d. nominal
   e. ratio
   f. ordinal
   g. nominal
   h. interval
   i. ratio
   j. interval
   k. ratio
   l. ordinal

41
   a. Inmates may not feel comfortable refusing participation, or may feel obligated to take advantage of the promised benefits. They may not feel truly free to refuse participation.
   b. Parents can provide consent on behalf of their children, but children are not competent to provide consent for themselves.
   c. All risks and benefits must be clearly outlined. Study participants must be informed of relevant aspects of the study in order to give appropriate consent.

43
   a. all children who take ski or snowboard lessons
   b. a group of these children
   c. the population mean age of children who take their first snowboard lesson
   d. the sample mean age of children who take their first snowboard lesson

e. X = the age of one child who takes his or her first ski or snowboard lesson

f.   values for X, such as 3, 7, and so on

45

a.   the clients of the insurance companies

b.   a group of the clients

c.   the mean health costs of the clients

d.   the mean health costs of the sample

e. X = the health costs of one client

f.   values for X, such as 34, 9, 82, and so on

47

a.   all the clients of this counselor

b.   a group of clients of this marriage counselor

c.   the proportion of all her clients who stay married

d.   the proportion of the sample of the counselor's clients who stay married

e. X = the number of couples who stay married

f.   yes, no

49

a.   all people (maybe in a certain geographic area, such as the United States)

b.   a group of the people

c.   the proportion of all people who will buy the product

d.   the proportion of the sample who will buy the product

e. X = the number of people who will buy it

f.   buy, not buy

51 a

53 quantitative discrete, 150

55 qualitative, Oakland A's

57 quantitative discrete, 11,234 students

59 qualitative, Crest

61 quantitative continuous, 47.3 years

63 b

65

a.   The survey was conducted using six similar flights.
The survey would not be a true representation of the entire population of air travelers.
Conducting the survey on a holiday weekend will not produce representative results.

b.   Conduct the survey during different times of the year.
Conduct the survey using flights to and from various locations.
Conduct the survey on different days of the week.

67 Answers will vary. Sample Answer: You could use a systematic sampling method. Stop the tenth person as they leave one of the buildings on campus at 9:50 in the morning. Then stop the tenth person as they leave a different building on campus at 1:50 in the afternoon.

69 Answers will vary. Sample Answer: Many people will not respond to mail surveys. If they do respond to the surveys, you can't be sure who is responding. In addition, mailing lists can be incomplete.

71 b

73 convenience; cluster; stratified ; systematic; simple random

75

a. qualitative

b. quantitative discrete

c. quantitative discrete

d. qualitative

77 Causality: The fact that two variables are related does not guarantee that one variable is influencing the other. We cannot assume that crime rate impacts education level or that education level impacts crime rate. Confounding: There are many factors that define a community other than education level and crime rate. Communities with high crime rates and high education levels may have other lurking variables that distinguish them from communities with lower crime rates and lower education levels. Because we cannot isolate these variables of interest, we cannot draw valid conclusions about the connection between education and crime. Possible lurking variables include police expenditures, unemployment levels, region, average age, and size.

79

a. Possible reasons: increased use of caller id, decreased use of landlines, increased use of private numbers, voice mail, privacy managers, hectic nature of personal schedules, decreased willingness to be interviewed

b. When a large number of people refuse to participate, then the sample may not have the same characteristics of the population. Perhaps the majority of people willing to participate are doing so because they feel strongly about the subject of the survey.

81

a.

# Flossing per Week	Frequency	Relative Frequency	Cumulative Relative Frequency
0	27	0.4500	0.4500
1	18	0.3000	0.7500
3	11	0.1833	0.9333
6	3	0.0500	0.9833
7	1	0.0167	1

Table 1.40

b. 5.00%

c. 93.33%

83 The sum of the travel times is 1,173.1. Divide the sum by 50 to calculate the mean value: 23.462. Because each state's travel time was measured to the nearest tenth, round this calculation to the nearest hundredth: 23.46.

85 b

87 Explanatory variable: amount of sleep
Response variable: performance measured in assigned tasks
Treatments: normal sleep and 27 hours of total sleep deprivation
Experimental Units: 19 professional drivers
Lurking variables: none – all drivers participated in both treatments
Random assignment: treatments were assigned in random order; this eliminated the effect of any "learning" that may take place during the first experimental session
Control/Placebo: completing the experimental session under normal sleep conditions
Blinding: researchers evaluating subjects' performance must not know which treatment is being applied at the time

89 You cannot assume that the numbers of complaints reflect the quality of the airlines. The airlines shown with the greatest number of complaints are the ones with the most passengers. You must consider the appropriateness of methods for presenting data; in this case displaying totals is misleading.

91 Answers will vary. Sample answer: The sample is not representative of the population of all college textbooks. Two reasons why it is not representative are that he only sampled seven subjects and he only investigated one textbook in each subject. There are several possible sources of bias in the study. The seven subjects that he investigated are all in mathematics and the sciences; there are many subjects in the humanities, social sciences, and other subject areas, (for example: literature, art, history, psychology, sociology, business) that he did not investigate at all. It may be that different subject areas exhibit different patterns of textbook availability, but his sample would not detect such results. He also looked only at the most popular textbook in each of the subjects he investigated. The availability of the most popular textbooks may differ from the availability of other textbooks in one of two ways:

- the most popular textbooks may be more readily available online, because more new copies are printed, and more students nationwide are selling back their used copies OR

- the most popular textbooks may be harder to find available online, because more student demand exhausts the supply more quickly.

In reality, many college students do not use the most popular textbook in their subject, and this study gives no useful information about the situation for those less popular textbooks. He could improve this study by:

- expanding the selection of subjects he investigates so that it is more representative of all subjects studied by college students, and

- expanding the selection of textbooks he investigates within each subject to include a mixed representation of both the most popular and less popular textbooks.

2 | DESCRIPTIVE STATISTICS

Figure 2.1 When you have large amounts of data, you will need to organize it in a way that makes sense. These ballots from an election are rolled together with similar ballots to keep them organized. (credit: William Greeson)

Introduction

Chapter Objectives

By the end of this chapter, the student should be able to:

- Display data graphically and interpret graphs: stemplots, histograms, and box plots.
- Recognize, describe, and calculate the measures of location of data: quartiles and percentiles.
- Recognize, describe, and calculate the measures of the center of data: mean, median, and mode.
- Recognize, describe, and calculate the measures of the spread of data: variance, standard deviation, and range.

Once you have collected data, what will you do with it? Data can be described and presented in many different formats. For example, suppose you are interested in buying a house in a particular area. You may have no clue about the house prices, so you might ask your real estate agent to give you a sample data set of prices. Looking at all the prices in the sample often is overwhelming. A better way might be to look at the median price and the variation of prices. The median and variation are just two ways that you will learn to describe data. Your agent might also provide you with a graph of the data.

In this chapter, you will study numerical and graphical ways to describe and display your data. This area of statistics is called **"Descriptive Statistics."** You will learn how to calculate, and even more importantly, how to interpret these measurements and graphs.

A statistical graph is a tool that helps you learn about the shape or distribution of a sample or a population. A graph can be a more effective way of presenting data than a mass of numbers because we can see where data clusters and where there are only a few data values. Newspapers and the Internet use graphs to show trends and to enable readers to compare facts and figures quickly. Statisticians often graph data first to get a picture of the data. Then, more formal tools may be applied.

Some of the types of graphs that are used to summarize and organize data are the dot plot, the bar graph, the histogram, the stem-and-leaf plot, the frequency polygon (a type of broken line graph), the pie chart, and the box plot. In this chapter, we will briefly look at stem-and-leaf plots, line graphs, and bar graphs, as well as frequency polygons, and time series graphs. Our emphasis will be on histograms and box plots.

NOTE

This book contains instructions for constructing a histogram and a box plot for the TI-83+ and TI-84 calculators. The **Texas Instruments (TI) website (http://education.ti.com/educationportal/sites/US/sectionHome/ support.html)** provides additional instructions for using these calculators.

2.1 | Stem-and-Leaf Graphs (Stemplots), Line Graphs, and Bar Graphs

One simple graph, the **stem-and-leaf graph** or **stemplot**, comes from the field of exploratory data analysis. It is a good choice when the data sets are small. To create the plot, divide each observation of data into a stem and a leaf. The leaf consists of a **final significant digit**. For example, 23 has stem two and leaf three. The number 432 has stem 43 and leaf two. Likewise, the number 5,432 has stem 543 and leaf two. The decimal 9.3 has stem nine and leaf three. Write the stems in a vertical line from smallest to largest. Draw a vertical line to the right of the stems. Then write the leaves in increasing order next to their corresponding stem.

Example 2.1

For Susan Dean's spring pre-calculus class, scores for the first exam were as follows (smallest to largest):
33; 42; 49; 49; 53; 55; 55; 61; 63; 67; 68; 68; 69; 69; 72; 73; 74; 78; 80; 83; 88; 88; 88; 90; 92; 94; 94; 94; 94; 96; 100

Stem	Leaf
3	3
4	2 9 9
5	3 5 5
6	1 3 7 8 8 9 9
7	2 3 4 8
8	0 3 8 8 8
9	0 2 4 4 4 4 6
10	0

Table 2.1 Stem-and-Leaf Graph

The stemplot shows that most scores fell in the 60s, 70s, 80s, and 90s. Eight out of the 31 scores or approximately 26% $\left(\frac{8}{31}\right)$ were in the 90s or 100, a fairly high number of As.

Try It Σ

2.1 For the Park City basketball team, scores for the last 30 games were as follows (smallest to largest):
32; 32; 33; 34; 38; 40; 42; 42; 43; 44; 46; 47; 47; 48; 48; 48; 49; 50; 50; 51; 52; 52; 52; 53; 54; 56; 57; 57; 60; 61
Construct a stem plot for the data.

The stemplot is a quick way to graph data and gives an exact picture of the data. You want to look for an overall pattern and any outliers. An **outlier** is an observation of data that does not fit the rest of the data. It is sometimes called an **extreme value.** When you graph an outlier, it will appear not to fit the pattern of the graph. Some outliers are due to mistakes (for example, writing down 50 instead of 500) while others may indicate that something unusual is happening. It takes some background information to explain outliers, so we will cover them in more detail later.

Example 2.2

The data are the distances (in kilometers) from a home to local supermarkets. Create a stemplot using the data:
1.1; 1.5; 2.3; 2.5; 2.7; 3.2; 3.3; 3.3; 3.5; 3.8; 4.0; 4.2; 4.5; 4.5; 4.7; 4.8; 5.5; 5.6; 6.5; 6.7; 12.3

Do the data seem to have any concentration of values?

The leaves are to the right of the decimal.

Solution 2.2

The value 12.3 may be an outlier. Values appear to concentrate at three and four kilometers.

Stem	Leaf
1	1 5
2	3 5 7
3	2 3 3 5 8
4	0 2 5 5 7 8
5	5 6
6	5 7
7	
8	
9	
10	
11	
12	3

Table 2.2

Try It Σ

2.2 The following data show the distances (in miles) from the homes of off-campus statistics students to the college. Create a stem plot using the data and identify any outliers:

0.5; 0.7; 1.1; 1.2; 1.2; 1.3; 1.3; 1.5; 1.5; 1.7; 1.7; 1.8; 1.9; 2.0; 2.2; 2.5; 2.6; 2.8; 2.8; 2.8; 3.5; 3.8; 4.4; 4.8; 4.9; 5.2; 5.5; 5.7; 5.8; 8.0

Example 2.3

A **side-by-side stem-and-leaf plot** allows a comparison of the two data sets in two columns. In a side-by-side stem-and-leaf plot, two sets of leaves share the same stem. The leaves are to the left and the right of the stems. **Table 2.4** and **Table 2.5** show the ages of presidents at their inauguration and at their death. Construct a side-by-side stem-and-leaf plot using this data.

Solution 2.3

Ages at Inauguration		Ages at Death
9 9 8 7 7 7 6 3 2	4	6 9
8 7 7 7 7 6 6 6 5 5 5 5 4 4 4 4 4 2 1 1 1 1 1 0	5	3 6 6 7 7 8
9 5 4 4 2 1 1 1 0	6	0 0 3 3 4 4 5 6 7 7 7 8
	7	0 0 1 1 1 4 7 8 8 9
	8	0 1 3 5 8
	9	0 0 3 3

Table 2.3

President	Age	President	Age	President	Age
Washington	57	Lincoln	52	Hoover	54
J. Adams	61	A. Johnson	56	F. Roosevelt	51
Jefferson	57	Grant	46	Truman	60
Madison	57	Hayes	54	Eisenhower	62
Monroe	58	Garfield	49	Kennedy	43
J. Q. Adams	57	Arthur	51	L. Johnson	55
Jackson	61	Cleveland	47	Nixon	56
Van Buren	54	B. Harrison	55	Ford	61
W. H. Harrison	68	Cleveland	55	Carter	52
Tyler	51	McKinley	54	Reagan	69
Polk	49	T. Roosevelt	42	G.H.W. Bush	64
Taylor	64	Taft	51	Clinton	47
Fillmore	50	Wilson	56	G. W. Bush	54
Pierce	48	Harding	55	Obama	47
Buchanan	65	Coolidge	51		

Table 2.4 Presidential Ages at Inauguration

President	Age	President	Age	President	Age
Washington	67	Lincoln	56	Hoover	90

Table 2.5 Presidential Age at Death

President	Age	President	Age	President	Age
J. Adams	90	A. Johnson	66	F. Roosevelt	63
Jefferson	83	Grant	63	Truman	88
Madison	85	Hayes	70	Eisenhower	78
Monroe	73	Garfield	49	Kennedy	46
J. Q. Adams	80	Arthur	56	L. Johnson	64
Jackson	78	Cleveland	71	Nixon	81
Van Buren	79	B. Harrison	67	Ford	93
W. H. Harrison	68	Cleveland	71	Reagan	93
Tyler	71	McKinley	58		
Polk	53	T. Roosevelt	60		
Taylor	65	Taft	72		
Fillmore	74	Wilson	67		
Pierce	64	Harding	57		
Buchanan	77	Coolidge	60		

Table 2.5 Presidential Age at Death

2.3 The table shows the number of wins and losses the Atlanta Hawks have had in 42 seasons. Create a side-by-side stem-and-leaf plot of these wins and losses.

Losses	Wins	Year	Losses	Wins	Year
34	48	1968–1969	41	41	1989–1990
34	48	1969–1970	39	43	1990–1991
46	36	1970–1971	44	38	1991–1992
46	36	1971–1972	39	43	1992–1993
36	46	1972–1973	25	57	1993–1994
47	35	1973–1974	40	42	1994–1995
51	31	1974–1975	36	46	1995–1996
53	29	1975–1976	26	56	1996–1997
51	31	1976–1977	32	50	1997–1998
41	41	1977–1978	19	31	1998–1999
36	46	1978–1979	54	28	1999–2000
32	50	1979–1980	57	25	2000–2001
51	31	1980–1981	49	33	2001–2002
40	42	1981–1982	47	35	2002–2003

Table 2.6

Losses	Wins	Year	Losses	Wins	Year
39	43	1982–1983	54	28	2003–2004
42	40	1983–1984	69	13	2004–2005
48	34	1984–1985	56	26	2005–2006
32	50	1985–1986	52	30	2006–2007
25	57	1986–1987	45	37	2007–2008
32	50	1987–1988	35	47	2008–2009
30	52	1988–1989	29	53	2009–2010

Table 2.6

Another type of graph that is useful for specific data values is a **line graph**. In the particular line graph shown in **Example 2.4**, the *x*-axis (horizontal axis) consists of **data values** and the *y*-axis (vertical axis) consists of **frequency points**. The frequency points are connected using line segments.

Example 2.4

In a survey, 40 mothers were asked how many times per week a teenager must be reminded to do his or her chores. The results are shown in **Table 2.7** and in **Figure 2.2**.

Number of times teenager is reminded	Frequency
0	2
1	5
2	8
3	14
4	7
5	4

Table 2.7

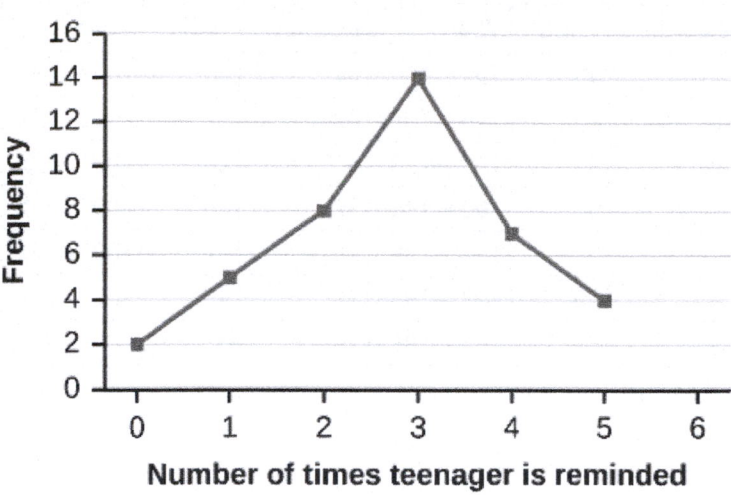

Figure 2.2

2.4 In a survey, 40 people were asked how many times per year they had their car in the shop for repairs. The results are shown in **Table 2.8**. Construct a line graph.

Number of times in shop	Frequency
0	7
1	10
2	14
3	9

Table 2.8

Bar graphs consist of bars that are separated from each other. The bars can be rectangles or they can be rectangular boxes (used in three-dimensional plots), and they can be vertical or horizontal. The **bar graph** shown in **Example 2.5** has age groups represented on the *x*-axis and proportions on the *y*-axis.

Example 2.5

By the end of 2011, Facebook had over 146 million users in the United States. **Table 2.8** shows three age groups, the number of users in each age group, and the proportion (%) of users in each age group. Construct a bar graph using this data.

Age groups	Number of Facebook users	Proportion (%) of Facebook users
13–25	65,082,280	45%
26–44	53,300,200	36%
45–64	27,885,100	19%

Table 2.9

Solution 2.5

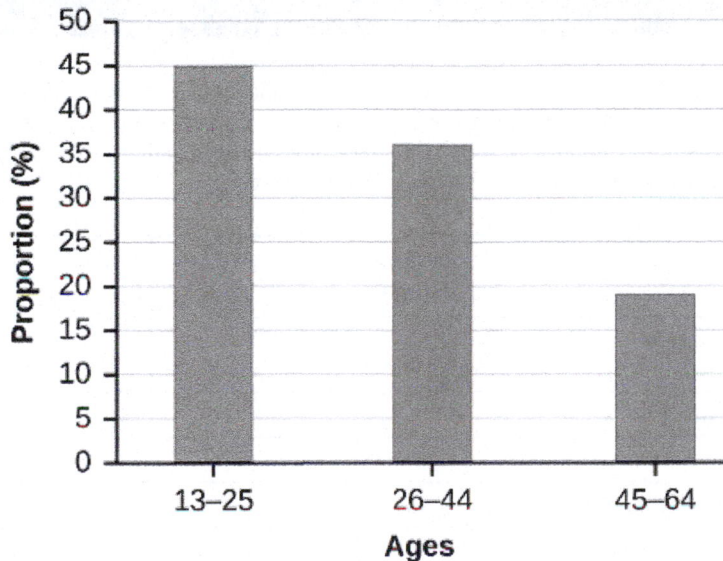

Figure 2.3

2.5 The population in Park City is made up of children, working-age adults, and retirees. **Table 2.10** shows the three age groups, the number of people in the town from each age group, and the proportion (%) of people in each age group. Construct a bar graph showing the proportions.

Age groups	Number of people	Proportion of population
Children	67,059	19%
Working-age adults	152,198	43%
Retirees	131,662	38%

Table 2.10

Example 2.6

The columns in **Table 2.10** contain: the race or ethnicity of students in U.S. Public Schools for the class of 2011, percentages for the Advanced Placement examine population for that class, and percentages for the overall student population. Create a bar graph with the student race or ethnicity (qualitative data) on the *x*-axis, and the Advanced Placement examinee population percentages on the *y*-axis.

Race/Ethnicity	AP Examinee Population	Overall Student Population
1 = Asian, Asian American or Pacific Islander	10.3%	5.7%
2 = Black or African American	9.0%	14.7%
3 = Hispanic or Latino	17.0%	17.6%
4 = American Indian or Alaska Native	0.6%	1.1%
5 = White	57.1%	59.2%
6 = Not reported/other	6.0%	1.7%

Table 2.11

Solution 2.6

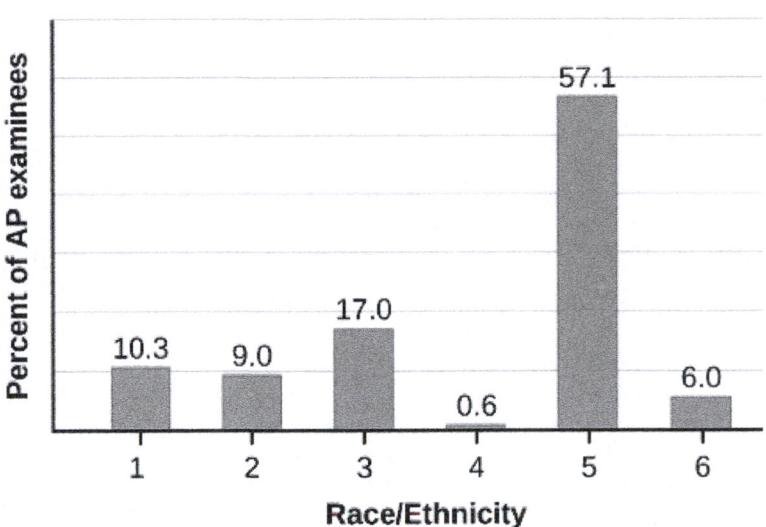

Figure 2.4

2.6 Park city is broken down into six voting districts. The table shows the percent of the total registered voter population that lives in each district as well as the percent total of the entire population that lives in each district. Construct a bar graph that shows the registered voter population by district.

District	Registered voter population	Overall city population
1	15.5%	19.4%
2	12.2%	15.6%
3	9.8%	9.0%
4	17.4%	18.5%
5	22.8%	20.7%
6	22.3%	16.8%

Table 2.12

2.2 | Histograms, Frequency Polygons, and Time Series Graphs

For most of the work you do in this book, you will use a histogram to display the data. One advantage of a histogram is that it can readily display large data sets. A rule of thumb is to use a histogram when the data set consists of 100 values or more.

A **histogram** consists of contiguous (adjoining) boxes. It has both a horizontal axis and a vertical axis. The horizontal axis is labeled with what the data represents (for instance, distance from your home to school). The vertical axis is labeled either **frequency** or **relative frequency** (or percent frequency or probability). The graph will have the same shape with either label. The histogram (like the stemplot) can give you the shape of the data, the center, and the spread of the data.

The relative frequency is equal to the frequency for an observed value of the data divided by the total number of data values in the sample.(Remember, frequency is defined as the number of times an answer occurs.) If:

- f = frequency
- n = total number of data values (or the sum of the individual frequencies), and
- RF = relative frequency,

then:

$$RF = \frac{f}{n}$$

For example, if three students in Mr. Ahab's English class of 40 students received from 90% to 100%, then, f = 3, n = 40, and $RF = \frac{f}{n} = \frac{3}{40}$ = 0.075. 7.5% of the students received 90–100%. 90–100% are quantitative measures.

To construct a histogram, first decide how many **bars** or **intervals**, also called classes, represent the data. Many histograms consist of five to 15 bars or classes for clarity. The number of bars needs to be chosen. Choose a starting point for the first interval to be less than the smallest data value. A **convenient starting point** is a lower value carried out to one more decimal place than the value with the most decimal places. For example, if the value with the most decimal places is 6.1 and this is the smallest value, a convenient starting point is 6.05 (6.1 – 0.05 = 6.05). We say that 6.05 has more precision. If the value with the most decimal places is 2.23 and the lowest value is 1.5, a convenient starting point is 1.495 (1.5 – 0.005 = 1.495). If the value with the most decimal places is 3.234 and the lowest value is 1.0, a convenient starting point is 0.9995 (1.0 – 0.0005 = 0.9995). If all the data happen to be integers and the smallest value is two, then a convenient starting point is 1.5 (2 – 0.5 = 1.5). Also, when the starting point and other boundaries are carried to one additional decimal place, no data value will fall on a boundary. The next two examples go into detail about how to construct a histogram using continuous data and how to create a histogram using discrete data.

Example 2.7

The following data are the heights (in inches to the nearest half inch) of 100 male semiprofessional soccer players. The heights are **continuous** data, since height is measured.
60; 60.5; 61; 61; 61.5
63.5; 63.5; 63.5
64; 64; 64; 64; 64; 64; 64; 64.5; 64.5; 64.5; 64.5; 64.5; 64.5; 64.5; 64.5
66; 66; 66; 66; 66; 66; 66; 66; 66; 66; 66.5; 66.5; 66.5; 66.5; 66.5; 66.5; 66.5; 66.5; 66.5; 66.5; 66.5; 67; 67; 67; 67; 67; 67; 67; 67; 67; 67; 67; 67; 67.5; 67.5; 67.5; 67.5; 67.5; 67.5; 67.5
68; 68; 69; 69; 69; 69; 69; 69; 69; 69; 69; 69.5; 69.5; 69.5; 69.5; 69.5
70; 70; 70; 70; 70; 70; 70.5; 70.5; 70.5; 71; 71; 71
72; 72; 72; 72.5; 72.5; 73; 73.5
74

The smallest data value is 60. Since the data with the most decimal places has one decimal (for instance, 61.5), we want our starting point to have two decimal places. Since the numbers 0.5, 0.05, 0.005, etc. are convenient numbers, use 0.05 and subtract it from 60, the smallest value, for the convenient starting point.

60 – 0.05 = 59.95 which is more precise than, say, 61.5 by one decimal place. The starting point is, then, 59.95.

The largest value is 74, so 74 + 0.05 = 74.05 is the ending value.

Next, calculate the width of each bar or class interval. To calculate this width, subtract the starting point from the ending value and divide by the number of bars (you must choose the number of bars you desire). Suppose you choose eight bars.

$$\frac{74.05 - 59.95}{8} = 1.76$$

NOTE

We will round up to two and make each bar or class interval two units wide. Rounding up to two is one way to prevent a value from falling on a boundary. Rounding to the next number is often necessary even if it goes against the standard rules of rounding. For this example, using 1.76 as the width would also work. A guideline that is followed by some for the width of a bar or class interval is to take the square root of the number of data values and then round to the nearest whole number, if necessary. For example, if there are 150 values of data, take the square root of 150 and round to 12 bars or intervals.

The boundaries are:

- 59.95
- 59.95 + 2 = 61.95
- 61.95 + 2 = 63.95
- 63.95 + 2 = 65.95
- 65.95 + 2 = 67.95
- 67.95 + 2 = 69.95
- 69.95 + 2 = 71.95
- 71.95 + 2 = 73.95
- 73.95 + 2 = 75.95

The heights 60 through 61.5 inches are in the interval 59.95–61.95. The heights that are 63.5 are in the interval 61.95–63.95. The heights that are 64 through 64.5 are in the interval 63.95–65.95. The heights 66 through 67.5 are in the interval 65.95–67.95. The heights 68 through 69.5 are in the interval 67.95–69.95. The heights 70 through 71 are in the interval 69.95–71.95. The heights 72 through 73.5 are in the interval 71.95–73.95. The height 74 is in the interval 73.95–75.95.

The following histogram displays the heights on the x-axis and relative frequency on the y-axis.

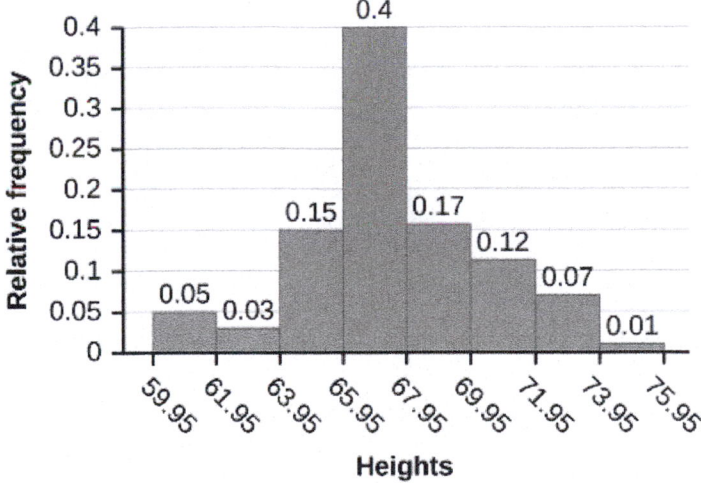

Figure 2.5

Try It Σ

2.7 The following data are the shoe sizes of 50 male students. The sizes are continuous data since shoe size is measured. Construct a histogram and calculate the width of each bar or class interval. Suppose you choose six bars.
9; 9; 9.5; 9.5; 10; 10; 10; 10; 10; 10; 10.5; 10.5; 10.5; 10.5; 10.5; 10.5; 10.5; 10.5

11; 11; 11; 11; 11; 11; 11; 11; 11; 11; 11; 11; 11; 11.5; 11.5; 11.5; 11.5; 11.5; 11.5; 11.5
12; 12; 12; 12; 12; 12; 12; 12.5; 12.5; 12.5; 12.5; 14

Example 2.8

The following data are the number of books bought by 50 part-time college students at ABC College. The number of books is **discrete data**, since books are counted.

1; 1; 1; 1; 1; 1; 1; 1; 1; 1; 1

2; 2; 2; 2; 2; 2; 2; 2; 2; 2

3; 3; 3; 3; 3; 3; 3; 3; 3; 3; 3; 3; 3; 3; 3; 3

4; 4; 4; 4; 4; 4

5; 5; 5; 5; 5

6; 6

Eleven students buy one book. Ten students buy two books. Sixteen students buy three books. Six students buy four books. Five students buy five books. Two students buy six books.

Because the data are integers, subtract 0.5 from 1, the smallest data value and add 0.5 to 6, the largest data value. Then the starting point is 0.5 and the ending value is 6.5.

Next, calculate the width of each bar or class interval. If the data are discrete and there are not too many different values, a width that places the data values in the middle of the bar or class interval is the most convenient. Since the data consist of the numbers 1, 2, 3, 4, 5, 6, and the starting point is 0.5, a width of one places the 1 in the middle of the interval from 0.5 to 1.5, the 2 in the middle of the interval from 1.5 to 2.5, the 3 in the middle of the interval from 2.5 to 3.5, the 4 in the middle of the interval from _____ to _____, the 5 in the middle of the interval from _____ to _____, and the _____ in the middle of the interval from _____ to _____ .

Solution 2.8

- 3.5 to 4.5

- 4.5 to 5.5

- 6

- 5.5 to 6.5

Calculate the number of bars as follows:

$$\frac{6.5 - 0.5}{\text{number of bars}} = 1$$

where 1 is the width of a bar. Therefore, bars = 6.

The following histogram displays the number of books on the x-axis and the frequency on the y-axis.

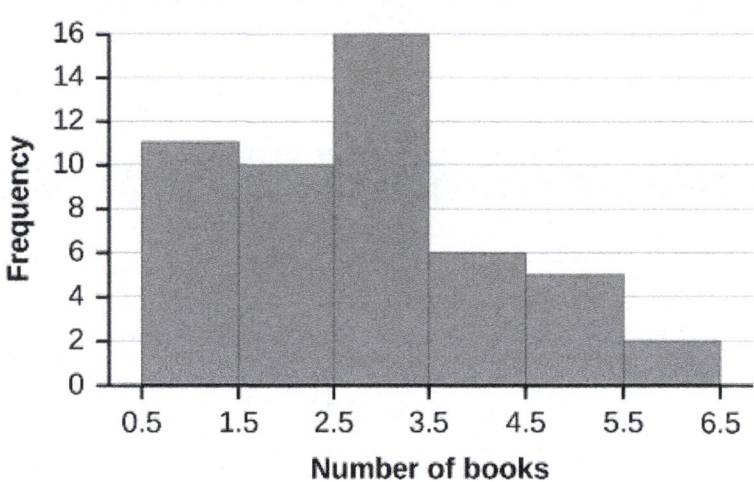

Figure 2.6

Using the TI-83, 83+, 84, 84+ Calculator

Go to **Appendix G**. There are calculator instructions for entering data and for creating a customized histogram. Create the histogram for **Example 2.8**.

- Press Y=. Press CLEAR to delete any equations.
- Press STAT 1:EDIT. If L1 has data in it, arrow up into the name L1, press CLEAR and then arrow down. If necessary, do the same for L2.
- Into L1, enter 1, 2, 3, 4, 5, 6.
- Into L2, enter 11, 10, 16, 6, 5, 2.
- Press WINDOW. Set Xmin = .5, Xscl = (6.5 – .5)/6, Ymin = –1, Ymax = 20, Yscl = 1, Xres = 1.
- Press 2nd Y=. Start by pressing 4:Plotsoff ENTER.
- Press 2nd Y=. Press 1:Plot1. Press ENTER. Arrow down to TYPE. Arrow to the 3rd picture (histogram). Press ENTER.
- Arrow down to Xlist: Enter L1 (2nd 1). Arrow down to Freq. Enter L2 (2nd 2).
- Press GRAPH.
- Use the TRACE key and the arrow keys to examine the histogram.

Try It Σ

2.8 The following data are the number of sports played by 50 student athletes. The number of sports is discrete data since sports are counted.

1; 1; 1; 1; 1; 1; 1; 1; 1; 1; 1; 1; 1; 1; 1; 1; 1; 1; 1; 1
2; 2; 2; 2; 2; 2; 2; 2; 2; 2; 2; 2; 2; 2; 2; 2; 2; 2; 2; 2; 2; 2
3; 3; 3; 3; 3; 3; 3; 3
20 student athletes play one sport. 22 student athletes play two sports. Eight student athletes play three sports.

Fill in the blanks for the following sentence. Since the data consist of the numbers 1, 2, 3, and the starting point is 0.5, a width of one places the 1 in the middle of the interval 0.5 to _____, the 2 in the middle of the interval from _____ to _____, and the 3 in the middle of the interval from _____ to _____.

Example 2.9

Using this data set, construct a histogram.

Number of Hours My Classmates Spent Playing Video Games on Weekends				
9.95	10	2.25	16.75	0
19.5	22.5	7.5	15	12.75
5.5	11	10	20.75	17.5
23	21.9	24	23.75	18
20	15	22.9	18.8	20.5

Table 2.13

Solution 2.9

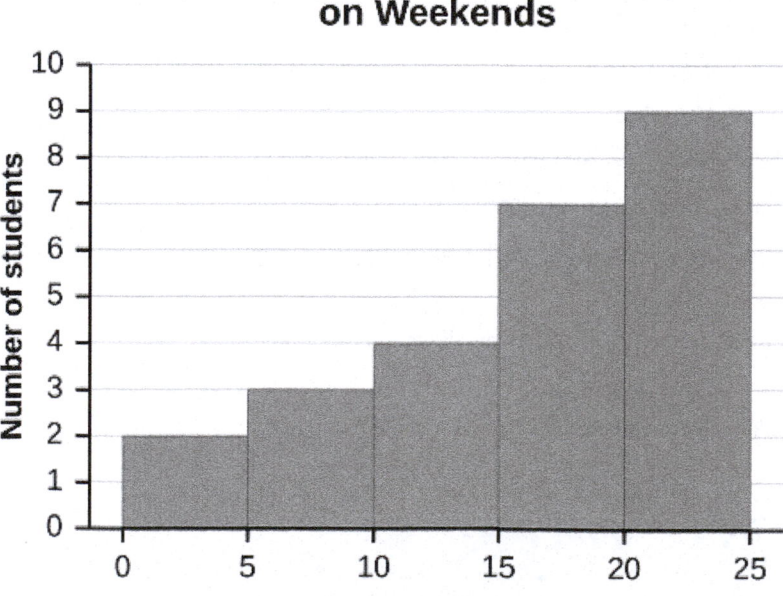

Figure 2.7

Some values in this data set fall on boundaries for the class intervals. A value is counted in a class interval if it falls on the left boundary, but not if it falls on the right boundary. Different researchers may set up histograms for the same data in different ways. There is more than one correct way to set up a histogram.

Try It Σ

2.9 The following data represent the number of employees at various restaurants in New York City. Using this data, create a histogram.

22; 35; 15; 26; 40; 28; 18; 20; 25; 34; 39; 42; 24; 22; 19; 27; 22; 34; 40; 20; 38; and 28
Use 10–19 as the first interval.

Collaborative Exercise

Count the money (bills and change) in your pocket or purse. Your instructor will record the amounts. As a class, construct a histogram displaying the data. Discuss how many intervals you think is appropriate. You may want to experiment with the number of intervals.

Frequency Polygons

Frequency polygons are analogous to line graphs, and just as line graphs make continuous data visually easy to interpret, so too do frequency polygons.

To construct a frequency polygon, first examine the data and decide on the number of intervals, or class intervals, to use on the x-axis and y-axis. After choosing the appropriate ranges, begin plotting the data points. After all the points are plotted, draw line segments to connect them.

Example 2.10

A frequency polygon was constructed from the frequency table below.

Frequency Distribution for Calculus Final Test Scores			
Lower Bound	**Upper Bound**	**Frequency**	**Cumulative Frequency**
49.5	59.5	5	5
59.5	69.5	10	15
69.5	79.5	30	45
79.5	89.5	40	85
89.5	99.5	15	100

Table 2.14

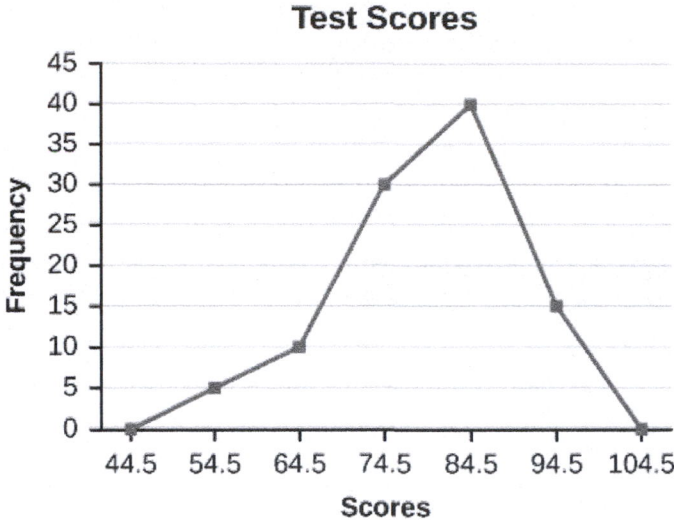

Figure 2.8

The first label on the *x*-axis is 44.5. This represents an interval extending from 39.5 to 49.5. Since the lowest test score is 54.5, this interval is used only to allow the graph to touch the *x*-axis. The point labeled 54.5 represents the next interval, or the first "real" interval from the table, and contains five scores. This reasoning is followed for each of the remaining intervals with the point 104.5 representing the interval from 99.5 to 109.5. Again, this interval contains no data and is only used so that the graph will touch the *x*-axis. Looking at the graph, we say that this distribution is skewed because one side of the graph does not mirror the other side.

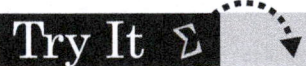

2.10 Construct a frequency polygon of U.S. Presidents' ages at inauguration shown in **Table 2.15**.

Age at Inauguration	Frequency
41.5–46.5	4
46.5–51.5	11
51.5–56.5	14
56.5–61.5	9
61.5–66.5	4
66.5–71.5	2

Table 2.15

Frequency polygons are useful for comparing distributions. This is achieved by overlaying the frequency polygons drawn for different data sets.

Example 2.11

We will construct an overlay frequency polygon comparing the scores from **Example 2.10** with the students' final numeric grade.

Frequency Distribution for Calculus Final Test Scores			
Lower Bound	Upper Bound	Frequency	Cumulative Frequency
49.5	59.5	5	5
59.5	69.5	10	15
69.5	79.5	30	45
79.5	89.5	40	85
89.5	99.5	15	100

Table 2.16

Frequency Distribution for Calculus Final Grades			
Lower Bound	Upper Bound	Frequency	Cumulative Frequency
49.5	59.5	10	10
59.5	69.5	10	20
69.5	79.5	30	50
79.5	89.5	45	95
89.5	99.5	5	100

Table 2.17

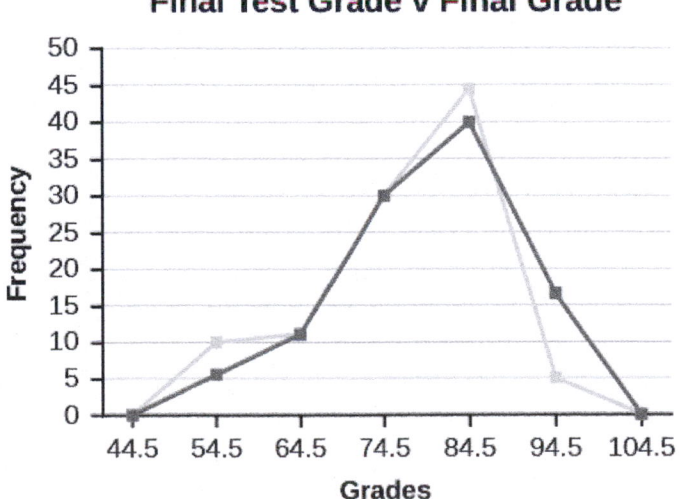

Figure 2.9

Suppose that we want to study the temperature range of a region for an entire month. Every day at noon we note the temperature and write this down in a log. A variety of statistical studies could be done with this data. We could find the mean or the median temperature for the month. We could construct a histogram displaying the number of days that temperatures reach a certain range of values. However, all of these methods ignore a portion of the data that we have collected.

One feature of the data that we may want to consider is that of time. Since each date is paired with the temperature reading for the day, we don't have to think of the data as being random. We can instead use the times given to impose a chronological order on the data. A graph that recognizes this ordering and displays the changing temperature as the month progresses is called a time series graph.

Constructing a Time Series Graph

To construct a time series graph, we must look at both pieces of our **paired data set**. We start with a standard Cartesian coordinate system. The horizontal axis is used to plot the date or time increments, and the vertical axis is used to plot the values of the variable that we are measuring. By doing this, we make each point on the graph correspond to a date and a measured quantity. The points on the graph are typically connected by straight lines in the order in which they occur.

Example 2.12

The following data shows the Annual Consumer Price Index, each month, for ten years. Construct a time series graph for the Annual Consumer Price Index data only.

Year	Jan	Feb	Mar	Apr	May	Jun	Jul
2003	181.7	183.1	184.2	183.8	183.5	183.7	183.9
2004	185.2	186.2	187.4	188.0	189.1	189.7	189.4
2005	190.7	191.8	193.3	194.6	194.4	194.5	195.4
2006	198.3	198.7	199.8	201.5	202.5	202.9	203.5
2007	202.416	203.499	205.352	206.686	207.949	208.352	208.299
2008	211.080	211.693	213.528	214.823	216.632	218.815	219.964
2009	211.143	212.193	212.709	213.240	213.856	215.693	215.351
2010	216.687	216.741	217.631	218.009	218.178	217.965	218.011
2011	220.223	221.309	223.467	224.906	225.964	225.722	225.922
2012	226.665	227.663	229.392	230.085	229.815	229.478	229.104

Table 2.18

Year	Aug	Sep	Oct	Nov	Dec	Annual
2003	184.6	185.2	185.0	184.5	184.3	184.0
2004	189.5	189.9	190.9	191.0	190.3	188.9
2005	196.4	198.8	199.2	197.6	196.8	195.3
2006	203.9	202.9	201.8	201.5	201.8	201.6
2007	207.917	208.490	208.936	210.177	210.036	207.342
2008	219.086	218.783	216.573	212.425	210.228	215.303
2009	215.834	215.969	216.177	216.330	215.949	214.537
2010	218.312	218.439	218.711	218.803	219.179	218.056
2011	226.545	226.889	226.421	226.230	225.672	224.939
2012	230.379	231.407	231.317	230.221	229.601	229.594

Table 2.19

Solution 2.12

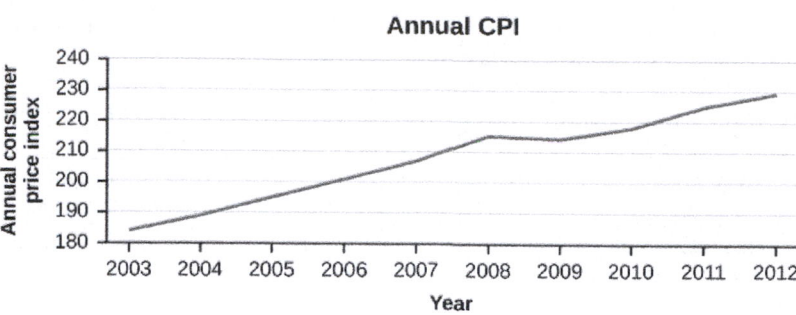

Figure 2.10

Try It Σ

2.12 The following table is a portion of a data set from www.worldbank.org. Use the table to construct a time series graph for CO_2 emissions for the United States.

CO2 Emissions			
	Ukraine	**United Kingdom**	**United States**
2003	352,259	540,640	5,681,664
2004	343,121	540,409	5,790,761
2005	339,029	541,990	5,826,394
2006	327,797	542,045	5,737,615
2007	328,357	528,631	5,828,697
2008	323,657	522,247	5,656,839
2009	272,176	474,579	5,299,563

Table 2.20

Uses of a Time Series Graph

Time series graphs are important tools in various applications of statistics. When recording values of the same variable over an extended period of time, sometimes it is difficult to discern any trend or pattern. However, once the same data points are displayed graphically, some features jump out. Time series graphs make trends easy to spot.

2.3 | Measures of the Location of the Data

The common measures of location are **quartiles** and **percentiles**

Quartiles are special percentiles. The first quartile, Q_1, is the same as the 25th percentile, and the third quartile, Q_3, is the same as the 75th percentile. The median, M, is called both the second quartile and the 50th percentile.

To calculate quartiles and percentiles, the data must be ordered from smallest to largest. Quartiles divide ordered data into quarters. Percentiles divide ordered data into hundredths. To score in the 90th percentile of an exam does not mean, necessarily, that you received 90% on a test. It means that 90% of test scores are the same or less than your score and 10% of the test scores are the same or greater than your test score.

Percentiles are useful for comparing values. For this reason, universities and colleges use percentiles extensively. One instance in which colleges and universities use percentiles is when SAT results are used to determine a minimum testing score that will be used as an acceptance factor. For example, suppose Duke accepts SAT scores at or above the 75th percentile. That translates into a score of at least 1220.

Percentiles are mostly used with very large populations. Therefore, if you were to say that 90% of the test scores are less (and not the same or less) than your score, it would be acceptable because removing one particular data value is not significant.

The **median** is a number that measures the "center" of the data. You can think of the median as the "middle value," but it does not actually have to be one of the observed values. It is a number that separates ordered data into halves. Half the values are the same number or smaller than the median, and half the values are the same number or larger. For example, consider the following data.
1; 11.5; 6; 7.2; 4; 8; 9; 10; 6.8; 8.3; 2; 2; 10; 1
Ordered from smallest to largest:
1; 1; 2; 2; 4; 6; 6.8; 7.2; 8; 8.3; 9; 10; 10; 11.5

Since there are 14 observations, the median is between the seventh value, 6.8, and the eighth value, 7.2. To find the median, add the two values together and divide by two.

$$\frac{6.8 + 7.2}{2} = 7$$

The median is seven. Half of the values are smaller than seven and half of the values are larger than seven.

Quartiles are numbers that separate the data into quarters. Quartiles may or may not be part of the data. To find the quartiles, first find the median or second quartile. The first quartile, Q_1, is the middle value of the lower half of the data, and the third quartile, Q_3, is the middle value, or median, of the upper half of the data. To get the idea, consider the same data set:
1; 1; 2; 2; 4; 6; 6.8; 7.2; 8; 8.3; 9; 10; 10; 11.5

The median or **second quartile** is seven. The lower half of the data are 1, 1, 2, 2, 4, 6, 6.8. The middle value of the lower half is two.
1; 1; 2; 2; 4; 6; 6.8

The number two, which is part of the data, is the **first quartile**. One-fourth of the entire sets of values are the same as or less than two and three-fourths of the values are more than two.

The upper half of the data is 7.2, 8, 8.3, 9, 10, 10, 11.5. The middle value of the upper half is nine.

The **third quartile**, Q_3, is nine. Three-fourths (75%) of the ordered data set are less than nine. One-fourth (25%) of the ordered data set are greater than nine. The third quartile is part of the data set in this example.

The **interquartile range** is a number that indicates the spread of the middle half or the middle 50% of the data. It is the difference between the third quartile (Q_3) and the first quartile (Q_1).

$IQR = Q_3 - Q_1$

The IQR can help to determine potential **outliers. A value is suspected to be a potential outlier if it is less than (1.5)(IQR) below the first quartile or more than (1.5)(IQR) above the third quartile**. Potential outliers always require further investigation.

NOTE

A potential outlier is a data point that is significantly different from the other data points. These special data points may be errors or some kind of abnormality or they may be a key to understanding the data.

Example 2.13

For the following 13 real estate prices, calculate the IQR and determine if any prices are potential outliers. Prices are in dollars.
389,950; 230,500; 158,000; 479,000; 639,000; 114,950; 5,500,000; 387,000; 659,000; 529,000; 575,000; 488,800; 1,095,000

Solution 2.13

Order the data from smallest to largest.
114,950; 158,000; 230,500; 387,000; 389,950; 479,000; 488,800; 529,000; 575,000; 639,000; 659,000; 1,095,000; 5,500,000

$M = 488,800$

$$Q_1 = \frac{230,500 + 387,000}{2} = 308,750$$

$$Q_3 = \frac{639,000 + 659,000}{2} = 649,000$$

$IQR = 649,000 - 308,750 = 340,250$

$(1.5)(IQR) = (1.5)(340,250) = 510,375$

$Q_1 - (1.5)(IQR) = 308,750 - 510,375 = -201,625$

$Q_3 + (1.5)(IQR) = 649,000 + 510,375 = 1,159,375$

No house price is less than −201,625. However, 5,500,000 is more than 1,159,375. Therefore, 5,500,000 is a potential **outlier**.

Try It Σ

2.13 For the following 11 salaries, calculate the *IQR* and determine if any salaries are outliers. The salaries are in dollars.

$33,000; $64,500; $28,000; $54,000; $72,000; $68,500; $69,000; $42,000; $54,000; $120,000; $40,500

Example 2.14

For the two data sets in the **test scores example**, find the following:

a. The interquartile range. Compare the two interquartile ranges.

b. Any outliers in either set.

Solution 2.14

The five number summary for the day and night classes is

	Minimum	Q_1	Median	Q_3	Maximum
Day	32	56	74.5	82.5	99
Night	25.5	78	81	89	98

Table 2.21

a. The IQR for the day group is $Q_3 - Q_1 = 82.5 - 56 = 26.5$
    The IQR for the night group is $Q_3 - Q_1 = 89 - 78 = 11$

    The interquartile range (the spread or variability) for the day class is larger than the night class *IQR*. This suggests more variation will be found in the day class's class test scores.

b. Day class outliers are found using the IQR times 1.5 rule. So,
    $Q_1 - IQR(1.5) = 56 - 26.5(1.5) = 16.25$

$Q_3 + IQR(1.5) = 82.5 + 26.5(1.5) = 122.25$

Since the minimum and maximum values for the day class are greater than 16.25 and less than 122.25, there are no outliers.

Night class outliers are calculated as:

$Q_1 - IQR (1.5) = 78 - 11(1.5) = 61.5$

$Q_3 + IQR(1.5) = 89 + 11(1.5) = 105.5$

For this class, any test score less than 61.5 is an outlier. Therefore, the scores of 45 and 25.5 are outliers. Since no test score is greater than 105.5, there is no upper end outlier.

Try It Σ

2.14 Find the interquartile range for the following two data sets and compare them.

Test Scores for Class *A*
69; 96; 81; 79; 65; 76; 83; 99; 89; 67; 90; 77; 85; 98; 66; 91; 77; 69; 80; 94
Test Scores for Class *B*
90; 72; 80; 92; 90; 97; 92; 75; 79; 68; 70; 80; 99; 95; 78; 73; 71; 68; 95; 100

Example 2.15

Fifty statistics students were asked how much sleep they get per school night (rounded to the nearest hour). The results were:

AMOUNT OF SLEEP PER SCHOOL NIGHT (HOURS)	FREQUENCY	RELATIVE FREQUENCY	CUMULATIVE RELATIVE FREQUENCY
4	2	0.04	0.04
5	5	0.10	0.14
6	7	0.14	0.28
7	12	0.24	0.52
8	14	0.28	0.80
9	7	0.14	0.94
10	3	0.06	1.00

Table 2.22

Find the 28th percentile. Notice the 0.28 in the "cumulative relative frequency" column. Twenty-eight percent of 50 data values is 14 values. There are 14 values less than the 28th percentile. They include the two 4s, the five 5s, and the seven 6s. The 28th percentile is between the last six and the first seven. **The 28th percentile is 6.5.**

Find the median. Look again at the "cumulative relative frequency" column and find 0.52. The median is the 50th percentile or the second quartile. 50% of 50 is 25. There are 25 values less than the median. They include the two 4s, the five 5s, the seven 6s, and eleven of the 7s. The median or 50th percentile is between the 25th, or seven, and 26th, or seven, values. **The median is seven.**

Find the third quartile. The third quartile is the same as the 75th percentile. You can "eyeball" this answer. If you look at the "cumulative relative frequency" column, you find 0.52 and 0.80. When you have all the fours, fives, sixes and sevens, you have 52% of the data. When you include all the 8s, you have 80% of the data. **The 75th percentile, then, must be an eight**. Another way to look at the problem is to find 75% of 50, which is 37.5, and round up to 38. The third quartile, Q_3, is the 38th value, which is an eight. You can check this answer by counting the values. (There are 37 values below the third quartile and 12 values above.)

Try It Σ

2.15 Forty bus drivers were asked how many hours they spend each day running their routes (rounded to the nearest hour). Find the 65th percentile.

Amount of time spent on route (hours)	Frequency	Relative Frequency	Cumulative Relative Frequency
2	12	0.30	0.30
3	14	0.35	0.65
4	10	0.25	0.90
5	4	0.10	1.00

Table 2.23

Example 2.16

Using **Table 2.22**:

a. Find the 80th percentile.

b. Find the 90th percentile.

c. Find the first quartile. What is another name for the first quartile?

Solution 2.16

Using the data from the frequency table, we have:

a. The 80th percentile is between the last eight and the first nine in the table (between the 40th and 41st values). Therefore, we need to take the mean of the 40th an 41st values. The 80th percentile $= \frac{8+9}{2} = 8.5$

b. The 90th percentile will be the 45th data value (location is 0.90(50) = 45) and the 45th data value is nine.

c. Q_1 is also the 25th percentile. The 25th percentile location calculation: $P_{25} = 0.25(50) = 12.5 \approx 13$ the 13th data value. Thus, the 25th percentile is six.

Try It Σ

2.16 Refer to the **Table 2.23**. Find the third quartile. What is another name for the third quartile?

Collaborative Exercise

Your instructor or a member of the class will ask everyone in class how many sweaters they own. Answer the following questions:

1. How many students were surveyed?

2. What kind of sampling did you do?

3. Construct two different histograms. For each, starting value = _____ ending value = _____.

4. Find the median, first quartile, and third quartile.

5. Construct a table of the data to find the following:

    a. the 10th percentile

    b. the 70th percentile

    c. the percent of students who own less than four sweaters

A Formula for Finding the *k*th Percentile

If you were to do a little research, you would find several formulas for calculating the kth percentile. Here is one of them.

k = the kth percentile. It may or may not be part of the data.

i = the index (ranking or position of a data value)

n = the total number of data

- Order the data from smallest to largest.

- Calculate $i = \frac{k}{100}(n + 1)$

- If i is an integer, then the kth percentile is the data value in the ith position in the ordered set of data.

- If i is not an integer, then round i up and round i down to the nearest integers. Average the two data values in these two positions in the ordered data set. This is easier to understand in an example.

Example 2.17

Listed are 29 ages for Academy Award winning best actors *in order from smallest to largest.*
18; 21; 22; 25; 26; 27; 29; 30; 31; 33; 36; 37; 41; 42; 47; 52; 55; 57; 58; 62; 64; 67; 69; 71; 72; 73; 74; 76; 77

a. Find the 70th percentile.

b. Find the 83rd percentile.

Solution 2.17

a. $k = 70$
i = the index
$n = 29$
$i = \frac{k}{100}(n + 1) = (\frac{70}{100})(29 + 1) = 21$. Twenty-one is an integer, and the data value in the 21st position in

b. the ordered data set is 64. The 70th percentile is 64 years.

$k = 83^{rd}$ percentile

i = the index

$n = 29$

$i = \frac{k}{100}(n+1) =)\frac{83}{100})(29+1) = 24.9$, which is NOT an integer. Round it down to 24 and up to 25. The

age in the 24th position is 71 and the age in the 25th position is 72. Average 71 and 72. The 83rd percentile is 71.5 years.

2.17 Listed are 29 ages for Academy Award winning best actors *in order from smallest to largest.*

18; 21; 22; 25; 26; 27; 29; 30; 31; 33; 36; 37; 41; 42; 47; 52; 55; 57; 58; 62; 64; 67; 69; 71; 72; 73; 74; 76; 77

Calculate the 20th percentile and the 55th percentile.

NOTE

 You can calculate percentiles using calculators and computers. There are a variety of online calculators.

A Formula for Finding the Percentile of a Value in a Data Set

- Order the data from smallest to largest.

- x = the number of data values counting from the bottom of the data list up to but not including the data value for which you want to find the percentile.

- y = the number of data values equal to the data value for which you want to find the percentile.

- n = the total number of data.

- Calculate $\frac{x + 0.5y}{n}(100)$. Then round to the nearest integer.

Example 2.18

Listed are 29 ages for Academy Award winning best actors *in order from smallest to largest.*
18; 21; 22; 25; 26; 27; 29; 30; 31; 33; 36; 37; 41; 42; 47; 52; 55; 57; 58; 62; 64; 67; 69; 71; 72; 73; 74; 76; 77

a. Find the percentile for 58.

b. Find the percentile for 25.

Solution 2.18

a. Counting from the bottom of the list, there are 18 data values less than 58. There is one value of 58.

$x = 18$ and $y = 1$. $\dfrac{x + 0.5y}{n}(100) = \dfrac{18 + 0.5(1)}{29}(100) = 63.80$. 58 is the 64th percentile.

b. Counting from the bottom of the list, there are three data values less than 25. There is one value of 25.

$x = 3$ and $y = 1$. $\dfrac{x + 0.5y}{n}(100) = \dfrac{3 + 0.5(1)}{29}(100) = 12.07$. Twenty-five is the 12th percentile.

Try It Σ

2.18 Listed are 30 ages for Academy Award winning best actors <u>in order from smallest to largest.</u>

18; 21; 22; 25; 26; 27; 29; 30; 31, 31; 33; 36; 37; 41; 42; 47; 52; 55; 57; 58; 62; 64; 67; 69; 71; 72; 73; 74; 76; 77
Find the percentiles for 47 and 31.

Interpreting Percentiles, Quartiles, and Median

A percentile indicates the relative standing of a data value when data are sorted into numerical order from smallest to largest. Percentages of data values are less than or equal to the pth percentile. For example, 15% of data values are less than or equal to the 15th percentile.

- Low percentiles always correspond to lower data values.

- High percentiles always correspond to higher data values.

A percentile may or may not correspond to a value judgment about whether it is "good" or "bad." The interpretation of whether a certain percentile is "good" or "bad" depends on the context of the situation to which the data applies. In some situations, a low percentile would be considered "good;" in other contexts a high percentile might be considered "good". In many situations, there is no value judgment that applies.

Understanding how to interpret percentiles properly is important not only when describing data, but also when calculating probabilities in later chapters of this text.

GUIDELINE

When writing the interpretation of a percentile in the context of the given data, the sentence should contain the following information.

- information about the context of the situation being considered

- the data value (value of the variable) that represents the percentile

- the percent of individuals or items with data values below the percentile

- the percent of individuals or items with data values above the percentile.

Example 2.19

On a timed math test, the first quartile for time it took to finish the exam was 35 minutes. Interpret the first quartile in the context of this situation.

Solution 2.19
- Twenty-five percent of students finished the exam in 35 minutes or less.
- Seventy-five percent of students finished the exam in 35 minutes or more.
- A low percentile could be considered good, as finishing more quickly on a timed exam is desirable. (If you take too long, you might not be able to finish.)

2.19 For the 100-meter dash, the third quartile for times for finishing the race was 11.5 seconds. Interpret the third quartile in the context of the situation.

Example 2.20

On a 20 question math test, the 70th percentile for number of correct answers was 16. Interpret the 70th percentile in the context of this situation.

Solution 2.20
- Seventy percent of students answered 16 or fewer questions correctly.
- Thirty percent of students answered 16 or more questions correctly.
- A higher percentile could be considered good, as answering more questions correctly is desirable.

2.20 On a 60 point written assignment, the 80th percentile for the number of points earned was 49. Interpret the 80th percentile in the context of this situation.

Example 2.21

At a community college, it was found that the 30th percentile of credit units that students are enrolled for is seven units. Interpret the 30th percentile in the context of this situation.

Solution 2.21
- Thirty percent of students are enrolled in seven or fewer credit units.
- Seventy percent of students are enrolled in seven or more credit units.
- In this example, there is no "good" or "bad" value judgment associated with a higher or lower percentile. Students attend community college for varied reasons and needs, and their course load varies according to their needs.

Try It Σ

2.21 During a season, the 40th percentile for points scored per player in a game is eight. Interpret the 40th percentile in the context of this situation.

Example 2.22

Sharpe Middle School is applying for a grant that will be used to add fitness equipment to the gym. The principal surveyed 15 anonymous students to determine how many minutes a day the students spend exercising. The results from the 15 anonymous students are shown.

0 minutes; 40 minutes; 60 minutes; 30 minutes; 60 minutes

10 minutes; 45 minutes; 30 minutes; 300 minutes; 90 minutes;

30 minutes; 120 minutes; 60 minutes; 0 minutes; 20 minutes

Determine the following five values.

Min = 0
Q_1 = 20
Med = 40
Q_3 = 60
Max = 300

If you were the principal, would you be justified in purchasing new fitness equipment? Since 75% of the students exercise for 60 minutes or less daily, and since the *IQR* is 40 minutes (60 − 20 = 40), we know that half of the students surveyed exercise between 20 minutes and 60 minutes daily. This seems a reasonable amount of time spent exercising, so the principal would be justified in purchasing the new equipment.

However, the principal needs to be careful. The value 300 appears to be a potential outlier.

$Q_3 + 1.5(IQR) = 60 + (1.5)(40) = 120.$

The value 300 is greater than 120 so it is a potential outlier. If we delete it and calculate the five values, we get the following values:

Min = 0
Q_1 = 20
Q_3 = 60
Max = 120

We still have 75% of the students exercising for 60 minutes or less daily and half of the students exercising between 20 and 60 minutes a day. However, 15 students is a small sample and the principal should survey more students to be sure of his survey results.

2.4 | Box Plots

Box plots (also called **box-and-whisker plots** or **box-whisker plots**) give a good graphical image of the concentration of the data. They also show how far the extreme values are from most of the data. A box plot is constructed from five values: the minimum value, the first quartile, the median, the third quartile, and the maximum value. We use these values to compare how close other data values are to them.

To construct a box plot, use a horizontal or vertical number line and a rectangular box. The smallest and largest data values label the endpoints of the axis. The first quartile marks one end of the box and the third quartile marks the other end of the box. Approximately **the middle 50 percent of the data fall inside the box.** The "whiskers" extend from the ends of the box to the smallest and largest data values. The median or second quartile can be between the first and third quartiles, or it can be one, or the other, or both. The box plot gives a good, quick picture of the data.

NOTE

You may encounter box-and-whisker plots that have dots marking outlier values. In those cases, the whiskers are not extending to the minimum and maximum values.

Consider, again, this dataset.

1; 1; 2; 2; 4; 6; 6.8; 7.2; 8; 8.3; 9; 10; 10; 11.5

The first quartile is two, the median is seven, and the third quartile is nine. The smallest value is one, and the largest value is 11.5. The following image shows the constructed box plot.

NOTE

See the calculator instructions on the **TI web site (http://education.ti.com/educationportal/sites/US/ sectionHome/support.html)** or in the appendix.

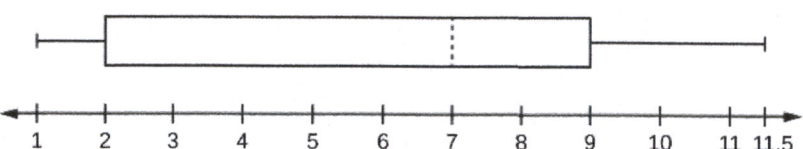

Figure 2.11

The two whiskers extend from the first quartile to the smallest value and from the third quartile to the largest value. The median is shown with a dashed line.

NOTE

It is important to start a box plot with a **scaled number line**. Otherwise the box plot may not be useful.

Example 2.23

The following data are the heights of 40 students in a statistics class.

59; 60; 61; 62; 62; 63; 63; 64; 64; 64; 65; 65; 65; 65; 65; 65; 65; 65; 65; 66; 66; 67; 67; 68; 68; 69; 70; 70; 70; 70; 70; 71; 71; 72; 72; 73; 74; 74; 75; 77

Construct a box plot with the following properties; the calculator intructions for the minimum and maximum values as well as the quartiles follow the example.

- Minimum value = 59

- Maximum value = 77

- Q1: First quartile = 64.5

- Q2: Second quartile or median= 66

- Q3: Third quartile = 70

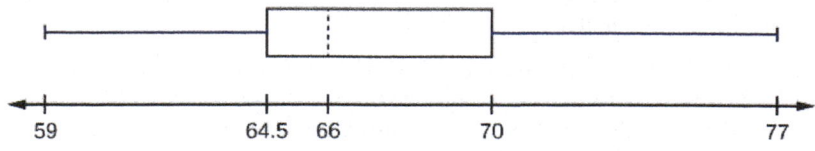

Figure 2.12

a. Each quarter has approximately 25% of the data.

b. The spreads of the four quarters are 64.5 – 59 = 5.5 (first quarter), 66 – 64.5 = 1.5 (second quarter), 70 – 66 = 4 (third quarter), and 77 – 70 = 7 (fourth quarter). So, the second quarter has the smallest spread and the fourth quarter has the largest spread.

c. Range = maximum value – the minimum value = 77 – 59 = 18

d. Interquartile Range: $IQR = Q3 - Q1 = 70 - 64.5 = 5.5$.

e. The interval 59–65 has more than 25% of the data so it has more data in it than the interval 66 through 70 which has 25% of the data.

f. The middle 50% (middle half) of the data has a range of 5.5 inches.

Using the TI-83, 83+, 84, 84+ Calculator

To find the minimum, maximum, and quartiles:

Enter data into the list editor (Pres STAT 1:EDIT). If you need to clear the list, arrow up to the name L1, press CLEAR, and then arrow down.

Put the data values into the list L1.

Press STAT and arrow to CALC. Press 1:1-VarStats. Enter L1.

Press ENTER.

Use the down and up arrow keys to scroll.

Smallest value = 59.

Largest value = 77.

Q_1: First quartile = 64.5.

Q_2: Second quartile or median = 66.

Q_3: Third quartile = 70.

To construct the box plot:

Press 4:Plotsoff. Press ENTER.

Arrow down and then use the right arrow key to go to the fifth picture, which is the box plot. Press ENTER.

Arrow down to Xlist: Press 2nd 1 for L1

Arrow down to Freq: Press ALPHA. Press 1.

Press Zoom. Press 9: ZoomStat.

Press TRACE, and use the arrow keys to examine the box plot.

Try It Σ

2.23 The following data are the number of pages in 40 books on a shelf. Construct a box plot using a graphing calculator, and state the interquartile range.

136; 140; 178; 190; 205; 215; 217; 218; 232; 234; 240; 255; 270; 275; 290; 301; 303; 315; 317; 318; 326; 333; 343; 349; 360; 369; 377; 388; 391; 392; 398; 400; 402; 405; 408; 422; 429; 450; 475; 512

For some sets of data, some of the largest value, smallest value, first quartile, median, and third quartile may be the same. For instance, you might have a data set in which the median and the third quartile are the same. In this case, the diagram would not have a dotted line inside the box displaying the median. The right side of the box would display both the third quartile and the median. For example, if the smallest value and the first quartile were both one, the median and the third quartile were both five, and the largest value was seven, the box plot would look like:

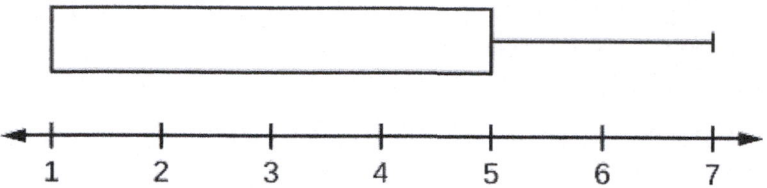

Figure 2.13

In this case, at least 25% of the values are equal to one. Twenty-five percent of the values are between one and five, inclusive. At least 25% of the values are equal to five. The top 25% of the values fall between five and seven, inclusive.

Example 2.24

Test scores for a college statistics class held during the day are:

99; 56; 78; 55.5; 32; 90; 80; 81; 56; 59; 45; 77; 84.5; 84; 70; 72; 68; 32; 79; 90

Test scores for a college statistics class held during the evening are:

98; 78; 68; 83; 81; 89; 88; 76; 65; 45; 98; 90; 80; 84.5; 85; 79; 78; 98; 90; 79; 81; 25.5

a. Find the smallest and largest values, the median, and the first and third quartile for the day class.

b. Find the smallest and largest values, the median, and the first and third quartile for the night class.

c. For each data set, what percentage of the data is between the smallest value and the first quartile? the first quartile and the median? the median and the third quartile? the third quartile and the largest value? What percentage of the data is between the first quartile and the largest value?

d. Create a box plot for each set of data. Use one number line for both box plots.

e. Which box plot has the widest spread for the middle 50% of the data (the data between the first and third quartiles)? What does this mean for that set of data in comparison to the other set of data?

Solution 2.24

a. Min = 32
   $Q_1 = 56$
   $M = 74.5$
   $Q_3 = 82.5$
   Max = 99

b. Min = 25.5
   $Q_1 = 78$
   $M = 81$
   $Q_3 = 89$
   Max = 98

c. Day class: There are six data values ranging from 32 to 56: 30%. There are six data values ranging from 56 to 74.5: 30%. There are five data values ranging from 74.5 to 82.5: 25%. There are five data values ranging from 82.5 to 99: 25%. There are 16 data values between the first quartile, 56, and the largest value, 99: 75%.

d. Night class:

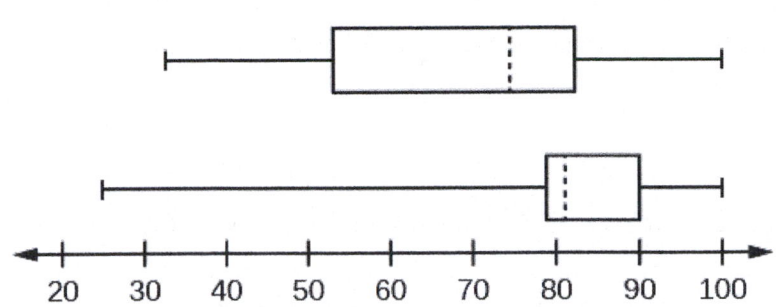

Figure 2.14

e. The first data set has the wider spread for the middle 50% of the data. The *IQR* for the first data set is greater than the *IQR* for the second set. This means that there is more variability in the middle 50% of the first data set.

Try It Σ

☞ **2.24** The following data set shows the heights in inches for the boys in a class of 40 students.

66; 66; 67; 67; 68; 68; 68; 68; 68; 69; 69; 69; 70; 71; 72; 72; 72; 73; 73; 74
The following data set shows the heights in inches for the girls in a class of 40 students.
61; 61; 62; 62; 63; 63; 63; 65; 65; 65; 66; 66; 66; 67; 68; 68; 68; 69; 69; 69
Construct a box plot using a graphing calculator for each data set, and state which box plot has the wider spread for the middle 50% of the data.

Example 2.25

Graph a box-and-whisker plot for the data values shown.

10; 10; 10; 15; 35; 75; 90; 95; 100; 175; 420; 490; 515; 515; 790

The five numbers used to create a box-and-whisker plot are:

Min: 10
Q_1: 15
Med: 95
Q_3: 490
Max: 790

The following graph shows the box-and-whisker plot.

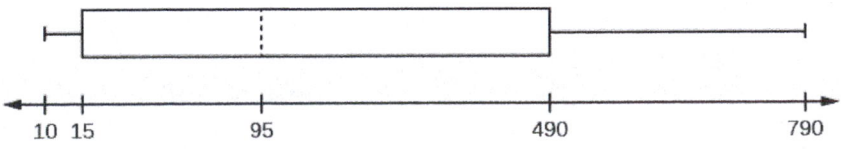

Figure 2.15

2.25 Follow the steps you used to graph a box-and-whisker plot for the data values shown.

0; 5; 5; 15; 30; 30; 45; 50; 50; 60; 75; 110; 140; 240; 330

2.5 | Measures of the Center of the Data

The "center" of a data set is also a way of describing location. The two most widely used measures of the "center" of the data are the **mean** (average) and the **median**. To calculate the **mean weight** of 50 people, add the 50 weights together and divide by 50. To find the **median weight** of the 50 people, order the data and find the number that splits the data into two equal parts. The median is generally a better measure of the center when there are extreme values or outliers because it is not affected by the precise numerical values of the outliers. The mean is the most common measure of the center.

NOTE

The words "mean" and "average" are often used interchangeably. The substitution of one word for the other is common practice. The technical term is "arithmetic mean" and "average" is technically a center location. However, in practice among non-statisticians, "average" is commonly accepted for "arithmetic mean."

When each value in the data set is not unique, the mean can be calculated by multiplying each distinct value by its frequency and then dividing the sum by the total number of data values. The letter used to represent the **sample mean** is an x with a bar over it (pronounced "x bar"): $\bar{x}$.

The Greek letter μ (pronounced "mew") represents the **population mean**. One of the requirements for the **sample mean** to be a good estimate of the **population mean** is for the sample taken to be truly random.

To see that both ways of calculating the mean are the same, consider the sample:
1; 1; 1; 2; 2; 3; 4; 4; 4; 4; 4

$$\bar{x} = \frac{1+1+1+2+2+3+4+4+4+4+4}{11} = 2.7$$

$$\bar{x} = \frac{3(1)+2(2)+1(3)+5(4)}{11} = 2.7$$

In the second example, the frequencies are 3(1) + 2(2) + 1(3) + 5(4).

You can quickly find the location of the median by using the expression $\frac{n+1}{2}$.

The letter n is the total number of data values in the sample. If n is an odd number, the median is the middle value of the ordered data (ordered smallest to largest). If n is an even number, the median is equal to the two middle values added together and divided by two after the data has been ordered. For example, if the total number of data values is 97, then $\frac{n+1}{2} = \frac{97+1}{2}$ = 49. The median is the 49th value in the ordered data. If the total number of data values is 100, then $\frac{n+1}{2} = \frac{100+1}{2}$ = 50.5. The median occurs midway between the 50th and 51st values. The location of the median and the value of the median are **not** the same. The upper case letter M is often used to represent the median. The next example illustrates the location of the median and the value of the median.

Example 2.26

AIDS data indicating the number of months a patient with AIDS lives after taking a new antibody drug are as follows (smallest to largest):
3; 4; 8; 8; 10; 11; 12; 13; 14; 15; 15; 16; 16; 17; 17; 18; 21; 22; 22; 24; 24; 25; 26; 26; 27; 27; 29; 29; 31; 32; 33;

33; 34; 34; 35; 37; 40; 44; 44; 47;
Calculate the mean and the median.

Solution 2.26

The calculation for the mean is:

$$\bar{x} = \frac{[3 + 4 + (8)(2) + 10 + 11 + 12 + 13 + 14 + (15)(2) + (16)(2) + ... + 35 + 37 + 40 + (44)(2) + 47]}{40} = 23.6$$

To find the median, M, first use the formula for the location. The location is:

$$\frac{n+1}{2} = \frac{40+1}{2} = 20.5$$

Starting at the smallest value, the median is located between the 20th and 21st values (the two 24s):
3; 4; 8; 8; 10; 11; 12; 13; 14; 15; 15; 16; 16; 17; 17; 18; 21; 22; 22; 24; 24; 25; 26; 26; 27; 27; 29; 29; 31; 32; 33;
33; 34; 34; 35; 37; 40; 44; 44; 47;

$$M = \frac{24 + 24}{2} = 24$$

Using the TI-83, 83+, 84, 84+ Calculator

To find the mean and the median:

Clear list L1. Pres STAT 4:ClrList. Enter 2nd 1 for list L1. Press ENTER.

Enter data into the list editor. Press STAT 1:EDIT.

Put the data values into list L1.

Press STAT and arrow to CALC. Press 1:1-VarStats. Press 2nd 1 for L1 and then ENTER.

Press the down and up arrow keys to scroll.

$\bar{x}$ = 23.6, M = 24

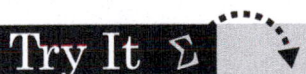

Try It Σ

2.26 The following data show the number of months patients typically wait on a transplant list before getting surgery. The data are ordered from smallest to largest. Calculate the mean and median.

3; 4; 5; 7; 7; 7; 7; 8; 8; 9; 9; 10; 10; 10; 10; 10; 11; 12; 12; 13; 14; 14; 15; 15; 17; 17; 18; 19; 19; 19; 21; 21; 22; 22; 23; 24; 24; 24; 24

Example 2.27

Suppose that in a small town of 50 people, one person earns $5,000,000 per year and the other 49 each earn $30,000. Which is the better measure of the "center": the mean or the median?

Solution 2.27

$$\bar{x} = \frac{5,000,000 + 49(30,000)}{50} = 129,400$$

M = 30,000

(There are 49 people who earn $30,000 and one person who earns $5,000,000.)

The median is a better measure of the "center" than the mean because 49 of the values are 30,000 and one is 5,000,000. The 5,000,000 is an outlier. The 30,000 gives us a better sense of the middle of the data.

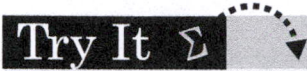

2.27 In a sample of 60 households, one house is worth $2,500,000. Half of the rest are worth $280,000, and all the others are worth $315,000. Which is the better measure of the "center": the mean or the median?

Another measure of the center is the mode. The **mode** is the most frequent value. There can be more than one mode in a data set as long as those values have the same frequency and that frequency is the highest. A data set with two modes is called bimodal.

Example 2.28

Statistics exam scores for 20 students are as follows:

50; 53; 59; 59; 63; 63; 72; 72; 72; 72; 72; 76; 78; 81; 83; 84; 84; 84; 90; 93

Find the mode.

Solution 2.28
The most frequent score is 72, which occurs five times. Mode = 72.

Try It Σ

2.28 The number of books checked out from the library from 25 students are as follows:

0; 0; 0; 1; 2; 3; 3; 4; 4; 5; 5; 7; 7; 7; 7; 8; 8; 8; 9; 10; 10; 11; 11; 12; 12
Find the mode.

Example 2.29

Five real estate exam scores are 430, 430, 480, 480, 495. The data set is bimodal because the scores 430 and 480 each occur twice.

When is the mode the best measure of the "center"? Consider a weight loss program that advertises a mean weight loss of six pounds the first week of the program. The mode might indicate that most people lose two pounds the first week, making the program less appealing.

NOTE

The mode can be calculated for qualitative data as well as for quantitative data. For example, if the data set is: red, red, red, green, green, yellow, purple, black, blue, the mode is red.

Statistical software will easily calculate the mean, the median, and the mode. Some graphing calculators can also make these calculations. In the real world, people make these calculations using software.

Try It Σ

2.29 Five credit scores are 680, 680, 700, 720, 720. The data set is bimodal because the scores 680 and 720 each occur twice. Consider the annual earnings of workers at a factory. The mode is $25,000 and occurs 150 times out of 301. The median is $50,000 and the mean is $47,500. What would be the best measure of the "center"?

The Law of Large Numbers and the Mean

The Law of Large Numbers says that if you take samples of larger and larger size from any population, then the mean $\bar{x}$ of the sample is very likely to get closer and closer to μ. This is discussed in more detail later in the text.

Sampling Distributions and Statistic of a Sampling Distribution

You can think of a **sampling distribution** as a **relative frequency distribution** with a great many samples. (See **Sampling and Data** for a review of relative frequency). Suppose thirty randomly selected students were asked the number of movies they watched the previous week. The results are in the **relative frequency table** shown below.

# of movies	Relative Frequency
0	$\frac{5}{30}$
1	$\frac{15}{30}$
2	$\frac{6}{30}$
3	$\frac{3}{30}$
4	$\frac{1}{30}$

Table 2.24

If you let the number of samples get very large (say, 300 million or more), the relative frequency table becomes a relative frequency distribution.

A **statistic** is a number calculated from a sample. Statistic examples include the mean, the median and the mode as well as others. The sample mean $\bar{x}$ is an example of a statistic which estimates the population mean μ.

Calculating the Mean of Grouped Frequency Tables

When only grouped data is available, you do not know the individual data values (we only know intervals and interval frequencies); therefore, you cannot compute an exact mean for the data set. What we must do is estimate the actual mean by calculating the mean of a frequency table. A frequency table is a data representation in which grouped data is displayed along with the corresponding frequencies. To calculate the mean from a grouped frequency table we can apply the basic definition of mean: $mean = \frac{data\ sum}{number\ of\ data\ values}$ We simply need to modify the definition to fit within the restrictions of a frequency table.

Since we do not know the individual data values we can instead find the midpoint of each interval. The midpoint is $\frac{lower\ boundary + upper\ boundary}{2}$. We can now modify the mean definition to be

$Mean\ of\ Frequency\ Table = \frac{\sum fm}{\sum f}$ where f = the frequency of the interval and m = the midpoint of the interval.

Example 2.30

A frequency table displaying professor Blount's last statistic test is shown. Find the best estimate of the class mean.

Grade Interval	Number of Students
50–56.5	1
56.5–62.5	0
62.5–68.5	4
68.5–74.5	4
74.5–80.5	2
80.5–86.5	3
86.5–92.5	4
92.5–98.5	1

Table 2.25

Solution 2.30

- Find the midpoints for all intervals

Grade Interval	Midpoint
50–56.5	53.25
56.5–62.5	59.5
62.5–68.5	65.5
68.5–74.5	71.5
74.5–80.5	77.5
80.5–86.5	83.5
86.5–92.5	89.5
92.5–98.5	95.5

Table 2.26

- Calculate the sum of the product of each interval frequency and midpoint. $\sum fm$

$$53.25(1) + 59.5(0) + 65.5(4) + 71.5(4) + 77.5(2) + 83.5(3) + 89.5(4) + 95.5(1) = 1460.25$$

- $\mu = \dfrac{\sum fm}{\sum f} = \dfrac{1460.25}{19} = 76.86$

2.30 Maris conducted a study on the effect that playing video games has on memory recall. As part of her study, she compiled the following data:

Hours Teenagers Spend on Video Games	Number of Teenagers
0–3.5	3
3.5–7.5	7
7.5–11.5	12
11.5–15.5	7
15.5–19.5	9

Table 2.27

What is the best estimate for the mean number of hours spent playing video games?

2.6 | Skewness and the Mean, Median, and Mode

Consider the following data set.
4; 5; 6; 6; 6; 7; 7; 7; 7; 7; 7; 8; 8; 8; 9; 10

This data set can be represented by following histogram. Each interval has width one, and each value is located in the middle of an interval.

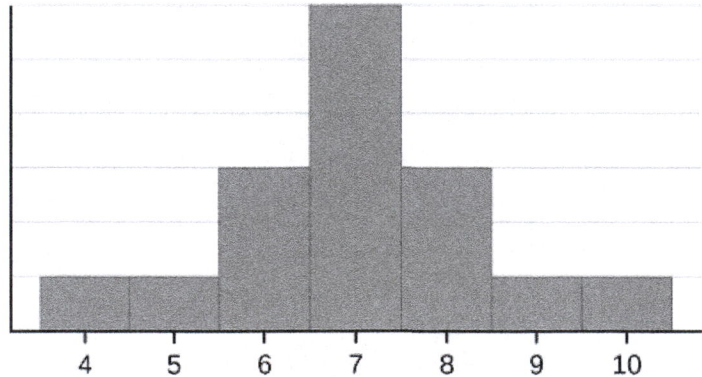

Figure 2.16

The histogram displays a **symmetrical** distribution of data. A distribution is symmetrical if a vertical line can be drawn at some point in the histogram such that the shape to the left and the right of the vertical line are mirror images of each other. The mean, the median, and the mode are each seven for these data. **In a perfectly symmetrical distribution, the mean and the median are the same.** This example has one mode (unimodal), and the mode is the same as the mean and median. In a symmetrical distribution that has two modes (bimodal), the two modes would be different from the mean and median.

The histogram for the data: 4; 5; 6; 6; 6; 7; 7; 7; 7; 8 is not symmetrical. The right-hand side seems "chopped off" compared to the left side. A distribution of this type is called **skewed to the left** because it is pulled out to the left.

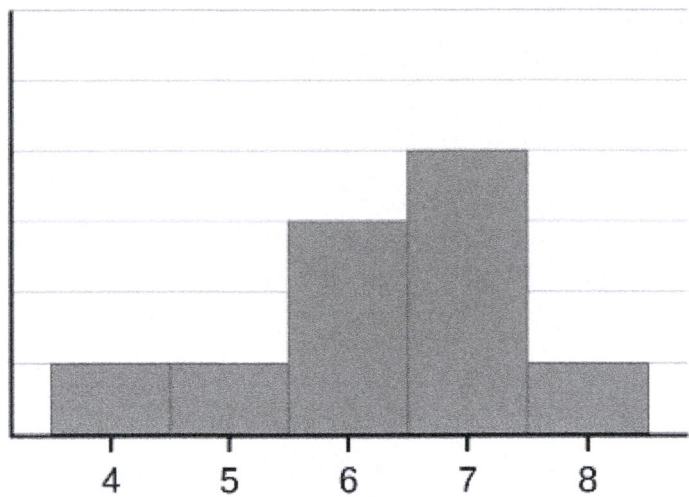

Figure 2.17

The mean is 6.3, the median is 6.5, and the mode is seven. **Notice that the mean is less than the median, and they are both less than the mode.** The mean and the median both reflect the skewing, but the mean reflects it more so.

The histogram for the data: 6; 7; 7; 7; 7; 8; 8; 8; 9; 10, is also not symmetrical. It is **skewed to the right**.

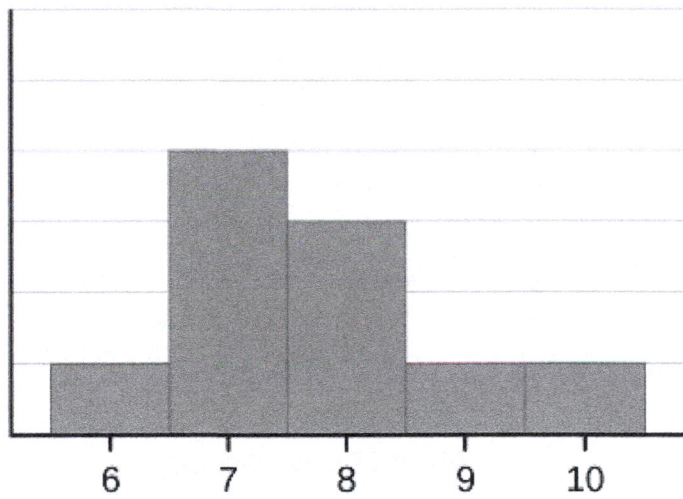

Figure 2.18

The mean is 7.7, the median is 7.5, and the mode is seven. Of the three statistics, **the mean is the largest, while the mode is the smallest**. Again, the mean reflects the skewing the most.

To summarize, generally if the distribution of data is skewed to the left, the mean is less than the median, which is often less than the mode. If the distribution of data is skewed to the right, the mode is often less than the median, which is less than the mean.

Skewness and symmetry become important when we discuss probability distributions in later chapters.

Example 2.31

Statistics are used to compare and sometimes identify authors. The following lists shows a simple random sample that compares the letter counts for three authors.

Terry: 7; 9; 3; 3; 3; 4; 1; 3; 2; 2

Davis: 3; 3; 3; 4; 1; 4; 3; 2; 3; 1

Maris: 2; 3; 4; 4; 4; 6; 6; 6; 8; 3

a. Make a dot plot for the three authors and compare the shapes.

b. Calculate the mean for each.

c. Calculate the median for each.

d. Describe any pattern you notice between the shape and the measures of center.

Solution 2.31

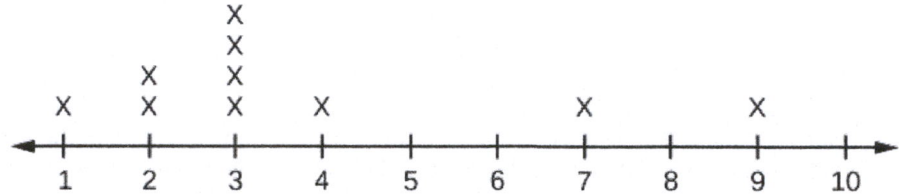

a.

Figure 2.19 Terry's distribution has a right (positive) skew.

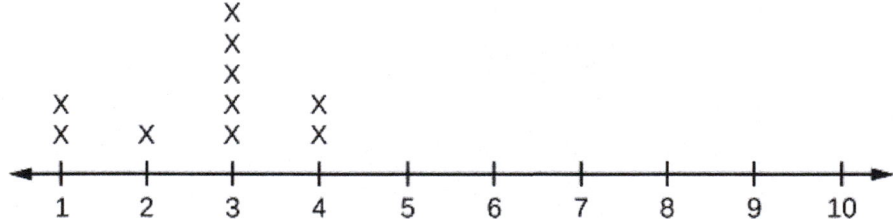

Figure 2.20 Davis' distribution has a left (negative) skew

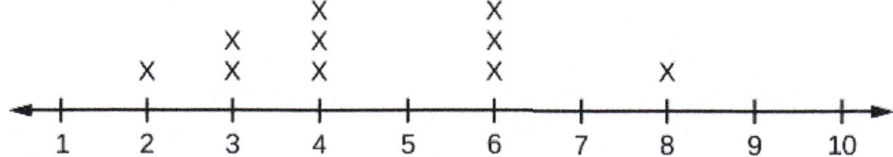

Figure 2.21 Maris' distribution is symmetrically shaped.

b. Terry's mean is 3.7, Davis' mean is 2.7, Maris' mean is 4.6.

c. Terry's median is three, Davis' median is three. Maris' median is four.

d. It appears that the median is always closest to the high point (the mode), while the mean tends to be farther out on the tail. In a symmetrical distribution, the mean and the median are both centrally located close to the high point of the distribution.

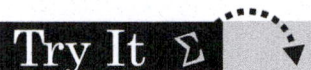

2.31 Discuss the mean, median, and mode for each of the following problems. Is there a pattern between the shape and measure of the center?

a.

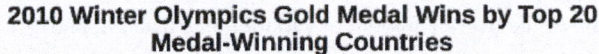

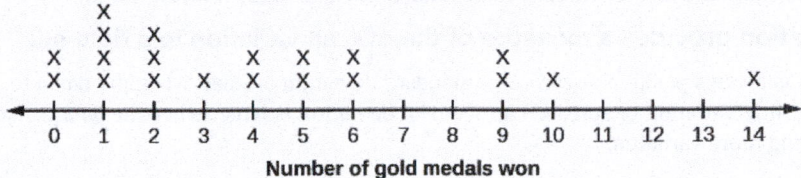

2010 Winter Olympics Gold Medal Wins by Top 20 Medal-Winning Countries

```
      X
      X   X
  X   X   X       X   X   X           X               X
  X   X   X   X   X   X   X           X   X
+---+---+---+---+---+---+---+---+---+---+---+---+---+---+-->
  0   1   2   3   4   5   6   7   8   9  10  11  12  13  14
```

Number of gold medals won

Figure 2.22

b.

The Ages Former U.S Presidents Died	
4	6 9
5	3 6 7 7 7 8
6	0 0 3 3 4 4 5 6 7 7 7 8
7	0 1 1 2 3 4 7 8 8 9
8	0 1 3 5 8
9	0 0 3 3
Key: 8|0 means 80.	

Table 2.28

c.

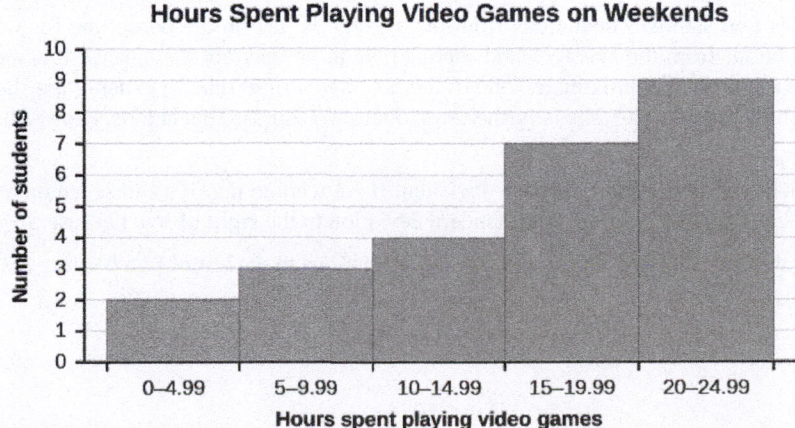

Hours Spent Playing Video Games on Weekends

Figure 2.23

2.7 | Measures of the Spread of the Data

An important characteristic of any set of data is the variation in the data. In some data sets, the data values are concentrated closely near the mean; in other data sets, the data values are more widely spread out from the mean. The most common measure of variation, or spread, is the standard deviation. The **standard deviation** is a number that measures how far data values are from their mean.

The standard deviation

- provides a numerical measure of the overall amount of variation in a data set, and
- can be used to determine whether a particular data value is close to or far from the mean.

The standard deviation provides a measure of the overall variation in a data set

The standard deviation is always positive or zero. The standard deviation is small when the data are all concentrated close to the mean, exhibiting little variation or spread. The standard deviation is larger when the data values are more spread out from the mean, exhibiting more variation.

Suppose that we are studying the amount of time customers wait in line at the checkout at supermarket A and supermarket B. the average wait time at both supermarkets is five minutes. At supermarket A, the standard deviation for the wait time is two minutes; at supermarket B the standard deviation for the wait time is four minutes.

Because supermarket B has a higher standard deviation, we know that there is more variation in the wait times at supermarket B. Overall, wait times at supermarket B are more spread out from the average; wait times at supermarket A are more concentrated near the average.

The standard deviation can be used to determine whether a data value is close to or far from the mean.

Suppose that Rosa and Binh both shop at supermarket A. Rosa waits at the checkout counter for seven minutes and Binh waits for one minute. At supermarket A, the mean waiting time is five minutes and the standard deviation is two minutes. The standard deviation can be used to determine whether a data value is close to or far from the mean.

Rosa waits for seven minutes:

- Seven is two minutes longer than the average of five; two minutes is equal to one standard deviation.
- Rosa's wait time of seven minutes is **two minutes longer than the average** of five minutes.
- Rosa's wait time of seven minutes is **one standard deviation above the average** of five minutes.

Binh waits for one minute.

- One is four minutes less than the average of five; four minutes is equal to two standard deviations.
- Binh's wait time of one minute is **four minutes less than the average** of five minutes.
- Binh's wait time of one minute is **two standard deviations below the average** of five minutes.
- A data value that is two standard deviations from the average is just on the borderline for what many statisticians would consider to be far from the average. Considering data to be far from the mean if it is more than two standard deviations away is more of an approximate "rule of thumb" than a rigid rule. In general, the shape of the distribution of the data affects how much of the data is further away than two standard deviations. (You will learn more about this in later chapters.)

The number line may help you understand standard deviation. If we were to put five and seven on a number line, seven is to the right of five. We say, then, that seven is **one** standard deviation to the **right** of five because $5 + (1)(2) = 7$.

If one were also part of the data set, then one is **two** standard deviations to the **left** of five because $5 + (-2)(2) = 1$.

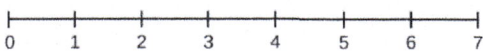

Figure 2.24

- In general, a **value = mean + (#ofSTDEV)(standard deviation)**
- where #ofSTDEVs = the number of standard deviations
- #ofSTDEV does not need to be an integer

- One is **two standard deviations less than the mean** of five because: $1 = 5 + (-2)(2)$.

The equation **value = mean + (#ofSTDEVs)(standard deviation)** can be expressed for a sample and for a population.

- **sample:** $x = \bar{x} + (\#\,of\,STDEV)(s)$

- **Population:** $x = \mu + (\#\,of\,STDEV)(\sigma)$

The lower case letter s represents the sample standard deviation and the Greek letter σ (sigma, lower case) represents the population standard deviation.

The symbol $\bar{x}$ is the sample mean and the Greek symbol μ is the population mean.

Calculating the Standard Deviation

If x is a number, then the difference "x – mean" is called its **deviation**. In a data set, there are as many deviations as there are items in the data set. The deviations are used to calculate the standard deviation. If the numbers belong to a population, in symbols a deviation is $x - \mu$. For sample data, in symbols a deviation is $x - \bar{x}$.

The procedure to calculate the standard deviation depends on whether the numbers are the entire population or are data from a sample. The calculations are similar, but not identical. Therefore the symbol used to represent the standard deviation depends on whether it is calculated from a population or a sample. The lower case letter s represents the sample standard deviation and the Greek letter σ (sigma, lower case) represents the population standard deviation. If the sample has the same characteristics as the population, then s should be a good estimate of σ.

To calculate the standard deviation, we need to calculate the variance first. The **variance** is the **average of the squares of the deviations** (the $x - \bar{x}$ values for a sample, or the $x - \mu$ values for a population). The symbol σ^2 represents the population variance; the population standard deviation σ is the square root of the population variance. The symbol s^2 represents the sample variance; the sample standard deviation s is the square root of the sample variance. You can think of the standard deviation as a special average of the deviations.

If the numbers come from a census of the entire **population** and not a sample, when we calculate the average of the squared deviations to find the variance, we divide by N, the number of items in the population. If the data are from a **sample** rather than a population, when we calculate the average of the squared deviations, we divide by $n - 1$, one less than the number of items in the sample.

Formulas for the Sample Standard Deviation

- $s = \sqrt{\dfrac{\Sigma(x - \bar{x})^2}{n - 1}}$ or $s = \sqrt{\dfrac{\Sigma f(x - \bar{x})^2}{n - 1}}$

- For the sample standard deviation, the denominator is $n - 1$, that is the sample size MINUS 1.

Formulas for the Population Standard Deviation

- $\sigma = \sqrt{\dfrac{\Sigma(x - \mu)^2}{N}}$ or $\sigma = \sqrt{\dfrac{\Sigma f(x - \mu)^2}{N}}$

- For the population standard deviation, the denominator is N, the number of items in the population.

In these formulas, f represents the frequency with which a value appears. For example, if a value appears once, f is one. If a value appears three times in the data set or population, f is three.

Sampling Variability of a Statistic

The statistic of a sampling distribution was discussed in **Descriptive Statistics: Measuring the Center of the Data**. How much the statistic varies from one sample to another is known as the **sampling variability of a statistic**. You typically measure the sampling variability of a statistic by its standard error. The **standard error of the mean** is an example of a standard error. It is a special standard deviation and is known as the standard deviation of the sampling distribution of the mean. You will cover the standard error of the mean in the chapter **The Central Limit Theorem** (not now). The notation for the standard error of the mean is $\frac{\sigma}{\sqrt{n}}$ where σ is the standard deviation of the population and n is the size of the sample.

NOTE

☞ **In practice, USE A CALCULATOR OR COMPUTER SOFTWARE TO CALCULATE THE STANDARD DEVIATION. If you are using a TI-83, 83+, 84+ calculator, you need to select the appropriate standard deviation** σ_x **or** s_x **from the summary statistics.** We will concentrate on using and interpreting the information that the standard deviation gives us. However you should study the following step-by-step example to help you understand how the standard deviation measures variation from the mean. (The calculator instructions appear at the end of this example.)

Example 2.32

In a fifth grade class, the teacher was interested in the average age and the sample standard deviation of the ages of her students. The following data are the ages for a SAMPLE of $n = 20$ fifth grade students. The ages are rounded to the nearest half year:

9; 9.5; 9.5; 10; 10; 10; 10; 10.5; 10.5; 10.5; 10.5; 11; 11; 11; 11; 11; 11; 11.5; 11.5; 11.5;

$$\bar{x} = \frac{9 + 9.5(2) + 10(4) + 10.5(4) + 11(6) + 11.5(3)}{20} = 10.525$$

The average age is 10.53 years, rounded to two places.

The variance may be calculated by using a table. Then the standard deviation is calculated by taking the square root of the variance. We will explain the parts of the table after calculating s.

Data	Freq.	Deviations	*Deviations*^{2}	(Freq.)(*Deviations*^{2})
x	f	$(x - \bar{x})$	$(x - \bar{x})^2$	$(f)(x - \bar{x})^2$
9	1	$9 - 10.525 = -1.525$	$(-1.525)^2 = 2.325625$	$1 \times 2.325625 = 2.325625$
9.5	2	$9.5 - 10.525 = -1.025$	$(-1.025)^2 = 1.050625$	$2 \times 1.050625 = 2.101250$
10	4	$10 - 10.525 = -0.525$	$(-0.525)^2 = 0.275625$	$4 \times 0.275625 = 1.1025$
10.5	4	$10.5 - 10.525 = -0.025$	$(-0.025)^2 = 0.000625$	$4 \times 0.000625 = 0.0025$
11	6	$11 - 10.525 = 0.475$	$(0.475)^2 = 0.225625$	$6 \times 0.225625 = 1.35375$
11.5	3	$11.5 - 10.525 = 0.975$	$(0.975)^2 = 0.950625$	$3 \times 0.950625 = 2.851875$
				The total is 9.7375

Table 2.29

The sample variance, s^2, is equal to the sum of the last column (9.7375) divided by the total number of data values minus one (20 – 1):

$$s^2 = \frac{9.7375}{20 - 1} = 0.5125$$

The **sample standard deviation** s is equal to the square root of the sample variance:

$s = \sqrt{0.5125} = 0.715891$, which is rounded to two decimal places, $s = 0.72$.

Typically, you do the calculation for the standard deviation on your calculator or computer. The intermediate results are not rounded. This is done for accuracy.

- For the following problems, recall that **value = mean + (#ofSTDEVs)(standard deviation)**. Verify the mean and standard deviation or a calculator or computer.

- For a sample: $x = \bar{x} + (\text{\#ofSTDEVs})(s)$

- For a population: $x = \mu + (\text{\#ofSTDEVs})(\sigma)$

- For this example, use $x = \bar{x} + (\text{\#ofSTDEVs})(s)$ because the data is from a sample

a. Verify the mean and standard deviation on your calculator or computer.

b. Find the value that is one standard deviation above the mean. Find ($\bar{x} + 1s$).

c. Find the value that is two standard deviations below the mean. Find ($\bar{x} - 2s$).

d. Find the values that are 1.5 standard deviations **from** (below and above) the mean.

Solution 2.32

a. ◦ Clear lists L1 and L2. Press STAT 4:ClrList. Enter 2nd 1 for L1, the comma (,), and 2nd 2 for L2.

◦ Enter data into the list editor. Press STAT 1:EDIT. If necessary, clear the lists by arrowing up into the name. Press CLEAR and arrow down.

◦ Put the data values (9, 9.5, 10, 10.5, 11, 11.5) into list L1 and the frequencies (1, 2, 4, 4, 6, 3) into list L2. Use the arrow keys to move around.

◦ Press STAT and arrow to CALC. Press 1:1-VarStats and enter L1 (2nd 1), L2 (2nd 2). Do not forget the comma. Press ENTER.

◦ $\bar{x} = 10.525$

◦ Use Sx because this is sample data (not a population): Sx=0.715891

b. ($\bar{x} + 1s$) = 10.53 + (1)(0.72) = 11.25

c. ($\bar{x} - 2s$) = 10.53 − (2)(0.72) = 9.09

d. ◦ ($\bar{x} - 1.5s$) = 10.53 − (1.5)(0.72) = 9.45

◦ ($\bar{x} + 1.5s$) = 10.53 + (1.5)(0.72) = 11.61

Try It Σ

☞ **2.32** On a baseball team, the ages of each of the players are as follows:

21; 21; 22; 23; 24; 24; 25; 25; 28; 29; 29; 31; 32; 33; 33; 34; 35; 36; 36; 36; 36; 38; 38; 38; 40

Use your calculator or computer to find the mean and standard deviation. Then find the value that is two standard deviations above the mean.

Explanation of the standard deviation calculation shown in the table

The deviations show how spread out the data are about the mean. The data value 11.5 is farther from the mean than is the data value 11 which is indicated by the deviations 0.97 and 0.47. A positive deviation occurs when the data value is greater than the mean, whereas a negative deviation occurs when the data value is less than the mean. The deviation is −1.525 for the data value nine. **If you add the deviations, the sum is always zero.** (For **Example 2.32**, there are $n = 20$ deviations.) So you cannot simply add the deviations to get the spread of the data. By squaring the deviations, you make them positive numbers, and the sum will also be positive. The variance, then, is the average squared deviation.

The variance is a squared measure and does not have the same units as the data. Taking the square root solves the problem. The standard deviation measures the spread in the same units as the data.

Notice that instead of dividing by $n = 20$, the calculation divided by $n - 1 = 20 - 1 = 19$ because the data is a sample. For the **sample** variance, we divide by the sample size minus one ($n - 1$). Why not divide by n? The answer has to do with the population variance. **The sample variance is an estimate of the population variance.** Based on the theoretical mathematics that lies behind these calculations, dividing by ($n - 1$) gives a better estimate of the population variance.

NOTE

☞ Your concentration should be on what the standard deviation tells us about the data. The standard deviation is a number which measures how far the data are spread from the mean. Let a calculator or computer do the arithmetic.

The standard deviation, s or σ, is either zero or larger than zero. When the standard deviation is zero, there is no spread; that is, the all the data values are equal to each other. The standard deviation is small when the data are all concentrated close to the mean, and is larger when the data values show more variation from the mean. When the standard deviation is a lot larger than zero, the data values are very spread out about the mean; outliers can make s or σ very large.

The standard deviation, when first presented, can seem unclear. By graphing your data, you can get a better "feel" for the deviations and the standard deviation. You will find that in symmetrical distributions, the standard deviation can be very helpful but in skewed distributions, the standard deviation may not be much help. The reason is that the two sides of a skewed distribution have different spreads. In a skewed distribution, it is better to look at the first quartile, the median, the third quartile, the smallest value, and the largest value. Because numbers can be confusing, **always graph your data**. Display your data in a histogram or a box plot.

Example 2.33

Use the following data (first exam scores) from Susan Dean's spring pre-calculus class:

33; 42; 49; 49; 53; 55; 55; 61; 63; 67; 68; 68; 69; 69; 72; 73; 74; 78; 80; 83; 88; 88; 88; 90; 92; 94; 94; 94; 94; 96; 100

a. Create a chart containing the data, frequencies, relative frequencies, and cumulative relative frequencies to three decimal places.

b. Calculate the following to one decimal place using a TI-83+ or TI-84 calculator:

  i. The sample mean

  ii. The sample standard deviation

  iii. The median

  iv. The first quartile

  v. The third quartile

  vi. *IQR*

c. Construct a box plot and a histogram on the same set of axes. Make comments about the box plot, the histogram, and the chart.

Solution 2.33
a. See **Table 2.30**

b.   i.   The sample mean = 73.5

  ii.   The sample standard deviation = 17.9

  iii.   The median = 73

  iv.   The first quartile = 61

  v.   The third quartile = 90

vi. $IQR = 90 - 61 = 29$

c.   The x-axis goes from 32.5 to 100.5; y-axis goes from –2.4 to 15 for the histogram. The number of intervals is five, so the width of an interval is $(100.5 - 32.5)$ divided by five, is equal to 13.6. Endpoints of the intervals are as follows: the starting point is 32.5, 32.5 + 13.6 = 46.1, 46.1 + 13.6 = 59.7, 59.7 + 13.6 = 73.3, 73.3 + 13.6 = 86.9, 86.9 + 13.6 = 100.5 = the ending value; No data values fall on an interval boundary.

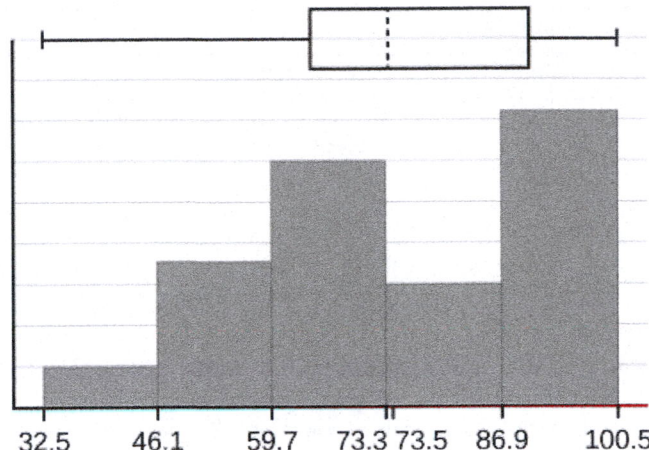

Figure 2.25

The long left whisker in the box plot is reflected in the left side of the histogram. The spread of the exam scores in the lower 50% is greater (73 – 33 = 40) than the spread in the upper 50% (100 – 73 = 27). The histogram, box plot, and chart all reflect this. There are a substantial number of A and B grades (80s, 90s, and 100). The histogram clearly shows this. The box plot shows us that the middle 50% of the exam scores ($IQR = 29$) are Ds, Cs, and Bs. The box plot also shows us that the lower 25% of the exam scores are Ds and Fs.

Data	Frequency	Relative Frequency	Cumulative Relative Frequency
33	1	0.032	0.032
42	1	0.032	0.064
49	2	0.065	0.129
53	1	0.032	0.161
55	2	0.065	0.226
61	1	0.032	0.258
63	1	0.032	0.29
67	1	0.032	0.322
68	2	0.065	0.387
69	2	0.065	0.452
72	1	0.032	0.484
73	1	0.032	0.516
74	1	0.032	0.548
78	1	0.032	0.580

Table 2.30

Data	Frequency	Relative Frequency	Cumulative Relative Frequency
80	1	0.032	0.612
83	1	0.032	0.644
88	3	0.097	0.741
90	1	0.032	0.773
92	1	0.032	0.805
94	4	0.129	0.934
96	1	0.032	0.966
100	1	0.032	0.998 (Why isn't this value 1?)

Table 2.30

Try It Σ

👉 **2.33** The following data show the different types of pet food stores in the area carry.
6; 6; 6; 6; 7; 7; 7; 7; 7; 8; 9; 9; 9; 9; 10; 10; 10; 10; 10; 11; 11; 11; 11; 12; 12; 12; 12; 12; 12;
Calculate the sample mean and the sample standard deviation to one decimal place using a TI-83+ or TI-84 calculator.

Standard deviation of Grouped Frequency Tables

Recall that for grouped data we do not know individual data values, so we cannot describe the typical value of the data with precision. In other words, we cannot find the exact mean, median, or mode. We can, however, determine the best estimate of

the measures of center by finding the mean of the grouped data with the formula: $Mean\ of\ Frequency\ Table = \dfrac{\sum fm}{\sum f}$

where $f =$ interval frequencies and $m =$ interval midpoints.

Just as we could not find the exact mean, neither can we find the exact standard deviation. Remember that standard deviation describes numerically the expected deviation a data value has from the mean. In simple English, the standard deviation allows us to compare how "unusual" individual data is compared to the mean.

Example 2.34

Find the standard deviation for the data in **Table 2.31**.

Class	Frequency, f	Midpoint, m	m^2	$\bar{x}^2$	fm^2	Standard Deviation
0–2	1	1	1	7.58	1	3.5
3–5	6	4	16	7.58	96	3.5
6–8	10	7	49	7.58	490	3.5
9–11	7	10	100	7.58	700	3.5
12–14	0	13	169	7.58	0	3.5
15–17	2	16	256	7.58	512	3.5

Table 2.31

For this data set, we have the mean, $\bar{x}$ = 7.58 and the standard deviation, s_x = 3.5. This means that a randomly selected data value would be expected to be 3.5 units from the mean. If we look at the first class, we see that the class midpoint is equal to one. This is almost two full standard deviations from the mean since 7.58 − 3.5 − 3.5 = 0.58. While the formula for calculating the standard deviation is not complicated, $s_x = \sqrt{\dfrac{f(m - \bar{x})^2}{n - 1}}$ where s_x = sample standard deviation, $\bar{x}$ = sample mean, the calculations are tedious. It is usually best to use technology when performing the calculations.

Try It Σ

☞ **2.34** Find the standard deviation for the data from the previous example

Class	Frequency, *f*
0–2	1
3–5	6
6–8	10
9–11	7
12–14	0
15–17	2

Table 2.32

First, press the **STAT** key and select **1:Edit**

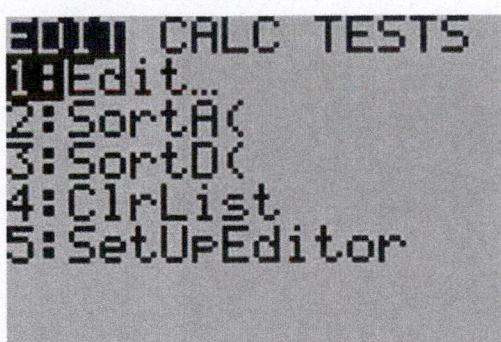

Figure 2.26

Input the midpoint values into **L1** and the frequencies into **L2**

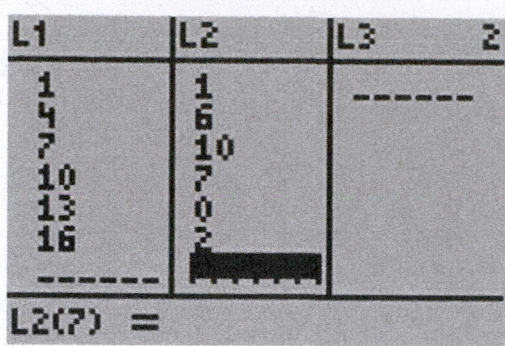

Figure 2.27

Select **STAT**, **CALC**, and **1: 1-Var Stats**

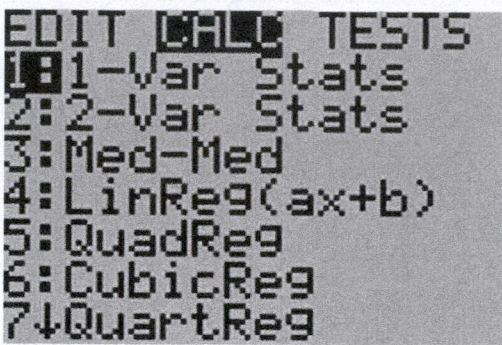

Figure 2.28

Select 2nd then **1** then , 2nd then **2 Enter**

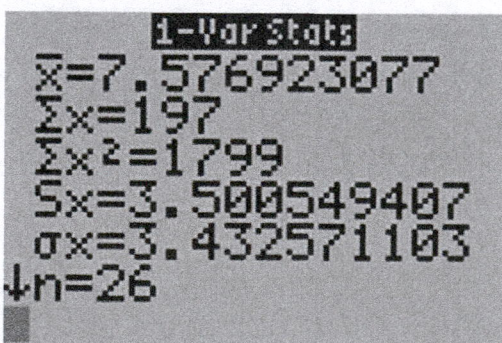

Figure 2.29

You will see displayed both a population standard deviation, σ_x, and the sample standard deviation, s_x.

Comparing Values from Different Data Sets

The standard deviation is useful when comparing data values that come from different data sets. If the data sets have different means and standard deviations, then comparing the data values directly can be misleading.

- For each data value, calculate how many standard deviations away from its mean the value is.
- Use the formula: value = mean + (#ofSTDEVs)(standard deviation); solve for #ofSTDEVs.

- $\# ofSTDEVs = \dfrac{value - mean}{standard\ deviation}$

- Compare the results of this calculation.

#ofSTDEVs is often called a "z-score"; we can use the symbol z. In symbols, the formulas become:

Sample	$x = \bar{x} + zs$	$z = \dfrac{x - \bar{x}}{s}$
Population	$x = \mu + z\sigma$	$z = \dfrac{x - \mu}{\sigma}$

Table 2.33

Example 2.35

Two students, John and Ali, from different high schools, wanted to find out who had the highest GPA when compared to his school. Which student had the highest GPA when compared to his school?

Student	GPA	School Mean GPA	School Standard Deviation
John	2.85	3.0	0.7
Ali	77	80	10

Table 2.34

Solution 2.35

For each student, determine how many standard deviations (#ofSTDEVs) his GPA is away from the average, for his school. Pay careful attention to signs when comparing and interpreting the answer.

$$z = \#\ of\ STDEVs = \dfrac{value\ -\ mean}{standard\ deviation} = \dfrac{x + \mu}{\sigma}$$

For John, $z = \# ofSTDEVs = \dfrac{2.85 - 3.0}{0.7} = -0.21$

For Ali, $z = \# ofSTDEVs = \dfrac{77 - 80}{10} = -0.3$

John has the better GPA when compared to his school because his GPA is 0.21 standard deviations **below** his school's mean while Ali's GPA is 0.3 standard deviations **below** his school's mean.

John's z-score of –0.21 is higher than Ali's z-score of –0.3. For GPA, higher values are better, so we conclude that John has the better GPA when compared to his school.

Try It Σ

2.35 Two swimmers, Angie and Beth, from different teams, wanted to find out who had the fastest time for the 50 meter freestyle when compared to her team. Which swimmer had the fastest time when compared to her team?

Swimmer	Time (seconds)	Team Mean Time	Team Standard Deviation
Angie	26.2	27.2	0.8
Beth	27.3	30.1	1.4

Table 2.35

The following lists give a few facts that provide a little more insight into what the standard deviation tells us about the distribution of the data.

For ANY data set, no matter what the distribution of the data is:

- At least 75% of the data is within two standard deviations of the mean.

- At least 89% of the data is within three standard deviations of the mean.

- At least 95% of the data is within 4.5 standard deviations of the mean.

- This is known as Chebyshev's Rule.

For data having a distribution that is BELL-SHAPED and SYMMETRIC:

- Approximately 68% of the data is within one standard deviation of the mean.

- Approximately 95% of the data is within two standard deviations of the mean.

- More than 99% of the data is within three standard deviations of the mean.

- This is known as the Empirical Rule.

- It is important to note that this rule only applies when the shape of the distribution of the data is bell-shaped and symmetric. We will learn more about this when studying the "Normal" or "Gaussian" probability distribution in later chapters.

2.8 | Descriptive Statistics

Stats **l.**ab

2.1 Descriptive Statistics

Class Time:

Names:

Student Learning Outcomes

- The student will construct a histogram and a box plot.
- The student will calculate univariate statistics.
- The student will examine the graphs to interpret what the data implies.

Collect the Data

Record the number of pairs of shoes you own.

1. Randomly survey 30 classmates about the number of pairs of shoes they own. Record their values.

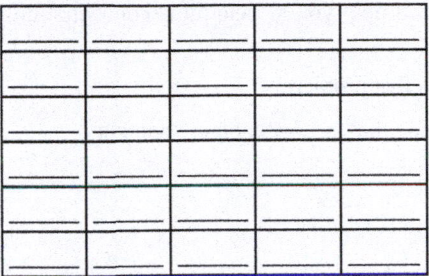

Table 2.36 Survey Results

2. Construct a histogram. Make five to six intervals. Sketch the graph using a ruler and pencil and scale the axes.

Figure 2.30

3. Calculate the following values.

   a. $\bar{x} = $ _____

   b. $s = $ _____

4. Are the data discrete or continuous? How do you know?

5. In complete sentences, describe the shape of the histogram.

6. Are there any potential outliers? List the value(s) that could be outliers. Use a formula to check the end values to determine if they are potential outliers.

Analyze the Data

1. Determine the following values.

    a. Min = _____

    b. M = _____

    c. Max = _____

    d. Q_1 = _____

    e. Q_3 = _____

    f. IQR = _____

2. Construct a box plot of data

3. What does the shape of the box plot imply about the concentration of data? Use complete sentences.

4. Using the box plot, how can you determine if there are potential outliers?

5. How does the standard deviation help you to determine concentration of the data and whether or not there are potential outliers?

6. What does the IQR represent in this problem?

7. Show your work to find the value that is 1.5 standard deviations:

    a. above the mean.

    b. below the mean.

KEY TERMS

Box plot a graph that gives a quick picture of the middle 50% of the data

First Quartile the value that is the median of the of the lower half of the ordered data set

Frequency the number of times a value of the data occurs

Frequency Polygon looks like a line graph but uses intervals to display ranges of large amounts of data

Frequency Table a data representation in which grouped data is displayed along with the corresponding frequencies

Histogram a graphical representation in x-y form of the distribution of data in a data set; x represents the data and y represents the frequency, or relative frequency. The graph consists of contiguous rectangles.

Interquartile Range or IQR, is the range of the middle 50 percent of the data values; the IQR is found by subtracting the first quartile from the third quartile.

Interval also called a class interval; an interval represents a range of data and is used when displaying large data sets

Mean a number that measures the central tendency of the data; a common name for mean is 'average.' The term 'mean' is a shortened form of 'arithmetic mean.' By definition, the mean for a sample (denoted by $\bar{x}$) is $\bar{x} = \dfrac{\text{Sum of all values in the sample}}{\text{Number of values in the sample}}$, and the mean for a population (denoted by μ) is $\mu = \dfrac{\text{Sum of all values in the population}}{\text{Number of values in the population}}$.

Median a number that separates ordered data into halves; half the values are the same number or smaller than the median and half the values are the same number or larger than the median. The median may or may not be part of the data.

Midpoint the mean of an interval in a frequency table

Mode the value that appears most frequently in a set of data

Outlier an observation that does not fit the rest of the data

Paired Data Set two data sets that have a one to one relationship so that:
- both data sets are the same size, and
- each data point in one data set is matched with exactly one point from the other set.

Percentile a number that divides ordered data into hundredths; percentiles may or may not be part of the data. The median of the data is the second quartile and the 50th percentile. The first and third quartiles are the 25th and the 75th percentiles, respectively.

Quartiles the numbers that separate the data into quarters; quartiles may or may not be part of the data. The second quartile is the median of the data.

Relative Frequency the ratio of the number of times a value of the data occurs in the set of all outcomes to the number of all outcomes

Skewed used to describe data that is not symmetrical; when the right side of a graph looks "chopped off" compared the left side, we say it is "skewed to the left." When the left side of the graph looks "chopped off" compared to the right side, we say the data is "skewed to the right." Alternatively: when the lower values of the data are more spread out, we say the data are skewed to the left. When the greater values are more spread out, the data are skewed to the right.

Standard Deviation a number that is equal to the square root of the variance and measures how far data values are from their mean; notation: s for sample standard deviation and σ for population standard deviation.

Variance mean of the squared deviations from the mean, or the square of the standard deviation; for a set of data, a deviation can be represented as $x - \bar{x}$ where x is a value of the data and $\bar{x}$ is the sample mean. The sample variance is equal to the sum of the squares of the deviations divided by the difference of the sample size and one.

CHAPTER REVIEW

2.1 Stem-and-Leaf Graphs (Stemplots), Line Graphs, and Bar Graphs

A **stem-and-leaf plot** is a way to plot data and look at the distribution. In a stem-and-leaf plot, all data values within a class are visible. The advantage in a stem-and-leaf plot is that all values are listed, unlike a histogram, which gives classes of data values. A **line graph** is often used to represent a set of data values in which a quantity varies with time. These graphs are useful for finding trends. That is, finding a general pattern in data sets including temperature, sales, employment, company profit or cost over a period of time. A **bar graph** is a chart that uses either horizontal or vertical bars to show comparisons among categories. One axis of the chart shows the specific categories being compared, and the other axis represents a discrete value. Some bar graphs present bars clustered in groups of more than one (grouped bar graphs), and others show the bars divided into subparts to show cumulative effect (stacked bar graphs). Bar graphs are especially useful when categorical data is being used.

2.2 Histograms, Frequency Polygons, and Time Series Graphs

A **histogram** is a graphic version of a frequency distribution. The graph consists of bars of equal width drawn adjacent to each other. The horizontal scale represents classes of quantitative data values and the vertical scale represents frequencies. The heights of the bars correspond to frequency values. Histograms are typically used for large, continuous, quantitative data sets. A frequency polygon can also be used when graphing large data sets with data points that repeat. The data usually goes on y-axis with the frequency being graphed on the x-axis. Time series graphs can be helpful when looking at large amounts of data for one variable over a period of time.

2.3 Measures of the Location of the Data

The values that divide a rank-ordered set of data into 100 equal parts are called percentiles. Percentiles are used to compare and interpret data. For example, an observation at the 50th percentile would be greater than 50 percent of the other obeservations in the set. Quartiles divide data into quarters. The first quartile (Q_1) is the 25th percentile,the second quartile (Q_2 or median) is 50th percentile, and the third quartile (Q_3) is the the 75th percentile. The interquartile range, or IQR, is the range of the middle 50 percent of the data values. The IQR is found by subtracting Q_1 from Q_3, and can help determine outliers by using the following two expressions.

- $Q_3 + IQR(1.5)$
- $Q_1 - IQR(1.5)$

2.4 Box Plots

Box plots are a type of graph that can help visually organize data. To graph a box plot the following data points must be calculated: the minimum value, the first quartile, the median, the third quartile, and the maximum value. Once the box plot is graphed, you can display and compare distributions of data.

2.5 Measures of the Center of the Data

The mean and the median can be calculated to help you find the "center" of a data set. The mean is the best estimate for the actual data set, but the median is the best measurement when a data set contains several outliers or extreme values. The mode will tell you the most frequently occuring datum (or data) in your data set. The mean, median, and mode are extremely helpful when you need to analyze your data, but if your data set consists of ranges which lack specific values, the mean may seem impossible to calculate. However, the mean can be approximated if you add the lower boundary with the upper boundary and divide by two to find the midpoint of each interval. Multiply each midpoint by the number of values found in the corresponding range. Divide the sum of these values by the total number of data values in the set.

2.6 Skewness and the Mean, Median, and Mode

Looking at the distribution of data can reveal a lot about the relationship between the mean, the median, and the mode. There are <u>three types of distributions</u>. A **right (or positive) skewed** distribution has a shape like **Figure 2.17**. A **left (or negative) skewed** distribution has a shape like **Figure 2.18**. A **symmetrical** distrubtion looks like **Figure 2.16**.

2.7 Measures of the Spread of the Data

The standard deviation can help you calculate the spread of data. There are different equations to use if are calculating the standard deviation of a sample or of a population.

- The Standard Deviation allows us to compare individual data or classes to the data set mean numerically.

- $s = \sqrt{\dfrac{\sum (x - \bar{x})^2}{n - 1}}$ or $s = \sqrt{\dfrac{\sum f(x - \bar{x})^2}{n - 1}}$ is the formula for calculating the standard deviation of a sample.

  To calculate the standard deviation of a population, we would use the population mean, μ, and the formula $\sigma = \sqrt{\dfrac{\sum (x - \mu)^2}{N}}$ or $\sigma = \sqrt{\dfrac{\sum f(x - \mu)^2}{N}}$.

FORMULA REVIEW

2.3 Measures of the Location of the Data

$i = \left(\dfrac{k}{100}\right)(n + 1)$

where i = the ranking or position of a data value,

k = the kth percentile,

n = total number of data.

Expression for finding the percentile of a data value:
$\left(\dfrac{x + 0.5y}{n}\right)(100)$

where x = the number of values counting from the bottom of the data list up to but not including the data value for which you want to find the percentile,

y = the number of data values equal to the data value for which you want to find the percentile,

n = total number of data

2.5 Measures of the Center of the Data

$\mu = \dfrac{\sum fm}{\sum f}$ Where f = interval frequencies and m = interval midpoints.

2.7 Measures of the Spread of the Data

$s_x = \sqrt{\dfrac{\sum fm^2}{n} - \bar{x}^2}$ where

s_x = sample standard deviation

$\bar{x}$ = sample mean

PRACTICE

2.1 Stem-and-Leaf Graphs (Stemplots), Line Graphs, and Bar Graphs

For each of the following data sets, create a stem plot and identify any outliers.

1. The miles per gallon rating for 30 cars are shown below (lowest to highest).
19, 19, 19, 20, 21, 21, 25, 25, 25, 26, 26, 28, 29, 31, 31, 32, 32, 33, 34, 35, 36, 37, 37, 38, 38, 38, 38, 41, 43, 43

2. The height in feet of 25 trees is shown below (lowest to highest).
25, 27, 33, 34, 34, 34, 35, 37, 37, 38, 39, 39, 39, 40, 41, 45, 46, 47, 49, 50, 50, 53, 53, 54, 54

3. The data are the prices of different laptops at an electronics store. Round each value to the nearest ten.
249, 249, 260, 265, 265, 280, 299, 299, 309, 319, 325, 326, 350, 350, 350, 365, 369, 389, 409, 459, 489, 559, 569, 570, 610

4. The data are daily high temperatures in a town for one month.
61, 61, 62, 64, 66, 67, 67, 67, 68, 69, 70, 70, 70, 71, 71, 72, 74, 74, 74, 75, 75, 75, 76, 76, 77, 78, 78, 79, 79, 95

For the next three exercises, use the data to construct a line graph.

5. In a survey, 40 people were asked how many times they visited a store before making a major purchase. The results are shown in **Table 2.37**.

Number of times in store	Frequency
1	4
2	10
3	16
4	6
5	4

Table 2.37

6. In a survey, several people were asked how many years it has been since they purchased a mattress. The results are shown in **Table 2.38**.

Years since last purchase	Frequency
0	2
1	8
2	13
3	22
4	16
5	9

Table 2.38

7. Several children were asked how many TV shows they watch each day. The results of the survey are shown in **Table 2.39**.

Number of TV Shows	Frequency
0	12
1	18
2	36
3	7
4	2

Table 2.39

8. The students in Ms. Ramirez's math class have birthdays in each of the four seasons. **Table 2.40** shows the four seasons, the number of students who have birthdays in each season, and the percentage (%) of students in each group. Construct a bar graph showing the number of students.

Seasons	Number of students	Proportion of population
Spring	8	24%
Summer	9	26%
Autumn	11	32%
Winter	6	18%

Table 2.40

9. Using the data from Mrs. Ramirez's math class supplied in **Exercise 2.8**, construct a bar graph showing the percentages.

10. David County has six high schools. Each school sent students to participate in a county-wide science competition. **Table 2.41** shows the percentage breakdown of competitors from each school, and the percentage of the entire student population of the county that goes to each school. Construct a bar graph that shows the population percentage of competitors from each school.

High School	Science competition population	Overall student population
Alabaster	28.9%	8.6%
Concordia	7.6%	23.2%
Genoa	12.1%	15.0%
Mocksville	18.5%	14.3%
Tynneson	24.2%	10.1%
West End	8.7%	28.8%

Table 2.41

11. Use the data from the David County science competition supplied in **Exercise 2.10**. Construct a bar graph that shows the county-wide population percentage of students at each school.

2.2 Histograms, Frequency Polygons, and Time Series Graphs

12. Sixty-five randomly selected car salespersons were asked the number of cars they generally sell in one week. Fourteen people answered that they generally sell three cars; nineteen generally sell four cars; twelve generally sell five cars; nine generally sell six cars; eleven generally sell seven cars. Complete the table.

Data Value (# cars)	Frequency	Relative Frequency	Cumulative Relative Frequency

Table 2.42

13. What does the frequency column in **Table 2.42** sum to? Why?

14. What does the relative frequency column in **Table 2.42** sum to? Why?

15. What is the difference between relative frequency and frequency for each data value in **Table 2.42**?

16. What is the difference between cumulative relative frequency and relative frequency for each data value?

17. To construct the histogram for the data in **Table 2.42**, determine appropriate minimum and maximum x and y values and the scaling. Sketch the histogram. Label the horizontal and vertical axes with words. Include numerical scaling.

Figure 2.31

18. Construct a frequency polygon for the following:

a.

Pulse Rates for Women	Frequency
60–69	12
70–79	14
80–89	11
90–99	1
100–109	1
110–119	0
120–129	1

Table 2.43

b.

Actual Speed in a 30 MPH Zone	Frequency
42–45	25
46–49	14
50–53	7
54–57	3
58–61	1

Table 2.44

c.

Tar (mg) in Nonfiltered Cigarettes	Frequency
10–13	1
14–17	0
18–21	15

Tar (mg) in Nonfiltered Cigarettes	Frequency
22–25	7
26–29	2

Table 2.45

19. Construct a frequency polygon from the frequency distribution for the 50 highest ranked countries for depth of hunger.

Depth of Hunger	Frequency
230–259	21
260–289	13
290–319	5
320–349	7
350–379	1
380–409	1
410–439	1

Table 2.46

20. Use the two frequency tables to compare the life expectancy of men and women from 20 randomly selected countries. Include an overlayed frequency polygon and discuss the shapes of the distributions, the center, the spread, and any outliers. What can we conclude about the life expectancy of women compared to men?

Life Expectancy at Birth – Women	Frequency
49–55	3
56–62	3
63–69	1
70–76	3
77–83	8
84–90	2

Table 2.47

Life Expectancy at Birth – Men	Frequency
49–55	3
56–62	3
63–69	1
70–76	1
77–83	7
84–90	5

21. Construct a times series graph for (a) the number of male births, (b) the number of female births, and (c) the total number of births.

Sex/Year	1855	1856	1857	1858	1859	1860	1861
Female	45,545	49,582	50,257	50,324	51,915	51,220	52,403
Male	47,804	52,239	53,158	53,694	54,628	54,409	54,606
Total	93,349	101,821	103,415	104,018	106,543	105,629	107,009

Table 2.49

Sex/Year	1862	1863	1864	1865	1866	1867	1868	1869
Female	51,812	53,115	54,959	54,850	55,307	55,527	56,292	55,033
Male	55,257	56,226	57,374	58,220	58,360	58,517	59,222	58,321
Total	107,069	109,341	112,333	113,070	113,667	114,044	115,514	113,354

Table 2.50

Sex/Year	1871	1870	1872	1871	1872	1827	1874	1875
Female	56,099	56,431	57,472	56,099	57,472	58,233	60,109	60,146
Male	60,029	58,959	61,293	60,029	61,293	61,467	63,602	63,432
Total	116,128	115,390	118,765	116,128	118,765	119,700	123,711	123,578

Table 2.51

22. The following data sets list full time police per 100,000 citizens along with homicides per 100,000 citizens for the city of Detroit, Michigan during the period from 1961 to 1973.

Year	1961	1962	1963	1964	1965	1966	1967
Police	260.35	269.8	272.04	272.96	272.51	261.34	268.89
Homicides	8.6	8.9	8.52	8.89	13.07	14.57	21.36

Table 2.52

Year	1968	1969	1970	1971	1972	1973
Police	295.99	319.87	341.43	356.59	376.69	390.19
Homicides	28.03	31.49	37.39	46.26	47.24	52.33

Table 2.53

a. Construct a double time series graph using a common *x*-axis for both sets of data.
b. Which variable increased the fastest? Explain.
c. Did Detroit's increase in police officers have an impact on the murder rate? Explain.

2.3 Measures of the Location of the Data

23. Listed are 29 ages for Academy Award winning best actors *in order from smallest to largest.*

18; 21; 22; 25; 26; 27; 29; 30; 31; 33; 36; 37; 41; 42; 47; 52; 55; 57; 58; 62; 64; 67; 69; 71; 72; 73; 74; 76; 77

    a. Find the 40th percentile.
    b. Find the 78th percentile.

24. Listed are 32 ages for Academy Award winning best actors *in order from smallest to largest.*

18; 18; 21; 22; 25; 26; 27; 29; 30; 31; 31; 33; 36; 37; 37; 41; 42; 47; 52; 55; 57; 58; 62; 64; 67; 69; 71; 72; 73; 74; 76; 77

    a. Find the percentile of 37.
    b. Find the percentile of 72.

25. Jesse was ranked 37th in his graduating class of 180 students. At what percentile is Jesse's ranking?

26.
    a. For runners in a race, a low time means a faster run. The winners in a race have the shortest running times. Is it more desirable to have a finish time with a high or a low percentile when running a race?
    b. The 20th percentile of run times in a particular race is 5.2 minutes. Write a sentence interpreting the 20th percentile in the context of the situation.
    c. A bicyclist in the 90th percentile of a bicycle race completed the race in 1 hour and 12 minutes. Is he among the fastest or slowest cyclists in the race? Write a sentence interpreting the 90th percentile in the context of the situation.

27.
    a. For runners in a race, a higher speed means a faster run. Is it more desirable to have a speed with a high or a low percentile when running a race?
    b. The 40th percentile of speeds in a particular race is 7.5 miles per hour. Write a sentence interpreting the 40th percentile in the context of the situation.

28. On an exam, would it be more desirable to earn a grade with a high or low percentile? Explain.

29. Mina is waiting in line at the Department of Motor Vehicles (DMV). Her wait time of 32 minutes is the 85th percentile of wait times. Is that good or bad? Write a sentence interpreting the 85th percentile in the context of this situation.

30. In a survey collecting data about the salaries earned by recent college graduates, Li found that her salary was in the 78th percentile. Should Li be pleased or upset by this result? Explain.

31. In a study collecting data about the repair costs of damage to automobiles in a certain type of crash tests, a certain model of car had $1,700 in damage and was in the 90th percentile. Should the manufacturer and the consumer be pleased or upset by this result? Explain and write a sentence that interprets the 90th percentile in the context of this problem.

32. The University of California has two criteria used to set admission standards for freshman to be admitted to a college in the UC system:
    a. Students' GPAs and scores on standardized tests (SATs and ACTs) are entered into a formula that calculates an "admissions index" score. The admissions index score is used to set eligibility standards intended to meet the goal of admitting the top 12% of high school students in the state. In this context, what percentile does the top 12% represent?
    b. Students whose GPAs are at or above the 96th percentile of all students at their high school are eligible (called eligible in the local context), even if they are not in the top 12% of all students in the state. What percentage of students from each high school are "eligible in the local context"?

33. Suppose that you are buying a house. You and your realtor have determined that the most expensive house you can afford is the 34th percentile. The 34th percentile of housing prices is $240,000 in the town you want to move to. In this town, can you afford 34% of the houses or 66% of the houses?
Use **Exercise 2.25** to calculate the following values:

34. First quartile = _____

35. Second quartile = median = 50th percentile = _____

36. Third quartile = _____

37. Interquartile range (*IQR*) = _____ – _____ = _____

38. 10th percentile = _____

39. 70th percentile = _____

2.4 Box Plots

Sixty-five randomly selected car salespersons were asked the number of cars they generally sell in one week. Fourteen people answered that they generally sell three cars; nineteen generally sell four cars; twelve generally sell five cars; nine generally sell six cars; eleven generally sell seven cars.

40. Construct a box plot below. Use a ruler to measure and scale accurately.

41. Looking at your box plot, does it appear that the data are concentrated together, spread out evenly, or concentrated in some areas, but not in others? How can you tell?

2.5 Measures of the Center of the Data

42. Find the mean for the following frequency tables.

a.

Grade	Frequency
49.5–59.5	2
59.5–69.5	3
69.5–79.5	8
79.5–89.5	12
89.5–99.5	5

Table 2.54

b.

Daily Low Temperature	Frequency
49.5–59.5	53
59.5–69.5	32
69.5–79.5	15
79.5–89.5	1
89.5–99.5	0

Table 2.55

c.

Points per Game	Frequency
49.5–59.5	14
59.5–69.5	32
69.5–79.5	15
79.5–89.5	23

Points per Game	Frequency
89.5–99.5	2

Table 2.56

Use the following information to answer the next three exercises: The following data show the lengths of boats moored in a marina. The data are ordered from smallest to largest: 16; 17; 19; 20; 20; 21; 23; 24; 25; 25; 25; 26; 26; 27; 27; 27; 28; 29; 30; 32; 33; 33; 34; 35; 37; 39; 40

43. Calculate the mean.

44. Identify the median.

45. Identify the mode.

Use the following information to answer the next three exercises: Sixty-five randomly selected car salespersons were asked the number of cars they generally sell in one week. Fourteen people answered that they generally sell three cars; nineteen generally sell four cars; twelve generally sell five cars; nine generally sell six cars; eleven generally sell seven cars. Calculate the following:

46. sample mean = $\bar{x}$ = _____

47. median = _____

48. mode = _____

2.6 Skewness and the Mean, Median, and Mode

Use the following information to answer the next three exercises: State whether the data are symmetrical, skewed to the left, or skewed to the right.

49. 1; 1; 1; 2; 2; 2; 2; 3; 3; 3; 3; 3; 3; 3; 3; 4; 4; 4; 5; 5

50. 16; 17; 19; 22; 22; 22; 22; 22; 23

51. 87; 87; 87; 87; 87; 88; 89; 89; 90; 91

52. When the data are skewed left, what is the typical relationship between the mean and median?

53. When the data are symmetrical, what is the typical relationship between the mean and median?

54. What word describes a distribution that has two modes?

55. Describe the shape of this distribution.

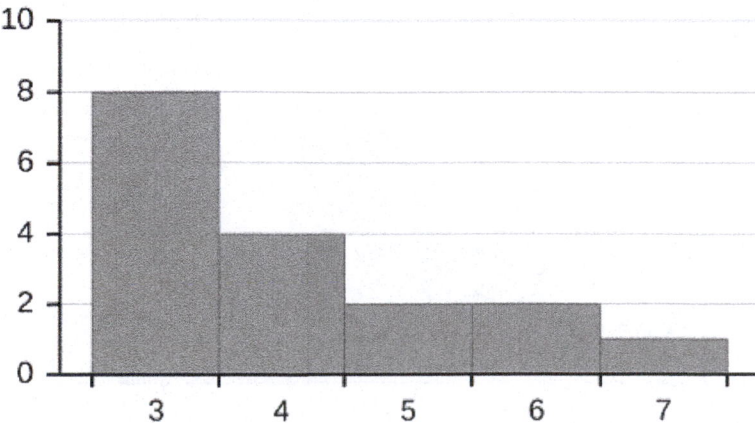

Figure 2.32

56. Describe the relationship between the mode and the median of this distribution.

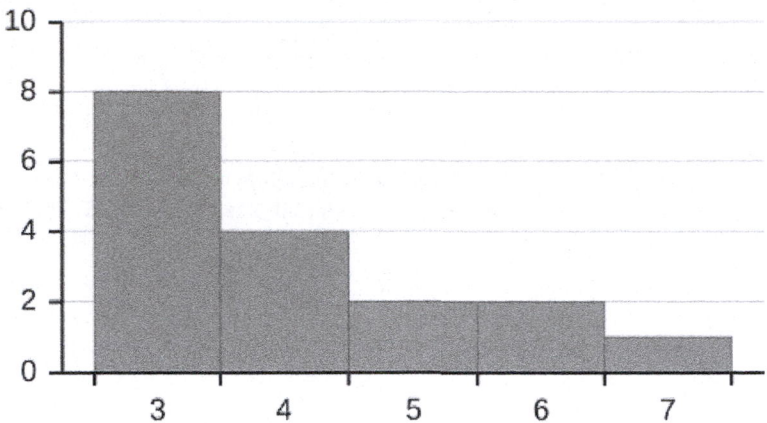

Figure 2.33

57. Describe the relationship between the mean and the median of this distribution.

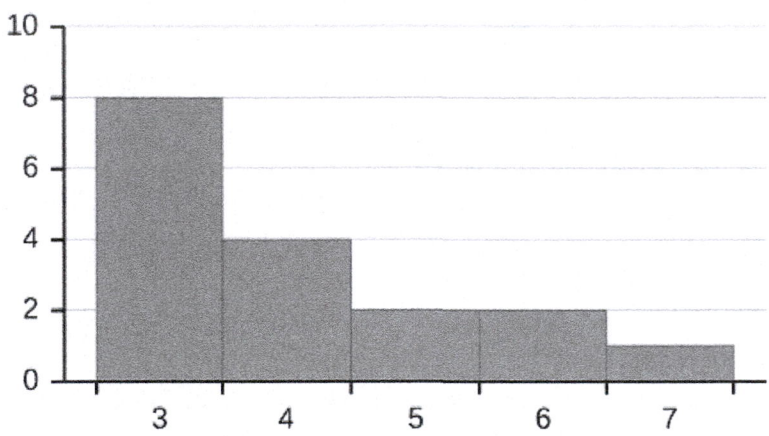

Figure 2.34

58. Describe the shape of this distribution.

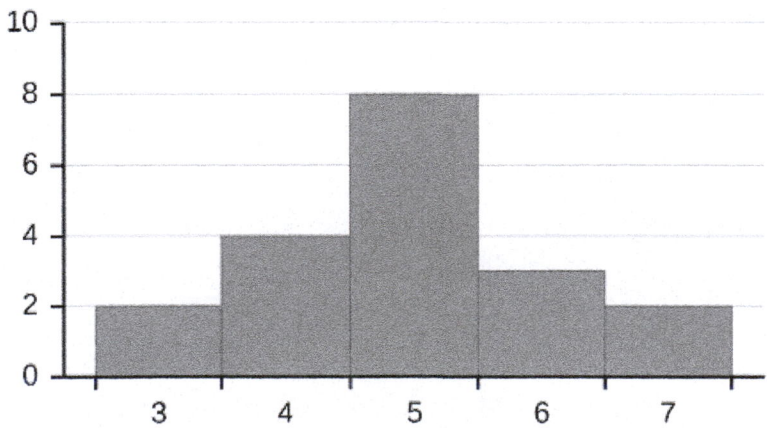

Figure 2.35

59. Describe the relationship between the mode and the median of this distribution.

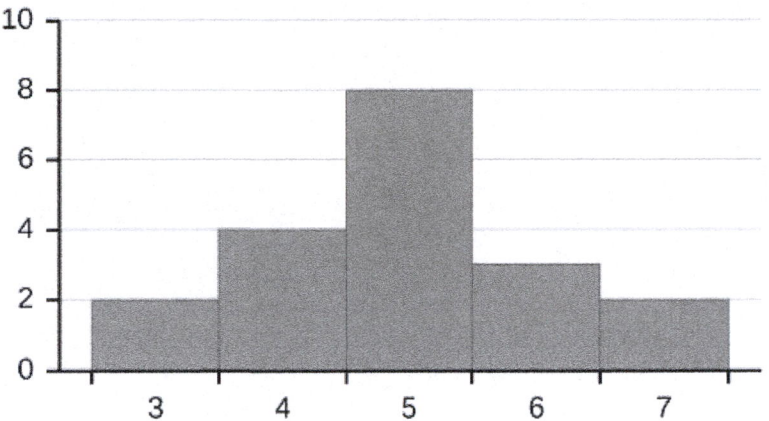

Figure 2.36

60. Are the mean and the median the exact same in this distribution? Why or why not?

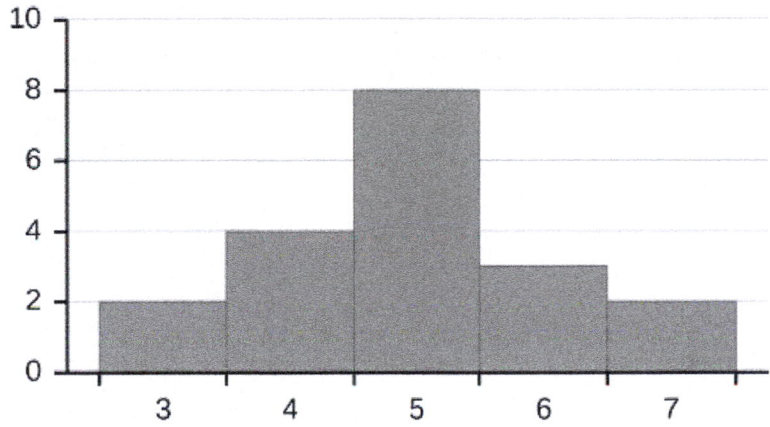

Figure 2.37

61. Describe the shape of this distribution.

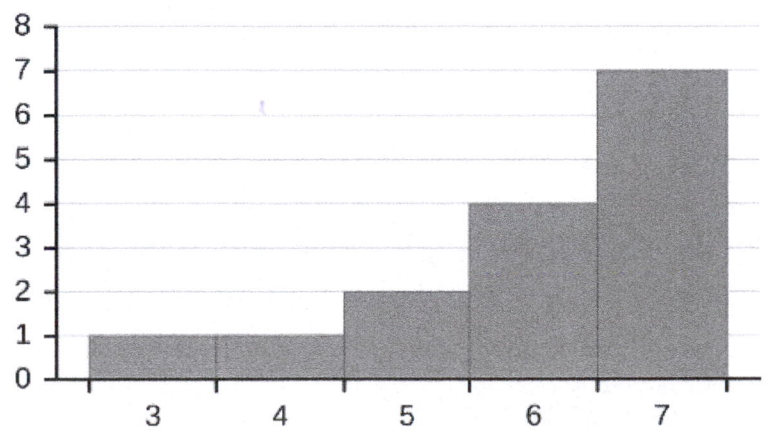

Figure 2.38

62. Describe the relationship between the mode and the median of this distribution.

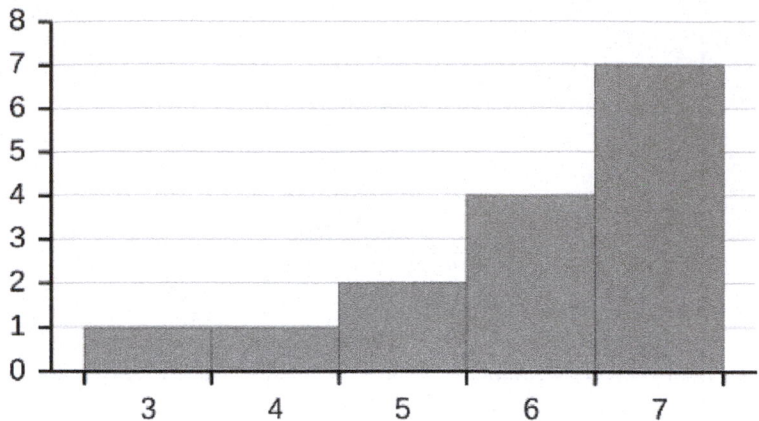

Figure 2.39

63. Describe the relationship between the mean and the median of this distribution.

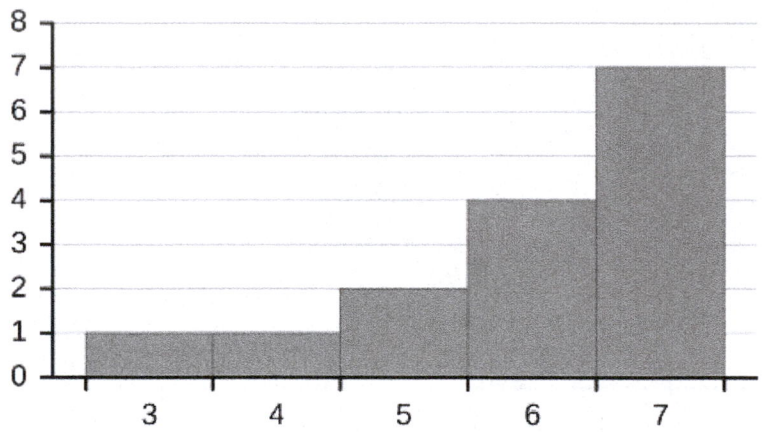

Figure 2.40

64. The mean and median for the data are the same.

3; 4; 5; 5; 6; 6; 6; 6; 7; 7; 7; 7; 7; 7; 7

Is the data perfectly symmetrical? Why or why not?

65. Which is the greatest, the mean, the mode, or the median of the data set?

11; 11; 12; 12; 12; 12; 13; 15; 17; 22; 22; 22

66. Which is the least, the mean, the mode, and the median of the data set?

56; 56; 56; 58; 59; 60; 62; 64; 64; 65; 67

67. Of the three measures, which tends to reflect skewing the most, the mean, the mode, or the median? Why?

68. In a perfectly symmetrical distribution, when would the mode be different from the mean and median?

2.7 Measures of the Spread of the Data

Use the following information to answer the next two exercises: The following data are the distances between 20 retail stores and a large distribution center. The distances are in miles.
29; 37; 38; 40; 58; 67; 68; 69; 76; 86; 87; 95; 96; 96; 99; 106; 112; 127; 145; 150

69. Use a graphing calculator or computer to find the standard deviation and round to the nearest tenth.

70. Find the value that is one standard deviation below the mean.

71. Two baseball players, Fredo and Karl, on different teams wanted to find out who had the higher batting average when compared to his team. Which baseball player had the higher batting average when compared to his team?

Baseball Player	Batting Average	Team Batting Average	Team Standard Deviation
Fredo	0.158	0.166	0.012
Karl	0.177	0.189	0.015

Table 2.57

72. Use **Table 2.57** to find the value that is three standard deviations:
a. above the mean
b. below the mean

Find the standard deviation for the following frequency tables using the formula. Check the calculations with the TI 83/84.

73. Find the standard deviation for the following frequency tables using the formula. Check the calculations with the TI 83/84.

a.

Grade	Frequency
49.5–59.5	2
59.5–69.5	3
69.5–79.5	8
79.5–89.5	12
89.5–99.5	5

Table 2.58

b.

Daily Low Temperature	Frequency
49.5–59.5	53
59.5–69.5	32
69.5–79.5	15
79.5–89.5	1
89.5–99.5	0

Table 2.59

c.

Points per Game	Frequency
49.5–59.5	14
59.5–69.5	32
69.5–79.5	15
79.5–89.5	23
89.5–99.5	2

HOMEWORK

2.1 Stem-and-Leaf Graphs (Stemplots), Line Graphs, and Bar Graphs

74. Student grades on a chemistry exam were: 77, 78, 76, 81, 86, 51, 79, 82, 84, 99
    a. Construct a stem-and-leaf plot of the data.
    b. Are there any potential outliers? If so, which scores are they? Why do you consider them outliers?

75. Table 2.61 contains the 2010 obesity rates in U.S. states and Washington, DC.

State	Percent (%)	State	Percent (%)	State	Percent (%)
Alabama	32.2	Kentucky	31.3	North Dakota	27.2
Alaska	24.5	Louisiana	31.0	Ohio	29.2
Arizona	24.3	Maine	26.8	Oklahoma	30.4
Arkansas	30.1	Maryland	27.1	Oregon	26.8
California	24.0	Massachusetts	23.0	Pennsylvania	28.6
Colorado	21.0	Michigan	30.9	Rhode Island	25.5
Connecticut	22.5	Minnesota	24.8	South Carolina	31.5
Delaware	28.0	Mississippi	34.0	South Dakota	27.3
Washington, DC	22.2	Missouri	30.5	Tennessee	30.8
Florida	26.6	Montana	23.0	Texas	31.0
Georgia	29.6	Nebraska	26.9	Utah	22.5
Hawaii	22.7	Nevada	22.4	Vermont	23.2
Idaho	26.5	New Hampshire	25.0	Virginia	26.0
Illinois	28.2	New Jersey	23.8	Washington	25.5
Indiana	29.6	New Mexico	25.1	West Virginia	32.5
Iowa	28.4	New York	23.9	Wisconsin	26.3
Kansas	29.4	North Carolina	27.8	Wyoming	25.1

Table 2.61

    a. Use a random number generator to randomly pick eight states. Construct a bar graph of the obesity rates of those eight states.
    b. Construct a bar graph for all the states beginning with the letter "A."
    c. Construct a bar graph for all the states beginning with the letter "M."

2.2 Histograms, Frequency Polygons, and Time Series Graphs

76. Suppose that three book publishers were interested in the number of fiction paperbacks adult consumers purchase per month. Each publisher conducted a survey. In the survey, adult consumers were asked the number of fiction paperbacks they had purchased the previous month. The results are as follows:

# of books	Freq.	Rel. Freq.
0	10	
1	12	
2	16	
3	12	
4	8	
5	6	
6	2	
8	2	

Table 2.62 Publisher A

# of books	Freq.	Rel. Freq.
0	18	
1	24	
2	24	
3	22	
4	15	
5	10	
7	5	
9	1	

Table 2.63 Publisher B

# of books	Freq.	Rel. Freq.
0–1	20	
2–3	35	
4–5	12	
6–7	2	
8–9	1	

Table 2.64 Publisher C

a. Find the relative frequencies for each survey. Write them in the charts.
b. Using either a graphing calculator, computer, or by hand, use the frequency column to construct a histogram for each publisher's survey. For Publishers A and B, make bar widths of one. For Publisher C, make bar widths of two.
c. In complete sentences, give two reasons why the graphs for Publishers A and B are not identical.
d. Would you have expected the graph for Publisher C to look like the other two graphs? Why or why not?
e. Make new histograms for Publisher A and Publisher B. This time, make bar widths of two.
f. Now, compare the graph for Publisher C to the new graphs for Publishers A and B. Are the graphs more similar or more different? Explain your answer.

77. Often, cruise ships conduct all on-board transactions, with the exception of gambling, on a cashless basis. At the end of the cruise, guests pay one bill that covers all onboard transactions. Suppose that 60 single travelers and 70 couples were surveyed as to their on-board bills for a seven-day cruise from Los Angeles to the Mexican Riviera. Following is a summary of the bills for each group.

Amount($)	Frequency	Rel. Frequency
51–100	5	
101–150	10	
151–200	15	
201–250	15	
251–300	10	
301–350	5	

Table 2.65 Singles

Amount($)	Frequency	Rel. Frequency
100–150	5	
201–250	5	
251–300	5	
301–350	5	
351–400	10	
401–450	10	
451–500	10	
501–550	10	
551–600	5	
601–650	5	

Table 2.66 Couples

a. Fill in the relative frequency for each group.
b. Construct a histogram for the singles group. Scale the x-axis by $50 widths. Use relative frequency on the y-axis.
c. Construct a histogram for the couples group. Scale the x-axis by $50 widths. Use relative frequency on the y-axis.
d. Compare the two graphs:
   i. List two similarities between the graphs.
   ii. List two differences between the graphs.
   iii. Overall, are the graphs more similar or different?
e. Construct a new graph for the couples by hand. Since each couple is paying for two individuals, instead of scaling the x-axis by $50, scale it by $100. Use relative frequency on the y-axis.
f. Compare the graph for the singles with the new graph for the couples:
   i. List two similarities between the graphs.
   ii. Overall, are the graphs more similar or different?
g. How did scaling the couples graph differently change the way you compared it to the singles graph?
h. Based on the graphs, do you think that individuals spend the same amount, more or less, as singles as they do person by person as a couple? Explain why in one or two complete sentences.

78. Twenty-five randomly selected students were asked the number of movies they watched the previous week. The results are as follows.

# of movies	Frequency	Relative Frequency	Cumulative Relative Frequency
0	5		
1	9		
2	6		
3	4		
4	1		

Table 2.67

    a. Construct a histogram of the data.
    b. Complete the columns of the chart.

Use the following information to answer the next two exercises: Suppose one hundred eleven people who shopped in a special t-shirt store were asked the number of t-shirts they own costing more than $19 each.

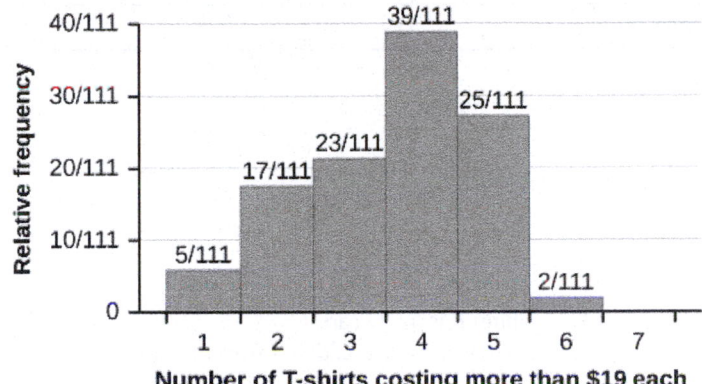

79. The percentage of people who own at most three t-shirts costing more than $19 each is approximately:
    a. 21
    b. 59
    c. 41
    d. Cannot be determined

80. If the data were collected by asking the first 111 people who entered the store, then the type of sampling is:
    a. cluster
    b. simple random
    c. stratified
    d. convenience

81. Following are the 2010 obesity rates by U.S. states and Washington, DC.

State	Percent (%)	State	Percent (%)	State	Percent (%)
Alabama	32.2	Kentucky	31.3	North Dakota	27.2
Alaska	24.5	Louisiana	31.0	Ohio	29.2
Arizona	24.3	Maine	26.8	Oklahoma	30.4
Arkansas	30.1	Maryland	27.1	Oregon	26.8

State	Percent (%)	State	Percent (%)	State	Percent (%)
California	24.0	Massachusetts	23.0	Pennsylvania	28.6
Colorado	21.0	Michigan	30.9	Rhode Island	25.5
Connecticut	22.5	Minnesota	24.8	South Carolina	31.5
Delaware	28.0	Mississippi	34.0	South Dakota	27.3
Washington, DC	22.2	Missouri	30.5	Tennessee	30.8
Florida	26.6	Montana	23.0	Texas	31.0
Georgia	29.6	Nebraska	26.9	Utah	22.5
Hawaii	22.7	Nevada	22.4	Vermont	23.2
Idaho	26.5	New Hampshire	25.0	Virginia	26.0
Illinois	28.2	New Jersey	23.8	Washington	25.5
Indiana	29.6	New Mexico	25.1	West Virginia	32.5
Iowa	28.4	New York	23.9	Wisconsin	26.3
Kansas	29.4	North Carolina	27.8	Wyoming	25.1

Table 2.68

Construct a bar graph of obesity rates of your state and the four states closest to your state. Hint: Label the x-axis with the states.

2.3 Measures of the Location of the Data

82. The median age for U.S. blacks currently is 30.9 years; for U.S. whites it is 42.3 years.
   a. Based upon this information, give two reasons why the black median age could be lower than the white median age.
   b. Does the lower median age for blacks necessarily mean that blacks die younger than whites? Why or why not?
   c. How might it be possible for blacks and whites to die at approximately the same age, but for the median age for whites to be higher?

83. Six hundred adult Americans were asked by telephone poll, "What do you think constitutes a middle-class income?" The results are in **Table 2.69**. Also, include left endpoint, but not the right endpoint.

Salary ($)	Relative Frequency
< 20,000	0.02
20,000–25,000	0.09
25,000–30,000	0.19
30,000–40,000	0.26
40,000–50,000	0.18
50,000–75,000	0.17
75,000–99,999	0.02
100,000+	0.01

Table 2.69

      a.   What percentage of the survey answered "not sure"?

      b.   What percentage think that middle-class is from $25,000 to $50,000?

      c.   Construct a histogram of the data.

           i.   Should all bars have the same width, based on the data? Why or why not?

          ii.   How should the <20,000 and the 100,000+ intervals be handled? Why?

      d.   Find the 40th and 80th percentiles

      e.   Construct a bar graph of the data

84. Given the following box plot:

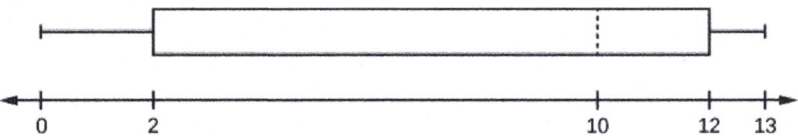

Figure 2.41

      a.   which quarter has the smallest spread of data? What is that spread?

      b.   which quarter has the largest spread of data? What is that spread?

      c.   find the interquartile range (*IQR*).

      d.   are there more data in the interval 5–10 or in the interval 10–13? How do you know this?

      e.   which interval has the fewest data in it? How do you know this?

           i.   0–2

          ii.   2–4

         iii.   10–12

         iv.   12–13

          v.   need more information

85. The following box plot shows the U.S. population for 1990, the latest available year.

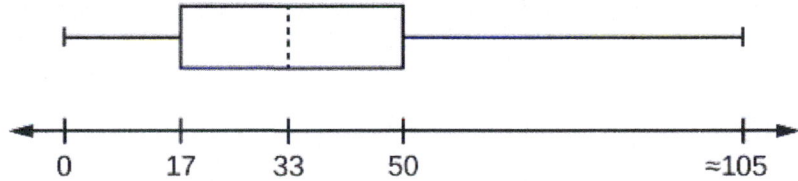

Figure 2.42

      a.   Are there fewer or more children (age 17 and under) than senior citizens (age 65 and over)? How do you know?

      b.   12.6% are age 65 and over. Approximately what percentage of the population are working age adults (above age 17 to age 65)?

2.4 Box Plots

86. In a survey of 20-year-olds in China, Germany, and the United States, people were asked the number of foreign countries they had visited in their lifetime. The following box plots display the results.

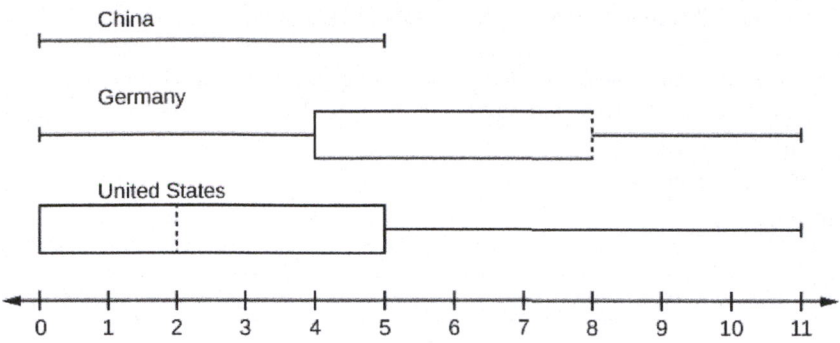

Figure 2.43
   a. In complete sentences, describe what the shape of each box plot implies about the distribution of the data collected.
   b. Have more Americans or more Germans surveyed been to over eight foreign countries?
   c. Compare the three box plots. What do they imply about the foreign travel of 20-year-old residents of the three countries when compared to each other?

87. Given the following box plot, answer the questions.

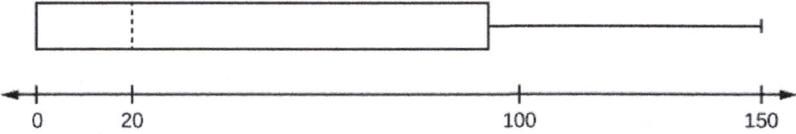

Figure 2.44
   a. Think of an example (in words) where the data might fit into the above box plot. In 2–5 sentences, write down the example.
   b. What does it mean to have the first and second quartiles so close together, while the second to third quartiles are far apart?

88. Given the following box plots, answer the questions.

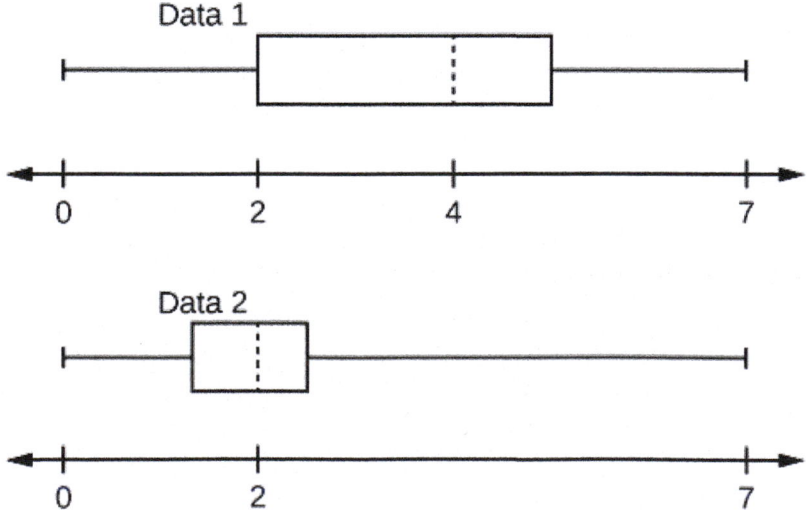

Figure 2.45
   a. In complete sentences, explain why each statement is false.
      i. **Data 1** has more data values above two than **Data 2** has above two.

     ii.   The data sets cannot have the same mode.

    iii.   For **Data 1**, there are more data values below four than there are above four.

  b. For which group, Data 1 or Data 2, is the value of "7" more likely to be an outlier? Explain why in complete sentences.

89. A survey was conducted of 130 purchasers of new BMW 3 series cars, 130 purchasers of new BMW 5 series cars, and 130 purchasers of new BMW 7 series cars. In it, people were asked the age they were when they purchased their car. The following box plots display the results.

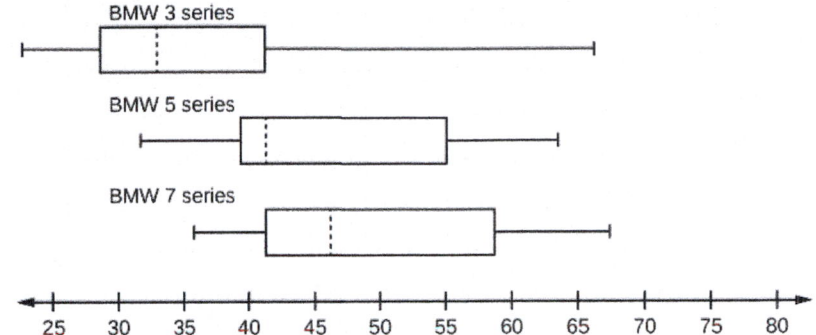

Figure 2.46

  a. In complete sentences, describe what the shape of each box plot implies about the distribution of the data collected for that car series.

  b. Which group is most likely to have an outlier? Explain how you determined that.

  c. Compare the three box plots. What do they imply about the age of purchasing a BMW from the series when compared to each other?

  d. Look at the BMW 5 series. Which quarter has the smallest spread of data? What is the spread?

  e. Look at the BMW 5 series. Which quarter has the largest spread of data? What is the spread?

  f. Look at the BMW 5 series. Estimate the interquartile range (IQR).

  g. Look at the BMW 5 series. Are there more data in the interval 31 to 38 or in the interval 45 to 55? How do you know this?

  h. Look at the BMW 5 series. Which interval has the fewest data in it? How do you know this?

     i.   31–35

    ii.   38–41

    iii.   41–64

90. Twenty-five randomly selected students were asked the number of movies they watched the previous week. The results are as follows:

# of movies	Frequency
0	5
1	9
2	6
3	4
4	1

Table 2.70

Construct a box plot of the data.

2.5 Measures of the Center of the Data

91. The most obese countries in the world have obesity rates that range from 11.4% to 74.6%. This data is summarized in the following table.

Percent of Population Obese	Number of Countries
11.4–20.45	29
20.45–29.45	13
29.45–38.45	4
38.45–47.45	0
47.45–56.45	2
56.45–65.45	1
65.45–74.45	0
74.45–83.45	1

Table 2.71

a. What is the best estimate of the average obesity percentage for these countries?
b. The United States has an average obesity rate of 33.9%. Is this rate above average or below?
c. How does the United States compare to other countries?

92. Table 2.72 gives the percent of children under five considered to be underweight. What is the best estimate for the mean percentage of underweight children?

Percent of Underweight Children	Number of Countries
16–21.45	23
21.45–26.9	4
26.9–32.35	9
32.35–37.8	7
37.8–43.25	6
43.25–48.7	1

Table 2.72

2.6 Skewness and the Mean, Median, and Mode

93. The median age of the U.S. population in 1980 was 30.0 years. In 1991, the median age was 33.1 years.
a. What does it mean for the median age to rise?
b. Give two reasons why the median age could rise.
c. For the median age to rise, is the actual number of children less in 1991 than it was in 1980? Why or why not?

2.7 Measures of the Spread of the Data

Use the following information to answer the next nine exercises: The population parameters below describe the full-time equivalent number of students (FTES) each year at Lake Tahoe Community College from 1976–1977 through 2004–2005.

- $\mu = 1000$ FTES
- median = 1,014 FTES

- $\sigma = 474$ FTES

- first quartile = 528.5 FTES

- third quartile = 1,447.5 FTES

- $n = 29$ years

94. A sample of 11 years is taken. About how many are expected to have a FTES of 1014 or above? Explain how you determined your answer.

95. 75% of all years have an FTES:
  a. at or below: _____
  b. at or above: _____

96. The population standard deviation = _____

97. What percent of the FTES were from 528.5 to 1447.5? How do you know?

98. What is the *IQR*? What does the *IQR* represent?

99. How many standard deviations away from the mean is the median?

Additional Information: The population FTES for 2005–2006 through 2010–2011 was given in an updated report. The data are reported here.

Year	2005–06	2006–07	2007–08	2008–09	2009–10	2010–11
Total FTES	1,585	1,690	1,735	1,935	2,021	1,890

Table 2.73

100. Calculate the mean, median, standard deviation, the first quartile, the third quartile and the *IQR*. Round to one decimal place.

101. Construct a box plot for the FTES for 2005–2006 through 2010–2011 and a box plot for the FTES for 1976–1977 through 2004–2005.

102. Compare the *IQR* for the FTES for 1976–77 through 2004–2005 with the *IQR* for the FTES for 2005-2006 through 2010–2011. Why do you suppose the *IQR*s are so different?

103. Three students were applying to the same graduate school. They came from schools with different grading systems. Which student had the best GPA when compared to other students at his school? Explain how you determined your answer.

Student	GPA	School Average GPA	School Standard Deviation
Thuy	2.7	3.2	0.8
Vichet	87	75	20
Kamala	8.6	8	0.4

Table 2.74

104. A music school has budgeted to purchase three musical instruments. They plan to purchase a piano costing $3,000, a guitar costing $550, and a drum set costing $600. The mean cost for a piano is $4,000 with a standard deviation of $2,500. The mean cost for a guitar is $500 with a standard deviation of $200. The mean cost for drums is $700 with a standard deviation of $100. Which cost is the lowest, when compared to other instruments of the same type? Which cost is the highest when compared to other instruments of the same type. Justify your answer.

105. An elementary school class ran one mile with a mean of 11 minutes and a standard deviation of three minutes. Rachel, a student in the class, ran one mile in eight minutes. A junior high school class ran one mile with a mean of nine minutes and a standard deviation of two minutes. Kenji, a student in the class, ran 1 mile in 8.5 minutes. A high school class ran one mile with a mean of seven minutes and a standard deviation of four minutes. Nedda, a student in the class, ran one mile in eight minutes.

a. Why is Kenji considered a better runner than Nedda, even though Nedda ran faster than he?
b. Who is the fastest runner with respect to his or her class? Explain why.

106. The most obese countries in the world have obesity rates that range from 11.4% to 74.6%. This data is summarized in **Table 14**.

Percent of Population Obese	Number of Countries
11.4–20.45	29
20.45–29.45	13
29.45–38.45	4
38.45–47.45	0
47.45–56.45	2
56.45–65.45	1
65.45–74.45	0
74.45–83.45	1

Table 2.75

What is the best estimate of the average obesity percentage for these countries? What is the standard deviation for the listed obesity rates? The United States has an average obesity rate of 33.9%. Is this rate above average or below? How "unusual" is the United States' obesity rate compared to the average rate? Explain.

107. Table 2.76 gives the percent of children under five considered to be underweight.

Percent of Underweight Children	Number of Countries
16–21.45	23
21.45–26.9	4
26.9–32.35	9
32.35–37.8	7
37.8–43.25	6
43.25–48.7	1

Table 2.76

What is the best estimate for the mean percentage of underweight children? What is the standard deviation? Which interval(s) could be considered unusual? Explain.

BRINGING IT TOGETHER: HOMEWORK

108. Santa Clara County, CA, has approximately 27,873 Japanese-Americans. Their ages are as follows:

Age Group	Percent of Community
0–17	18.9
18–24	8.0
25–34	22.8
35–44	15.0
45–54	13.1
55–64	11.9
65+	10.3

Table 2.77

a. Construct a histogram of the Japanese-American community in Santa Clara County, CA. The bars will **not** be the same width for this example. Why not? What impact does this have on the reliability of the graph?
b. What percentage of the community is under age 35?
c. Which box plot most resembles the information above?

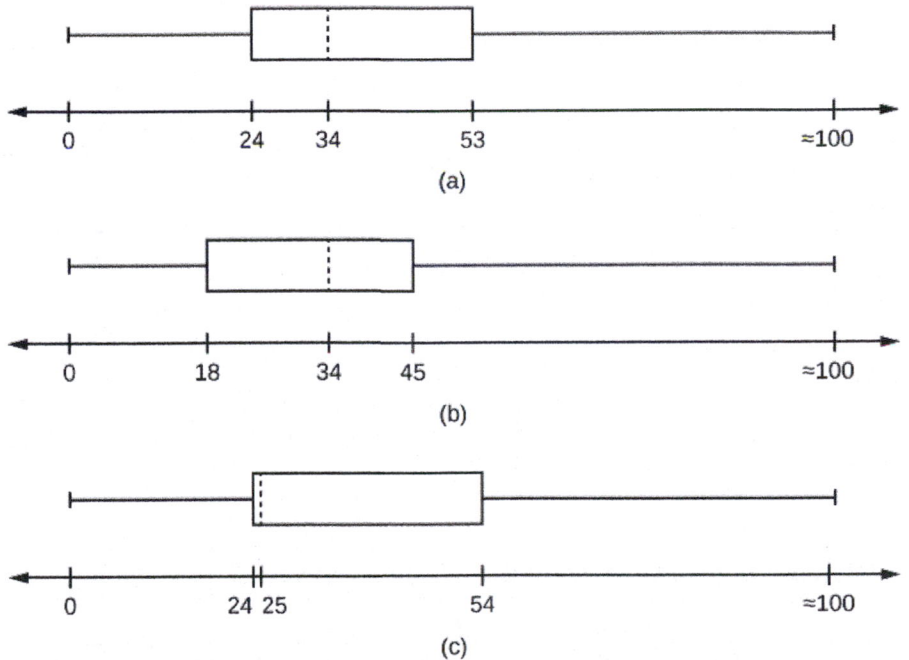

Figure 2.47

109. Javier and Ercilia are supervisors at a shopping mall. Each was given the task of estimating the mean distance that shoppers live from the mall. They each randomly surveyed 100 shoppers. The samples yielded the following information.

	Javier	Ercilia
$\bar{x}$	6.0 miles	6.0 miles
s	4.0 miles	7.0 miles

Table 2.78

a. How can you determine which survey was correct ?
b. Explain what the difference in the results of the surveys implies about the data.
c. If the two histograms depict the distribution of values for each supervisor, which one depicts Ercilia's sample? How do you know?

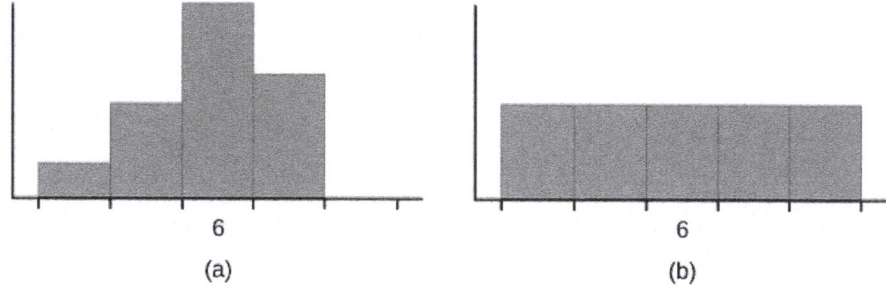

Figure 2.48

d. If the two box plots depict the distribution of values for each supervisor, which one depicts Ercilia's sample? How do you know?

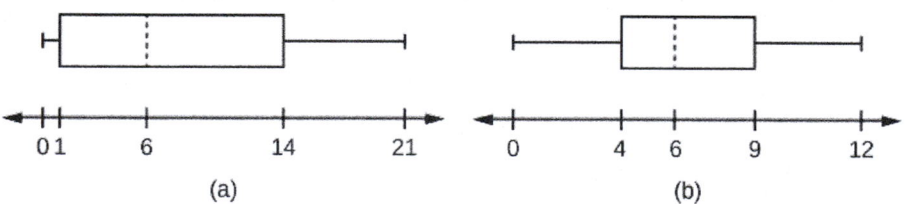

Figure 2.49

Use the following information to answer the next three exercises: We are interested in the number of years students in a particular elementary statistics class have lived in California. The information in the following table is from the entire section.

Number of years	Frequency	Number of years	Frequency
7	1	22	1
14	3	23	1
15	1	26	1
18	1	40	2
19	4	42	2
			Total = 20

Number of years	Frequency	Number of years	Frequency
20	3		
			Total = 20

Table 2.79

110. What is the *IQR*?
   a. 8
   b. 11
   c. 15
   d. 35

111. What is the mode?
   a. 19
   b. 19.5
   c. 14 and 20
   d. 22.65

112. Is this a sample or the entire population?
   a. sample
   b. entire population
   c. neither

113. Twenty-five randomly selected students were asked the number of movies they watched the previous week. The results are as follows:

# of movies	Frequency
0	5
1	9
2	6
3	4
4	1

Table 2.80

   a. Find the sample mean $\bar{x}$.
   b. Find the approximate sample standard deviation, *s*.

114. Forty randomly selected students were asked the number of pairs of sneakers they owned. Let *X* = the number of pairs of sneakers owned. The results are as follows:

X	Frequency
1	2
2	5
3	8
4	12

Table 2.81

X	Frequency
5	12
6	0
7	1

Table 2.81

a. Find the sample mean $\bar{x}$
b. Find the sample standard deviation, s
c. Construct a histogram of the data.
d. Complete the columns of the chart.
e. Find the first quartile.
f. Find the median.
g. Find the third quartile.
h. Construct a box plot of the data.
i. What percent of the students owned at least five pairs?
j. Find the 40th percentile.
k. Find the 90th percentile.
l. Construct a line graph of the data
m. Construct a stemplot of the data

115. Following are the published weights (in pounds) of all of the team members of the San Francisco 49ers from a previous year.

177; 205; 210; 210; 232; 205; 185; 185; 178; 210; 206; 212; 184; 174; 185; 242; 188; 212; 215; 247; 241; 223; 220; 260; 245; 259; 278; 270; 280; 295; 275; 285; 290; 272; 273; 280; 285; 286; 200; 215; 185; 230; 250; 241; 190; 260; 250; 302; 265; 290; 276; 228; 265

a. Organize the data from smallest to largest value.
b. Find the median.
c. Find the first quartile.
d. Find the third quartile.
e. Construct a box plot of the data.
f. The middle 50% of the weights are from _____ to _____.
g. If our population were all professional football players, would the above data be a sample of weights or the population of weights? Why?
h. If our population included every team member who ever played for the San Francisco 49ers, would the above data be a sample of weights or the population of weights? Why?
i. Assume the population was the San Francisco 49ers. Find:
    i. the population mean, μ.
    ii. the population standard deviation, σ.
    iii. the weight that is two standard deviations below the mean.
    iv. When Steve Young, quarterback, played football, he weighed 205 pounds. How many standard deviations above or below the mean was he?
j. That same year, the mean weight for the Dallas Cowboys was 240.08 pounds with a standard deviation of 44.38 pounds. Emmit Smith weighed in at 209 pounds. With respect to his team, who was lighter, Smith or Young? How did you determine your answer?

116. One hundred teachers attended a seminar on mathematical problem solving. The attitudes of a representative sample of 12 of the teachers were measured before and after the seminar. A positive number for change in attitude indicates that a teacher's attitude toward math became more positive. The 12 change scores are as follows:

3; 8; –1; 2; 0; 5; –3; 1; –1; 6; 5; –2

    a. What is the mean change score?
    b. What is the standard deviation for this population?
    c. What is the median change score?
    d. Find the change score that is 2.2 standard deviations below the mean.

117. Refer to **Figure 2.50** determine which of the following are true and which are false. Explain your solution to each part in complete sentences.

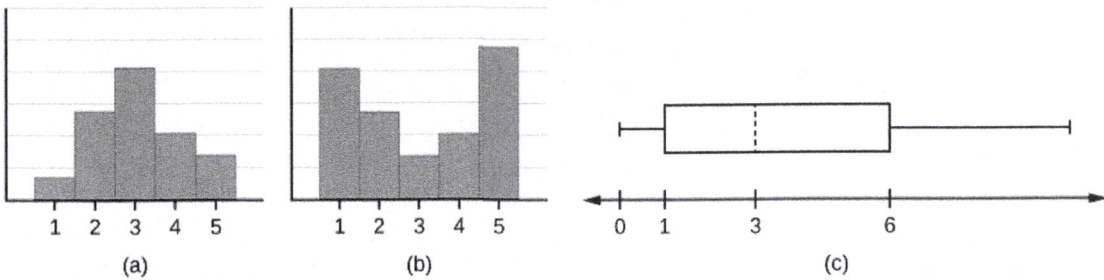

Figure 2.50
    a. The medians for all three graphs are the same.
    b. We cannot determine if any of the means for the three graphs is different.
    c. The standard deviation for graph b is larger than the standard deviation for graph a.
    d. We cannot determine if any of the third quartiles for the three graphs is different.

118. In a recent issue of the *IEEE Spectrum*, 84 engineering conferences were announced. Four conferences lasted two days. Thirty-six lasted three days. Eighteen lasted four days. Nineteen lasted five days. Four lasted six days. One lasted seven days. One lasted eight days. One lasted nine days. Let X = the length (in days) of an engineering conference.

    a. Organize the data in a chart.
    b. Find the median, the first quartile, and the third quartile.
    c. Find the 65th percentile.
    d. Find the 10th percentile.
    e. Construct a box plot of the data.
    f. The middle 50% of the conferences last from _____ days to _____ days.
    g. Calculate the sample mean of days of engineering conferences.
    h. Calculate the sample standard deviation of days of engineering conferences.
    i. Find the mode.
    j. If you were planning an engineering conference, which would you choose as the length of the conference: mean; median; or mode? Explain why you made that choice.
    k. Give two reasons why you think that three to five days seem to be popular lengths of engineering conferences.

119. A survey of enrollment at 35 community colleges across the United States yielded the following figures:

6414; 1550; 2109; 9350; 21828; 4300; 5944; 5722; 2825; 2044; 5481; 5200; 5853; 2750; 10012; 6357; 27000; 9414; 7681; 3200; 17500; 9200; 7380; 18314; 6557; 13713; 17768; 7493; 2771; 2861; 1263; 7285; 28165; 5080; 11622

    a. Organize the data into a chart with five intervals of equal width. Label the two columns "Enrollment" and "Frequency."
    b. Construct a histogram of the data.
    c. If you were to build a new community college, which piece of information would be more valuable: the mode or the mean?
    d. Calculate the sample mean.
    e. Calculate the sample standard deviation.
    f. A school with an enrollment of 8000 would be how many standard deviations away from the mean?

Use the following information to answer the next two exercises. X = the number of days per week that 100 clients use a particular exercise facility.

x	Frequency
0	3
1	12
2	33
3	28
4	11
5	9
6	4

Table 2.82

120. The 80th percentile is _____
   a. 5
   b. 80
   c. 3
   d. 4

121. The number that is 1.5 standard deviations BELOW the mean is approximately _____
   a. 0.7
   b. 4.8
   c. −2.8
   d. Cannot be determined

122. Suppose that a publisher conducted a survey asking adult consumers the number of fiction paperback books they had purchased in the previous month. The results are summarized in the **Table 2.83**.

# of books	Freq.	Rel. Freq.
0	18	
1	24	
2	24	
3	22	
4	15	
5	10	
7	5	
9	1	

Table 2.83

   a. Are there any outliers in the data? Use an appropriate numerical test involving the *IQR* to identify outliers, if any, and clearly state your conclusion.
   b. If a data value is identified as an outlier, what should be done about it?
   c. Are any data values further than two standard deviations away from the mean? In some situations, statisticians may use this criteria to identify data values that are unusual, compared to the other data values. (Note that this criteria is most appropriate to use for data that is mound-shaped and symmetric, rather than for skewed data.)
   d. Do parts a and c of this problem give the same answer?
   e. Examine the shape of the data. Which part, a or c, of this question gives a more appropriate result for this data?

    f.   Based on the shape of the data which is the most appropriate measure of center for this data: mean, median or mode?

REFERENCES

2.1 Stem-and-Leaf Graphs (Stemplots), Line Graphs, and Bar Graphs

Burbary, Ken. *Facebook Demographics Revisited – 2001 Statistics*, 2011. Available online at http://www.kenburbary.com/2011/03/facebook-demographics-revisited-2011-statistics-2/ (accessed August 21, 2013).

"9th Annual AP Report to the Nation." CollegeBoard, 2013. Available online at http://apreport.collegeboard.org/goals-and-findings/promoting-equity (accessed September 13, 2013).

"Overweight and Obesity: Adult Obesity Facts." Centers for Disease Control and Prevention. Available online at http://www.cdc.gov/obesity/data/adult.html (accessed September 13, 2013).

2.2 Histograms, Frequency Polygons, and Time Series Graphs

Data on annual homicides in Detroit, 1961–73, from Gunst & Mason's book 'Regression Analysis and its Application', Marcel Dekker

"Timeline: Guide to the U.S. Presidents: Information on every president's birthplace, political party, term of office, and more." Scholastic, 2013. Available online at http://www.scholastic.com/teachers/article/timeline-guide-us-presidents (accessed April 3, 2013).

"Presidents." Fact Monster. Pearson Education, 2007. Available online at http://www.factmonster.com/ipka/A0194030.html (accessed April 3, 2013).

"Food Security Statistics." Food and Agriculture Organization of the United Nations. Available online at http://www.fao.org/economic/ess/ess-fs/en/ (accessed April 3, 2013).

"Consumer Price Index." United States Department of Labor: Bureau of Labor Statistics. Available online at http://data.bls.gov/pdq/SurveyOutputServlet (accessed April 3, 2013).

"CO_2 emissions (kt)." The World Bank, 2013. Available online at http://databank.worldbank.org/data/home.aspx (accessed April 3, 2013).

"Births Time Series Data." General Register Office For Scotland, 2013. Available online at http://www.gro-scotland.gov.uk/statistics/theme/vital-events/births/time-series.html (accessed April 3, 2013).

"Demographics: Children under the age of 5 years underweight." Indexmundi. Available online at http://www.indexmundi.com/g/r.aspx?t=50&v=2224&aml=en (accessed April 3, 2013).

Gunst, Richard, Robert Mason. *Regression Analysis and Its Application: A Data-Oriented Approach*. CRC Press: 1980.

"Overweight and Obesity: Adult Obesity Facts." Centers for Disease Control and Prevention. Available online at http://www.cdc.gov/obesity/data/adult.html (accessed September 13, 2013).

2.3 Measures of the Location of the Data

Cauchon, Dennis, Paul Overberg. "Census data shows minorities now a majority of U.S. births." USA Today, 2012. Available online at http://usatoday30.usatoday.com/news/nation/story/2012-05-17/minority-birthscensus/55029100/1 (accessed April 3, 2013).

Data from the United States Department of Commerce: United States Census Bureau. Available online at http://www.census.gov/ (accessed April 3, 2013).

"1990 Census." United States Department of Commerce: United States Census Bureau. Available online at http://www.census.gov/main/www/cen1990.html (accessed April 3, 2013).

Data from *San Jose Mercury News*.

Data from *Time Magazine*; survey by Yankelovich Partners, Inc.

2.4 Box Plots

Data from *West Magazine*.

2.5 Measures of the Center of the Data

Data from The World Bank, available online at http://www.worldbank.org (accessed April 3, 2013).

"Demographics: Obesity – adult prevalence rate." Indexmundi. Available online at http://www.indexmundi.com/g/r.aspx?t=50&v=2228&l=en (accessed April 3, 2013).

2.7 Measures of the Spread of the Data

Data from Microsoft Bookshelf.

King, Bill."Graphically Speaking." Institutional Research, Lake Tahoe Community College. Available online at http://www.ltcc.edu/web/about/institutional-research (accessed April 3, 2013).

SOLUTIONS

1

Stem	Leaf
1	9 9 9
2	0 1 1 5 5 5 6 6 8 9
3	1 1 2 2 3 4 5 6 7 7 8 8 8 8
4	1 3 3

Table 2.84

3

Stem	Leaf
2	5 5 6 7 7 8
3	0 0 1 2 3 3 5 5 5 7 7 9
4	1 6 9
5	6 7 7
6	1

Table 2.85

5

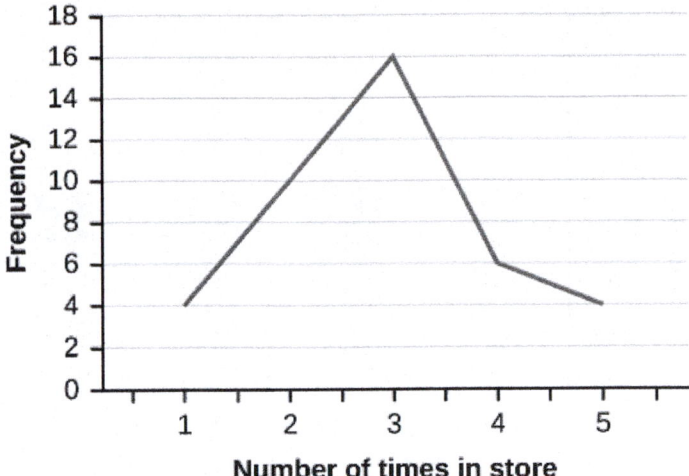

Figure 2.51

7

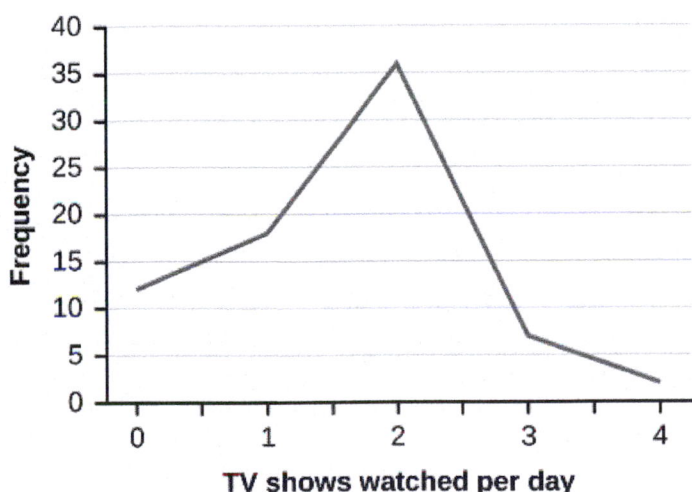

Figure 2.52

9

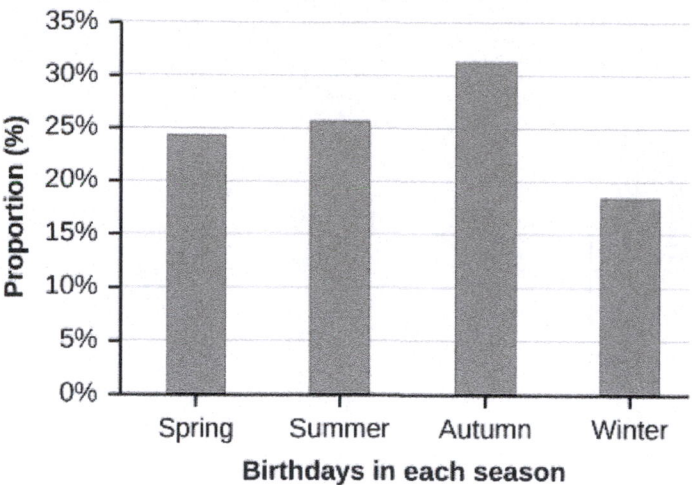

Figure 2.53

11

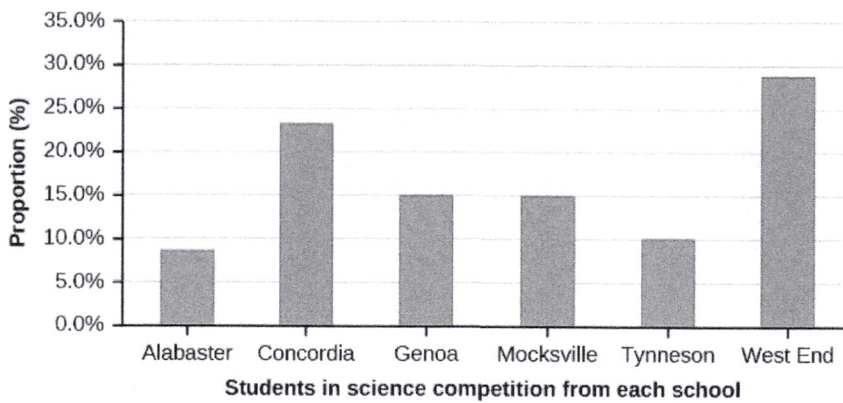

Figure 2.54

13 65

15 The relative frequency shows the *proportion* of data points that have each value. The frequency tells the *number* of data points that have each value.

17 Answers will vary. One possible histogram is shown:

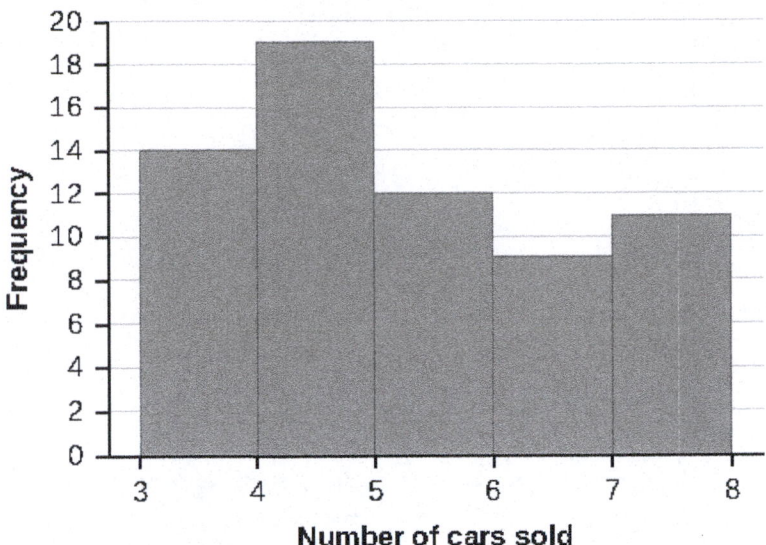

Figure 2.55

19 Find the midpoint for each class. These will be graphed on the *x*-axis. The frequency values will be graphed on the *y*-axis values.

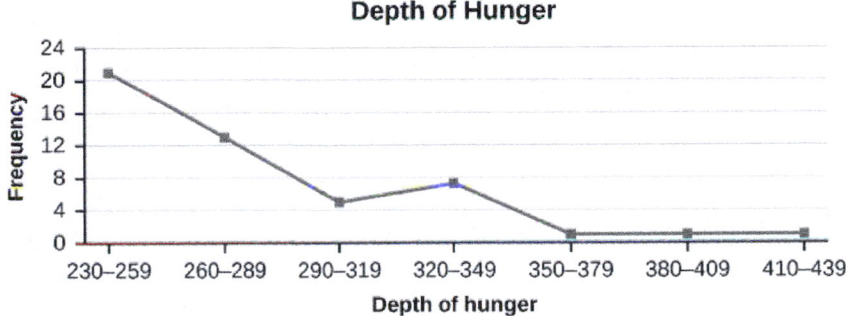

Figure 2.56

21

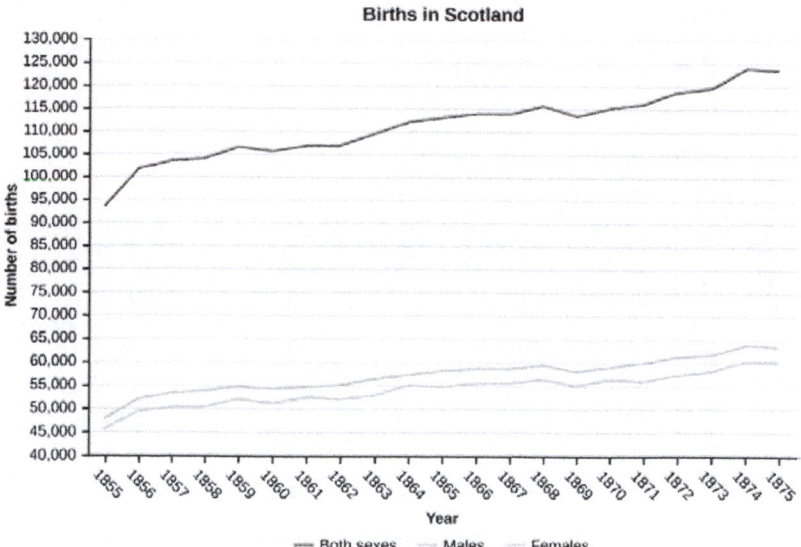

Figure 2.57

23

a. The 40th percentile is 37 years.

b. The 78th percentile is 70 years.

25 Jesse graduated 37th out of a class of 180 students. There are 180 − 37 = 143 students ranked below Jesse. There is one rank of 37. $x = 143$ and $y = 1$. $\frac{x + 0.5y}{n}(100) = \frac{143 + 0.5(1)}{180}(100) = 79.72$. Jesse's rank of 37 puts him at the 80th percentile.

27

a. For runners in a race it is more desirable to have a high percentile for speed. A high percentile means a higher speed which is faster.

b. 40% of runners ran at speeds of 7.5 miles per hour or less (slower). 60% of runners ran at speeds of 7.5 miles per hour or more (faster).

29 When waiting in line at the DMV, the 85th percentile would be a long wait time compared to the other people waiting. 85% of people had shorter wait times than Mina. In this context, Mina would prefer a wait time corresponding to a lower percentile. 85% of people at the DMV waited 32 minutes or less. 15% of people at the DMV waited 32 minutes or longer.

31 The manufacturer and the consumer would be upset. This is a large repair cost for the damages, compared to the other cars in the sample. INTERPRETATION: 90% of the crash tested cars had damage repair costs of $1700 or less; only 10% had damage repair costs of $1700 or more.

33 You can afford 34% of houses. 66% of the houses are too expensive for your budget. INTERPRETATION: 34% of houses cost $240,000 or less. 66% of houses cost $240,000 or more.

35 4

37 6 − 4 = 2

39 6

41 More than 25% of salespersons sell four cars in a typical week. You can see this concentration in the box plot because the first quartile is equal to the median. The top 25% and the bottom 25% are spread out evenly; the whiskers have the same length.

43 Mean: $16 + 17 + 19 + 20 + 20 + 21 + 23 + 24 + 25 + 25 + 25 + 26 + 26 + 27 + 27 + 27 + 28 + 29 + 30 + 32 + 33 + 33 + 34 + 35 + 37 + 39 + 40 = 738$; $\frac{738}{27} = 27.33$

45 The most frequent lengths are 25 and 27, which occur three times. Mode = 25, 27

47 4

49 The data are symmetrical. The median is 3 and the mean is 2.85. They are close, and the mode lies close to the middle of the data, so the data are symmetrical.

51 The data are skewed right. The median is 87.5 and the mean is 88.2. Even though they are close, the mode lies to the left of the middle of the data, and there are many more instances of 87 than any other number, so the data are skewed right.

53 When the data are symmetrical, the mean and median are close or the same.

55 The distribution is skewed right because it looks pulled out to the right.

57 The mean is 4.1 and is slightly greater than the median, which is four.

59 The mode and the median are the same. In this case, they are both five.

61 The distribution is skewed left because it looks pulled out to the left.

63 The mean and the median are both six.

65 The mode is 12, the median is 13.5, and the mean is 15.1. The mean is the largest.

67 The mean tends to reflect skewing the most because it is affected the most by outliers.

69 $s = 34.5$

71 For Fredo: $z = \frac{0.158 - 0.166}{0.012} = -0.67$ For Karl: $z = \frac{0.177 - 0.189}{0.015} = -0.8$ Fredo's z-score of -0.67 is higher than Karl's z-score of -0.8. For batting average, higher values are better, so Fredo has a better batting average compared to his team.

73

a. $s_x = \sqrt{\dfrac{\sum fm^2}{n} - \bar{x}^2} = \sqrt{\dfrac{193157.45}{30} - 79.5^2} = 10.88$

b. $s_x = \sqrt{\dfrac{\sum fm^2}{n} - \bar{x}^2} = \sqrt{\dfrac{380945.3}{101} - 60.94^2} = 7.62$

c. $s_x = \sqrt{\dfrac{\sum fm^2}{n} - \bar{x}^2} = \sqrt{\dfrac{440051.5}{86} - 70.66^2} = 11.14$

75

a. Example solution for using the random number generator for the TI-84+ to generate a simple random sample of 8 states. Instructions are as follows.

Number the entries in the table 1–51 (Includes Washington, DC; Numbered vertically)

Press MATH

Arrow over to PRB

Press 5:randInt(

Enter 51,1,8)

Eight numbers are generated (use the right arrow key to scroll through the numbers). The numbers correspond to the numbered states (for this example: {47 21 9 23 51 13 25 4}. If any numbers are repeated, generate a different number by using 5:randInt(51,1)). Here, the states (and Washington DC) are {Arkansas, Washington DC, Idaho, Maryland, Michigan, Mississippi, Virginia, Wyoming}.

Corresponding percents are {30.1, 22.2, 26.5, 27.1, 30.9, 34.0, 26.0, 25.1}.

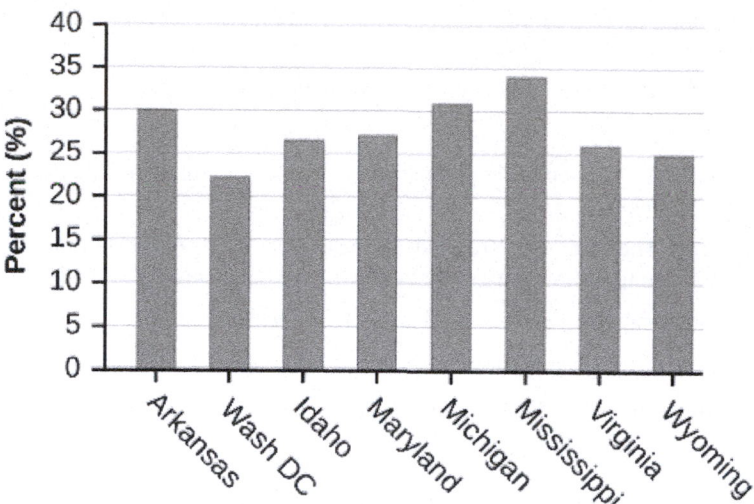

Figure 2.58

b.

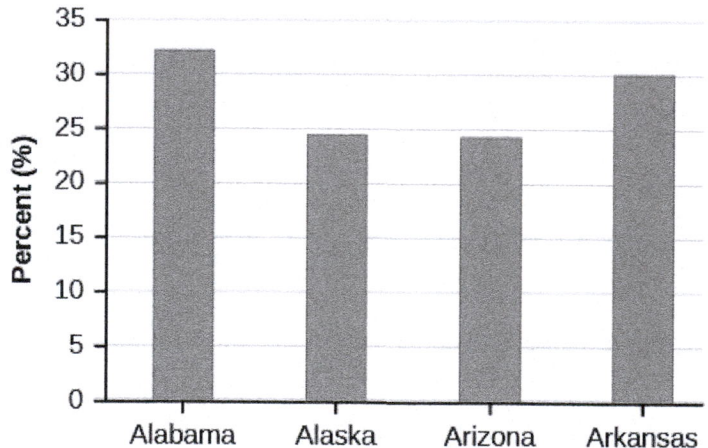

Figure 2.59

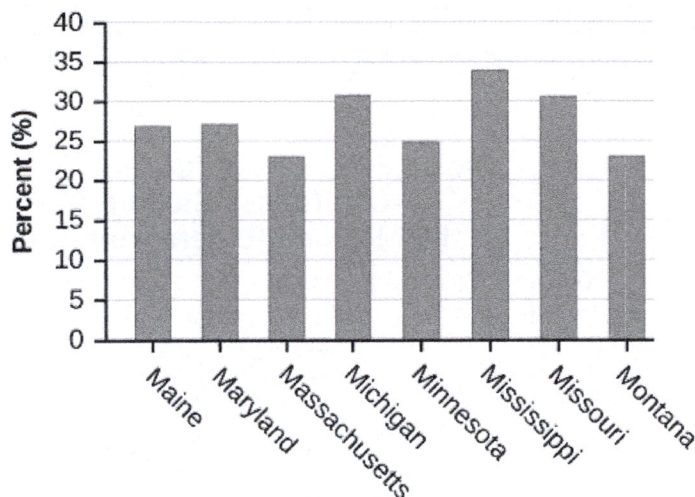

c.

Figure 2.60

77

Amount($)	Frequency	Relative Frequency
51–100	5	0.08
101–150	10	0.17
151–200	15	0.25
201–250	15	0.25
251–300	10	0.17
301–350	5	0.08

Table 2.86 Singles

Amount($)	Frequency	Relative Frequency
100–150	5	0.07
201–250	5	0.07
251–300	5	0.07
301–350	5	0.07
351–400	10	0.14
401–450	10	0.14
451–500	10	0.14
501–550	10	0.14
551–600	5	0.07
601–650	5	0.07

Table 2.87 Couples

a. See **Table 2.86** and **Table 2.87**.

b. In the following histogram data values that fall on the right boundary are counted in the class interval, while values that fall on the left boundary are not counted (with the exception of the first interval where both boundary values are included).

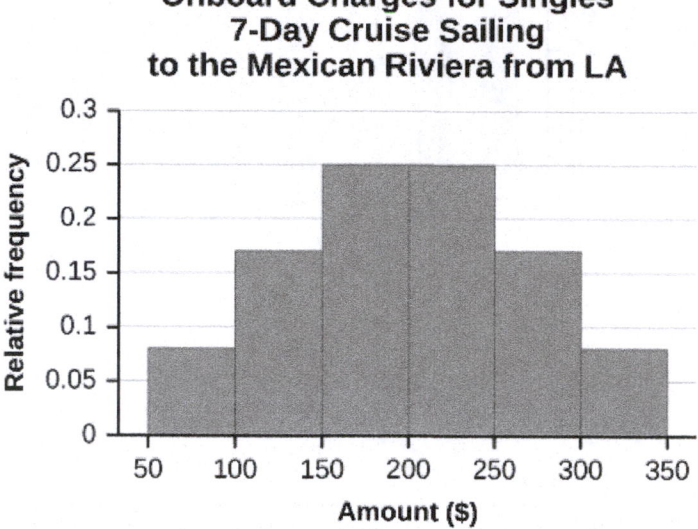

Figure 2.61

c. In the following histogram, the data values that fall on the right boundary are counted in the class interval, while values that fall on the left boundary are not counted (with the exception of the first interval where values on both boundaries are included).

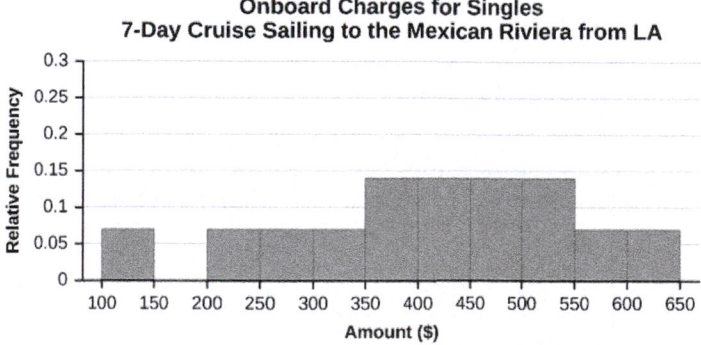

Figure 2.62

d. Compare the two graphs:

    i. Answers may vary. Possible answers include:

        - Both graphs have a single peak.
        - Both graphs use class intervals with width equal to $50.

    ii. Answers may vary. Possible answers include:

        - The couples graph has a class interval with no values.
        - It takes almost twice as many class intervals to display the data for couples.

      iii. Answers may vary. Possible answers include: The graphs are more similar than different because the overall patterns for the graphs are the same.

  e. Check student's solution.

  f. Compare the graph for the Singles with the new graph for the Couples:

      i.   ▪ Both graphs have a single peak.

            ▪ Both graphs display 6 class intervals.

            ▪ Both graphs show the same general pattern.

      ii. Answers may vary. Possible answers include: Although the width of the class intervals for couples is double that of the class intervals for singles, the graphs are more similar than they are different.

  g. Answers may vary. Possible answers include: You are able to compare the graphs interval by interval. It is easier to compare the overall patterns with the new scale on the Couples graph. Because a couple represents two individuals, the new scale leads to a more accurate comparison.

  h. Answers may vary. Possible answers include: Based on the histograms, it seems that spending does not vary much from singles to individuals who are part of a couple. The overall patterns are the same. The range of spending for couples is approximately double the range for individuals.

79 c

81 Answers will vary.

83

  a. $1 - (0.02+0.09+0.19+0.26+0.18+0.17+0.02+0.01) = 0.06$

  b. $0.19+0.26+0.18 = 0.63$

  c. Check student's solution.

  d. 40th percentile will fall between 30,000 and 40,000

      80th percentile will fall between 50,000 and 75,000

  e. Check student's solution.

85

  a. more children; the left whisker shows that 25% of the population are children 17 and younger. The right whisker shows that 25% of the population are adults 50 and older, so adults 65 and over represent less than 25%.

  b. 62.4%

87

  a. Answers will vary. Possible answer: State University conducted a survey to see how involved its students are in community service. The box plot shows the number of community service hours logged by participants over the past year.

  b. Because the first and second quartiles are close, the data in this quarter is very similar. There is not much variation in the values. The data in the third quarter is much more variable, or spread out. This is clear because the second quartile is so far away from the third quartile.

89

  a. Each box plot is spread out more in the greater values. Each plot is skewed to the right, so the ages of the top 50% of buyers are more variable than the ages of the lower 50%.

  b. The BMW 3 series is most likely to have an outlier. It has the longest whisker.

  c. Comparing the median ages, younger people tend to buy the BMW 3 series, while older people tend to buy the BMW 7 series. However, this is not a rule, because there is so much variability in each data set.

  d. The second quarter has the smallest spread. There seems to be only a three-year difference between the first quartile and the median.

e. The third quarter has the largest spread. There seems to be approximately a 14-year difference between the median and the third quartile.

f. *IQR* ~ 17 years

g. There is not enough information to tell. Each interval lies within a quarter, so we cannot tell exactly where the data in that quarter is concentrated.

h. The interval from 31 to 35 years has the fewest data values. Twenty-five percent of the values fall in the interval 38 to 41, and 25% fall between 41 and 64. Since 25% of values fall between 31 and 38, we know that fewer than 25% fall between 31 and 35.

92 The mean percentage, $\bar{x} = \dfrac{1328.65}{50} = 26.75$

94 The median value is the middle value in the ordered list of data values. The median value of a set of 11 will be the 6th number in order. Six years will have totals at or below the median.

96 474 FTES

98 919

100

- mean = 1,809.3

- median = 1,812.5

- standard deviation = 151.2

- first quartile = 1,690

- third quartile = 1,935

- *IQR* = 245

102 Hint: Think about the number of years covered by each time period and what happened to higher education during those periods.

104 For pianos, the cost of the piano is 0.4 standard deviations BELOW the mean. For guitars, the cost of the guitar is 0.25 standard deviations ABOVE the mean. For drums, the cost of the drum set is 1.0 standard deviations BELOW the mean. Of the three, the drums cost the lowest in comparison to the cost of other instruments of the same type. The guitar costs the most in comparison to the cost of other instruments of the same type.

106

- $\bar{x} = 23.32$

- Using the TI 83/84, we obtain a standard deviation of: $s_x = 12.95$.

- The obesity rate of the United States is 10.58% higher than the average obesity rate.

- Since the standard deviation is 12.95, we see that 23.32 + 12.95 = 36.27 is the obesity percentage that is one standard deviation from the mean. The United States obesity rate is slightly less than one standard deviation from the mean. Therefore, we can assume that the United States, while 34% obese, does not hav e an unusually high percentage of obese people.

108

a. For graph, check student's solution.

b. 49.7% of the community is under the age of 35.

c. Based on the information in the table, graph (a) most closely represents the data.

110 a

112 b

113

a. 1.48

  b. 1.12

115

  a. 174; 177; 178; 184; 185; 185; 185; 185; 188; 190; 200; 205; 205; 206; 210; 210; 210; 212; 212; 215; 215; 220; 223; 228; 230; 232; 241; 241; 242; 245; 247; 250; 250; 259; 260; 260; 265; 265; 270; 272; 273; 275; 276; 278; 280; 280; 285; 285; 286; 290; 290; 295; 302

  b. 241

  c. 205.5

  d. 272.5

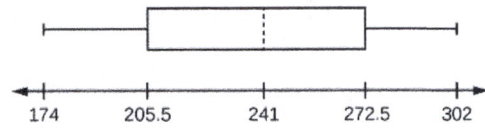

  e.

  f. 205.5, 272.5

  g. sample

  h. population

  i.   i.   236.34

      ii. 37.50

      iii. 161.34

      iv. 0.84 std. dev. below the mean

  j. Young

117

  a. True

  b. True

  c. True

  d. False

119

  a.

Enrollment	Frequency
1000-5000	10
5000-10000	16
10000-15000	3
15000-20000	3
20000-25000	1
25000-30000	2

Table 2.88

  b. Check student's solution.

  c. mode

  d. 8628.74

  e. 6943.88

f. −0.09

121 a

3 | PROBABILITY TOPICS

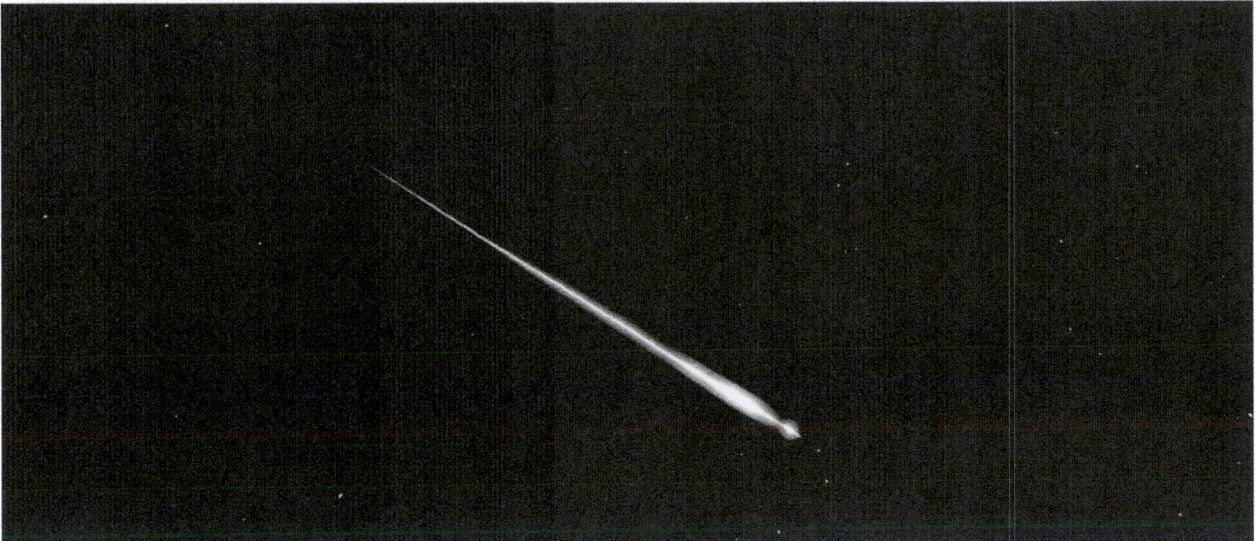

Figure 3.1 Meteor showers are rare, but the probability of them occurring can be calculated. (credit: Navicore/flickr)

Introduction

Chapter Objectives
By the end of this chapter, the student should be able to: • Understand and use the terminology of probability.   • Determine whether two events are mutually exclusive and whether two events are independent.   • Calculate probabilities using the Addition Rules and Multiplication Rules.   • Construct and interpret Contingency Tables.   • Construct and interpret Venn Diagrams.   • Construct and interpret Tree Diagrams.

It is often necessary to "guess" about the outcome of an event in order to make a decision. Politicians study polls to guess their likelihood of winning an election. Teachers choose a particular course of study based on what they think students can comprehend. Doctors choose the treatments needed for various diseases based on their assessment of likely results. You may have visited a casino where people play games chosen because of the belief that the likelihood of winning is good. You may have chosen your course of study based on the probable availability of jobs.

You have, more than likely, used probability. In fact, you probably have an intuitive sense of probability. Probability deals with the chance of an event occurring. Whenever you weigh the odds of whether or not to do your homework or to study for an exam, you are using probability. In this chapter, you will learn how to solve probability problems using a systematic approach.

Collaborative Exercise

Your instructor will survey your class. Count the number of students in the class today.

- Raise your hand if you have any change in your pocket or purse. Record the number of raised hands.

- Raise your hand if you rode a bus within the past month. Record the number of raised hands.

- Raise your hand if you answered "yes" to BOTH of the first two questions. Record the number of raised hands.

Use the class data as estimates of the following probabilities. P(change) means the probability that a randomly chosen person in your class has change in his/her pocket or purse. P(bus) means the probability that a randomly chosen person in your class rode a bus within the last month and so on. Discuss your answers.

- Find P(change).

- Find P(bus).

- Find P(change AND bus). Find the probability that a randomly chosen student in your class has change in his/her pocket or purse and rode a bus within the last month.

- Find P(change|bus). Find the probability that a randomly chosen student has change given that he or she rode a bus within the last month. Count all the students that rode a bus. From the group of students who rode a bus, count those who have change. The probability is equal to those who have change and rode a bus divided by those who rode a bus.

3.1 | Terminology

Probability is a measure that is associated with how certain we are of outcomes of a particular experiment or activity. An **experiment** is a planned operation carried out under controlled conditions. If the result is not predetermined, then the experiment is said to be a **chance** experiment. Flipping one fair coin twice is an example of an experiment.

A result of an experiment is called an **outcome**. The **sample space** of an experiment is the set of all possible outcomes. Three ways to represent a sample space are: to list the possible outcomes, to create a tree diagram, or to create a Venn diagram. The uppercase letter S is used to denote the sample space. For example, if you flip one fair coin, $S = \{H, T\}$ where H = heads and T = tails are the outcomes.

An **event** is any combination of outcomes. Upper case letters like A and B represent events. For example, if the experiment is to flip one fair coin, event A might be getting at most one head. The probability of an event A is written $P(A)$.

The **probability** of any outcome is the **long-term relative frequency** of that outcome. **Probabilities are between zero and one, inclusive** (that is, zero and one and all numbers between these values). $P(A) = 0$ means the event A can never happen. $P(A) = 1$ means the event A always happens. $P(A) = 0.5$ means the event A is equally likely to occur or not to occur. For example, if you flip one fair coin repeatedly (from 20 to 2,000 to 20,000 times) the relative frequency of heads approaches 0.5 (the probability of heads).

Equally likely means that each outcome of an experiment occurs with equal probability. For example, if you toss a **fair**, six-sided die, each face (1, 2, 3, 4, 5, or 6) is as likely to occur as any other face. If you toss a fair coin, a Head (H) and a Tail (T) are equally likely to occur. If you randomly guess the answer to a true/false question on an exam, you are equally likely to select a correct answer or an incorrect answer.

To calculate the probability of an event A when all outcomes in the sample space are equally likely, count the number of outcomes for event A and divide by the total number of outcomes in the sample space. For example, if you toss a fair dime and a fair nickel, the sample space is $\{HH, TH, HT, TT\}$ where T = tails and H = heads. The sample space has four outcomes. A = getting one head. There are two outcomes that meet this condition $\{HT, TH\}$, so $P(A) = \frac{2}{4} = 0.5$.

Suppose you roll one fair six-sided die, with the numbers $\{1, 2, 3, 4, 5, 6\}$ on its faces. Let event E = rolling a number that is at least five. There are two outcomes $\{5, 6\}$. $P(E) = \frac{2}{6}$. If you were to roll the die only a few times, you would not be surprised if your observed results did not match the probability. If you were to roll the die a very large number of times, you would expect that, overall, $\frac{2}{6}$ of the rolls would result in an outcome of "at least five". You would not expect exactly $\frac{2}{6}$.

The long-term relative frequency of obtaining this result would approach the theoretical probability of $\frac{2}{6}$ as the number of repetitions grows larger and larger.

This important characteristic of probability experiments is known as the **law of large numbers** which states that as the number of repetitions of an experiment is increased, the relative frequency obtained in the experiment tends to become closer and closer to the theoretical probability. Even though the outcomes do not happen according to any set pattern or order, overall, the long-term observed relative frequency will approach the theoretical probability. (The word **empirical** is often used instead of the word observed.)

It is important to realize that in many situations, the outcomes are not equally likely. A coin or die may be **unfair**, or **biased**. Two math professors in Europe had their statistics students test the Belgian one Euro coin and discovered that in 250 trials, a head was obtained 56% of the time and a tail was obtained 44% of the time. The data seem to show that the coin is not a fair coin; more repetitions would be helpful to draw a more accurate conclusion about such bias. Some dice may be biased. Look at the dice in a game you have at home; the spots on each face are usually small holes carved out and then painted to make the spots visible. Your dice may or may not be biased; it is possible that the outcomes may be affected by the slight weight differences due to the different numbers of holes in the faces. Gambling casinos make a lot of money depending on outcomes from rolling dice, so casino dice are made differently to eliminate bias. Casino dice have flat faces; the holes are completely filled with paint having the same density as the material that the dice are made out of so that each face is equally likely to occur. Later we will learn techniques to use to work with probabilities for events that are not equally likely.

"OR" Event:

An outcome is in the event A OR B if the outcome is in A or is in B or is in both A and B. For example, let A = {1, 2, 3, 4, 5} and B = {4, 5, 6, 7, 8}. A OR B = {1, 2, 3, 4, 5, 6, 7, 8}. Notice that 4 and 5 are NOT listed twice.

"AND" Event:

An outcome is in the event A AND B if the outcome is in both A and B at the same time. For example, let A and B be {1, 2, 3, 4, 5} and {4, 5, 6, 7, 8}, respectively. Then A AND B = {4, 5}.

The **complement** of event A is denoted A' (read "A prime"). A' consists of all outcomes that are **NOT** in A. Notice that $P(A) + P(A') = 1$. For example, let S = {1, 2, 3, 4, 5, 6} and let A = {1, 2, 3, 4}. Then, A' = {5, 6}. $P(A) = \frac{4}{6}$, $P(A') = \frac{2}{6}$, and

$$P(A) + P(A') = \frac{4}{6} + \frac{2}{6} = 1$$

The **conditional probability** of A given B is written P(A|B). P(A|B) is the probability that event A will occur given that the event B has already occurred. **A conditional reduces the sample space**. We calculate the probability of A from the reduced sample space B. The formula to calculate P(A|B) is $P(A|B) = \frac{P(A \text{ AND } B)}{P(B)}$ where P(B) is greater than zero.

For example, suppose we toss one fair, six-sided die. The sample space S = {1, 2, 3, 4, 5, 6}. Let A = face is 2 or 3 and B = face is even (2, 4, 6). To calculate P(A|B), we count the number of outcomes 2 or 3 in the sample space B = {2, 4, 6}. Then we divide that by the number of outcomes B (rather than S).

We get the same result by using the formula. Remember that S has six outcomes.

$$P(A|B) = \frac{P(A \text{ AND } B)}{P(B)} = \frac{\frac{\text{(the number of outcomes that are 2 or 3 and even in } S)}{6}}{\frac{\text{(the number of outcomes that are even in } S)}{6}} = \frac{\frac{1}{6}}{\frac{3}{6}} = \frac{1}{3}$$

Understanding Terminology and Symbols

It is important to read each problem carefully to think about and understand what the events are. Understanding the wording is the first very important step in solving probability problems. Reread the problem several times if necessary. Clearly identify the event of interest. Determine whether there is a condition stated in the wording that would indicate that the probability is conditional; carefully identify the condition, if any.

Example 3.1

The sample space S is the whole numbers starting at one and less than 20.

a. $S = $ _____

Let event A = the even numbers and event B = numbers greater than 13.

b. $A = $ _____, $B = $ _____

c. $P(A) = $ _____, $P(B) = $ _____

d. A AND $B = $ _____, A OR $B = $ _____

e. $P(A$ AND $B) = $ _____, $P(A$ OR $B) = $ _____

f. $A' = $ _____, $P(A') = $ _____

g. $P(A) + P(A') = $ _____

h. $P(A|B) = $ _____, $P(B|A) = $ _____; are the probabilities equal?

Solution 3.1

a. S = {1, 2, 3, 4, 5, 6, 7, 8, 9, 10, 11, 12, 13, 14, 15, 16, 17, 18, 19}

b. A = {2, 4, 6, 8, 10, 12, 14, 16, 18}, B = {14, 15, 16, 17, 18, 19}

c. $P(A) = \frac{9}{19}$, $P(B) = \frac{6}{19}$

d. A AND B = {14,16,18}, A OR B = 2, 4, 6, 8, 10, 12, 14, 15, 16, 17, 18, 19}

e. $P(A$ AND $B) = \frac{3}{19}$, $P(A$ OR $B) = \frac{12}{19}$

f. A' = 1, 3, 5, 7, 9, 11, 13, 15, 17, 19; $P(A') = \frac{10}{19}$

g. $P(A) + P(A') = 1$ ($\frac{9}{19} + \frac{10}{19} = 1$)

h. $P(A|B) = \frac{P(A \text{ AND } B)}{P(B)} = \frac{3}{6}$, $P(B|A) = \frac{P(A \text{ AND } B)}{P(A)} = \frac{3}{9}$, No

Try It Σ

3.1 The sample space S is the ordered pairs of two whole numbers, the first from one to three and the second from one to four (Example: (1, 4)).

a. $S = $ _____

Let event A = the sum is even and event B = the first number is prime.

b. $A = $ _____, $B = $ _____

c. $P(A) = $ _____, $P(B) = $ _____

d. A AND $B = $ _____, A OR $B = $ _____

e. $P(A$ AND $B) = $ _____, $P(A$ OR $B) = $ _____

f. $B' = $ _____, $P(B') = $ _____

g. $P(A) + P(A') = $ _____

h. $P(A|B) = $ _____, $P(B|A) = $ _____; are the probabilities equal?

Example 3.2

A fair, six-sided die is rolled. Describe the sample space S, identify each of the following events with a subset of S and compute its probability (an outcome is the number of dots that show up).

a. Event T = the outcome is two.

b. Event A = the outcome is an even number.

c. Event B = the outcome is less than four.

d. The complement of A.

e. A GIVEN B

f. B GIVEN A

g. A AND B

h. A OR B

i. A OR B'

j. Event N = the outcome is a prime number.

k. Event I = the outcome is seven.

Solution 3.2

a. $T = \{2\}$, $P(T) = \frac{1}{6}$

b. $A = \{2, 4, 6\}$, $P(A) = \frac{1}{2}$

c. $B = \{1, 2, 3\}$, $P(B) = \frac{1}{2}$

d. $A' = \{1, 3, 5\}$, $P(A') = \frac{1}{2}$

e. $A|B = \{2\}$, $P(A|B) = \frac{1}{3}$

f. $B|A = \{2\}$, $P(B|A) = \frac{1}{3}$

g. A AND $B = \{2\}$, $P(A$ AND $B) = \frac{1}{6}$

h. A OR $B = \{1, 2, 3, 4, 6\}$, $P(A$ OR $B) = \frac{5}{6}$

i. A OR $B' = \{2, 4, 5, 6\}$, $P(A$ OR $B') = \frac{2}{3}$

j. $N = \{2, 3, 5\}$, $P(N) = \frac{1}{2}$

k. A six-sided die does not have seven dots. $P(7) = 0$.

Example 3.3

Table 3.1 describes the distribution of a random sample S of 100 individuals, organized by gender and whether they are right- or left-handed.

	Right-handed	Left-handed
Males	43	9
Females	44	4

Table 3.1

Let's denote the events M = the subject is male, F = the subject is female, R = the subject is right-handed, L = the subject is left-handed. Compute the following probabilities:

a. $P(M)$

b. $P(F)$

c. $P(R)$

d. $P(L)$

e. $P(M \text{ AND } R)$

f. $P(F \text{ AND } L)$

g. $P(M \text{ OR } F)$

h. $P(M \text{ OR } R)$

i. $P(F \text{ OR } L)$

j. $P(M')$

k. $P(R|M)$

l. $P(F|L)$

m. $P(L|F)$

Solution 3.3

a. $P(M) = 0.52$

b. $P(F) = 0.48$

c. $P(R) = 0.87$

d. $P(L) = 0.13$

e. $P(M \text{ AND } R) = 0.43$

f. $P(F \text{ AND } L) = 0.04$

g. $P(M \text{ OR } F) = 1$

h. $P(M \text{ OR } R) = 0.96$

i. $P(F \text{ OR } L) = 0.57$

j. $P(M') = 0.48$

k. $P(R|M) = 0.8269$ (rounded to four decimal places)

l. $P(F|L) = 0.3077$ (rounded to four decimal places)

m. $P(L|F) = 0.0833$

3.2 | Independent and Mutually Exclusive Events

Independent and mutually exclusive do **not** mean the same thing.

Independent Events

Two events are independent if the following are true:

- $P(A|B) = P(A)$
- $P(B|A) = P(B)$
- $P(A \text{ AND } B) = P(A)P(B)$

Two events A and B are **independent** if the knowledge that one occurred does not affect the chance the other occurs. For example, the outcomes of two roles of a fair die are independent events. The outcome of the first roll does not change the probability for the outcome of the second roll. To show two events are independent, you must show **only one** of the above conditions. If two events are NOT independent, then we say that they are **dependent**.

Sampling may be done **with replacement** or **without replacement**.

- **With replacement**: If each member of a population is replaced after it is picked, then that member has the possibility of being chosen more than once. When sampling is done with replacement, then events are considered to be independent, meaning the result of the first pick will not change the probabilities for the second pick.

- **Without replacement**: When sampling is done without replacement, each member of a population may be chosen only once. In this case, the probabilities for the second pick are affected by the result of the first pick. The events are considered to be dependent or not independent.

If it is not known whether A and B are independent or dependent, **assume they are dependent until you can show otherwise**.

Example 3.4

You have a fair, well-shuffled deck of 52 cards. It consists of four suits. The suits are clubs, diamonds, hearts and spades. There are 13 cards in each suit consisting of 1, 2, 3, 4, 5, 6, 7, 8, 9, 10, J (jack), Q (queen), K (king) of that suit.

a. Sampling with replacement:
Suppose you pick three cards with replacement. The first card you pick out of the 52 cards is the Q of spades. You put this card back, reshuffle the cards and pick a second card from the 52-card deck. It is the ten of clubs. You put this card back, reshuffle the cards and pick a third card from the 52-card deck. This time, the card is the Q of spades again. Your picks are {Q of spades, ten of clubs, Q of spades}. You have picked the Q of spades twice. You pick each card from the 52-card deck.

b. Sampling without replacement:
Suppose you pick three cards without replacement. The first card you pick out of the 52 cards is the K of hearts. You put this card aside and pick the second card from the 51 cards remaining in the deck. It is the three of diamonds. You put this card aside and pick the third card from the remaining 50 cards in the deck. The third card is the J of spades. Your picks are {K of hearts, three of diamonds, J of spades}. Because you have picked the cards without replacement, you cannot pick the same card twice.

Try It Σ

3.4 You have a fair, well-shuffled deck of 52 cards. It consists of four suits. The suits are clubs, diamonds, hearts and spades. There are 13 cards in each suit consisting of 1, 2, 3, 4, 5, 6, 7, 8, 9, 10, J (jack), Q (queen), K (king) of that suit. Three cards are picked at random.

a. Suppose you know that the picked cards are Q of spades, K of hearts and Q of spades. Can you decide if the sampling was with or without replacement?

b. Suppose you know that the picked cards are Q of spades, K of hearts, and J of spades. Can you decide if the sampling was with or without replacement?

Example 3.5

You have a fair, well-shuffled deck of 52 cards. It consists of four suits. The suits are clubs, diamonds, hearts, and spades. There are 13 cards in each suit consisting of 1, 2, 3, 4, 5, 6, 7, 8, 9, 10, J (jack), Q (queen), and K (king) of that suit. S = spades, H = Hearts, D = Diamonds, C = Clubs.

a. Suppose you pick four cards, but do not put any cards back into the deck. Your cards are QS, $1D$, $1C$, QD.

b. Suppose you pick four cards and put each card back before you pick the next card. Your cards are KH, $7D$, $6D$, KH.

Which of a. or b. did you sample with replacement and which did you sample without replacement?

Solution 3.5

a. Without replacement; b. With replacement

Try It Σ

3.5 You have a fair, well-shuffled deck of 52 cards. It consists of four suits. The suits are clubs, diamonds, hearts, and spades. There are 13 cards in each suit consisting of 1, 2, 3, 4, 5, 6, 7, 8, 9, 10, J (jack), Q (queen), and K (king) of that suit. S = spades, H = Hearts, D = Diamonds, C = Clubs. Suppose that you sample four cards without replacement. Which of the following outcomes are possible? Answer the same question for sampling with replacement.

a. *QS*, 1*D*, 1*C*, *QD*

b. *KH*, 7*D*, 6*D*, *KH*

c. *QS*, 7*D*, 6*D*, *KS*

Mutually Exclusive Events

A and B are **mutually exclusive** events if they cannot occur at the same time. This means that A and B do not share any outcomes and $P(A \text{ AND } B) = 0$.

For example, suppose the sample space S = {1, 2, 3, 4, 5, 6, 7, 8, 9, 10}. Let A = {1, 2, 3, 4, 5}, B = {4, 5, 6, 7, 8}, and C = {7, 9}. $A \text{ AND } B$ = {4, 5}. $P(A \text{ AND } B) = \frac{2}{10}$ and is not equal to zero. Therefore, A and B are not mutually exclusive. A and C do not have any numbers in common so $P(A \text{ AND } C) = 0$. Therefore, A and C are mutually exclusive.

If it is not known whether A and B are mutually exclusive, **assume they are not until you can show otherwise**. The following examples illustrate these definitions and terms.

Example 3.6

Flip two fair coins. (This is an experiment.)

The sample space is {*HH*, *HT*, *TH*, *TT*} where *T* = tails and *H* = heads. The outcomes are *HH*, *HT*, *TH*, and *TT*. The outcomes HT and TH are different. The *HT* means that the first coin showed heads and the second coin showed tails. The *TH* means that the first coin showed tails and the second coin showed heads.

- Let *A* = the event of getting **at most one tail**. (At most one tail means zero or one tail.) Then *A* can be written as {*HH*, *HT*, *TH*}. The outcome *HH* shows zero tails. *HT* and *TH* each show one tail.

- Let *B* = the event of getting all tails. *B* can be written as {*TT*}. *B* is the **complement** of A, so *B* = *A*'. Also, $P(A) + P(B) = P(A) + P(A') = 1$.

- The probabilities for *A* and for *B* are $P(A) = \frac{3}{4}$ and $P(B) = \frac{1}{4}$.

- Let *C* = the event of getting all heads. *C* = {*HH*}. Since *B* = {*TT*}, $P(B \text{ AND } C) = 0$. *B* and *C* are mutually exclusive. (*B* and *C* have no members in common because you cannot have all tails and all heads at the same time.)

- Let *D* = event of getting **more than one** tail. *D* = {*TT*}. $P(D) = \frac{1}{4}$

- Let *E* = event of getting a head on the first roll. (This implies you can get either a head or tail on the second roll.) *E* = {*HT*, *HH*}. $P(E) = \frac{2}{4}$

- Find the probability of getting **at least one** (one or two) tail in two flips. Let *F* = event of getting at least one tail in two flips. *F* = {*HT*, *TH*, *TT*}. $P(F) = \frac{3}{4}$

Try It Σ

3.6 Draw two cards from a standard 52-card deck with replacement. Find the probability of getting at least one black card.

Example 3.7

Flip two fair coins. Find the probabilities of the events.

a. Let *F* = the event of getting at most one tail (zero or one tail).

b. Let *G* = the event of getting two faces that are the same.

c. Let *H* = the event of getting a head on the first flip followed by a head or tail on the second flip.

d. Are *F* and *G* mutually exclusive?

e. Let *J* = the event of getting all tails. Are *J* and *H* mutually exclusive?

Solution 3.7

Look at the sample space in **Example 3.6**.

a. Zero (0) or one (1) tails occur when the outcomes *HH*, *TH*, *HT* show up. $P(F) = \frac{3}{4}$

b. Two faces are the same if *HH* or *TT* show up. $P(G) = \frac{2}{4}$

  c. A head on the first flip followed by a head or tail on the second flip occurs when *HH* or *HT* show up. $P(H) = \frac{2}{4}$

  d. *F* and *G* share *HH* so *P*(*F* AND *G*) is not equal to zero (0). *F* and *G* are not mutually exclusive.

  e. Getting all tails occurs when tails shows up on both coins (*TT*). *H*'s outcomes are *HH* and *HT*.

J and *H* have nothing in common so *P*(*J* AND *H*) = 0. *J* and *H* are mutually exclusive.

Try It Σ

3.7 A box has two balls, one white and one red. We select one ball, put it back in the box, and select a second ball (sampling with replacement). Find the probability of the following events:

a. Let *F* = the event of getting the white ball twice.

b. Let *G* = the event of getting two balls of different colors.

c. Let *H* = the event of getting white on the first pick.

d. Are *F* and *G* mutually exclusive?

e. Are *G* and *H* mutually exclusive?

Example 3.8

Roll one fair, six-sided die. The sample space is {1, 2, 3, 4, 5, 6}. Let event *A* = a face is odd. Then *A* = {1, 3, 5}. Let event *B* = a face is even. Then *B* = {2, 4, 6}.

- Find the complement of *A*, *A*'. The complement of *A*, *A*', is *B* because *A* and *B* together make up the sample space. $P(A) + P(B) = P(A) + P(A') = 1$. Also, $P(A) = \frac{3}{6}$ and $P(B) = \frac{3}{6}$.

- Let event *C* = odd faces larger than two. Then *C* = {3, 5}. Let event *D* = all even faces smaller than five. Then *D* = {2, 4}. *P*(*C* AND *D*) = 0 because you cannot have an odd and even face at the same time. Therefore, *C* and *D* are mutually exclusive events.

- Let event *E* = all faces less than five. *E* = {1, 2, 3, 4}.

Are *C* and *E* mutually exclusive events? (Answer yes or no.) Why or why not?

Solution 3.8

No. *C* = {3, 5} and *E* = {1, 2, 3, 4}. $P(C \text{ AND } E) = \frac{1}{6}$. To be mutually exclusive, *P*(*C* AND *E*) must be zero.

- Find *P*(*C*|*A*). This is a conditional probability. Recall that the event *C* is {3, 5} and event *A* is {1, 3, 5}. To find *P*(*C*|*A*), find the probability of *C* using the sample space *A*. You have reduced the sample space from the original sample space {1, 2, 3, 4, 5, 6} to {1, 3, 5}. So, $P(C|A) = \frac{2}{3}$.

Try It Σ

3.8 Let event *A* = learning Spanish. Let event *B* = learning German. Then *A* AND *B* = learning Spanish and German. Suppose *P*(*A*) = 0.4 and *P*(*B*) = 0.2. *P*(*A* AND *B*) = 0.08. Are events *A* and *B* independent? Hint: You must show ONE of the following:

- $P(A|B) = P(A)$

- $P(B|A)$

- $P(A \text{ AND } B) = P(A)P(B)$

Example 3.9

Let event G = taking a math class. Let event H = taking a science class. Then, G AND H = taking a math class and a science class. Suppose $P(G) = 0.6$, $P(H) = 0.5$, and $P(G \text{ AND } H) = 0.3$. Are G and H independent?

If G and H are independent, then you must show **ONE** of the following:

- $P(G|H) = P(G)$

- $P(H|G) = P(H)$

- $P(G \text{ AND } H) = P(G)P(H)$

NOTE

The choice you make depends on the information you have. You could choose any of the methods here because you have the necessary information.

a. Show that $P(G|H) = P(G)$.

Solution 3.9

$P(G|H) = \dfrac{P(G \text{ AND } H)}{P(H)} = \dfrac{0.3}{0.5} = 0.6 = P(G)$

b. Show $P(G \text{ AND } H) = P(G)P(H)$.

Solution 3.9

$P(G)P(H) = (0.6)(0.5) = 0.3 = P(G \text{ AND } H)$

Since G and H are independent, knowing that a person is taking a science class does not change the chance that he or she is taking a math class. If the two events had not been independent (that is, they are dependent) then knowing that a person is taking a science class would change the chance he or she is taking math. For practice, show that $P(H|G) = P(H)$ to show that G and H are independent events.

Try It

3.9 In a bag, there are six red marbles and four green marbles. The red marbles are marked with the numbers 1, 2, 3, 4, 5, and 6. The green marbles are marked with the numbers 1, 2, 3, and 4.

- R = a red marble

- G = a green marble

- O = an odd-numbered marble

- The sample space is $S = \{R1, R2, R3, R4, R5, R6, G1, G2, G3, G4\}$.

S has ten outcomes. What is $P(G \text{ AND } O)$?

Example 3.10

Let event C = taking an English class. Let event D = taking a speech class.

Suppose $P(C) = 0.75$, $P(D) = 0.3$, $P(C|D) = 0.75$ and $P(C \text{ AND } D) = 0.225$.

Justify your answers to the following questions numerically.

   a. Are C and D independent?

   b. Are C and D mutually exclusive?

   c. What is $P(D|C)$?

Solution 3.10

   a. Yes, because $P(C|D) = P(C)$.

   b. No, because $P(C \text{ AND } D)$ is not equal to zero.

   c. $P(D|C) = \dfrac{P(C \text{ AND } D)}{P(C)} = \dfrac{0.225}{0.75} = 0.3$

Try It Σ

3.10 A student goes to the library. Let events B = the student checks out a book and D = the student checks out a DVD. Suppose that $P(B) = 0.40$, $P(D) = 0.30$ and $P(B \text{ AND } D) = 0.20$.

   a. Find $P(B|D)$.

   b. Find $P(D|B)$.

   c. Are B and D independent?

   d. Are B and D mutually exclusive?

Example 3.11

In a box there are three red cards and five blue cards. The red cards are marked with the numbers 1, 2, and 3, and the blue cards are marked with the numbers 1, 2, 3, 4, and 5. The cards are well-shuffled. You reach into the box (you cannot see into it) and draw one card.

Let R = red card is drawn, B = blue card is drawn, E = even-numbered card is drawn.

The sample space $S = R1, R2, R3, B1, B2, B3, B4, B5$. S has eight outcomes.

- $P(R) = \frac{3}{8}$. $P(B) = \frac{5}{8}$. $P(R \text{ AND } B) = 0$. (You cannot draw one card that is both red and blue.)

- $P(E) = \frac{3}{8}$. (There are three even-numbered cards, $R2$, $B2$, and $B4$.)

- $P(E|B) = \frac{2}{5}$. (There are five blue cards: $B1$, $B2$, $B3$, $B4$, and $B5$. Out of the blue cards, there are two even cards; $B2$ and $B4$.)

- $P(B|E) = \frac{2}{3}$. (There are three even-numbered cards: $R2$, $B2$, and $B4$. Out of the even-numbered cards, to are blue; $B2$ and $B4$.)

- The events R and B are mutually exclusive because $P(R \text{ AND } B) = 0$.

- Let G = card with a number greater than 3. G = {$B4$, $B5$}. $P(G)$ = $\frac{2}{8}$. Let H = blue card numbered between one and four, inclusive. H = {$B1$, $B2$, $B3$, $B4$}. $P(G|H)$ = $\frac{1}{4}$. (The only card in H that has a number greater than three is $B4$.) Since $\frac{2}{8}$ = $\frac{1}{4}$, $P(G)$ = $P(G|H)$, which means that G and H are independent.

Try It Σ

3.11 In a basketball arena,

- 70% of the fans are rooting for the home team.
- 25% of the fans are wearing blue.
- 20% of the fans are wearing blue and are rooting for the away team.
- Of the fans rooting for the away team, 67% are wearing blue.

Let A be the event that a fan is rooting for the away team.
Let B be the event that a fan is wearing blue.
Are the events of rooting for the away team and wearing blue independent? Are they mutually exclusive?

Example 3.12

In a particular college class, 60% of the students are female. Fifty percent of all students in the class have long hair. Forty-five percent of the students are female and have long hair. Of the female students, 75% have long hair. Let F be the event that a student is female. Let L be the event that a student has long hair. One student is picked randomly. Are the events of being female and having long hair independent?

- The following probabilities are given in this example:
- $P(F)$ = 0.60; $P(L)$ = 0.50
- $P(F \text{ AND } L)$ = 0.45
- $P(L|F)$ = 0.75

NOTE

The choice you make depends on the information you have. You could use the first or last condition on the list for this example. You do not know $P(F|L)$ yet, so you cannot use the second condition.

Solution 1

Check whether $P(F \text{ AND } L)$ = $P(F)P(L)$. We are given that $P(F \text{ AND } L)$ = 0.45, but $P(F)P(L)$ = (0.60)(0.50) = 0.30. The events of being female and having long hair are not independent because $P(F \text{ AND } L)$ does not equal $P(F)P(L)$.

Solution 2

Check whether $P(L|F)$ equals $P(L)$. We are given that $P(L|F)$ = 0.75, but $P(L)$ = 0.50; they are not equal. The events of being female and having long hair are not independent.

Interpretation of Results

The events of being female and having long hair are not independent; knowing that a student is female changes the probability that a student has long hair.

Try It Σ

3.12 Mark is deciding which route to take to work. His choices are I = the Interstate and F = Fifth Street.

- $P(I) = 0.44$ and $P(F) = 0.55$

- $P(I \text{ AND } F) = 0$ because Mark will take only one route to work.

What is the probability of $P(I \text{ OR } F)$?

Example 3.13

a. Toss one fair coin (the coin has two sides, H and T). The outcomes are _____. Count the outcomes. There are ____ outcomes.

b. Toss one fair, six-sided die (the die has 1, 2, 3, 4, 5 or 6 dots on a side). The outcomes are _____. Count the outcomes. There are ____ outcomes.

c. Multiply the two numbers of outcomes. The answer is _____.

d. If you flip one fair coin and follow it with the toss of one fair, six-sided die, the answer in three is the number of outcomes (size of the sample space). What are the outcomes? (Hint: Two of the outcomes are $H1$ and $T6$.)

e. Event A = heads (H) on the coin followed by an even number (2, 4, 6) on the die.
$A = \{$_____$\}$. Find $P(A)$.

f. Event B = heads on the coin followed by a three on the die. $B = \{$_____$\}$. Find $P(B)$.

g. Are A and B mutually exclusive? (Hint: What is $P(A \text{ AND } B)$? If $P(A \text{ AND } B) = 0$, then A and B are mutually exclusive.)

h. Are A and B independent? (Hint: Is $P(A \text{ AND } B) = P(A)P(B)$? If $P(A \text{ AND } B) = P(A)P(B)$, then A and B are independent. If not, then they are dependent).

Solution 3.13

a. H and T; 2

b. 1, 2, 3, 4, 5, 6; 6

c. $2(6) = 12$

d. $T1, T2, T3, T4, T5, T6, H1, H2, H3, H4, H5, H6$

e. $A = \{H2, H4, H6\}$; $P(A) = \frac{3}{12}$

f. $B = \{H3\}$; $P(B) = \frac{1}{12}$

g. Yes, because $P(A \text{ AND } B) = 0$

h. $P(A \text{ AND } B) = 0.P(A)P(B) = \left(\frac{3}{12}\right)\left(\frac{1}{12}\right)$. $P(A \text{ AND } B)$ does not equal $P(A)P(B)$, so A and B are dependent.

Try It Σ

3.13 A box has two balls, one white and one red. We select one ball, put it back in the box, and select a second ball (sampling with replacement). Let T be the event of getting the white ball twice, F the event of picking the white ball first, S the event of picking the white ball in the second drawing.

a. Compute $P(T)$.

b. Compute $P(T|F)$.

c. Are T and F independent?.

d. Are F and S mutually exclusive?

e. Are F and S independent?

3.3 | Two Basic Rules of Probability

When calculating probability, there are two rules to consider when determining if two events are independent or dependent and if they are mutually exclusive or not.

The Multiplication Rule

If A and B are two events defined on a **sample space**, then: $P(A \text{ AND } B) = P(B)P(A|B)$.

This rule may also be written as: $P(A|B) = \dfrac{P(A \text{ AND } B)}{P(B)}$

(The probability of A given B equals the probability of A and B divided by the probability of B.)

If A and B are **independent**, then $P(A|B) = P(A)$. Then $P(A \text{ AND } B) = P(A|B)P(B)$ becomes $P(A \text{ AND } B) = P(A)P(B)$.

The Addition Rule

If A and B are defined on a sample space, then: $P(A \text{ OR } B) = P(A) + P(B) - P(A \text{ AND } B)$.

If A and B are **mutually exclusive**, then $P(A \text{ AND } B) = 0$. Then $P(A \text{ OR } B) = P(A) + P(B) - P(A \text{ AND } B)$ becomes $P(A \text{ OR } B) = P(A) + P(B)$.

Example 3.14

Klaus is trying to choose where to go on vacation. His two choices are: A = New Zealand and B = Alaska

- Klaus can only afford one vacation. The probability that he chooses A is $P(A) = 0.6$ and the probability that he chooses B is $P(B) = 0.35$.

- $P(A \text{ AND } B) = 0$ because Klaus can only afford to take one vacation

- Therefore, the probability that he chooses either New Zealand or Alaska is $P(A \text{ OR } B) = P(A) + P(B) = 0.6 + 0.35 = 0.95$. Note that the probability that he does not choose to go anywhere on vacation must be 0.05.

Example 3.15

Carlos plays college soccer. He makes a goal 65% of the time he shoots. Carlos is going to attempt two goals in a row in the next game. A = the event Carlos is successful on his first attempt. $P(A) = 0.65$. B = the event Carlos is successful on his second attempt. $P(B) = 0.65$. Carlos tends to shoot in streaks. The probability that he makes the second goal **GIVEN** that he made the first goal is 0.90.

a. What is the probability that he makes both goals?

Solution 3.15

a. The problem is asking you to find $P(A \text{ AND } B) = P(B \text{ AND } A)$. Since $P(B|A) = 0.90$: $P(B \text{ AND } A) = P(B|A)P(A) = (0.90)(0.65) = 0.585$

Carlos makes the first and second goals with probability 0.585.

b. What is the probability that Carlos makes either the first goal or the second goal?

Solution 3.15

b. The problem is asking you to find $P(A$ OR $B)$.

$P(A$ OR $B) = P(A) + P(B) - P(A$ AND $B) = 0.65 + 0.65 - 0.585 = 0.715$

Carlos makes either the first goal or the second goal with probability 0.715.

c. Are A and B independent?

Solution 3.15

c. No, they are not, because $P(B$ AND $A) = 0.585$.

$P(B)P(A) = (0.65)(0.65) = 0.423$

$0.423 \neq 0.585 = P(B$ AND $A)$

So, $P(B$ AND $A)$ is **not** equal to $P(B)P(A)$.

d. Are A and B mutually exclusive?

Solution 3.15

d. No, they are not because $P(A$ and $B) = 0.585$.

To be mutually exclusive, $P(A$ AND $B)$ must equal zero.

Try It Σ

3.15 Helen plays basketball. For free throws, she makes the shot 75% of the time. Helen must now attempt two free throws. C = the event that Helen makes the first shot. $P(C) = 0.75$. D = the event Helen makes the second shot. $P(D) = 0.75$. The probability that Helen makes the second free throw given that she made the first is 0.85. What is the probability that Helen makes both free throws?

Example 3.16

A community swim team has **150** members. **Seventy-five** of the members are advanced swimmers. **Forty-seven** of the members are intermediate swimmers. The remainder are novice swimmers. **Forty** of the advanced swimmers practice four times a week. **Thirty** of the intermediate swimmers practice four times a week. **Ten** of the novice swimmers practice four times a week. Suppose one member of the swim team is chosen randomly.

a. What is the probability that the member is a novice swimmer?

Solution 3.16

a. $\frac{28}{150}$

b. What is the probability that the member practices four times a week?

Solution 3.16

b. $\frac{80}{150}$

c. What is the probability that the member is an advanced swimmer and practices four times a week?

Solution 3.16

c. $\frac{40}{150}$

d. What is the probability that a member is an advanced swimmer and an intermediate swimmer? Are being an advanced swimmer and an intermediate swimmer mutually exclusive? Why or why not?

Solution 3.16

d. P(advanced AND intermediate) = 0, so these are mutually exclusive events. A swimmer cannot be an advanced swimmer and an intermediate swimmer at the same time.

e. Are being a novice swimmer and practicing four times a week independent events? Why or why not?

Solution 3.16

e. No, these are not independent events.
P(novice AND practices four times per week) = 0.0667
P(novice)P(practices four times per week) = 0.0996
$0.0667 \neq 0.0996$

Try It Σ

3.16 A school has 200 seniors of whom 140 will be going to college next year. Forty will be going directly to work. The remainder are taking a gap year. Fifty of the seniors going to college play sports. Thirty of the seniors going directly to work play sports. Five of the seniors taking a gap year play sports. What is the probability that a senior is taking a gap year?

Example 3.17

Felicity attends Modesto JC in Modesto, CA. The probability that Felicity enrolls in a math class is 0.2 and the probability that she enrolls in a speech class is 0.65. The probability that she enrolls in a math class GIVEN that she enrolls in speech class is 0.25.

Let: M = math class, S = speech class, $M|S$ = math given speech

a. What is the probability that Felicity enrolls in math and speech?
   Find $P(M \text{ AND } S) = P(M|S)P(S)$.

b. What is the probability that Felicity enrolls in math or speech classes?
   Find $P(M \text{ OR } S) = P(M) + P(S) - P(M \text{ AND } S)$.

c. Are M and S independent? Is $P(M|S) = P(M)$?

d. Are M and S mutually exclusive? Is $P(M \text{ AND } S) = 0$?

Solution 3.17
a. 0.1625, b. 0.6875, c. No, d. No

Try It Σ

3.17 A student goes to the library. Let events B = the student checks out a book and D = the student check out a DVD. Suppose that $P(B) = 0.40$, $P(D) = 0.30$ and $P(D|B) = 0.5$.

a. Find $P(B \text{ AND } D)$.

b. Find $P(B \text{ OR } D)$.

Example 3.18

Studies show that about one woman in seven (approximately 14.3%) who live to be 90 will develop breast cancer. Suppose that of those women who develop breast cancer, a test is negative 2% of the time. Also suppose that in the general population of women, the test for breast cancer is negative about 85% of the time. Let B = woman develops breast cancer and let N = tests negative. Suppose one woman is selected at random.

a. What is the probability that the woman develops breast cancer? What is the probability that woman tests negative?

Solution 3.18
a. $P(B) = 0.143$; $P(N) = 0.85$

b. Given that the woman has breast cancer, what is the probability that she tests negative?

Solution 3.18
b. $P(N|B) = 0.02$

c. What is the probability that the woman has breast cancer AND tests negative?

Solution 3.18
c. $P(B \text{ AND } N) = P(B)P(N|B) = (0.143)(0.02) = 0.0029$

d. What is the probability that the woman has breast cancer or tests negative?

Solution 3.18
d. $P(B \text{ OR } N) = P(B) + P(N) - P(B \text{ AND } N) = 0.143 + 0.85 - 0.0029 = 0.9901$

e. Are having breast cancer and testing negative independent events?

Solution 3.18
e. No. $P(N) = 0.85$; $P(N|B) = 0.02$. So, $P(N|B)$ does not equal $P(N)$.

f. Are having breast cancer and testing negative mutually exclusive?

Solution 3.18
f. No. $P(B \text{ AND } N) = 0.0029$. For B and N to be mutually exclusive, $P(B \text{ AND } N)$ must be zero.

3.18 A school has 200 seniors of whom 140 will be going to college next year. Forty will be going directly to work. The remainder are taking a gap year. Fifty of the seniors going to college play sports. Thirty of the seniors going directly to work play sports. Five of the seniors taking a gap year play sports. What is the probability that a senior is going to college and plays sports?

Example 3.19

Refer to the information in **Example 3.18**. P = tests positive.

  a. Given that a woman develops breast cancer, what is the probability that she tests positive. Find $P(P|B) = 1 - P(N|B)$.

  b. What is the probability that a woman develops breast cancer and tests positive. Find $P(B \text{ AND } P) = P(P|B)P(B)$.

  c. What is the probability that a woman does not develop breast cancer. Find $P(B') = 1 - P(B)$.

  d. What is the probability that a woman tests positive for breast cancer. Find $P(P) = 1 - P(N)$.

Solution 3.19
a. 0.98; b. 0.1401; c. 0.857; d. 0.15

Try It Σ

3.19 A student goes to the library. Let events B = the student checks out a book and D = the student checks out a DVD. Suppose that $P(B) = 0.40$, $P(D) = 0.30$ and $P(D|B) = 0.5$.

  a. Find $P(B')$.

  b. Find $P(D \text{ AND } B)$.

  c. Find $P(B|D)$.

  d. Find $P(D \text{ AND } B')$.

  e. Find $P(D|B')$.

3.4 | Contingency Tables

A **contingency table** provides a way of portraying data that can facilitate calculating probabilities. The table helps in determining conditional probabilities quite easily. The table displays sample values in relation to two different variables that may be dependent or contingent on one another. Later on, we will use contingency tables again, but in another manner.

Example 3.20

Suppose a study of speeding violations and drivers who use cell phones produced the following fictional data:

	Speeding violation in the last year	No speeding violation in the last year	Total
Cell phone user	25	280	305
Not a cell phone user	45	405	450
Total	70	685	755

Table 3.2

The total number of people in the sample is 755. The row totals are 305 and 450. The column totals are 70 and 685. Notice that 305 + 450 = 755 and 70 + 685 = 755.

Calculate the following probabilities using the table.

a. Find P(Person is a car phone user).

Solution 3.20

a. $\dfrac{\text{number of car phone users}}{\text{total number in study}} = \dfrac{305}{755}$

b. Find P(person had no violation in the last year).

Solution 3.20

b. $\dfrac{\text{number that had no violation}}{\text{total number in study}} = \dfrac{685}{755}$

c. Find P(Person had no violation in the last year AND was a car phone user).

Solution 3.20

c. $\dfrac{280}{755}$

d. Find P(Person is a car phone user OR person had no violation in the last year).

Solution 3.20

d. $\left(\dfrac{305}{755} + \dfrac{685}{755}\right) - \dfrac{280}{755} = \dfrac{710}{755}$

e. Find P(Person is a car phone user GIVEN person had a violation in the last year).

Solution 3.20

e. $\dfrac{25}{70}$ (The sample space is reduced to the number of persons who had a violation.)

f. Find P(Person had no violation last year GIVEN person was not a car phone user)

Solution 3.20

f. $\dfrac{405}{450}$ (The sample space is reduced to the number of persons who were not car phone users.)

Try It Σ

3.20 **Table 3.3** shows the number of athletes who stretch before exercising and how many had injuries within the past year.

	Injury in last year	No injury in last year	Total
Stretches	55	295	350
Does not stretch	231	219	450
Total	286	514	800

Table 3.3

a. What is P(athlete stretches before exercising)?

b. What is P(athlete stretches before exercising|no injury in the last year)?

Example 3.21

Table 3.4 shows a random sample of 100 hikers and the areas of hiking they prefer.

Sex	The Coastline	Near Lakes and Streams	On Mountain Peaks	Total
Female	18	16	___	45
Male	___	___	14	55
Total	___	41	___	___

Table 3.4 Hiking Area Preference

a. Complete the table.

Solution 3.21

a.

Sex	The Coastline	Near Lakes and Streams	On Mountain Peaks	Total
Female	18	16	**11**	45
Male	**16**	**25**	14	55
Total	**34**	41	**25**	**100**

Table 3.5 Hiking Area Preference

b. Are the events "being female" and "preferring the coastline" independent events?

Let F = being female and let C = preferring the coastline.

1. Find $P(F \text{ AND } C)$.

2. Find $P(F)P(C)$

Are these two numbers the same? If they are, then F and C are independent. If they are not, then F and C are not independent.

Solution 3.21

b.

1. $P(F \text{ AND } C) = \frac{18}{100} = 0.18$

2. $P(F)P(C) = \left(\frac{45}{100}\right)\left(\frac{34}{100}\right) = (0.45)(0.34) = 0.153$

$P(F \text{ AND } C) \neq P(F)P(C)$, so the events F and C are not independent.

c. Find the probability that a person is male given that the person prefers hiking near lakes and streams. Let M = being male, and let L = prefers hiking near lakes and streams.

1. What word tells you this is a conditional?

2. Fill in the blanks and calculate the probability: $P(\_\_|\_\_) = \_\_$.

3. Is the sample space for this problem all 100 hikers? If not, what is it?

Solution 3.21

c.

1. The word 'given' tells you that this is a conditional.

2. $P(M|L) = \frac{25}{41}$

3. No, the sample space for this problem is the 41 hikers who prefer lakes and streams.

d. Find the probability that a person is female or prefers hiking on mountain peaks. Let F = being female, and let P = prefers mountain peaks.

1. Find $P(F)$.

2. Find $P(P)$.

3. Find $P(F \text{ AND } P)$.

4. Find $P(F \text{ OR } P)$.

Solution 3.21

d.

1. $P(F) = \frac{45}{100}$

2. $P(P) = \frac{25}{100}$

3. $P(F \text{ AND } P) = \frac{11}{100}$

4. $P(F \text{ OR } P) = \frac{45}{100} + \frac{25}{100} - \frac{11}{100} = \frac{59}{100}$

Try It Σ

3.21 **Table 3.6** shows a random sample of 200 cyclists and the routes they prefer. Let *M* = males and *H* = hilly path.

Gender	Lake Path	Hilly Path	Wooded Path	Total
Female	45	38	27	110
Male	26	52	12	90
Total	71	90	39	200

Table 3.6

a. Out of the males, what is the probability that the cyclist prefers a hilly path?

b. Are the events "being male" and "preferring the hilly path" independent events?

Example 3.22

Muddy Mouse lives in a cage with three doors. If Muddy goes out the first door, the probability that he gets caught by Alissa the cat is $\frac{1}{5}$ and the probability he is not caught is $\frac{4}{5}$. If he goes out the second door, the probability he gets caught by Alissa is $\frac{1}{4}$ and the probability he is not caught is $\frac{3}{4}$. The probability that Alissa catches Muddy coming out of the third door is $\frac{1}{2}$ and the probability she does not catch Muddy is $\frac{1}{2}$. It is equally likely that Muddy will choose any of the three doors so the probability of choosing each door is $\frac{1}{3}$.

Caught or Not	Door One	Door Two	Door Three	Total
Caught	$\frac{1}{15}$	$\frac{1}{12}$	$\frac{1}{6}$	____
Not Caught	$\frac{4}{15}$	$\frac{3}{12}$	$\frac{1}{6}$	____
Total	____	____	____	1

Table 3.7 Door Choice

- The first entry $\frac{1}{15} = \left(\frac{1}{5}\right)\left(\frac{1}{3}\right)$ is *P*(Door One AND Caught)

- The entry $\frac{4}{15} = \left(\frac{4}{5}\right)\left(\frac{1}{3}\right)$ is *P*(Door One AND Not Caught)

Verify the remaining entries.

a. Complete the probability contingency table. Calculate the entries for the totals. Verify that the lower-right corner entry is 1.

Solution 3.22

a.

Caught or Not	Door One	Door Two	Door Three	Total
Caught	$\frac{1}{15}$	$\frac{1}{12}$	$\frac{1}{6}$	$\frac{19}{60}$
Not Caught	$\frac{4}{15}$	$\frac{3}{12}$	$\frac{1}{6}$	$\frac{41}{60}$
Total	$\frac{5}{15}$	$\frac{4}{12}$	$\frac{2}{6}$	1

Table 3.8 Door Choice

b. What is the probability that Alissa does not catch Muddy?

Solution 3.22

b. $\frac{41}{60}$

c. What is the probability that Muddy chooses Door One OR Door Two given that Muddy is caught by Alissa?

Solution 3.22

c. $\frac{9}{19}$

Example 3.23

Table 3.9 contains the number of crimes per 100,000 inhabitants from 2008 to 2011 in the U.S.

Year	Robbery	Burglary	Rape	Vehicle	Total
2008	145.7	732.1	29.7	314.7	
2009	133.1	717.7	29.1	259.2	
2010	119.3	701	27.7	239.1	
2011	113.7	702.2	26.8	229.6	
Total					

Table 3.9 United States Crime Index Rates Per 100,000 Inhabitants 2008–2011

TOTAL each column and each row. Total data = 4,520.7

  a. Find P(2009 AND Robbery).
  b. Find P(2010 AND Burglary).
  c. Find P(2010 OR Burglary).
  d. Find P(2011|Rape).
  e. Find P(Vehicle|2008).

Solution 3.23

a. 0.0294, b. 0.1551, c. 0.7165, d. 0.2365, e. 0.2575

3.23 **Table 3.10** relates the weights and heights of a group of individuals participating in an observational study.

Weight/Height	Tall	Medium	Short	Totals
Obese	18	28	14	
Normal	20	51	28	
Underweight	12	25	9	
Totals				

Table 3.10

a. Find the total for each row and column

b. Find the probability that a randomly chosen individual from this group is Tall.

c. Find the probability that a randomly chosen individual from this group is Obese and Tall.

d. Find the probability that a randomly chosen individual from this group is Tall given that the idividual is Obese.

e. Find the probability that a randomly chosen individual from this group is Obese given that the individual is Tall.

f. Find the probability a randomly chosen individual from this group is Tall and Underweight.

g. Are the events Obese and Tall independent?

3.5 | Tree and Venn Diagrams

Sometimes, when the probability problems are complex, it can be helpful to graph the situation. Tree diagrams and Venn diagrams are two tools that can be used to visualize and solve conditional probabilities.

Tree Diagrams

A **tree diagram** is a special type of graph used to determine the outcomes of an experiment. It consists of "branches" that are labeled with either frequencies or probabilities. Tree diagrams can make some probability problems easier to visualize and solve. The following example illustrates how to use a tree diagram.

Example 3.24

In an urn, there are 11 balls. Three balls are red (*R*) and eight balls are blue (*B*). Draw two balls, one at a time, **with replacement**. "With replacement" means that you put the first ball back in the urn before you select the second ball. The tree diagram using frequencies that show all the possible outcomes follows.

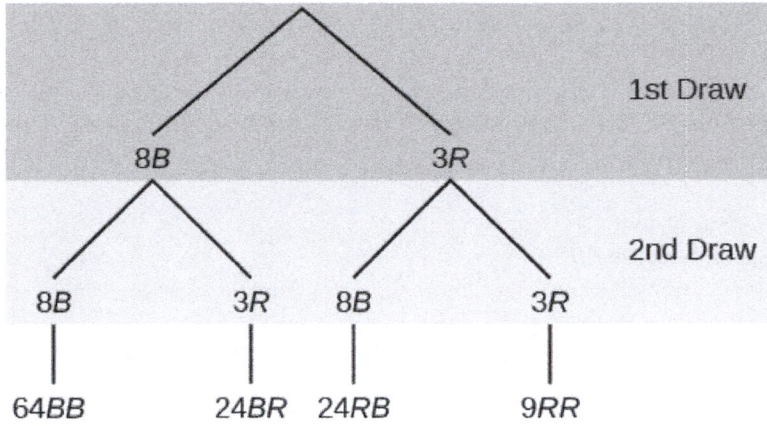

Figure 3.2 Total = 64 + 24 + 24 + 9 = 121

The first set of branches represents the first draw. The second set of branches represents the second draw. Each of the outcomes is distinct. In fact, we can list each red ball as $R1$, $R2$, and $R3$ and each blue ball as $B1$, $B2$, $B3$, $B4$, $B5$, $B6$, $B7$, and $B8$. Then the nine RR outcomes can be written as:

$R1R1$; $R1R2$; $R1R3$; $R2R1$; $R2R2$; $R2R3$; $R3R1$; $R3R2$; $R3R3$

The other outcomes are similar.

There are a total of 11 balls in the urn. Draw two balls, one at a time, with replacement. There are 11(11) = 121 outcomes, the size of the **sample space**.

a. List the 24 BR outcomes: $B1R1$, $B1R2$, $B1R3$, ...

Solution 3.24
a. $B1R1$; $B1R2$; $B1R3$; $B2R1$; $B2R2$; $B2R3$; $B3R1$; $B3R2$; $B3R3$; $B4R1$; $B4R2$; $B4R3$; $B5R1$; $B5R2$; $B5R3$; $B6R1$; $B6R2$; $B6R3$; $B7R1$; $B7R2$; $B7R3$; $B8R1$; $B8R2$; $B8R3$

b. Using the tree diagram, calculate $P(RR)$.

Solution 3.24
b. $P(RR) = \left(\frac{3}{11}\right)\left(\frac{3}{11}\right) = \frac{9}{121}$

c. Using the tree diagram, calculate $P(RB \text{ OR } BR)$.

Solution 3.24
c. $P(RB \text{ OR } BR) = \left(\frac{3}{11}\right)\left(\frac{8}{11}\right) + \left(\frac{8}{11}\right)\left(\frac{3}{11}\right) = \frac{48}{121}$

d. Using the tree diagram, calculate $P(R \text{ on 1st draw AND } B \text{ on 2nd draw})$.

Solution 3.24
d. $P(R \text{ on 1st draw AND } B \text{ on 2nd draw}) = P(RB) = \left(\frac{3}{11}\right)\left(\frac{8}{11}\right) = \frac{24}{121}$

e. Using the tree diagram, calculate $P(R \text{ on 2nd draw GIVEN } B \text{ on 1st draw})$.

Solution 3.24

e. $P(R$ on 2nd draw GIVEN B on 1st draw$) = P(R$ on 2nd$|B$ on 1st$) = \frac{24}{88} = \frac{3}{11}$

This problem is a conditional one. The sample space has been reduced to those outcomes that already have a blue on the first draw. There are 24 + 64 = 88 possible outcomes (24 *BR* and 64 *BB*). Twenty-four of the 88 possible outcomes are *BR*. $\frac{24}{88} = \frac{3}{11}$.

f. Using the tree diagram, calculate $P(BB)$.

Solution 3.24

f. $P(BB) = \frac{64}{121}$

g. Using the tree diagram, calculate $P(B$ on the 2nd draw given R on the first draw$)$.

Solution 3.24

g. $P(B$ on 2nd draw$|R$ on 1st draw$) = \frac{8}{11}$

There are 9 + 24 outcomes that have *R* on the first draw (9 *RR* and 24 *RB*). The sample space is then 9 + 24 = 33. 24 of the 33 outcomes have *B* on the second draw. The probability is then $\frac{24}{33}$.

Try It Σ

3.24 In a standard deck, there are 52 cards. 12 cards are face cards (event *F*) and 40 cards are not face cards (event *N*). Draw two cards, one at a time, with replacement. All possible outcomes are shown in the tree diagram as frequencies. Using the tree diagram, calculate $P(FF)$.

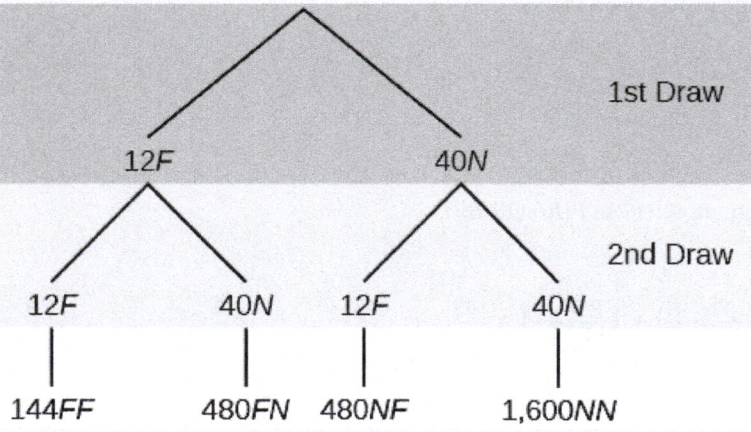

Figure 3.3

Example 3.25

An urn has three red marbles and eight blue marbles in it. Draw two marbles, one at a time, this time without replacement, from the urn. **"Without replacement"** means that you do not put the first ball back before you select the second marble. Following is a tree diagram for this situation. The branches are labeled with probabilities instead of frequencies. The numbers at the ends of the branches are calculated by multiplying the numbers on the two corresponding branches, for example, $\left(\frac{3}{11}\right)\left(\frac{2}{10}\right) = \frac{6}{110}$.

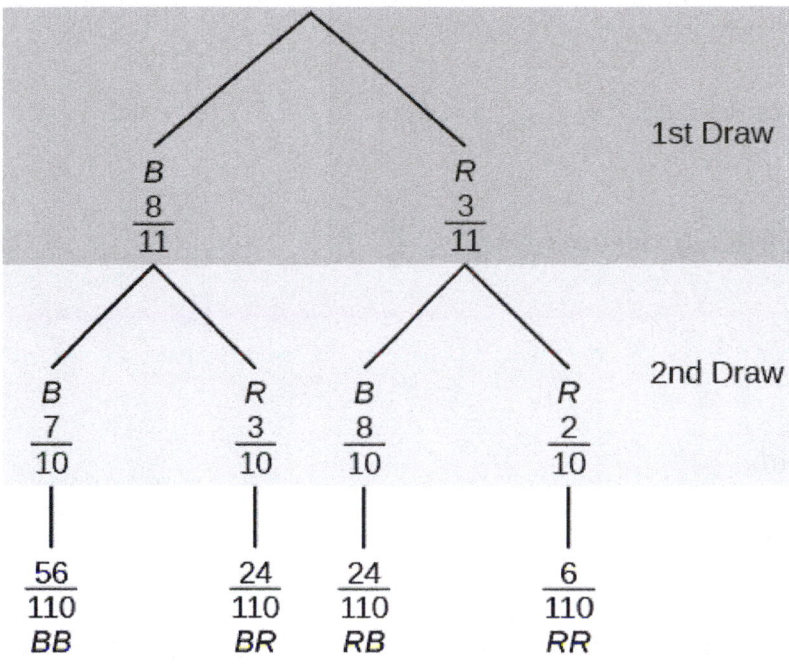

Figure 3.4 Total = $\frac{56 + 24 + 24 + 6}{110} = \frac{110}{110} = 1$

NOTE

If you draw a red on the first draw from the three red possibilities, there are two red marbles left to draw on the second draw. You do not put back or replace the first marble after you have drawn it. You draw **without replacement**, so that on the second draw there are ten marbles left in the urn.

Calculate the following probabilities using the tree diagram.

a. $P(RR) = $ _____

Solution 3.25

a. $P(RR) = \left(\frac{3}{11}\right)\left(\frac{2}{10}\right) = \frac{6}{110}$

b. Fill in the blanks:

$P(RB \text{ OR } BR) = \left(\frac{3}{11}\right)\left(\frac{8}{10}\right) + (\_\_\_)(\_\_\_) = \frac{48}{110}$

Solution 3.25

b. $P(RB \text{ OR } BR) = \left(\frac{3}{11}\right)\left(\frac{8}{10}\right) + \left(\frac{8}{11}\right)\left(\frac{3}{10}\right) = \frac{48}{110}$

c. $P(R \text{ on 2nd}|B \text{ on 1st}) =$

Solution 3.25

c. $P(R \text{ on 2nd}|B \text{ on 1st}) = \frac{3}{10}$

d. Fill in the blanks.

$P(R \text{ on 1st AND } B \text{ on 2nd}) = P(RB) = (\underline{\quad})(\underline{\quad}) = \frac{24}{100}$

Solution 3.25

d. $P(R \text{ on 1st AND } B \text{ on 2nd}) = P(RB) = \left(\frac{3}{11}\right)\left(\frac{8}{10}\right) = \frac{24}{100}$

e. Find $P(BB)$.

Solution 3.25

e. $P(BB) = \left(\frac{8}{11}\right)\left(\frac{7}{10}\right)$

f. Find $P(B \text{ on 2nd}|R \text{ on 1st})$.

Solution 3.25

f. Using the tree diagram, $P(B \text{ on 2nd}|R \text{ on 1st}) = P(R|B) = \frac{8}{10}$.

If we are using probabilities, we can label the tree in the following general way.

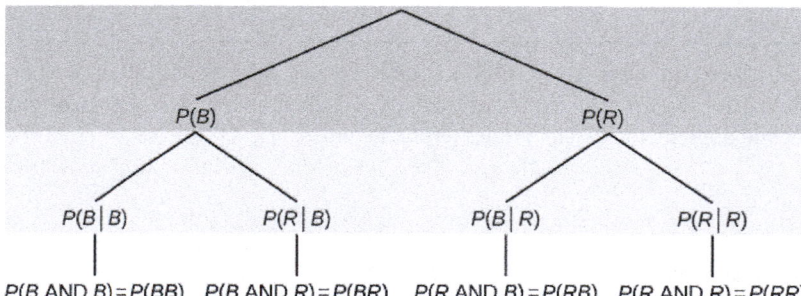

- $P(R|R)$ here means $P(R \text{ on 2nd}|R \text{ on 1st})$
- $P(B|R)$ here means $P(B \text{ on 2nd}|R \text{ on 1st})$
- $P(R|B)$ here means $P(R \text{ on 2nd}|B \text{ on 1st})$
- $P(B|B)$ here means $P(B \text{ on 2nd}|B \text{ on 1st})$

Try It Σ

3.25 In a standard deck, there are 52 cards. Twelve cards are face cards (*F*) and 40 cards are not face cards (*N*). Draw two cards, one at a time, without replacement. The tree diagram is labeled with all possible probabilities.

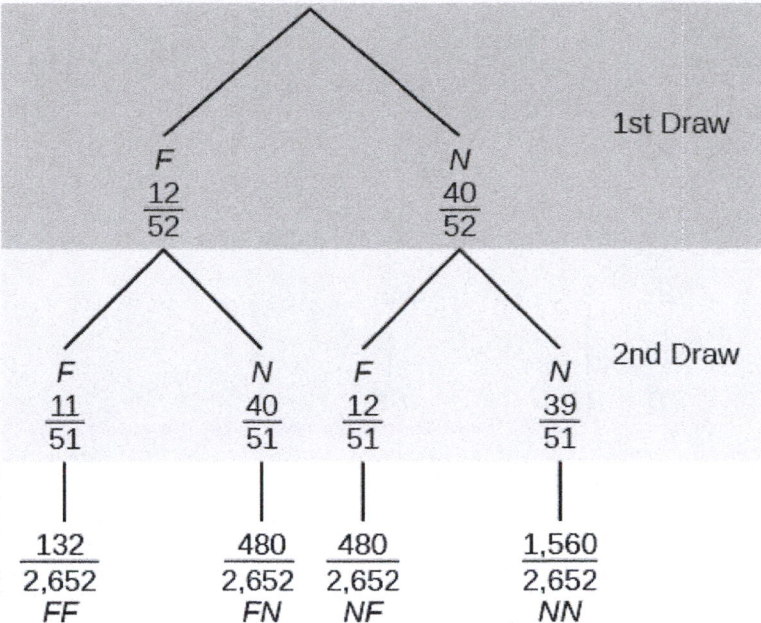

Figure 3.5

a. Find $P(FN \text{ OR } NF)$.

b. Find $P(N|F)$.

c. Find P(at most one face card).
    Hint: "At most one face card" means zero or one face card.

d. Find P(at least on face card).
    Hint: "At least one face card" means one or two face cards.

Example 3.26

A litter of kittens available for adoption at the Humane Society has four tabby kittens and five black kittens. A family comes in and randomly selects two kittens (without replacement) for adoption.

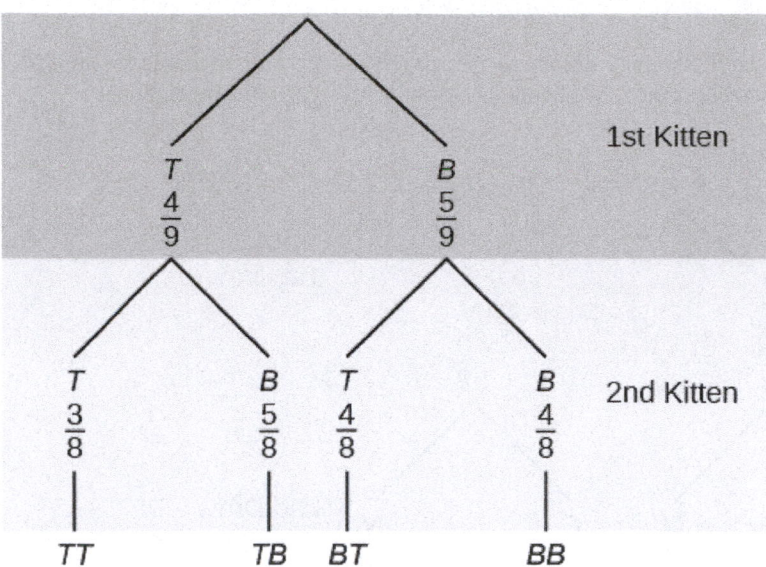

a. What is the probability that both kittens are tabby?

    a. $\left(\frac{1}{2}\right)\left(\frac{1}{2}\right)$ b. $\left(\frac{4}{9}\right)\left(\frac{4}{9}\right)$ c. $\left(\frac{4}{9}\right)\left(\frac{3}{8}\right)$ d. $\left(\frac{4}{9}\right)\left(\frac{5}{9}\right)$

b. What is the probability that one kitten of each coloring is selected?

    a. $\left(\frac{4}{9}\right)\left(\frac{5}{9}\right)$ b. $\left(\frac{4}{9}\right)\left(\frac{5}{8}\right)$ c. $\left(\frac{4}{9}\right)\left(\frac{5}{9}\right)+\left(\frac{5}{9}\right)\left(\frac{4}{9}\right)$ d. $\left(\frac{4}{9}\right)\left(\frac{5}{8}\right)+\left(\frac{5}{9}\right)\left(\frac{4}{8}\right)$

c. What is the probability that a tabby is chosen as the second kitten when a black kitten was chosen as the first?

d. What is the probability of choosing two kittens of the same color?

Solution 3.26

a. c, b. d, c. $\frac{4}{8}$, d. $\frac{32}{72}$

3.26 Suppose there are four red balls and three yellow balls in a box. Three balls are drawn from the box without replacement. What is the probability that one ball of each coloring is selected?

Venn Diagram

A **Venn diagram** is a picture that represents the outcomes of an experiment. It generally consists of a box that represents the sample space S together with circles or ovals. The circles or ovals represent events.

Example 3.27

Suppose an experiment has the outcomes 1, 2, 3, ... , 12 where each outcome has an equal chance of occurring. Let event A = {1, 2, 3, 4, 5, 6} and event B = {6, 7, 8, 9}. Then A AND B = {6} and A OR B = {1, 2, 3, 4, 5, 6, 7, 8, 9}. The Venn diagram is as follows:

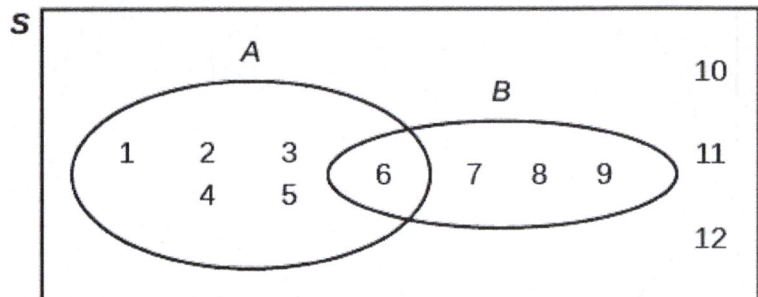

Figure 3.6

3.27 Suppose an experiment has outcomes black, white, red, orange, yellow, green, blue, and purple, where each outcome has an equal chance of occurring. Let event C = {green, blue, purple} and event P = {red, yellow, blue}. Then C AND P = {blue} and C OR P = {green, blue, purple, red, yellow}. Draw a Venn diagram representing this situation.

Example 3.28

Flip two fair coins. Let A = tails on the first coin. Let B = tails on the second coin. Then A = {TT, TH} and B = {TT, HT}. Therefore, A AND B = {TT}. A OR B = {TH, TT, HT}.

The sample space when you flip two fair coins is X = {HH, HT, TH, TT}. The outcome HH is in NEITHER A NOR B. The Venn diagram is as follows:

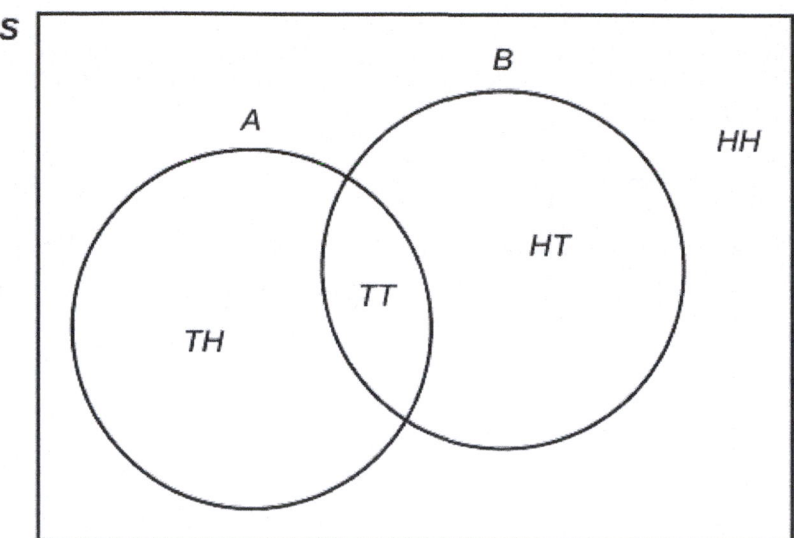

Figure 3.7

3.28 Roll a fair, six-sided die. Let A = a prime number of dots is rolled. Let B = an odd number of dots is rolled. Then $A = \{2, 3, 5\}$ and $B = \{1, 3, 5\}$. Therefore, A AND $B = \{3, 5\}$. A OR $B = \{1, 2, 3, 5\}$. The sample space for rolling a fair die is $S = \{1, 2, 3, 4, 5, 6\}$. Draw a Venn diagram representing this situation.

Example 3.29

Forty percent of the students at a local college belong to a club and **50%** work part time. **Five percent** of the students work part time and belong to a club. Draw a Venn diagram showing the relationships. Let C = student belongs to a club and PT = student works part time.

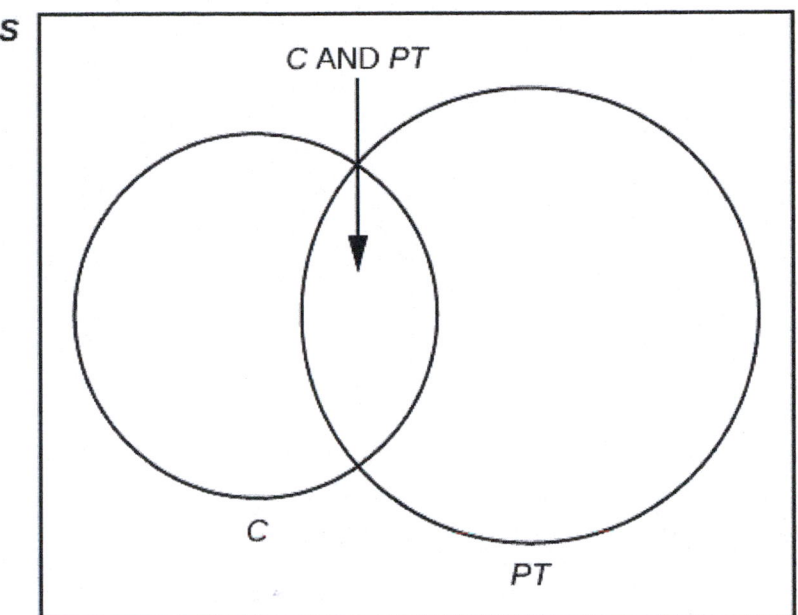

Figure 3.8

If a student is selected at random, find

- the probability that the student belongs to a club. $P(C) = 0.40$
- the probability that the student works part time. $P(PT) = 0.50$
- the probability that the student belongs to a club AND works part time. $P(C \text{ AND } PT) = 0.05$
- the probability that the student belongs to a club **given** that the student works part time.
$$P(C|PT) = \frac{P(C \text{ AND } PT)}{P(PT)} = \frac{0.05}{0.50} = 0.1$$
- the probability that the student belongs to a club **OR** works part time. $P(C \text{ OR } PT) = P(C) + P(PT) - P(C \text{ AND } PT) = 0.40 + 0.50 - 0.05 = 0.85$

Try It Σ

3.29 Fifty percent of the workers at a factory work a second job, 25% have a spouse who also works, 5% work a second job and have a spouse who also works. Draw a Venn diagram showing the relationships. Let W = works a second job and S = spouse also works.

Example 3.30

A person with type O blood and a negative Rh factor (Rh-) can donate blood to any person with any blood type. Four percent of African Americans have type O blood and a negative RH factor, 5–10% of African Americans have the Rh- factor, and 51% have type O blood.

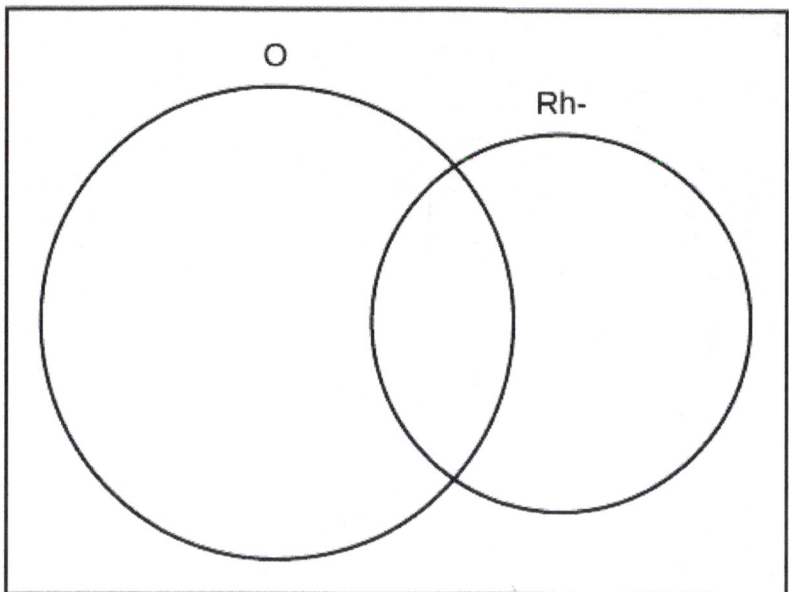

Figure 3.9

The "O" circle represents the African Americans with type O blood. The "Rh-" oval represents the African Americans with the Rh- factor.

We will take the average of 5% and 10% and use 7.5% as the percent of African Americans who have the Rh-factor. Let O = African American with Type O blood and R = African American with Rh- factor.

   a. $P(O)$ = _____

   b. $P(R)$ = _____

   c. $P(O$ AND $R)$ = _____

   d. $P(O$ OR $R)$ = _____

   e. In the Venn Diagram, describe the overlapping area using a complete sentence.

   f. In the Venn Diagram, describe the area in the rectangle but outside both the circle and the oval using a complete sentence.

Solution 3.30
a. 0.51; b. 0.075; c. 0.04; d. 0.545; e. The area represents the African Americans that have type O blood and the Rh- factor. f. The area represents the African Americans that have neither type O blood nor the Rh- factor.

3.30 In a bookstore, the probability that the customer buys a novel is 0.6, and the probability that the customer buys a non-fiction book is 0.4. Suppose that the probability that the customer buys both is 0.2.

   a. Draw a Venn diagram representing the situation.

   b. Find the probability that the customer buys either a novel or anon-fiction book.

   c. In the Venn diagram, describe the overlapping area using a complete sentence.

   d. Suppose that some customers buy only compact disks. Draw an oval in your Venn diagram representing this event.

3.6 | Probability Topics

Stats Lab

3.1 Probability Topics

Class time:

Names:

Student Learning Outcomes

- The student will use theoretical and empirical methods to estimate probabilities.
- The student will appraise the differences between the two estimates.
- The student will demonstrate an understanding of long-term relative frequencies.

Do the Experiment

Count out 40 mixed-color M&Ms® which is approximately one small bag's worth. Record the number of each color in **Table 3.11**. Use the information from this table to complete **Table 3.12**. Next, put the M&Ms in a cup. The experiment is to pick two M&Ms, one at a time. Do **not** look at them as you pick them. The first time through, replace the first M&M before picking the second one. Record the results in the "With Replacement" column of **Table 3.13**. Do this 24 times. The second time through, after picking the first M&M, do **not** replace it before picking the second one. Then, pick the second one. Record the results in the "Without Replacement" column section of **Table 3.14**. After you record the pick, put **both** M&Ms back. Do this a total of 24 times, also. Use the data from **Table 3.14** to calculate the empirical probability questions. Leave your answers in unreduced fractional form. Do **not** multiply out any fractions.

Color	Quantity
Yellow (Y)	
Green (G)	
Blue (BL)	
Brown (B)	
Orange (O)	
Red (R)	

Table 3.11 Population

	With Replacement	Without Replacement
$P(\text{2 reds})$		
$P(R_1B_2 \text{ OR } B_1R_2)$		
$P(R_1 \text{ AND } G_2)$		
$P(G_2|R_1)$		
$P(\text{no yellows})$		
$P(\text{doubles})$		
$P(\text{no doubles})$		

Table 3.12 Theoretical Probabilities

NOTE

G_2 = green on second pick; R_1 = red on first pick; B_1 = brown on first pick; B_2 = brown on second pick; doubles = both picks are the same colour.

With Replacement	Without Replacement
(__ , __) (__ , __)	(__ , __) (__ , __)
(__ , __) (__ , __)	(__ , __) (__ , __)
(__ , __) (__ , __)	(__ , __) (__ , __)
(__ , __) (__ , __)	(__ , __) (__ , __)
(__ , __) (__ , __)	(__ , __) (__ , __)
(__ , __) (__ , __)	(__ , __) (__ , __)
(__ , __) (__ , __)	(__ , __) (__ , __)
(__ , __) (__ , __)	(__ , __) (__ , __)
(__ , __) (__ , __)	(__ , __) (__ , __)
(__ , __) (__ , __)	(__ , __) (__ , __)
(__ , __) (__ , __)	(__ , __) (__ , __)
(__ , __) (__ , __)	(__ , __) (__ , __)

Table 3.13 Empirical Results

	With Replacement	Without Replacement
P(2 reds)		
$P(R_1B_2$ OR $B_1R_2)$		
$P(R_1$ AND $G_2)$		
$P(G_2\|R_1)$		
P(no yellows)		
P(doubles)		
P(no doubles)		

Table 3.14 Empirical Probabilities

Discussion Questions

1. Why are the "With Replacement" and "Without Replacement" probabilities different?

2. Convert P(no yellows) to decimal format for both Theoretical "With Replacement" and for Empirical "With Replacement". Round to four decimal places.

   a. Theoretical "With Replacement": P(no yellows) = _____

   b. Empirical "With Replacement": P(no yellows) = _____

   c. Are the decimal values "close"? Did you expect them to be closer together or farther apart? Why?

3. If you increased the number of times you picked two M&Ms to 240 times, why would empirical probability values change?

4. Would this change (see part 3) cause the empirical probabilities and theoretical probabilities to be closer together or farther apart? How do you know?

5. Explain the differences in what $P(G_1 \text{ AND } R_2)$ and $P(R_1|G_2)$ represent. Hint: Think about the sample space for each probability.

KEY TERMS

Conditional Probability the likelihood that an event will occur given that another event has already occurred

contingency table the method of displaying a frequency distribution as a table with rows and columns to show how two variables may be dependent (contingent) upon each other; the table provides an easy way to calculate conditional probabilities.

Dependent Events If two events are NOT independent, then we say that they are dependent.

Equally Likely Each outcome of an experiment has the same probability.

Event a subset of the set of all outcomes of an experiment; the set of all outcomes of an experiment is called a **sample space** and is usually denoted by S. An event is an arbitrary subset in S. It can contain one outcome, two outcomes, no outcomes (empty subset), the entire sample space, and the like. Standard notations for events are capital letters such as A, B, C, and so on.

Experiment a planned activity carried out under controlled conditions

Independent Events The occurrence of one event has no effect on the probability of the occurrence of another event. Events A and B are independent if one of the following is true:

1. $P(A|B) = P(A)$
2. $P(B|A) = P(B)$
3. $P(A \text{ AND } B) = P(A)P(B)$

Mutually Exclusive Two events are mutually exclusive if the probability that they both happen at the same time is zero. If events A and B are mutually exclusive, then $P(A \text{ AND } B) = 0$.

Outcome a particular result of an experiment

Probability a number between zero and one, inclusive, that gives the likelihood that a specific event will occur; the foundation of statistics is given by the following 3 axioms (by A.N. Kolmogorov, 1930's): Let S denote the sample space and A and B are two events in S. Then:

- $0 \le P(A) \le 1$
- If A and B are any two mutually exclusive events, then $P(A \text{ OR } B) = P(A) + P(B)$.
- $P(S) = 1$

Sample Space the set of all possible outcomes of an experiment

Sampling with Replacement If each member of a population is replaced after it is picked, then that member has the possibility of being chosen more than once.

Sampling without Replacement When sampling is done without replacement, each member of a population may be chosen only once.

The AND Event An outcome is in the event A AND B if the outcome is in both A AND B at the same time.

The Complement Event The complement of event A consists of all outcomes that are NOT in A.

The Conditional Probability of A GIVEN B $P(A|B)$ is the probability that event A will occur given that the event B has already occurred.

The Conditional Probability of One Event Given Another Event $P(A|B)$ is the probability that event A will occur given that the event B has already occurred.

The Or Event An outcome is in the event A OR B if the outcome is in A or is in B or is in both A and B.

The OR of Two Events An outcome is in the event A OR B if the outcome is in A, is in B, or is in both A and B.

Tree Diagram the useful visual representation of a sample space and events in the form of a "tree" with branches marked by possible outcomes together with associated probabilities (frequencies, relative frequencies)

Venn Diagram the visual representation of a sample space and events in the form of circles or ovals showing their intersections

CHAPTER REVIEW

3.1 Terminology

In this module we learned the basic terminology of probability. The set of all possible outcomes of an experiment is called the sample space. Events are subsets of the sample space, and they are assigned a probability that is a number between zero and one, inclusive.

3.2 Independent and Mutually Exclusive Events

Two events A and B are independent if the knowledge that one occurred does not affect the chance the other occurs. If two events are not independent, then we say that they are dependent.

In sampling with replacement, each member of a population is replaced after it is picked, so that member has the possibility of being chosen more than once, and the events are considered to be independent. In sampling without replacement, each member of a population may be chosen only once, and the events are considered not to be independent. When events do not share outcomes, they are mutually exclusive of each other.

3.3 Two Basic Rules of Probability

The multiplication rule and the addition rule are used for computing the probability of A and B, as well as the probability of A or B for two given events A, B defined on the sample space. In sampling with replacement each member of a population is replaced after it is picked, so that member has the possibility of being chosen more than once, and the events are considered to be independent. In sampling without replacement, each member of a population may be chosen only once, and the events are considered to be not independent. The events A and B are mutually exclusive events when they do not have any outcomes in common.

3.4 Contingency Tables

There are several tools you can use to help organize and sort data when calculating probabilities. Contingency tables help display data and are particularly useful when calculating probabilites that have multiple dependent variables.

3.5 Tree and Venn Diagrams

A tree diagram use branches to show the different outcomes of experiments and makes complex probability questions easy to visualize.

A Venn diagram is a picture that represents the outcomes of an experiment. It generally consists of a box that represents the sample space S together with circles or ovals. The circles or ovals represent events. A Venn diagram is especially helpful for visualizing the OR event, the AND event, and the complement of an event and for understanding conditional probabilities.

FORMULA REVIEW

3.1 Terminology

A and B are events

$P(S) = 1$ where S is the sample space

$0 \le P(A) \le 1$

$P(A|B) = \dfrac{P(A \text{ AND } B)}{P(B)}$

3.2 Independent and Mutually Exclusive Events

If A and B are independent, $P(A \text{ AND } B) = P(A)P(B)$, $P(A|B) = P(A)$ and $P(B|A) = P(B)$.

If A and B are mutually exclusive, $P(A \text{ OR } B) = P(A) + P(B)$ and $P(A \text{ AND } B) = 0$.

3.3 Two Basic Rules of Probability

The multiplication rule: $P(A \text{ AND } B) = P(A|B)P(B)$

The addition rule: $P(A \text{ OR } B) = P(A) + P(B) - P(A \text{ AND } B)$

PRACTICE

3.1 Terminology

1. In a particular college class, there are male and female students. Some students have long hair and some students have short hair. Write the **symbols** for the probabilities of the events for parts a through j. (Note that you cannot find numerical answers here. You were not given enough information to find any probability values yet; concentrate on understanding the symbols.)

- Let F be the event that a student is female.
- Let M be the event that a student is male.
- Let S be the event that a student has short hair.
- Let L be the event that a student has long hair.

    a. The probability that a student does not have long hair.
    b. The probability that a student is male or has short hair.
    c. The probability that a student is a female and has long hair.
    d. The probability that a student is male, given that the student has long hair.
    e. The probability that a student has long hair, given that the student is male.
    f. Of all the female students, the probability that a student has short hair.
    g. Of all students with long hair, the probability that a student is female.
    h. The probability that a student is female or has long hair.
    i. The probability that a randomly selected student is a male student with short hair.
    j. The probability that a student is female.

Use the following information to answer the next four exercises. A box is filled with several party favors. It contains 12 hats, 15 noisemakers, ten finger traps, and five bags of confetti.

Let H = the event of getting a hat.
Let N = the event of getting a noisemaker.
Let F = the event of getting a finger trap.
Let C = the event of getting a bag of confetti.

2. Find $P(H)$.

3. Find $P(N)$.

4. Find $P(F)$.

5. Find $P(C)$.

Use the following information to answer the next six exercises. A jar of 150 jelly beans contains 22 red jelly beans, 38 yellow, 20 green, 28 purple, 26 blue, and the rest are orange.

Let B = the event of getting a blue jelly bean
Let G = the event of getting a green jelly bean.
Let O = the event of getting an orange jelly bean.
Let P = the event of getting a purple jelly bean.
Let R = the event of getting a red jelly bean.
Let Y = the event of getting a yellow jelly bean.

6. Find $P(B)$.

7. Find $P(G)$.

8. Find $P(P)$.

9. Find $P(R)$.

10. Find $P(Y)$.

11. Find $P(O)$.

Use the following information to answer the next six exercises. There are 23 countries in North America, 12 countries in South America, 47 countries in Europe, 44 countries in Asia, 54 countries in Africa, and 14 in Oceania (Pacific Ocean region).
Let A = the event that a country is in Asia.
Let E = the event that a country is in Europe.
Let F = the event that a country is in Africa.
Let N = the event that a country is in North America.
Let O = the event that a country is in Oceania.
Let S = the event that a country is in South America.

12. Find $P(A)$.

13. Find $P(E)$.

14. Find $P(F)$.

15. Find $P(N)$.

16. Find $P(O)$.

17. Find $P(S)$.

18. What is the probability of drawing a red card in a standard deck of 52 cards?

19. What is the probability of drawing a club in a standard deck of 52 cards?

20. What is the probability of rolling an even number of dots with a fair, six-sided die numbered one through six?

21. What is the probability of rolling a prime number of dots with a fair, six-sided die numbered one through six?

Use the following information to answer the next two exercises. You see a game at a local fair. You have to throw a dart at a color wheel. Each section on the color wheel is equal in area.

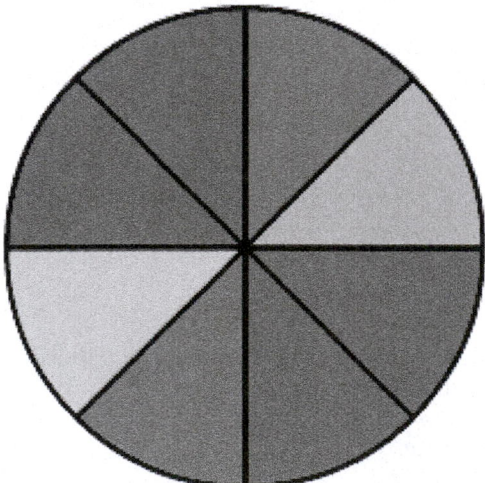

Figure 3.10

Let B = the event of landing on blue.
Let R = the event of landing on red.
Let G = the event of landing on green.
Let Y = the event of landing on yellow.

22. If you land on Y, you get the biggest prize. Find $P(Y)$.

23. If you land on red, you don't get a prize. What is $P(R)$?

Use the following information to answer the next ten exercises. On a baseball team, there are infielders and outfielders. Some players are great hitters, and some players are not great hitters.

Let *I* = the event that a player in an infielder.

Let *O* = the event that a player is an outfielder.

Let *H* = the event that a player is a great hitter.

Let *N* = the event that a player is not a great hitter.

24. Write the symbols for the probability that a player is not an outfielder.

25. Write the symbols for the probability that a player is an outfielder or is a great hitter.

26. Write the symbols for the probability that a player is an infielder and is not a great hitter.

27. Write the symbols for the probability that a player is a great hitter, given that the player is an infielder.

28. Write the symbols for the probability that a player is an infielder, given that the player is a great hitter.

29. Write the symbols for the probability that of all the outfielders, a player is not a great hitter.

30. Write the symbols for the probability that of all the great hitters, a player is an outfielder.

31. Write the symbols for the probability that a player is an infielder or is not a great hitter.

32. Write the symbols for the probability that a player is an outfielder and is a great hitter.

33. Write the symbols for the probability that a player is an infielder.

34. What is the word for the set of all possible outcomes?

35. What is conditional probability?

36. A shelf holds 12 books. Eight are fiction and the rest are nonfiction. Each is a different book with a unique title. The fiction books are numbered one to eight. The nonfiction books are numbered one to four. Randomly select one book

Let *F* = event that book is fiction

Let *N* = event that book is nonfiction

What is the sample space?

37. What is the sum of the probabilities of an event and its complement?

Use the following information to answer the next two exercises. You are rolling a fair, six-sided number cube. Let *E* = the event that it lands on an even number. Let *M* = the event that it lands on a multiple of three.

38. What does $P(E|M)$ mean in words?

39. What does $P(E \text{ OR } M)$ mean in words?

3.2 Independent and Mutually Exclusive Events

40. *E* and *F* are mutually exclusive events. $P(E) = 0.4$; $P(F) = 0.5$. Find $P(E|F)$.

41. *J* and *K* are independent events. $P(J|K) = 0.3$. Find $P(J)$.

42. *U* and *V* are mutually exclusive events. $P(U) = 0.26$; $P(V) = 0.37$. Find:
   a. $P(U \text{ AND } V) =$
   b. $P(U|V) =$
   c. $P(U \text{ OR } V) =$

43. *Q* and *R* are independent events. $P(Q) = 0.4$ and $P(Q \text{ AND } R) = 0.1$. Find $P(R)$.

3.3 Two Basic Rules of Probability

Use the following information to answer the next ten exercises. Forty-eight percent of all Californians registered voters prefer life in prison without parole over the death penalty for a person convicted of first degree murder. Among Latino California registered voters, 55% prefer life in prison without parole over the death penalty for a person convicted of first degree murder. 37.6% of all Californians are Latino.

In this problem, let:

- C = Californians (registered voters) preferring life in prison without parole over the death penalty for a person convicted of first degree murder.
- L = Latino Californians

Suppose that one Californian is randomly selected.

44. Find $P(C)$.

45. Find $P(L)$.

46. Find $P(C|L)$.

47. In words, what is $C|L$?

48. Find $P(L$ AND $C)$.

49. In words, what is L AND C?

50. Are L and C independent events? Show why or why not.

51. Find $P(L$ OR $C)$.

52. In words, what is L OR C?

53. Are L and C mutually exclusive events? Show why or why not.

3.4 Contingency Tables

Use the following information to answer the next four exercises. **Table 3.15** shows a random sample of musicians and how they learned to play their instruments.

Gender	Self-taught	Studied in School	Private Instruction	Total
Female	12	38	22	72
Male	19	24	15	58
Total	31	62	37	130

Table 3.15

54. Find P(musician is a female).

55. Find P(musician is a male AND had private instruction).

56. Find P(musician is a female OR is self taught).

57. Are the events "being a female musician" and "learning music in school" mutually exclusive events?

3.5 Tree and Venn Diagrams

58. The probability that a man develops some form of cancer in his lifetime is 0.4567. The probability that a man has at least one false positive test result (meaning the test comes back for cancer when the man does not have it) is 0.51. Let: C = a man develops cancer in his lifetime; P = man has at least one false positive. Construct a tree diagram of the situation.

BRINGING IT TOGETHER: PRACTICE

Use the following information to answer the next seven exercises. An article in the *New England Journal of Medicine*, reported about a study of smokers in California and Hawaii. In one part of the report, the self-reported ethnicity and smoking levels per day were given. Of the people smoking at most ten cigarettes per day, there were 9,886 African Americans, 2,745 Native Hawaiians, 12,831 Latinos, 8,378 Japanese Americans, and 7,650 Whites. Of the people smoking 11 to 20 cigarettes per day, there were 6,514 African Americans, 3,062 Native Hawaiians, 4,932 Latinos, 10,680 Japanese Americans, and 9,877 Whites. Of the people smoking 21 to 30 cigarettes per day, there were 1,671 African Americans, 1,419 Native Hawaiians, 1,406 Latinos, 4,715 Japanese Americans, and 6,062 Whites. Of the people smoking at least 31 cigarettes per day, there were 759 African Americans, 788 Native Hawaiians, 800 Latinos, 2,305 Japanese Americans, and 3,970 Whites.

59. Complete the table using the data provided. Suppose that one person from the study is randomly selected. Find the probability that person smoked 11 to 20 cigarettes per day.

Smoking Level	African American	Native Hawaiian	Latino	Japanese Americans	White	TOTALS
1–10						
11–20						
21–30						
31+						
TOTALS						

Table 3.16 Smoking Levels by Ethnicity

60. Suppose that one person from the study is randomly selected. Find the probability that person smoked 11 to 20 cigarettes per day.

61. Find the probability that the person was Latino.

62. In words, explain what it means to pick one person from the study who is "Japanese American **AND** smokes 21 to 30 cigarettes per day." Also, find the probability.

63. In words, explain what it means to pick one person from the study who is "Japanese American **OR** smokes 21 to 30 cigarettes per day." Also, find the probability.

64. In words, explain what it means to pick one person from the study who is "Japanese American **GIVEN** that person smokes 21 to 30 cigarettes per day." Also, find the probability.

65. Prove that smoking level/day and ethnicity are dependent events.

HOMEWORK

3.1 Terminology

66.

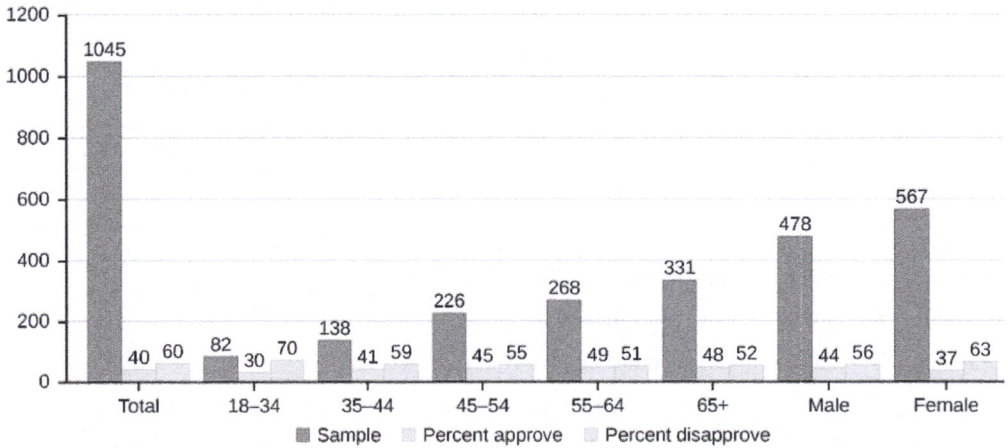

Figure 3.11 The graph in **Figure 3.11** displays the sample sizes and percentages of people in different age and gender groups who were polled concerning their approval of Mayor Ford's actions in office. The total number in the sample of all the age groups is 1,045.

a. Define three events in the graph.
b. Describe in words what the entry 40 means.
c. Describe in words the complement of the entry in question 2.
d. Describe in words what the entry 30 means.
e. Out of the males and females, what percent are males?
f. Out of the females, what percent disapprove of Mayor Ford?
g. Out of all the age groups, what percent approve of Mayor Ford?
h. Find $P(\text{Approve}|\text{Male})$.
i. Out of the age groups, what percent are more than 44 years old?
j. Find $P(\text{Approve}|\text{Age} < 35)$.

67. Explain what is wrong with the following statements. Use complete sentences.
a. If there is a 60% chance of rain on Saturday and a 70% chance of rain on Sunday, then there is a 130% chance of rain over the weekend.
b. The probability that a baseball player hits a home run is greater than the probability that he gets a successful hit.

3.2 Independent and Mutually Exclusive Events

Use the following information to answer the next 12 exercises. The graph shown is based on more than 170,000 interviews done by Gallup that took place from January through December 2012. The sample consists of employed Americans 18 years of age or older. The Emotional Health Index Scores are the sample space. We randomly sample one Emotional Health Index Score.

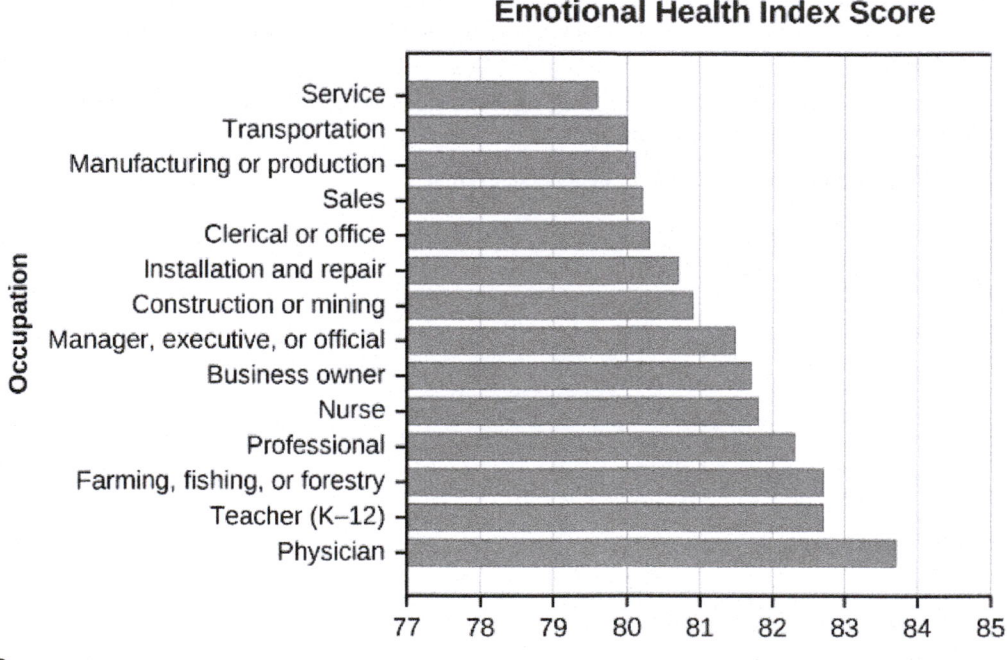

Figure 3.12

68. Find the probability that an Emotional Health Index Score is 82.7.

69. Find the probability that an Emotional Health Index Score is 81.0.

70. Find the probability that an Emotional Health Index Score is more than 81?

71. Find the probability that an Emotional Health Index Score is between 80.5 and 82?

72. If we know an Emotional Health Index Score is 81.5 or more, what is the probability that it is 82.7?

73. What is the probability that an Emotional Health Index Score is 80.7 or 82.7?

74. What is the probability that an Emotional Health Index Score is less than 80.2 given that it is already less than 81.

75. What occupation has the highest emotional index score?

76. What occupation has the lowest emotional index score?

77. What is the range of the data?

78. Compute the average EHIS.

79. If all occupations are equally likely for a certain individual, what is the probability that he or she will have an occupation with lower than average EHIS?

3.3 Two Basic Rules of Probability

80. On February 28, 2013, a Field Poll Survey reported that 61% of California registered voters approved of allowing two people of the same gender to marry and have regular marriage laws apply to them. Among 18 to 39 year olds (California registered voters), the approval rating was 78%. Six in ten California registered voters said that the upcoming Supreme Court's ruling about the constitutionality of California's Proposition 8 was either very or somewhat important to them. Out of those CA registered voters who support same-sex marriage, 75% say the ruling is important to them.

In this problem, let:

- C = California registered voters who support same-sex marriage.
- B = California registered voters who say the Supreme Court's ruling about the constitutionality of California's Proposition 8 is very or somewhat important to them
- A = California registered voters who are 18 to 39 years old.

    a. Find $P(C)$.
    b. Find $P(B)$.
    c. Find $P(C|A)$.
    d. Find $P(B|C)$.
    e. In words, what is $C|A$?
    f. In words, what is $B|C$?
    g. Find $P(C \text{ AND } B)$.
    h. In words, what is $C \text{ AND } B$?
    i. Find $P(C \text{ OR } B)$.
    j. Are C and B mutually exclusive events? Show why or why not.

81. After Rob Ford, the mayor of Toronto, announced his plans to cut budget costs in late 2011, the Forum Research polled 1,046 people to measure the mayor's popularity. Everyone polled expressed either approval or disapproval. These are the results their poll produced:
- In early 2011, 60 percent of the population approved of Mayor Ford's actions in office.
- In mid-2011, 57 percent of the population approved of his actions.
- In late 2011, the percentage of popular approval was measured at 42 percent.

    a. What is the sample size for this study?
    b. What proportion in the poll disapproved of Mayor Ford, according to the results from late 2011?
    c. How many people polled responded that they approved of Mayor Ford in late 2011?
    d. What is the probability that a person supported Mayor Ford, based on the data collected in mid-2011?
    e. What is the probability that a person supported Mayor Ford, based on the data collected in early 2011?

Use the following information to answer the next three exercises. The casino game, roulette, allows the gambler to bet on the probability of a ball, which spins in the roulette wheel, landing on a particular color, number, or range of numbers. The table used to place bets contains of 38 numbers, and each number is assigned to a color and a range.

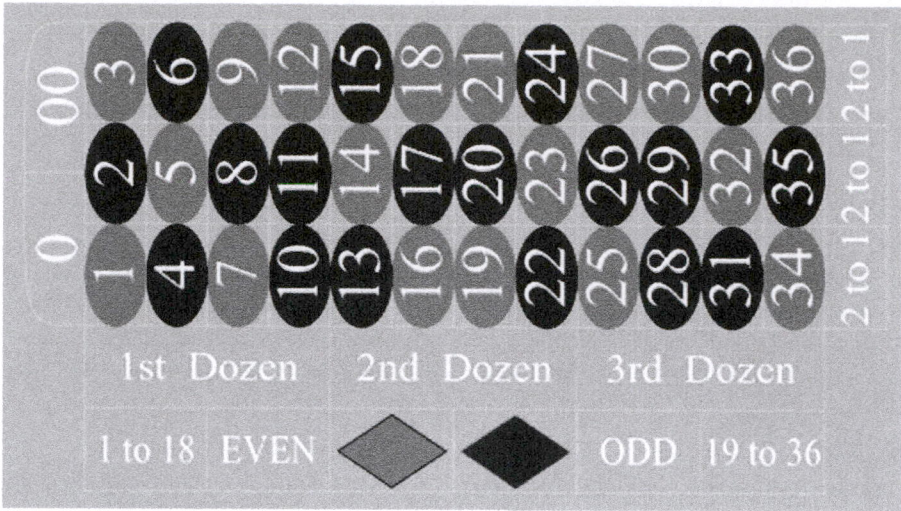

Figure 3.13 (credit: film8ker/wikibooks)

82.
  a. List the sample space of the 38 possible outcomes in roulette.
  b. You bet on red. Find *P*(red).
  c. You bet on -1st 12- (1st Dozen). Find *P*(-1st 12-).
  d. You bet on an even number. Find *P*(even number).
  e. Is getting an odd number the complement of getting an even number? Why?
  f. Find two mutually exclusive events.
  g. Are the events Even and 1st Dozen independent?

83. Compute the probability of winning the following types of bets:
  a. Betting on two lines that touch each other on the table as in 1-2-3-4-5-6
  b. Betting on three numbers in a line, as in 1-2-3
  c. Betting on one number
  d. Betting on four numbers that touch each other to form a square, as in 10-11-13-14
  e. Betting on two numbers that touch each other on the table, as in 10-11 or 10-13
  f. Betting on 0-00-1-2-3
  g. Betting on 0-1-2; or 0-00-2; or 00-2-3

84. Compute the probability of winning the following types of bets:
  a. Betting on a color
  b. Betting on one of the dozen groups
  c. Betting on the range of numbers from 1 to 18
  d. Betting on the range of numbers 19–36
  e. Betting on one of the columns
  f. Betting on an even or odd number (excluding zero)

85. Suppose that you have eight cards. Five are green and three are yellow. The five green cards are numbered 1, 2, 3, 4, and 5. The three yellow cards are numbered 1, 2, and 3. The cards are well shuffled. You randomly draw one card.
  - *G* = card drawn is green
  - *E* = card drawn is even-numbered
      a. List the sample space.
      b. $P(G) = $ _____
      c. $P(G|E) = $ _____
      d. $P(G \text{ AND } E) = $ _____
      e. $P(G \text{ OR } E) = $ _____
      f. Are *G* and *E* mutually exclusive? Justify your answer numerically.

86. Roll two fair dice. Each die has six faces.
  a. List the sample space.

    b. Let A be the event that either a three or four is rolled first, followed by an even number. Find $P(A)$.

    c. Let B be the event that the sum of the two rolls is at most seven. Find $P(B)$.

    d. In words, explain what "$P(A|B)$" represents. Find $P(A|B)$.

    e. Are A and B mutually exclusive events? Explain your answer in one to three complete sentences, including numerical justification.

    f. Are A and B independent events? Explain your answer in one to three complete sentences, including numerical justification.

87. A special deck of cards has ten cards. Four are green, three are blue, and three are red. When a card is picked, its color of it is recorded. An experiment consists of first picking a card and then tossing a coin.

    a. List the sample space.

    b. Let A be the event that a blue card is picked first, followed by landing a head on the coin toss. Find $P(A)$.

    c. Let B be the event that a red or green is picked, followed by landing a head on the coin toss. Are the events A and B mutually exclusive? Explain your answer in one to three complete sentences, including numerical justification.

    d. Let C be the event that a red or blue is picked, followed by landing a head on the coin toss. Are the events A and C mutually exclusive? Explain your answer in one to three complete sentences, including numerical justification.

88. An experiment consists of first rolling a die and then tossing a coin.

    a. List the sample space.

    b. Let A be the event that either a three or a four is rolled first, followed by landing a head on the coin toss. Find $P(A)$.

    c. Let B be the event that the first and second tosses land on heads. Are the events A and B mutually exclusive? Explain your answer in one to three complete sentences, including numerical justification.

89. An experiment consists of tossing a nickel, a dime, and a quarter. Of interest is the side the coin lands on.

    a. List the sample space.

    b. Let A be the event that there are at least two tails. Find $P(A)$.

    c. Let B be the event that the first and second tosses land on heads. Are the events A and B mutually exclusive? Explain your answer in one to three complete sentences, including justification.

90. Consider the following scenario:

Let $P(C) = 0.4$.

Let $P(D) = 0.5$.

Let $P(C|D) = 0.6$.

    a. Find $P(C \text{ AND } D)$.

    b. Are C and D mutually exclusive? Why or why not?

    c. Are C and D independent events? Why or why not?

    d. Find $P(C \text{ OR } D)$.

    e. Find $P(D|C)$.

91. Y and Z are independent events.

    a. Rewrite the basic Addition Rule $P(Y \text{ OR } Z) = P(Y) + P(Z) - P(Y \text{ AND } Z)$ using the information that Y and Z are independent events.

    b. Use the rewritten rule to find $P(Z)$ if $P(Y \text{ OR } Z) = 0.71$ and $P(Y) = 0.42$.

92. G and H are mutually exclusive events. $P(G) = 0.5$ $P(H) = 0.3$

    a. Explain why the following statement MUST be false: $P(H|G) = 0.4$.

    b. Find $P(H \text{ OR } G)$.

    c. Are G and H independent or dependent events? Explain in a complete sentence.

93. Approximately 281,000,000 people over age five live in the United States. Of these people, 55,000,000 speak a language other than English at home. Of those who speak another language at home, 62.3% speak Spanish.

Let: E = speaks English at home; E' = speaks another language at home; S = speaks Spanish;

Finish each probability statement by matching the correct answer.

Probability Statements	Answers
a. $P(E') =$	i. 0.8043

Probability Statements	Answers
b. $P(E)$ =	ii. 0.623
c. $P(S$ and $E')$ =	iii. 0.1957
d. $P(S|E')$ =	iv. 0.1219

Table 3.17

94. 1994, the U.S. government held a lottery to issue 55,000 Green Cards (permits for non-citizens to work legally in the U.S.). Renate Deutsch, from Germany, was one of approximately 6.5 million people who entered this lottery. Let G = won green card.

  a. What was Renate's chance of winning a Green Card? Write your answer as a probability statement.
  b. In the summer of 1994, Renate received a letter stating she was one of 110,000 finalists chosen. Once the finalists were chosen, assuming that each finalist had an equal chance to win, what was Renate's chance of winning a Green Card? Write your answer as a conditional probability statement. Let F = was a finalist.
  c. Are G and F independent or dependent events? Justify your answer numerically and also explain why.
  d. Are G and F mutually exclusive events? Justify your answer numerically and explain why.

95. Three professors at George Washington University did an experiment to determine if economists are more selfish than other people. They dropped 64 stamped, addressed envelopes with $10 cash in different classrooms on the George Washington campus. 44% were returned overall. From the economics classes 56% of the envelopes were returned. From the business, psychology, and history classes 31% were returned.

Let: R = money returned; E = economics classes; O = other classes

  a. Write a probability statement for the overall percent of money returned.
  b. Write a probability statement for the percent of money returned out of the economics classes.
  c. Write a probability statement for the percent of money returned out of the other classes.
  d. Is money being returned independent of the class? Justify your answer numerically and explain it.
  e. Based upon this study, do you think that economists are more selfish than other people? Explain why or why not. Include numbers to justify your answer.

96. The following table of data obtained from www.baseball-almanac.com shows hit information for four players. Suppose that one hit from the table is randomly selected.

Name	Single	Double	Triple	Home Run	Total Hits
Babe Ruth	1,517	506	136	714	2,873
Jackie Robinson	1,054	273	54	137	1,518
Ty Cobb	3,603	174	295	114	4,189
Hank Aaron	2,294	624	98	755	3,771
Total	8,471	1,577	583	1,720	12,351

Table 3.18

Are "the hit being made by Hank Aaron" and "the hit being a double" independent events?

  a. Yes, because P(hit by Hank Aaron|hit is a double) = P(hit by Hank Aaron)
  b. No, because P(hit by Hank Aaron|hit is a double) $\neq$ P(hit is a double)
  c. No, because P(hit is by Hank Aaron|hit is a double) $\neq$ P(hit by Hank Aaron)
  d. Yes, because P(hit is by Hank Aaron|hit is a double) = P(hit is a double)

97. United Blood Services is a blood bank that serves more than 500 hospitals in 18 states. According to their website, a person with type O blood and a negative Rh factor (Rh-) can donate blood to any person with any bloodtype. Their data show that 43% of people have type O blood and 15% of people have Rh- factor; 52% of people have type O or Rh- factor.

    a. Find the probability that a person has both type O blood and the Rh- factor.
    b. Find the probability that a person does NOT have both type O blood and the Rh- factor.

98. At a college, 72% of courses have final exams and 46% of courses require research papers. Suppose that 32% of courses have a research paper and a final exam. Let F be the event that a course has a final exam. Let R be the event that a course requires a research paper.
    a. Find the probability that a course has a final exam or a research project.
    b. Find the probability that a course has NEITHER of these two requirements.

99. In a box of assorted cookies, 36% contain chocolate and 12% contain nuts. Of those, 8% contain both chocolate and nuts. Sean is allergic to both chocolate and nuts.
    a. Find the probability that a cookie contains chocolate or nuts (he can't eat it).
    b. Find the probability that a cookie does not contain chocolate or nuts (he can eat it).

100. A college finds that 10% of students have taken a distance learning class and that 40% of students are part time students. Of the part time students, 20% have taken a distance learning class. Let D = event that a student takes a distance learning class and E = event that a student is a part time student
    a. Find $P(D \text{ AND } E)$.
    b. Find $P(E|D)$.
    c. Find $P(D \text{ OR } E)$.
    d. Using an appropriate test, show whether D and E are independent.
    e. Using an appropriate test, show whether D and E are mutually exclusive.

3.4 Contingency Tables

*Use the information in the **Table 3.19** to answer the next eight exercises.* The table shows the political party affiliation of each of 67 members of the US Senate in June 2012, and when they are up for reelection.

Up for reelection:	Democratic Party	Republican Party	Other	Total
November 2014	20	13	0	
November 2016	10	24	0	
Total				

Table 3.19

101. What is the probability that a randomly selected senator has an "Other" affiliation?

102. What is the probability that a randomly selected senator is up for reelection in November 2016?

103. What is the probability that a randomly selected senator is a Democrat and up for reelection in November 2016?

104. What is the probability that a randomly selected senator is a Republican or is up for reelection in November 2014?

105. Suppose that a member of the US Senate is randomly selected. Given that the randomly selected senator is up for reelection in November 2016, what is the probability that this senator is a Democrat?

106. Suppose that a member of the US Senate is randomly selected. What is the probability that the senator is up for reelection in November 2014, knowing that this senator is a Republican?

107. The events "Republican" and "Up for reelection in 2016" are _____
    a. mutually exclusive.
    b. independent.
    c. both mutually exclusive and independent.
    d. neither mutually exclusive nor independent.

108. The events "Other" and "Up for reelection in November 2016" are _____
    a. mutually exclusive.
    b. independent.
    c. both mutually exclusive and independent.

d. neither mutually exclusive nor independent.

109. Table 3.20 gives the number of suicides estimated in the U.S. for a recent year by age, race (black or white), and sex. We are interested in possible relationships between age, race, and sex. We will let suicide victims be our population.

Race and Sex	1–14	15–24	25–64	over 64	TOTALS
white, male	210	3,360	13,610		22,050
white, female	80	580	3,380		4,930
black, male	10	460	1,060		1,670
black, female	0	40	270		330
all others					
TOTALS	310	4,650	18,780		29,760

Table 3.20

Do not include "all others" for parts f and g.

a. Fill in the column for the suicides for individuals over age 64.
b. Fill in the row for all other races.
c. Find the probability that a randomly selected individual was a white male.
d. Find the probability that a randomly selected individual was a black female.
e. Find the probability that a randomly selected individual was black
f. Find the probability that a randomly selected individual was male.
g. Out of the individuals over age 64, find the probability that a randomly selected individual was a black or white male.

Use the following information to answer the next two exercises. The table of data obtained from *www.baseball-almanac.com* shows hit information for four well known baseball players. Suppose that one hit from the table is randomly selected.

NAME	Single	Double	Triple	Home Run	TOTAL HITS
Babe Ruth	1,517	506	136	714	2,873
Jackie Robinson	1,054	273	54	137	1,518
Ty Cobb	3,603	174	295	114	4,189
Hank Aaron	2,294	624	98	755	3,771
TOTAL	8,471	1,577	583	1,720	12,351

Table 3.21

110. Find *P*(hit was made by Babe Ruth).
a. $\frac{1518}{2873}$
b. $\frac{2873}{12351}$
c. $\frac{583}{12351}$
d. $\frac{4189}{12351}$

111. Find *P*(hit was made by Ty Cobb|The hit was a Home Run).

a. $\dfrac{4189}{12351}$

b. $\dfrac{114}{1720}$

c. $\dfrac{1720}{4189}$

d. $\dfrac{114}{12351}$

112. Table 3.22 identifies a group of children by one of four hair colors, and by type of hair.

Hair Type	Brown	Blond	Black	Red	Totals
Wavy	20		15	3	43
Straight	80	15		12	
Totals		20			215

Table 3.22

a. Complete the table.
b. What is the probability that a randomly selected child will have wavy hair?
c. What is the probability that a randomly selected child will have either brown or blond hair?
d. What is the probability that a randomly selected child will have wavy brown hair?
e. What is the probability that a randomly selected child will have red hair, given that he or she has straight hair?
f. If B is the event of a child having brown hair, find the probability of the complement of B.
g. In words, what does the complement of B represent?

113. In a previous year, the weights of the members of the **San Francisco 49ers** and the **Dallas Cowboys** were published in the *San Jose Mercury News*. The factual data were compiled into the following table.

Shirt#	≤ 210	211–250	251–290	> 290
1–33	21	5	0	0
34–66	6	18	7	4
66–99	6	12	22	5

Table 3.23

For the following, suppose that you randomly select one player from the 49ers or Cowboys.

a. Find the probability that his shirt number is from 1 to 33.
b. Find the probability that he weighs at most 210 pounds.
c. Find the probability that his shirt number is from 1 to 33 AND he weighs at most 210 pounds.
d. Find the probability that his shirt number is from 1 to 33 OR he weighs at most 210 pounds.
e. Find the probability that his shirt number is from 1 to 33 GIVEN that he weighs at most 210 pounds.

3.5 Tree and Venn Diagrams

Use the following information to answer the next two exercises. This tree diagram shows the tossing of an unfair coin followed by drawing one bead from a cup containing three red (R), four yellow (Y) and five blue (B) beads. For the coin, $P(H) = \frac{2}{3}$ and $P(T) = \frac{1}{3}$ where H is heads and T is tails.

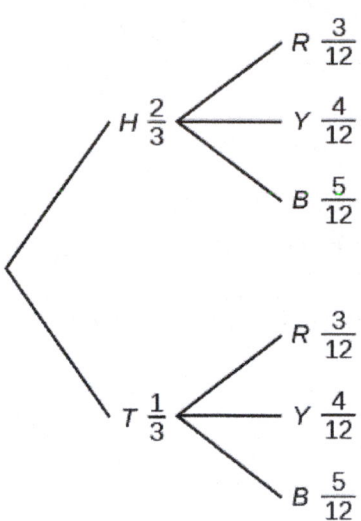

Figure 3.14

114. Find P(tossing a Head on the coin AND a Red bead)

    a. $\frac{2}{3}$

    b. $\frac{5}{15}$

    c. $\frac{6}{36}$

    d. $\frac{5}{36}$

115. Find P(Blue bead).

    a. $\frac{15}{36}$

    b. $\frac{10}{36}$

    c. $\frac{10}{12}$

    d. $\frac{6}{36}$

116. A box of cookies contains three chocolate and seven butter cookies. Miguel randomly selects a cookie and eats it. Then he randomly selects another cookie and eats it. (How many cookies did he take?)

    a. Draw the tree that represents the possibilities for the cookie selections. Write the probabilities along each branch of the tree.

    b. Are the probabilities for the flavor of the SECOND cookie that Miguel selects independent of his first selection? Explain.

    c. For each complete path through the tree, write the event it represents and find the probabilities.

    d. Let S be the event that both cookies selected were the same flavor. Find $P(S)$.

    e. Let T be the event that the cookies selected were different flavors. Find $P(T)$ by two different methods: by using the complement rule and by using the branches of the tree. Your answers should be the same with both methods.

    f. Let U be the event that the second cookie selected is a butter cookie. Find $P(U)$.

BRINGING IT TOGETHER: HOMEWORK

117. A previous year, the weights of the members of the **San Francisco 49ers** and the **Dallas Cowboys** were published in the *San Jose Mercury News*. The factual data are compiled into **Table 3.24**.

Shirt#	≤ 210	211–250	251–290	290≤
1–33	21	5	0	0
34–66	6	18	7	4
66–99	6	12	22	5

Table 3.24

For the following, suppose that you randomly select one player from the 49ers or Cowboys.

If having a shirt number from one to 33 and weighing at most 210 pounds were independent events, then what should be true about $P(\text{Shirt\# 1–33} | \leq 210 \text{ pounds})$?

118. The probability that a male develops some form of cancer in his lifetime is 0.4567. The probability that a male has at least one false positive test result (meaning the test comes back for cancer when the man does not have it) is 0.51. Some of the following questions do not have enough information for you to answer them. Write "not enough information" for those answers. Let C = a man develops cancer in his lifetime and P = man has at least one false positive.
   a. $P(C) = $ _____
   b. $P(P|C) = $ _____
   c. $P(P|C') = $ _____
   d. If a test comes up positive, based upon numerical values, can you assume that man has cancer? Justify numerically and explain why or why not.

119. Given events G and H: $P(G) = 0.43$; $P(H) = 0.26$; $P(H \text{ AND } G) = 0.14$
   a. Find $P(H \text{ OR } G)$.
   b. Find the probability of the complement of event ($H \text{ AND } G$).
   c. Find the probability of the complement of event ($H \text{ OR } G$).

120. Given events J and K: $P(J) = 0.18$; $P(K) = 0.37$; $P(J \text{ OR } K) = 0.45$
   a. Find $P(J \text{ AND } K)$.
   b. Find the probability of the complement of event ($J \text{ AND } K$).
   c. Find the probability of the complement of event ($J \text{ AND } K$).

Use the following information to answer the next two exercises. Suppose that you have eight cards. Five are green and three are yellow. The cards are well shuffled.

121. Suppose that you randomly draw two cards, one at a time, **with replacement**.
Let G_1 = first card is green
Let G_2 = second card is green
   a. Draw a tree diagram of the situation.
   b. Find $P(G_1 \text{ AND } G_2)$.
   c. Find $P(\text{at least one green})$.
   d. Find $P(G_2|G_1)$.
   e. Are G_2 and G_1 independent events? Explain why or why not.

122. Suppose that you randomly draw two cards, one at a time, **without replacement**.
G_1 = first card is green
G_2 = second card is green
   a. Draw a tree diagram of the situation.
   b. Find $P(G_1 \text{ AND } G_2)$.
   c. Find $P(\text{at least one green})$.
   d. Find $P(G_2|G_1)$.
   e. Are G_2 and G_1 independent events? Explain why or why not.

Use the following information to answer the next two exercises. The percent of licensed U.S. drivers (from a recent year) that are female is 48.60. Of the females, 5.03% are age 19 and under; 81.36% are age 20–64; 13.61% are age 65 or over. Of the licensed U.S. male drivers, 5.04% are age 19 and under; 81.43% are age 20–64; 13.53% are age 65 or over.

123. Complete the following.
   a. Construct a table or a tree diagram of the situation.

b. Find *P*(driver is female).
c. Find *P*(driver is age 65 or over|driver is female).
d. Find *P*(driver is age 65 or over AND female).
e. In words, explain the difference between the probabilities in part c and part d.
f. Find *P*(driver is age 65 or over).
g. Are being age 65 or over and being female mutually exclusive events? How do you know?

124. Suppose that 10,000 U.S. licensed drivers are randomly selected.
a. How many would you expect to be male?
b. Using the table or tree diagram, construct a contingency table of gender versus age group.
c. Using the contingency table, find the probability that out of the age 20–64 group, a randomly selected driver is female.

125. Approximately 86.5% of Americans commute to work by car, truck, or van. Out of that group, 84.6% drive alone and 15.4% drive in a carpool. Approximately 3.9% walk to work and approximately 5.3% take public transportation.
a. Construct a table or a tree diagram of the situation. Include a branch for all other modes of transportation to work.
b. Assuming that the walkers walk alone, what percent of all commuters travel alone to work?
c. Suppose that 1,000 workers are randomly selected. How many would you expect to travel alone to work?
d. Suppose that 1,000 workers are randomly selected. How many would you expect to drive in a carpool?

126. When the Euro coin was introduced in 2002, two math professors had their statistics students test whether the Belgian one Euro coin was a fair coin. They spun the coin rather than tossing it and found that out of 250 spins, 140 showed a head (event *H*) while 110 showed a tail (event *T*). On that basis, they claimed that it is not a fair coin.
a. Based on the given data, find *P*(*H*) and *P*(*T*).
b. Use a tree to find the probabilities of each possible outcome for the experiment of tossing the coin twice.
c. Use the tree to find the probability of obtaining exactly one head in two tosses of the coin.
d. Use the tree to find the probability of obtaining at least one head.

127. *Use the following information to answer the next two exercises.* The following are real data from Santa Clara County, CA. As of a certain time, there had been a total of 3,059 documented cases of AIDS in the county. They were grouped into the following categories:

	Homosexual/Bisexual	IV Drug User*	Heterosexual Contact	Other	Totals
Female	0	70	136	49	____
Male	2,146	463	60	135	____
Totals	____	____	____	____	____

Table 3.25 * includes homosexual/bisexual IV drug users

Suppose a person with AIDS in Santa Clara County is randomly selected.

a. Find *P*(Person is female).
b. Find *P*(Person has a risk factor heterosexual contact).
c. Find *P*(Person is female OR has a risk factor of IV drug user).
d. Find *P*(Person is female AND has a risk factor of homosexual/bisexual).
e. Find *P*(Person is male AND has a risk factor of IV drug user).
f. Find *P*(Person is female GIVEN person got the disease from heterosexual contact).
g. Construct a Venn diagram. Make one group females and the other group heterosexual contact.

128. Answer these questions using probability rules. Do NOT use the contingency table. Three thousand fifty-nine cases of AIDS had been reported in Santa Clara County, CA, through a certain date. Those cases will be our population. Of those cases, 6.4% obtained the disease through heterosexual contact and 7.4% are female. Out of the females with the disease, 53.3% got the disease from heterosexual contact.
a. Find *P*(Person is female).
b. Find *P*(Person obtained the disease through heterosexual contact).
c. Find *P*(Person is female GIVEN person got the disease from heterosexual contact)

    d. Construct a Venn diagram representing this situation. Make one group females and the other group heterosexual contact. Fill in all values as probabilities.

REFERENCES

3.1 Terminology

"Countries List by Continent." Worldatlas, 2013. Available online at http://www.worldatlas.com/cntycont.htm (accessed May 2, 2013).

3.2 Independent and Mutually Exclusive Events

Lopez, Shane, Preety Sidhu. "U.S. Teachers Love Their Lives, but Struggle in the Workplace." Gallup Wellbeing, 2013. http://www.gallup.com/poll/161516/teachers-love-lives-struggle-workplace.aspx (accessed May 2, 2013).

Data from Gallup. Available online at www.gallup.com/ (accessed May 2, 2013).

3.3 Two Basic Rules of Probability

DiCamillo, Mark, Mervin Field. "The File Poll." Field Research Corporation. Available online at http://www.field.com/fieldpollonline/subscribers/Rls2443.pdf (accessed May 2, 2013).

Rider, David, "Ford support plummeting, poll suggests," The Star, September 14, 2011. Available online at http://www.thestar.com/news/gta/2011/09/14/ford_support_plummeting_poll_suggests.html (accessed May 2, 2013).

"Mayor's Approval Down." News Release by Forum Research Inc. Available online at http://www.forumresearch.com/forms/News Archives/News Releases/74209_TO_Issues_-_Mayoral_Approval_%28Forum_Research%29%2820130320%29.pdf (accessed May 2, 2013).

"Roulette." Wikipedia. Available online at http://en.wikipedia.org/wiki/Roulette (accessed May 2, 2013).

Shin, Hyon B., Robert A. Kominski. "Language Use in the United States: 2007." United States Census Bureau. Available online at http://www.census.gov/hhes/socdemo/language/data/acs/ACS-12.pdf (accessed May 2, 2013).

Data from the Baseball-Almanac, 2013. Available online at www.baseball-almanac.com (accessed May 2, 2013).

Data from U.S. Census Bureau.

Data from the Wall Street Journal.

Data from The Roper Center: Public Opinion Archives at the University of Connecticut. Available online at http://www.ropercenter.uconn.edu/ (accessed May 2, 2013).

Data from Field Research Corporation. Available online at www.field.com/fieldpollonline (accessed May 2,2 013).

3.4 Contingency Tables

"Blood Types." American Red Cross, 2013. Available online at http://www.redcrossblood.org/learn-about-blood/blood-types (accessed May 3, 2013).

Data from the National Center for Health Statistics, part of the United States Department of Health and Human Services.

Data from United States Senate. Available online at www.senate.gov (accessed May 2, 2013).

Haiman, Christopher A., Daniel O. Stram, Lynn R. Wilkens, Malcom C. Pike, Laurence N. Kolonel, Brien E. Henderson, and Loïc Le Marchand. "Ethnic and Racial Differences in the Smoking-Related Risk of Lung Cancer." The New England Journal of Medicine, 2013. Available online at http://www.nejm.org/doi/full/10.1056/NEJMoa033250 (accessed May 2, 2013).

"Human Blood Types." Unite Blood Services, 2011. Available online at http://www.unitedbloodservices.org/learnMore.aspx (accessed May 2, 2013).

Samuel, T. M. "Strange Facts about RH Negative Blood." eHow Health, 2013. Available online at http://www.ehow.com/facts_5552003_strange-rh-negative-blood.html (accessed May 2, 2013).

"United States: Uniform Crime Report – State Statistics from 1960–2011." The Disaster Center. Available online at http://www.disastercenter.com/crime/ (accessed May 2, 2013).

3.5 Tree and Venn Diagrams

Data from Clara County Public H.D.

Data from the American Cancer Society.

Data from The Data and Story Library, 1996. Available online at http://lib.stat.cmu.edu/DASL/ (accessed May 2, 2013).

Data from the Federal Highway Administration, part of the United States Department of Transportation.

Data from the United States Census Bureau, part of the United States Department of Commerce.

Data from USA Today.

"Environment." The World Bank, 2013. Available online at http://data.worldbank.org/topic/environment (accessed May 2, 2013).

"Search for Datasets." Roper Center: Public Opinion Archives, University of Connecticut., 2013. Available online at http://www.ropercenter.uconn.edu/data_access/data/search_for_datasets.html (accessed May 2, 2013).

SOLUTIONS

1
 a. $P(L') = P(S)$
 b. $P(M \text{ OR } S)$
 c. $P(F \text{ AND } L)$
 d. $P(M|L)$
 e. $P(L|M)$
 f. $P(S|F)$
 g. $P(F|L)$
 h. $P(F \text{ OR } L)$
 i. $P(M \text{ AND } S)$
 j. $P(F)$

3 $P(N) = \frac{15}{42} = \frac{5}{14} = 0.36$

5 $P(C) = \frac{5}{42} = 0.12$

7 $P(G) = \frac{20}{150} = \frac{2}{15} = 0.13$

9 $P(R) = \frac{22}{150} = \frac{11}{75} = 0.15$

11 $P(O) = \frac{150 - 22 - 38 - 20 - 28 - 26}{150} = \frac{16}{150} = \frac{8}{75} = 0.11$

13 $P(E) = \frac{47}{194} = 0.24$

15 $P(N) = \frac{23}{194} = 0.12$

17 $P(S) = \frac{12}{194} = \frac{6}{97} = 0.06$

19 $\frac{13}{52} = \frac{1}{4} = 0.25$

21 $\frac{3}{6} = \frac{1}{2} = 0.5$

23 $P(R) = \frac{4}{8} = 0.5$

25 $P(O \text{ OR } H)$

27 $P(H|I)$

29 $P(N|O)$

31 $P(I \text{ OR } N)$

33 $P(I)$

35 The likelihood that an event will occur given that another event has already occurred.

37 1

39 the probability of landing on an even number or a multiple of three

41 $P(J) = 0.3$

43 $P(Q \text{ AND } R) = P(Q)P(R)\ 0.1 = (0.4)P(R)\ P(R) = 0.25$

45 0.376

47 $C|L$ means, given the person chosen is a Latino Californian, the person is a registered voter who prefers life in prison without parole for a person convicted of first degree murder.

49 L AND C is the event that the person chosen is a Latino California registered voter who prefers life without parole over the death penalty for a person convicted of first degree murder.

51 0.6492

53 No, because $P(L \text{ AND } C)$ does not equal 0.

55 $P(\text{musician is a male AND had private instruction}) = \frac{15}{130} = \frac{3}{26} = 0.12$

57 $P(\text{being a female musician AND learning music in school}) = \frac{38}{130} = \frac{19}{65} = 0.29$ $P(\text{being a female musician})P(\text{learning music in school}) = \left(\frac{72}{130}\right)\left(\frac{62}{130}\right) = \frac{4,464}{16,900} = \frac{1,116}{4,225} = 0.26$ No, they are not independent because $P(\text{being a female musician AND learning music in school})$ is not equal to $P(\text{being a female musician})P(\text{learning music in school})$.

58

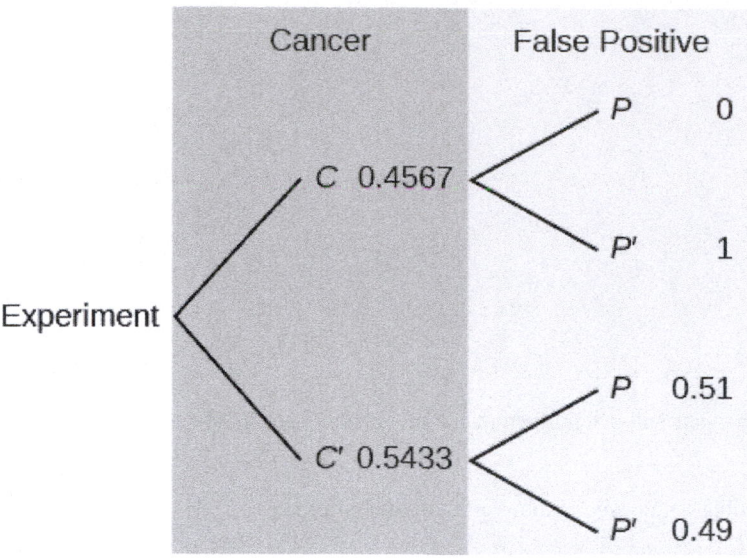

Figure 3.15

60 $\dfrac{35,065}{100,450}$

62 To pick one person from the study who is Japanese American AND smokes 21 to 30 cigarettes per day means that the person has to meet both criteria: both Japanese American and smokes 21 to 30 cigarettes. The sample space should include everyone in the study. The probability is $\dfrac{4,715}{100,450}$.

64 To pick one person from the study who is Japanese American given that person smokes 21-30 cigarettes per day, means that the person must fulfill both criteria and the sample space is reduced to those who smoke 21-30 cigarettes per day. The probability is $\dfrac{4715}{15,273}$.

67
  a. You can't calculate the joint probability knowing the probability of both events occurring, which is not in the information given; the probabilities should be multiplied, not added; and probability is never greater than 100%
  b. A home run by definition is a successful hit, so he has to have at least as many successful hits as home runs.

69 0

71 0.3571

73 0.2142

75 Physician (83.7)

77 83.7 − 79.6 = 4.1

79 P(Occupation < 81.3) = 0.5

81
  a. The Forum Research surveyed 1,046 Torontonians.
  b. 58%
  c. 42% of 1,046 = 439 (rounding to the nearest integer)
  d. 0.57
  e. 0.60.

83

a. P(Betting on two line that touch each other on the table) = $\frac{6}{38}$

b. P(Betting on three numbers in a line) = $\frac{3}{38}$

c. P(Bettting on one number) = $\frac{1}{38}$

d. P(Betting on four number that touch each other to form a square) = $\frac{4}{38}$

e. P(Betting on two number that touch each other on the table) = $\frac{2}{38}$

f. P(Betting on 0-00-1-2-3) = $\frac{5}{38}$

g. P(Betting on 0-1-2; or 0-00-2; or 00-2-3) = $\frac{3}{38}$

85

a. $\{G1, G2, G3, G4, G5, Y1, Y2, Y3\}$

b. $\frac{5}{8}$

c. $\frac{2}{3}$

d. $\frac{2}{8}$

e. $\frac{6}{8}$

f. No, because $P(G$ AND $E)$ does not equal 0.

87

> **NOTE**
> _____
>
> The coin toss is independent of the card picked first.

a. $\{(G,H)\ (G,T)\ (B,H)\ (B,T)\ (R,H)\ (R,T)\}$

b. $P(A) = P$(blue)P(head) = $\left(\frac{3}{10}\right)\left(\frac{1}{2}\right) = \frac{3}{20}$

c. Yes, A and B are mutually exclusive because they cannot happen at the same time; you cannot pick a card that is both blue and also (red or green). $P(A$ AND $B) = 0$

d. No, A and C are not mutually exclusive because they can occur at the same time. In fact, C includes all of the outcomes of A; if the card chosen is blue it is also (red or blue). $P(A$ AND $C) = P(A) = \frac{3}{20}$

89

a. $S = \{(HHH), (HHT), (HTH), (HTT), (THH), (THT), (TTH), (TTT)\}$

b. $\frac{4}{8}$

c. Yes, because if A has occurred, it is impossible to obtain two tails. In other words, $P(A$ AND $B) = 0$.

91

a. If Y and Z are independent, then $P(Y$ AND $Z) = P(Y)P(Z)$, so $P(Y$ OR $Z) = P(Y) + P(Z) - P(Y)P(Z)$.

b. 0.5

93 iii; i; iv; ii

95

a. $P(R) = 0.44$

b. $P(R|E) = 0.56$

c. $P(R|O) = 0.31$

d. No, whether the money is returned is not independent of which class the money was placed in. There are several ways to justify this mathematically, but one is that the money placed in economics classes is not returned at the same overall rate; $P(R|E) \neq P(R)$.

e. No, this study definitely does not support that notion; *in fact*, it suggests the opposite. The money placed in the economics classrooms was returned at a higher rate than the money place in all classes collectively; $P(R|E) > P(R)$.

97

a. $P(\text{type O OR Rh-}) = P(\text{type O}) + P(\text{Rh-}) - P(\text{type O AND Rh-})$

$0.52 = 0.43 + 0.15 - P(\text{type O AND Rh-})$; solve to find $P(\text{type O AND Rh-}) = 0.06$

6% of people have type O, Rh- blood

b. $P(\text{NOT(type O AND Rh-)}) = 1 - P(\text{type O AND Rh-}) = 1 - 0.06 = 0.94$

94% of people do not have type O, Rh- blood

99

a. Let C = be the event that the cookie contains chocolate. Let N = the event that the cookie contains nuts.

b. $P(C \text{ OR } N) = P(C) + P(N) - P(C \text{ AND } N) = 0.36 + 0.12 - 0.08 = 0.40$

c. $P(\text{NEITHER chocolate NOR nuts}) = 1 - P(C \text{ OR } N) = 1 - 0.40 = 0.60$

101 0

103 $\frac{10}{67}$

105 $\frac{10}{34}$

107 d

109

a.

Race and Sex	1–14	15–24	25–64	over 64	TOTALS
white, male	210	3,360	13,610	4,870	22,050
white, female	80	580	3,380	890	4,930
black, male	10	460	1,060	140	1,670
black, female	0	40	270	20	330
all others				100	
TOTALS	310	4,650	18,780	6,020	29,760

Table 3.26

b.

Race and Sex	1–14	15–24	25–64	over 64	TOTALS
white, male	210	3,360	13,610	4,870	22,050
white, female	80	580	3,380	890	4,930
black, male	10	460	1,060	140	1,670
black, female	0	40	270	20	330
all others	10	210	460	100	780
TOTALS	310	4,650	18,780	6,020	29,760

Table 3.27

c. $\dfrac{22{,}050}{29{,}760}$

d. $\dfrac{330}{29{,}760}$

e. $\dfrac{2{,}000}{29{,}760}$

f. $\dfrac{23{,}720}{29{,}760}$

g. $\dfrac{5{,}010}{6{,}020}$

111 b

113

a. $\dfrac{26}{106}$

b. $\dfrac{33}{106}$

c. $\dfrac{21}{106}$

d. $\left(\dfrac{26}{106}\right) + \left(\dfrac{33}{106}\right) - \left(\dfrac{21}{106}\right) = \left(\dfrac{38}{106}\right)$

e. $\dfrac{21}{33}$

115 a

118

a. $P(C) = 0.4567$

b. not enough information

c. not enough information

d. No, because over half (0.51) of men have at least one false positive text

120

a. $P(J \text{ OR } K) = P(J) + P(K) - P(J \text{ AND } K)$; $0.45 = 0.18 + 0.37 - P(J \text{ AND } K)$; solve to find $P(J \text{ AND } K) = 0.10$

b. $P(\text{NOT } (J \text{ AND } K)) = 1 - P(J \text{ AND } K) = 1 - 0.10 = 0.90$

c. $P(\text{NOT } (J \text{ OR } K)) = 1 - P(J \text{ OR } K) = 1 - 0.45 = 0.55$

121

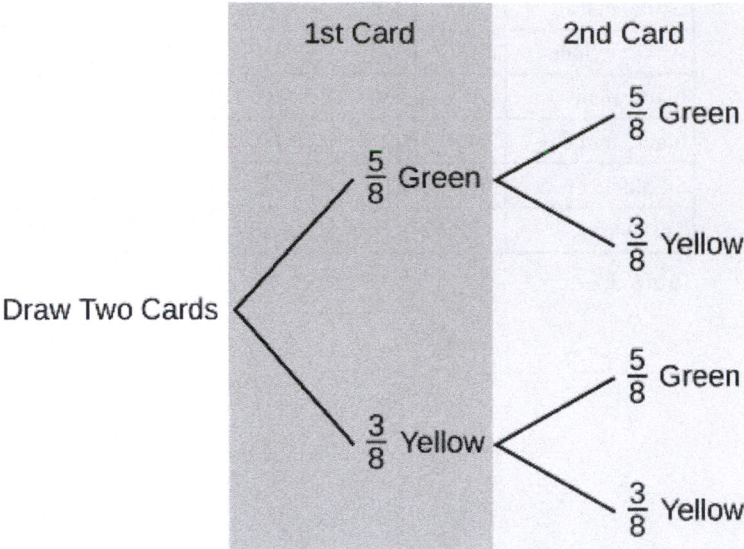

a.

Figure 3.16

b. $P(GG) = \left(\frac{5}{8}\right)\left(\frac{5}{8}\right) = \frac{25}{64}$

c. $P(\text{at least one green}) = P(GG) + P(GY) + P(YG) = \frac{25}{64} + \frac{15}{64} + \frac{15}{64} = \frac{55}{64}$

d. $P(G|G) = \frac{5}{8}$

e. Yes, they are independent because the first card is placed back in the bag before the second card is drawn; the composition of cards in the bag remains the same from draw one to draw two.

123

a.

	<20	20–64	>64	Totals
Female	0.0244	0.3954	0.0661	0.486
Male	0.0259	0.4186	0.0695	0.514
Totals	0.0503	0.8140	0.1356	1

Table 3.28

b. $P(F) = 0.486$

c. $P(>64|F) = 0.1361$

d. $P(>64 \text{ and } F) = P(F) \, P(>64|F) = (0.486)(0.1361) = 0.0661$

e. $P(>64|F)$ is the percentage of female drivers who are 65 or older and $P(>64 \text{ and } F)$ is the percentage of drivers who are female and 65 or older.

f. $P(>64) = P(>64 \text{ and } F) + P(>64 \text{ and } M) = 0.1356$

g. No, being female and 65 or older are not mutually exclusive because they can occur at the same time $P(>64 \text{ and } F) = 0.0661$.

125

	Car, Truck or Van	Walk	Public Transportation	Other	Totals
Alone	0.7318				
Not Alone	0.1332				
Totals	0.8650	0.0390	0.0530	0.0430	1

Table 3.29

b. If we assume that all walkers are alone and that none from the other two groups travel alone (which is a big assumption) we have: $P(\text{Alone}) = 0.7318 + 0.0390 = 0.7708$.

c. Make the same assumptions as in (b) we have: $(0.7708)(1{,}000) = 771$

d. $(0.1332)(1{,}000) = 133$

127 The completed contingency table is as follows:

	Homosexual/Bisexual	IV Drug User*	Heterosexual Contact	Other	Totals
Female	0	70	136	49	**255**
Male	2,146	463	60	135	**2,804**
Totals	**2,146**	**533**	**196**	**184**	**3,059**

Table 3.30 * includes homosexual/bisexual IV drug users

a. $\dfrac{255}{3059}$

b. $\dfrac{196}{3059}$

c. $\dfrac{718}{3059}$

d. 0

e. $\dfrac{463}{3059}$

f. $\dfrac{136}{196}$

g.

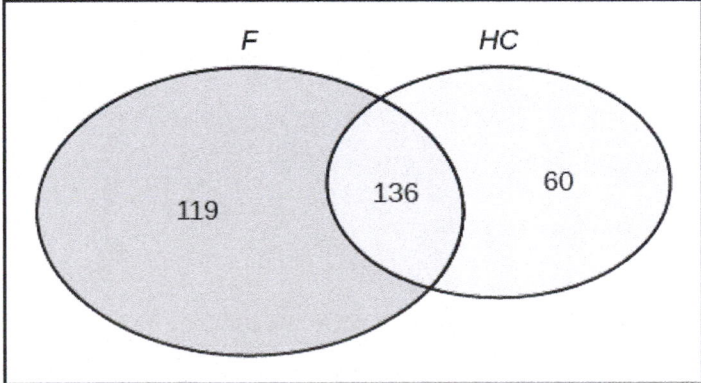

Figure 3.17

4 | DISCRETE RANDOM VARIABLES

Figure 4.1 You can use probability and discrete random variables to calculate the likelihood of lightning striking the ground five times during a half-hour thunderstorm. (Credit: Leszek Leszczynski)

Introduction

Chapter Objectives
By the end of this chapter, the student should be able to: • Recognize and understand discrete probability distribution functions, in general. • Calculate and interpret expected values. • Recognize the binomial probability distribution and apply it appropriately. • Recognize the Poisson probability distribution and apply it appropriately. • Recognize the geometric probability distribution and apply it appropriately. • Recognize the hypergeometric probability distribution and apply it appropriately. • Classify discrete word problems by their distributions.

A student takes a ten-question, true-false quiz. Because the student had such a busy schedule, he or she could not study and guesses randomly at each answer. What is the probability of the student passing the test with at least a 70%?

Small companies might be interested in the number of long-distance phone calls their employees make during the peak time of the day. Suppose the average is 20 calls. What is the probability that the employees make more than 20 long-distance phone calls during the peak time?

These two examples illustrate two different types of probability problems involving discrete random variables. Recall that discrete data are data that you can count. A **random variable** describes the outcomes of a statistical experiment in words. The values of a random variable can vary with each repetition of an experiment.

Random Variable Notation

Upper case letters such as X or Y denote a random variable. Lower case letters like x or y denote the value of a random variable. If X **is a random variable, then** X **is written in words, and** x **is given as a number.**

For example, let X = the number of heads you get when you toss three fair coins. The sample space for the toss of three fair coins is *TTT*; *THH*; *HTH*; *HHT*; *HTT*; *THT*; *TTH*; *HHH*. Then, x = 0, 1, 2, 3. X is in words and x is a number. Notice that for this example, the x values are countable outcomes. Because you can count the possible values that X can take on and the outcomes are random (the x values 0, 1, 2, 3), X is a discrete random variable.

Collaborative Exercise

Toss a coin ten times and record the number of heads. After all members of the class have completed the experiment (tossed a coin ten times and counted the number of heads), fill in **Table 4.1**. Let X = the number of heads in ten tosses of the coin.

x	Frequency of x	Relative Frequency of x

Table 4.1

a. Which value(s) of x occurred most frequently?

b. If you tossed the coin 1,000 times, what values could x take on? Which value(s) of x do you think would occur most frequently?

c. What does the relative frequency column sum to?

4.1 | Probability Distribution Function (PDF) for a Discrete Random Variable

A discrete **probability distribution function** has two characteristics:

1. Each probability is between zero and one, inclusive.

2. The sum of the probabilities is one.

Example 4.1

A child psychologist is interested in the number of times a newborn baby's crying wakes its mother after midnight. For a random sample of 50 mothers, the following information was obtained. Let X = the number of times per week a newborn baby's crying wakes its mother after midnight. For this example, x = 0, 1, 2, 3, 4, 5.

$P(x)$ = probability that X takes on a value x.

x	P(x)
0	$P(x = 0) = \frac{2}{50}$
1	$P(x = 1) = \frac{11}{50}$
2	$P(x = 2) = \frac{23}{50}$
3	$P(x = 3) = \frac{9}{50}$
4	$P(x = 4) = \frac{4}{50}$
5	$P(x = 5) = \frac{1}{50}$

Table 4.2

X takes on the values 0, 1, 2, 3, 4, 5. This is a discrete PDF because:

a. Each $P(x)$ is between zero and one, inclusive.

b. The sum of the probabilities is one, that is,

$$\frac{2}{50} + \frac{11}{50} + \frac{23}{50} + \frac{9}{50} + \frac{4}{50} + \frac{1}{50} = 1$$

Try It Σ

4.1 A hospital researcher is interested in the number of times the average post-op patient will ring the nurse during a 12-hour shift. For a random sample of 50 patients, the following information was obtained. Let X = the number of times a patient rings the nurse during a 12-hour shift. For this exercise, x = 0, 1, 2, 3, 4, 5. $P(x)$ = the probability that X takes on value x. Why is this a discrete probability distribution function (two reasons)?

X	P(x)
0	$P(x = 0) = \frac{4}{50}$
1	$P(x = 1) = \frac{8}{50}$
2	$P(x = 2) = \frac{16}{50}$

Table 4.3

X	P(x)
3	$P(x = 3) = \frac{14}{50}$
4	$P(x = 4) = \frac{6}{50}$
5	$P(x = 5) = \frac{2}{50}$

Table 4.3

Example 4.2

Suppose Nancy has classes **three days** a week. She attends classes three days a week **80%** of the time, **two days** **15%** of the time, **one day 4%** of the time, and **no days 1%** of the time. Suppose one week is randomly selected.

a. Let X = the number of days Nancy _____.

Solution 4.2
a. Let X = the number of days Nancy attends class per week.

b. X takes on what values?

Solution 4.2
b. 0, 1, 2, and 3

c. Suppose one week is randomly chosen. Construct a probability distribution table (called a PDF table) like the one in **Example 4.1**. The table should have two columns labeled x and $P(x)$. What does the $P(x)$ column sum to?

Solution 4.2
c.

x	P(x)
0	0.01
1	0.04
2	0.15
3	0.80

Table 4.4

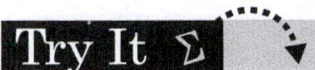

Try It

4.2 Jeremiah has basketball practice two days a week. Ninety percent of the time, he attends both practices. Eight percent of the time, he attends one practice. Two percent of the time, he does not attend either practice. What is X and what values does it take on?

4.2 | Mean or Expected Value and Standard Deviation

The **expected value** is often referred to as the **"long-term" average or mean**. This means that over the long term of doing an experiment over and over, you would **expect** this average.

You toss a coin and record the result. What is the probability that the result is heads? If you flip a coin two times, does probability tell you that these flips will result in one heads and one tail? You might toss a fair coin ten times and record nine heads. As you learned in **Section 3.**, probability does not describe the short-term results of an experiment. It gives information about what can be expected in the long term. To demonstrate this, Karl Pearson once tossed a fair coin 24,000 times! He recorded the results of each toss, obtaining heads 12,012 times. **In his experiment, Pearson illustrated the Law of Large Numbers**.

The Law of Large Numbers states that, as the number of trials in a probability experiment increases, the difference between the theoretical probability of an event and the relative frequency approaches zero **(the theoretical probability and the relative frequency get closer and closer together)**. When evaluating the long-term results of statistical experiments, we often want to know the "average" outcome. This "long-term average" is known as the **mean** or **expected value** of the experiment and is denoted by the Greek letter μ. In other words, after conducting many trials of an experiment, you would expect this average value.

NOTE

To find the expected value or long term average, μ, simply multiply each value of the random variable by its probability and add the products.

Example 4.3

A men's soccer team plays soccer zero, one, or two days a week. The probability that they play zero days is 0.2, the probability that they play one day is 0.5, and the probability that they play two days is 0.3. Find the long-term average or expected value, μ, of the number of days per week the men's soccer team plays soccer.

To do the problem, first let the random variable X = the number of days the men's soccer team plays soccer per week. X takes on the values 0, 1, 2. Construct a PDF table adding a column $x*P(x)$. In this column, you will multiply each x value by its probability.

x	P(x)	x*P(x)
0	0.2	(0)(0.2) = 0
1	0.5	(1)(0.5) = 0.5
2	0.3	(2)(0.3) = 0.6

Table 4.5 Expected Value Table This table is called an expected value table. The table helps you calculate the expected value or long-term average.

Add the last column $x*P(x)$ to find the long term average or expected value: $(0)(0.2) + (1)(0.5) + (2)(0.3) = 0 + 0.5 + 0.6 = 1.1$.

The expected value is 1.1. The men's soccer team would, on the average, expect to play soccer 1.1 days per week. The number 1.1 is the long-term average or expected value if the men's soccer team plays soccer week after week after week. We say $\mu = 1.1$.

Example 4.4

Find the expected value of the number of times a newborn baby's crying wakes its mother after midnight. The expected value is the expected number of times per week a newborn baby's crying wakes its mother after midnight. Calculate the standard deviation of the variable as well.

x	P(x)	x*P(x)	$(x-\mu)^2 \cdot P(x)$
0	$P(x = 0) = \frac{2}{50}$	$(0)\left(\frac{2}{50}\right) = 0$	$(0 - 2.1)^2 \cdot 0.04 = 0.1764$
1	$P(x = 1) = \left(\frac{11}{50}\right)$	$(1)\left(\frac{11}{50}\right) = \frac{11}{50}$	$(1 - 2.1)^2 \cdot 0.22 = 0.2662$
2	$P(x = 2) = \frac{23}{50}$	$(2)\left(\frac{23}{50}\right) = \frac{46}{50}$	$(2 - 2.1)^2 \cdot 0.46 = 0.0046$
3	$P(x = 3) = \frac{9}{50}$	$(3)\left(\frac{9}{50}\right) = \frac{27}{50}$	$(3 - 2.1)^2 \cdot 0.18 = 0.1458$

Table 4.6 You expect a newborn to wake its mother after midnight 2.1 times per week, on the average.

x	P(x)	x*P(x)	$(x - \mu)^2 \cdot P(x)$
4	$P(x = 4) = \frac{4}{50}$	$(4)\left(\frac{4}{50}\right) = \frac{16}{50}$	$(4 - 2.1)^2 \cdot 0.08 = 0.2888$
5	$P(x = 5) = \frac{1}{50}$	$(5)\left(\frac{1}{50}\right) = \frac{5}{50}$	$(5 - 2.1)^2 \cdot 0.02 = 0.1682$

Table 4.6 You expect a newborn to wake its mother after midnight 2.1 times per week, on the average.

Add the values in the third column of the table to find the expected value of X:

μ = Expected Value = $\frac{105}{50}$ = 2.1

Use μ to complete the table. The fourth column of this table will provide the values you need to calculate the standard deviation. For each value x, multiply the square of its deviation by its probability. (Each deviation has the format $x - \mu$).

Add the values in the fourth column of the table:

0.1764 + 0.2662 + 0.0046 + 0.1458 + 0.2888 + 0.1682 = 1.05

The standard deviation of X is the square root of this sum: $\sigma = \sqrt{1.05} \approx 1.0247$

Try It Σ

4.4 A hospital researcher is interested in the number of times the average post-op patient will ring the nurse during a 12-hour shift. For a random sample of 50 patients, the following information was obtained. What is the expected value?

x	P(x)
0	$P(x = 0) = \frac{4}{50}$
1	$P(x = 1) = \frac{8}{50}$
2	$P(x = 2) = \frac{16}{50}$
3	$P(x = 3) = \frac{14}{50}$
4	$P(x = 4) = \frac{6}{50}$
5	$P(x = 5) = \frac{2}{50}$

Table 4.7

Example 4.5

Suppose you play a game of chance in which five numbers are chosen from 0, 1, 2, 3, 4, 5, 6, 7, 8, 9. A computer randomly selects five numbers from zero to nine with replacement. You pay $2 to play and could profit $100,000 if you match all five numbers in order (you get your $2 back plus $100,000). Over the long term, what is your **expected** profit of playing the game?

To do this problem, set up an expected value table for the amount of money you can profit.

Let X = the amount of money you profit. The values of x are not 0, 1, 2, 3, 4, 5, 6, 7, 8, 9. Since you are interested in your profit (or loss), the values of x are 100,000 dollars and −2 dollars.

To win, you must get all five numbers correct, in order. The probability of choosing one correct number is $\frac{1}{10}$ because there are ten numbers. You may choose a number more than once. The probability of choosing all five numbers correctly and in order is

$$\left(\frac{1}{10}\right)\left(\frac{1}{10}\right)\left(\frac{1}{10}\right)\left(\frac{1}{10}\right)\left(\frac{1}{10}\right) = (1)(10^{-5}) = 0.00001.$$

Therefore, the probability of winning is 0.00001 and the probability of losing is

$$1 - 0.00001 = 0.99999.$$

The expected value table is as follows:

	x	$P(x)$	$x*P(x)$
Loss	−2	0.99999	(−2)(0.99999) = −1.99998
Profit	100,000	0.00001	(100000)(0.00001) = 1

Table 4.8 Add the last column. −1.99998 + 1 = −0.99998

Since −0.99998 is about −1, you would, on average, expect to lose approximately $1 for each game you play. However, each time you play, you either lose $2 or profit $100,000. The $1 is the average or expected LOSS per game after playing this game over and over.

Try It Σ

4.5 You are playing a game of chance in which four cards are drawn from a standard deck of 52 cards. You guess the suit of each card before it is drawn. The cards are replaced in the deck on each draw. You pay $1 to play. If you guess the right suit every time, you get your money back and $256. What is your expected profit of playing the game over the long term?

Example 4.6

Suppose you play a game with a biased coin. You play each game by tossing the coin once. $P(\text{heads}) = \frac{2}{3}$ and $P(\text{tails}) = \frac{1}{3}$. If you toss a head, you pay $6. If you toss a tail, you win $10. If you play this game many times, will you come out ahead?

a. Define a random variable X.

Solution 4.6

a. X = amount of profit

b. Complete the following expected value table.

	x	___	___
WIN	10	$\frac{1}{3}$	___
LOSE	___	___	$\frac{-12}{3}$

Table 4.9

Solution 4.6

b.

	x	$P(x)$	$xP(x)$
WIN	10	$\frac{1}{3}$	$\frac{10}{3}$
LOSE	–6	$\frac{2}{3}$	$\frac{-12}{3}$

Table 4.10

c. What is the expected value, μ? Do you come out ahead?

Solution 4.6

c. Add the last column of the table. The expected value $\mu = \frac{-2}{3}$. You lose, on average, about 67 cents each time

you play the game so you do not come out ahead.

Try It Σ

4.6 Suppose you play a game with a spinner. You play each game by spinning the spinner once. $P(\text{red}) = \frac{2}{5}$, $P(\text{blue})$ = $\frac{2}{5}$, and $P(\text{green}) = \frac{1}{5}$. If you land on red, you pay \$10. If you land on blue, you don't pay or win anything. If you land on green, you win \$10. Complete the following expected value table.

	x	P(x)	
Red			$\frac{20}{5}$
Blue		$\frac{2}{5}$	
Green	10		

Table 4.11

Like data, probability distributions have standard deviations. To calculate the standard deviation (σ) of a probability distribution, find each deviation from its expected value, square it, multiply it by its probability, add the products, and take the square root. To understand how to do the calculation, look at the table for the number of days per week a men's soccer team plays soccer. To find the standard deviation, add the entries in the column labeled $(x - \mu)^2 P(x)$ and take the square root.

x	P(x)	x*P(x)	$(x - \mu)^2 P(x)$
0	0.2	(0)(0.2) = 0	$(0 - 1.1)^2(0.2) = 0.242$
1	0.5	(1)(0.5) = 0.5	$(1 - 1.1)^2(0.5) = 0.005$
2	0.3	(2)(0.3) = 0.6	$(2 - 1.1)^2(0.3) = 0.243$

Table 4.12

Add the last column in the table. 0.242 + 0.005 + 0.243 = 0.490. The standard deviation is the square root of 0.49, or $\sigma = \sqrt{0.49} = 0.7$

Generally for probability distributions, we use a calculator or a computer to calculate μ and σ to reduce roundoff error. For some probability distributions, there are short-cut formulas for calculating μ and σ.

Example 4.7

Toss a fair, six-sided die twice. Let X = the number of faces that show an even number. Construct a table like **Table 4.11** and calculate the mean μ and standard deviation σ of X.

Solution 4.7

Tossing one fair six-sided die twice has the same sample space as tossing two fair six-sided dice. The sample space has 36 outcomes:

(1, 1)	(1, 2)	(1, 3)	(1, 4)	(1, 5)	(1, 6)
(2, 1)	(2, 2)	(2, 3)	(2, 4)	(2, 5)	(2, 6)
(3, 1)	(3, 2)	(3, 3)	(3, 4)	(3, 5)	(3, 6)
(4, 1)	(4, 2)	(4, 3)	(4, 4)	(4, 5)	(4, 6)
(5, 1)	(5, 2)	(5, 3)	(5, 4)	(5, 5)	(5, 6)
(6, 1)	(6, 2)	(6, 3)	(6, 4)	(6, 5)	(6, 6)

Table 4.13

Use the sample space to complete the following table:

x	$P(x)$	$xP(x)$	$(x - \mu)^2 \cdot P(x)$
0	$\frac{9}{36}$	0	$(0 - 1)^2 \cdot \frac{9}{36} = \frac{9}{36}$
1	$\frac{18}{36}$	$\frac{18}{36}$	$(1 - 1)^2 \cdot \frac{18}{36} = 0$
2	$\frac{9}{36}$	$\frac{18}{36}$	$(1 - 1)^2 \cdot \frac{9}{36} = \frac{9}{36}$

Table 4.14 Calculating μ and σ.

Add the values in the third column to find the expected value: $\mu = \frac{36}{36} = 1$. Use this value to complete the fourth column.

Add the values in the fourth column and take the square root of the sum: $\sigma = \sqrt{\frac{18}{36}} \approx 0.7071$.

Example 4.8

On May 11, 2013 at 9:30 PM, the probability that moderate seismic activity (one moderate earthquake) would occur in the next 48 hours in Iran was about 21.42%. Suppose you make a bet that a moderate earthquake will occur in Iran during this period. If you win the bet, you win $50. If you lose the bet, you pay $20. Let X = the amount of profit from a bet.

P(win) = P(one moderate earthquake will occur) = 21.42%

P(loss) = P(one moderate earthquake will *not* occur) = 100% − 21.42%

If you bet many times, will you come out ahead? Explain your answer in a complete sentence using numbers. What is the standard deviation of X? Construct a table similar to **Table 4.12** and **Table 4.12** to help you answer these questions.

Solution 4.8

	x	P(x)	x(Px)	$(x-\mu)^2P(x)$
win	50	0.2142	10.71	$[50-(-5.006)]^2(0.2142) = 648.0964$
loss	–20	0.7858	–15.716	$[-20-(-5.006)]^2(0.7858) = 176.6636$

Table 4.15

Mean = Expected Value = 10.71 + (–15.716) = –5.006.

If you make this bet many times under the same conditions, your long term outcome will be an average *loss* of $5.01 per bet.

Standard Deviation = $\sqrt{648.0964 + 176.6636} \approx 28.7186$

4.8 On May 11, 2013 at 9:30 PM, the probability that moderate seismic activity (one moderate earthquake) would occur in the next 48 hours in Japan was about 1.08%. As in **Example 4.8**, you bet that a moderate earthquake will occur in Japan during this period. If you win the bet, you win $100. If you lose the bet, you pay $10. Let X = the amount of profit from a bet. Find the mean and standard deviation of X.

Some of the more common discrete probability functions are binomial, geometric, hypergeometric, and Poisson. Most elementary courses do not cover the geometric, hypergeometric, and Poisson. Your instructor will let you know if he or she wishes to cover these distributions.

A probability distribution function is a pattern. You try to fit a probability problem into a **pattern** or distribution in order to perform the necessary calculations. These distributions are tools to make solving probability problems easier. Each distribution has its own special characteristics. Learning the characteristics enables you to distinguish among the different distributions.

4.3 | Binomial Distribution

There are three characteristics of a binomial experiment.

1. There are a fixed number of trials. Think of trials as repetitions of an experiment. The letter n denotes the number of trials.

2. There are only two possible outcomes, called "success" and "failure," for each trial. The letter p denotes the probability of a success on one trial, and q denotes the probability of a failure on one trial. $p + q = 1$.

3. The n trials are independent and are repeated using identical conditions. Because the n trials are independent, the outcome of one trial does not help in predicting the outcome of another trial. Another way of saying this is that for each individual trial, the probability, p, of a success and probability, q, of a failure remain the same. For example, randomly guessing at a true-false statistics question has only two outcomes. If a success is guessing correctly, then a failure is guessing incorrectly. Suppose Joe always guesses correctly on any statistics true-false question with probability p = 0.6. Then, q = 0.4. This means that for every true-false statistics question Joe answers, his probability of success (p = 0.6) and his probability of failure (q = 0.4) remain the same.

The outcomes of a binomial experiment fit a **binomial probability distribution**. The random variable X = the number of successes obtained in the n independent trials.

The mean, μ, and variance, σ^2, for the binomial probability distribution are $\mu = np$ and $\sigma^2 = npq$. The standard deviation, σ, is then $\sigma = \sqrt{npq}$.

Any experiment that has characteristics two and three and where $n = 1$ is called a **Bernoulli Trial** (named after Jacob Bernoulli who, in the late 1600s, studied them extensively). A binomial experiment takes place when the number of successes is counted in one or more Bernoulli Trials.

Example 4.9

At ABC College, the withdrawal rate from an elementary physics course is 30% for any given term. This implies that, for any given term, 70% of the students stay in the class for the entire term. A "success" could be defined as an individual who withdrew. The random variable X = the number of students who withdraw from the randomly selected elementary physics class.

Try It Σ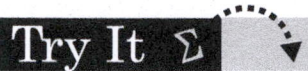

4.9 The state health board is concerned about the amount of fruit available in school lunches. Forty-eight percent of schools in the state offer fruit in their lunches every day. This implies that 52% do not. What would a "success" be in this case?

Example 4.10

Suppose you play a game that you can only either win or lose. The probability that you win any game is 55%, and the probability that you lose is 45%. Each game you play is independent. If you play the game 20 times, write the function that describes the probability that you win 15 of the 20 times. Here, if you define X as the number of wins, then X takes on the values 0, 1, 2, 3, ..., 20. The probability of a success is $p = 0.55$. The probability of a failure is $q = 0.45$. The number of trials is $n = 20$. The probability question can be stated mathematically as $P(x = 15)$.

Try It Σ

4.10 A trainer is teaching a dolphin to do tricks. The probability that the dolphin successfully performs the trick is 35%, and the probability that the dolphin does not successfully perform the trick is 65%. Out of 20 attempts, you want to find the probability that the dolphin succeeds 12 times. State the probability question mathematically.

Example 4.11

A fair coin is flipped 15 times. Each flip is independent. What is the probability of getting more than ten heads? Let X = the number of heads in 15 flips of the fair coin. X takes on the values 0, 1, 2, 3, ..., 15. Since the coin is fair, $p = 0.5$ and $q = 0.5$. The number of trials is $n = 15$. State the probability question mathematically.

Solution 4.11
$P(x > 10)$

Try It Σ

4.11 A fair, six-sided die is rolled ten times. Each roll is independent. You want to find the probability of rolling a one more than three times. State the probability question mathematically.

Example 4.12

Approximately 70% of statistics students do their homework in time for it to be collected and graded. Each student does homework independently. In a statistics class of 50 students, what is the probability that at least 40 will do their homework on time? Students are selected randomly.

a. This is a binomial problem because there is only a success or a _____, there are a fixed number of trials, and the probability of a success is 0.70 for each trial.

Solution 4.12
a. failure

b. If we are interested in the number of students who do their homework on time, then how do we define X?

Solution 4.12
b. X = the number of statistics students who do their homework on time

c. What values does x take on?

Solution 4.12
c. 0, 1, 2, …, 50

d. What is a "failure," in words?

Solution 4.12
d. Failure is defined as a student who does not complete his or her homework on time.

The probability of a success is $p = 0.70$. The number of trials is $n = 50$.

e. If $p + q = 1$, then what is q?

Solution 4.12
e. $q = 0.30$

f. The words "at least" translate as what kind of inequality for the probability question $P(x$ _____ 40).

Solution 4.12
f. greater than or equal to ($\geq$)
The probability question is $P(x \geq 40)$.

Try It Σ

4.12 Sixty-five percent of people pass the state driver's exam on the first try. A group of 50 individuals who have taken the driver's exam is randomly selected. Give two reasons why this is a binomial problem.

Notation for the Binomial: *B* = Binomial Probability Distribution Function

$X \sim B(n, p)$

Read this as "X is a random variable with a binomial distribution." The parameters are n and p; n = number of trials, p = probability of a success on each trial.

Example 4.13

It has been stated that about 41% of adult workers have a high school diploma but do not pursue any further education. If 20 adult workers are randomly selected, find the probability that at most 12 of them have a high school diploma but do not pursue any further education. How many adult workers do you expect to have a high school diploma but do not pursue any further education?

Let X = the number of workers who have a high school diploma but do not pursue any further education.

X takes on the values 0, 1, 2, ..., 20 where $n = 20$, $p = 0.41$, and $q = 1 - 0.41 = 0.59$. $X \sim B(20, 0.41)$

Find $P(x \le 12)$. $P(x \le 12) = 0.9738$. (calculator or computer)

 Using the TI-83, 83+, 84, 84+ Calculator

Go into 2nd DISTR. The syntax for the instructions are as follows:

To calculate (x = value): binompdf(n, p, number) if "number" is left out, the result is the binomial probability table.
To calculate $P(x \le$ value): binomcdf(n, p, number) if "number" is left out, the result is the cumulative binomial probability table.

For this problem: After you are in 2nd DISTR, arrow down to binomcdf. Press ENTER. Enter 20,0.41,12). The result is $P(x \le 12) = 0.9738$.

NOTE

If you want to find $P(x = 12)$, use the pdf (binompdf). If you want to find $P(x > 12)$, use 1 - binomcdf(20,0.41,12).

The probability that at most 12 workers have a high school diploma but do not pursue any further education is 0.9738.

The graph of $X \sim B(20, 0.41)$ is as follows:

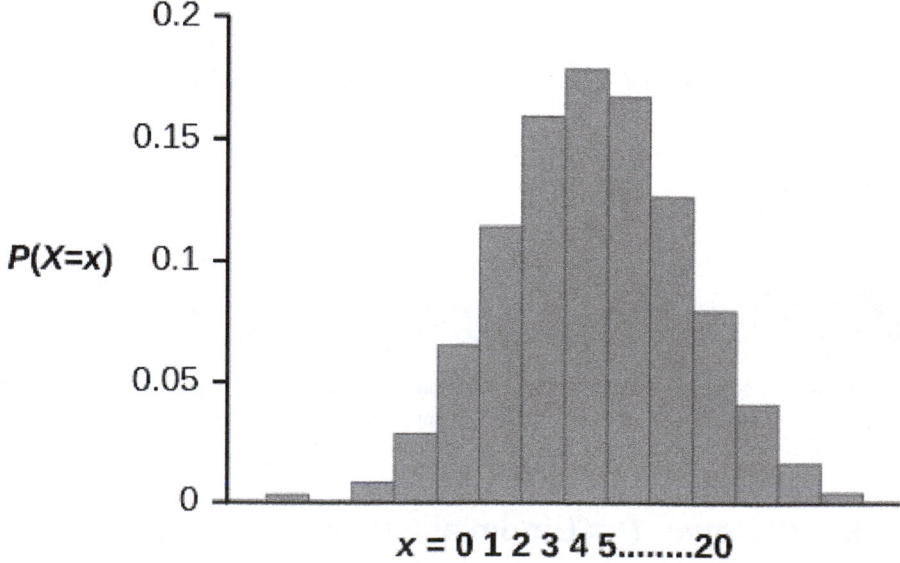

Figure 4.2

The y-axis contains the probability of x, where X = the number of workers who have only a high school diploma.

The number of adult workers that you expect to have a high school diploma but not pursue any further education is the mean, $\mu = np = (20)(0.41) = 8.2$.

The formula for the variance is $\sigma^2 = npq$. The standard deviation is $\sigma = \sqrt{npq}$.

$\sigma = \sqrt{(20)(0.41)(0.59)} = 2.20$.

Try It Σ

☞ **4.13** About 32% of students participate in a community volunteer program outside of school. If 30 students are selected at random, find the probability that at most 14 of them participate in a community volunteer program outside of school. Use the TI-83+ or TI-84 calculator to find the answer.

Example 4.14

In the 2013 *Jerry's Artarama* art supplies catalog, there are 560 pages. Eight of the pages feature signature artists. Suppose we randomly sample 100 pages. Let X = the number of pages that feature signature artists.

a.   What values does x take on?

b.   What is the probability distribution? Find the following probabilities:

   i.   the probability that two pages feature signature artists

   ii.   the probability that at most six pages feature signature artists

   iii.   the probability that more than three pages feature signature artists.

c.   Using the formulas, calculate the (i) mean and (ii) standard deviation.

Solution 4.14

a. $x = 0, 1, 2, 3, 4, 5, 6, 7, 8$

b. $X \sim B\left(100, \frac{8}{560}\right)$

   i. $P(x = 2) = \text{binompdf}\left(100, \frac{8}{560}, 2\right) = 0.2466$

   ii. $P(x \le 6) = \text{binomcdf}\left(100, \frac{8}{560}, 6\right) = 0.9994$

   iii. $P(x > 3) = 1 - P(x \le 3) = 1 - \text{binomcdf}\left(100, \frac{8}{560}, 3\right) = 1 - 0.9443 = 0.0557$

c.   i. $\text{Mean} = np = (100)\left(\frac{8}{560}\right) = \frac{800}{560} \approx 1.4286$

   ii. $\text{Standard Deviation} = \sqrt{npq} = \sqrt{(100)\left(\frac{8}{560}\right)\left(\frac{552}{560}\right)} \approx 1.1867$

Try It Σ

4.14 According to a Gallup poll, 60% of American adults prefer saving over spending. Let X = the number of American adults out of a random sample of 50 who prefer saving to spending.

a. What is the probability distribution for X?

b. Use your calculator to find the following probabilities:

　　i. the probability that 25 adults in the sample prefer saving over spending

　　ii. the probability that at most 20 adults prefer saving

　　iii. the probability that more than 30 adults prefer saving

c. Using the formulas, calculate the (i) mean and (ii) standard deviation of X.

Example 4.15

The lifetime risk of developing pancreatic cancer is about one in 78 (1.28%). Suppose we randomly sample 200 people. Let X = the number of people who will develop pancreatic cancer.

a. What is the probability distribution for X?

b. Using the formulas, calculate the (i) mean and (ii) standard deviation of X.

c. Use your calculator to find the probability that at most eight people develop pancreatic cancer

d. Is it more likely that five or six people will develop pancreatic cancer? Justify your answer numerically.

Solution 4.15

a. $X \sim B(200, 0.0128)$

b. 　i. Mean = np = 200(0.0128) = 2.56

　　ii. Standard Deviation = $\sqrt{npq} = \sqrt{(200)(0.0128)(0.9872)} \approx 1.5897$

c. Using the TI-83, 83+, 84 calculator with instructions as provided in **Example 4.13**:
$P(x \le 8)$ = binomcdf(200, 0.0128, 8) = 0.9988

d. $P(x = 5)$ = binompdf(200, 0.0128, 5) = 0.0707
$P(x = 6)$ = binompdf(200, 0.0128, 6) = 0.0298
So $P(x = 5) > P(x = 6)$; it is more likely that five people will develop cancer than six.

Try It Σ

4.15 During the 2013 regular NBA season, DeAndre Jordan of the Los Angeles Clippers had the highest field goal completion rate in the league. DeAndre scored with 61.3% of his shots. Suppose you choose a random sample of 80 shots made by DeAndre during the 2013 season. Let X = the number of shots that scored points.

a. What is the probability distribution for X?

b. Using the formulas, calculate the (i) mean and (ii) standard deviation of X.

c. Use your calculator to find the probability that DeAndre scored with 60 of these shots.

d. Find the probability that DeAndre scored with more than 50 of these shots.

Example 4.16

The following example illustrates a problem that is **not** binomial. It violates the condition of independence. ABC College has a student advisory committee made up of ten staff members and six students. The committee wishes to choose a chairperson and a recorder. What is the probability that the chairperson and recorder are both students? The names of all committee members are put into a box, and two names are drawn **without replacement**. The first name drawn determines the chairperson and the second name the recorder. There are two trials. However, the trials are not independent because the outcome of the first trial affects the outcome of the second trial. The probability of a student on the first draw is $\frac{6}{16}$. The probability of a student on the second draw is $\frac{5}{15}$, when the first draw selects a student. The probability is $\frac{6}{15}$, when the first draw selects a staff member. The probability of drawing a student's name changes for each of the trials and, therefore, violates the condition of independence.

4.16 A lacrosse team is selecting a captain. The names of all the seniors are put into a hat, and the first three that are drawn will be the captains. The names are not replaced once they are drawn (one person cannot be two captains). You want to see if the captains all play the same position. State whether this is binomial or not and state why.

4.4 | Geometric Distribution

There are three main characteristics of a geometric experiment.

1. There are one or more Bernoulli trials with all failures except the last one, which is a success. In other words, you keep repeating what you are doing until the first success. Then you stop. For example, you throw a dart at a bullseye until you hit the bullseye. The first time you hit the bullseye is a "success" so you stop throwing the dart. It might take six tries until you hit the bullseye. You can think of the trials as failure, failure, failure, failure, failure, success, STOP.

2. In theory, the number of trials could go on forever. There must be at least one trial.

3. The probability, p, of a success and the probability, q, of a failure is the same for each trial. $p + q = 1$ and $q = 1 - p$. For example, the probability of rolling a three when you throw one fair die is $\frac{1}{6}$. This is true no matter how many times you roll the die. Suppose you want to know the probability of getting the first three on the fifth roll. On rolls one through four, you do not get a face with a three. The probability for each of the rolls is $q = \frac{5}{6}$, the probability of a failure. The probability of getting a three on the fifth roll is $\left(\frac{5}{6}\right)\left(\frac{5}{6}\right)\left(\frac{5}{6}\right)\left(\frac{5}{6}\right)\left(\frac{1}{6}\right) = 0.0804$

X = the number of independent trials until the first success.

Example 4.17

You play a game of chance that you can either win or lose (there are no other possibilities) **until** you lose. Your probability of losing is $p = 0.57$. What is the probability that it takes five games until you lose? Let X = the number of games you play until you lose (includes the losing game). Then X takes on the values 1, 2, 3, ... (could go on indefinitely). The probability question is $P(x = 5)$.

Try It

4.17 You throw darts at a board until you hit the center area. Your probability of hitting the center area is $p = 0.17$. You want to find the probability that it takes eight throws until you hit the center. What values does X take on?

Example 4.18

A safety engineer feels that 35% of all industrial accidents in her plant are caused by failure of employees to follow instructions. She decides to look at the accident reports (selected randomly and replaced in the pile after reading) **until** she finds one that shows an accident caused by failure of employees to follow instructions. On average, how many reports would the safety engineer **expect** to look at until she finds a report showing an accident caused by employee failure to follow instructions? What is the probability that the safety engineer will have to examine at least three reports until she finds a report showing an accident caused by employee failure to follow instructions?

Let X = the number of accidents the safety engineer must examine **until** she finds a report showing an accident caused by employee failure to follow instructions. X takes on the values 1, 2, 3, ... The first question asks you to find the **expected value** or the mean. The second question asks you to find $P(x \geq 3)$. ("At least" translates to a "greater than or equal to" symbol).

Try It

4.18 An instructor feels that 15% of students get below a C on their final exam. She decides to look at final exams (selected randomly and replaced in the pile after reading) until she finds one that shows a grade below a C. We want to know the probability that the instructor will have to examine at least ten exams until she finds one with a grade below a C. What is the probability question stated mathematically?

Example 4.19

Suppose that you are looking for a student at your college who lives within five miles of you. You know that 55% of the 25,000 students do live within five miles of you. You randomly contact students from the college **until** one says he or she lives within five miles of you. What is the probability that you need to contact four people?

This is a geometric problem because you may have a number of failures before you have the one success you desire. Also, the probability of a success stays the same each time you ask a student if he or she lives within five miles of you. There is no definite number of trials (number of times you ask a student).

a. Let X = the number of _____ you must ask _____ one says yes.

Solution 4.19

a. Let X = the number of **students** you must ask **until** one says yes.

b. What values does X take on?

Solution 4.19
b. 1, 2, 3, ..., (total number of students)

c. What are p and q?

Solution 4.19
c. $p = 0.55$; $q = 0.45$

d. The probability question is $P(\rule{1.5cm}{0.15mm})$.

Solution 4.19
d. $P(x = 4)$

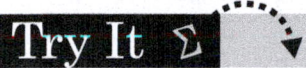

Try It Σ

4.19 You need to find a store that carries a special printer ink. You know that of the stores that carry printer ink, 10% of them carry the special ink. You randomly call each store until one has the ink you need. What are p and q?

Notation for the Geometric: G = Geometric Probability Distribution Function

$X \sim G(p)$

Read this as "X is a random variable with a **geometric distribution**." The parameter is p; p = the probability of a success for each trial.

Example 4.20

Assume that the probability of a defective computer component is 0.02. Components are randomly selected. Find the probability that the first defect is caused by the seventh component tested. How many components do you expect to test until one is found to be defective?

Let X = the number of computer components tested until the first defect is found.

X takes on the values 1, 2, 3, ... where $p = 0.02$. $X \sim G(0.02)$

Find $P(x = 7)$. $P(x = 7) = 0.0177$.

 Using the TI-83, 83+, 84, 84+ Calculator

To find the probability that $x = 7$,

- Enter 2nd, DISTR
- Scroll down and select geometpdf(
- Press ENTER
- Enter 0.02, 7); press ENTER to see the result: $P(x = 7) = 0.0177$

To find the probability that $x \le 7$, follow the same instructions EXCEPT select E:geometcdf(as the distribution function.

The probability that the seventh component is the first defect is 0.0177.

The graph of $X \sim G(0.02)$ is:

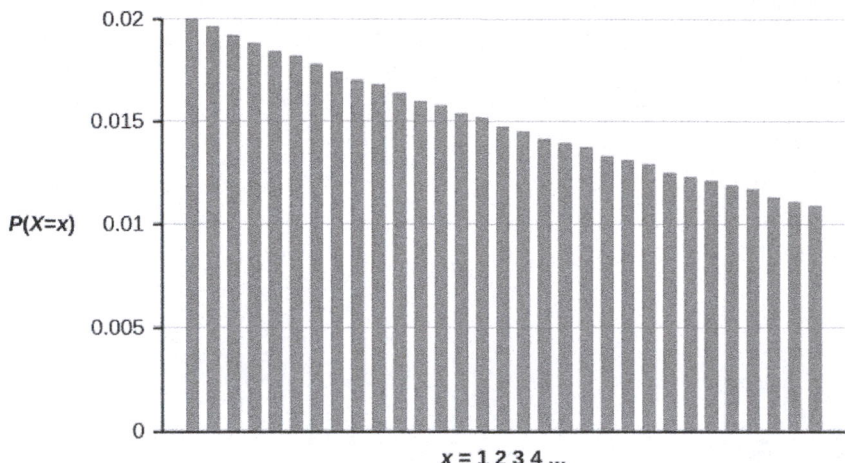

Figure 4.3

The y-axis contains the probability of x, where $X =$ the number of computer components tested.

The number of components that you would expect to test until you find the first defective one is the mean, $\mu = 50$.

The formula for the mean is $\mu = \frac{1}{p} = \frac{1}{0.02} = 50$

The formula for the variance is $\sigma^2 = \left(\frac{1}{p}\right)\left(\frac{1}{p} - 1\right) = \left(\frac{1}{0.02}\right)\left(\frac{1}{0.02} - 1\right) = 2{,}450$

The standard deviation is $\sigma = \sqrt{\left(\frac{1}{p}\right)\left(\frac{1}{p} - 1\right)} = \sqrt{\left(\frac{1}{0.02}\right)\left(\frac{1}{0.02} - 1\right)} = 49.5$

Try It Σ

☞ **4.20** The probability of a defective steel rod is 0.01. Steel rods are selected at random. Find the probability that the first defect occurs on the ninth steel rod. Use the TI-83+ or TI-84 calculator to find the answer.

Example 4.21

The lifetime risk of developing pancreatic cancer is about one in 78 (1.28%). Let $X =$ the number of people you ask until one says he or she has pancreatic cancer. Then X is a discrete random variable with a geometric distribution: $X \sim G\left(\frac{1}{78}\right)$ or $X \sim G(0.0128)$.

  a. What is the probability of that you ask ten people before one says he or she has pancreatic cancer?

  b. What is the probability that you must ask 20 people?

  c. Find the (i) mean and (ii) standard deviation of X.

Solution 4.21

a. $P(x = 10) = \text{geometpdf}(0.0128, 10) = 0.0114$

b. $P(x = 20) = \text{geometpdf}(0.0128, 20) = 0.01$

c.   i.   Mean $= \mu = \dfrac{1}{p} = \dfrac{1}{0.0128} = 78$

ii.   Standard Deviation $= \sigma = \sqrt{\dfrac{1-p}{p^2}} = \sqrt{\dfrac{1-0.0128}{0.0128^2}} \approx 77.6234$

Try It Σ

4.21 The literacy rate for a nation measures the proportion of people age 15 and over who can read and write. The literacy rate for women in Afghanistan is 12%. Let X = the number of Afghani women you ask until one says that she is literate.

a. What is the probability distribution of X?

b. What is the probability that you ask five women before one says she is literate?

c. What is the probability that you must ask ten women?

d. Find the (i) mean and (ii) standard deviation of X.

4.5 | Hypergeometric Distribution

There are five characteristics of a hypergeometric experiment.

1. You take samples from **two** groups.

2. You are concerned with a group of interest, called the first group.

3. You sample **without replacement** from the combined groups. For example, you want to choose a softball team from a combined group of 11 men and 13 women. The team consists of ten players.

4. Each pick is **not** independent, since sampling is without replacement. In the softball example, the probability of picking a woman first is $\dfrac{13}{24}$. The probability of picking a man second is $\dfrac{11}{23}$ if a woman was picked first. It is $\dfrac{10}{23}$ if a man was picked first. The probability of the second pick depends on what happened in the first pick.

5. You are **not** dealing with Bernoulli Trials.

The outcomes of a hypergeometric experiment fit a **hypergeometric probability** distribution. The random variable X = the number of items from the group of interest.

Example 4.22

A candy dish contains 100 jelly beans and 80 gumdrops. Fifty candies are picked at random. What is the probability that 35 of the 50 are gumdrops? The two groups are jelly beans and gumdrops. Since the probability question asks for the probability of picking gumdrops, the group of interest (first group) is gumdrops. The size of the group of interest (first group) is 80. The size of the second group is 100. The size of the sample is 50 (jelly beans or gumdrops). Let X = the number of gumdrops in the sample of 50. X takes on the values x = 0, 1, 2, ..., 50. What is the probability statement written mathematically?

Solution 4.22
$P(x = 35)$

4.22 A bag contains letter tiles. Forty-four of the tiles are vowels, and 56 are consonants. Seven tiles are picked at random. You want to know the probability that four of the seven tiles are vowels. What is the group of interest, the size of the group of interest, and the size of the sample?

Example 4.23

Suppose a shipment of 100 DVD players is known to have ten defective players. An inspector randomly chooses 12 for inspection. He is interested in determining the probability that, among the 12 players, at most two are defective. The two groups are the 90 non-defective DVD players and the 10 defective DVD players. The group of interest (first group) is the defective group because the probability question asks for the probability of at most two defective DVD players. The size of the sample is 12 DVD players. (They may be non-defective or defective.) Let X = the number of defective DVD players in the sample of 12. X takes on the values 0, 1, 2, ..., 10. X may not take on the values 11 or 12. The sample size is 12, but there are only 10 defective DVD players. Write the probability statement mathematically.

Solution 4.23
$P(x \le 2)$

4.23 A gross of eggs contains 144 eggs. A particular gross is known to have 12 cracked eggs. An inspector randomly chooses 15 for inspection. She wants to know the probability that, among the 15, at most three are cracked. What is X, and what values does it take on?

Example 4.24

You are president of an on-campus special events organization. You need a committee of seven students to plan a special birthday party for the president of the college. Your organization consists of 18 women and 15 men. You are interested in the number of men on your committee. If the members of the committee are randomly selected, what is the probability that your committee has more than four men?

This is a hypergeometric problem because you are choosing your committee from two groups (men and women).

a. Are you choosing with or without replacement?

Solution 4.24
a. without

b. What is the group of interest?

Solution 4.24
b. the men

c. How many are in the group of interest?

Solution 4.24
c. 15 men

d. How many are in the other group?

Solution 4.24
d. 18 women

e. Let X = _____ on the committee. What values does X take on?

Solution 4.24
e. Let X = **the number of men** on the committee. x = 0, 1, 2, …, 7.

f. The probability question is $P($_____$)$.

Solution 4.24
f. $P(x > 4)$

Try It Σ

4.24 A palette has 200 milk cartons. Of the 200 cartons, it is known that ten of them have leaked and cannot be sold. A stock clerk randomly chooses 18 for inspection. He wants to know the probability that among the 18, no more than two are leaking. Give five reasons why this is a hypergeometric problem.

Notation for the Hypergeometric: H = Hypergeometric Probability Distribution Function

$X \sim H(r, b, n)$

Read this as "X is a random variable with a hypergeometric distribution." The parameters are r, b, and n; r = the size of the group of interest (first group), b = the size of the second group, n = the size of the chosen sample.

Example 4.25

A school site committee is to be chosen randomly from six men and five women. If the committee consists of four members chosen randomly, what is the probability that two of them are men? How many men do you expect to be on the committee?

Let X = the number of men on the committee of four. The men are the group of interest (first group).

X takes on the values 0, 1, 2, 3, 4, where *r* = 6, *b* = 5, and *n* = 4. *X* ~ *H*(6, 5, 4)

Find *P*(*x* = 2). *P*(*x* = 2) = 0.4545 (calculator or computer)

NOTE

 Currently, the TI-83+ and TI-84 do not have hypergeometric probability functions. There are a number of computer packages, including Microsoft Excel, that do.

The probability that there are two men on the committee is about 0.45.

The graph of *X* ~ *H*(6, 5, 4) is:

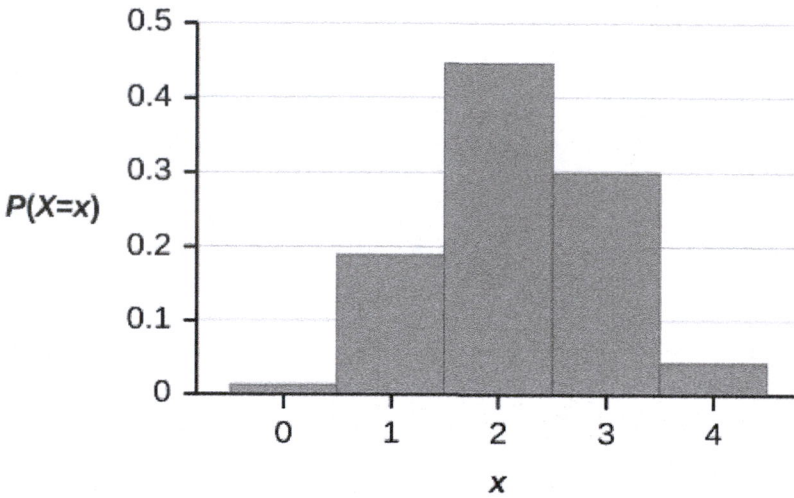

Figure 4.4

The *y*-axis contains the probability of *X*, where *X* = the number of men on the committee.

You would expect *m* = 2.18 (about two) men on the committee.

The formula for the mean is $\mu = \frac{nr}{r+b} = \frac{(4)(6)}{6+5} = 2.18$

4.25 An intramural basketball team is to be chosen randomly from 15 boys and 12 girls. The team has ten slots. You want to know the probability that eight of the players will be boys. What is the group of interest and the sample?

4.6 | Poisson Distribution

There are two main characteristics of a Poisson experiment.

1. The **Poisson probability distribution** gives the probability of a number of events occurring in a **fixed interval** of time or space if these events happen with a known average rate and independently of the time since the last event. For example, a book editor might be interested in the number of words spelled incorrectly in a particular book. It might be that, on the average, there are five words spelled incorrectly in 100 pages. The interval is the 100 pages.

2. The Poisson distribution may be used to approximate the binomial if the probability of success is "small" (such as 0.01) and the number of trials is "large" (such as 1,000). You will verify the relationship in the homework exercises. n is the number of trials, and p is the probability of a "success."

The random variable X = the number of occurrences in the interval of interest.

Example 4.26

The average number of loaves of bread put on a shelf in a bakery in a half-hour period is 12. Of interest is the number of loaves of bread put on the shelf in five minutes. The time interval of interest is five minutes. What is the probability that the number of loaves, selected randomly, put on the shelf in five minutes is three?

Let X = the number of loaves of bread put on the shelf in five minutes. If the average number of loaves put on the shelf in 30 minutes (half-hour) is 12, **then the average number of loaves put on the shelf in five minutes is** $\left(\frac{5}{30}\right)(12)$ = 2 loaves of bread.

The probability question asks you to find $P(x = 3)$.

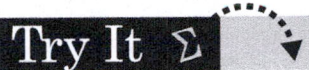

4.26 The average number of fish caught in an hour is eight. Of interest is the number of fish caught in 15 minutes. The time interval of interest is 15 minutes. What is the average number of fish caught in 15 minutes?

Example 4.27

A bank expects to receive six bad checks per day, on average. What is the probability of the bank getting fewer than five bad checks on any given day? Of interest is the number of checks the bank receives in one day, so the time interval of interest is one day. Let X = the number of bad checks the bank receives in one day. If the bank expects to receive six bad checks per day then the average is six checks per day. Write a mathematical statement for the probability question.

Solution 4.27
$P(x < 5)$

4.27 An electronics store expects to have ten returns per day on average. The manager wants to know the probability of the store getting fewer than eight returns on any given day. State the probability question mathematically.

Example 4.28

You notice that a news reporter says "uh," on average, two times per broadcast. What is the probability that the news reporter says "uh" more than two times per broadcast.

This is a Poisson problem because you are interested in knowing the number of times the news reporter says "uh" during a broadcast.

a. What is the interval of interest?

Solution 4.28
a. one broadcast

b. What is the average number of times the news reporter says "uh" during one broadcast?

Solution 4.28
b. 2

c. Let $X =$ _____. What values does X take on?

Solution 4.28
c. Let $X =$ the number of times the news reporter says "uh" during one broadcast.
$x = 0, 1, 2, 3, ...$

d. The probability question is $P($_____$)$.

Solution 4.28
d. $P(x > 2)$

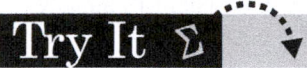

4.28 An emergency room at a particular hospital gets an average of five patients per hour. A doctor wants to know the probability that the ER gets more than five patients per hour. Give the reason why this would be a Poisson distribution.

Notation for the Poisson: P = Poisson Probability Distribution Function

$X \sim P(\mu)$

Read this as "X is a random variable with a Poisson distribution." The parameter is μ (or λ); μ (or λ) = the mean for the interval of interest.

Example 4.29

Leah's answering machine receives about six telephone calls between 8 a.m. and 10 a.m. What is the probability that Leah receives more than one call **in the next 15 minutes?**

Let $X =$ the number of calls Leah receives in 15 minutes. (The **interval of interest** is 15 minutes or $\frac{1}{4}$ hour.)

$x = 0, 1, 2, 3, ...$

If Leah receives, on the average, six telephone calls in two hours, and there are eight 15 minute intervals in two hours, then Leah receives

$\left(\frac{1}{8}\right)(6) = 0.75$ calls in 15 minutes, on average. So, $\mu = 0.75$ for this problem.

$X \sim P(0.75)$

Find $P(x > 1)$. $P(x > 1) = 0.1734$ (calculator or computer)

Using the TI-83, 83+, 84, 84+ Calculator

- Press 1 – and then press 2nd DISTR.
- Arrow down to poissoncdf. Press ENTER.
- Enter (.75,1).
- The result is $P(x > 1) = 0.1734$.

NOTE

 The TI calculators use λ (lambda) for the mean.

The probability that Leah receives more than one telephone call in the next 15 minutes is about 0.1734: $P(x > 1) = 1 - \text{poissoncdf}(0.75, 1)$.

The graph of $X \sim P(0.75)$ is:

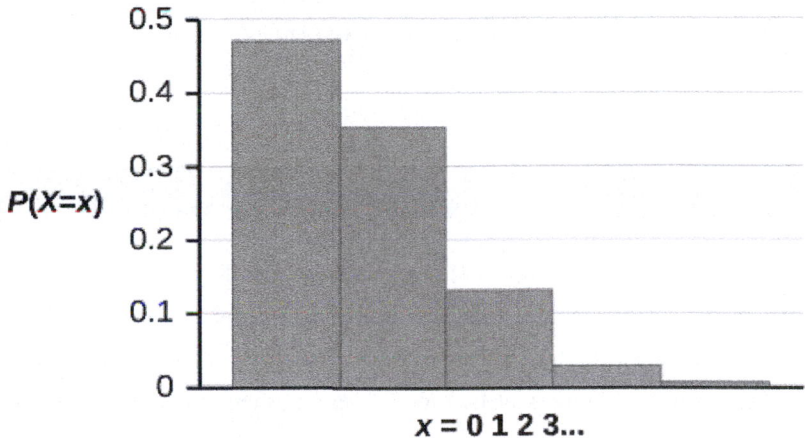

Figure 4.5

The y-axis contains the probability of x where X = the number of calls in 15 minutes.

 **Try It**

☞ **4.29** A customer service center receives about ten emails every half-hour. What is the probability that the customer service center receives more than four emails in the next six minutes? Use the TI-83+ or TI-84 calculator to find the answer.

Example 4.30

According to Baydin, an email management company, an email user gets, on average, 147 emails per day. Let X = the number of emails an email user receives per day. The discrete random variable X takes on the values $x = 0$, 1, 2 …. The random variable X has a Poisson distribution: $X \sim P(147)$. The mean is 147 emails.

a. What is the probability that an email user receives exactly 160 emails per day?

b. What is the probability that an email user receives at most 160 emails per day?

c. What is the standard deviation?

Solution 4.30

a. $P(x = 160) = \text{poissonpdf}(147, 160) \approx 0.0180$

b. $P(x \leq 160) = \text{poissoncdf}(147, 160) \approx 0.8666$

c. Standard Deviation $= \sigma = \sqrt{\mu} = \sqrt{147} \approx 12.1244$

Try It Σ

4.30 According to a recent poll by the Pew Internet Project, girls between the ages of 14 and 17 send an average of 187 text messages each day. Let X = the number of texts that a girl aged 14 to 17 sends per day. The discrete random variable X takes on the values $x = 0, 1, 2 \ldots$. The random variable X has a Poisson distribution: $X \sim P(187)$. The mean is 187 text messages.

a. What is the probability that a teen girl sends exactly 175 texts per day?

b. What is the probability that a teen girl sends at most 150 texts per day?

c. What is the standard deviation?

Example 4.31

Text message users receive or send an average of 41.5 text messages per day.

a. How many text messages does a text message user receive or send per hour?

b. What is the probability that a text message user receives or sends two messages per hour?

c. What is the probability that a text message user receives or sends more than two messages per hour?

Solution 4.31

a. Let X = the number of texts that a user sends or receives in one hour. The average number of texts received per hour is $\frac{41.5}{24} \approx 1.7292$.

b. $X \sim P(1.7292)$, so $P(x = 2) = \text{poissonpdf}(1.7292, 2) \approx 0.2653$

c. $P(x > 2) = 1 - P(x \leq 2) = 1 - \text{poissoncdf}(1.7292, 2) \approx 1 - 0.7495 = 0.2505$

Try It Σ

4.31 Atlanta's Hartsfield-Jackson International Airport is the busiest airport in the world. On average there are 2,500 arrivals and departures each day.

a. How many airplanes arrive and depart the airport per hour?

b. What is the probability that there are exactly 100 arrivals and departures in one hour?

c. What is the probability that there are at most 100 arrivals and departures in one hour?

Example 4.32

On May 13, 2013, starting at 4:30 PM, the probability of low seismic activity for the next 48 hours in Alaska was reported as about 1.02%. Use this information for the next 200 days to find the probability that there will be low seismic activity in ten of the next 200 days. Use both the binomial and Poisson distributions to calculate the probabilities. Are they close?

Solution 4.32

Let X = the number of days with low seismic activity.

Using the binomial distribution:

- $P(x = 10) = \text{binompdf}(200, .0102, 10) \approx 0.000039$

Using the Poisson distribution:

- Calculate $\mu = np = 200(0.0102) \approx 2.04$

- $P(x = 10) = \text{poissonpdf}(2.04, 10) \approx 0.000045$

We expect the approximation to be good because n is large (greater than 20) and p is small (less than 0.05). The results are close—both probabilities reported are almost 0.

4.32 On May 13, 2013, starting at 4:30 PM, the probability of moderate seismic activity for the next 48 hours in the Kuril Islands off the coast of Japan was reported at about 1.43%. Use this information for the next 100 days to find the probability that there will be low seismic activity in five of the next 100 days. Use both the binomial and Poisson distributions to calculate the probabilities. Are they close?

4.7 | Discrete Distribution (Playing Card Experiment)

Stats Lab

4.1 Discrete Distribution (Playing Card Experiment)

Class Time:

Names:

Student Learning Outcomes

- The student will compare empirical data and a theoretical distribution to determine if an everyday experiment fits a discrete distribution.
- The student will demonstrate an understanding of long-term probabilities.

Supplies

- One full deck of playing cards

Procedure

The experimental procedure is to pick one card from a deck of shuffled cards.

1. The theoretical probability of picking a diamond from a deck is ___$\frac{1}{4}$___.
2. Shuffle a deck of cards.
3. Pick one card from it.
4. Record whether it was a diamond or not a diamond.
5. Put the card back and reshuffle.
6. Do this a total of ten times.
7. Record the number of diamonds picked.
8. Let X = number of diamonds. Theoretically, $X \sim B(\underline{\quad},\underline{\quad})$

Organize the Data

1. Record the number of diamonds picked for your class in **Table 4.16**. Then calculate the relative frequency.

x	Frequency	Relative Frequency
0	_____	_____
1	_____	_____
2	_____	_____
3	_____	_____
4	_____	_____
5	_____	_____
6	_____	_____
7	_____	_____
8	_____	_____
9	_____	_____
10	_____	_____

Table 4.16

2. Calculate the following:

    a. $\overline{x}$ = _____

    b. $s =$ _____

3. Construct a histogram of the empirical data.

Figure 4.6

Theoretical Distribution

a. Build the theoretical PDF chart based on the distribution in the **Procedure** section.

x	P(x)
0	
1	
2	
3	
4	
5	
6	
7	
8	
9	
10	

Table 4.17

b. Calculate the following:

    a. $\mu =$ _____

    b. $\sigma =$ _____

c. Construct a histogram of the theoretical distribution.

This is a blank graph template. The x-axis is labeled Number of diamonds.
The y-axis is labeled Probability.

Figure 4.7

Using the Data

NOTE

RF = relative frequency

Use the table from the **Theoretical Distribution** section to calculate the following answers. Round your answers to four decimal places.

- $P(x = 3) = $ _____
- $P(1 < x < 4) = $ _____
- $P(x \geq 8) = $ _____

Use the data from the **Organize the Data** section to calculate the following answers. Round your answers to four decimal places.

- $RF(x = 3) = $ _____
- $RF(1 < x < 4) = $ _____
- $RF(x \geq 8) = $ _____

Discussion Questions

For questions 1 and 2, think about the shapes of the two graphs, the probabilities, the relative frequencies, the means, and the standard deviations.

1. Knowing that data vary, describe three similarities between the graphs and distributions of the theoretical and empirical distributions. Use complete sentences.

2. Describe the three most significant differences between the graphs or distributions of the theoretical and empirical distributions.

3. Using your answers from questions 1 and 2, does it appear that the data fit the theoretical distribution? In complete sentences, explain why or why not.

4. Suppose that the experiment had been repeated 500 times. Would you expect **Table 4.16** or **Table 4.17** to change, and how would it change? Why? Why wouldn't the other table change?

4.8 | Discrete Distribution (Lucky Dice Experiment)

Stats Lab

4.2 Discrete Distribution (Lucky Dice Experiment)

Class Time:

Names:

Student Learning Outcomes

- The student will compare empirical data and a theoretical distribution to determine if a Tet gambling game fits a discrete distribution.
- The student will demonstrate an understanding of long-term probabilities.

Supplies

- one "Lucky Dice" game or three regular dice

Procedure

Round answers to relative frequency and probability problems to four decimal places.

1. The experimental procedure is to bet on one object. Then, roll three Lucky Dice and count the number of matches. The number of matches will decide your profit.
2. What is the theoretical probability of one die matching the object?
3. Choose one object to place a bet on. Roll the three Lucky Dice. Count the number of matches.
4. Let X = number of matches. Theoretically, $X \sim B(_____,_____)$
5. Let Y = profit per game.

Organize the Data

In **Table 4.18**, fill in the y value that corresponds to each x value. Next, record the number of matches picked for your class. Then, calculate the relative frequency.

1. Complete the table.

x	y	Frequency	Relative Frequency
0			
1			
2			
3			

Table 4.18

2. Calculate the following:

   a. $\bar{x}$ = _____

   b. s_x = _____

   c. $\bar{y}$ = _____

   d. s_y = _____

3. Explain what $\bar{x}$ represents.

4. Explain what $\bar{y}$ represents.

5. Based upon the experiment:

   a. What was the average profit per game?

   b. Did this represent an average win or loss per game?

   c. How do you know? Answer in complete sentences.

6. Construct a histogram of the empirical data.

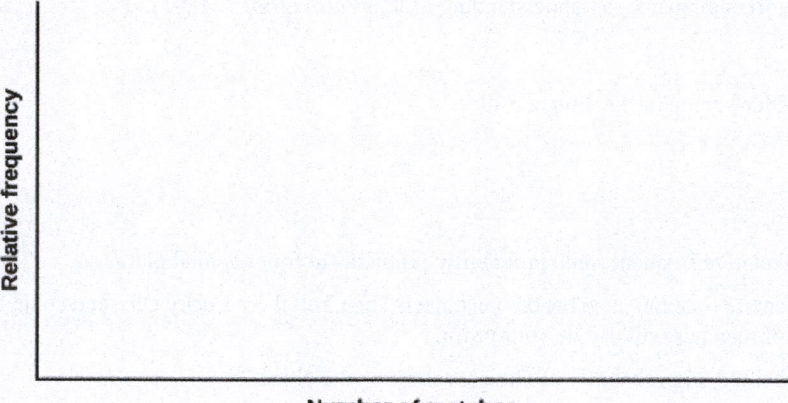

Figure 4.8

Theoretical Distribution

Build the theoretical PDF chart for x and y based on the distribution from the **Procedure** section.

1.

x	y	P(x) = P(y)
0		
1		
2		
3		

Table 4.19

2. Calculate the following:

   a. $\mu_x =$ _____

   b. $\sigma_x =$ _____

   c. $\mu_x =$ _____

3. Explain what μ_x represents.

4. Explain what μ_y represents.

5. Based upon theory:

   a. What was the expected profit per game?

   b. Did the expected profit represent an average win or loss per game?

    c. How do you know? Answer in complete sentences.

6. Construct a histogram of the theoretical distribution.

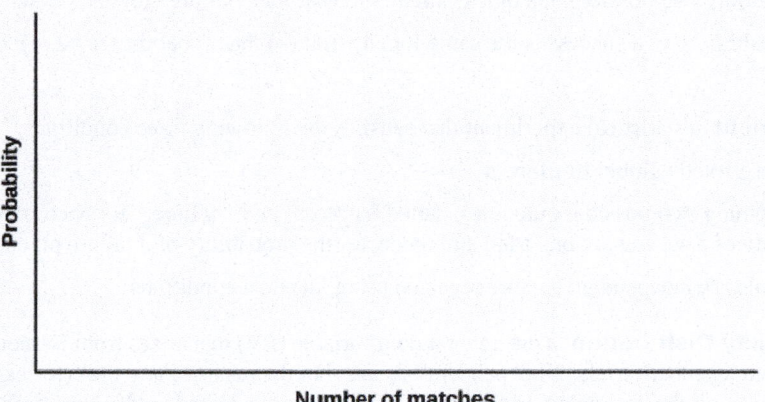

Figure 4.9

Use the Data

> **NOTE**
> _____
>
> *RF* = relative frequency

Use the data from the **Theoretical Distribution** section to calculate the following answers. Round your answers to four decimal places.

1. $P(x = 3)$ = _____

2. $P(0 < x < 3)$ = _____

3. $P(x \geq 2)$ = _____

Use the data from the **Organize the Data** section to calculate the following answers. Round your answers to four decimal places.

1. $RF(x = 3)$ = _____

2. $RF(0 < x < 3)$ = _____

3. $RF(x \geq 2)$ = _____

Discussion Question

For questions 1 and 2, consider the graphs, the probabilities, the relative frequencies, the means, and the standard deviations.

1. Knowing that data vary, describe three similarities between the graphs and distributions of the theoretical and empirical distributions. Use complete sentences.

2. Describe the three most significant differences between the graphs or distributions of the theoretical and empirical distributions.

3. Thinking about your answers to questions 1 and 2, does it appear that the data fit the theoretical distribution? In complete sentences, explain why or why not.

4. Suppose that the experiment had been repeated 500 times. Would you expect **Table 4.18** or **Table 4.19** to change, and how would it change? Why? Why wouldn't the other table change?

KEY TERMS

Bernoulli Trials an experiment with the following characteristics:

1. There are only two possible outcomes called "success" and "failure" for each trial.

2. The probability p of a success is the same for any trial (so the probability $q = 1 - p$ of a failure is the same for any trial).

Binomial Experiment a statistical experiment that satisfies the following three conditions:

1. There are a fixed number of trials, n.

2. There are only two possible outcomes, called "success" and, "failure," for each trial. The letter p denotes the probability of a success on one trial, and q denotes the probability of a failure on one trial.

3. The n trials are independent and are repeated using identical conditions.

Binomial Probability Distribution a discrete random variable (RV) that arises from Bernoulli trials; there are a fixed number, n, of independent trials. "Independent" means that the result of any trial (for example, trial one) does not affect the results of the following trials, and all trials are conducted under the same conditions. Under these circumstances the binomial RV X is defined as the number of successes in n trials. The notation is: $X \sim B(n, p)$. The mean is $\mu = np$ and the standard deviation is $\sigma = \sqrt{npq}$. The probability of exactly x successes in n trials is

$$P(X = x) = \binom{n}{x} p^x q^{n-x}.$$

Expected Value expected arithmetic average when an experiment is repeated many times; also called the mean. Notations: μ. For a discrete random variable (RV) with probability distribution function $P(x)$, the definition can also be written in the form $\mu = \sum xP(x)$.

Geometric Distribution a discrete random variable (RV) that arises from the Bernoulli trials; the trials are repeated until the first success. The geometric variable X is defined as the number of trials until the first success. Notation: $X \sim G(p)$. The mean is $\mu = \frac{1}{p}$ and the standard deviation is $\sigma = \sqrt{\frac{1}{p}\left(\frac{1}{p} - 1\right)}$. The probability of exactly x failures before the first success is given by the formula: $P(X = x) = p(1-p)^{x-1}$.

Geometric Experiment a statistical experiment with the following properties:

1. There are one or more Bernoulli trials with all failures except the last one, which is a success.

2. In theory, the number of trials could go on forever. There must be at least one trial.

3. The probability, p, of a success and the probability, q, of a failure do not change from trial to trial.

Hypergeometric Experiment a statistical experiment with the following properties:

1. You take samples from two groups.

2. You are concerned with a group of interest, called the first group.

3. You sample without replacement from the combined groups.

4. Each pick is not independent, since sampling is without replacement.

5. You are not dealing with Bernoulli Trials.

Hypergeometric Probability a discrete random variable (RV) that is characterized by:

1. A fixed number of trials.

2. The probability of success is not the same from trial to trial.

We sample from two groups of items when we are interested in only one group. X is defined as the number of successes out of the total number of items chosen. Notation: $X \sim H(r, b, n)$, where r = the number of items in the group of interest, b = the number of items in the group not of interest, and n = the number of items chosen.

Mean a number that measures the central tendency; a common name for mean is 'average.' The term 'mean' is a shortened form of 'arithmetic mean.' By definition, the mean for a sample (detonated by $\bar{x}$) is

$\bar{x} = \dfrac{\text{Sum of all values in the sample}}{\text{Number of values in the sample}}$ and the mean for a population (denoted by μ) is $\mu = $
$\dfrac{\text{Sum of all values in the population}}{\text{Number of values in the population}}$.

Mean of a Probability Distribution the long-term average of many trials of a statistical experiment

Poisson Probability Distribution a discrete random variable (RV) that counts the number of times a certain event will occur in a specific interval; characteristics of the variable:

- The probability that the event occurs in a given interval is the same for all intervals.

- The events occur with a known mean and independently of the time since the last event.

The distribution is defined by the mean μ of the event in the interval. Notation: $X \sim P(\mu)$. The mean is $\mu = np$. The

standard deviation is $\sigma = \sqrt{\mu}$. The probability of having exactly x successes in r trials is $P(X = x) = (e^{-\mu})\dfrac{\mu^x}{x!}$.

The Poisson distribution is often used to approximate the binomial distribution, when n is "large" and p is "small" (a general rule is that n should be greater than or equal to 20 and p should be less than or equal to 0.05).

Probability Distribution Function (PDF) a mathematical description of a discrete random variable (RV), given either in the form of an equation (formula) or in the form of a table listing all the possible outcomes of an experiment and the probability associated with each outcome.

Random Variable (RV) a characteristic of interest in a population being studied; common notation for variables are upper case Latin letters X, Y, Z,...; common notation for a specific value from the domain (set of all possible values of a variable) are lower case Latin letters x, y, and z. For example, if X is the number of children in a family, then x represents a specific integer 0, 1, 2, 3,.... Variables in statistics differ from variables in intermediate algebra in the two following ways.

- The domain of the random variable (RV) is not necessarily a numerical set; the domain may be expressed in words; for example, if X = hair color then the domain is {black, blond, gray, green, orange}.

- We can tell what specific value x the random variable X takes only after performing the experiment.

Standard Deviation of a Probability Distribution a number that measures how far the outcomes of a statistical experiment are from the mean of the distribution

The Law of Large Numbers As the number of trials in a probability experiment increases, the difference between the theoretical probability of an event and the relative frequency probability approaches zero.

CHAPTER REVIEW

4.1 Probability Distribution Function (PDF) for a Discrete Random Variable

The characteristics of a probability distribution function (PDF) for a discrete random variable are as follows:

1. Each probability is between zero and one, inclusive (*inclusive* means to include zero and one).

2. The sum of the probabilities is one.

4.2 Mean or Expected Value and Standard Deviation

The expected value, or mean, of a discrete random variable predicts the long-term results of a statistical experiment that has been repeated many times. The standard deviation of a probability distribution is used to measure the variability of possible outcomes.

4.3 Binomial Distribution

A statistical experiment can be classified as a binomial experiment if the following conditions are met:

1. There are a fixed number of trials, n.

2. There are only two possible outcomes, called "success" and, "failure" for each trial. The letter p denotes the probability of a success on one trial and q denotes the probability of a failure on one trial.

3. The n trials are independent and are repeated using identical conditions.

The outcomes of a binomial experiment fit a binomial probability distribution. The random variable X = the number of successes obtained in the n independent trials. The mean of X can be calculated using the formula $\mu = np$, and the standard deviation is given by the formula $\sigma = \sqrt{npq}$.

4.4 Geometric Distribution

There are three characteristics of a geometric experiment:

1. There are one or more Bernoulli trials with all failures except the last one, which is a success.

2. In theory, the number of trials could go on forever. There must be at least one trial.

3. The probability, p, of a success and the probability, q, of a failure are the same for each trial.

In a geometric experiment, define the discrete random variable X as the number of independent trials until the first success. We say that X has a geometric distribution and write $X \sim G(p)$ where p is the probability of success in a single trial.

The mean of the geometric distribution $X \sim G(p)$ is $\mu = \sqrt{\dfrac{1-p}{p^2}} = \sqrt{\dfrac{1}{p}\left(\dfrac{1}{p} - 1\right)}$.

4.5 Hypergeometric Distribution

A **hypergeometric experiment** is a statistical experiment with the following properties:

1. You take samples from two groups.

2. You are concerned with a group of interest, called the first group.

3. You sample without replacement from the combined groups.

4. Each pick is not independent, since sampling is without replacement.

5. You are not dealing with Bernoulli Trials.

The outcomes of a hypergeometric experiment fit a hypergeometric probability distribution. The random variable X = the number of items from the group of interest. The distribution of X is denoted $X \sim H(r, b, n)$, where r = the size of the group of interest (first group), b = the size of the second group, and n = the size of the chosen sample. It follows that $n \le r + b$. The mean of X is $\mu = \dfrac{nr}{r+b}$ and the standard deviation is $\sigma = \sqrt{\dfrac{rbn(r+b-n)}{(r+b)^2(r+b-1)}}$.

4.6 Poisson Distribution

A **Poisson probability distribution** of a discrete random variable gives the probability of a number of events occurring in a fixed interval of time or space, if these events happen at a known average rate and independently of the time since the last event. The Poisson distribution may be used to approximate the binomial, if the probability of success is "small" (less than or equal to 0.05) and the number of trials is "large" (greater than or equal to 20).

FORMULA REVIEW

4.2 Mean or Expected Value and Standard Deviation

Mean or Expected Value: $\mu = \displaystyle\sum_{x \in X} xP(x)$

Standard Deviation: $\sigma = \sqrt{\displaystyle\sum_{x \in X} (x - \mu)^2 P(x)}$

4.3 Binomial Distribution

$X \sim B(n, p)$ means that the discrete random variable X has a binomial probability distribution with n trials and probability of success p.

X = the number of successes in n independent trials

n = the number of independent trials

X takes on the values $x = 0, 1, 2, 3, ..., n$

p = the probability of a success for any trial

q = the probability of a failure for any trial

$p + q = 1$

$q = 1 - p$

The mean of X is $\mu = np$. The standard deviation of X is $\sigma = \sqrt{npq}$.

4.4 Geometric Distribution

$X \sim G(p)$ means that the discrete random variable X has a geometric probability distribution with probability of success in a single trial p.

X = the number of independent trials until the first success

X takes on the values $x = 1, 2, 3, ...$

p = the probability of a success for any trial

q = the probability of a failure for any trial $p + q = 1$
$q = 1 - p$

The mean is $\mu = \frac{1}{p}$.

The standard deviation is $\sigma = \sqrt{\frac{1 - p}{p^2}} = \sqrt{\frac{1}{p}\left(\frac{1}{p} - 1\right)}$.

4.5 Hypergeometric Distribution

$X \sim H(r, b, n)$ means that the discrete random variable X has a hypergeometric probability distribution with r = the size of the group of interest (first group), b = the size of the second group, and n = the size of the chosen sample.

X = the number of items from the group of interest that are in the chosen sample, and X may take on the values $x = 0$, 1, ..., up to the size of the group of interest. (The minimum value for X may be larger than zero in some instances.)

$n \leq r + b$

The mean of X is given by the formula $\mu = \frac{nr}{r + b}$ and the

standard deviation is $= \sqrt{\frac{rbn(r + b - n)}{(r + b)^2(r + b - 1)}}$.

4.6 Poisson Distribution

$X \sim P(\mu)$ means that X has a Poisson probability distribution where X = the number of occurrences in the interval of interest.

X takes on the values $x = 0, 1, 2, 3, ...$

The mean μ is typically given.

The variance is $\sigma^2 = \mu$, and the standard deviation is $\sigma = \sqrt{\mu}$.

When $P(\mu)$ is used to approximate a binomial distribution, $\mu = np$ where n represents the number of independent trials and p represents the probability of success in a single trial.

PRACTICE

4.1 Probability Distribution Function (PDF) for a Discrete Random Variable

Use the following information to answer the next five exercises: A company wants to evaluate its attrition rate, in other words, how long new hires stay with the company. Over the years, they have established the following probability distribution.

Let X = the number of years a new hire will stay with the company.

Let $P(x)$ = the probability that a new hire will stay with the company x years.

1. Complete **Table 4.20** using the data provided.

x	P(x)
0	0.12
1	0.18
2	0.30
3	0.15
4	
5	0.10
6	0.05

Table 4.20

2. $P(x = 4) =$ _____

3. $P(x \geq 5) =$ _____

4. On average, how long would you expect a new hire to stay with the company?

5. What does the column "$P(x)$" sum to?

Use the following information to answer the next six exercises: A baker is deciding how many batches of muffins to make to sell in his bakery. He wants to make enough to sell every one and no fewer. Through observation, the baker has established a probability distribution.

x	P(x)
1	0.15
2	0.35
3	0.40
4	0.10

Table 4.21

6. Define the random variable X.

7. What is the probability the baker will sell more than one batch? $P(x > 1) =$ _____

8. What is the probability the baker will sell exactly one batch? $P(x = 1) =$ _____

9. On average, how many batches should the baker make?

Use the following information to answer the next four exercises: Ellen has music practice three days a week. She practices for all of the three days 85% of the time, two days 8% of the time, one day 4% of the time, and no days 3% of the time. One week is selected at random.

10. Define the random variable X.

11. Construct a probability distribution table for the data.

12. We know that for a probability distribution function to be discrete, it must have two characteristics. One is that the sum of the probabilities is one. What is the other characteristic?

Use the following information to answer the next five exercises: Javier volunteers in community events each month. He does not do more than five events in a month. He attends exactly five events 35% of the time, four events 25% of the time, three events 20% of the time, two events 10% of the time, one event 5% of the time, and no events 5% of the time.

13. Define the random variable X.

14. What values does x take on?

15. Construct a PDF table.

16. Find the probability that Javier volunteers for less than three events each month. $P(x < 3) =$ _____

17. Find the probability that Javier volunteers for at least one event each month. $P(x > 0) =$ _____

4.2 Mean or Expected Value and Standard Deviation

18. Complete the expected value table.

x	P(x)	x*P(x)
0	0.2	
1	0.2	
2	0.4	
3	0.2	

Table 4.22

19. Find the expected value from the expected value table.

x	P(x)	x*P(x)
2	0.1	2(0.1) = 0.2
4	0.3	4(0.3) = 1.2
6	0.4	6(0.4) = 2.4
8	0.2	8(0.2) = 1.6

Table 4.23

20. Find the standard deviation.

x	P(x)	x*P(x)	$(x - \mu)^2 P(x)$
2	0.1	2(0.1) = 0.2	$(2{-}5.4)^2(0.1) = 1.156$
4	0.3	4(0.3) = 1.2	$(4{-}5.4)^2(0.3) = 0.588$
6	0.4	6(0.4) = 2.4	$(6{-}5.4)^2(0.4) = 0.144$
8	0.2	8(0.2) = 1.6	$(8{-}5.4)^2(0.2) = 1.352$

Table 4.24

21. Identify the mistake in the probability distribution table.

x	P(x)	x*P(x)
1	0.15	0.15
2	0.25	0.50
3	0.30	0.90
4	0.20	0.80
5	0.15	0.75

Table 4.25

22. Identify the mistake in the probability distribution table.

x	P(x)	x*P(x)
1	0.15	0.15
2	0.25	0.40
3	0.25	0.65
4	0.20	0.85
5	0.15	1

Table 4.26

Use the following information to answer the next five exercises: A physics professor wants to know what percent of physics majors will spend the next several years doing post-graduate research. He has the following probability distribution.

x	P(x)	x*P(x)
1	0.35	
2	0.20	
3	0.15	
4		
5	0.10	

x	P(x)	x*P(x)
6	0.05	

Table 4.27

23. Define the random variable X.

24. Define $P(x)$, or the probability of x.

25. Find the probability that a physics major will do post-graduate research for four years. $P(x = 4) =$ _____

26. FInd the probability that a physics major will do post-graduate research for at most three years. $P(x \le 3) =$ _____

27. On average, how many years would you expect a physics major to spend doing post-graduate research?

Use the following information to answer the next seven exercises: A ballet instructor is interested in knowing what percent of each year's class will continue on to the next, so that she can plan what classes to offer. Over the years, she has established the following probability distribution.

- Let $X =$ the number of years a student will study ballet with the teacher.

- Let $P(x) =$ the probability that a student will study ballet x years.

28. Complete **Table 4.28** using the data provided.

x	P(x)	x*P(x)
1	0.10	
2	0.05	
3	0.10	
4		
5	0.30	
6	0.20	
7	0.10	

Table 4.28

29. In words, define the random variable X.

30. $P(x = 4) =$ _____

31. $P(x < 4) =$ _____

32. On average, how many years would you expect a child to study ballet with this teacher?

33. What does the column "$P(x)$" sum to and why?

34. What does the column "$x*P(x)$" sum to and why?

35. You are playing a game by drawing a card from a standard deck and replacing it. If the card is a face card, you win $30. If it is not a face card, you pay $2. There are 12 face cards in a deck of 52 cards. What is the expected value of playing the game?

36. You are playing a game by drawing a card from a standard deck and replacing it. If the card is a face card, you win $30. If it is not a face card, you pay $2. There are 12 face cards in a deck of 52 cards. Should you play the game?

4.3 Binomial Distribution

Use the following information to answer the next eight exercises: The Higher Education Research Institute at UCLA collected data from 203,967 incoming first-time, full-time freshmen from 270 four-year colleges and universities in the U.S.

71.3% of those students replied that, yes, they believe that same-sex couples should have the right to legal marital status. Suppose that you randomly pick eight first-time, full-time freshmen from the survey. You are interested in the number that believes that same sex-couples should have the right to legal marital status.

37. In words, define the random variable X.

38. $X \sim$ _____(_____,_____)

39. What values does the random variable X take on?

40. Construct the probability distribution function (PDF).

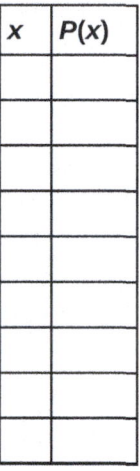

x	P(x)

Table 4.29

41. On average (μ), how many would you expect to answer yes?

42. What is the standard deviation (σ)?

43. What is the probability that at most five of the freshmen reply "yes"?

44. What is the probability that at least two of the freshmen reply "yes"?

4.4 Geometric Distribution

Use the following information to answer the next six exercises: The Higher Education Research Institute at UCLA collected data from 203,967 incoming first-time, full-time freshmen from 270 four-year colleges and universities in the U.S. 71.3% of those students replied that, yes, they believe that same-sex couples should have the right to legal marital status. Suppose that you randomly select freshman from the study until you find one who replies "yes." You are interested in the number of freshmen you must ask.

45. In words, define the random variable X.

46. $X \sim$ _____(_____,_____)

47. What values does the random variable X take on?

48. Construct the probability distribution function (PDF). Stop at $x = 6$.

x	P(x)
1	
2	
3	
4	

x	P(x)
5	
6	

Table 4.30

49. On average (μ), how many freshmen would you expect to have to ask until you found one who replies "yes?"

50. What is the probability that you will need to ask fewer than three freshmen?

4.5 Hypergeometric Distribution

Use the following information to answer the next five exercises: Suppose that a group of statistics students is divided into two groups: business majors and non-business majors. There are 16 business majors in the group and seven non-business majors in the group. A random sample of nine students is taken. We are interested in the number of business majors in the sample.

51. In words, define the random variable X.

52. $X \sim$ _____(_____,_____)

53. What values does X take on?

54. Find the standard deviation.

55. On average (μ), how many would you expect to be business majors?

4.6 Poisson Distribution

Use the following information to answer the next six exercises: On average, a clothing store gets 120 customers per day.

56. Assume the event occurs independently in any given day. Define the random variable X.

57. What values does X take on?

58. What is the probability of getting 150 customers in one day?

59. What is the probability of getting 35 customers in the first four hours? Assume the store is open 12 hours each day.

60. What is the probability that the store will have more than 12 customers in the first hour?

61. What is the probability that the store will have fewer than 12 customers in the first two hours?

62. Which type of distribution can the Poisson model be used to approximate? When would you do this?

Use the following information to answer the next six exercises: On average, eight teens in the U.S. die from motor vehicle injuries per day. As a result, states across the country are debating raising the driving age.

63. Assume the event occurs independently in any given day. In words, define the random variable X.

64. $X \sim$ _____(_____,_____)

65. What values does X take on?

66. For the given values of the random variable X, fill in the corresponding probabilities.

67. Is it likely that there will be no teens killed from motor vehicle injuries on any given day in the U.S? Justify your answer numerically.

68. Is it likely that there will be more than 20 teens killed from motor vehicle injuries on any given day in the U.S.? Justify your answer numerically.

HOMEWORK

4.1 Probability Distribution Function (PDF) for a Discrete Random Variable

69. Suppose that the PDF for the number of years it takes to earn a Bachelor of Science (B.S.) degree is given in **Table 4.31**.

x	P(x)
3	0.05
4	0.40
5	0.30
6	0.15
7	0.10

Table 4.31

   a. In words, define the random variable X.
   b. What does it mean that the values zero, one, and two are not included for x in the PDF?

4.2 Mean or Expected Value and Standard Deviation

70. A theater group holds a fund-raiser. It sells 100 raffle tickets for $5 apiece. Suppose you purchase four tickets. The prize is two passes to a Broadway show, worth a total of $150.
   a. What are you interested in here?
   b. In words, define the random variable X.
   c. List the values that X may take on.
   d. Construct a PDF.
   e. If this fund-raiser is repeated often and you always purchase four tickets, what would be your expected average winnings per raffle?

71. A game involves selecting a card from a regular 52-card deck and tossing a coin. The coin is a fair coin and is equally likely to land on heads or tails.
   • If the card is a face card, and the coin lands on Heads, you win $6
   • If the card is a face card, and the coin lands on Tails, you win $2
   • If the card is not a face card, you lose $2, no matter what the coin shows.

   a. Find the expected value for this game (expected net gain or loss).
   b. Explain what your calculations indicate about your long-term average profits and losses on this game.
   c. Should you play this game to win money?

72. You buy a lottery ticket to a lottery that costs $10 per ticket. There are only 100 tickets available to be sold in this lottery. In this lottery there are one $500 prize, two $100 prizes, and four $25 prizes. Find your expected gain or loss.

73. Complete the PDF and answer the questions.

x	P(x)	xP(x)
0	0.3	
1	0.2	
2		
3	0.4	

Table 4.32

   a. Find the probability that $x = 2$.

    b. Find the expected value.

74. Suppose that you are offered the following "deal." You roll a die. If you roll a six, you win \$10. If you roll a four or five, you win \$5. If you roll a one, two, or three, you pay \$6.

    a. What are you ultimately interested in here (the value of the roll or the money you win)?

    b. In words, define the Random Variable X.

    c. List the values that X may take on.

    d. Construct a PDF.

    e. Over the long run of playing this game, what are your expected average winnings per game?

    f. Based on numerical values, should you take the deal? Explain your decision in complete sentences.

75. A venture capitalist, willing to invest \$1,000,000, has three investments to choose from. The first investment, a software company, has a 10% chance of returning \$5,000,000 profit, a 30% chance of returning \$1,000,000 profit, and a 60% chance of losing the million dollars. The second company, a hardware company, has a 20% chance of returning \$3,000,000 profit, a 40% chance of returning \$1,000,000 profit, and a 40% chance of losing the million dollars. The third company, a biotech firm, has a 10% chance of returning \$6,000,000 profit, a 70% of no profit or loss, and a 20% chance of losing the million dollars.

    a. Construct a PDF for each investment.

    b. Find the expected value for each investment.

    c. Which is the safest investment? Why do you think so?

    d. Which is the riskiest investment? Why do you think so?

    e. Which investment has the highest expected return, on average?

76. Suppose that 20,000 married adults in the United States were randomly surveyed as to the number of children they have. The results are compiled and are used as theoretical probabilities. Let X = the number of children married people have.

x	P(x)	xP(x)
0	0.10	
1	0.20	
2	0.30	
3		
4	0.10	
5	0.05	
6 (or more)	0.05	

Table 4.33

    a. Find the probability that a married adult has three children.

    b. In words, what does the expected value in this example represent?

    c. Find the expected value.

    d. Is it more likely that a married adult will have two to three children or four to six children? How do you know?

77. Suppose that the PDF for the number of years it takes to earn a Bachelor of Science (B.S.) degree is given as in **Table 4.34**.

x	P(x)
3	0.05
4	0.40
5	0.30

x	P(x)
6	0.15
7	0.10

Table 4.34

On average, how many years do you expect it to take for an individual to earn a B.S.?

78. People visiting video rental stores often rent more than one DVD at a time. The probability distribution for DVD rentals per customer at Video To Go is given in the following table. There is a five-video limit per customer at this store, so nobody ever rents more than five DVDs.

x	P(x)
0	0.03
1	0.50
2	0.24
3	
4	0.70
5	0.04

Table 4.35

a. Describe the random variable X in words.
b. Find the probability that a customer rents three DVDs.
c. Find the probability that a customer rents at least four DVDs.
d. Find the probability that a customer rents at most two DVDs.
   Another shop, Entertainment Headquarters, rents DVDs and video games. The probability distribution for DVD rentals per customer at this shop is given as follows. They also have a five-DVD limit per customer.

x	P(x)
0	0.35
1	0.25
2	0.20
3	0.10
4	0.05
5	0.05

Table 4.36

e. At which store is the expected number of DVDs rented per customer higher?
f. If Video to Go estimates that they will have 300 customers next week, how many DVDs do they expect to rent next week? Answer in sentence form.
g. If Video to Go expects 300 customers next week, and Entertainment HQ projects that they will have 420 customers, for which store is the expected number of DVD rentals for next week higher? Explain.
h. Which of the two video stores experiences more variation in the number of DVD rentals per customer? How do you know that?

79. A "friend" offers you the following "deal." For a $10 fee, you may pick an envelope from a box containing 100 seemingly identical envelopes. However, each envelope contains a coupon for a free gift.
- Ten of the coupons are for a free gift worth $6.
- Eighty of the coupons are for a free gift worth $8.
- Six of the coupons are for a free gift worth $12.
- Four of the coupons are for a free gift worth $40.

Based upon the financial gain or loss over the long run, should you play the game?

    a. Yes, I expect to come out ahead in money.
    b. No, I expect to come out behind in money.
    c. It doesn't matter. I expect to break even.

80. Florida State University has 14 statistics classes scheduled for its Summer 2013 term. One class has space available for 30 students, eight classes have space for 60 students, one class has space for 70 students, and four classes have space for 100 students.
    a. What is the average class size assuming each class is filled to capacity?
    b. Space is available for 980 students. Suppose that each class is filled to capacity and select a statistics student at random. Let the random variable X equal the size of the student's class. Define the PDF for X.
    c. Find the mean of X.
    d. Find the standard deviation of X.

81. In a lottery, there are 250 prizes of $5, 50 prizes of $25, and ten prizes of $100. Assuming that 10,000 tickets are to be issued and sold, what is a fair price to charge to break even?

4.3 Binomial Distribution

82. According to a recent article the average number of babies born with significant hearing loss (deafness) is approximately two per 1,000 babies in a healthy baby nursery. The number climbs to an average of 30 per 1,000 babies in an intensive care nursery.

Suppose that 1,000 babies from healthy baby nurseries were randomly surveyed. Find the probability that exactly two babies were born deaf.

Use the following information to answer the next four exercises. Recently, a nurse commented that when a patient calls the medical advice line claiming to have the flu, the chance that he or she truly has the flu (and not just a nasty cold) is only about 4%. Of the next 25 patients calling in claiming to have the flu, we are interested in how many actually have the flu.

83. Define the random variable and list its possible values.

84. State the distribution of X.

85. Find the probability that at least four of the 25 patients actually have the flu.

86. On average, for every 25 patients calling in, how many do you expect to have the flu?

87. People visiting video rental stores often rent more than one DVD at a time. The probability distribution for DVD rentals per customer at Video To Go is given **Table 4.37**. There is five-video limit per customer at this store, so nobody ever rents more than five DVDs.

x	P(x)
0	0.03
1	0.50
2	0.24
3	
4	0.07
5	0.04

a. Describe the random variable X in words.
b. Find the probability that a customer rents three DVDs.
c. Find the probability that a customer rents at least four DVDs.
d. Find the probability that a customer rents at most two DVDs.

88. A school newspaper reporter decides to randomly survey 12 students to see if they will attend Tet (Vietnamese New Year) festivities this year. Based on past years, she knows that 18% of students attend Tet festivities. We are interested in the number of students who will attend the festivities.
a. In words, define the random variable X.
b. List the values that X may take on.
c. Give the distribution of X. $X \sim$ _____(_____,_____)
d. How many of the 12 students do we expect to attend the festivities?
e. Find the probability that at most four students will attend.
f. Find the probability that more than two students will attend.

Use the following information to answer the next two exercises: The probability that the San Jose Sharks will win any given game is 0.3694 based on a 13-year win history of 382 wins out of 1,034 games played (as of a certain date). An upcoming monthly schedule contains 12 games.

89. The expected number of wins for that upcoming month is:
a. 1.67
b. 12
c. $\frac{382}{1043}$
d. 4.43

Let X = the number of games won in that upcoming month.

90. What is the probability that the San Jose Sharks win six games in that upcoming month?
a. 0.1476
b. 0.2336
c. 0.7664
d. 0.8903

91. What is the probability that the San Jose Sharks win at least five games in that upcoming month
a. 0.3694
b. 0.5266
c. 0.4734
d. 0.2305

92. A student takes a ten-question true-false quiz, but did not study and randomly guesses each answer. Find the probability that the student passes the quiz with a grade of at least 70% of the questions correct.

93. A student takes a 32-question multiple-choice exam, but did not study and randomly guesses each answer. Each question has three possible choices for the answer. Find the probability that the student guesses **more than** 75% of the questions correctly.

94. Six different colored dice are rolled. Of interest is the number of dice that show a one.
a. In words, define the random variable X.
b. List the values that X may take on.
c. Give the distribution of X. $X \sim$ _____(_____,_____)
d. On average, how many dice would you expect to show a one?
e. Find the probability that all six dice show a one.
f. Is it more likely that three or that four dice will show a one? Use numbers to justify your answer numerically.

95. More than 96 percent of the very largest colleges and universities (more than 15,000 total enrollments) have some online offerings. Suppose you randomly pick 13 such institutions. We are interested in the number that offer distance learning courses.
a. In words, define the random variable X.
b. List the values that X may take on.
c. Give the distribution of X. $X \sim$ _____(_____,_____)
d. On average, how many schools would you expect to offer such courses?

    e. Find the probability that at most ten offer such courses.

    f. Is it more likely that 12 or that 13 will offer such courses? Use numbers to justify your answer numerically and answer in a complete sentence.

96. Suppose that about 85% of graduating students attend their graduation. A group of 22 graduating students is randomly chosen.

    a. In words, define the random variable X.

    b. List the values that X may take on.

    c. Give the distribution of X. $X \sim$ _____(_____,_____)

    d. How many are expected to attend their graduation?

    e. Find the probability that 17 or 18 attend.

    f. Based on numerical values, would you be surprised if all 22 attended graduation? Justify your answer numerically.

97. At The Fencing Center, 60% of the fencers use the foil as their main weapon. We randomly survey 25 fencers at The Fencing Center. We are interested in the number of fencers who do **not** use the foil as their main weapon.

    a. In words, define the random variable X.

    b. List the values that X may take on.

    c. Give the distribution of X. $X \sim$ _____(_____,_____)

    d. How many are expected to **not** to use the foil as their main weapon?

    e. Find the probability that six do **not** use the foil as their main weapon.

    f. Based on numerical values, would you be surprised if all 25 did **not** use foil as their main weapon? Justify your answer numerically.

98. Approximately 8% of students at a local high school participate in after-school sports all four years of high school. A group of 60 seniors is randomly chosen. Of interest is the number who participated in after-school sports all four years of high school.

    a. In words, define the random variable X.

    b. List the values that X may take on.

    c. Give the distribution of X. $X \sim$ _____(_____,_____)

    d. How many seniors are expected to have participated in after-school sports all four years of high school?

    e. Based on numerical values, would you be surprised if none of the seniors participated in after-school sports all four years of high school? Justify your answer numerically.

    f. Based upon numerical values, is it more likely that four or that five of the seniors participated in after-school sports all four years of high school? Justify your answer numerically.

99. The chance of an IRS audit for a tax return with over $25,000 in income is about 2% per year. We are interested in the expected number of audits a person with that income has in a 20-year period. Assume each year is independent.

    a. In words, define the random variable X.

    b. List the values that X may take on.

    c. Give the distribution of X. $X \sim$ _____(_____,_____)

    d. How many audits are expected in a 20-year period?

    e. Find the probability that a person is not audited at all.

    f. Find the probability that a person is audited more than twice.

100. It has been estimated that only about 30% of California residents have adequate earthquake supplies. Suppose you randomly survey 11 California residents. We are interested in the number who have adequate earthquake supplies.

    a. In words, define the random variable X.

    b. List the values that X may take on.

    c. Give the distribution of X. $X \sim$ _____(_____,_____)

    d. What is the probability that at least eight have adequate earthquake supplies?

    e. Is it more likely that none or that all of the residents surveyed will have adequate earthquake supplies? Why?

    f. How many residents do you expect will have adequate earthquake supplies?

101. There are two similar games played for Chinese New Year and Vietnamese New Year. In the Chinese version, fair dice with numbers 1, 2, 3, 4, 5, and 6 are used, along with a board with those numbers. In the Vietnamese version, fair dice with pictures of a gourd, fish, rooster, crab, crayfish, and deer are used. The board has those six objects on it, also. We will play with bets being $1. The player places a bet on a number or object. The "house" rolls three dice. If none of the dice show the number or object that was bet, the house keeps the $1 bet. If one of the dice shows the number or object bet (and the other two do not show it), the player gets back his or her $1 bet, plus $1 profit. If two of the dice show the number or object bet (and the third die does not show it), the player gets back his or her $1 bet, plus $2 profit. If all three dice show the number or object bet, the player gets back his or her $1 bet, plus $3 profit. Let X = number of matches and Y = profit per game.

    a. In words, define the random variable X.
    b. List the values that X may take on.
    c. Give the distribution of X. $X \sim$ _____(_____,_____)
    d. List the values that Y may take on. Then, construct one PDF table that includes both X and Y and their probabilities.
    e. Calculate the average expected matches over the long run of playing this game for the player.
    f. Calculate the average expected earnings over the long run of playing this game for the player.
    g. Determine who has the advantage, the player or the house.

102. According to The World Bank, only 9% of the population of Uganda had access to electricity as of 2009. Suppose we randomly sample 150 people in Uganda. Let X = the number of people who have access to electricity.
    a. What is the probability distribution for X?
    b. Using the formulas, calculate the mean and standard deviation of X.
    c. Use your calculator to find the probability that 15 people in the sample have access to electricity.
    d. Find the probability that at most ten people in the sample have access to electricity.
    e. Find the probability that more than 25 people in the sample have access to electricity.

103. The literacy rate for a nation measures the proportion of people age 15 and over that can read and write. The literacy rate in Afghanistan is 28.1%. Suppose you choose 15 people in Afghanistan at random. Let X = the number of people who are literate.
    a. Sketch a graph of the probability distribution of X.
    b. Using the formulas, calculate the (i) mean and (ii) standard deviation of X.
    c. Find the probability that more than five people in the sample are literate. Is it is more likely that three people or four people are literate.

4.4 Geometric Distribution

104. A consumer looking to buy a used red Miata car will call dealerships until she finds a dealership that carries the car. She estimates the probability that any independent dealership will have the car will be 28%. We are interested in the number of dealerships she must call.
    a. In words, define the random variable X.
    b. List the values that X may take on.
    c. Give the distribution of X. $X \sim$ _____(_____,_____)
    d. On average, how many dealerships would we expect her to have to call until she finds one that has the car?
    e. Find the probability that she must call at most four dealerships.
    f. Find the probability that she must call three or four dealerships.

105. Suppose that the probability that an adult in America will watch the Super Bowl is 40%. Each person is considered independent. We are interested in the number of adults in America we must survey until we find one who will watch the Super Bowl.
    a. In words, define the random variable X.
    b. List the values that X may take on.
    c. Give the distribution of X. $X \sim$ _____(_____,_____)
    d. How many adults in America do you expect to survey until you find one who will watch the Super Bowl?
    e. Find the probability that you must ask seven people.
    f. Find the probability that you must ask three or four people.

106. It has been estimated that only about 30% of California residents have adequate earthquake supplies. Suppose we are interested in the number of California residents we must survey until we find a resident who does **not** have adequate earthquake supplies.
    a. In words, define the random variable X.
    b. List the values that X may take on.
    c. Give the distribution of X. $X \sim$ _____(_____,_____)
    d. What is the probability that we must survey just one or two residents until we find a California resident who does not have adequate earthquake supplies?
    e. What is the probability that we must survey at least three California residents until we find a California resident who does not have adequate earthquake supplies?

f. How many California residents do you expect to need to survey until you find a California resident who **does not** have adequate earthquake supplies?

g. How many California residents do you expect to need to survey until you find a California resident who **does** have adequate earthquake supplies?

107. In one of its Spring catalogs, L.L. Bean® advertised footwear on 29 of its 192 catalog pages. Suppose we randomly survey 20 pages. We are interested in the number of pages that advertise footwear. Each page may be picked more than once.

a. In words, define the random variable X.

b. List the values that X may take on.

c. Give the distribution of X. $X \sim$ _____(_____,_____)

d. How many pages do you expect to advertise footwear on them?

e. Is it probable that all twenty will advertise footwear on them? Why or why not?

f. What is the probability that fewer than ten will advertise footwear on them?

g. Reminder: A page may be picked more than once. We are interested in the number of pages that we must randomly survey until we find one that has footwear advertised on it. Define the random variable X and give its distribution.

h. What is the probability that you only need to survey at most three pages in order to find one that advertises footwear on it?

i. How many pages do you expect to need to survey in order to find one that advertises footwear?

108. Suppose that you are performing the probability experiment of rolling one fair six-sided die. Let F be the event of rolling a four or a five. You are interested in how many times you need to roll the die in order to obtain the first four or five as the outcome.

• p = probability of success (event F occurs)

• q = probability of failure (event F does not occur)

a. Write the description of the random variable X.

b. What are the values that X can take on?

c. Find the values of p and q.

d. Find the probability that the first occurrence of event F (rolling a four or five) is on the second trial.

109. Ellen has music practice three days a week. She practices for all of the three days 85% of the time, two days 8% of the time, one day 4% of the time, and no days 3% of the time. One week is selected at random. What values does X take on?

110. The World Bank records the prevalence of HIV in countries around the world. According to their data, "Prevalence of HIV refers to the percentage of people ages 15 to 49 who are infected with HIV."[1] In South Africa, the prevalence of HIV is 17.3%. Let X = the number of people you test until you find a person infected with HIV.

a. Sketch a graph of the distribution of the discrete random variable X.

b. What is the probability that you must test 30 people to find one with HIV?

c. What is the probability that you must ask ten people?

d. Find the (i) mean and (ii) standard deviation of the distribution of X.

111. According to a recent Pew Research poll, 75% of millenials (people born between 1981 and 1995) have a profile on a social networking site. Let X = the number of millenials you ask until you find a person without a profile on a social networking site.

a. Describe the distribution of X.

b. Find the (i) mean and (ii) standard deviation of X.

c. What is the probability that you must ask ten people to find one person without a social networking site?

d. What is the probability that you must ask 20 people to find one person without a social networking site?

e. What is the probability that you must ask *at most* five people?

4.5 Hypergeometric Distribution

---

1. "Prevalence of HIV, total (% of populations ages 15-49)," The World Bank, 2013. Available online at http://data.worldbank.org/indicator/
SH.DYN.AIDS.ZS?order=wbapi_data_value_2011+wbapi_data_value+wbapi_data_value-last&sort=desc (accessed May 15, 2013).

112. A group of Martial Arts students is planning on participating in an upcoming demonstration. Six are students of Tae Kwon Do; seven are students of Shotokan Karate. Suppose that eight students are randomly picked to be in the first demonstration. We are interested in the number of Shotokan Karate students in that first demonstration.
   a. In words, define the random variable X.
   b. List the values that X may take on.
   c. Give the distribution of X. X ~ _____(_____,_____)
   d. How many Shotokan Karate students do we expect to be in that first demonstration?

113. In one of its Spring catalogs, L.L. Bean® advertised footwear on 29 of its 192 catalog pages. Suppose we randomly survey 20 pages. We are interested in the number of pages that advertise footwear. Each page may be picked at most once.

   a. In words, define the random variable X.
   b. List the values that X may take on.
   c. Give the distribution of X. X ~ _____(_____,_____)
   d. How many pages do you expect to advertise footwear on them?
   e. Calculate the standard deviation.

114. Suppose that a technology task force is being formed to study technology awareness among instructors. Assume that ten people will be randomly chosen to be on the committee from a group of 28 volunteers, 20 who are technically proficient and eight who are not. We are interested in the number on the committee who are **not** technically proficient.
   a. In words, define the random variable X.
   b. List the values that X may take on.
   c. Give the distribution of X. X ~ _____(_____,_____)
   d. How many instructors do you expect on the committee who are **not** technically proficient?
   e. Find the probability that at least five on the committee are not technically proficient.
   f. Find the probability that at most three on the committee are not technically proficient.

115. Suppose that nine Massachusetts athletes are scheduled to appear at a charity benefit. The nine are randomly chosen from eight volunteers from the Boston Celtics and four volunteers from the New England Patriots. We are interested in the number of Patriots picked.
   a. In words, define the random variable X.
   b. List the values that X may take on.
   c. Give the distribution of X. X ~ _____(_____,_____)
   d. Are you choosing the nine athletes with or without replacement?

116. A bridge hand is defined as 13 cards selected at random and without replacement from a deck of 52 cards. In a standard deck of cards, there are 13 cards from each suit: hearts, spades, clubs, and diamonds. What is the probability of being dealt a hand that does not contain a heart?
   a. What is the group of interest?
   b. How many are in the group of interest?
   c. How many are in the other group?
   d. Let X = _____. What values does X take on?
   e. The probability question is P(_____).
   f. Find the probability in question.
   g. Find the (i) mean and (ii) standard deviation of X.

4.6 Poisson Distribution

117. The switchboard in a Minneapolis law office gets an average of 5.5 incoming phone calls during the noon hour on Mondays. Experience shows that the existing staff can handle up to six calls in an hour. Let X = the number of calls received at noon.
   a. Find the mean and standard deviation of X.
   b. What is the probability that the office receives at most six calls at noon on Monday?
   c. Find the probability that the law office receives six calls at noon. What does this mean to the law office staff who get, on average, 5.5 incoming phone calls at noon?
   d. What is the probability that the office receives more than eight calls at noon?

118. The maternity ward at Dr. Jose Fabella Memorial Hospital in Manila in the Philippines is one of the busiest in the world with an average of 60 births per day. Let X = the number of births in an hour.
   a. Find the mean and standard deviation of X.
   b. Sketch a graph of the probability distribution of X.

    c.   What is the probability that the maternity ward will deliver three babies in one hour?

    d.   What is the probability that the maternity ward will deliver at most three babies in one hour?

    e.   What is the probability that the maternity ward will deliver more than five babies in one hour?

119. A manufacturer of Christmas tree light bulbs knows that 3% of its bulbs are defective. Find the probability that a string of 100 lights contains at most four defective bulbs using both the binomial and Poisson distributions.

120. The average number of children a Japanese woman has in her lifetime is 1.37. Suppose that one Japanese woman is randomly chosen.

    a.   In words, define the random variable X.

    b.   List the values that X may take on.

    c.   Give the distribution of X. $X \sim$ _____(_____,_____)

    d.   Find the probability that she has no children.

    e.   Find the probability that she has fewer children than the Japanese average.

    f.   Find the probability that she has more children than the Japanese average.

121. The average number of children a Spanish woman has in her lifetime is 1.47. Suppose that one Spanish woman is randomly chosen.

    a.   In words, define the Random Variable X.

    b.   List the values that X may take on.

    c.   Give the distribution of X. $X \sim$ _____(_____,_____)

    d.   Find the probability that she has no children.

    e.   Find the probability that she has fewer children than the Spanish average.

    f.   Find the probability that she has more children than the Spanish average .

122. Fertile, female cats produce an average of three litters per year. Suppose that one fertile, female cat is randomly chosen. In one year, find the probability she produces:

    a.   In words, define the random variable X.

    b.   List the values that X may take on.

    c.   Give the distribution of X. $X \sim$ _____

    d.   Find the probability that she has no litters in one year.

    e.   Find the probability that she has at least two litters in one year.

    f.   Find the probability that she has exactly three litters in one year.

123. The chance of having an extra fortune in a fortune cookie is about 3%. Given a bag of 144 fortune cookies, we are interested in the number of cookies with an extra fortune. Two distributions may be used to solve this problem, but only use one distribution to solve the problem.

    a.   In words, define the random variable X.

    b.   List the values that X may take on.

    c.   Give the distribution of X. $X \sim$ _____(_____,_____)

    d.   How many cookies do we expect to have an extra fortune?

    e.   Find the probability that none of the cookies have an extra fortune.

    f.   Find the probability that more than three have an extra fortune.

    g.   As n increases, what happens involving the probabilities using the two distributions? Explain in complete sentences.

124. According to the South Carolina Department of Mental Health web site, for every 200 U.S. women, the average number who suffer from anorexia is one. Out of a randomly chosen group of 600 U.S. women determine the following.

    a.   In words, define the random variable X.

    b.   List the values that X may take on.

    c.   Give the distribution of$X$. $X \sim$ _____(_____,_____)

    d.   How many are expected to suffer from anorexia?

    e.   Find the probability that no one suffers from anorexia.

    f.   Find the probability that more than four suffer from anorexia.

125. The chance of an IRS audit for a tax return with over $25,000 in income is about 2% per year. Suppose that 100 people with tax returns over $25,000 are randomly picked. We are interested in the number of people audited in one year. Use a Poisson distribution to anwer the following questions.

    a.   In words, define the random variable X.

    b.   List the values that X may take on.

    c. Give the distribution of X. $X \sim$ _____(_____,_____)

    d. How many are expected to be audited?

    e. Find the probability that no one was audited.

    f. Find the probability that at least three were audited.

126. Approximately 8% of students at a local high school participate in after-school sports all four years of high school. A group of 60 seniors is randomly chosen. Of interest is the number that participated in after-school sports all four years of high school.

    a. In words, define the random variable X.

    b. List the values that X may take on.

    c. Give the distribution of X. $X \sim$ _____(_____,_____)

    d. How many seniors are expected to have participated in after-school sports all four years of high school?

    e. Based on numerical values, would you be surprised if none of the seniors participated in after-school sports all four years of high school? Justify your answer numerically.

    f. Based on numerical values, is it more likely that four or that five of the seniors participated in after-school sports all four years of high school? Justify your answer numerically.

127. On average, Pierre, an amateur chef, drops three pieces of egg shell into every two cake batters he makes. Suppose that you buy one of his cakes.

    a. In words, define the random variable X.

    b. List the values that X may take on.

    c. Give the distribution of X. $X \sim$ _____(_____,_____)

    d. On average, how many pieces of egg shell do you expect to be in the cake?

    e. What is the probability that there will not be any pieces of egg shell in the cake?

    f. Let's say that you buy one of Pierre's cakes each week for six weeks. What is the probability that there will not be any egg shell in any of the cakes?

    g. Based upon the average given for Pierre, is it possible for there to be seven pieces of shell in the cake? Why?

Use the following information to answer the next two exercises: The average number of times per week that Mrs. Plum's cats wake her up at night because they want to play is ten. We are interested in the number of times her cats wake her up each week.

128. In words, the random variable $X =$ _____

    a. the number of times Mrs. Plum's cats wake her up each week.

    b. the number of times Mrs. Plum's cats wake her up each hour.

    c. the number of times Mrs. Plum's cats wake her up each night.

    d. the number of times Mrs. Plum's cats wake her up.

129. Find the probability that her cats will wake her up no more than five times next week.

    a. 0.5000

    b. 0.9329

    c. 0.0378

    d. 0.0671

REFERENCES

4.2 Mean or Expected Value and Standard Deviation

Class Catalogue at the Florida State University. Available online at https://apps.oti.fsu.edu/RegistrarCourseLookup/SearchFormLegacy (accessed May 15, 2013).

"World Earthquakes: Live Earthquake News and Highlights," World Earthquakes, 2012. http://www.world-earthquakes.com/index.php?option=ethq_prediction (accessed May 15, 2013).

4.3 Binomial Distribution

"Access to electricity (% of population)," The World Bank, 2013. Available online at http://data.worldbank.org/indicator/EG.ELC.ACCS.ZS?order=wbapi_data_value_2009%20wbapi_data_value%20wbapi_data_value-first&sort=asc (accessed May 15, 2015).

"Distance Education." Wikipedia. Available online at http://en.wikipedia.org/wiki/Distance_education (accessed May 15, 2013).

"NBA Statistics – 2013," ESPN NBA, 2013. Available online at http://espn.go.com/nba/statistics/_/seasontype/2 (accessed May 15, 2013).

Newport, Frank. "Americans Still Enjoy Saving Rather than Spending: Few demographic differences seen in these views other than by income," GALLUP® Economy, 2013. Available online at http://www.gallup.com/poll/162368/americans-enjoy-saving-rather-spending.aspx (accessed May 15, 2013).

Pryor, John H., Linda DeAngelo, Laura Palucki Blake, Sylvia Hurtado, Serge Tran. *The American Freshman: National Norms Fall 2011*. Los Angeles: Cooperative Institutional Research Program at the Higher Education Research Institute at UCLA, 2011. Also available online at http://heri.ucla.edu/PDFs/pubs/TFS/Norms/Monographs/TheAmericanFreshman2011.pdf (accessed May 15, 2013).

"The World FactBook," Central Intelligence Agency. Available online at https://www.cia.gov/library/publications/the-world-factbook/geos/af.html (accessed May 15, 2013).

"What are the key statistics about pancreatic cancer?" American Cancer Society, 2013. Available online at http://www.cancer.org/cancer/pancreaticcancer/detailedguide/pancreatic-cancer-key-statistics (accessed May 15, 2013).

4.4 Geometric Distribution

"Millennials: A Portrait of Generation Next," PewResearchCenter. Available online at http://www.pewsocialtrends.org/files/2010/10/millennials-confident-connected-open-to-change.pdf (accessed May 15, 2013).

"Millennials: Confident. Connected. Open to Change." Executive Summary by PewResearch Social & Demographic Trends, 2013. Available online at http://www.pewsocialtrends.org/2010/02/24/millennials-confident-connected-open-to-change/ (accessed May 15, 2013).

"Prevalence of HIV, total (% of populations ages 15-49)," The World Bank, 2013. Available online at http://data.worldbank.org/indicator/SH.DYN.AIDS.ZS?order=wbapi_data_value_2011+wbapi_data_value+wbapi_data_value-last&sort=desc (accessed May 15, 2013).

Pryor, John H., Linda DeAngelo, Laura Palucki Blake, Sylvia Hurtado, Serge Tran. *The American Freshman: National Norms Fall 2011*. Los Angeles: Cooperative Institutional Research Program at the Higher Education Research Institute at UCLA, 2011. Also available online at http://heri.ucla.edu/PDFs/pubs/TFS/Norms/Monographs/TheAmericanFreshman2011.pdf (accessed May 15, 2013).

"Summary of the National Risk and Vulnerability Assessment 2007/8: A profile of Afghanistan," The European Union and ICON-Institute. Available online at http://ec.europa.eu/europeaid/where/asia/documents/afgh_brochure_summary_en.pdf (accessed May 15, 2013).

"The World FactBook," Central Intelligence Agency. Available online at https://www.cia.gov/library/publications/the-world-factbook/geos/af.html (accessed May 15, 2013).

"UNICEF reports on Female Literacy Centers in Afghanistan established to teach women and girls basic resading [sic] and writing skills," UNICEF Television. Video available online at http://www.unicefusa.org/assets/video/afghan-female-literacy-centers.html (accessed May 15, 2013).

4.6 Poisson Distribution

"ATL Fact Sheet," Department of Aviation at the Hartsfield-Jackson Atlanta International Airport, 2013. Available online at http://www.atlanta-airport.com/Airport/ATL/ATL_FactSheet.aspx (accessed May 15, 2013).

Center for Disease Control and Prevention. "Teen Drivers: Fact Sheet," Injury Prevention & Control: Motor Vehicle Safety, October 2, 2012. Available online at http://www.cdc.gov/Motorvehiclesafety/Teen_Drivers/teendrivers_factsheet.html (accessed May 15, 2013).

"Children and Childrearing," Ministry of Health, Labour, and Welfare. Available online at http://www.mhlw.go.jp/english/policy/children/children-childrearing/index.html (accessed May 15, 2013).

"Eating Disorder Statistics," South Carolina Department of Mental Health, 2006. Available online at http://www.state.sc.us/dmh/anorexia/statistics.htm (accessed May 15, 2013).

"Giving Birth in Manila: The maternity ward at the Dr Jose Fabella Memorial Hospital in Manila, the busiest in the Philippines, where there is an average of 60 births a day," theguardian, 2013. Available online at http://www.theguardian.com/world/gallery/2011/jun/08/philippines-health#/?picture=375471900&index=2 (accessed May 15, 2013).

"How Americans Use Text Messaging," Pew Internet, 2013. Available online at http://pewinternet.org/Reports/2011/Cell-Phone-Texting-2011/Main-Report.aspx (accessed May 15, 2013).

Lenhart, Amanda. "Teens, Smartphones & Testing: Texting volum is up while the frequency of voice calling is down. About one in four teens say they own smartphones," Pew Internet, 2012. Available online at http://www.pewinternet.org/~/media/Files/Reports/2012/PIP_Teens_Smartphones_and_Texting.pdf (accessed May 15, 2013).

"One born every minute: the maternity unit where mothers are THREE to a bed," MailOnline. Available online at http://www.dailymail.co.uk/news/article-2001422/Busiest-maternity-ward-planet-averages-60-babies-day-mothers-bed.html (accessed May 15, 2013).

Vanderkam, Laura. "Stop Checking Your Email, Now." CNNMoney, 2013. Available online at http://management.fortune.cnn.com/2012/10/08/stop-checking-your-email-now/ (accessed May 15, 2013).

"World Earthquakes: Live Earthquake News and Highlights," World Earthquakes, 2012. http://www.world-earthquakes.com/index.php?option=ethq_prediction (accessed May 15, 2013).

SOLUTIONS

1

x	P(x)
0	0.12
1	0.18
2	0.30
3	0.15
4	0.10
5	0.10
6	0.05

Table 4.38

3 $0.10 + 0.05 = 0.15$

5 1

7 $0.35 + 0.40 + 0.10 = 0.85$

9 $1(0.15) + 2(0.35) + 3(0.40) + 4(0.10) = 0.15 + 0.70 + 1.20 + 0.40 = 2.45$

11

x	P(x)
0	0.03
1	0.04
2	0.08
3	0.85

Table 4.39

13 Let X = the number of events Javier volunteers for each month.

15

x	P(x)
0	0.05
1	0.05
2	0.10
3	0.20
4	0.25
5	0.35

Table 4.40

17 $1 - 0.05 = 0.95$

19 $0.2 + 1.2 + 2.4 + 1.6 = 5.4$

21 The values of $P(x)$ do not sum to one.

23 Let X = the number of years a physics major will spend doing post-graduate research.

25 $1 - 0.35 - 0.20 - 0.15 - 0.10 - 0.05 = 0.15$

27 $1(0.35) + 2(0.20) + 3(0.15) + 4(0.15) + 5(0.10) + 6(0.05) = 0.35 + 0.40 + 0.45 + 0.60 + 0.50 + 0.30 = 2.6$ years

29 X is the number of years a student studies ballet with the teacher.

31 $0.10 + 0.05 + 0.10 = 0.25$

33 The sum of the probabilities sum to one because it is a probability distribution.

35 $-2\left(\frac{40}{52}\right) + 30\left(\frac{12}{52}\right) = -1.54 + 6.92 = 5.38$

37 X = the number that reply "yes"

39 0, 1, 2, 3, 4, 5, 6, 7, 8

41 5.7

43 0.4151

45 X = the number of freshmen selected from the study until one replied "yes" that same-sex couples should have the right to legal marital status.

47 1,2,…

49 1.4

51 X = the number of business majors in the sample.

53 2, 3, 4, 5, 6, 7, 8, 9

55 6.26

57 0, 1, 2, 3, 4, …

59 0.0485

61 0.0214

63 X = the number of U.S. teens who die from motor vehicle injuries per day.

65 0, 1, 2, 3, 4, ...

67 No

71 The variable of interest is X, or the gain or loss, in dollars. The face cards jack, queen, and king. There are (3)(4) = 12 face cards and 52 – 12 = 40 cards that are not face cards. We first need to construct the probability distribution for X. We use the card and coin events to determine the probability for each outcome, but we use the monetary value of X to determine the expected value.

Card Event	X net gain/loss	$P(X)$
Face Card and Heads	6	$\left(\frac{12}{52}\right)\left(\frac{1}{2}\right) = \left(\frac{6}{52}\right)$
Face Card and Tails	2	$\left(\frac{12}{52}\right)\left(\frac{1}{2}\right) = \left(\frac{6}{52}\right)$
(Not Face Card) and (H or T)	–2	$\left(\frac{40}{52}\right)(1) = \left(\frac{40}{52}\right)$

Table 4.41

- Expected value = $(6)\left(\frac{6}{52}\right) + (2)\left(\frac{6}{52}\right) + (-2)\left(\frac{40}{52}\right) = -\frac{32}{52}$

- Expected value = –$0.62, rounded to the nearest cent

- If you play this game repeatedly, over a long string of games, you would expect to lose 62 cents per game, on average.

- You should not play this game to win money because the expected value indicates an expected average loss.

73
a. 0.1
b. 1.6

75

a.

Software Company	
x	$P(x)$
5,000,000	0.10
1,000,000	0.30
–1,000,000	0.60

Table 4.42

Hardware Company

x	P(x)
3,000,000	0.20
1,000,000	0.40
−1,000,00	0.40

Table 4.43

Biotech Firm

x	P(x)
6,00,000	0.10
0	0.70
−1,000,000	0.20

Table 4.44

b. $200,000; $600,000; $400,000

c. third investment because it has the lowest probability of loss

d. first investment because it has the highest probability of loss

e. second investment

77 4.85 years

79 b

81 Let X = the amount of money to be won on a ticket. The following table shows the PDF for X.

x	P(x)
0	0.969
5	$\frac{250}{10,000} = 0.025$
25	$\frac{50}{10,000} = 0.005$
100	$\frac{10}{10,000} = 0.001$

Table 4.45

Calculate the expected value of X. $0(0.969) + 5(0.025) + 25(0.005) + 100(0.001) = 0.35$ A fair price for a ticket is $0.35. Any price over $0.35 will enable the lottery to raise money.

83 X = the number of patients calling in claiming to have the flu, who actually have the flu. X = 0, 1, 2, ...25

85 0.0165

87

a. X = the number of DVDs a Video to Go customer rents

b. 0.12

   c. 0.11

   d. 0.77

89 d. 4.43

91 c

93

- X = number of questions answered correctly

- $X \sim B\left(32, \frac{1}{3}\right)$

- We are interested in MORE THAN 75% of 32 questions correct. 75% of 32 is 24. We want to find $P(x > 24)$. The event "more than 24" is the complement of "less than or equal to 24."

- Using your calculator's distribution menu: $1 - \text{binomcdf}\left(32, \frac{1}{3}, \; 24\right)$

- $P(x > 24) = 0$

- The probability of getting more than 75% of the 32 questions correct when randomly guessing is very small and practically zero.

95

   a. X = the number of college and universities that offer online offerings.

   b. 0, 1, 2, …, 13

   c. $X \sim B(13, 0.96)$

   d. 12.48

   e. 0.0135

   f. $P(x = 12) = 0.3186$ $P(x = 13) = 0.5882$ More likely to get 13.

97

   a. X = the number of fencers who do **not** use the foil as their main weapon

   b. 0, 1, 2, 3,... 25

   c. $X \sim B(25, 0.40)$

   d. 10

   e. 0.0442

   f. The probability that all 25 not use the foil is almost zero. Therefore, it would be very surprising.

99

   a. X = the number of audits in a 20-year period

   b. 0, 1, 2, …, 20

   c. $X \sim B(20, 0.02)$

   d. 0.4

   e. 0.6676

   f. 0.0071

101

   1. X = the number of matches

   2. 0, 1, 2, 3

   3. $X \sim B\left(3, \frac{1}{6}\right)$

   4. In dollars: −1, 1, 2, 3

5. $\frac{1}{2}$

6. Multiply each Y value by the corresponding X probability from the PDF table. The answer is −0.0787. You lose about eight cents, on average, per game.

7. The house has the advantage.

103

a. $X \sim B(15, 0.281)$

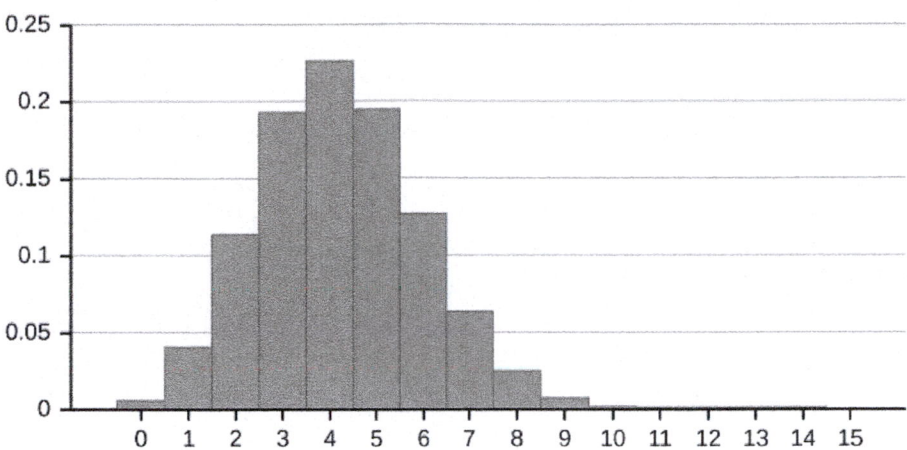

Figure 4.10

b.   i.   Mean = $\mu = np = 15(0.281) = 4.215$

    ii.   Standard Deviation = $\sigma = \sqrt{npq} = \sqrt{15(0.281)(0.719)} = 1.7409$

c. $P(x > 5) = 1 - P(x \le 5) = 1 - \text{binomcdf}(15, 0.281, 5) = 1 - 0.7754 = 0.2246$
$P(x = 3) = \text{binompdf}(15, 0.281, 3) = 0.1927$
$P(x = 4) = \text{binompdf}(15, 0.281, 4) = 0.2259$
It is more likely that four people are literate that three people are.

105

a. X = the number of adults in America who are surveyed until one says he or she will watch the Super Bowl.

b. $X \sim G(0.40)$

c. 2.5

d. 0.0187

e. 0.2304

107

a. X = the number of pages that advertise footwear

b. X takes on the values 0, 1, 2, ..., 20

c. $X \sim B(20, \frac{29}{192})$

d. 3.02

e. No

f. 0.9997

g. X = the number of pages we must survey until we find one that advertises footwear. $X \sim G(\frac{29}{192})$

h. 0.3881

i. 6.6207 pages

109 0, 1, 2, and 3

111

a. $X \sim G(0.25)$

b. i. Mean $= \mu = \frac{1}{p} = \frac{1}{0.25} = 4$

ii. Standard Deviation $= \sigma = \sqrt{\frac{1-p}{p^2}} = \sqrt{\frac{1-0.25}{0.25^2}} \approx 3.4641$

c. $P(x = 10) = \text{geometpdf}(0.25, 10) = 0.0188$

d. $P(x = 20) = \text{geometpdf}(0.25, 20) = 0.0011$

e. $P(x \leq 5) = \text{geometcdf}(0.25, 5) = 0.7627$

113

a. X = the number of pages that advertise footwear

b. 0, 1, 2, 3, ..., 20

c. $X \sim H(29, 163, 20)$; $r = 29$, $b = 163$, $n = 20$

d. 3.03

e. 1.5197

115

a. X = the number of Patriots picked

b. 0, 1, 2, 3, 4

c. $X \sim H(4, 8, 9)$

d. Without replacement

117

a. $X \sim P(5.5)$; $\mu = 5.5$; $\sigma = \sqrt{5.5} \approx 2.3452$

b. $P(x \leq 6) = \text{poissoncdf}(5.5, 6) \approx 0.6860$

c. There is a 15.7% probability that the law staff will receive more calls than they can handle.

d. $P(x > 8) = 1 - P(x \leq 8) = 1 - \text{poissoncdf}(5.5, 8) \approx 1 - 0.8944 = 0.1056$

119 Let X = the number of defective bulbs in a string. Using the Poisson distribution:

- $\mu = np = 100(0.03) = 3$

- $X \sim P(3)$

- $P(x \leq 4) = \text{poissoncdf}(3, 4) \approx 0.8153$

Using the binomial distribution:

- $X \sim B(100, 0.03)$

- $P(x \leq 4) = \text{binomcdf}(100, 0.03, 4) \approx 0.8179$

The Poisson approximation is very good—the difference between the probabilities is only 0.0026.

121

a. X = the number of children for a Spanish woman

b. 0, 1, 2, 3,...

c. $X \sim P(1.47)$

d. 0.2299

    e. 0.5679

    f. 0.4321

123

    a. X = the number of fortune cookies that have an extra fortune

    b. 0, 1, 2, 3,... 144

    c. $X \sim B(144, 0.03)$ or $P(4.32)$

    d. 4.32

    e. 0.0124 or 0.0133

    f. 0.6300 or 0.6264

    g. As n gets larger, the probabilities get closer together.

125

    a. X = the number of people audited in one year

    b. 0, 1, 2, ..., 100

    c. $X \sim P(2)$

    d. 2

    e. 0.1353

    f. 0.3233

127

    a. X = the number of shell pieces in one cake

    b. 0, 1, 2, 3,...

    c. $X \sim P(1.5)$

    d. 1.5

    e. 0.2231

    f. 0.0001

    g. Yes

129 d

5 | CONTINUOUS RANDOM VARIABLES

Figure 5.1 The heights of these radish plants are continuous random variables. (Credit: Rev Stan)

Introduction

Chapter Objectives
By the end of this chapter, the student should be able to: • Recognize and understand continuous probability density functions in general.   • Recognize the uniform probability distribution and apply it appropriately.   • Recognize the exponential probability distribution and apply it appropriately.

Continuous random variables have many applications. Baseball batting averages, IQ scores, the length of time a long distance telephone call lasts, the amount of money a person carries, the length of time a computer chip lasts, and SAT scores are just a few. The field of reliability depends on a variety of continuous random variables.

NOTE

The values of discrete and continuous random variables can be ambiguous. For example, if X is equal to the number of miles (to the nearest mile) you drive to work, then X is a discrete random variable. You count the miles. If X is the distance you drive to work, then you measure values of X and X is a continuous random variable. For a second example, if X is equal to the number of books in a backpack, then X is a discrete random variable. If X is the weight of a book, then X is a continuous random variable because weights are measured. How the random variable is defined is very important.

Properties of Continuous Probability Distributions

The graph of a continuous probability distribution is a curve. Probability is represented by area under the curve.

The curve is called the **probability density function** (abbreviated as **pdf**). We use the symbol $f(x)$ to represent the curve. $f(x)$ is the function that corresponds to the graph; we use the density function $f(x)$ to draw the graph of the probability distribution.

Area under the curve is given by a different function called the **cumulative distribution function** (abbreviated as **cdf**). The cumulative distribution function is used to evaluate probability as area.

- The outcomes are measured, not counted.
- The entire area under the curve and above the x-axis is equal to one.
- Probability is found for intervals of x values rather than for individual x values.
- $P(c < x < d)$ is the probability that the random variable X is in the interval between the values c and d. $P(c < x < d)$ is the area under the curve, above the x-axis, to the right of c and the left of d.
- $P(x = c) = 0$ The probability that x takes on any single individual value is zero. The area below the curve, above the x-axis, and between $x = c$ and $x = c$ has no width, and therefore no area (area = 0). Since the probability is equal to the area, the probability is also zero.
- $P(c < x < d)$ is the same as $P(c \le x \le d)$ because probability is equal to area.

We will find the area that represents probability by using geometry, formulas, technology, or probability tables. In general, calculus is needed to find the area under the curve for many probability density functions. When we use formulas to find the area in this textbook, the formulas were found by using the techniques of integral calculus. However, because most students taking this course have not studied calculus, we will not be using calculus in this textbook.

There are many continuous probability distributions. When using a continuous probability distribution to model probability, the distribution used is selected to model and fit the particular situation in the best way.

In this chapter and the next, we will study the uniform distribution, the exponential distribution, and the normal distribution. The following graphs illustrate these distributions.

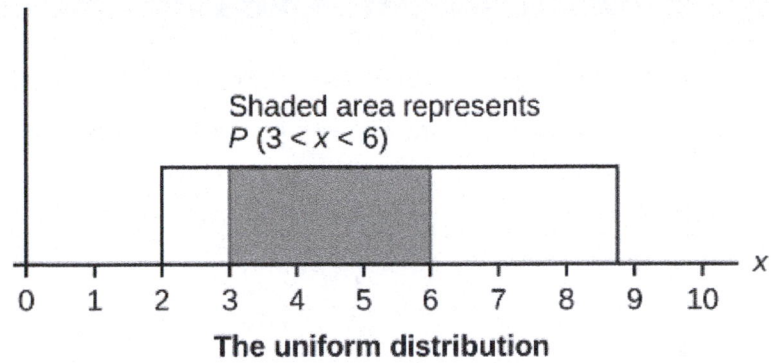

Figure 5.2 The graph shows a Uniform Distribution with the area between $x = 3$ and $x = 6$ shaded to represent the probability that the value of the random variable X is in the interval between three and six.

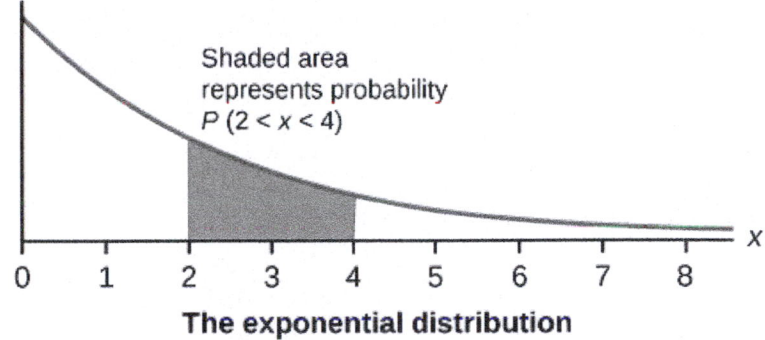

Figure 5.3 The graph shows an Exponential Distribution with the area between $x = 2$ and $x = 4$ shaded to represent the probability that the value of the random variable X is in the interval between two and four.

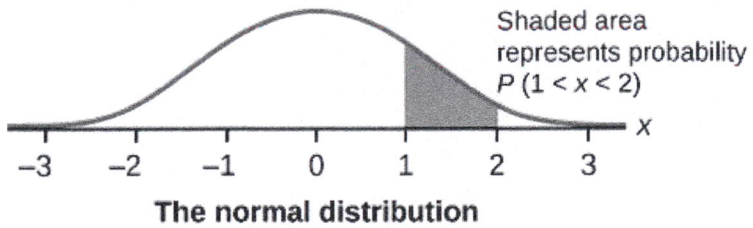

Figure 5.4 The graph shows the Standard Normal Distribution with the area between $x = 1$ and $x = 2$ shaded to represent the probability that the value of the random variable X is in the interval between one and two.

5.1 | Continuous Probability Functions

We begin by defining a continuous probability density function. We use the function notation $f(x)$. Intermediate algebra may have been your first formal introduction to functions. In the study of probability, the functions we study are special. We define the function $f(x)$ so that the area between it and the x-axis is equal to a probability. Since the maximum probability is one, the maximum area is also one. **For continuous probability distributions, PROBABILITY = AREA.**

Example 5.1

Consider the function $f(x) = \frac{1}{20}$ for $0 \le x \le 20$. x = a real number. The graph of $f(x) = \frac{1}{20}$ is a horizontal line. However, since $0 \le x \le 20$, $f(x)$ is restricted to the portion between $x = 0$ and $x = 20$, inclusive.

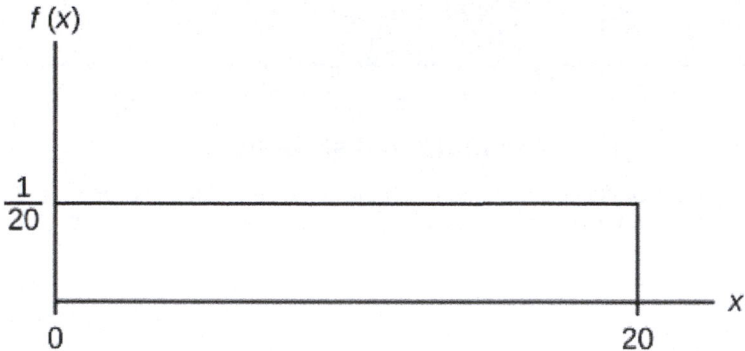

Figure 5.5

$f(x) = \frac{1}{20}$ **for** $0 \le x \le 20$.

The graph of $f(x) = \frac{1}{20}$ is a horizontal line segment when $0 \le x \le 20$.

The area between $f(x) = \frac{1}{20}$ where $0 \le x \le 20$ and the x-axis is the area of a rectangle with base = 20 and height $= \frac{1}{20}$.

$$\text{AREA} = 20\left(\frac{1}{20}\right) = 1$$

Suppose we want to find the area between $f(x) = \frac{1}{20}$ **and the x-axis where** $0 < x < 2$.

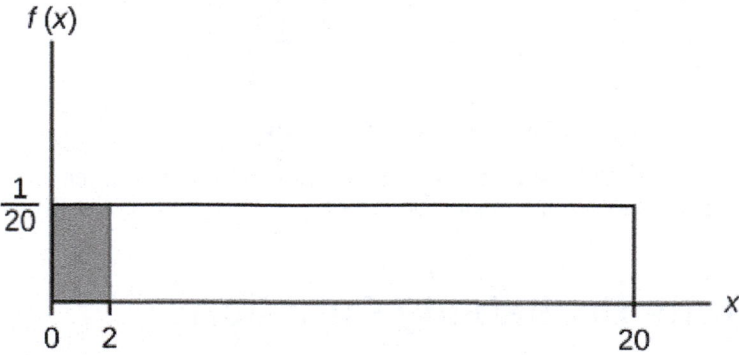

Figure 5.6

$$\text{AREA} = (2 - 0)\left(\frac{1}{20}\right) = 0.1$$

$(2 - 0) = 2 =$ base of a rectangle

area of a rectangle = (base)(height).

The area corresponds to a probability. The probability that x is between zero and two is 0.1, which can be written mathematically as $P(0 < x < 2) = P(x < 2) = 0.1$.

Suppose we want to find the area between $f(x) = \frac{1}{20}$ and the x-axis where $4 < x < 15$.

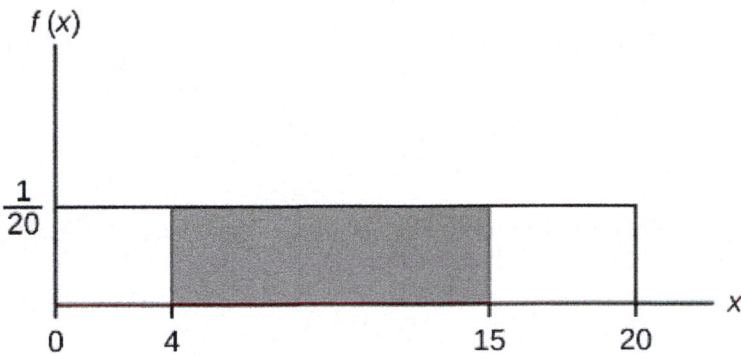

Figure 5.7

$\text{AREA} = (15 - 4)\left(\frac{1}{20}\right) = 0.55$

$\text{AREA} = (15 - 4)\left(\frac{1}{20}\right) = 0.55$

$(15 - 4) = 11 = \text{the base of a rectangle}$

The area corresponds to the probability $P(4 < x < 15) = 0.55$.

Suppose we want to find $P(x = 15)$. On an x-y graph, $x = 15$ is a vertical line. A vertical line has no width (or zero width). Therefore, $P(x = 15) = (\text{base})(\text{height}) = (0)\left(\frac{1}{20}\right) = 0$

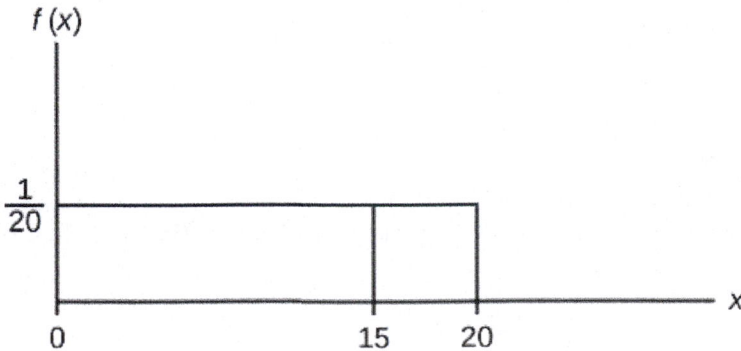

Figure 5.8

$P(X \leq x)$ (can be written as $P(X < x)$ for continuous distributions) is called the cumulative distribution function or CDF. Notice the "less than or equal to" symbol. We can use the CDF to calculate $P(X > x)$. The CDF gives "area to the left" and $P(X > x)$ gives "area to the right." We calculate $P(X > x)$ for continuous distributions as follows: $P(X > x) = 1 - P(X < x)$.

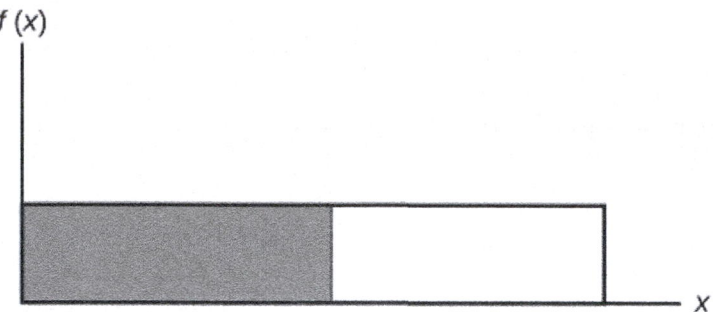

Figure 5.9

Label the graph with $f(x)$ and x. Scale the x and y axes with the maximum x and y values. $f(x) = \frac{1}{20}$, $0 \le x \le 20$.

To calculate the probability that x is between two values, look at the following graph. Shade the region between $x = 2.3$ and $x = 12.7$. Then calculate the shaded area of a rectangle.

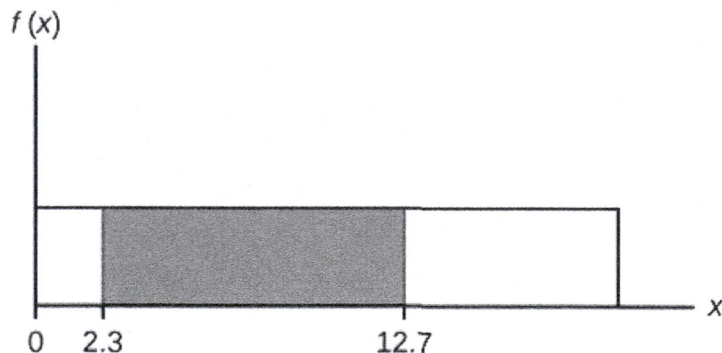

Figure 5.10

$$P(2.3 < x < 12.7) = (\text{base})(\text{height}) = (12.7 - 2.3)\left(\frac{1}{20}\right) = 0.52$$

Try It Σ

5.1 Consider the function $f(x) = \frac{1}{8}$ for $0 \le x \le 8$. Draw the graph of $f(x)$ and find $P(2.5 < x < 7.5)$.

5.2 | The Uniform Distribution

The uniform distribution is a continuous probability distribution and is concerned with events that are equally likely to occur. When working out problems that have a uniform distribution, be careful to note if the data is inclusive or exclusive.

Example 5.2

The data in **Table 5.1** are 55 smiling times, in seconds, of an eight-week-old baby.

10.4	19.6	18.8	13.9	17.8	16.8	21.6	17.9	12.5	11.1	4.9
12.8	14.8	22.8	20.0	15.9	16.3	13.4	17.1	14.5	19.0	22.8
1.3	0.7	8.9	11.9	10.9	7.3	5.9	3.7	17.9	19.2	9.8
5.8	6.9	2.6	5.8	21.7	11.8	3.4	2.1	4.5	6.3	10.7
8.9	9.4	9.4	7.6	10.0	3.3	6.7	7.8	11.6	13.8	18.6

Table 5.1

The sample mean = 11.49 and the sample standard deviation = 6.23.

We will assume that the smiling times, in seconds, follow a uniform distribution between zero and 23 seconds, inclusive. This means that any smiling time from zero to and including 23 seconds is **equally likely**. The histogram that could be constructed from the sample is an empirical distribution that closely matches the theoretical uniform distribution.

Let X = length, in seconds, of an eight-week-old baby's smile.

The notation for the uniform distribution is

$X \sim U(a, b)$ where a = the lowest value of x and b = the highest value of x.

The probability density function is $f(x) = \dfrac{1}{b-a}$ for $a \leq x \leq b$.

For this example, $X \sim U(0, 23)$ and $f(x) = \dfrac{1}{23-0}$ for $0 \leq X \leq 23$.

Formulas for the theoretical mean and standard deviation are

$$\mu = \frac{a+b}{2} \text{ and } \sigma = \sqrt{\frac{(b-a)^2}{12}}$$

For this problem, the theoretical mean and standard deviation are

$$\mu = \frac{0+23}{2} = 11.50 \text{ seconds and } \sigma = \sqrt{\frac{(23-0)^2}{12}} = 6.64 \text{ seconds.}$$

Notice that the theoretical mean and standard deviation are close to the sample mean and standard deviation in this example.

Try It Σ

5.2 The data that follow are the number of passengers on 35 different charter fishing boats. The sample mean = 7.9 and the sample standard deviation = 4.33. The data follow a uniform distribution where all values between and including zero and 14 are equally likely. State the values of a and b. Write the distribution in proper notation, and calculate the theoretical mean and standard deviation.

1	12	4	10	4	14	11
7	11	4	13	2	4	6
3	10	0	12	6	9	10
5	13	4	10	14	12	11
6	10	11	0	11	13	2

Table 5.2

Example 5.3

a. Refer to **Example 5.2**. What is the probability that a randomly chosen eight-week-old baby smiles between two and 18 seconds?

Solution 5.3

a. Find $P(2 < x < 18)$.

$P(2 < x < 18) = \text{(base)(height)} = (18 - 2)\left(\frac{1}{23}\right) = \left(\frac{16}{23}\right)$.

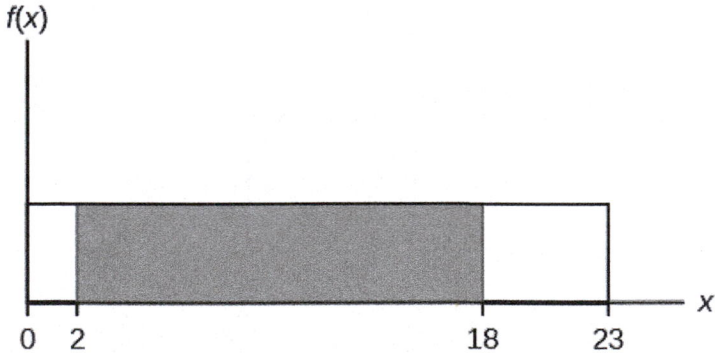

Figure 5.11

b. Find the 90th percentile for an eight-week-old baby's smiling time.

Solution 5.3

b. Ninety percent of the smiling times fall below the 90th percentile, k, so $P(x < k) = 0.90$

$P(x < k) = 0.90$

$\text{(base)(height)} = 0.90$

$$(k - 0)\left(\frac{1}{23}\right) = 0.90$$

$$k = (23)(0.90) = 20.7$$

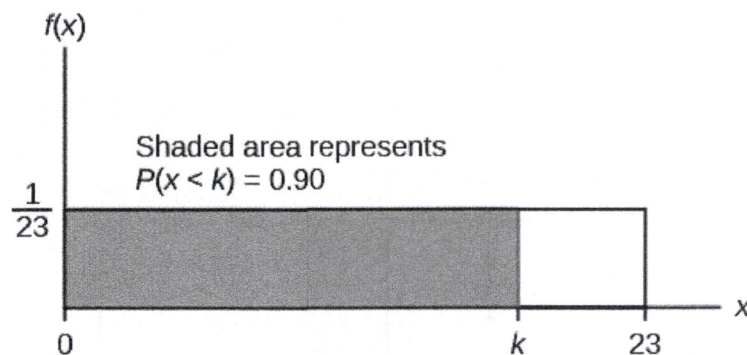

Figure 5.12

c. Find the probability that a random eight-week-old baby smiles more than 12 seconds **KNOWING** that the baby smiles **MORE THAN EIGHT SECONDS**.

Solution 5.3

c. This probability question is a **conditional**. You are asked to find the probability that an eight-week-old baby smiles more than 12 seconds when you **already know** the baby has smiled for more than eight seconds.

Find $P(x > 12 | x > 8)$ There are two ways to do the problem. **For the first way**, use the fact that this is a **conditional** and changes the sample space. The graph illustrates the new sample space. You already know the baby smiled more than eight seconds.

Write a new $f(x)$: $f(x) = \dfrac{1}{23 - 8} = \dfrac{1}{15}$

for $8 < x < 23$

$$P(x > 12 | x > 8) = (23 - 12)\left(\frac{1}{15}\right) = \left(\frac{11}{15}\right)$$

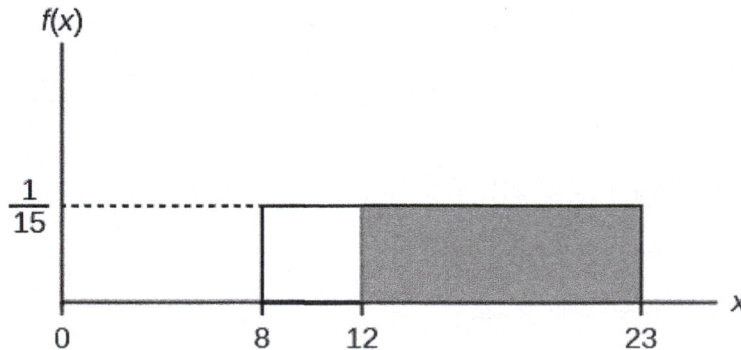

Figure 5.13

For the second way, use the conditional formula from **Probability Topics** with the original distribution $X \sim U$ (0, 23):

$$P(A|B) = \frac{P(A \text{ AND } B)}{P(B)}$$

For this problem, A is $(x > 12)$ and B is $(x > 8)$.

So, $P(x > 12|x > 8) = \frac{(x > 12 \text{ AND } x > 8)}{P(x > 8)} = \frac{P(x > 12)}{P(x > 8)} = \frac{\frac{11}{23}}{\frac{15}{23}} = \frac{11}{15}$

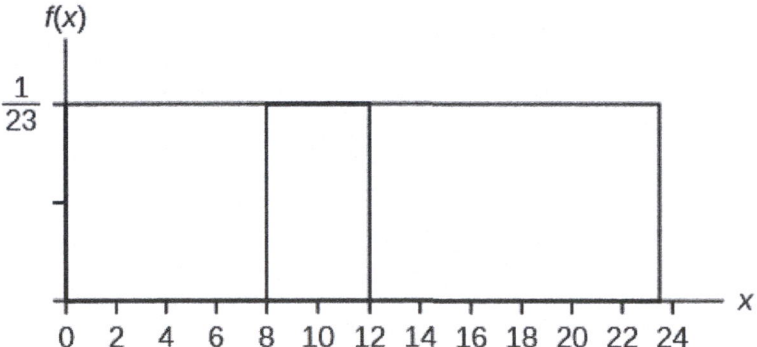

Figure 5.14

Try It Σ

5.3 A distribution is given as $X \sim U(0, 20)$. What is $P(2 < x < 18)$? Find the 90th percentile.

Example 5.4

The amount of time, in minutes, that a person must wait for a bus is uniformly distributed between zero and 15 minutes, inclusive.

a. What is the probability that a person waits fewer than 12.5 minutes?

Solution 5.4

a. Let X = the number of minutes a person must wait for a bus. $a = 0$ and $b = 15$. $X \sim U(0, 15)$. Write the probability density function. $f(x) = \frac{1}{15 - 0} = \frac{1}{15}$ for $0 \leq x \leq 15$.

Find $P(x < 12.5)$. Draw a graph.

$$P(x < k) = (\text{base})(\text{height}) = (12.5 - 0)\left(\frac{1}{15}\right) = 0.8333$$

The probability a person waits less than 12.5 minutes is 0.8333.

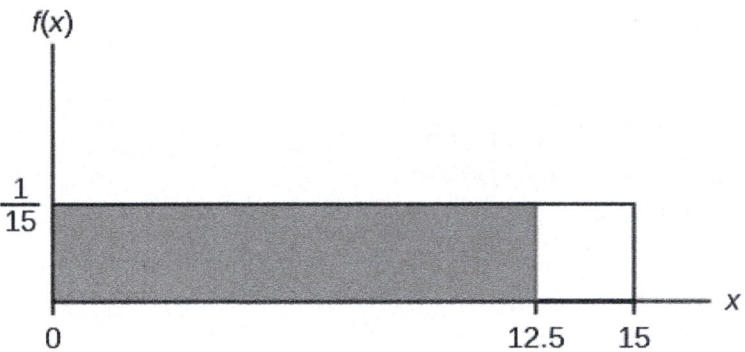

Figure 5.15

b. On the average, how long must a person wait? Find the mean, μ, and the standard deviation, σ.

Solution 5.4

b. $\mu = \dfrac{a + b}{2} = \dfrac{15 + 0}{2}$ = 7.5. On the average, a person must wait 7.5 minutes.

$\sigma = \sqrt{\dfrac{(b - a)^2}{12}} = \sqrt{\dfrac{(15 - 0)^2}{12}}$ = 4.3. The Standard deviation is 4.3 minutes.

c. Ninety percent of the time, the time a person must wait falls below what value?

This asks for the 90th percentile.

Solution 5.4

c. Find the 90th percentile. Draw a graph. Let k = the 90th percentile.

$P(x < k) = (\text{base})(\text{height}) = (k - 0)(\frac{1}{15})$

$0.90 = (k)\left(\frac{1}{15}\right)$

$k = (0.90)(15) = 13.5$

k is sometimes called a critical value.

The 90th percentile is 13.5 minutes. Ninety percent of the time, a person must wait at most 13.5 minutes.

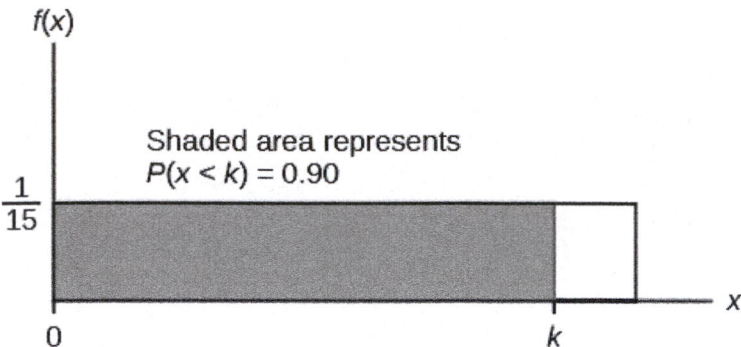

Figure 5.16

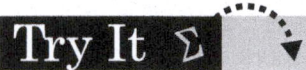

5.4 The total duration of baseball games in the major league in the 2011 season is uniformly distributed between 447 hours and 521 hours inclusive.

a. Find *a* and *b* and describe what they represent.

b. Write the distribution.

c. Find the mean and the standard deviation.

d. What is the probability that the duration of games for a team for the 2011 season is between 480 and 500 hours?

e. What is the 65th percentile for the duration of games for a team for the 2011 season?

Example 5.5

Suppose the time it takes a nine-year old to eat a donut is between 0.5 and 4 minutes, inclusive. Let X = the time, in minutes, it takes a nine-year old child to eat a donut. Then $X \sim U(0.5, 4)$.

a. The probability that a randomly selected nine-year old child eats a donut in at least two minutes is _____.

Solution 5.5
a. 0.5714

b. Find the probability that a different nine-year old child eats a donut in more than two minutes given that the child has already been eating the donut for more than 1.5 minutes.

The second question has a **conditional probability**. You are asked to find the probability that a nine-year old child eats a donut in more than two minutes given that the child has already been eating the donut for more than 1.5 minutes. Solve the problem two different ways (see **Example 5.2**). You must reduce the sample space. **First way**: Since you know the child has already been eating the donut for more than 1.5 minutes, you are no longer starting at *a* = 0.5 minutes. Your starting point is 1.5 minutes.

Write a new $f(x)$:

$$f(x) = \frac{1}{4 - 1.5} = \frac{2}{5} \text{ for } 1.5 \le x \le 4.$$

Find $P(x > 2 | x > 1.5)$. Draw a graph.

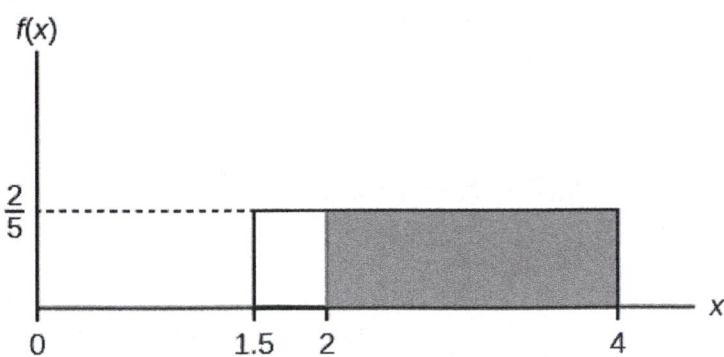

Figure 5.17

$P(x > 2|x > 1.5) = (\text{base})(\text{new height}) = (4 - 2)\left(\frac{2}{5}\right) = ?$

Solution 5.5

b. $\frac{4}{5}$

The probability that a nine-year old child eats a donut in more than two minutes given that the child has already been eating the donut for more than 1.5 minutes is $\frac{4}{5}$.

Second way: Draw the original graph for $X \sim U (0.5, 4)$. Use the conditional formula

$P(x > 2|x > 1.5) = \dfrac{P(x > 2 \text{ AND } x > 1.5)}{P(x > 1.5)} = \dfrac{P(x > 2)}{P(x > 1.5)} = \dfrac{\frac{2}{3.5}}{\frac{2.5}{3.5}} = 0.8 = \frac{4}{5}$

Try It Σ

5.5 Suppose the time it takes a student to finish a quiz is uniformly distributed between six and 15 minutes, inclusive. Let $X = $ the time, in minutes, it takes a student to finish a quiz. Then $X \sim U (6, 15)$.

Find the probability that a randomly selected student needs at least eight minutes to complete the quiz. Then find the probability that a different student needs at least eight minutes to finish the quiz given that she has already taken more than seven minutes.

Example 5.6

Ace Heating and Air Conditioning Service finds that the amount of time a repairman needs to fix a furnace is uniformly distributed between 1.5 and four hours. Let $x = $ the time needed to fix a furnace. Then $x \sim U (1.5, 4)$.

a. Find the probability that a randomly selected furnace repair requires more than two hours.

b. Find the probability that a randomly selected furnace repair requires less than three hours.

c. Find the 30th percentile of furnace repair times.

d. The longest 25% of furnace repair times take at least how long? (In other words: find the minimum time for the longest 25% of repair times.) What percentile does this represent?

e. Find the mean and standard deviation

Solution 5.6

a. To find $f(x)$: $f(x) = \dfrac{1}{4 - 1.5} = \dfrac{1}{2.5}$ so $f(x) = 0.4$

$P(x > 2) = (\text{base})(\text{height}) = (4 - 2)(0.4) = 0.8$

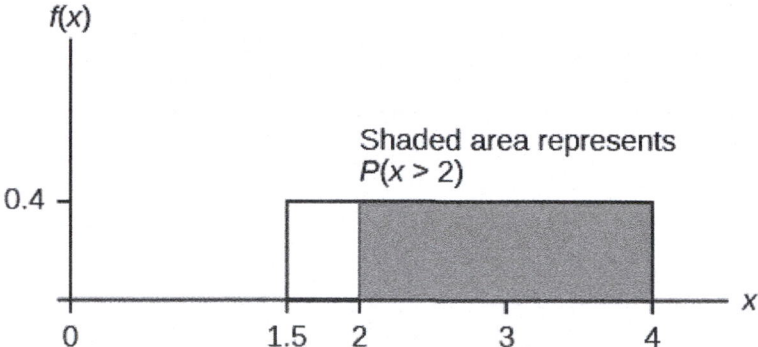

Figure 5.18 Uniform Distribution between 1.5 and four with shaded area between two and four representing the probability that the repair time x is greater than two

Solution 5.6

b. $P(x < 3) = (\text{base})(\text{height}) = (3 - 1.5)(0.4) = 0.6$

The graph of the rectangle showing the entire distribution would remain the same. However the graph should be shaded between $x = 1.5$ and $x = 3$. Note that the shaded area starts at $x = 1.5$ rather than at $x = 0$; since $X \sim U(1.5, 4)$, x can not be less than 1.5.

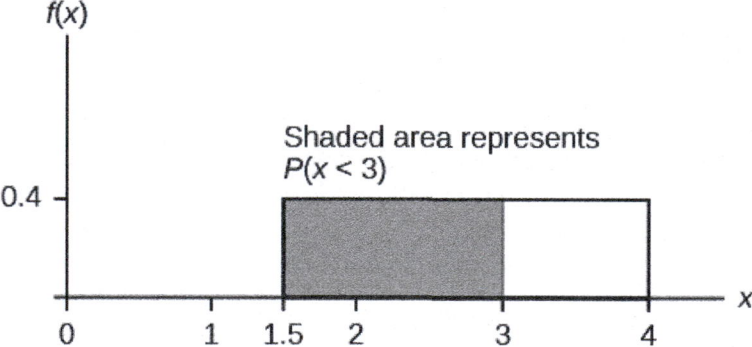

Figure 5.19 Uniform Distribution between 1.5 and four with shaded area between 1.5 and three representing the probability that the repair time x is less than three

Solution 5.6

c.

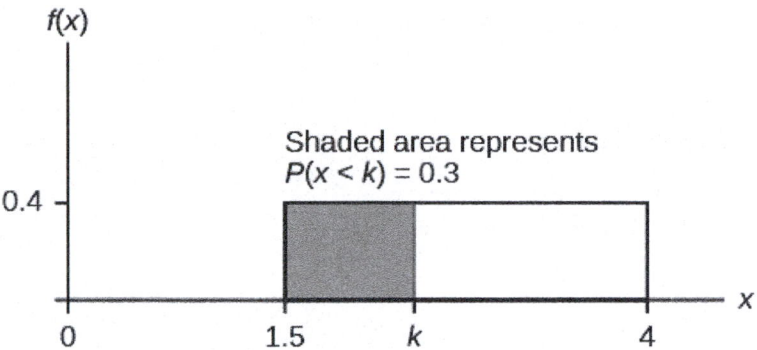

Figure 5.20 Uniform Distribution between 1.5 and 4 with an area of 0.30 shaded to the left, representing the shortest 30% of repair times.

$P (x < k) = 0.30$

$P(x < k) = (\text{base})(\text{height}) = (k - 1.5)(0.4)$

0.3 = (k − 1.5) (0.4); Solve to find k:

$0.75 = k - 1.5$, obtained by dividing both sides by 0.4

k = 2.25 , obtained by adding 1.5 to both sides

The 30th percentile of repair times is 2.25 hours. 30% of repair times are 2.5 hours or less.

Solution 5.6

d.

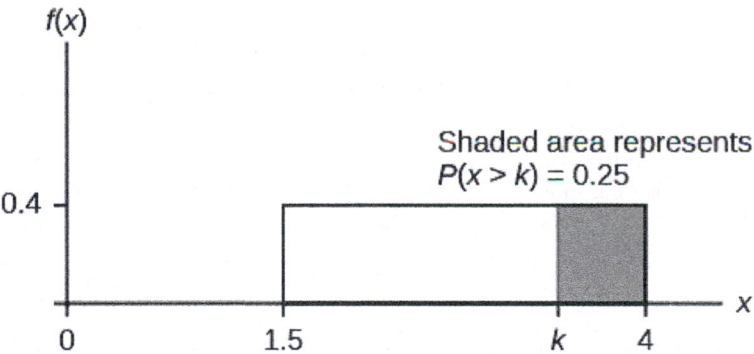

Figure 5.21 Uniform Distribution between 1.5 and 4 with an area of 0.25 shaded to the right representing the longest 25% of repair times.

$P(x > k) = 0.25$

$P(x > k) = (\text{base})(\text{height}) = (4 - k)(0.4)$

0.25 = (4 − k)(0.4); Solve for k:

$0.625 = 4 - k$,

obtained by dividing both sides by 0.4

$-3.375 = -k$,

obtained by subtracting four from both sides: **k = 3.375**

The longest 25% of furnace repairs take at least 3.375 hours (3.375 hours or longer).

Note: Since 25% of repair times are 3.375 hours or longer, that means that 75% of repair times are 3.375 hours or less. 3.375 hours is the **75th percentile** of furnace repair times.

Solution 5.6

e. $\mu = \frac{a+b}{2}$ and $\sigma = \sqrt{\frac{(b-a)^2}{12}}$

$\mu = \frac{1.5+4}{2} = 2.75$ hours and $\sigma = \sqrt{\frac{(4-1.5)^2}{12}} = 0.7217$ hours

Try It Σ

5.6 The amount of time a service technician needs to change the oil in a car is uniformly distributed between 11 and 21 minutes. Let X = the time needed to change the oil on a car.

a. Write the random variable X in words. X = _____.

b. Write the distribution.

c. Graph the distribution.

d. Find $P(x > 19)$.

e. Find the 50th percentile.

5.3 | The Exponential Distribution

The **exponential distribution** is often concerned with the amount of time until some specific event occurs. For example, the amount of time (beginning now) until an earthquake occurs has an exponential distribution. Other examples include the length, in minutes, of long distance business telephone calls, and the amount of time, in months, a car battery lasts. It can be shown, too, that the value of the change that you have in your pocket or purse approximately follows an exponential distribution.

Values for an exponential random variable occur in the following way. There are fewer large values and more small values. For example, the amount of money customers spend in one trip to the supermarket follows an exponential distribution. There are more people who spend small amounts of money and fewer people who spend large amounts of money.

The exponential distribution is widely used in the field of reliability. Reliability deals with the amount of time a product lasts.

Example 5.7

Let X = amount of time (in minutes) a postal clerk spends with his or her customer. The time is known to have an exponential distribution with the average amount of time equal to four minutes.

X is a **continuous random variable** since time is measured. It is given that μ = 4 minutes. To do any calculations, you must know m, the decay parameter.

$m = \frac{1}{\mu}$. Therefore, $m = \frac{1}{4} = 0.25$.

The standard deviation, σ, is the same as the mean. $\mu = \sigma$

The distribution notation is $X \sim Exp(m)$. Therefore, $X \sim Exp(0.25)$.

The probability density function is $f(x) = me^{-mx}$. The number $e = 2.71828182846...$ It is a number that is used often in mathematics. Scientific calculators have the key "e^x." If you enter one for x, the calculator will display the value e.

The curve is:

$f(x) = 0.25e^{-0.25x}$ where x is at least zero and $m = 0.25$.

For example, $f(5) = 0.25e^{-(0.25)(5)} = 0.072$. The postal clerk spends five minutes with the customers.

The graph is as follows:

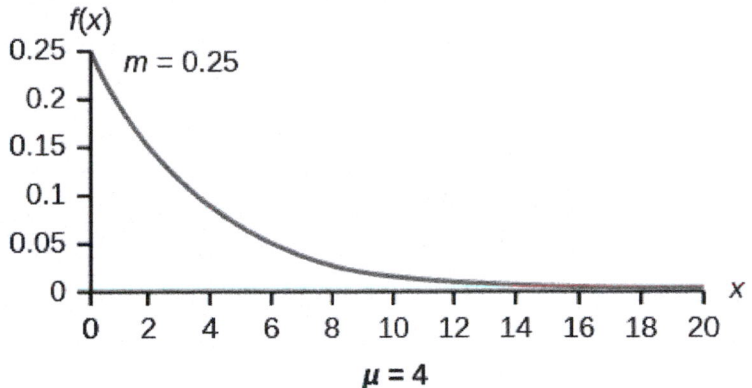

Figure 5.22

Notice the graph is a declining curve. When $x = 0$,

$f(x) = 0.25e^{(-0.25)(0)} = (0.25)(1) = 0.25 = m$. The maximum value on the y-axis is m.

Try It Σ

5.7 The amount of time spouses shop for anniversary cards can be modeled by an exponential distribution with the average amount of time equal to eight minutes. Write the distribution, state the probability density function, and graph the distribution.

Example 5.8

a. Using the information in **Exercise 5.0**, find the probability that a clerk spends four to five minutes with a randomly selected customer.

Solution 5.8

a. Find $P(4 < x < 5)$.
The **cumulative distribution function (CDF)** gives the area to the left.
$P(x < x) = 1 - e^{-mx}$
$P(x < 5) = 1 - e(-0.25)(5) = 0.7135$ and $P(x < 4) = 1 - e^{(-0.25)(4)} = 0.6321$

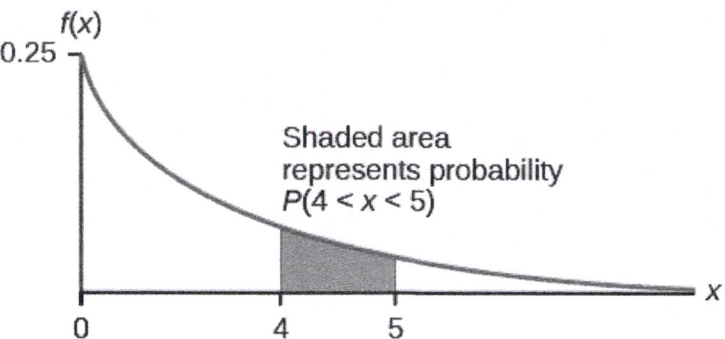

Figure 5.23

NOTE

You can do these calculations easily on a calculator.

The probability that a postal clerk spends four to five minutes with a randomly selected customer is $P(4 < x < 5)$ = $P(x < 5) - P(x < 4) = 0.7135 - 0.6321 = 0.0814$.

 Using the TI-83, 83+, 84, 84+ Calculator

On the home screen, enter $(1 - e^{(-0.25*5)}) - (1 - e^{(-0.25*4)})$ or enter $e^{(-0.25*4)} - e^{(-0.25*5)}$.

b. Half of all customers are finished within how long? (Find the 50th percentile)

Solution 5.8

b. Find the 50th percentile.

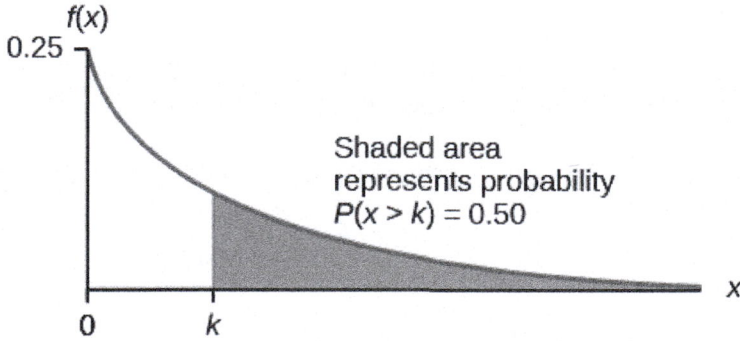

Figure 5.24

$P(x < k) = 0.50$, $k = 2.8$ minutes (calculator or computer)

Half of all customers are finished within 2.8 minutes.

You can also do the calculation as follows:

$P(x < k) = 0.50$ and $P(x < k) = 1 - e^{-0.25k}$

Therefore, $0.50 = 1 - e^{-0.25k}$ and $e^{-0.25k} = 1 - 0.50 = 0.5$

Take natural logs: $ln(e^{-0.25k}) = ln(0.50)$. So, $-0.25k = ln(0.50)$

Solve for k: $k = \dfrac{ln(0.50)}{-0.25} = 2.8$ minutes. The calculator simplifies the calculation for percentile k. See the following two notes.

NOTE

A formula for the percentile k is $k = \dfrac{ln(1 - AreaToTheLeft)}{-m}$ where ln is the natural log.

 Using the TI-83, 83+, 84, 84+ Calculator

On the home screen, enter $ln(1 - 0.50)/-0.25$. Press the (-) for the negative.

c. Which is larger, the mean or the median?

Solution 5.8

c. From part b, the median or 50th percentile is 2.8 minutes. The theoretical mean is four minutes. The mean is larger.

 **Try It**

5.8 The number of days ahead travelers purchase their airline tickets can be modeled by an exponential distribution with the average amount of time equal to 15 days. Find the probability that a traveler will purchase a ticket fewer than ten days in advance. How many days do half of all travelers wait?

Collaborative Exercise

Have each class member count the change he or she has in his or her pocket or purse. Your instructor will record the amounts in dollars and cents. Construct a histogram of the data taken by the class. Use five intervals. Draw a smooth curve through the bars. The graph should look approximately exponential. Then calculate the mean.

Let X = the amount of money a student in your class has in his or her pocket or purse.

The distribution for X is approximately exponential with mean, μ = _____ and m = _____. The standard deviation, σ = _____.

Draw the appropriate exponential graph. You should label the x– and y–axes, the decay rate, and the mean. Shade the area that represents the probability that one student has less than $.40 in his or her pocket or purse. (Shade $P(x < 0.40)$).

Example 5.9

On the average, a certain computer part lasts ten years. The length of time the computer part lasts is exponentially distributed.

a. What is the probability that a computer part lasts more than 7 years?

Solution 5.9

a. Let x = the amount of time (in years) a computer part lasts.

$\mu = 10$ so $m = \frac{1}{\mu} = \frac{1}{10} = 0.1$

Find $P(x > 7)$. Draw the graph.

$P(x > 7) = 1 - P(x < 7)$.

Since $P(X < x) = 1 - e^{-mx}$ then $P(X > x) = 1 - (1 - e^{-mx}) = e^{-mx}$

$P(x > 7) = e^{(-0.1)(7)} = 0.4966$. The probability that a computer part lasts more than seven years is 0.4966.

Using the TI-83, 83+, 84, 84+ Calculator

On the home screen, enter e^(-.1*7).

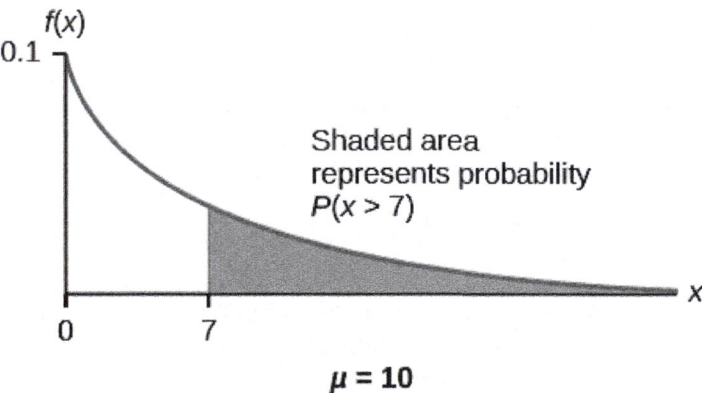

$\mu = 10$

Figure 5.25

b. On the average, how long would five computer parts last if they are used one after another?

Solution 5.9

b. On the average, one computer part lasts ten years. Therefore, five computer parts, if they are used one right after the other would last, on the average, $(5)(10) = 50$ years.

c. Eighty percent of computer parts last at most how long?

Solution 5.9

c. Find the 80th percentile. Draw the graph. Let k = the 80th percentile.

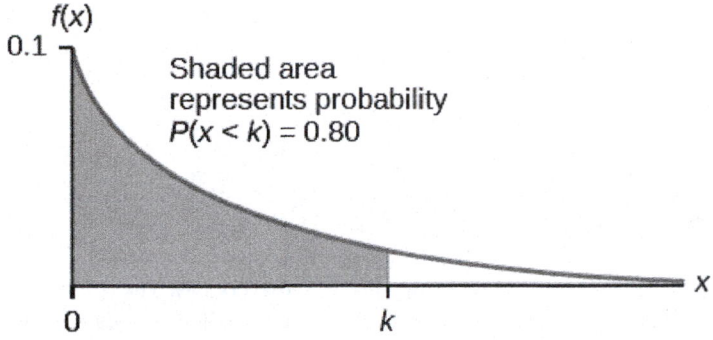

Figure 5.26

Solve for k: $k = \dfrac{\ln(1 - 0.80)}{-0.1} = 16.1$ years

Eighty percent of the computer parts last at most 16.1 years.

 Using the TI-83, 83+, 84, 84+ Calculator

On the home screen, enter $\dfrac{\ln(1 - 0.80)}{-0.1}$

d. What is the probability that a computer part lasts between nine and 11 years?

Solution 5.9

d. Find $P(9 < x < 11)$. Draw the graph.

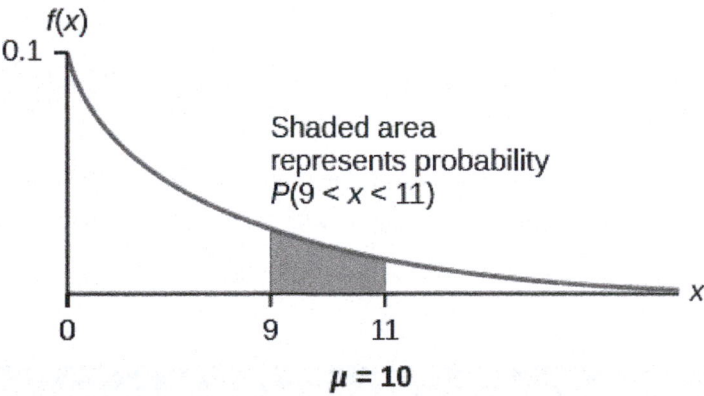

Figure 5.27

$P(9 < x < 11) = P(x < 11) - P(x < 9) = (1 - e^{(-0.1)(11)}) - (1 - e^{(-0.1)(9)}) = 0.6671 - 0.5934 = 0.0737$. The probability that a computer part lasts between nine and 11 years is 0.0737.

Using the TI-83, 83+, 84, 84+ Calculator

On the home screen, enter $e^{\wedge}(-0.1*9) - e^{\wedge}(-0.1*11)$.

Try It Σ

5.9 On average, a pair of running shoes can last 18 months if used every day. The length of time running shoes last is exponentially distributed. What is the probability that a pair of running shoes last more than 15 months? On average, how long would six pairs of running shoes last if they are used one after the other? Eighty percent of running shoes last at most how long if used every day?

Example 5.10

Suppose that the length of a phone call, in minutes, is an exponential random variable with decay parameter = $\frac{1}{12}$. If another person arrives at a public telephone just before you, find the probability that you will have to wait more than five minutes. Let X = the length of a phone call, in minutes.

What is m, μ, and σ? The probability that you must wait more than five minutes is _____ .

Solution 5.10

- $m = \frac{1}{12}$

- $\mu = 12$

- $\sigma = 12$

$P(x > 5) = 0.6592$

Try It Σ

5.10 Suppose that the distance, in miles, that people are willing to commute to work is an exponential random variable with a decay parameter $\frac{1}{20}$. Let X = the distance people are willing to commute in miles. What is m, μ, and σ? What is the probability that a person is willing to commute more than 25 miles?

Example 5.11

The time spent waiting between events is often modeled using the exponential distribution. For example, suppose that an average of 30 customers per hour arrive at a store and the time between arrivals is exponentially distributed.

  a. On average, how many minutes elapse between two successive arrivals?

  b. When the store first opens, how long on average does it take for three customers to arrive?

  c. After a customer arrives, find the probability that it takes less than one minute for the next customer to arrive.

   d. After a customer arrives, find the probability that it takes more than five minutes for the next customer to arrive.

   e. Seventy percent of the customers arrive within how many minutes of the previous customer?

   f. Is an exponential distribution reasonable for this situation?

Solution 5.11

   a. Since we expect 30 customers to arrive per hour (60 minutes), we expect on average one customer to arrive every two minutes on average.

   b. Since one customer arrives every two minutes on average, it will take six minutes on average for three customers to arrive.

   c. Let X = the time between arrivals, in minutes. By part a, $\mu = 2$, so $m = \frac{1}{2} = 0.5$.

Therefore, $X \sim Exp(0.5)$.

The cumulative distribution function is $P(X < x) = 1 - e^{(-0.5x)^e}$.

Therefore $P(X < 1) = 1 - e^{(-0.5)(1)} \approx 0.3935$.

      $1 - e^{\wedge}(-0.5) \approx 0.3935$

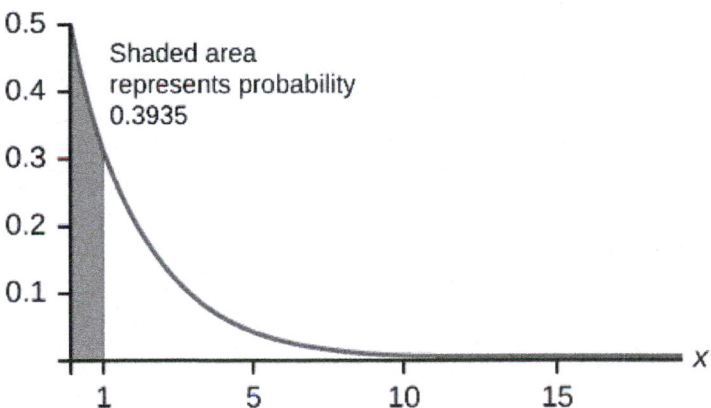

Figure 5.28

   d. $P(X > 5) = 1 - P(X < 5) = 1 - (1 - e^{(-5)(0.5)}) = e^{-2.5} \approx 0.0821$.

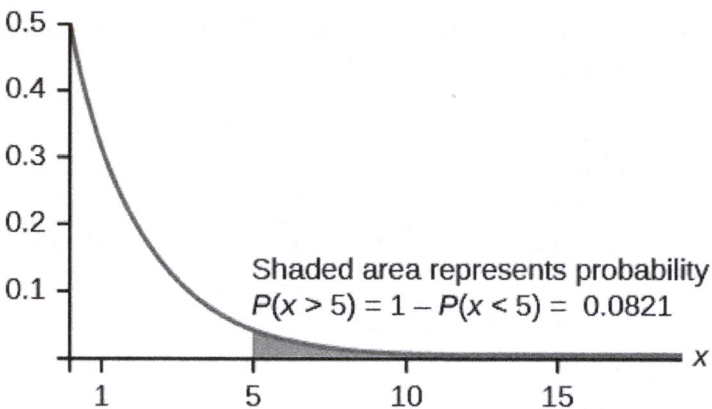

Figure 5.29

$$1 - (1 - e^\wedge(-5*0.5)) \text{ or } e^\wedge(-5*0.5)$$

e. We want to solve $0.70 = P(X < x)$ for x.

Substituting in the cumulative distribution function gives $0.70 = 1 - e^{-0.5x}$, so that $e^{-0.5x} = 0.30$. Converting this to logarithmic form gives $-0.5x = ln(0.30)$, or $x = \dfrac{ln(0.30)}{-0.5} \approx 2.41$ minutes.

Thus, seventy percent of customers arrive within 2.41 minutes of the previous customer.

You are finding the 70th percentile k so you can use the formula $k = \dfrac{ln(1 - Area\_To\_The\_Left\_Of\_k)}{(-m)}$

$$k = \frac{ln(1 - 0.70)}{(-0.5)} \approx 2.41 \text{ minutes}$$

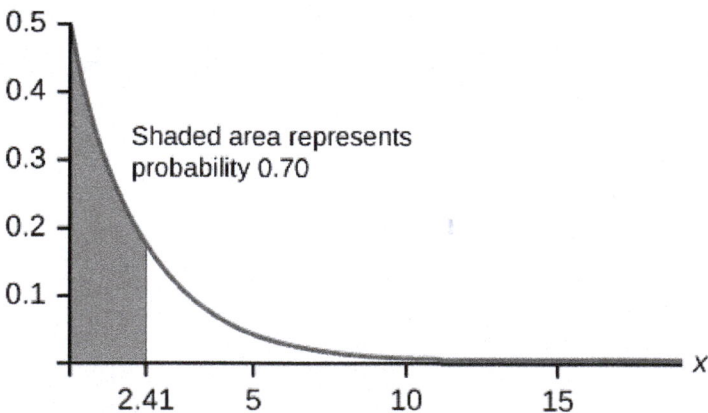

Figure 5.30

f. This model assumes that a single customer arrives at a time, which may not be reasonable since people might shop in groups, leading to several customers arriving at the same time. It also assumes that the flow of customers does not change throughout the day, which is not valid if some times of the day are busier than others.

Try It Σ

5.11 Suppose that on a certain stretch of highway, cars pass at an average rate of five cars per minute. Assume that the duration of time between successive cars follows the exponential distribution.

a. On average, how many seconds elapse between two successive cars?

b. After a car passes by, how long on average will it take for another seven cars to pass by?

c. Find the probability that after a car passes by, the next car will pass within the next 20 seconds.

d. Find the probability that after a car passes by, the next car will not pass for at least another 15 seconds.

Memorylessness of the Exponential Distribution

In **Example 5.7** recall that the amount of time between customers is exponentially distributed with a mean of two minutes ($X \sim Exp$ (0.5)). Suppose that five minutes have elapsed since the last customer arrived. Since an unusually long amount of time has now elapsed, it would seem to be more likely for a customer to arrive within the next minute. With the exponential distribution, this is not the case–the additional time spent waiting for the next customer does not depend on how much time has already elapsed since the last customer. This is referred to as the **memoryless property**. Specifically, the **memoryless property** says that

$P(X > r + t \mid X > r) = P(X > t)$ for all $r \geq 0$ and $t \geq 0$

For example, if five minutes has elapsed since the last customer arrived, then the probability that more than one minute will elapse before the next customer arrives is computed by using $r = 5$ and $t = 1$ in the foregoing equation.

$P(X > 5 + 1 \mid X > 5) = P(X > 1) = e^{(-0.5)(1)} \approx 0.6065.$

This is the same probability as that of waiting more than one minute for a customer to arrive after the previous arrival.

The exponential distribution is often used to model the longevity of an electrical or mechanical device. In **Example 5.9**, the lifetime of a certain computer part has the exponential distribution with a mean of ten years ($X \sim Exp(0.1)$). The **memoryless property** says that knowledge of what has occurred in the past has no effect on future probabilities. In this case it means that an old part is not any more likely to break down at any particular time than a brand new part. In other words, the part stays as good as new until it suddenly breaks. For example, if the part has already lasted ten years, then the probability that it lasts another seven years is $P(X > 17|X > 10) = P(X > 7) = 0.4966.$

Example 5.12

Refer to **Example 5.7** where the time a postal clerk spends with his or her customer has an exponential distribution with a mean of four minutes. Suppose a customer has spent four minutes with a postal clerk. What is the probability that he or she will spend at least an additional three minutes with the postal clerk?

The decay parameter of X is $m = \frac{1}{4} = 0.25$, so $X \sim Exp(0.25)$.

The cumulative distribution function is $P(X < x) = 1 - e^{-0.25x}$.

We want to find $P(X > 7|X > 4)$. The **memoryless property** says that $P(X > 7|X > 4) = P(X > 3)$, so we just need to find the probability that a customer spends more than three minutes with a postal clerk.

This is $P(X > 3) = 1 - P(X < 3) = 1 - (1 - e^{-0.25 \cdot 3}) = e^{-0.75} \approx 0.4724.$

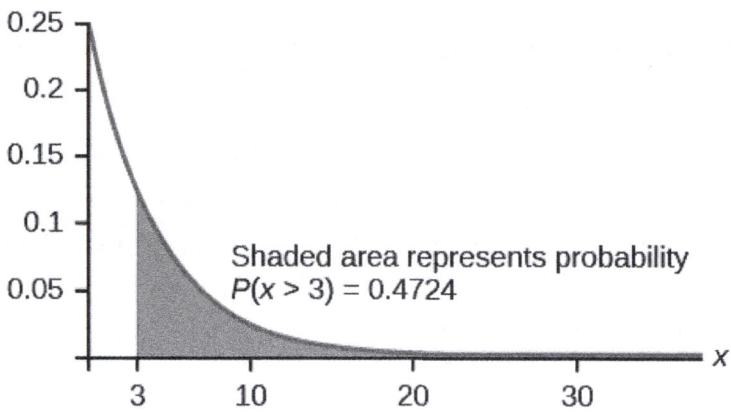

Figure 5.31

 Using the TI-83, 83+, 84, 84+ Calculator

1–(1–e^(–0.25*2)) = e^(–0.25*2).

5.12 Suppose that the longevity of a light bulb is exponential with a mean lifetime of eight years. If a bulb has already lasted 12 years, find the probability that it will last a total of over 19 years.

Relationship between the Poisson and the Exponential Distribution

There is an interesting relationship between the exponential distribution and the Poisson distribution. Suppose that the time that elapses between two successive events follows the exponential distribution with a mean of μ units of time. Also assume that these times are independent, meaning that the time between events is not affected by the times between previous events. If these assumptions hold, then the number of events per unit time follows a Poisson distribution with mean $\lambda = 1/\mu$. Recall from the chapter on **Discrete Random Variables** that if X has the Poisson distribution with mean λ, then $P(X = k) = \frac{\lambda^k e^{-\lambda}}{k!}$. Conversely, if the number of events per unit time follows a Poisson distribution, then the amount of time between events follows the exponential distribution. ($k! = k*(k-1*)(k–2)*(k-3)\ldots3*2*1$)

 Using the TI-83, 83+, 84, 84+ Calculator

Suppose X has the Poisson distribution with mean λ. Compute $P(X = k)$ by entering 2nd, VARS(DISTR), C: poissonpdf(λ, k). To compute $P(X \leq k)$, enter 2nd, VARS (DISTR), D:poissoncdf(λ, k).

Example 5.13

At a police station in a large city, calls come in at an average rate of four calls per minute. Assume that the time that elapses from one call to the next has the exponential distribution. Take note that we are concerned only with the rate at which calls come in, and we are ignoring the time spent on the phone. We must also assume that the times spent between calls are independent. This means that a particularly long delay between two calls does not mean that there will be a shorter waiting period for the next call. We may then deduce that the total number of calls received during a time period has the Poisson distribution.

a. Find the average time between two successive calls.

b. Find the probability that after a call is received, the next call occurs in less than ten seconds.

c. Find the probability that exactly five calls occur within a minute.

d. Find the probability that less than five calls occur within a minute.

e. Find the probability that more than 40 calls occur in an eight-minute period.

Solution 5.13

a. On average there are four calls occur per minute, so 15 seconds, or $\frac{15}{60}$ = 0.25 minutes occur between successive calls on average.

b. Let T = time elapsed between calls. From part a, μ = 0.25, so $m = \frac{1}{0.25}$ = 4. Thus, $T \sim Exp(4)$.

The cumulative distribution function is $P(T < t) = 1 - e^{-4t}$.
The probability that the next call occurs in less than ten seconds (ten seconds = 1/6 minute) is

$$P\left(T < \frac{1}{6}\right) = 1 - e^{-4\frac{1}{6}} \approx 0.4866.$$

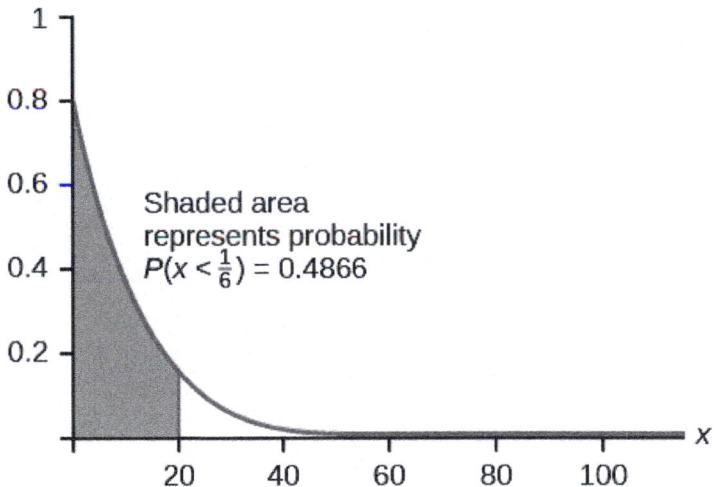

Shaded area represents probability $P(x < \frac{1}{6})$ = 0.4866

Figure 5.32

c. Let X = the number of calls per minute. As previously stated, the number of calls per minute has a Poisson distribution, with a mean of four calls per minute.

Therefore, $X \sim Poisson(4)$, and so $P(X = 5) = \frac{4^5 e^{-4}}{5!} \approx 0.1563$. (5! = (5)(4)(3)(2)(1))

poissonpdf(4, 5) = 0.1563.

d. Keep in mind that X must be a whole number, so $P(X < 5) = P(X \le 4)$.
To compute this, we could take $P(X = 0) + P(X = 1) + P(X = 2) + P(X = 3) + P(X = 4)$.
Using technology, we see that $P(X \le 4) = 0.6288$.

poisssoncdf(4, 4) = 0.6288

e. Let Y = the number of calls that occur during an eight minute period.
    Since there is an average of four calls per minute, there is an average of (8)(4) = 32 calls during each eight
    minute period.
    Hence, $Y \sim Poisson(32)$. Therefore, $P(Y > 40) = 1 - P(Y \le 40) = 1 - 0.9294 = 0.0707$.

    1 – poissoncdf(32, 40). = 0.0707

Try It Σ

5.13 In a small city, the number of automobile accidents occur with a Poisson distribution at an average of three per
week.

a. Calculate the probability that there are at most 2 accidents occur in any given week.

b. What is the probability that there is at least two weeks between any 2 accidents?

5.4 | Continuous Distribution

Stats Lab

5.1 Continuous Distribution

Class Time:

Names:

Student Learning Outcomes

- The student will compare and contrast empirical data from a random number generator with the uniform distribution.

Collect the Data

Use a random number generator to generate 50 values between zero and one (inclusive). List them in **Table 5.3**. Round the numbers to four decimal places or set the calculator MODE to four places.

1. Complete the table.

Table 5.3

2. Calculate the following:

   a. $\bar{x}$ = _____

   b. s = _____

   c. first quartile = _____

   d. third quartile = _____

   e. median = _____

Organize the Data

1. Construct a histogram of the empirical data. Make eight bars.

Figure 5.33

2. Construct a histogram of the empirical data. Make five bars.

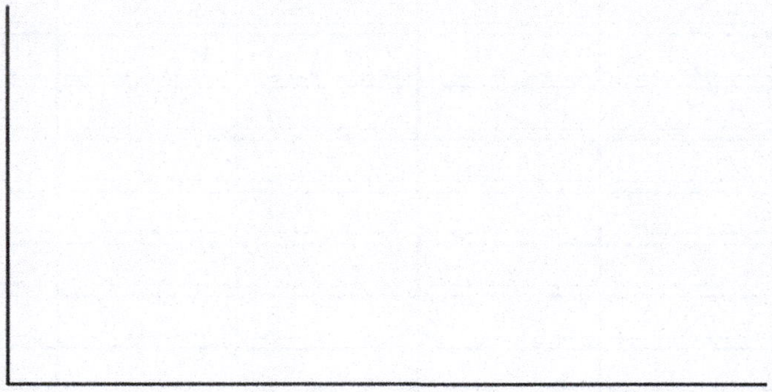

Figure 5.34

Describe the Data

1. In two to three complete sentences, describe the shape of each graph. (Keep it simple. Does the graph go straight across, does it have a V shape, does it have a hump in the middle or at either end, and so on. One way to help you determine a shape is to draw a smooth curve roughly through the top of the bars.)

2. Describe how changing the number of bars might change the shape.

Theoretical Distribution

1. In words, X = _____.

2. The theoretical distribution of X is $X \sim U(0,1)$.

3. In theory, based upon the distribution $X \sim U(0,1)$, complete the following.

    a. μ = _____

    b. σ = _____

    c. first quartile = _____

    d. third quartile = _____

    e. median = _____

4. Are the empirical values (the data) in the section titled **Collect the Data** close to the corresponding theoretical values? Why or why not?

Plot the Data

1. Construct a box plot of the data. Be sure to use a ruler to scale accurately and draw straight edges.
2. Do you notice any potential outliers? If so, which values are they? Either way, justify your answer numerically. (Recall that any DATA that are less than $Q_1 - 1.5(IQR)$ or more than $Q_3 + 1.5(IQR)$ are potential outliers. *IQR* means interquartile range.)

Compare the Data

1. For each of the following parts, use a complete sentence to comment on how the value obtained from the data compares to the theoretical value you expected from the distribution in the section titled **Theoretical Distribution**.
    a. minimum value: _____
    b. first quartile: _____
    c. median: _____
    d. third quartile: _____
    e. maximum value: _____
    f. width of *IQR*: _____
    g. overall shape: _____
2. Based on your comments in the section titled **Collect the Data**, how does the box plot fit or not fit what you would expect of the distribution in the section titled **Theoretical Distribution**?

Discussion Question

1. Suppose that the number of values generated was 500, not 50. How would that affect what you would expect the empirical data to be and the shape of its graph to look like?

KEY TERMS

Conditional Probability the likelihood that an event will occur given that another event has already occurred.

decay parameter The decay parameter describes the rate at which probabilities decay to zero for increasing values of x. It is the value m in the probability density function $f(x) = me^{(-mx)}$ of an exponential random variable. It is also equal to $m = \frac{1}{\mu}$, where μ is the mean of the random variable.

Exponential Distribution a continuous random variable (RV) that appears when we are interested in the intervals of time between some random events, for example, the length of time between emergency arrivals at a hospital; the notation is $X \sim Exp(m)$. The mean is $\mu = \frac{1}{m}$ and the standard deviation is $\sigma = \frac{1}{m}$. The probability density function is $f(x) = me^{-mx}$, $x \geq 0$ and the cumulative distribution function is $P(X \leq x) = 1 - e^{-mx}$.

memoryless property For an exponential random variable X, the memoryless property is the statement that knowledge of what has occurred in the past has no effect on future probabilities. This means that the probability that X exceeds $x + k$, given that it has exceeded x, is the same as the probability that X would exceed k if we had no knowledge about it. In symbols we say that $P(X > x + k | X > x) = P(X > k)$.

Poisson distribution If there is a known average of λ events occurring per unit time, and these events are independent of each other, then the number of events X occurring in one unit of time has the Poisson distribution. The probability of k events occurring in one unit time is equal to $P(X = k) = \frac{\lambda^k e^{-\lambda}}{k!}$.

Uniform Distribution a continuous random variable (RV) that has equally likely outcomes over the domain, $a < x < b$; it is often referred as the **rectangular distribution** because the graph of the pdf has the form of a rectangle. Notation: $X \sim U(a,b)$. The mean is $\mu = \frac{a+b}{2}$ and the standard deviation is $\sigma = \sqrt{\frac{(b-a)^2}{12}}$. The probability density function is $f(x) = \frac{1}{b-a}$ for $a < x < b$ or $a \leq x \leq b$. The cumulative distribution is $P(X \leq x) = \frac{x-a}{b-a}$.

CHAPTER REVIEW

5.1 Continuous Probability Functions

The probability density function (pdf) is used to describe probabilities for continuous random variables. The area under the density curve between two points corresponds to the probability that the variable falls between those two values. In other words, the area under the density curve between points a and b is equal to $P(a < x < b)$. The cumulative distribution function (cdf) gives the probability as an area. If X is a continuous random variable, the probability density function (pdf), $f(x)$, is used to draw the graph of the probability distribution. The total area under the graph of $f(x)$ is one. The area under the graph of $f(x)$ and between values a and b gives the probability $P(a < x < b)$.

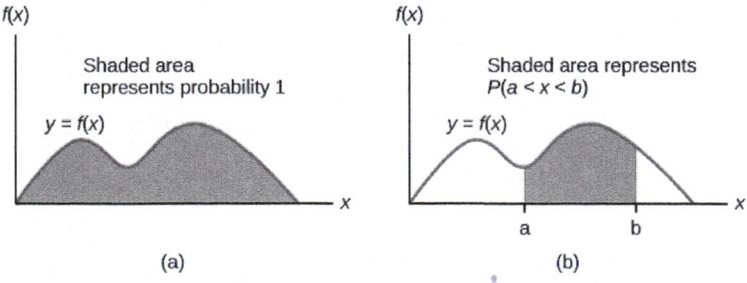

(a)   (b)

Figure 5.35

The cumulative distribution function (cdf) of X is defined by $P(X \leq x)$. It is a function of x that gives the probability that the random variable is less than or equal to x.

5.2 The Uniform Distribution

If X has a uniform distribution where $a < x < b$ or $a \le x \le b$, then X takes on values between a and b (may include a and b). All values x are equally likely. We write $X \sim U(a, b)$. The mean of X is $\mu = \frac{a+b}{2}$. The standard deviation of X is $\sigma = \sqrt{\frac{(b-a)^2}{12}}$. The probability density function of X is $f(x) = \frac{1}{b-a}$ for $a \le x \le b$. The cumulative distribution function of X is $P(X \le x) = \frac{x-a}{b-a}$. X is continuous.

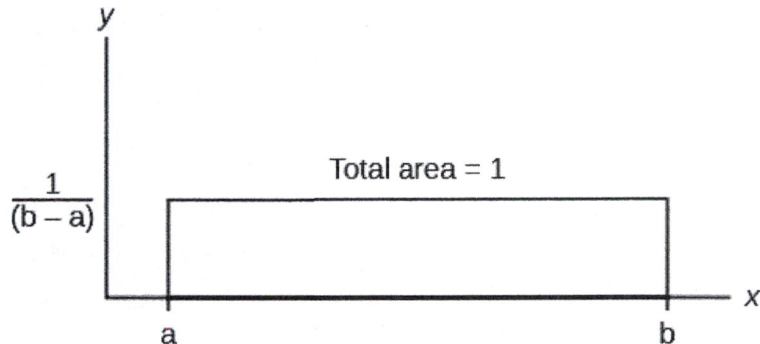

Figure 5.36

The probability $P(c < X < d)$ may be found by computing the area under $f(x)$, between c and d. Since the corresponding area is a rectangle, the area may be found simply by multiplying the width and the height.

5.3 The Exponential Distribution

If X has an **exponential distribution** with mean μ, then the **decay parameter** is $m = \frac{1}{\mu}$, and we write $X \sim Exp(m)$ where $x \ge 0$ and $m > 0$. The probability density function of X is $f(x) = me^{-mx}$ (or equivalently $f(x) = \frac{1}{\mu}e^{-x/\mu}$. The cumulative distribution function of X is $P(X \le x) = 1 - e^{-mx}$.

The exponential distribution has the **memoryless property**, which says that future probabilities do not depend on any past information. Mathematically, it says that $P(X > x + k | X > x) = P(X > k)$.

If T represents the waiting time between events, and if $T \sim Exp(\lambda)$, then the number of events X per unit time follows the Poisson distribution with mean λ. The probability density function of PX is $(X = k) = \frac{\lambda^k e^{-k}}{k!}$. This may be computed using a TI-83, 83+, 84, 84+ calculator with the command poissonpdf(λ, k). The cumulative distribution function $P(X \le k)$ may be computed using the TI-83, 83+,84, 84+ calculator with the command poissoncdf(λ, k).

FORMULA REVIEW

5.1 Continuous Probability Functions

Probability density function (pdf) $f(x)$:

- $f(x) \ge 0$

- The total area under the curve $f(x)$ is one.

Cumulative distribution function (cdf): $P(X \le x)$

5.2 The Uniform Distribution

X = a real number between a and b (in some instances, X can take on the values a and b). a = smallest X; b = largest X

$X \sim U(a, b)$

The mean is $\mu = \frac{a+b}{2}$

The standard deviation is $\sigma = \sqrt{\frac{(b-a)^2}{12}}$

Probability density function: $f(x) = \frac{1}{b-a}$ for $a \le X \le b$

Area to the Left of x: $P(X < x) = (x-a)\left(\frac{1}{b-a}\right)$

Area to the Right of x: $P(X > x) = (b-x)\left(\frac{1}{b-a}\right)$

Area Between c and d: $P(c < x < d) = (base)(height) = (d-c)\left(\frac{1}{b-a}\right)$

Uniform: $X \sim U(a, b)$ where $a < x < b$

- pdf: $f(x) = \frac{1}{b-a}$ for $a \le x \le b$

- cdf: $P(X \le x) = \frac{x-a}{b-a}$

- mean $\mu = \frac{a+b}{2}$

- standard deviation $\sigma = \sqrt{\frac{(b-a)^2}{12}}$

- $P(c < X < d) = (d-c)\left(\frac{1}{b-a}\right)$

5.3 The Exponential Distribution

Exponential: $X \sim Exp(m)$ where m = the decay parameter

- pdf: $f(x) = me^{(-mx)}$ where $x \ge 0$ and $m > 0$

- cdf: $P(X \le x) = 1 - e^{(-mx)}$

- mean $\mu = \frac{1}{m}$

- standard deviation $\sigma = \mu$

- percentile k: $k = \frac{ln(1 - AreaToTheLeftOfk)}{(-m)}$

- Additionally
    - $P(X > x) = e^{(-mx)}$
    - $P(a < X < b) = e^{(-ma)} - e^{(-mb)}$

- Memoryless Property: $P(X > x + k | X > x) = P(X > k)$

- Poisson probability: $P(X = k) = \frac{\lambda^k e^{-k}}{k!}$ with mean λ

- $k! = k*(k-1)*(k-2)*(k-3)\dots3*2*1$

PRACTICE

5.1 Continuous Probability Functions

1. Which type of distribution does the graph illustrate?

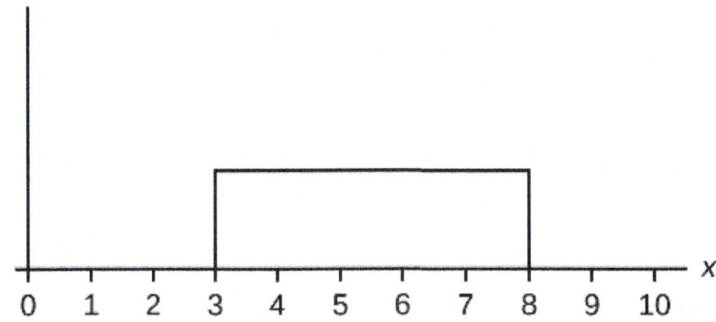

Figure 5.37

2. Which type of distribution does the graph illustrate?

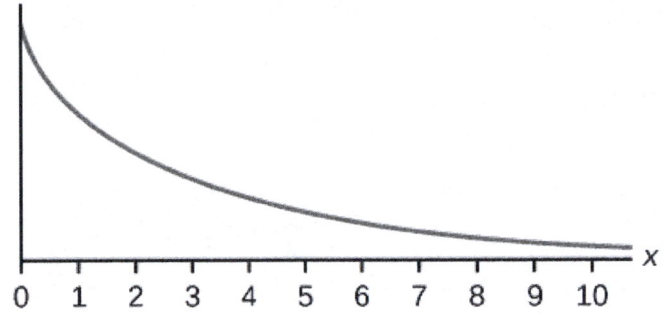

Figure 5.38

3. Which type of distribution does the graph illustrate?

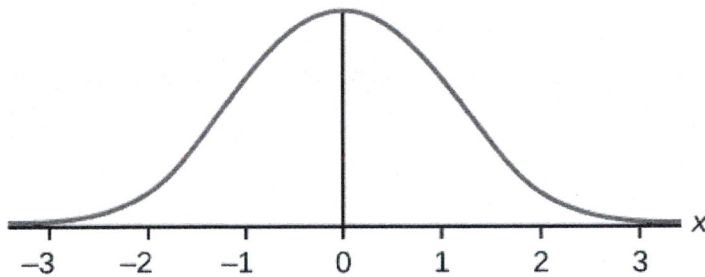

Figure 5.39

4. What does the shaded area represent? $P(\_\_\_ < x < \_\_\_)$

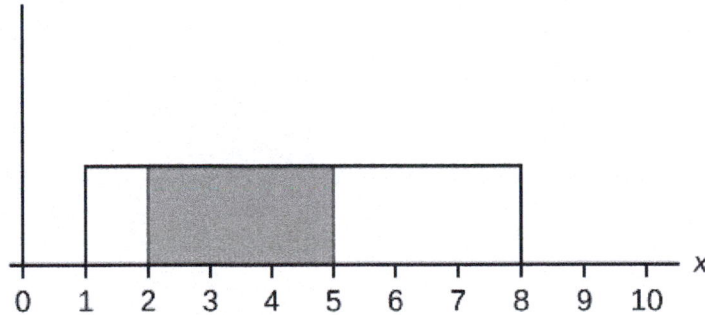

Figure 5.40

5. What does the shaded area represent? $P(\_\_\_ < x < \_\_\_)$

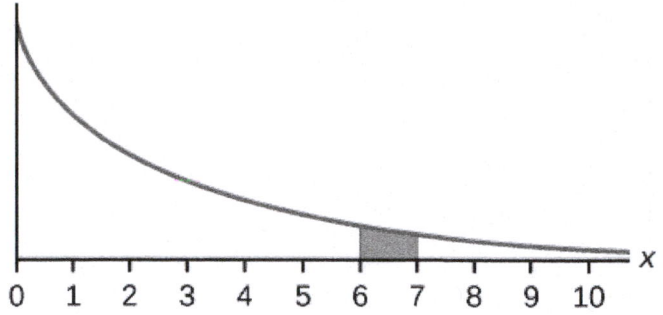

Figure 5.41

6. For a continuous probablity distribution, $0 \le x \le 15$. What is $P(x > 15)$?

7. What is the area under $f(x)$ if the function is a continuous probability density function?

8. For a continuous probability distribution, $0 \le x \le 10$. What is $P(x = 7)$?

9. A **continuous** probability function is restricted to the portion between $x = 0$ and 7. What is $P(x = 10)$?

10. $f(x)$ for a continuous probability function is $\frac{1}{5}$, and the function is restricted to $0 \le x \le 5$. What is $P(x < 0)$?

11. $f(x)$, a continuous probability function, is equal to $\frac{1}{12}$, and the function is restricted to $0 \le x \le 12$. What is $P(0 < x < 12)$?

12. Find the probability that x falls in the shaded area.

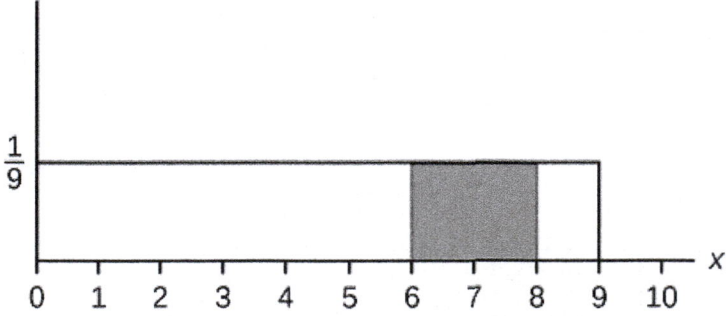

Figure 5.42

13. Find the probability that x falls in the shaded area.

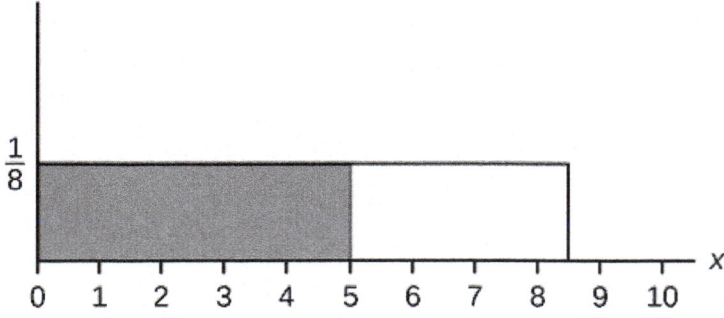

Figure 5.43

14. Find the probability that x falls in the shaded area.

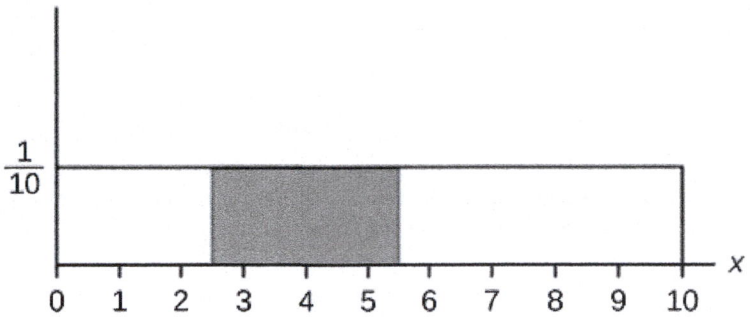

Figure 5.44

15. $f(x)$, a continuous probability function, is equal to $\frac{1}{3}$ and the function is restricted to $1 \leq x \leq 4$. Describe $P\left(x > \frac{3}{2}\right)$.

5.2 The Uniform Distribution

Use the following information to answer the next ten questions. The data that follow are the square footage (in 1,000 feet squared) of 28 homes.

1.5	2.4	3.6	2.6	1.6	2.4	2.0
3.5	2.5	1.8	2.4	2.5	3.5	4.0
2.6	1.6	2.2	1.8	3.8	2.5	1.5
2.8	1.8	4.5	1.9	1.9	3.1	1.6

Table 5.4

The sample mean = 2.50 and the sample standard deviation = 0.8302.

The distribution can be written as $X \sim U(1.5, 4.5)$.

16. What type of distribution is this?

17. In this distribution, outcomes are equally likely. What does this mean?

18. What is the height of $f(x)$ for the continuous probability distribution?

19. What are the constraints for the values of x?

20. Graph $P(2 < x < 3)$.

21. What is $P(2 < x < 3)$?

22. What is $P(x < 3.5 | x < 4)$?

23. What is $P(x = 1.5)$?

24. What is the 90th percentile of square footage for homes?

25. Find the probability that a randomly selected home has more than 3,000 square feet given that you already know the house has more than 2,000 square feet.

Use the following information to answer the next eight exercises. A distribution is given as $X \sim U(0, 12)$.

26. What is a? What does it represent?

27. What is b? What does it represent?

28. What is the probability density function?

29. What is the theoretical mean?

30. What is the theoretical standard deviation?

31. Draw the graph of the distribution for $P(x > 9)$.

32. Find $P(x > 9)$.

33. Find the 40th percentile.

Use the following information to answer the next eleven exercises. The age of cars in the staff parking lot of a suburban college is uniformly distributed from six months (0.5 years) to 9.5 years.

34. What is being measured here?

35. In words, define the random variable X.

36. Are the data discrete or continuous?

37. The interval of values for x is _____.

38. The distribution for X is _____.

39. Write the probability density function.

40. Graph the probability distribution.
   a. Sketch the graph of the probability distribution.

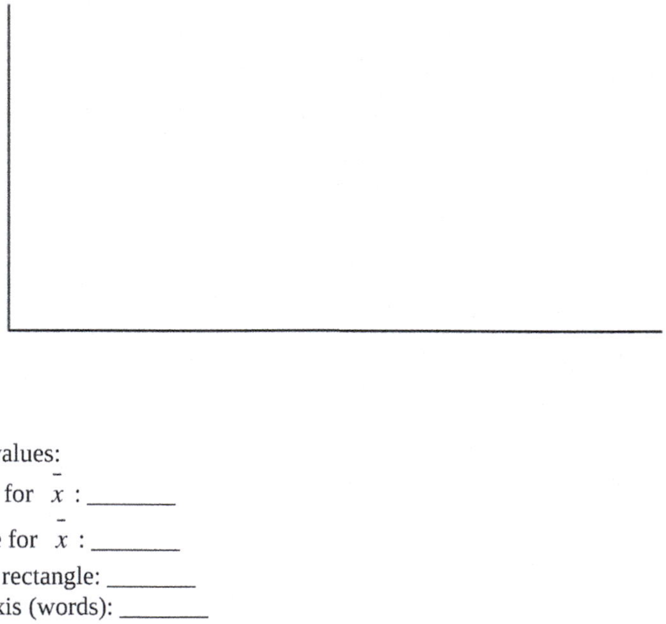

   Figure 5.45
   b. Identify the following values:
       i. Lowest value for $\bar{x}$: _____
       ii. Highest value for $\bar{x}$: _____
       iii. Height of the rectangle: _____
       iv. Label for x-axis (words): _____
       v. Label for y-axis (words): _____

41. Find the average age of the cars in the lot.

42. Find the probability that a randomly chosen car in the lot was less than four years old.
   a. Sketch the graph, and shade the area of interest.

Figure 5.46

   b. Find the probability. $P(x < 4) =$ _____

43. Considering only the cars less than 7.5 years old, find the probability that a randomly chosen car in the lot was less than four years old.

   a. Sketch the graph, shade the area of interest.

Figure 5.47

   b. Find the probability. $P(x < 4 | x < 7.5) =$ _____

44. What has changed in the previous two problems that made the solutions different?

45. Find the third quartile of ages of cars in the lot. This means you will have to find the value such that $\frac{3}{4}$, or 75%, of the cars are at most (less than or equal to) that age.

   a. Sketch the graph, and shade the area of interest.

Figure 5.48

b. Find the value k such that $P(x < k) = 0.75$.

c. The third quartile is _____

5.3 The Exponential Distribution

Use the following information to answer the next ten exercises. A customer service representative must spend different amounts of time with each customer to resolve various concerns. The amount of time spent with each customer can be modeled by the following distribution: $X \sim Exp(0.2)$

46. What type of distribution is this?

47. Are outcomes equally likely in this distribution? Why or why not?

48. What is m? What does it represent?

49. What is the mean?

50. What is the standard deviation?

51. State the probability density function.

52. Graph the distribution.

53. Find $P(2 < x < 10)$.

54. Find $P(x > 6)$.

55. Find the 70th percentile.

Use the following information to answer the next seven exercises. A distribution is given as $X \sim Exp(0.75)$.

56. What is m?

57. What is the probability density function?

58. What is the cumulative distribution function?

59. Draw the distribution.

60. Find $P(x < 4)$.

61. Find the 30th percentile.

62. Find the median.

63. Which is larger, the mean or the median?

Use the following information to answer the next 16 exercises. Carbon-14 is a radioactive element with a half-life of about 5,730 years. Carbon-14 is said to decay exponentially. The decay rate is 0.000121. We start with one gram of carbon-14. We are interested in the time (years) it takes to decay carbon-14.

64. What is being measured here?

65. Are the data discrete or continuous?

66. In words, define the random variable X.

67. What is the decay rate (m)?

68. The distribution for X is _____.

69. Find the amount (percent of one gram) of carbon-14 lasting less than 5,730 years. This means, find $P(x < 5,730)$.

    a. Sketch the graph, and shade the area of interest.

    Figure 5.49

    b. Find the probability. $P(x < 5,730) =$ _____

70. Find the percentage of carbon-14 lasting longer than 10,000 years.

    a. Sketch the graph, and shade the area of interest.

    Figure 5.50

    b. Find the probability. $P(x > 10,000) =$ _____

71. Thirty percent (30%) of carbon-14 will decay within how many years?

    a. Sketch the graph, and shade the area of interest.

Figure 5.51
b. Find the value k such that $P(x < k) = 0.30$.

HOMEWORK

5.1 Continuous Probability Functions

For each probability and percentile problem, draw the picture.

72. Consider the following experiment. You are one of 100 people enlisted to take part in a study to determine the percent of nurses in America with an R.N. (registered nurse) degree. You ask nurses if they have an R.N. degree. The nurses answer "yes" or "no." You then calculate the percentage of nurses with an R.N. degree. You give that percentage to your supervisor.
   a. What part of the experiment will yield discrete data?
   b. What part of the experiment will yield continuous data?

73. When age is rounded to the nearest year, do the data stay continuous, or do they become discrete? Why?

5.2 The Uniform Distribution

For each probability and percentile problem, draw the picture.

74. Births are approximately uniformly distributed between the 52 weeks of the year. They can be said to follow a uniform distribution from one to 53 (spread of 52 weeks).
   a. $X \sim$ _____
   b. Graph the probability distribution.
   c. $f(x) =$ _____
   d. $\mu =$ _____
   e. $\sigma =$ _____
   f. Find the probability that a person is born at the exact moment week 19 starts. That is, find $P(x = 19) =$ _____
   g. $P(2 < x < 31) =$ _____
   h. Find the probability that a person is born after week 40.
   i. $P(12 < x | x < 28) =$ _____
   j. Find the 70th percentile.
   k. Find the minimum for the upper quarter.

75. A random number generator picks a number from one to nine in a uniform manner.
   a. $X \sim$ _____
   b. Graph the probability distribution.
   c. $f(x) =$ _____
   d. $\mu =$ _____
   e. $\sigma =$ _____
   f. $P(3.5 < x < 7.25) =$ _____

    g. $P(x > 5.67)$

    h. $P(x > 5 | x > 3) =$ _____

    i. Find the 90th percentile.

76. According to a study by Dr. John McDougall of his live-in weight loss program at St. Helena Hospital, the people who follow his program lose between six and 15 pounds a month until they approach trim body weight. Let's suppose that the weight loss is uniformly distributed. We are interested in the weight loss of a randomly selected individual following the program for one month.

    a. Define the random variable. $X =$ _____

    b. $X \sim$ _____

    c. Graph the probability distribution.

    d. $f(x) =$ _____

    e. $\mu =$ _____

    f. $\sigma =$ _____

    g. Find the probability that the individual lost more than ten pounds in a month.

    h. Suppose it is known that the individual lost more than ten pounds in a month. Find the probability that he lost less than 12 pounds in the month.

    i. $P(7 < x < 13 | x > 9) =$ _____. State this in a probability question, similarly to parts g and h, draw the picture, and find the probability.

77. A subway train on the Red Line arrives every eight minutes during rush hour. We are interested in the length of time a commuter must wait for a train to arrive. The time follows a uniform distribution.

    a. Define the random variable. $X =$ _____

    b. $X \sim$ _____

    c. Graph the probability distribution.

    d. $f(x) =$ _____

    e. $\mu =$ _____

    f. $\sigma =$ _____

    g. Find the probability that the commuter waits less than one minute.

    h. Find the probability that the commuter waits between three and four minutes.

    i. Sixty percent of commuters wait more than how long for the train? State this in a probability question, similarly to parts g and h, draw the picture, and find the probability.

78. The age of a first grader on September 1 at Garden Elementary School is uniformly distributed from 5.8 to 6.8 years. We randomly select one first grader from the class.

    a. Define the random variable. $X =$ _____

    b. $X \sim$ _____

    c. Graph the probability distribution.

    d. $f(x) =$ _____

    e. $\mu =$ _____

    f. $\sigma =$ _____

    g. Find the probability that she is over 6.5 years old.

    h. Find the probability that she is between four and six years old.

    i. Find the 70th percentile for the age of first graders on September 1 at Garden Elementary School.

Use the following information to answer the next three exercises. The Sky Train from the terminal to the rental–car and long–term parking center is supposed to arrive every eight minutes. The waiting times for the train are known to follow a uniform distribution.

79. What is the average waiting time (in minutes)?

    a. zero

    b. two

    c. three

    d. four

80. Find the 30th percentile for the waiting times (in minutes).

    a. two

    b. 2.4

    c. 2.75

    d. three

81. The probability of waiting more than seven minutes given a person has waited more than four minutes is?
  a. 0.125
  b. 0.25
  c. 0.5
  d. 0.75

82. The time (in minutes) until the next bus departs a major bus depot follows a distribution with $f(x) = \frac{1}{20}$ where x goes

from 25 to 45 minutes.
  a. Define the random variable. $X =$ _____
  b. $X \sim$ _____
  c. Graph the probability distribution.
  d. The distribution is _____ (name of distribution). It is _____ (discrete or continuous).
  e. $\mu =$ _____
  f. $\sigma =$ _____
  g. Find the probability that the time is at most 30 minutes. Sketch and label a graph of the distribution. Shade the area of interest. Write the answer in a probability statement.
  h. Find the probability that the time is between 30 and 40 minutes. Sketch and label a graph of the distribution. Shade the area of interest. Write the answer in a probability statement.
  i. $P(25 < x < 55) =$ _____. State this in a probability statement, similarly to parts g and h, draw the picture, and find the probability.
  j. Find the 90th percentile. This means that 90% of the time, the time is less than _____ minutes.
  k. Find the 75th percentile. In a complete sentence, state what this means. (See part j.)
  l. Find the probability that the time is more than 40 minutes given (or knowing that) it is at least 30 minutes.

83. Suppose that the value of a stock varies each day from $16 to $25 with a uniform distribution.
  a. Find the probability that the value of the stock is more than $19.
  b. Find the probability that the value of the stock is between $19 and $22.
  c. Find the upper quartile - 25% of all days the stock is above what value? Draw the graph.
  d. Given that the stock is greater than $18, find the probability that the stock is more than $21.

84. A fireworks show is designed so that the time between fireworks is between one and five seconds, and follows a uniform distribution.
  a. Find the average time between fireworks.
  b. Find probability that the time between fireworks is greater than four seconds.

85. The number of miles driven by a truck driver falls between 300 and 700, and follows a uniform distribution.
  a. Find the probability that the truck driver goes more than 650 miles in a day.
  b. Find the probability that the truck drivers goes between 400 and 650 miles in a day.
  c. At least how many miles does the truck driver travel on the furthest 10% of days?

5.3 The Exponential Distribution

86. Suppose that the length of long distance phone calls, measured in minutes, is known to have an exponential distribution with the average length of a call equal to eight minutes.
  a. Define the random variable. $X =$ _____.
  b. Is X continuous or discrete?
  c. $X \sim$ _____
  d. $\mu =$ _____
  e. $\sigma =$ _____
  f. Draw a graph of the probability distribution. Label the axes.
  g. Find the probability that a phone call lasts less than nine minutes.
  h. Find the probability that a phone call lasts more than nine minutes.
  i. Find the probability that a phone call lasts between seven and nine minutes.
  j. If 25 phone calls are made one after another, on average, what would you expect the total to be? Why?

87. Suppose that the useful life of a particular car battery, measured in months, decays with parameter 0.025. We are interested in the life of the battery.
  a. Define the random variable. $X =$ _____.
  b. Is X continuous or discrete?

c. $X \sim$ _____
d. On average, how long would you expect one car battery to last?
e. On average, how long would you expect nine car batteries to last, if they are used one after another?
f. Find the probability that a car battery lasts more than 36 months.
g. Seventy percent of the batteries last at least how long?

88. The percent of persons (ages five and older) in each state who speak a language at home other than English is approximately exponentially distributed with a mean of 9.848. Suppose we randomly pick a state.
a. Define the random variable. $X =$ _____.
b. Is X continuous or discrete?
c. $X \sim$ _____
d. $\mu =$ _____
e. $\sigma =$ _____
f. Draw a graph of the probability distribution. Label the axes.
g. Find the probability that the percent is less than 12.
h. Find the probability that the percent is between eight and 14.
i. The percent of all individuals living in the United States who speak a language at home other than English is 13.8.

  i. Why is this number different from 9.848%?
  ii. What would make this number higher than 9.848%?

89. The time (in years) **after** reaching age 60 that it takes an individual to retire is approximately exponentially distributed with a mean of about five years. Suppose we randomly pick one retired individual. We are interested in the time after age 60 to retirement.
a. Define the random variable. $X =$ _____.
b. Is X continuous or discrete?
c. $X \sim =$ _____
d. $\mu =$ _____
e. $\sigma =$ _____
f. Draw a graph of the probability distribution. Label the axes.
g. Find the probability that the person retired after age 70.
h. Do more people retire before age 65 or after age 65?
i. In a room of 1,000 people over age 80, how many do you expect will NOT have retired yet?

90. The cost of all maintenance for a car during its first year is approximately exponentially distributed with a mean of $150.
a. Define the random variable. $X =$ _____.
b. $X \sim =$ _____
c. $\mu =$ _____
d. $\sigma =$ _____
e. Draw a graph of the probability distribution. Label the axes.
f. Find the probability that a car required over $300 for maintenance during its first year.

Use the following information to answer the next three exercises. The average lifetime of a certain new cell phone is three years. The manufacturer will replace any cell phone failing within two years of the date of purchase. The lifetime of these cell phones is known to follow an exponential distribution.

91. The decay rate is:
a. 0.3333
b. 0.5000
c. 2
d. 3

92. What is the probability that a phone will fail within two years of the date of purchase?
a. 0.8647
b. 0.4866
c. 0.2212
d. 0.9997

93. What is the median lifetime of these phones (in years)?

    a. 0.1941
    b. 1.3863
    c. 2.0794
    d. 5.5452

94. Let $X \sim Exp(0.1)$.
    a. decay rate = _____
    b. $\mu =$ _____
    c. Graph the probability distribution function.
    d. On the graph, shade the area corresponding to $P(x < 6)$ and find the probability.
    e. Sketch a new graph, shade the area corresponding to $P(3 < x < 6)$ and find the probability.
    f. Sketch a new graph, shade the area corresponding to $P(x < 7)$ and find the probability.
    g. Sketch a new graph, shade the area corresponding to the 40th percentile and find the value.
    h. Find the average value of x.

95. Suppose that the longevity of a light bulb is exponential with a mean lifetime of eight years.
    a. Find the probability that a light bulb lasts less than one year.
    b. Find the probability that a light bulb lasts between six and ten years.
    c. Seventy percent of all light bulbs last at least how long?
    d. A company decides to offer a warranty to give refunds to light bulbs whose lifetime is among the lowest two percent of all bulbs. To the nearest month, what should be the cutoff lifetime for the warranty to take place?
    e. If a light bulb has lasted seven years, what is the probability that it fails within the 8th year.

96. At a 911 call center, calls come in at an average rate of one call every two minutes. Assume that the time that elapses from one call to the next has the exponential distribution.
    a. On average, how much time occurs between five consecutive calls?
    b. Find the probability that after a call is received, it takes more than three minutes for the next call to occur.
    c. Ninety-percent of all calls occur within how many minutes of the previous call?
    d. Suppose that two minutes have elapsed since the last call. Find the probability that the next call will occur within the next minute.
    e. Find the probability that less than 20 calls occur within an hour.

97. In major league baseball, a no-hitter is a game in which a pitcher, or pitchers, doesn't give up any hits throughout the game. No-hitters occur at a rate of about three per season. Assume that the duration of time between no-hitters is exponential.
    a. What is the probability that an entire season elapses with a single no-hitter?
    b. If an entire season elapses without any no-hitters, what is the probability that there are no no-hitters in the following season?
    c. What is the probability that there are more than 3 no-hitters in a single season?

98. During the years 1998–2012, a total of 29 earthquakes of magnitude greater than 6.5 have occurred in Papua New Guinea. Assume that the time spent waiting between earthquakes is exponential.
    a. What is the probability that the next earthquake occurs within the next three months?
    b. Given that six months has passed without an earthquake in Papua New Guinea, what is the probability that the next three months will be free of earthquakes?
    c. What is the probability of zero earthquakes occurring in 2014?
    d. What is the probability that at least two earthquakes will occur in 2014?

99. According to the American Red Cross, about one out of nine people in the U.S. have Type B blood. Suppose the blood types of people arriving at a blood drive are independent. In this case, the number of Type B blood types that arrive roughly follows the Poisson distribution.
    a. If 100 people arrive, how many on average would be expected to have Type B blood?
    b. What is the probability that over 10 people out of these 100 have type B blood?
    c. What is the probability that more than 20 people arrive before a person with type B blood is found?

100. A web site experiences traffic during normal working hours at a rate of 12 visits per hour. Assume that the duration between visits has the exponential distribution.
    a. Find the probability that the duration between two successive visits to the web site is more than ten minutes.
    b. The top 25% of durations between visits are at least how long?
    c. Suppose that 20 minutes have passed since the last visit to the web site. What is the probability that the next visit will occur within the next 5 minutes?

    d. Find the probability that less than 7 visits occur within a one-hour period.

101. At an urgent care facility, patients arrive at an average rate of one patient every seven minutes. Assume that the duration between arrivals is exponentially distributed.

    a. Find the probability that the time between two successive visits to the urgent care facility is less than 2 minutes.

    b. Find the probability that the time between two successive visits to the urgent care facility is more than 15 minutes.

    c. If 10 minutes have passed since the last arrival, what is the probability that the next person will arrive within the next five minutes?

    d. Find the probability that more than eight patients arrive during a half-hour period.

REFERENCES

5.2 The Uniform Distribution

McDougall, John A. The McDougall Program for Maximum Weight Loss. Plume, 1995.

5.3 The Exponential Distribution

Data from the United States Census Bureau.

Data from World Earthquakes, 2013. Available online at http://www.world-earthquakes.com/ (accessed June 11, 2013).

"No-hitter." Baseball-Reference.com, 2013. Available online at http://www.baseball-reference.com/bullpen/No-hitter (accessed June 11, 2013).

Zhou, Rick. "Exponential Distribution lecture slides." Available online at www.public.iastate.edu/~riczw/stat330s11/lecture/lec13.pdf (accessed June 11, 2013).

SOLUTIONS

1 Uniform Distribution

3 Normal Distribution

5 $P(6 < x < 7)$

7 one

9 zero

11 one

13 0.625

15 The probability is equal to the area from $x = \frac{3}{2}$ to $x = 4$ above the x-axis and up to $f(x) = \frac{1}{3}$.

17 It means that the value of x is just as likely to be any number between 1.5 and 4.5.

19 $1.5 \le x \le 4.5$

21 0.3333

23 zero

25 0.6

27 b is 12, and it represents the highest value of x.

29 six

31

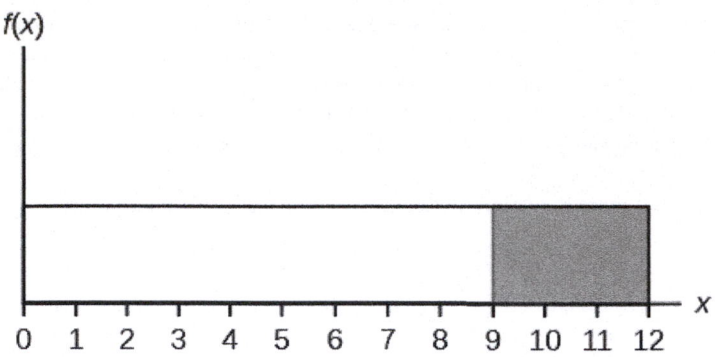

Figure 5.52

33 4.8

35 X = The age (in years) of cars in the staff parking lot

37 0.5 to 9.5

39 $f(x) = \frac{1}{9}$ where x is between 0.5 and 9.5, inclusive.

41 $\mu = 5$

43

a. Check student's solution.

b. $\frac{3.5}{7}$

45

a. Check student's solution.

b. $k = 7.25$

c. 7.25

47 No, outcomes are not equally likely. In this distribution, more people require a little bit of time, and fewer people require a lot of time, so it is more likely that someone will require less time.

49 five

51 $f(x) = 0.2e^{-0.2x}$

53 0.5350

55 6.02

57 $f(x) = 0.75e^{-0.75x}$

59

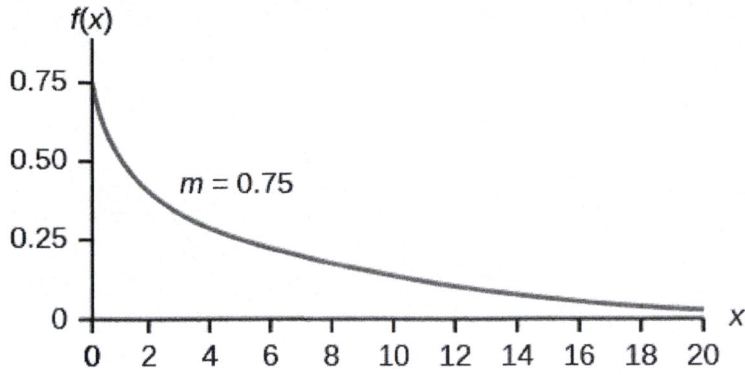

Figure 5.53

61 0.4756

63 The mean is larger. The mean is $\frac{1}{m} = \frac{1}{0.75} \approx 1.33$, which is greater than 0.9242.

65 continuous

67 $m = 0.000121$

69
a. Check student's solution
b. $P(x < 5,730) = 0.5001$

71
a. Check student's solution.
b. $k = 2947.73$

73 Age is a measurement, regardless of the accuracy used.

75
a. $X \sim U(1, 9)$
b. Check student's solution.
c. $f(x) = \frac{1}{8}$ where $1 \le x \le 9$
d. five
e. 2.3
f. $\frac{15}{32}$
g. $\frac{333}{800}$
h. $\frac{2}{3}$
i. 8.2

77
a. X represents the length of time a commuter must wait for a train to arrive on the Red Line.
b. $X \sim U(0, 8)$

c. $f(x) = \frac{1}{8}$ where $\le x \le 8$

d.   four

e.   2.31

f. $\frac{1}{8}$

g. $\frac{1}{8}$

h.   3.2

79 d

81 b

83

a.   The probability density function of X is $\frac{1}{25 - 16} = \frac{1}{9}$.

$P(X > 19) = (25 - 19) \left(\frac{1}{9}\right) = \frac{6}{9} = \frac{2}{3}$.

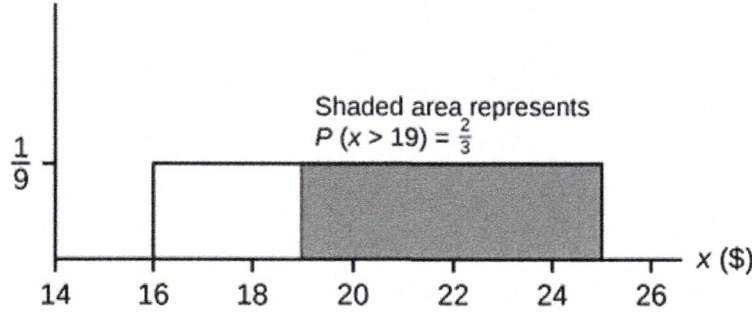

Figure 5.54

b. $P(19 < X < 22) = (22 - 19) \left(\frac{1}{9}\right) = \frac{3}{9} = \frac{1}{3}$.

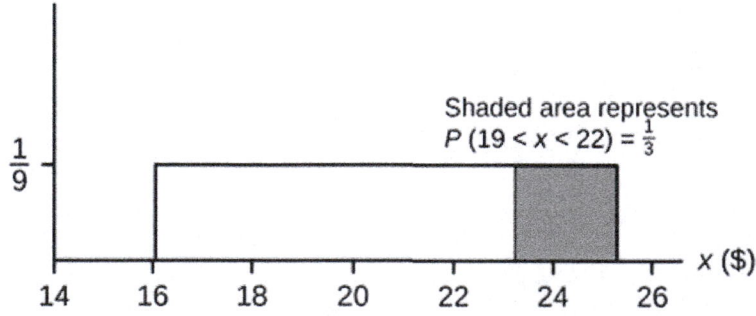

Figure 5.55

c.   The area must be 0.25, and $0.25 = (\text{width}) \left(\frac{1}{9}\right)$, so width = $(0.25)(9) = 2.25$. Thus, the value is $25 - 2.25 = 22.75$.

d.   This is a conditional probability question. $P(x > 21 | x > 18)$. You can do this two ways:

○ Draw the graph where a is now 18 and b is still 25. The height is $\dfrac{1}{(25-18)} = \dfrac{1}{7}$

So, $P(x > 21 | x > 18) = (25-21)\left(\dfrac{1}{7}\right) = 4/7$.

○ Use the formula: $P(x > 21 | x > 18) = \dfrac{P(x > 21 \text{ AND } x > 18)}{P(x > 18)}$

$= \dfrac{P(x > 21)}{P(x > 18)} = \dfrac{(25-21)}{(25-18)} = \dfrac{4}{7}$.

85

a. $P(X > 650) = \dfrac{700 - 650}{700 - 300} = \dfrac{500}{400} = \dfrac{1}{8} = 0.125$.

b. $P(400 < X < 650) = \dfrac{700 - 650}{700 - 300} = \dfrac{250}{400} = 0.625$

c. $0.10 = \dfrac{\text{width}}{700 - 300}$, so width = 400(0.10) = 40. Since 700 − 40 = 660, the drivers travel at least 660 miles on the furthest 10% of days.

87

a. X = the useful life of a particular car battery, measured in months.

b. X is continuous.

c. $X \sim Exp(0.025)$

d. 40 months

e. 360 months

f. 0.4066

g. 14.27

89

a. X = the time (in years) after reaching age 60 that it takes an individual to retire

b. X is continuous.

c. $X \sim Exp\left(\dfrac{1}{5}\right)$

d. five

e. five

f. Check student's solution.

g. 0.1353

h. before

i. 18.3

91 a

93 c

95 Let T = the life time of a light bulb. The decay parameter is $m = 1/8$, and $T \sim Exp(1/8)$. The cumulative distribution function is $P(T < t) = 1 - e^{-\frac{t}{8}}$

a. Therefore, $P(T < 1) = 1 - e^{-\frac{1}{8}} \approx 0.1175$.

b. We want to find $P(6 < t < 10)$.
    To do this, $P(6 < t < 10) - P(t < 6)$

$$= = \left(1 - e^{-\frac{1}{8}*10}\right) - \left(1 - e^{-\frac{1}{8}*6}\right) \approx 0.7135 - 0.5276 = 0.1859$$

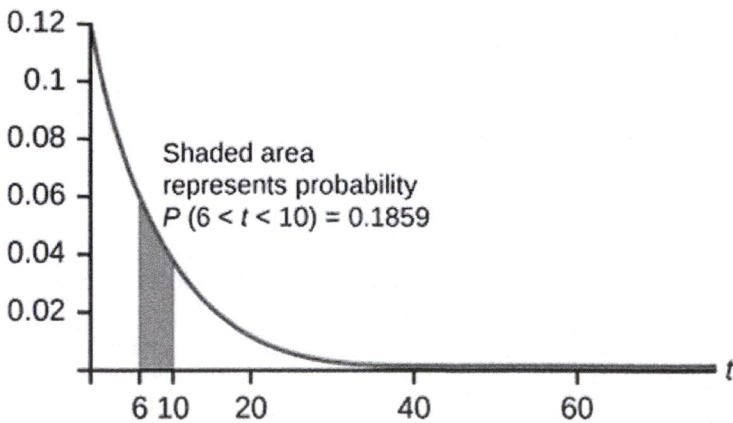

Figure 5.56

c. We want to find $0.70 = P(T > t) = 1 - \left(1 - e^{-\frac{t}{8}}\right) = e^{-\frac{t}{8}}$.

Solving for t, $e^{-\frac{t}{8}} = 0.70$, so $-\frac{t}{8} = ln(0.70)$, and $t = -8ln(0.70) \approx 2.85$ years.

Or use $t = \dfrac{ln(\text{area\_to\_the\_right})}{(-m)} = \dfrac{ln(0.70)}{-\frac{1}{8}} \approx 2.85$ years .

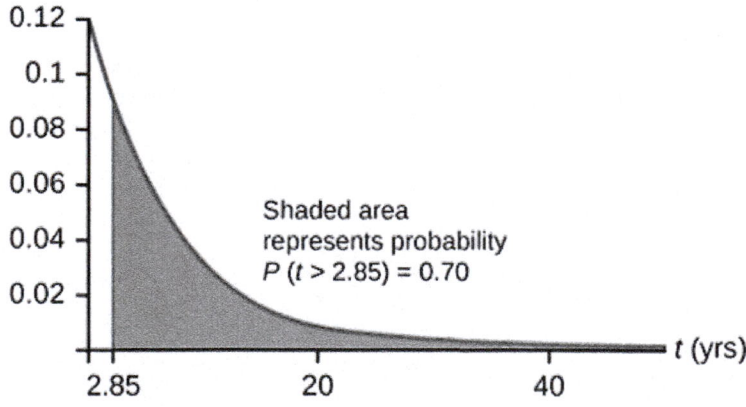

Figure 5.57

d. We want to find $0.02 = P(T < t) = 1 - e^{-\frac{t}{8}}$.

Solving for t, $e^{-\frac{t}{8}} = 0.98$, so $-\frac{t}{8} = ln(0.98)$, and $t = -8ln(0.98) \approx 0.1616$ years, or roughly two months.

The warranty should cover light bulbs that last less than 2 months.

Or use $\dfrac{\ln(\text{area\_to\_the\_right})}{(-m)} = \dfrac{\ln(1-0.2)}{-\frac{1}{8}} = 0.1616$.

e. We must find $P(T < 8 | T > 7)$.

Notice that by the rule of complement events, $P(T < 8 | T > 7) = 1 - P(T > 8 | T > 7)$.

By the memoryless property $(P(X > r + t | X > r) = P(X > t))$.

So $P(T > 8 | T > 7) = P(T > 1) = 1 - \left(1 - e^{-\frac{1}{8}}\right) = e^{-\frac{1}{8}} \approx 0.8825$

Therefore, $P(T < 8 | T > 7) = 1 - 0.8825 = 0.1175$.

97 Let X = the number of no-hitters throughout a season. Since the duration of time between no-hitters is exponential, the <u>number</u> of no-hitters <u>per season</u> is Poisson with mean $\lambda = 3$.

Therefore, $(X = 0) = \dfrac{3^0 e^{-3}}{0!} = e^{-3} \approx 0.0498$

You could let T = duration of time between no-hitters. Since the time is exponential and there are 3 no-hitters per season, then the time between no-hitters is $\frac{1}{3}$ season. For the exponential, $\mu = \frac{1}{3}$.

Therefore, $m = \dfrac{1}{\mu} = 3$ and $T \sim Exp(3)$.

---

a. The desired probability is $P(T > 1) = 1 - P(T < 1) = 1 - (1 - e^{-3}) = e^{-3} \approx 0.0498$.

b. Let T = duration of time between no-hitters. We find $P(T > 2 | T > 1)$, and by the **memoryless property** this is simply $P(T > 1)$, which we found to be 0.0498 in part a.

c. Let X = the <u>number</u> of no-hitters is a season. Assume that X is Poisson with mean $\lambda = 3$. Then $P(X > 3) = 1 - P(X \le 3)$ = 0.3528.

99

a. $\dfrac{100}{9} = 11.11$

b. $P(X > 10) = 1 - P(X \le 10) = 1 - \text{Poissoncdf}(11.11, 10) \approx 0.5532$.

c. The number of people with Type B blood encountered roughly follows the Poisson distribution, so the number of people X who arrive between successive Type B arrivals is roughly exponential with mean $\mu = 9$ and $m = \dfrac{1}{9}$. The cumulative distribution function of X is $P(X < x) = 1 - e^{-\frac{x}{9}}$. Thus hus, $P(X > 20) = 1 - P(X \le 20) = 1 - \left(1 - e^{-\frac{20}{9}}\right) \approx 0.1084$.

NOTE

We could also deduce that each person arriving has a 8/9 chance of not having Type B blood. So the probability that none of the first 20 people arrive have Type B blood is $\left(\dfrac{8}{9}\right)^{20} \approx 0.0948$. (The geometric distribution is more appropriate than the exponential because the number of people between Type B people is discrete instead of continuous.)

101 Let T = duration (in minutes) between successive visits. Since patients arrive at a rate of one patient every seven minutes, $\mu = 7$ and the decay constant is $m = \frac{1}{7}$. The cdf is $P(T < t) = 1 - e^{\frac{t}{7}}$

a. $P(T < 2) = 1 - \left[1 - e^{-\frac{2}{7}}\right] \approx 0.2485$.

b. $P(T > 15) = 1 - P(T < 15) = 1 - \left(1 - e^{-\frac{15}{7}}\right) \approx e^{-\frac{15}{7}} \approx 0.1173$.

c. $P(T > 15 | T > 10) = P(T > 5) = 1 - \left(1 - e^{-\frac{5}{7}}\right) = e^{-\frac{5}{7}} \approx 0.4895$.

d. Let X = # of patients arriving during a half-hour period. Then X has the Poisson distribution with a mean of $\frac{30}{7}$, $X \sim$ Poisson $\left(\frac{30}{7}\right)$. Find $P(X > 8) = 1 - P(X \le 8) \approx 0.0311$.

6 | THE NORMAL DISTRIBUTION

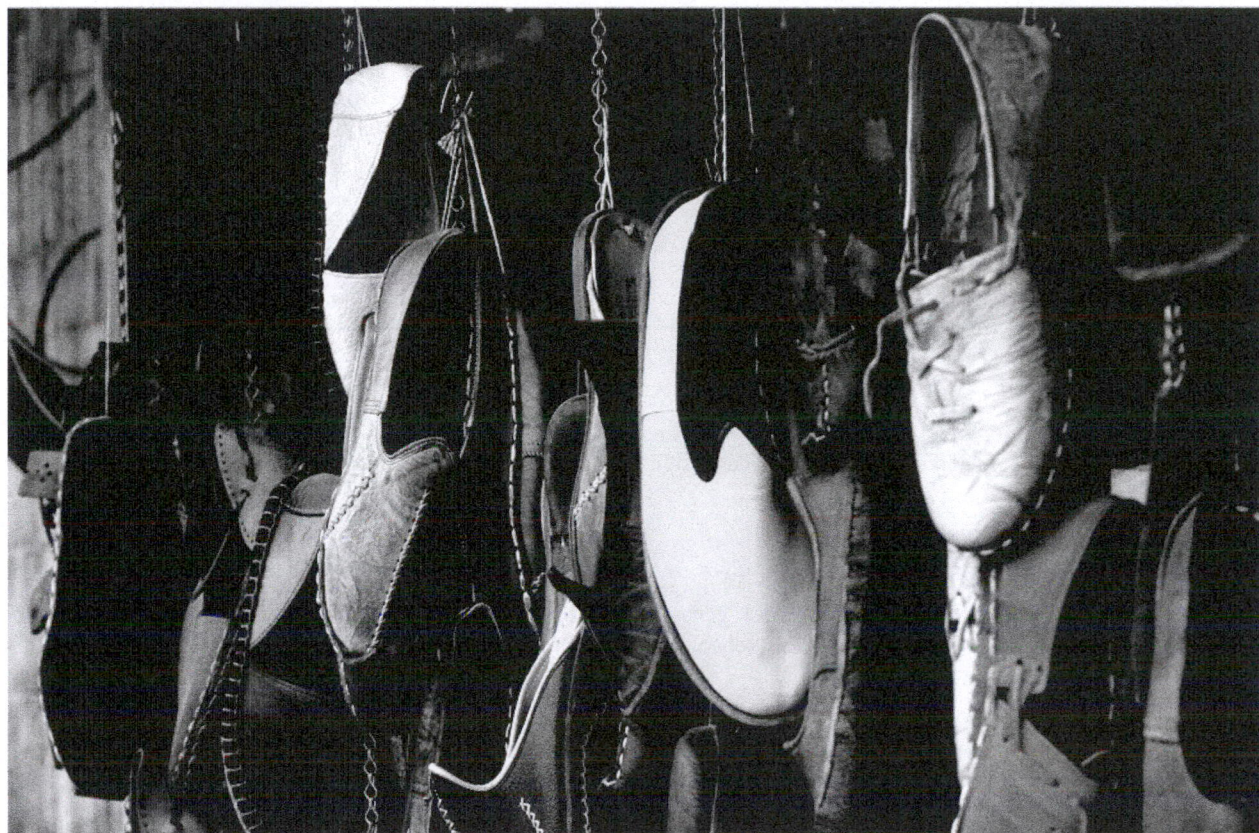

Figure 6.1 If you ask enough people about their shoe size, you will find that your graphed data is shaped like a bell curve and can be described as normally distributed. (credit: Ömer Ünlü)

Introduction

Chapter Objectives

By the end of this chapter, the student should be able to:

- Recognize the normal probability distribution and apply it appropriately.
- Recognize the standard normal probability distribution and apply it appropriately.
- Compare normal probabilities by converting to the standard normal distribution.

The normal, a continuous distribution, is the most important of all the distributions. It is widely used and even more widely abused. Its graph is bell-shaped. You see the bell curve in almost all disciplines. Some of these include psychology, business, economics, the sciences, nursing, and, of course, mathematics. Some of your instructors may use the normal distribution to help determine your grade. Most IQ scores are normally distributed. Often real-estate prices fit a normal distribution. The normal distribution is extremely important, but it cannot be applied to everything in the real world.

In this chapter, you will study the normal distribution, the standard normal distribution, and applications associated with them.

The normal distribution has two parameters (two numerical descriptive measures), the mean (μ) and the standard deviation (σ). If X is a quantity to be measured that has a normal distribution with mean (μ) and standard deviation (σ), we designate this by writing

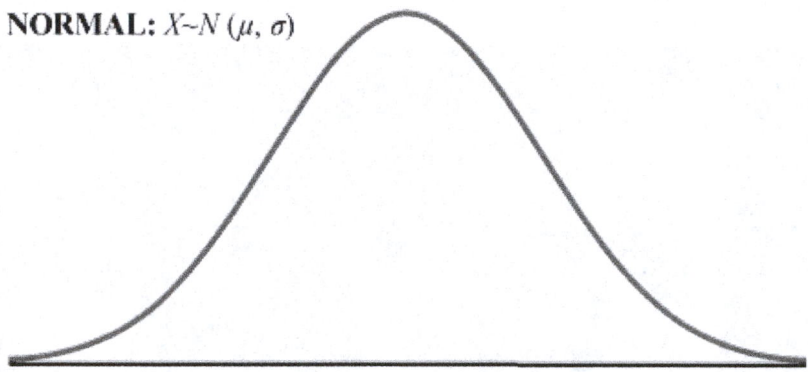

Figure 6.2

The probability density function is a rather complicated function. **Do not memorize it**. It is not necessary.

$$f(x) = \frac{1}{\sigma \cdot \sqrt{2 \cdot \pi}} \cdot e^{-\frac{1}{2} \cdot \left(\frac{x - \mu}{\sigma}\right)^2}$$

The cumulative distribution function is $P(X < x)$. It is calculated either by a calculator or a computer, or it is looked up in a table. Technology has made the tables virtually obsolete. For that reason, as well as the fact that there are various table formats, we are not including table instructions.

The curve is symmetrical about a vertical line drawn through the mean, μ. In theory, the mean is the same as the median, because the graph is symmetric about μ. As the notation indicates, the normal distribution depends only on the mean and the standard deviation. Since the area under the curve must equal one, a change in the standard deviation, σ, causes a change in the shape of the curve; the curve becomes fatter or skinnier depending on σ. A change in μ causes the graph to shift to the left or right. This means there are an infinite number of normal probability distributions. One of special interest is called the **standard normal distribution**.

Collaborative Exercise

Your instructor will record the heights of both men and women in your class, separately. Draw histograms of your data. Then draw a smooth curve through each histogram. Is each curve somewhat bell-shaped? Do you think that if you had recorded 200 data values for men and 200 for women that the curves would look bell-shaped? Calculate the mean for each data set. Write the means on the x-axis of the appropriate graph below the peak. Shade the approximate area that represents the probability that one randomly chosen male is taller than 72 inches. Shade the approximate area that represents the probability that one randomly chosen female is shorter than 60 inches. If the total area under each curve is one, does either probability appear to be more than 0.5?

6.1 | The Standard Normal Distribution

The **standard normal distribution** is a normal distribution of **standardized values called z-scores. A z-score is measured in units of the standard deviation.** For example, if the mean of a normal distribution is five and the standard deviation is two, the value 11 is three standard deviations above (or to the right of) the mean. The calculation is as follows:

$x = \mu + (z)(\sigma) = 5 + (3)(2) = 11$

The z-score is three.

The mean for the standard normal distribution is zero, and the standard deviation is one. The transformation $z = \frac{x - \mu}{\sigma}$ produces the distribution $Z \sim N(0, 1)$. The value x comes from a normal distribution with mean μ and standard deviation σ.

Z-Scores

If X is a normally distributed random variable and $X \sim N(\mu, \sigma)$, then the z-score is:

$$z = \frac{x - \mu}{\sigma}$$

The z-score tells you how many standard deviations the value x is above (to the right of) or below (to the left of) the mean, μ. Values of x that are larger than the mean have positive z-scores, and values of x that are smaller than the mean have negative z-scores. If x equals the mean, then x has a z-score of zero.

Example 6.1

Suppose $X \sim N(5, 6)$. This says that x is a normally distributed random variable with mean $\mu = 5$ and standard deviation $\sigma = 6$. Suppose $x = 17$. Then:

$$z = \frac{x - \mu}{\sigma} = \frac{17 - 5}{6} = 2$$

This means that $x = 17$ is **two standard deviations** (2σ) above or to the right of the mean $\mu = 5$. The standard deviation is $\sigma = 6$.

Notice that: $5 + (2)(6) = 17$ (The pattern is $\mu + z\sigma = x$)

Now suppose $x = 1$. Then: $z = \frac{x - \mu}{\sigma} = \frac{1 - 5}{6} = -0.67$ (rounded to two decimal places)

This means that $x = 1$ is 0.67 standard deviations (-0.67σ) below or to the left of the mean $\mu = 5$. Notice that: $5 + (-0.67)(6)$ is approximately equal to one (This has the pattern $\mu + (-0.67)\sigma = 1$)

Summarizing, when z is positive, x is above or to the right of μ and when z is negative, x is to the left of or below μ. Or, when z is positive, x is greater than μ, and when z is negative x is less than μ.

Try It Σ

6.1 What is the z-score of x, when $x = 1$ and $X \sim N(12,3)$?

Example 6.2

Some doctors believe that a person can lose five pounds, on the average, in a month by reducing his or her fat intake and by exercising consistently. Suppose weight loss has a normal distribution. Let X = the amount of weight lost(in pounds) by a person in a month. Use a standard deviation of two pounds. $X \sim N(5, 2)$. Fill in the blanks.

a. Suppose a person **lost** ten pounds in a month. The z-score when $x = 10$ pounds is $z = 2.5$ (verify). This z-score tells you that $x = 10$ is _____ standard deviations to the _____ (right or left) of the mean _____ (What is the mean?).

Solution 6.2

a. This z-score tells you that $x = 10$ is **2.5** standard deviations to the **right** of the mean **five**.

b. Suppose a person **gained** three pounds (a negative weight loss). Then $z =$ _____. This z-score tells you that $x = –3$ is _____ standard deviations to the _____ (right or left) of the mean.

Solution 6.2

b. $z = –4$. This z-score tells you that $x = –3$ is **four** standard deviations to the **left** of the mean.

Suppose the random variables X and Y have the following normal distributions: $X \sim N(5, 6)$ and $Y \sim N(2, 1)$. If $x = 17$, then $z = 2$. (This was previously shown.) If $y = 4$, what is z?

$$z = \frac{y - \mu}{\sigma} = \frac{4 - 2}{1} = 2 \text{ where } \mu = 2 \text{ and } \sigma = 1.$$

The z-score for $y = 4$ is $z = 2$. This means that four is $z = 2$ standard deviations to the right of the mean. Therefore, $x = 17$ and $y = 4$ are both two (of **their own**) standard deviations to the right of **their** respective means.

The z-score allows us to compare data that are scaled differently. To understand the concept, suppose $X \sim N(5, 6)$ represents weight gains for one group of people who are trying to gain weight in a six week period and $Y \sim N(2, 1)$ measures the same weight gain for a second group of people. A negative weight gain would be a weight loss. Since $x = 17$ and $y = 4$ are each two standard deviations to the right of their means, they represent the same, standardized weight gain **relative to their means**.

6.2 Fill in the blanks.

Jerome averages 16 points a game with a standard deviation of four points. $X \sim N(16,4)$. Suppose Jerome scores ten points in a game. The z–score when $x = 10$ is $–1.5$. This score tells you that $x = 10$ is _____ standard deviations to the _____(right or left) of the mean_____(What is the mean?).

The Empirical Rule

If X is a random variable and has a normal distribution with mean μ and standard deviation σ, then the **Empirical Rule** says the following:

- About 68% of the x values lie between $–1\sigma$ and $+1\sigma$ of the mean μ (within one standard deviation of the mean).

- About 95% of the x values lie between $–2\sigma$ and $+2\sigma$ of the mean μ (within two standard deviations of the mean).

- About 99.7% of the x values lie between $–3\sigma$ and $+3\sigma$ of the mean μ (within three standard deviations of the mean). Notice that almost all the x values lie within three standard deviations of the mean.

- The z-scores for $+1\sigma$ and $–1\sigma$ are $+1$ and $–1$, respectively.

- The z-scores for $+2\sigma$ and $–2\sigma$ are $+2$ and $–2$, respectively.

- The z-scores for $+3\sigma$ and $–3\sigma$ are $+3$ and $–3$ respectively.

The empirical rule is also known as the 68-95-99.7 rule.

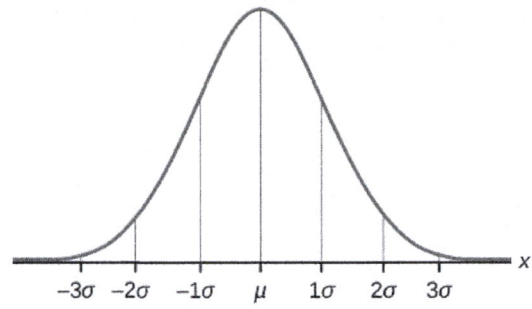

Figure 6.3

Example 6.3

The mean height of 15 to 18-year-old males from Chile from 2009 to 2010 was 170 cm with a standard deviation of 6.28 cm. Male heights are known to follow a normal distribution. Let X = the height of a 15 to 18-year-old male from Chile in 2009 to 2010. Then $X \sim N(170, 6.28)$.

a. Suppose a 15 to 18-year-old male from Chile was 168 cm tall from 2009 to 2010. The z-score when $x = 168$ cm is z = _____. This z-score tells you that $x = 168$ is _____ standard deviations to the _____ (right or left) of the mean _____ (What is the mean?).

Solution 6.3
a. −0.32, 0.32, left, 170

b. Suppose that the height of a 15 to 18-year-old male from Chile from 2009 to 2010 has a z-score of $z = 1.27$. What is the male's height? The z-score ($z = 1.27$) tells you that the male's height is _____ standard deviations to the _____ (right or left) of the mean.

Solution 6.3
b. 177.98, 1.27, right

Try It Σ

6.3 Use the information in **Example 6.3** to answer the following questions.

a. Suppose a 15 to 18-year-old male from Chile was 176 cm tall from 2009 to 2010. The z-score when $x = 176$ cm is z = _____. This z-score tells you that $x = 176$ cm is _____ standard deviations to the _____ (right or left) of the mean _____ (What is the mean?).

b. Suppose that the height of a 15 to 18-year-old male from Chile from 2009 to 2010 has a z-score of $z = -2$. What is the male's height? The z-score ($z = -2$) tells you that the male's height is _____ standard deviations to the _____ (right or left) of the mean.

Example 6.4

From 1984 to 1985, the mean height of 15 to 18-year-old males from Chile was 172.36 cm, and the standard deviation was 6.34 cm. Let Y = the height of 15 to 18-year-old males from 1984 to 1985. Then $Y \sim N(172.36, 6.34)$.

The mean height of 15 to 18-year-old males from Chile from 2009 to 2010 was 170 cm with a standard deviation of 6.28 cm. Male heights are known to follow a normal distribution. Let X = the height of a 15 to 18-year-old male from Chile in 2009 to 2010. Then $X \sim N(170, 6.28)$.

Find the z-scores for $x = 160.58$ cm and $y = 162.85$ cm. Interpret each z-score. What can you say about $x = 160.58$ cm and $y = 162.85$ cm?

Solution 6.4
The z-score for $x = 160.58$ is $z = -1.5$.
The z-score for $y = 162.85$ is $z = -1.5$.
Both $x = 160.58$ and $y = 162.85$ deviate the same number of standard deviations from their respective means and in the same direction.

Try It Σ

6.4 In 2012, 1,664,479 students took the SAT exam. The distribution of scores in the verbal section of the SAT had a mean $\mu = 496$ and a standard deviation $\sigma = 114$. Let X = a SAT exam verbal section score in 2012. Then $X \sim N(496, 114)$.

Find the z-scores for $x_1 = 325$ and $x_2 = 366.21$. Interpret each z-score. What can you say about $x_1 = 325$ and $x_2 = 366.21$?

Example 6.5

Suppose x has a normal distribution with mean 50 and standard deviation 6.

- About 68% of the x values lie between $-1\sigma = (-1)(6) = -6$ and $1\sigma = (1)(6) = 6$ of the mean 50. The values 50 − 6 = 44 and 50 + 6 = 56 are within one standard deviation of the mean 50. The z-scores are −1 and +1 for 44 and 56, respectively.

- About 95% of the x values lie between $-2\sigma = (-2)(6) = -12$ and $2\sigma = (2)(6) = 12$. The values 50 − 12 = 38 and 50 + 12 = 62 are within two standard deviations of the mean 50. The z-scores are −2 and +2 for 38 and 62, respectively.

- About 99.7% of the x values lie between $-3\sigma = (-3)(6) = -18$ and $3\sigma = (3)(6) = 18$ of the mean 50. The values 50 − 18 = 32 and 50 + 18 = 68 are within three standard deviations of the mean 50. The z-scores are −3 and +3 for 32 and 68, respectively.

Try It Σ

6.5 Suppose X has a normal distribution with mean 25 and standard deviation five. Between what values of x do 68% of the values lie?

Example 6.6

From 1984 to 1985, the mean height of 15 to 18-year-old males from Chile was 172.36 cm, and the standard deviation was 6.34 cm. Let Y = the height of 15 to 18-year-old males in 1984 to 1985. Then $Y \sim N(172.36, 6.34)$.

a. About 68% of the y values lie between what two values? These values are _____. The z-scores are _____, respectively.

b. About 95% of the *y* values lie between what two values? These values are _____. The *z*-scores are _____ respectively.

c. About 99.7% of the *y* values lie between what two values? These values are _____. The *z*-scores are _____, respectively.

Solution 6.6

a. About 68% of the values lie between 166.02 and 178.7. The *z*-scores are –1 and 1.

b. About 95% of the values lie between 159.68 and 185.04. The *z*-scores are –2 and 2.

c. About 99.7% of the values lie between 153.34 and 191.38. The *z*-scores are –3 and 3.

Try It Σ

6.6 The scores on a college entrance exam have an approximate normal distribution with mean, μ = 52 points and a standard deviation, σ = 11 points.

a. About 68% of the *y* values lie between what two values? These values are _____. The *z*-scores are _____, respectively.

b. About 95% of the *y* values lie between what two values? These values are _____. The *z*-scores are _____, respectively.

c. About 99.7% of the *y* values lie between what two values? These values are _____. The *z*-scores are _____, respectively.

6.2 | Using the Normal Distribution

The shaded area in the following graph indicates the area to the left of *x*. This area is represented by the probability $P(X < x)$. Normal tables, computers, and calculators provide or calculate the probability $P(X < x)$.

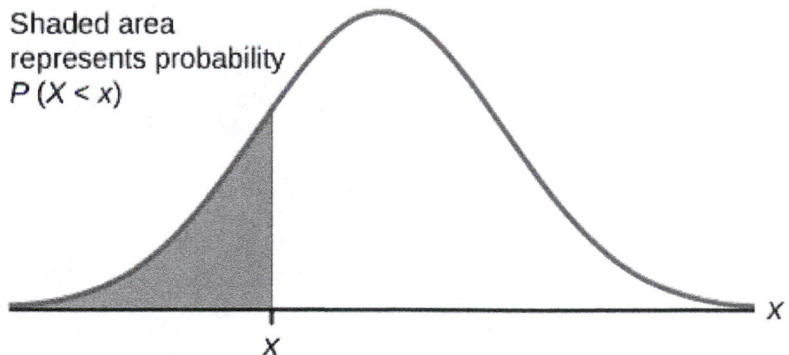

Figure 6.4

The area to the right is then $P(X > x) = 1 - P(X < x)$. Remember, $P(X < x) = $ **Area to the left** of the vertical line through *x*. $P(X < x) = 1 - P(X < x) = $ **Area to the right** of the vertical line through *x*. $P(X < x)$ is the same as $P(X \leq x)$ and $P(X > x)$ is the same as $P(X \geq x)$ for continuous distributions.

Calculations of Probabilities

Probabilities are calculated using technology. There are instructions given as necessary for the TI-83+ and TI-84 calculators.

NOTE

To calculate the probability, use the probability tables provided in **Appendix H** without the use of technology. The tables include instructions for how to use them.

Example 6.7

If the area to the left is 0.0228, then the area to the right is $1 - 0.0228 = 0.9772$.

Try It Σ

6.7 If the area to the left of x is 0.012, then what is the area to the right?

Example 6.8

The final exam scores in a statistics class were normally distributed with a mean of 63 and a standard deviation of five.

a. Find the probability that a randomly selected student scored more than 65 on the exam.

Solution 6.8

a. Let X = a score on the final exam. $X \sim N(63, 5)$, where $\mu = 63$ and $\sigma = 5$

Draw a graph.

Then, find $P(x > 65)$.

$P(x > 65) = 0.3446$

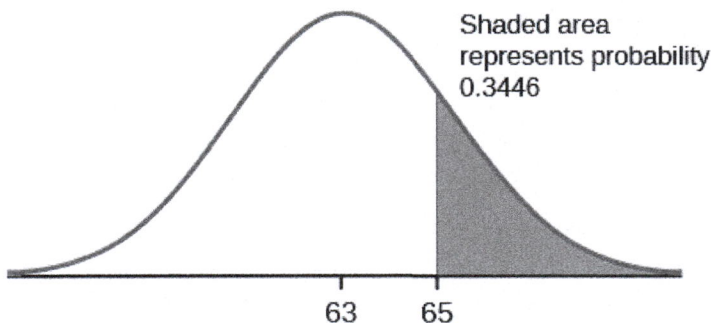

Figure 6.5

The probability that any student selected at random scores more than 65 is 0.3446.

Using the TI-83, 83+, 84, 84+ Calculator

Go into 2nd DISTR.

After pressing 2nd DISTR, press 2:normalcdf.

The syntax for the instructions are as follows:

normalcdf(lower value, upper value, mean, standard deviation) For this problem: normalcdf(65,1E99,63,5) = 0.3446. You get 1E99 (= 10^{99}) by pressing 1, the EE key (a 2nd key) and then 99. Or, you can enter 10^99 instead. The number 10^{99} is way out in the right tail of the normal curve. We are calculating the area between 65 and 10^{99}. In some instances, the lower number of the area might be –1E99 (= -10^{99}). The number -10^{99} is way out in the left tail of the normal curve.

HISTORICAL NOTE

The TI probability program calculates a z-score and then the probability from the z-score. Before technology, the z-score was looked up in a standard normal probability table (because the math involved is too cumbersome) to find the probability. In this example, a standard normal table with area to the left of the z-score was used. You calculate the z-score and look up the area to the left. The probability is the area to the right.

$$z = \frac{65 - 63}{5} = 0.4$$

Area to the left is 0.6554.

$P(x > 65) = P(z > 0.4) = 1 - 0.6554 = 0.3446$

 Using the TI-83, 83+, 84, 84+ Calculator

Calculate the z-score:

*Press 2nd Distr
*Press 3:invNorm(
*Enter the area to the left of z followed by)
*Press ENTER.
For this Example, the steps are
2nd Distr
3:invNorm(.6554) ENTER
The answer is 0.3999 which rounds to 0.4.

b. Find the probability that a randomly selected student scored less than 85.

Solution 6.8

b. Draw a graph.

Then find $P(x < 85)$, and shade the graph.

Using a computer or calculator, find $P(x < 85) = 1$.

normalcdf(0,85,63,5) = 1 (rounds to one)

The probability that one student scores less than 85 is approximately one (or 100%).

c. Find the 90th percentile (that is, find the score *k* that has 90% of the scores below *k* and 10% of the scores above *k*).

Solution 6.8

c. Find the 90th percentile. For each problem or part of a problem, draw a new graph. Draw the *x*-axis. Shade the area that corresponds to the 90th percentile.

Let *k* = the 90th percentile. The variable *k* is located on the *x*-axis. $P(x < k)$ is the area to the left of *k*. The 90th percentile *k* separates the exam scores into those that are the same or lower than *k* and those that are the same or higher. Ninety percent of the test scores are the same or lower than *k*, and ten percent are the same or higher. The variable *k* is often called a **critical value**.

k = 69.4

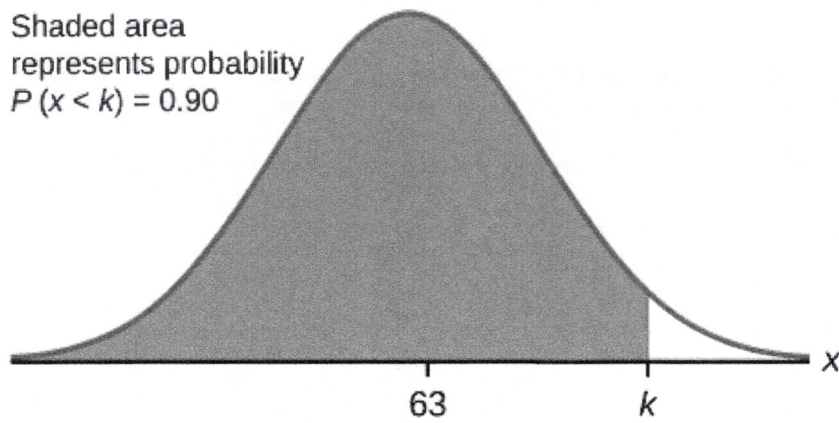

Figure 6.6

The 90th percentile is 69.4. This means that 90% of the test scores fall at or below 69.4 and 10% fall at or above. To get this answer on the calculator, follow this step:

 Using the TI-83, 83+, 84, 84+ Calculator

invNorm in 2nd DISTR. invNorm(area to the left, mean, standard deviation)
For this problem, invNorm(0.90,63,5) = 69.4

d. Find the 70th percentile (that is, find the score *k* such that 70% of scores are below *k* and 30% of the scores are above *k*).

Solution 6.8

d. Find the 70th percentile.

Draw a new graph and label it appropriately. *k* = 65.6

The 70th percentile is 65.6. This means that 70% of the test scores fall at or below 65.5 and 30% fall at or above.

invNorm(0.70,63,5) = 65.6

Try It

6.8 The golf scores for a school team were normally distributed with a mean of 68 and a standard deviation of three. Find the probability that a randomly selected golfer scored less than 65.

Example 6.9

A personal computer is used for office work at home, research, communication, personal finances, education, entertainment, social networking, and a myriad of other things. Suppose that the average number of hours a household personal computer is used for entertainment is two hours per day. Assume the times for entertainment are normally distributed and the standard deviation for the times is half an hour.

a. Find the probability that a household personal computer is used for entertainment between 1.8 and 2.75 hours per day.

Solution 6.9

a. Let X = the amount of time (in hours) a household personal computer is used for entertainment. $X \sim N(2, 0.5)$ where $\mu = 2$ and $\sigma = 0.5$.

Find $P(1.8 < x < 2.75)$.

The probability for which you are looking is the area **between** $x = 1.8$ and $x = 2.75$. $P(1.8 < x < 2.75) = 0.5886$

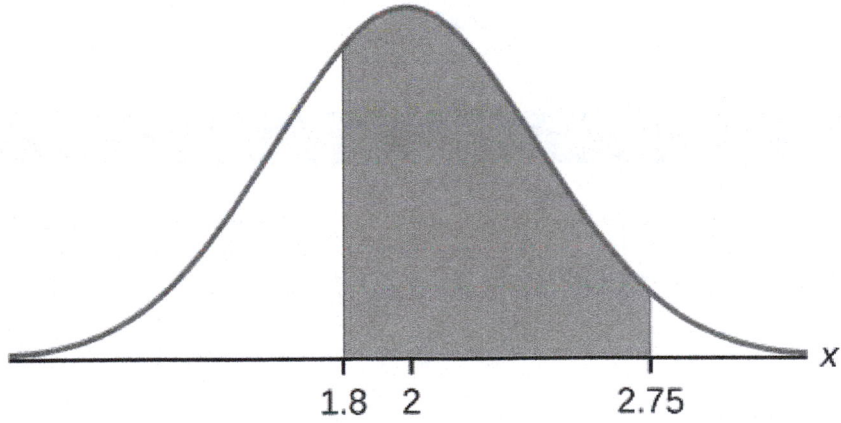

Figure 6.7

normalcdf(1.8,2.75,2,0.5) = 0.5886

The probability that a household personal computer is used between 1.8 and 2.75 hours per day for entertainment is 0.5886.

b. Find the maximum number of hours per day that the bottom quartile of households uses a personal computer for entertainment.

Solution 6.9

b. To find the maximum number of hours per day that the bottom quartile of households uses a personal computer for entertainment, **find the 25th percentile**, k, where $P(x < k) = 0.25$.

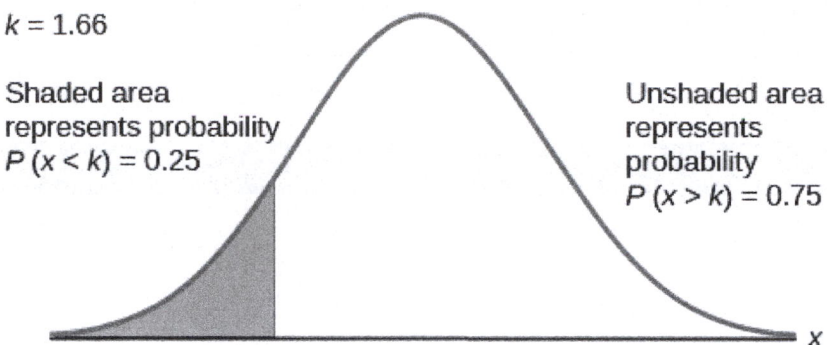

Figure 6.8

invNorm(0.25,2,0.5) = 1.66

The maximum number of hours per day that the bottom quartile of households uses a personal computer for entertainment is 1.66 hours.

 Try It Σ

6.9 The golf scores for a school team were normally distributed with a mean of 68 and a standard deviation of three. Find the probability that a golfer scored between 66 and 70.

Example 6.10

There are approximately one billion smartphone users in the world today. In the United States the ages 13 to 55+ of smartphone users approximately follow a normal distribution with approximate mean and standard deviation of 36.9 years and 13.9 years, respectively.

a. Determine the probability that a random smartphone user in the age range 13 to 55+ is between 23 and 64.7 years old.

Solution 6.10
a. normalcdf(23,64.7,36.9,13.9) = 0.8186

b. Determine the probability that a randomly selected smartphone user in the age range 13 to 55+ is at most 50.8 years old.

Solution 6.10
b. normalcdf(-10^{99},50.8,36.9,13.9) = 0.8413

c. Find the 80th percentile of this distribution, and interpret it in a complete sentence.

Solution 6.10
c.
invNorm(0.80,36.9,13.9) = 48.6

The 80th percentile is 48.6 years.

80% of the smartphone users in the age range 13 – 55+ are 48.6 years old or less.

Try It Σ

6.10 Use the information in **Example 6.10** to answer the following questions.

a. Find the 30th percentile, and interpret it in a complete sentence.

b. What is the probability that the age of a randomly selected smartphone user in the range 13 to 55+ is less than 27 years old.

Example 6.11

There are approximately one billion smartphone users in the world today. In the United States the ages 13 to 55+ of smartphone users approximately follow a normal distribution with approximate mean and standard deviation of 36.9 years and 13.9 years respectively. Using this information, answer the following questions (round answers to one decimal place).

a. Calculate the interquartile range (*IQR*).

Solution 6.11

a.

$IQR = Q_3 - Q_1$

Calculate Q_3 = 75th percentile and Q_1 = 25th percentile.

invNorm(0.75,36.9,13.9) = Q_3 = 46.2754

invNorm(0.25,36.9,13.9) = Q_1 = 27.5246

$IQR = Q_3 - Q_1$ = 18.7508

b. Forty percent of the ages that range from 13 to 55+ are at least what age?

Solution 6.11

b.

Find *k* where $P(x > k)$ = 0.40 ("At least" translates to "greater than or equal to.")

0.40 = the area to the right.

Area to the left = 1 – 0.40 = 0.60.

The area to the left of *k* = 0.60.

invNorm(0.60,36.9,13.9) = 40.4215.

k = 40.42.

Forty percent of the ages that range from 13 to 55+ are at least 40.42 years.

Try It Σ

6.11 Two thousand students took an exam. The scores on the exam have an approximate normal distribution with a mean μ = 81 points and standard deviation σ = 15 points.

a. Calculate the first- and third-quartile scores for this exam.

b. The middle 50% of the exam scores are between what two values?

Example 6.12

A citrus farmer who grows mandarin oranges finds that the diameters of mandarin oranges harvested on his farm follow a normal distribution with a mean diameter of 5.85 cm and a standard deviation of 0.24 cm.

a. Find the probability that a randomly selected mandarin orange from this farm has a diameter larger than 6.0 cm. Sketch the graph.

Solution 6.12

a. normalcdf(6,10^99,5.85,0.24) = 0.2660

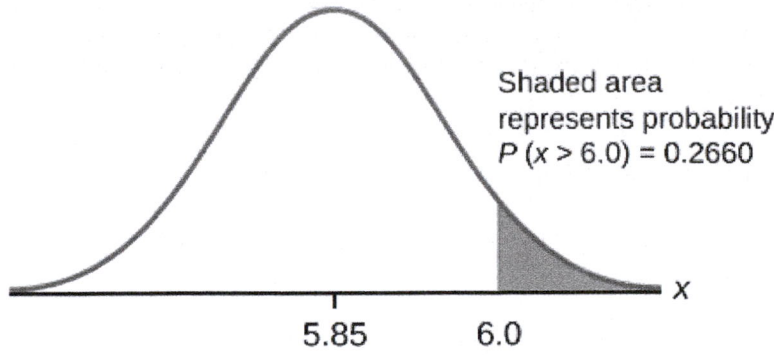

Shaded area
represents probability
$P(x > 6.0) = 0.2660$

Figure 6.9

b. The middle 20% of mandarin oranges from this farm have diameters between _____ and _____.

Solution 6.12
b.
$1 - 0.20 = 0.80$
The tails of the graph of the normal distribution each have an area of 0.40.

Find $k1$, the 40th percentile, and $k2$, the 60th percentile (0.40 + 0.20 = 0.60).
$k1 = $ invNorm(0.40,5.85,0.24) = 5.79 cm
$k2 = $ invNorm(0.60,5.85,0.24) = 5.91 cm

c. Find the 90th percentile for the diameters of mandarin oranges, and interpret it in a complete sentence.

Solution 6.12
c. 6.16: Ninety percent of the diameter of the mandarin oranges is at most 6.15 cm.

Try It Σ

6.12 Using the information from **Example 6.12**, answer the following:

a.   The middle 45% of mandarin oranges from this farm are between _____ and _____.

b.   Find the 16th percentile and interpret it in a complete sentence.

6.3 | Normal Distribution (Lap Times)

Stats Lab

6.1 Normal Distribution (Lap Times)

Class Time:

Names:

Student Learning Outcome

- The student will compare and contrast empirical data and a theoretical distribution to determine if Terry Vogel's lap times fit a continuous distribution.

Directions

Round the relative frequencies and probabilities to four decimal places. Carry all other decimal answers to two places.

Collect the Data

1. Use the data from **Appendix C**. Use a stratified sampling method by lap (races 1 to 20) and a random number generator to pick six lap times from each stratum. Record the lap times below for laps two to seven.

Table 6.1

2. Construct a histogram. Make five to six intervals. Sketch the graph using a ruler and pencil. Scale the axes.

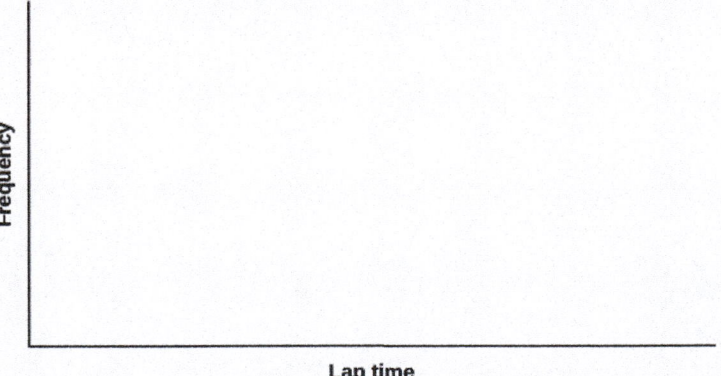

Figure 6.10

3. Calculate the following:

   a. $\bar{x}$ = _____

   b. s = _____

4. Draw a smooth curve through the tops of the bars of the histogram. Write one to two complete sentences to describe the general shape of the curve. (Keep it simple. Does the graph go straight across, does it have a v-shape, does it have a hump in the middle or at either end, and so on?)

Analyze the Distribution

Using your sample mean, sample standard deviation, and histogram to help, what is the approximate theoretical distribution of the data?

- $X \sim$ _____(_____,_____)
- How does the histogram help you arrive at the approximate distribution?

Describe the Data

Use the data you collected to complete the following statements.

- The *IQR* goes from _____ to _____.
- *IQR* = _____. (*IQR* = $Q_3 - Q_1$)
- The 15th percentile is _____.
- The 85th percentile is _____.
- The median is _____.
- The empirical probability that a randomly chosen lap time is more than 130 seconds is _____.
- Explain the meaning of the 85th percentile of this data.

Theoretical Distribution

Using the theoretical distribution, complete the following statements. You should use a normal approximation based on your sample data.

- The *IQR* goes from _____ to _____.
- *IQR* = _____.
- The 15th percentile is _____.
- The 85th percentile is _____.
- The median is _____.
- The probability that a randomly chosen lap time is more than 130 seconds is _____.
- Explain the meaning of the 85th percentile of this distribution.

Discussion Questions

Do the data from the section titled **Collect the Data** give a close approximation to the theoretical distribution in the section titled **Analyze the Distribution**? In complete sentences and comparing the result in the sections titled **Describe the Data** and **Theoretical Distribution**, explain why or why not.

6.4 | Normal Distribution (Pinkie Length)

Stats Lab

6.2 Normal Distribution (Pinkie Length)

Class Time:

Names:

Student Learning Outcomes

- The student will compare empirical data and a theoretical distribution to determine if data from the experiment follow a continuous distribution.

Collect the Data

Measure the length of your pinky finger (in centimeters).

1. Randomly survey 30 adults for their pinky finger lengths. Round the lengths to the nearest 0.5 cm.

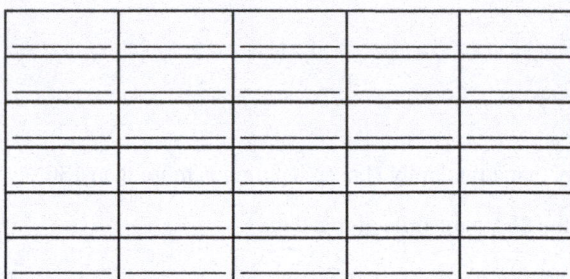

Table 6.2

2. Construct a histogram. Make five to six intervals. Sketch the graph using a ruler and pencil. Scale the axes.

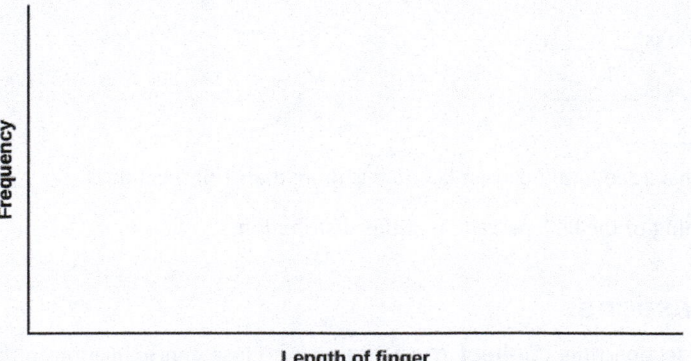

Figure 6.11

3. Calculate the following.

    a. $\bar{x}$ = _____

    b. s = _____

4. Draw a smooth curve through the top of the bars of the histogram. Write one to two complete sentences to describe the general shape of the curve. (Keep it simple. Does the graph go straight across, does it have a v-shape, does it have a hump in the middle or at either end, and so on?)

Analyze the Distribution

Using your sample mean, sample standard deviation, and histogram, what was the approximate theoretical distribution of the data you collected?

- $X \sim$ _____(_____,_____)

- How does the histogram help you arrive at the approximate distribution?

Describe the Data

Using the data you collected complete the following statements. (Hint: order the data)

REMEMBER

$(IQR = Q_3 - Q_1)$

- $IQR =$ _____
- The 15th percentile is _____.
- The 85th percentile is _____.
- Median is _____.
- What is the theoretical probability that a randomly chosen pinky length is more than 6.5 cm?
- Explain the meaning of the 85th percentile of this data.

Theoretical Distribution

Using the theoretical distribution, complete the following statements. Use a normal approximation based on the sample mean and standard deviation.

- $IQR =$ _____
- The 15th percentile is _____.
- The 85th percentile is _____.
- Median is _____.
- What is the theoretical probability that a randomly chosen pinky length is more than 6.5 cm?
- Explain the meaning of the 85th percentile of this data.

Discussion Questions

Do the data you collected give a close approximation to the theoretical distribution? In complete sentences and comparing the results in the sections titled **Describe the Data** and **Theoretical Distribution**, explain why or why not.

KEY TERMS

Normal Distribution
a continuous random variable (RV) with pdf $f(x) = \frac{1}{\sigma\sqrt{2\pi}} e^{\frac{-(x-m)^2}{2\sigma^2}}$, where μ is the mean of the distribution and σ is the standard deviation; notation: $X \sim N(\mu, \sigma)$. If $\mu = 0$ and $\sigma = 1$, the RV is called the **standard normal distribution**.

Standard Normal Distribution a continuous random variable (RV) $X \sim N(0, 1)$; when X follows the standard normal distribution, it is often noted as $Z \sim N(0, 1)$.

z-score the linear transformation of the form $z = \frac{x - \mu}{\sigma}$; if this transformation is applied to any normal distribution $X \sim N(\mu, \sigma)$ the result is the standard normal distribution $Z \sim N(0,1)$. If this transformation is applied to any specific value x of the RV with mean μ and standard deviation σ, the result is called the z-score of x. The z-score allows us to compare data that are normally distributed but scaled differently.

CHAPTER REVIEW

6.1 The Standard Normal Distribution

A z-score is a standardized value. Its distribution is the standard normal, $Z \sim N(0, 1)$. The mean of the z-scores is zero and the standard deviation is one. If z is the z-score for a value x from the normal distribution $N(\mu, \sigma)$ then z tells you how many standard deviations x is above (greater than) or below (less than) μ.

6.2 Using the Normal Distribution

The normal distribution, which is continuous, is the most important of all the probability distributions. Its graph is bell-shaped. This bell-shaped curve is used in almost all disciplines. Since it is a continuous distribution, the total area under the curve is one. The parameters of the normal are the mean μ and the standard deviation σ. A special normal distribution, called the standard normal distribution is the distribution of z-scores. Its mean is zero, and its standard deviation is one.

FORMULA REVIEW

6.0 Introduction

$X \sim N(\mu, \sigma)$

μ = the mean; σ = the standard deviation

6.1 The Standard Normal Distribution

$Z \sim N(0, 1)$

z = a standardized value (z-score)

mean = 0; standard deviation = 1

To find the Kth percentile of X when the z-scores is known:
$k = \mu + (z)\sigma$

z-score: $z = \frac{x-\mu}{\sigma}$

Z = the random variable for z-scores

$Z \sim N(0, 1)$

6.2 Using the Normal Distribution

Normal Distribution: $X \sim N(\mu, \sigma)$ where μ is the mean and σ is the standard deviation.

Standard Normal Distribution: $Z \sim N(0, 1)$.

Calculator function for probability: normalcdf (lower x value of the area, upper x value of the area, mean, standard deviation)

Calculator function for the kth percentile: k = invNorm (area to the left of k, mean, standard deviation)

PRACTICE

6.1 The Standard Normal Distribution

1. A bottle of water contains 12.05 fluid ounces with a standard deviation of 0.01 ounces. Define the random variable X in words. $X =$ _____.

2. A normal distribution has a mean of 61 and a standard deviation of 15. What is the median?

3. $X \sim N(1, 2)$

$\sigma =$ _____

4. A company manufactures rubber balls. The mean diameter of a ball is 12 cm with a standard deviation of 0.2 cm. Define the random variable X in words. $X =$ _____.

5. $X \sim N(-4, 1)$

What is the median?

6. $X \sim N(3, 5)$

$\sigma =$ _____

7. $X \sim N(-2, 1)$

$\mu =$ _____

8. What does a z-score measure?

9. What does standardizing a normal distribution do to the mean?

10. Is $X \sim N(0, 1)$ a standardized normal distribution? Why or why not?

11. What is the z-score of $x = 12$, if it is two standard deviations to the right of the mean?

12. What is the z-score of $x = 9$, if it is 1.5 standard deviations to the left of the mean?

13. What is the z-score of $x = -2$, if it is 2.78 standard deviations to the right of the mean?

14. What is the z-score of $x = 7$, if it is 0.133 standard deviations to the left of the mean?

15. Suppose $X \sim N(2, 6)$. What value of x has a z-score of three?

16. Suppose $X \sim N(8, 1)$. What value of x has a z-score of -2.25?

17. Suppose $X \sim N(9, 5)$. What value of x has a z-score of -0.5?

18. Suppose $X \sim N(2, 3)$. What value of x has a z-score of -0.67?

19. Suppose $X \sim N(4, 2)$. What value of x is 1.5 standard deviations to the left of the mean?

20. Suppose $X \sim N(4, 2)$. What value of x is two standard deviations to the right of the mean?

21. Suppose $X \sim N(8, 9)$. What value of x is 0.67 standard deviations to the left of the mean?

22. Suppose $X \sim N(-1, 2)$. What is the z-score of $x = 2$?

23. Suppose $X \sim N(12, 6)$. What is the z-score of $x = 2$?

24. Suppose $X \sim N(9, 3)$. What is the z-score of $x = 9$?

25. Suppose a normal distribution has a mean of six and a standard deviation of 1.5. What is the z-score of $x = 5.5$?

26. In a normal distribution, $x = 5$ and $z = -1.25$. This tells you that $x = 5$ is _____ standard deviations to the _____ (right or left) of the mean.

27. In a normal distribution, $x = 3$ and $z = 0.67$. This tells you that $x = 3$ is _____ standard deviations to the _____ (right or left) of the mean.

28. In a normal distribution, $x = -2$ and $z = 6$. This tells you that $x = -2$ is _____ standard deviations to the _____ (right or left) of the mean.

29. In a normal distribution, $x = -5$ and $z = -3.14$. This tells you that $x = -5$ is _____ standard deviations to the _____ (right or left) of the mean.

30. In a normal distribution, $x = 6$ and $z = -1.7$. This tells you that $x = 6$ is _____ standard deviations to the _____ (right or left) of the mean.

31. About what percent of x values from a normal distribution lie within one standard deviation (left and right) of the mean of that distribution?

32. About what percent of the x values from a normal distribution lie within two standard deviations (left and right) of the mean of that distribution?

33. About what percent of x values lie between the second and third standard deviations (both sides)?

34. Suppose $X \sim N(15, 3)$. Between what x values does 68.27% of the data lie? The range of x values is centered at the mean of the distribution (i.e., 15).

35. Suppose $X \sim N(-3, 1)$. Between what x values does 95.45% of the data lie? The range of x values is centered at the mean of the distribution(i.e., -3).

36. Suppose $X \sim N(-3, 1)$. Between what x values does 34.14% of the data lie?

37. About what percent of x values lie between the mean and three standard deviations?

38. About what percent of x values lie between the mean and one standard deviation?

39. About what percent of x values lie between the first and second standard deviations from the mean (both sides)?

40. About what percent of x values lie betwween the first and third standard deviations(both sides)?
Use the following information to answer the next two exercises: The life of Sunshine CD players is normally distributed with mean of 4.1 years and a standard deviation of 1.3 years. A CD player is guaranteed for three years. We are interested in the length of time a CD player lasts.

41. Define the random variable X in words. $X =$ _____.

42. $X \sim$ _____(_____,_____)

6.2 Using the Normal Distribution

43. How would you represent the area to the left of one in a probability statement?

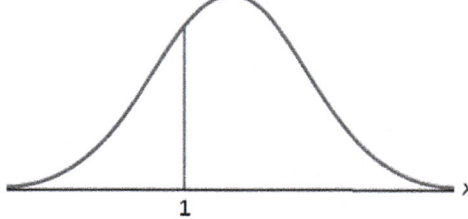

Figure 6.12

44. What is the area to the right of one?

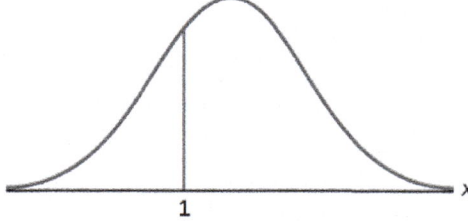

Figure 6.13

45. Is $P(x < 1)$ equal to $P(x \le 1)$? Why?

46. How would you represent the area to the left of three in a probability statement?

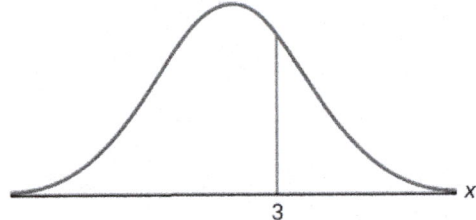

Figure 6.14

47. What is the area to the right of three?

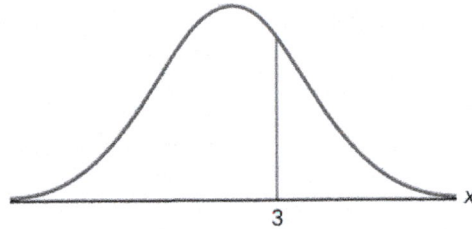

Figure 6.15

48. If the area to the left of x in a normal distribution is 0.123, what is the area to the right of x?

49. If the area to the right of x in a normal distribution is 0.543, what is the area to the left of x?
Use the following information to answer the next four exercises:

$X \sim N(54, 8)$

50. Find the probability that $x > 56$.

51. Find the probability that $x < 30$.

52. Find the 80th percentile.

53. Find the 60th percentile.

54. $X \sim N(6, 2)$

Find the probability that x is between three and nine.

55. $X \sim N(-3, 4)$

Find the probability that x is between one and four.

56. $X \sim N(4, 5)$

Find the maximum of x in the bottom quartile.

57. *Use the following information to answer the next three exercise:* The life of Sunshine CD players is normally distributed with a mean of 4.1 years and a standard deviation of 1.3 years. A CD player is guaranteed for three years. We are interested in the length of time a CD player lasts. Find the probability that a CD player will break down during the guarantee period.

    a.   Sketch the situation. Label and scale the axes. Shade the region corresponding to the probability.

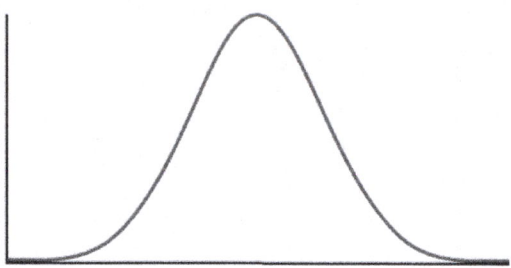

Figure 6.16

    b. $P(0 < x <$ _____$) =$ _____ (Use zero for the minimum value of x.)

58. Find the probability that a CD player will last between 2.8 and six years.
    a. Sketch the situation. Label and scale the axes. Shade the region corresponding to the probability.

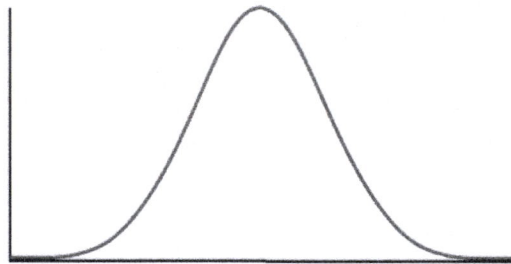

Figure 6.17

    b. $P($_____ $< x <$ _____$) =$ _____

59. Find the 70th percentile of the distribution for the time a CD player lasts.
    a. Sketch the situation. Label and scale the axes. Shade the region corresponding to the lower 70%.

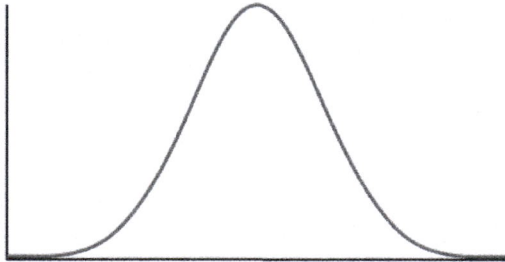

Figure 6.18

    b. $P(x < k) =$ _____ Therefore, $k =$ _____

HOMEWORK

6.1 The Standard Normal Distribution

Use the following information to answer the next two exercises: The patient recovery time from a particular surgical procedure is normally distributed with a mean of 5.3 days and a standard deviation of 2.1 days.

60. What is the median recovery time?
   a. 2.7
   b. 5.3
   c. 7.4
   d. 2.1

61. What is the z-score for a patient who takes ten days to recover?
   a. 1.5
   b. 0.2
   c. 2.2
   d. 7.3

62. The length of time to find it takes to find a parking space at 9 A.M. follows a normal distribution with a mean of five minutes and a standard deviation of two minutes. If the mean is significantly greater than the standard deviation, which of the following statements is true?
   I.   The data cannot follow the uniform distribution.
   II. The data cannot follow the exponential distribution..
   III. The data cannot follow the normal distribution.

   a. I only
   b. II only
   c. III only
   d. I, II, and III

63. The heights of the 430 National Basketball Association players were listed on team rosters at the start of the 2005–2006 season. The heights of basketball players have an approximate normal distribution with mean, $\mu = 79$ inches and a standard deviation, $\sigma = 3.89$ inches. For each of the following heights, calculate the z-score and interpret it using complete sentences.

   a. 77 inches
   b. 85 inches
   c. If an NBA player reported his height had a z-score of 3.5, would you believe him? Explain your answer.

64. The systolic blood pressure (given in millimeters) of males has an approximately normal distribution with mean $\mu = 125$ and standard deviation $\sigma = 14$. Systolic blood pressure for males follows a normal distribution.
   a. Calculate the z-scores for the male systolic blood pressures 100 and 150 millimeters.
   b. If a male friend of yours said he thought his systolic blood pressure was 2.5 standard deviations below the mean, but that he believed his blood pressure was between 100 and 150 millimeters, what would you say to him?

65. Kyle's doctor told him that the z-score for his systolic blood pressure is 1.75. Which of the following is the best interpretation of this standardized score? The systolic blood pressure (given in millimeters) of males has an approximately normal distribution with mean $\mu = 125$ and standard deviation $\sigma = 14$. If X = a systolic blood pressure score then $X \sim N (125, 14)$.
   a. Which answer(s) **is/are** correct?
        i.   Kyle's systolic blood pressure is 175.
        ii. Kyle's systolic blood pressure is 1.75 times the average blood pressure of men his age.
        iii. Kyle's systolic blood pressure is 1.75 above the average systolic blood pressure of men his age.
        iv. Kyles's systolic blood pressure is 1.75 standard deviations above the average systolic blood pressure for men.
   b. Calculate Kyle's blood pressure.

66. Height and weight are two measurements used to track a child's development. The World Health Organization measures child development by comparing the weights of children who are the same height and the same gender. In 2009, weights for all 80 cm girls in the reference population had a mean $\mu = 10.2$ kg and standard deviation $\sigma = 0.8$ kg. Weights are normally distributed. $X \sim N(10.2, 0.8)$. Calculate the z-scores that correspond to the following weights and interpret them.
   a. 11 kg
   b. 7.9 kg
   c. 12.2 kg

67. In 2005, 1,475,623 students heading to college took the SAT. The distribution of scores in the math section of the SAT follows a normal distribution with mean $\mu = 520$ and standard deviation $\sigma = 115$.
   a. Calculate the z-score for an SAT score of 720. Interpret it using a complete sentence.

    b.   What math SAT score is 1.5 standard deviations above the mean? What can you say about this SAT score?

    c.   For 2012, the SAT math test had a mean of 514 and standard deviation 117. The ACT math test is an alternate to the SAT and is approximately normally distributed with mean 21 and standard deviation 5.3. If one person took the SAT math test and scored 700 and a second person took the ACT math test and scored 30, who did better with respect to the test they took?

6.2 Using the Normal Distribution

Use the following information to answer the next two exercises: The patient recovery time from a particular surgical procedure is normally distributed with a mean of 5.3 days and a standard deviation of 2.1 days.

68. What is the probability of spending more than two days in recovery?

    a. 0.0580
    b. 0.8447
    c. 0.0553
    d. 0.9420

69. The 90th percentile for recovery times is?

    a. 8.89
    b. 7.07
    c. 7.99
    d. 4.32

Use the following information to answer the next three exercises: The length of time it takes to find a parking space at 9 A.M. follows a normal distribution with a mean of five minutes and a standard deviation of two minutes.

70. Based upon the given information and numerically justified, would you be surprised if it took less than one minute to find a parking space?

    a. Yes
    b. No
    c. Unable to determine

71. Find the probability that it takes at least eight minutes to find a parking space.

    a. 0.0001
    b. 0.9270
    c. 0.1862
    d. 0.0668

72. Seventy percent of the time, it takes more than how many minutes to find a parking space?

    a. 1.24
    b. 2.41
    c. 3.95
    d. 6.05

73. According to a study done by De Anza students, the height for Asian adult males is normally distributed with an average of 66 inches and a standard deviation of 2.5 inches. Suppose one Asian adult male is randomly chosen. Let X = height of the individual.

    a. $X \sim$ _____(_____,_____)
    b. Find the probability that the person is between 65 and 69 inches. Include a sketch of the graph, and write a probability statement.
    c. Would you expect to meet many Asian adult males over 72 inches? Explain why or why not, and justify your answer numerically.
    d. The middle 40% of heights fall between what two values? Sketch the graph, and write the probability statement.

74. IQ is normally distributed with a mean of 100 and a standard deviation of 15. Suppose one individual is randomly chosen. Let X = IQ of an individual.

    a. $X \sim$ _____(_____,_____)
    b. Find the probability that the person has an IQ greater than 120. Include a sketch of the graph, and write a probability statement.
    c. MENSA is an organization whose members have the top 2% of all IQs. Find the minimum IQ needed to qualify for the MENSA organization. Sketch the graph, and write the probability statement.
    d. The middle 50% of IQs fall between what two values? Sketch the graph and write the probability statement.

75. The percent of fat calories that a person in America consumes each day is normally distributed with a mean of about 36 and a standard deviation of 10. Suppose that one individual is randomly chosen. Let X = percent of fat calories.
  a. $X \sim$ _____(_____,_____)
  b. Find the probability that the percent of fat calories a person consumes is more than 40. Graph the situation. Shade in the area to be determined.
  c. Find the maximum number for the lower quarter of percent of fat calories. Sketch the graph and write the probability statement.

76. Suppose that the distance of fly balls hit to the outfield (in baseball) is normally distributed with a mean of 250 feet and a standard deviation of 50 feet.
  a. If X = distance in feet for a fly ball, then $X \sim$ _____(_____,_____)
  b. If one fly ball is randomly chosen from this distribution, what is the probability that this ball traveled fewer than 220 feet? Sketch the graph. Scale the horizontal axis X. Shade the region corresponding to the probability. Find the probability.
  c. Find the 80th percentile of the distribution of fly balls. Sketch the graph, and write the probability statement.

77. In China, four-year-olds average three hours a day unsupervised. Most of the unsupervised children live in rural areas, considered safe. Suppose that the standard deviation is 1.5 hours and the amount of time spent alone is normally distributed. We randomly select one Chinese four-year-old living in a rural area. We are interested in the amount of time the child spends alone per day.
  a. In words, define the random variable X.
  b. $X \sim$ _____(_____,_____)
  c. Find the probability that the child spends less than one hour per day unsupervised. Sketch the graph, and write the probability statement.
  d. What percent of the children spend over ten hours per day unsupervised?
  e. Seventy percent of the children spend at least how long per day unsupervised?

78. In the 1992 presidential election, Alaska's 40 election districts averaged 1,956.8 votes per district for President Clinton. The standard deviation was 572.3. (There are only 40 election districts in Alaska.) The distribution of the votes per district for President Clinton was bell-shaped. Let X = number of votes for President Clinton for an election district.
  a. State the approximate distribution of X.
  b. Is 1,956.8 a population mean or a sample mean? How do you know?
  c. Find the probability that a randomly selected district had fewer than 1,600 votes for President Clinton. Sketch the graph and write the probability statement.
  d. Find the probability that a randomly selected district had between 1,800 and 2,000 votes for President Clinton.
  e. Find the third quartile for votes for President Clinton.

79. Suppose that the duration of a particular type of criminal trial is known to be normally distributed with a mean of 21 days and a standard deviation of seven days.
  a. In words, define the random variable X.
  b. $X \sim$ _____(_____,_____)
  c. If one of the trials is randomly chosen, find the probability that it lasted at least 24 days. Sketch the graph and write the probability statement.
  d. Sixty percent of all trials of this type are completed within how many days?

80. Terri Vogel, an amateur motorcycle racer, averages 129.71 seconds per 2.5 mile lap (in a seven-lap race) with a standard deviation of 2.28 seconds. The distribution of her race times is normally distributed. We are interested in one of her randomly selected laps.
  a. In words, define the random variable X.
  b. $X \sim$ _____(_____,_____)
  c. Find the percent of her laps that are completed in less than 130 seconds.
  d. The fastest 3% of her laps are under _____.
  e. The middle 80% of her laps are from _____ seconds to _____ seconds.

81. Thuy Dau, Ngoc Bui, Sam Su, and Lan Voung conducted a survey as to how long customers at Lucky claimed to wait in the checkout line until their turn. Let X = time in line. **Table 6.3** displays the ordered real data (in minutes):

0.50	4.25	5	6	7.25
1.75	4.25	5.25	6	7.25
2	4.25	5.25	6.25	7.25
2.25	4.25	5.5	6.25	7.75
2.25	4.5	5.5	6.5	8
2.5	4.75	5.5	6.5	8.25
2.75	4.75	5.75	6.5	9.5
3.25	4.75	5.75	6.75	9.5
3.75	5	6	6.75	9.75
3.75	5	6	6.75	10.75

Table 6.3

a. Calculate the sample mean and the sample standard deviation.
b. Construct a histogram.
c. Draw a smooth curve through the midpoints of the tops of the bars.
d. In words, describe the shape of your histogram and smooth curve.
e. Let the sample mean approximate μ and the sample standard deviation approximate σ. The distribution of X can then be approximated by $X \sim$ _____(_____,_____)
f. Use the distribution in part e to calculate the probability that a person will wait fewer than 6.1 minutes.
g. Determine the cumulative relative frequency for waiting less than 6.1 minutes.
h. Why aren't the answers to part f and part g exactly the same?
i. Why are the answers to part f and part g as close as they are?
j. If only ten customers has been surveyed rather than 50, do you think the answers to part f and part g would have been closer together or farther apart? Explain your conclusion.

82. Suppose that Ricardo and Anita attend different colleges. Ricardo's GPA is the same as the average GPA at his school. Anita's GPA is 0.70 standard deviations above her school average. In complete sentences, explain why each of the following statements may be false.
a. Ricardo's actual GPA is lower than Anita's actual GPA.
b. Ricardo is not passing because his z-score is zero.
c. Anita is in the 70th percentile of students at her college.

83. Table 6.4 shows a sample of the maximum capacity (maximum number of spectators) of sports stadiums. The table does not include horse-racing or motor-racing stadiums.

40,000	40,000	45,050	45,500	46,249	48,134
49,133	50,071	50,096	50,466	50,832	51,100
51,500	51,900	52,000	52,132	52,200	52,530
52,692	53,864	54,000	55,000	55,000	55,000
55,000	55,000	55,000	55,082	57,000	58,008
59,680	60,000	60,000	60,492	60,580	62,380
62,872	64,035	65,000	65,050	65,647	66,000
66,161	67,428	68,349	68,976	69,372	70,107
70,585	71,594	72,000	72,922	73,379	74,500

| 75,025 | 76,212 | 78,000 | 80,000 | 80,000 | 82,300 |

Table 6.4

a. Calculate the sample mean and the sample standard deviation for the maximum capacity of sports stadiums (the data).
b. Construct a histogram.
c. Draw a smooth curve through the midpoints of the tops of the bars of the histogram.
d. In words, describe the shape of your histogram and smooth curve.
e. Let the sample mean approximate μ and the sample standard deviation approximate σ. The distribution of X can then be approximated by $X \sim$ _____(_____,_____).
f. Use the distribution in part e to calculate the probability that the maximum capacity of sports stadiums is less than 67,000 spectators.
g. Determine the cumulative relative frequency that the maximum capacity of sports stadiums is less than 67,000 spectators. Hint: Order the data and count the sports stadiums that have a maximum capacity less than 67,000. Divide by the total number of sports stadiums in the sample.
h. Why aren't the answers to part f and part g exactly the same?

84. An expert witness for a paternity lawsuit testifies that the length of a pregnancy is normally distributed with a mean of 280 days and a standard deviation of 13 days. An alleged father was out of the country from 240 to 306 days before the birth of the child, so the pregnancy would have been less than 240 days or more than 306 days long if he was the father. The birth was uncomplicated, and the child needed no medical intervention. What is the probability that he was NOT the father? What is the probability that he could be the father? Calculate the z-scores first, and then use those to calculate the probability.

85. A NUMMI assembly line, which has been operating since 1984, has built an average of 6,000 cars and trucks a week. Generally, 10% of the cars were defective coming off the assembly line. Suppose we draw a random sample of $n = 100$ cars. Let X represent the number of defective cars in the sample. What can we say about X in regard to the 68-95-99.7 empirical rule (one standard deviation, two standard deviations and three standard deviations from the mean are being referred to)? Assume a normal distribution for the defective cars in the sample.

86. We flip a coin 100 times ($n = 100$) and note that it only comes up heads 20% ($p = 0.20$) of the time. The mean and standard deviation for the number of times the coin lands on heads is $\mu = 20$ and $\sigma = 4$ (verify the mean and standard deviation). Solve the following:
a. There is about a 68% chance that the number of heads will be somewhere between ___ and ___.
b. There is about a ____chance that the number of heads will be somewhere between 12 and 28.
c. There is about a _____ chance that the number of heads will be somewhere between eight and 32.

87. A $1 scratch off lotto ticket will be a winner one out of five times. Out of a shipment of $n = 190$ lotto tickets, find the probability for the lotto tickets that there are
a. somewhere between 34 and 54 prizes.
b. somewhere between 54 and 64 prizes.
c. more than 64 prizes.

88. Facebook provides a variety of statistics on its Web site that detail the growth and popularity of the site.

On average, 28 percent of 18 to 34 year olds check their Facebook profiles before getting out of bed in the morning. Suppose this percentage follows a normal distribution with a standard deviation of five percent.

a. Find the probability that the percent of 18 to 34-year-olds who check Facebook before getting out of bed in the morning is at least 30.
b. Find the 95th percentile, and express it in a sentence.

REFERENCES

6.1 The Standard Normal Distribution

"Blood Pressure of Males and Females." StatCruch, 2013. Available online at http://www.statcrunch.com/5.0/viewreport.php?reportid=11960 (accessed May 14, 2013).

"The Use of Epidemiological Tools in Conflict-affected populations: Open-access educational resources for policy-makers: Calculation of z-scores." London School of Hygiene and Tropical Medicine, 2009. Available online at http://conflict.lshtm.ac.uk/page_125.htm (accessed May 14, 2013).

"2012 College-Bound Seniors Total Group Profile Report." CollegeBoard, 2012. Available online at http://media.collegeboard.com/digitalServices/pdf/research/TotalGroup-2012.pdf (accessed May 14, 2013).

"Digest of Education Statistics: ACT score average and standard deviations by sex and race/ethnicity and percentage of ACT test takers, by selected composite score ranges and planned fields of study: Selected years, 1995 through 2009." National Center for Education Statistics. Available online at http://nces.ed.gov/programs/digest/d09/tables/dt09_147.asp (accessed May 14, 2013).

Data from the *San Jose Mercury News*.

Data from *The World Almanac and Book of Facts*.

"List of stadiums by capacity." Wikipedia. Available online at https://en.wikipedia.org/wiki/List_of_stadiums_by_capacity (accessed May 14, 2013).

Data from the National Basketball Association. Available online at www.nba.com (accessed May 14, 2013).

6.2 Using the Normal Distribution

"Naegele's rule." Wikipedia. Available online at http://en.wikipedia.org/wiki/Naegele's_rule (accessed May 14, 2013).

"403: NUMMI." Chicago Public Media & Ira Glass, 2013. Available online at http://www.thisamericanlife.org/radio-archives/episode/403/nummi (accessed May 14, 2013).

"Scratch-Off Lottery Ticket Playing Tips." WinAtTheLottery.com, 2013. Available online at http://www.winatthelottery.com/public/department40.cfm (accessed May 14, 2013).

"Smart Phone Users, By The Numbers." Visual.ly, 2013. Available online at http://visual.ly/smart-phone-users-numbers (accessed May 14, 2013).

"Facebook Statistics." Statistics Brain. Available online at http://www.statisticbrain.com/facebook-statistics/(accessed May 14, 2013).

SOLUTIONS

1 ounces of water in a bottle

3 2

5 –4

7 –2

9 The mean becomes zero.

11 $z = 2$

13 $z = 2.78$

15 $x = 20$

17 $x = 6.5$

19 $x = 1$

21 $x = 1.97$

23 $z = -1.67$

25 $z \approx -0.33$

27 0.67, right

29 3.14, left

31 about 68%

33 about 4%

35 between –5 and –1

37 about 50%

39 about 27%

41 The lifetime of a Sunshine CD player measured in years.

43 $P(x < 1)$

45 Yes, because they are the same in a continuous distribution: $P(x = 1) = 0$

47 $1 - P(x < 3)$ or $P(x > 3)$

49 $1 - 0.543 = 0.457$

51 0.0013

53 56.03

55 0.1186

57
a. Check student's solution.
b. 3, 0.1979

59
a. Check student's solution.
b. 0.70, 4.78 years

61 c

63
a. Use the z-score formula. $z = -0.5141$. The height of 77 inches is 0.5141 standard deviations below the mean. An NBA player whose height is 77 inches is shorter than average.
b. Use the z-score formula. $z = 1.5424$. The height 85 inches is 1.5424 standard deviations above the mean. An NBA player whose height is 85 inches is taller than average.
c. Height = 79 + 3.5(3.89) = 90.67 inches, which is over 7.7 feet tall. There are very few NBA players this tall so the answer is no, not likely.

65
a. iv
b. Kyle's blood pressure is equal to 125 + (1.75)(14) = 149.5.

67 Let X = an SAT math score and Y = an ACT math score.
a. $X = 720$ $\frac{720 - 520}{15} = 1.74$ The exam score of 720 is 1.74 standard deviations above the mean of 520.
b. $z = 1.5$
   The math SAT score is $520 + 1.5(115) \approx 692.5$. The exam score of 692.5 is 1.5 standard deviations above the mean of 520.
c. $\frac{X - \mu}{\sigma} = \frac{700 - 514}{117} \approx 1.59$, the z-score for the SAT. $\frac{Y - \mu}{\sigma} = \frac{30 - 21}{5.3} \approx 1.70$, the z-scores for the ACT. With respect to the test they took, the person who took the ACT did better (has the higher z-score).

69 c

71 d

73

a. $X \sim N(66, 2.5)$

b. 0.5404

c. No, the probability that an Asian male is over 72 inches tall is 0.0082

75

a. $X \sim N(36, 10)$

b. The probability that a person consumes more than 40% of their calories as fat is 0.3446.

c. Approximately 25% of people consume less than 29.26% of their calories as fat.

77

a. X = number of hours that a Chinese four-year-old in a rural area is unsupervised during the day.

b. $X \sim N(3, 1.5)$

c. The probability that the child spends less than one hour a day unsupervised is 0.0918.

d. The probability that a child spends over ten hours a day unsupervised is less than 0.0001.

e. 2.21 hours

79

a. X = the distribution of the number of days a particular type of criminal trial will take

b. $X \sim N(21, 7)$

c. The probability that a randomly selected trial will last more than 24 days is 0.3336.

d. 22.77

81

a. mean = 5.51, s = 2.15

b. Check student's solution.

c. Check student's solution.

d. Check student's solution.

e. $X \sim N(5.51, 2.15)$

f. 0.6029

g. The cumulative frequency for less than 6.1 minutes is 0.64.

h. The answers to part f and part g are not exactly the same, because the normal distribution is only an approximation to the real one.

i. The answers to part f and part g are close, because a normal distribution is an excellent approximation when the sample size is greater than 30.

j. The approximation would have been less accurate, because the smaller sample size means that the data does not fit normal curve as well.

83

1. mean = 60,136
    s = 10,468

2. Answers will vary.

3. Answers will vary.

4. Answers will vary.

5. $X \sim N(60136, 10468)$

6. 0.7440

7. The cumulative relative frequency is 43/60 = 0.717.

8. The answers for part f and part g are not the same, because the normal distribution is only an approximation.

85

$n = 100$; $p = 0.1$; $q = 0.9$

$\mu = np = (100)(0.10) = 10$

$\sigma = \sqrt{npq} = \sqrt{(100)(0.1)(0.9)} = 3$

   i. $z = \pm 1$: $x_1 = \mu + z\sigma = 10 + 1(3) = 13$ and $x2 = \mu - z\sigma = 10 - 1(3) = 7$. 68% of the defective cars will fall between seven and 13.

  ii. $z = \pm 2$: $x_1 = \mu + z\sigma = 10 + 2(3) = 16$ and $x2 = \mu - z\sigma = 10 - 2(3) = 4$. 95 % of the defective cars will fall between four and 16

 iii. $z = \pm 3$: $x_1 = \mu + z\sigma = 10 + 3(3) = 19$ and $x2 = \mu - z\sigma = 10 - 3(3) = 1$. 99.7% of the defective cars will fall between one and 19.

87

$n = 190$; $p = \dfrac{1}{5} = 0.2$; $q = 0.8$

$\mu = np = (190)(0.2) = 38$

$\sigma = \sqrt{npq} = \sqrt{(190)(0.2)(0.8)} = 5.5136$

  a.   For this problem: $P(34 < x < 54) = \text{normalcdf}(34,54,48,5.5136) = 0.7641$

  b.   For this problem: $P(54 < x < 64) = \text{normalcdf}(54,64,48,5.5136) = 0.0018$

  c.   For this problem: $P(x > 64) = \text{normalcdf}(64,10^{99},48,5.5136) = 0.0000012$ (approximately 0)

7 | THE CENTRAL LIMIT THEOREM

Figure 7.1 If you want to figure out the distribution of the change people carry in their pockets, using the central limit theorem and assuming your sample is large enough, you will find that the distribution is normal and bell-shaped. (credit: John Lodder)

Introduction

Chapter Objectives
By the end of this chapter, the student should be able to: • Recognize central limit theorem problems. • Classify continuous word problems by their distributions. • Apply and interpret the central limit theorem for means. • Apply and interpret the central limit theorem for sums.

Why are we so concerned with means? Two reasons are: they give us a middle ground for comparison, and they are easy to calculate. In this chapter, you will study means and the **central limit theorem**.

The **central limit theorem** (clt for short) is one of the most powerful and useful ideas in all of statistics. There are two alternative forms of the theorem, and both alternatives are concerned with drawing finite samples size n from a population with a known mean, μ, and a known standard deviation, σ. The first alternative says that if we collect samples of size n with a "large enough n," calculate each sample's mean, and create a histogram of those means, then the resulting histogram will tend to have an approximate normal bell shape. The second alternative says that if we again collect samples of size n that are "large enough," calculate the sum of each sample and create a histogram, then the resulting histogram will again tend to have a normal bell-shape.

In either case, it does not matter what the distribution of the original population is, or whether you even need to know it. The important fact is that the distribution of sample means and the sums tend to follow the normal distribution.

The size of the sample, n, that is required in order to be "large enough" depends on the original population from which the samples are drawn (the sample size should be at least 30 or the data should come from a normal distribution). If the original population is far from normal, then more observations are needed for the sample means or sums to be normal. **Sampling is done with replacement.**

Collaborative Exercise

Suppose eight of you roll one fair die ten times, seven of you roll two fair dice ten times, nine of you roll five fair dice ten times, and 11 of you roll ten fair dice ten times.

Each time a person rolls more than one die, he or she calculates the sample **mean** of the faces showing. For example, one person might roll five fair dice and get 2, 2, 3, 4, 6 on one roll.

The mean is $\frac{2+2+3+4+6}{5}$ = 3.4. The 3.4 is one mean when five fair dice are rolled. This same person would roll the five dice nine more times and calculate nine more means for a total of ten means.

Your instructor will pass out the dice to several people. Roll your dice ten times. For each roll, record the faces, and find the mean. Round to the nearest 0.5.

Your instructor (and possibly you) will produce one graph (it might be a histogram) for one die, one graph for two dice, one graph for five dice, and one graph for ten dice. Since the "mean" when you roll one die is just the face on the die, what distribution do these **means** appear to be representing?

Draw the graph for the means using two dice. Do the sample means show any kind of pattern?

Draw the graph for the means using five dice. Do you see any pattern emerging?

Finally, draw the graph for the means using ten dice. Do you see any pattern to the graph? What can you conclude as you increase the number of dice?

As the number of dice rolled increases from one to two to five to ten, the following is happening:

1. The mean of the sample means remains approximately the same.

2. The spread of the sample means (the standard deviation of the sample means) gets smaller.

3. The graph appears steeper and thinner.

You have just demonstrated the central limit theorem (clt).

The central limit theorem tells you that as you increase the number of dice, **the sample means tend toward a normal distribution (the sampling distribution).**

7.1 | The Central Limit Theorem for Sample Means (Averages)

Suppose X is a random variable with a distribution that may be known or unknown (it can be any distribution). Using a subscript that matches the random variable, suppose:

a. μ_X = the mean of X

b. σ_X = the standard deviation of X

If you draw random samples of size n, then as n increases, the random variable $\overline{X}$ which consists of sample means, tends to be **normally distributed** and

$$\overline{X} \sim N\left(\mu_x, \frac{\sigma_X}{\sqrt{n}}\right).$$

The **central limit theorem** for sample means says that if you keep drawing larger and larger samples (such as rolling one, two, five, and finally, ten dice) and **calculating their means,** the sample means form their own **normal distribution** (the sampling distribution). The normal distribution has the same mean as the original distribution and a variance that equals the original variance divided by, the sample size. The variable n is the number of values that are averaged together, not the number of times the experiment is done.

To put it more formally, if you draw random samples of size n, the distribution of the random variable $\overline{X}$, which consists of sample means, is called the **sampling distribution of the mean**. The sampling distribution of the mean approaches a normal distribution as n, the **sample size**, increases.

The random variable $\overline{X}$ has a different z-score associated with it from that of the random variable X. The mean $\overline{x}$ is the value of $\overline{X}$ in one sample.

$$z = \frac{\overline{x} - \mu_x}{\left(\frac{\sigma_x}{\sqrt{n}}\right)}$$

μ_X is the average of both X and $\overline{X}$.

$\sigma \overline{x} = \frac{\sigma_x}{\sqrt{n}}$ = standard deviation of $\overline{X}$ and is called the **standard error of the mean.**

 Using the TI-83, 83+, 84, 84+ Calculator

To find probabilities for means on the calculator, follow these steps.

2nd DISTR
2:normalcdf

$$normalcdf\left(lower\ value\ of\ the\ area,\ upper\ value\ of\ the\ area,\ mean,\ \frac{standard\ \ deviation}{\sqrt{sample\ \ size}}\right)$$

where:

- *mean* is the mean of the original distribution
- *standard deviation* is the standard deviation of the original distribution
- *sample size* = n

Example 7.1

An unknown distribution has a mean of 90 and a standard deviation of 15. Samples of size $n = 25$ are drawn randomly from the population.

a. Find the probability that the **sample mean** is between 85 and 92.

Solution 7.1

a. Let X = one value from the original unknown population. The probability question asks you to find a probability for the **sample mean**.

Let $\bar{X}$ = the mean of a sample of size 25. Since $\mu_X = 90$, $\sigma_X = 15$, and $n = 25$,

$$\bar{X} \sim N\left(90, \frac{15}{\sqrt{25}}\right).$$

Find $P(85 < \bar{x} < 92)$. Draw a graph.

$P(85 < \bar{x} < 92) = 0.6997$

The probability that the sample mean is between 85 and 92 is 0.6997.

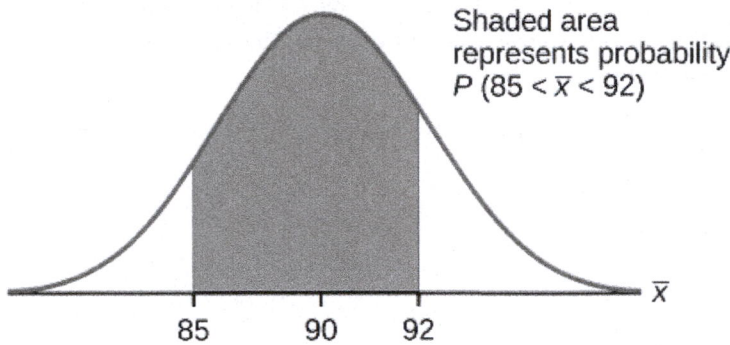

Figure 7.2

 Using the TI-83, 83+, 84, 84+ Calculator

`normalcdf`(lower value, upper value, mean, standard error of the mean)

The parameter list is abbreviated (lower value, upper value, μ, $\frac{\sigma}{\sqrt{n}}$)

`normalcdf`$(85,92,90, \frac{15}{\sqrt{25}}) = 0.6997$

b. Find the value that is two standard deviations above the expected value, 90, of the sample mean.

Solution 7.1

b. To find the value that is two standard deviations above the expected value 90, use the formula:

value $= \mu_x + (\text{\#ofTSDEVs})\left(\frac{\sigma_x}{\sqrt{n}}\right)$

value $= 90 + 2 \left(\frac{15}{\sqrt{25}}\right) = 96$

The value that is two standard deviations above the expected value is 96.

The standard error of the mean is $\frac{\sigma_x}{\sqrt{n}} = \frac{15}{\sqrt{25}} = 3$. Recall that the standard error of the mean is a description of how far (on average) that the sample mean will be from the population mean in repeated simple random samples of size n.

Try It Σ

7.1 An unknown distribution has a mean of 45 and a standard deviation of eight. Samples of size $n = 30$ are drawn randomly from the population. Find the probability that the sample mean is between 42 and 50.

Example 7.2

The length of time, in hours, it takes an "over 40" group of people to play one soccer match is normally distributed with a **mean of two hours** and a **standard deviation of 0.5 hours**. A **sample of size $n = 50$** is drawn randomly from the population. Find the probability that the **sample mean** is between 1.8 hours and 2.3 hours.

Solution 7.2

Let X = the time, in hours, it takes to play one soccer match.

The probability question asks you to find a probability for the **sample mean time, in hours**, it takes to play one soccer match.

Let $\bar{X}$ = the **mean** time, in hours, it takes to play one soccer match.

If $\mu_X =$ _____, $\sigma_X =$ _____, and $n =$ _____, then $X \sim N(\_\_\_\_\_, \_\_\_\_\_)$ by the **central limit theorem for means**.

$\mu_X = 2$, $\sigma_X = 0.5$, $n = 50$, and $X \sim N\left(2, \dfrac{0.5}{\sqrt{50}}\right)$

Find $P(1.8 < \bar{x} < 2.3)$. Draw a graph.

$P(1.8 < \bar{x} < 2.3) = 0.9977$

$\texttt{normalcdf}\left(1.8, 2.3, 2, \dfrac{.5}{\sqrt{50}}\right) = 0.9977$

The probability that the mean time is between 1.8 hours and 2.3 hours is 0.9977.

Try It Σ

7.2 The length of time taken on the SAT for a group of students is normally distributed with a mean of 2.5 hours and a standard deviation of 0.25 hours. A sample size of $n = 60$ is drawn randomly from the population. Find the probability that the sample mean is between two hours and three hours.

 Using the TI-83, 83+, 84, 84+ Calculator

To find percentiles for means on the calculator, follow these steps.

2nd DIStR
3:invNorm

$k = \text{invNorm}\left(\text{area to the left of } k, \text{ mean}, \dfrac{standard\ \ deviation}{\sqrt{sample\ \ size}}\right)$

where:

- k = the kth percentile
- *mean* is the mean of the original distribution
- *standard deviation* is the standard deviation of the original distribution
- *sample size* = n

Example 7.3

In a recent study reported Oct. 29, 2012 on the Flurry Blog, the mean age of tablet users is 34 years. Suppose the standard deviation is 15 years. Take a sample of size n = 100.

a. What are the mean and standard deviation for the sample mean ages of tablet users?

b. What does the distribution look like?

c. Find the probability that the sample mean age is more than 30 years (the reported mean age of tablet users in this particular study).

d. Find the 95th percentile for the sample mean age (to one decimal place).

Solution 7.3

a. Since the sample mean tends to target the population mean, we have $\mu_{\bar{x}} = \mu = 34$. The sample standard deviation is given by $\sigma_{\bar{x}} = \dfrac{\sigma}{\sqrt{n}} = \dfrac{15}{\sqrt{100}} = \dfrac{15}{10} = 1.5$

b. The central limit theorem states that for large sample sizes(n), the sampling distribution will be approximately normal.

c. The probability that the sample mean age is more than 30 is given by $P(X > 30) =$ $\texttt{normalcdf}(30,\text{E}99,34,1.5) = 0.9962$

d. Let k = the 95th percentile.

$k = \text{invNorm}\left(0.95,34,\dfrac{15}{\sqrt{100}}\right) = 36.5$

Try It Σ

7.3 In an article on Flurry Blog, a gaming marketing gap for men between the ages of 30 and 40 is identified. You are researching a startup game targeted at the 35-year-old demographic. Your idea is to develop a strategy game that can be played by men from their late 20s through their late 30s. Based on the article's data, industry research shows that the average strategy player is 28 years old with a standard deviation of 4.8 years. You take a sample of 100 randomly selected gamers. If your target market is 29- to 35-year-olds, should you continue with your development strategy?

Example 7.4

The mean number of minutes for app engagement by a tablet user is 8.2 minutes. Suppose the standard deviation is one minute. Take a sample of 60.

a. What are the mean and standard deviation for the sample mean number of app engagement by a tablet user?

b. What is the standard error of the mean?

c. Find the 90th percentile for the sample mean time for app engagement for a tablet user. Interpret this value in a complete sentence.

d. Find the probability that the sample mean is between eight minutes and 8.5 minutes.

Solution 7.4

a. $\mu_{\bar{x}} = \mu = 8.2 \; \sigma_{\bar{x}} = \frac{\sigma}{\sqrt{n}} = \frac{1}{\sqrt{60}} = 0.13$

b. This allows us to calculate the probability of sample means of a particular distance from the mean, in repeated samples of size 60.

c. Let k = the 90th percentile

$k = \mathtt{invNorm}\left(0.90, 8.2, \frac{1}{\sqrt{60}}\right) = 8.37$. This values indicates that 90 percent of the average app engagement time for table users is less than 8.37 minutes.

d. $P(8 < \bar{x} < 8.5) = \mathtt{normalcdf}\left(8, 8.5, 8.2, \frac{1}{\sqrt{60}}\right) = 0.9293$

Try It Σ

7.4 Cans of a cola beverage claim to contain 16 ounces. The amounts in a sample are measured and the statistics are $n = 34$, $\bar{x} = 16.01$ ounces. If the cans are filled so that $\mu = 16.00$ ounces (as labeled) and $\sigma = 0.143$ ounces, find the probability that a sample of 34 cans will have an average amount greater than 16.01 ounces. Do the results suggest that cans are filled with an amount greater than 16 ounces?

7.2 | The Central Limit Theorem for Sums

Suppose X is a random variable with a distribution that may be **known or unknown** (it can be any distribution) and suppose:

a. μ_X = the mean of X

b. σ_X = the standard deviation of X

If you draw random samples of size n, then as n increases, the random variable ΣX consisting of sums tends to be **normally distributed** and $\Sigma X \sim N((n)(\mu_X), (\sqrt{n})(\sigma_X))$.

The central limit theorem for sums says that if you keep drawing larger and larger samples and taking their sums, the sums form their own normal distribution (the sampling distribution), which approaches a normal distribution as the sample size increases. **The normal distribution has a mean equal to the original mean multiplied by the sample size and a standard deviation equal to the original standard deviation multiplied by the square root of the sample size.**

The random variable ΣX has the following z-score associated with it:

a. Σx is one sum.

b. $z = \frac{\Sigma x - (n)(\mu_X)}{(\sqrt{n})(\sigma_X)}$

    i. $(n)(\mu_X)$ = the mean of ΣX

    ii. $(\sqrt{n})(\sigma_X)$ = standard deviation of ΣX

Using the TI-83, 83+, 84, 84+ Calculator

To find probabilities for sums on the calculator, follow these steps.

2nd DISTR

2:normalcdf

normalcdf(lower value of the area, upper value of the area, (n)(mean), ($\sqrt{n}$)(standard deviation))

where:

- *mean* is the mean of the original distribution
- *standard deviation* is the standard deviation of the original distribution
- *sample size* = n

Example 7.5

An unknown distribution has a mean of 90 and a standard deviation of 15. A sample of size 80 is drawn randomly from the population.

a.   Find the probability that the sum of the 80 values (or the total of the 80 values) is more than 7,500.

b.   Find the sum that is 1.5 standard deviations above the mean of the sums.

Solution 7.5

Let X = one value from the original unknown population. The probability question asks you to find a probability for **the sum (or total of) 80 values.**

ΣX = the sum or total of 80 values. Since $\mu_X = 90$, $\sigma_X = 15$, and $n = 80$, $\Sigma X \sim N((80)(90),$ ($\sqrt{80}$)(15))

- mean of the sums = $(n)(\mu_X) = (80)(90) = 7{,}200$

- standard deviation of the sums = $(\sqrt{n})(\sigma_X) = (\sqrt{80})(15)$

- sum of 80 values = $\Sigma x = 7{,}500$

a. Find $P(\Sigma x > 7{,}500)$

$P(\Sigma x > 7{,}500) = 0.0127$

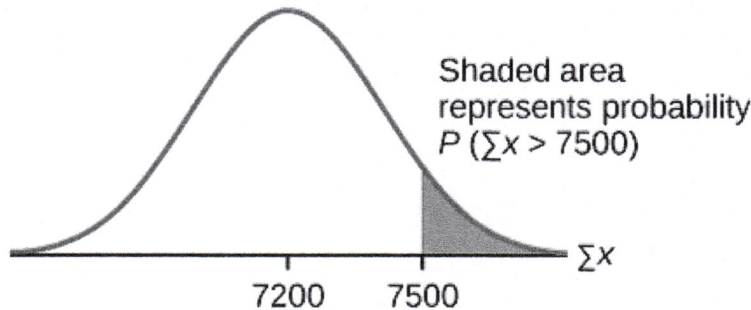

Figure 7.3

Using the TI-83, 83+, 84, 84+ Calculator

normalcdf(lower value, upper value, mean of sums, stdev of sums)

The parameter list is abbreviated(lower, upper, $(n)(\mu_X,\ (\sqrt{n})(\sigma_X))$

normalcdf $(7500, 1E99, (80)(90), (\sqrt{80})(15)) = 0.0127$

REMINDER

$1E99 = 10^{99}$.

Press the **EE** key for E.

b. Find Σx where $z = 1.5$.

$\Sigma x = (n)(\mu_X) + (z)(\sqrt{n})(\sigma_X) = (80)(90) + (1.5)(\sqrt{80})(15) = 7{,}401.2$

Try It Σ

7.5 An unknown distribution has a mean of 45 and a standard deviation of eight. A sample size of 50 is drawn randomly from the population. Find the probability that the sum of the 50 values is more than 2,400.

Using the TI-83, 83+, 84, 84+ Calculator

To find percentiles for sums on the calculator, follow these steps.

2nd DIStR
3:invNorm
k = invNorm (area to the left of k, (n)(mean), $(\sqrt{n})$ (standard deviation))

where:

- k is the kth **percentile**
- *mean* is the mean of the original distribution
- *standard deviation* is the standard deviation of the original distribution
- *sample size* = n

Example 7.6

In a recent study reported Oct. 29, 2012 on the Flurry Blog, the mean age of tablet users is 34 years. Suppose the standard deviation is 15 years. The sample of size is 50.

a. What are the mean and standard deviation for the sum of the ages of tablet users? What is the distribution?

b. Find the probability that the sum of the ages is between 1,500 and 1,800 years.

c. Find the 80th percentile for the sum of the 50 ages.

Solution 7.6

a. $\mu_{\Sigma x} = n\mu_x = 50(34) = 1{,}700$ and $\sigma_{\Sigma x} = \sqrt{n}\, \sigma_x = (\sqrt{50})(15) = 106.01$

The distribution is normal for sums by the central limit theorem.

b. $P(1500 < \Sigma x < 1800) = \texttt{normalcdf}\,(1{,}500,\ 1{,}800,\ (50)(34),\ (\sqrt{50})(15)) = 0.7974$

c. Let $k =$ the 80th percentile.
    $k = \texttt{invNorm}(0.80,(50)(34),(\sqrt{50})(15)) = 1{,}789.3$

Try It Σ

7.6 In a recent study reported Oct.29, 2012 on the Flurry Blog, the mean age of tablet users is 35 years. Suppose the standard deviation is ten years. The sample size is 39.

a. What are the mean and standard deviation for the sum of the ages of tablet users? What is the distribution?

b. Find the probability that the sum of the ages is between 1,400 and 1,500 years.

c. Find the 90th percentile for the sum of the 39 ages.

Example 7.7

The mean number of minutes for app engagement by a tablet user is 8.2 minutes. Suppose the standard deviation is one minute. Take a sample of size 70.

a. What are the mean and standard deviation for the sums?

b. Find the 95th percentile for the sum of the sample. Interpret this value in a complete sentence.

c. Find the probability that the sum of the sample is at least ten hours.

Solution 7.7

a. $\mu_{\Sigma x} = n\mu_x = 70(8.2) = 574$ minutes and $\sigma_{\Sigma x} = (\sqrt{n})(\sigma_x) = (\sqrt{70})(1) = 8.37$ minutes

b. Let $k =$ the 95th percentile.
    $k = \text{invNorm}\,(0.95,(70)(8.2),(\sqrt{70})(1)) = 587.76$ minutes

    Ninety five percent of the app engagement times are at most 587.76 minutes.

c. ten hours = 600 minutes
    $P(\Sigma x \geq 600) = \texttt{normalcdf}(600,\text{E}99,(70)(8.2),(\sqrt{70})(1)) = 0.0009$

Try It Σ

7.7 The mean number of minutes for app engagement by a table use is 8.2 minutes. Suppose the standard deviation is one minute. Take a sample size of 70.

a. What is the probability that the sum of the sample is between seven hours and ten hours? What does this mean in context of the problem?

b. Find the 84th and 16th percentiles for the sum of the sample. Interpret these values in context.

7.3 | Using the Central Limit Theorem

It is important for you to understand when to use the **central limit theorem**. If you are being asked to find the probability of the mean, use the clt for the mean. If you are being asked to find the probability of a sum or total, use the clt for sums. This also applies to percentiles for means and sums.

> **NOTE**
> _____
>
> If you are being asked to find the probability of an **individual** value, do **not** use the clt. **Use the distribution of its random variable.**

Examples of the Central Limit Theorem

Law of Large Numbers

The **law of large numbers** says that if you take samples of larger and larger size from any population, then the mean $\bar{x}$ of the sample tends to get closer and closer to μ. From the central limit theorem, we know that as n gets larger and larger, the sample means follow a normal distribution. The larger n gets, the smaller the standard deviation gets. (Remember that the standard deviation for $\bar{X}$ is $\frac{\sigma}{\sqrt{n}}$.) This means that the sample mean $\bar{x}$ must be close to the population mean μ. We can say that μ is the value that the sample means approach as n gets larger. The central limit theorem illustrates the law of large numbers.

Central Limit Theorem for the Mean and Sum Examples

Example 7.8

A study involving stress is conducted among the students on a college campus. **The stress scores follow a uniform distribution** with the lowest stress score equal to one and the highest equal to five. Using a sample of 75 students, find:

   a. The probability that the **mean stress score** for the 75 students is less than two.

   b. The 90th percentile for the **mean stress score** for the 75 students.

   c. The probability that the **total of the 75 stress scores** is less than 200.

   d. The 90th percentile for the **total stress score** for the 75 students.

Let X = one stress score.

Problems a and b ask you to find a probability or a percentile for a **mean**. Problems c and d ask you to find a probability or a percentile for a **total or sum**. The sample size, n, is equal to 75.

Since the individual stress scores follow a uniform distribution, $X \sim U(1, 5)$ where $a = 1$ and $b = 5$ (See **Continuous Random Variables** for an explanation on the uniform distribution).

$$\mu_X = \frac{a+b}{2} = \frac{1+5}{2} = 3$$

$$\sigma_X = \sqrt{\frac{(b-a)^2}{12}} = \sqrt{\frac{(5-1)^2}{12}} = 1.15$$

For problems 1. and 2., let $\bar{X}$ = the mean stress score for the 75 students. Then,

$$\bar{X} \sim N\left(3, \frac{1.15}{\sqrt{75}}\right) \text{ where } n = 75.$$

a. Find $P(\bar{x} < 2)$. Draw the graph.

Solution 7.8

a. $P(\bar{x} < 2) = 0$

The probability that the mean stress score is less than two is about zero.

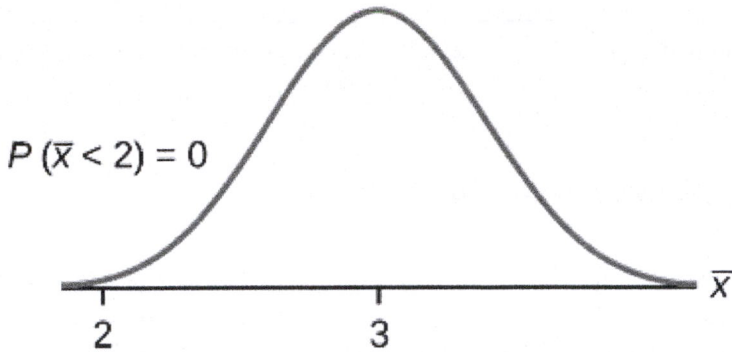

$$P(\overline{X} < 2) = 0$$

Figure 7.4

$\texttt{normalcdf}\left(1,2,3,\frac{1.15}{\sqrt{75}}\right) = 0$

REMINDER

The smallest stress score is one.

b. Find the 90$^\text{th}$ percentile for the mean of 75 stress scores. Draw a graph.

Solution 7.8

b. Let k = the 90$^\text{th}$ precentile.

Find k, where $P(\bar{x} < k) = 0.90$.

$k = 3.2$

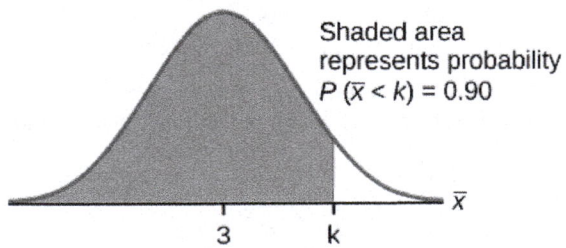

Shaded area represents probability $P(\overline{x} < k) = 0.90$

Figure 7.5

The 90$^\text{th}$ percentile for the mean of 75 scores is about 3.2. This tells us that 90% of all the means of 75 stress scores are at most 3.2, and that 10% are at least 3.2.

$\texttt{invNorm}\left(0.90,3,\frac{1.15}{\sqrt{75}}\right) = 3.2$

For problems c and d, let ΣX = the sum of the 75 stress scores. Then, $\Sigma X \sim N[(75)(3), (\sqrt{75})\,(1.15)]$

c. Find $P(\Sigma x < 200)$. Draw the graph.

Solution 7.8

c. The mean of the sum of 75 stress scores is $(75)(3) = 225$

The standard deviation of the sum of 75 stress scores is $(\sqrt{75})(1.15) = 9.96$

$P(\Sigma x < 200) = 0$

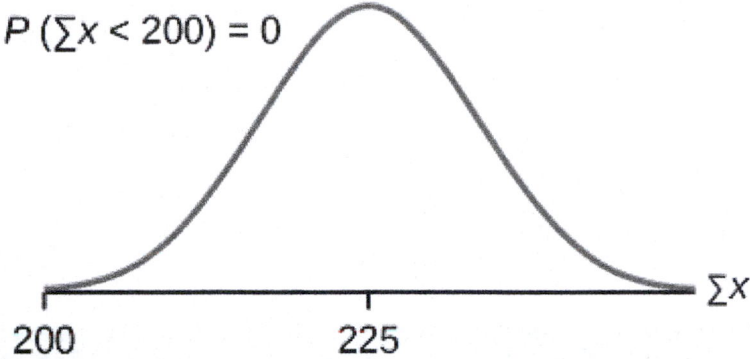

Figure 7.6

The probability that the total of 75 scores is less than 200 is about zero.

normalcdf $(75, 200, (75)(3), (\sqrt{75})(1.15))$.

REMINDER

The smallest total of 75 stress scores is 75, because the smallest single score is one.

---

d. Find the 90th percentile for the total of 75 stress scores. Draw a graph.

Solution 7.8

d. Let k = the 90th percentile.

Find k where $P(\Sigma x < k) = 0.90$.

$k = 237.8$

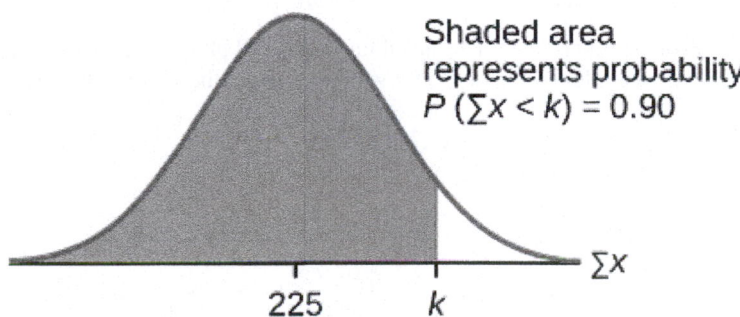

Figure 7.7

The 90th percentile for the sum of 75 scores is about 237.8. This tells us that 90% of all the sums of 75 scores are no more than 237.8 and 10% are no less than 237.8.

`invNorm`$(0.90,(75)(3), (\sqrt{75}) (1.15)) = 237.8$

Try It Σ

7.8 Use the information in **Example 7.8**, but use a sample size of 55 to answer the following questions.

a. Find $P(\bar{x} < 7)$.

b. Find $P(\Sigma x > 170)$.

c. Find the 80th percentile for the mean of 55 scores.

d. Find the 85th percentile for the sum of 55 scores.

Example 7.9

Suppose that a market research analyst for a cell phone company conducts a study of their customers who exceed the time allowance included on their basic cell phone contract; the analyst finds that for those people who exceed the time included in their basic contract, the **excess time used** follows an **exponential distribution** with a mean of 22 minutes.

Consider a random sample of 80 customers who exceed the time allowance included in their basic cell phone contract.

Let X = the excess time used by one INDIVIDUAL cell phone customer who exceeds his contracted time allowance.

$X \sim Exp\left(\frac{1}{22}\right)$. From previous chapters, we know that $\mu = 22$ and $\sigma = 22$.

Let $\bar{X}$ = the mean excess time used by a sample of $n = 80$ customers who exceed their contracted time allowance.

$\bar{X} \sim N\left(22, \frac{22}{\sqrt{80}}\right)$ by the central limit theorem for sample means

Using the clt to find probability

a. Find the probability that the mean excess time used by the 80 customers in the sample is longer than 20 minutes. This is asking us to find $P(\bar{x} > 20)$. Draw the graph.

b. Suppose that one customer who exceeds the time limit for his cell phone contract is randomly selected. Find the probability that this individual customer's excess time is longer than 20 minutes. This is asking us to find $P(x > 20)$.

c. Explain why the probabilities in parts a and b are different.

Solution 7.9

a. Find: $P(\bar{x} > 20)$

$P(\bar{x} > 20) = 0.79199$ using `normalcdf`$\left(20,1E99,22,\frac{22}{\sqrt{80}}\right)$

The probability is 0.7919 that the mean excess time used is more than 20 minutes, for a sample of 80 customers who exceed their contracted time allowance.

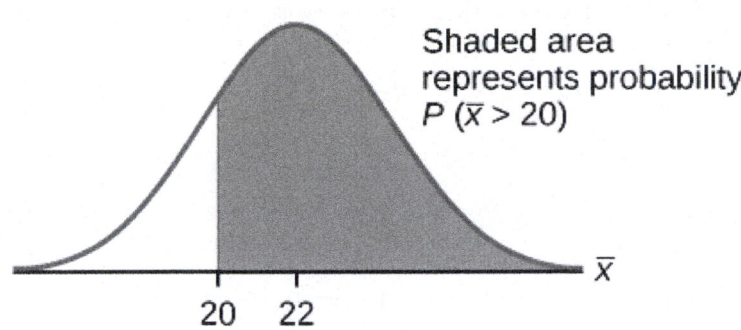

Figure 7.8

REMINDER

 1E99 = 10^{99} and –1E99 = -10^{99}. Press the **EE** key for E. Or just use 10^{99} instead of 1E99.

b. Find $P(x > 20)$. Remember to use the exponential distribution for an **individual**: $X \sim Exp\left(\frac{1}{22}\right)$.

$$P(x > 20) = e^{\left(-\left(\frac{1}{22}\right)(20)\right)} \text{ or } e^{(-0.04545(20))} = 0.4029$$

c. 1. $P(x > 20) = 0.4029$ but $P(\bar{x} > 20) = 0.7919$

2. The probabilities are not equal because we use different distributions to calculate the probability for individuals and for means.

3. When asked to find the probability of an individual value, use the stated distribution of its random variable; do not use the clt. Use the clt with the normal distribution when you are being asked to find the probability for a mean.

Using the clt to find percentiles Find the 95th percentile for the **sample mean excess time** for samples of 80 customers who exceed their basic contract time allowances. Draw a graph.

Solution 7.9

Let k = the 95th percentile. Find k where $P(\bar{x} < k) = 0.95$

$k = 26.0$ using $\mathtt{invNorm}\left(0.95, 22, \frac{22}{\sqrt{80}}\right) = 26.0$

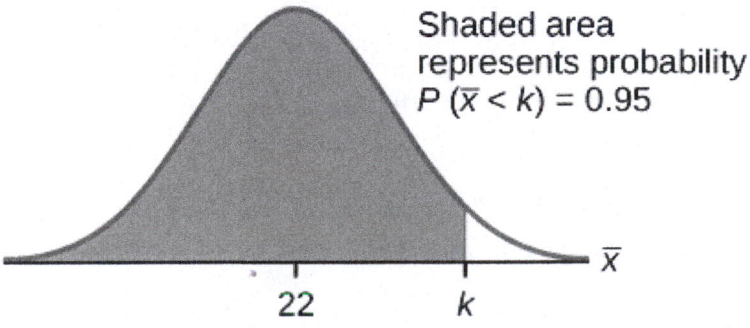

Figure 7.9

The 95th percentile for the **sample mean excess time used** is about 26.0 minutes for random samples of 80 customers who exceed their contractual allowed time.

Ninety five percent of such samples would have means under 26 minutes; only five percent of such samples would have means above 26 minutes.

Try It Σ

7.9 Use the information in **Example 7.9**, but change the sample size to 144.

a.   Find $P(20 < \bar{x} < 30)$.

b.   Find $P(\Sigma x$ is at least 3,000).

c.   Find the 75th percentile for the sample mean excess time of 144 customers.

d.   Find the 85th percentile for the sum of 144 excess times used by customers.

Example 7.10

In the United States, someone is sexually assaulted every two minutes, on average, according to a number of studies. Suppose the standard deviation is 0.5 minutes and the sample size is 100.

a.   Find the median, the first quartile, and the third quartile for the sample mean time of sexual assaults in the United States.

b.   Find the median, the first quartile, and the third quartile for the sum of sample times of sexual assaults in the United States.

c.   Find the probability that a sexual assault occurs on the average between 1.75 and 1.85 minutes.

d.   Find the value that is two standard deviations above the sample mean.

e.   Find the IQR for the sum of the sample times.

Solution 7.10

a.   We have, $\mu_x = \mu = 2$ and $\sigma_x = \frac{\sigma}{\sqrt{n}} = \frac{0.5}{10} = 0.05$. Therefore:

   1.   50th percentile $= \mu_x = \mu = 2$

   2.   25th percentile $= invNorm(0.25,2,0.05) = 1.97$

   3.   75th percentile $= invNorm(0.75,2,0.05) = 2.03$

b.   We have $\mu_{\Sigma x} = n(\mu_x) = 100(2) = 200$ and $\sigma_{\mu x} = \sqrt{n}\,(\sigma_x) = 10(0.5) = 5$. Therefore

   1.   50th percentile $= \mu_{\Sigma x} = n(\mu_x) = 100(2) = 200$

   2.   25th percentile $= invNorm(0.25,200,5) = 196.63$

   3.   75th percentile $= invNorm(0.75,200,5) = 203.37$

c. $P(1.75 < \bar{x} < 1.85) =$ `normalcdf`$(1.75,1.85,2,0.05) = 0.0013$

d.   Using the z-score equation, $z = \dfrac{\bar{x} - \mu_{\bar{x}}}{\sigma_{\bar{x}}}$, and solving for x, we have $x = 2(0.05) + 2 = 2.1$

e. The *IQR* is 75th percentile – 25th percentile = 203.37 – 196.63 = 6.74

Try It Σ

7.10 Based on data from the National Health Survey, women between the ages of 18 and 24 have an average systolic blood pressures (in mm Hg) of 114.8 with a standard deviation of 13.1. Systolic blood pressure for women between the ages of 18 to 24 follow a normal distribution.

a. If one woman from this population is randomly selected, find the probability that her systolic blood pressure is greater than 120.

b. If 40 women from this population are randomly selected, find the probability that their mean systolic blood pressure is greater than 120.

c. If the sample were four women between the ages of 18 to 24 and we did not know the original distribution, could the central limit theorem be used?

Example 7.11

A study was done about violence against prostitutes and the symptoms of the posttraumatic stress that they developed. The age range of the prostitutes was 14 to 61. The mean age was 30.9 years with a standard deviation of nine years.

a. In a sample of 25 prostitutes, what is the probability that the mean age of the prostitutes is less than 35?

b. Is it likely that the mean age of the sample group could be more than 50 years? Interpret the results.

c. In a sample of 49 prostitutes, what is the probability that the sum of the ages is no less than 1,600?

d. Is it likely that the sum of the ages of the 49 prostitutes is at most 1,595? Interpret the results.

e. Find the 95th percentile for the sample mean age of 65 prostitutes. Interpret the results.

f. Find the 90th percentile for the sum of the ages of 65 prostitutes. Interpret the results.

Solution 7.11

a. $P(\bar{x} < 35) = \texttt{normalcdf}(-E99,35,30.9,1.8) = 0.9886$

b. $P(\bar{x} > 50) = \texttt{normalcdf}(50, E99,30.9,1.8) \approx 0$. For this sample group, it is almost impossible for the group's average age to be more than 50. However, it is still possible for an individual in this group to have an age greater than 50.

c. $P(\Sigma x \geq 1,600) = \texttt{normalcdf}(1600,E99,1514.10,63) = 0.0864$

d. $P(\Sigma x \leq 1,595) = \texttt{normalcdf}(-E99,1595,1514.10,63) = 0.9005$. This means that there is a 90% chance that the sum of the ages for the sample group $n = 49$ is at most 1595.

e. The 95th percentile $= \texttt{invNorm}(0.95,30.9,1.1) = 32.7$. This indicates that 95% of the prostitutes in the sample of 65 are younger than 32.7 years, on average.

f. The 90th percentile $= \texttt{invNorm}(0.90,2008.5,72.56) = 2101.5$. This indicates that 90% of the prostitutes in the sample of 65 have a sum of ages less than 2,101.5 years.

7.11 According to Boeing data, the 757 airliner carries 200 passengers and has doors with a mean height of 72 inches. Assume for a certain population of men we have a mean of 69.0 inches and a standard deviation of 2.8 inches.

a. What mean doorway height would allow 95% of men to enter the aircraft without bending?

b. Assume that half of the 200 passengers are men. What mean doorway height satisfies the condition that there is a 0.95 probability that this height is greater than the mean height of 100 men?

c. For engineers designing the 757, which result is more relevant: the height from part a or part b? Why?

HISTORICAL NOTE

: Normal Approximation to the Binomial

Historically, being able to compute binomial probabilities was one of the most important applications of the central limit theorem. Binomial probabilities with a small value for n(say, 20) were displayed in a table in a book. To calculate the probabilities with large values of n, you had to use the binomial formula, which could be very complicated. Using the **normal approximation to the binomial** distribution simplified the process. To compute the normal approximation to the binomial distribution, take a simple random sample from a population. You must meet the conditions for a **binomial distribution**:

- there are a certain number n of independent trials

- the outcomes of any trial are success or failure

- each trial has the same probability of a success p

Recall that if X is the binomial random variable, then $X \sim B(n, p)$. The shape of the binomial distribution needs to be similar to the shape of the normal distribution. To ensure this, the quantities np and nq must both be greater than five ($np > 5$ and $nq > 5$; the approximation is better if they are both greater than or equal to 10). Then the binomial can be approximated by the normal distribution with mean $\mu = np$ and standard deviation $\sigma = \sqrt{npq}$. Remember that $q = 1 - p$. In order to get the best approximation, add 0.5 to x or subtract 0.5 from x (use $x + 0.5$ or $x - 0.5$). The number 0.5 is called the **continuity correction factor** and is used in the following example.

Example 7.12

Suppose in a local Kindergarten through 12th grade (K - 12) school district, 53 percent of the population favor a charter school for grades K through 5. A simple random sample of 300 is surveyed.

a. Find the probability that **at least 150** favor a charter school.

b. Find the probability that **at most 160** favor a charter school.

c. Find the probability that **more than 155** favor a charter school.

d. Find the probability that **fewer than 147** favor a charter school.

e. Find the probability that **exactly 175** favor a charter school.

Let X = the number that favor a charter school for grades K trough 5. $X \sim B(n, p)$ where n = 300 and p = 0.53. Since $np > 5$ and $nq > 5$, use the normal approximation to the binomial. The formulas for the mean and standard deviation are $\mu = np$ and $\sigma = \sqrt{npq}$. The mean is 159 and the standard deviation is 8.6447. The random variable for the normal distribution is Y. $Y \sim N(159, 8.6447)$. See **The Normal Distribution** for help with calculator instructions.

For part a, you **include 150** so $P(X \geq 150)$ has normal approximation $P(Y \geq 149.5) = 0.8641$.

normalcdf(149.5,10^99,159,8.6447) = 0.8641.

For part b, you **include 160** so $P(X \leq 160)$ has normal appraximation $P(Y \leq 160.5) = 0.5689$.

normalcdf(0,160.5,159,8.6447) = 0.5689

For part c, you **exclude 155** so $P(X > 155)$ has normal approximation $P(y > 155.5) = 0.6572$.

normalcdf(155.5,10^99,159,8.6447) = 0.6572.

For part d, you **exclude 147** so $P(X < 147)$ has normal approximation $P(Y < 146.5) = 0.0741$.

normalcdf(0,146.5,159,8.6447) = 0.0741

For part e,$P(X = 175)$ has normal approximation $P(174.5 < Y < 175.5) = 0.0083$.

normalcdf(174.5,175.5,159,8.6447) = 0.0083

Because of calculators and computer software that let you calculate binomial probabilities for large values of n easily, it is not necessary to use the the normal approximation to the binomial distribution, provided that you have access to these technology tools. Most school labs have Microsoft Excel, an example of computer software that calculates binomial probabilities. Many students have access to the TI-83 or 84 series calculators, and they easily calculate probabilities for the binomial distribution. If you type in "binomial probability distribution calculation" in an Internet browser, you can find at least one online calculator for the binomial.

For **Example 7.10**, the probabilities are calculated using the following binomial distribution: ($n = 300$ and $p = 0.53$). Compare the binomial and normal distribution answers. See **Discrete Random Variables** for help with calculator instructions for the binomial.

$P(X \geq 150)$:1 - binomialcdf(300,0.53,149) = 0.8641

$P(X \leq 160)$:binomialcdf(300,0.53,160) = 0.5684

$P(X > 155)$:1 - binomialcdf(300,0.53,155) = 0.6576

$P(X < 147)$:binomialcdf(300,0.53,146) = 0.0742

$P(X = 175)$:(You use the binomial pdf.)binomialpdf(300,0.53,175) = 0.0083

Try It Σ

7.12 In a city, 46 percent of the population favor the incumbent, Dawn Morgan, for mayor. A simple random sample of 500 is taken. Using the continuity correction factor, find the probability that at least 250 favor Dawn Morgan for mayor.

7.4 | Central Limit Theorem (Pocket Change)

Stats Lab

7.1 Central Limit Theorem (Pocket Change)

Class Time:

Names:

Student Learning Outcomes

- The student will demonstrate and compare properties of the central limit theorem.

NOTE

This lab works best when sampling from several classes and combining data.

Collect the Data

1. Count the change in your pocket. (Do not include bills.)
2. Randomly survey 30 classmates. Record the values of the change in **Table 7.1**.

Table 7.1

3. Construct a histogram. Make five to six intervals. Sketch the graph using a ruler and pencil. Scale the axes.

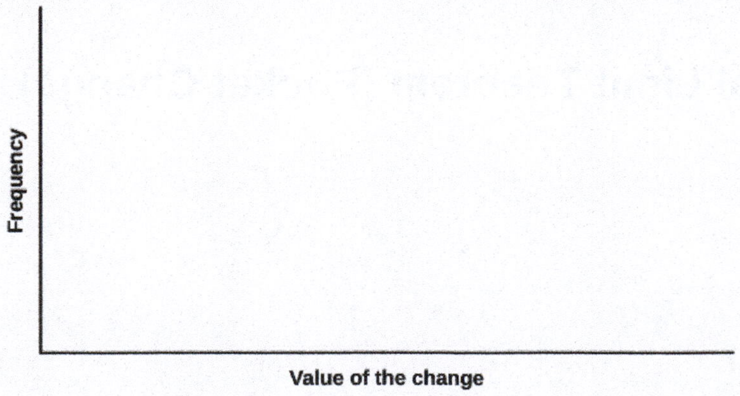

Figure 7.10

4. Calculate the following ($n = 1$; surveying one person at a time):

   a. $\bar{x} = $ _____

   b. $s = $ _____

5. Draw a smooth curve through the tops of the bars of the histogram. Use one to two complete sentences to describe the general shape of the curve.

Collecting Averages of Pairs

Repeat steps one through five of the section **Collect the Data.** with one exception. Instead of recording the change of 30 classmates, record the average change of 30 pairs.

1. Randomly survey 30 **pairs** of classmates.
2. Record the values of the average of their change in **Table 7.2**.

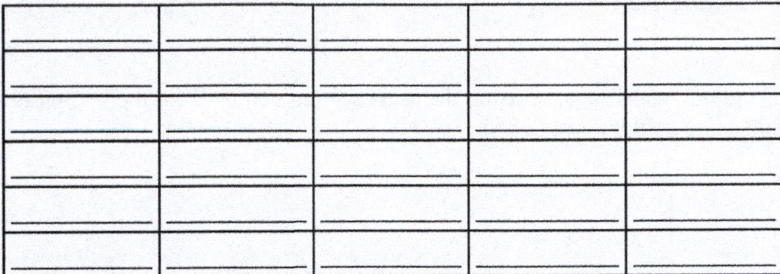

Table 7.2

3. Construct a histogram. Scale the axes using the same scaling you used for the section titled **Collect the Data**. Sketch the graph using a ruler and a pencil.

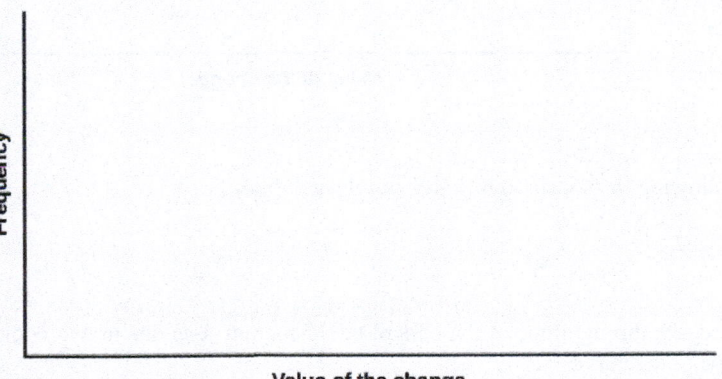

Figure 7.11

4. Calculate the following ($n = 2$; surveying two people at a time):

    a. $\bar{x} = $ _____

    b. $s = $ _____

5. Draw a smooth curve through tops of the bars of the histogram. Use one to two complete sentences to describe the general shape of the curve.

Collecting Averages of Groups of Five

Repeat steps one through five (of the section titled **Collect the Data**) with one exception. Instead of recording the change of 30 classmates, record the average change of 30 groups of five.

1. Randomly survey 30 **groups of five** classmates.
2. Record the values of the average of their change.

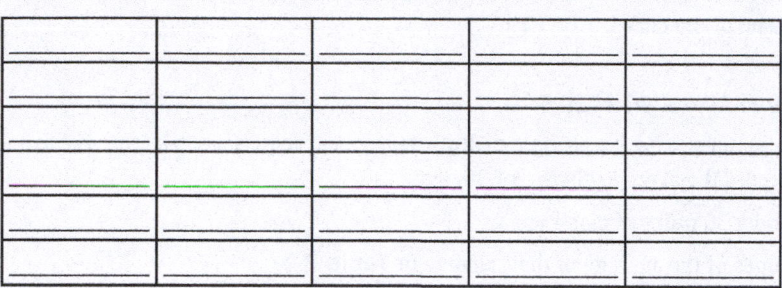

Table 7.3

3. Construct a histogram. Scale the axes using the same scaling you used for the section titled **Collect the Data**. Sketch the graph using a ruler and a pencil.

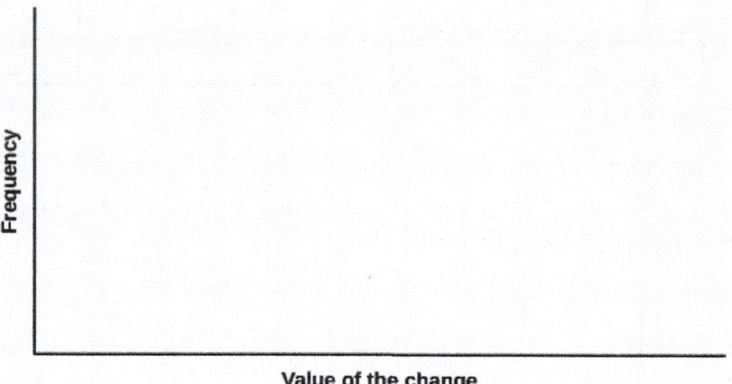

Value of the change

Figure 7.12

4. Calculate the following ($n = 5$; surveying five people at a time):

   a. $\bar{x} =$ _____

   b. $s =$ _____

5. Draw a smooth curve through tops of the bars of the histogram. Use one to two complete sentences to describe the general shape of the curve.

Discussion Questions

1. Why did the shape of the distribution of the data change, as n changed? Use one to two complete sentences to explain what happened.

2. In the section titled **Collect the Data**, what was the approximate distribution of the data? $X \sim$ _____(_____,_____)

3. In the section titled **Collecting Averages of Groups of Five**, what was the approximate distribution of the averages? $\bar{X} \sim$ _____(_____,_____)

4. In one to two complete sentences, explain any differences in your answers to the previous two questions.

7.5 | Central Limit Theorem (Cookie Recipes)

Stats Lab

7.2 Central Limit Theorem (Cookie Recipes)

Class Time:

Names:

Student Learning Outcomes

- The student will demonstrate and compare properties of the central limit theorem.

Given

X = length of time (in days) that a cookie recipe lasted at the Olmstead Homestead. (Assume that each of the different recipes makes the same quantity of cookies.)

Recipe #	X	Recipe #	X	Recipe #	X	Recipe #	X
1	1	16	2	31	3	46	2
2	5	17	2	32	4	47	2
3	2	18	4	33	5	48	11
4	5	19	6	34	6	49	5
5	6	20	1	35	6	50	5
6	1	21	6	36	1	51	4
7	2	22	5	37	1	52	6
8	6	23	2	38	2	53	5
9	5	24	5	39	1	54	1
10	2	25	1	40	6	55	1
11	5	26	6	41	1	56	2
12	1	27	4	42	6	57	4
13	1	28	1	43	2	58	3
14	3	29	6	44	6	59	6
15	2	30	2	45	2	60	5

Table 7.4

Calculate the following:

a. μ_x = _____

b. σ_x = _____

Collect the Data

Use a random number generator to randomly select four samples of size $n = 5$ from the given population. Record your samples in **Table 7.5**. Then, for each sample, calculate the mean to the nearest tenth. Record them in the spaces provided. Record the sample means for the rest of the class.

1. Complete the table:

	Sample 1	Sample 2	Sample 3	Sample 4	Sample means from other groups:
Means:	$\bar{x}$ = _____	$\bar{x}$ = _____	$\bar{x}$ = _____	$\bar{x}$ = _____	

Table 7.5

2. Calculate the following:

    a. $\bar{x}$ = _____

    b. $s\bar{x}$ = _____

3. Again, use a random number generator to randomly select four samples from the population. This time, make the samples of size $n = 10$. Record the samples in **Table 7.6**. As before, for each sample, calculate the mean to the nearest tenth. Record them in the spaces provided. Record the sample means for the rest of the class.

	Sample 1	Sample 2	Sample 3	Sample 4	Sample means from other groups
Means:	$\bar{x}$ = _____	$\bar{x}$ = _____	$\bar{x}$ = _____	$\bar{x}$ = _____	

Table 7.6

4. Calculate the following:

    a. $\bar{x}$ = _____

    b. $s\bar{x}$ = _____

5. For the original population, construct a histogram. Make intervals with a bar width of one day. Sketch the graph using a ruler and pencil. Scale the axes.

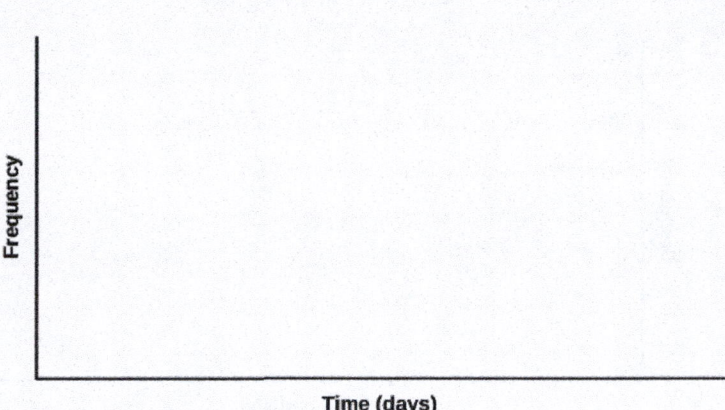

Figure 7.13

6. Draw a smooth curve through the tops of the bars of the histogram. Use one to two complete sentences to describe the general shape of the curve.

Repeat the Procedure for *n* = 5

1. For the sample of *n* = 5 days averaged together, construct a histogram of the averages (your means together with the means of the other groups). Make intervals with bar widths of $\frac{1}{2}$ a day. Sketch the graph using a ruler and pencil. Scale the axes.

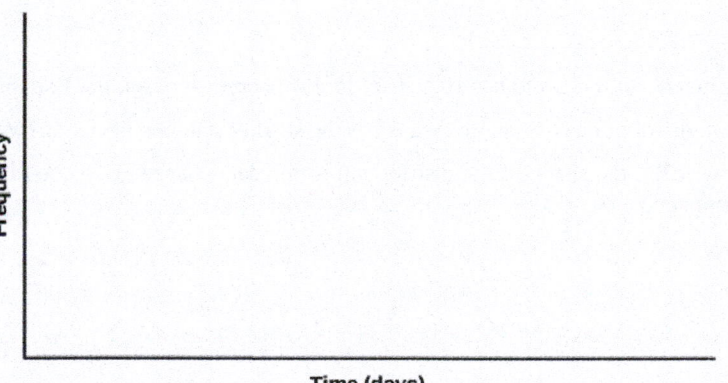

Figure 7.14

2. Draw a smooth curve through the tops of the bars of the histogram. Use one to two complete sentences to describe the general shape of the curve.

Repeat the Procedure for *n* = 10

1. For the sample of *n* = 10 days averaged together, construct a histogram of the averages (your means together with the means of the other groups). Make intervals with bar widths of $\frac{1}{2}$ a day. Sketch the graph using a ruler and pencil. Scale the axes.

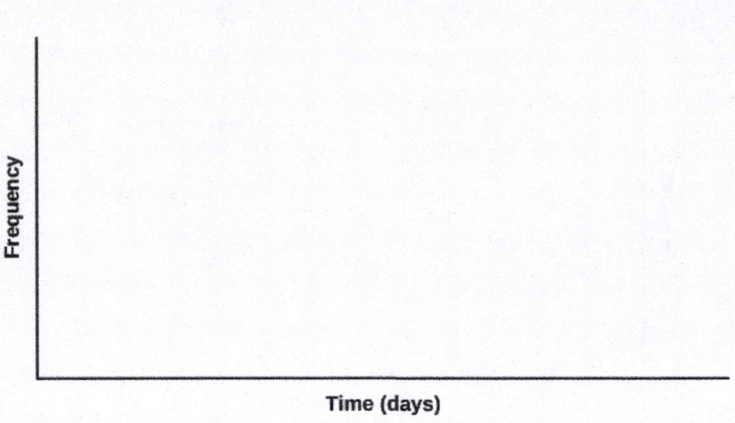

Figure 7.15

2. Draw a smooth curve through the tops of the bars of the histogram. Use one to two complete sentences to describe the general shape of the curve.

Discussion Questions

1. Compare the three histograms you have made, the one for the population and the two for the sample means. In three to five sentences, describe the similarities and differences.

2. State the theoretical (according to the clt) distributions for the sample means.

   a. $n = 5$: $\bar{x}$ ~ _____(_____,_____)

   b. $n = 10$: $\bar{x}$ ~ _____(_____,_____)

3. Are the sample means for $n = 5$ and $n = 10$ "close" to the theoretical mean, μ_x? Explain why or why not.

4. Which of the two distributions of sample means has the smaller standard deviation? Why?

5. As n changed, why did the shape of the distribution of the data change? Use one to two complete sentences to explain what happened.

KEY TERMS

Average a number that describes the central tendency of the data; there are a number of specialized averages, including the arithmetic mean, weighted mean, median, mode, and geometric mean.

Central Limit Theorem Given a random variable (RV) with known mean μ and known standard deviation, σ, we are sampling with size n, and we are interested in two new RVs: the sample mean, $\bar{X}$, and the sample sum, ΣX. If the size (n) of the sample is sufficiently large, then $\bar{X} \sim N(\mu, \frac{\sigma}{\sqrt{n}})$ and $\Sigma X \sim N(n\mu, (\sqrt{n})(\sigma))$. If the size ($n$) of the sample is sufficiently large, then the distribution of the sample means and the distribution of the sample sums will approximate a normal distributions regardless of the shape of the population. The mean of the sample means will equal the population mean, and the mean of the sample sums will equal n times the population mean. The standard deviation of the distribution of the sample means, $\frac{\sigma}{\sqrt{n}}$, is called the standard error of the mean.

Exponential Distribution a continuous random variable (RV) that appears when we are interested in the intervals of time between some random events, for example, the length of time between emergency arrivals at a hospital, notation: $X \sim Exp(m)$. The mean is $\mu = \frac{1}{m}$ and the standard deviation is $\sigma = \frac{1}{m}$. The probability density function is $f(x) = me^{-mx}$, $x \geq 0$ and the cumulative distribution function is $P(X \leq x) = 1 - e^{-mx}$.

Mean a number that measures the central tendency; a common name for mean is "average." The term "mean" is a shortened form of "arithmetic mean." By definition, the mean for a sample (denoted by $\bar{x}$) is $\bar{x} = \frac{\text{Sum of all values in the sample}}{\text{Number of values in the sample}}$, and the mean for a population (denoted by μ) is $\mu = \frac{\text{Sum of all values in the population}}{\text{Number of values in the population}}$.

Normal Distribution a continuous random variable (RV) with pdf $f(x) = \frac{1}{\sigma\sqrt{2\pi}} e^{\frac{-(x-\mu)^2}{2\sigma^2}}$, where μ is the mean of the distribution and σ is the standard deviation; notation: $X \sim N(\mu, \sigma)$. If $\mu = 0$ and $\sigma = 1$, the RV is called a **standard normal distribution**.

Normal Distribution a continuous random variable (RV) with pdf $f(x) = \frac{1}{\sigma\sqrt{2\pi}} e^{\frac{-(x-\mu)^2}{2\sigma^2}}$, where μ is the mean of the distribution and σ is the standard deviation.; notation: $X \sim N(\mu, \sigma)$. If $\mu = 0$ and $\sigma = 1$, the RV is called the **standard normal distribution**.

Sampling Distribution Given simple random samples of size n from a given population with a measured characteristic such as mean, proportion, or standard deviation for each sample, the probability distribution of all the measured characteristics is called a sampling distribution.

Standard Error of the Mean the standard deviation of the distribution of the sample means, or $\frac{\sigma}{\sqrt{n}}$.

Uniform Distribution a continuous random variable (RV) that has equally likely outcomes over the domain, $a < x < b$; often referred as the **Rectangular Distribution** because the graph of the pdf has the form of a rectangle. Notation: $X \sim U(a, b)$. The mean is $\mu = \frac{a+b}{2}$ and the standard deviation is $\sigma = \sqrt{\frac{(b-a)^2}{12}}$. The probability density function is $f(x) = \frac{1}{b-a}$ for $a < x < b$ or $a \leq x \leq b$. The cumulative distribution is $P(X \leq x) = \frac{x-a}{b-a}$.

CHAPTER REVIEW

7.1 The Central Limit Theorem for Sample Means (Averages)

In a population whose distribution may be known or unknown, if the size (n) of samples is sufficiently large, the distribution of the sample means will be approximately normal. The mean of the sample means will equal the population mean. The standard deviation of the distribution of the sample means, called the standard error of the mean, is equal to the population standard deviation divided by the square root of the sample size (n).

7.2 The Central Limit Theorem for Sums

The central limit theorem tells us that for a population with any distribution, the distribution of the sums for the sample means approaches a normal distribution as the sample size increases. In other words, if the sample size is large enough, the distribution of the sums can be approximated by a normal distribution even if the original population is not normally distributed. Additionally, if the original population has a mean of μ_X and a standard deviation of σ_X, the mean of the sums is $n\mu_X$ and the standard deviation is $(\sqrt{n})(\sigma_X)$ where n is the sample size.

7.3 Using the Central Limit Theorem

The central limit theorem can be used to illustrate the law of large numbers. The law of large numbers states that the larger the sample size you take from a population, the closer the sample mean $\bar{x}$ gets to μ.

FORMULA REVIEW

7.1 The Central Limit Theorem for Sample Means (Averages)

The Central Limit Theorem for Sample Means: $\bar{X} \sim N\left(\mu_x, \frac{\sigma_x}{\sqrt{n}}\right)$

The Mean $\bar{X}$: μ_x

Central Limit Theorem for Sample Means z-score and standard error of the mean: $z = \frac{\bar{x} - \mu_x}{\left(\frac{\sigma_x}{\sqrt{n}}\right)}$

Standard Error of the Mean (Standard Deviation ($\bar{X}$)): $\frac{\sigma_x}{\sqrt{n}}$

7.2 The Central Limit Theorem for Sums

The Central Limit Theorem for Sums: $\sum X \sim N[(n)(\mu_x), (\sqrt{n})(\sigma_x)]$

Mean for Sums ($\sum X$): $(n)(\mu_x)$

The Central Limit Theorem for Sums z-score and standard deviation for sums:

z for the sample mean $= \frac{\sum x - (n)(\mu_X)}{(\sqrt{n})(\sigma_X)}$

Standard deviation for Sums ($\sum X$): $(\sqrt{n})(\sigma_x)$

PRACTICE

7.1 The Central Limit Theorem for Sample Means (Averages)

Use the following information to answer the next six exercises: Yoonie is a personnel manager in a large corporation. Each month she must review 16 of the employees. From past experience, she has found that the reviews take her approximately four hours each to do with a population standard deviation of 1.2 hours. Let X be the random variable representing the time it takes her to complete one review. Assume X is normally distributed. Let $\bar{X}$ be the random variable representing the mean time to complete the 16 reviews. Assume that the 16 reviews represent a random set of reviews.

1. What is the mean, standard deviation, and sample size?

2. Complete the distributions.

a. $X \sim$ _____(_____,_____)

b. $\overline{X} \sim$ _____(_____,_____)

3. Find the probability that **one** review will take Yoonie from 3.5 to 4.25 hours. Sketch the graph, labeling and scaling the horizontal axis. Shade the region corresponding to the probability.

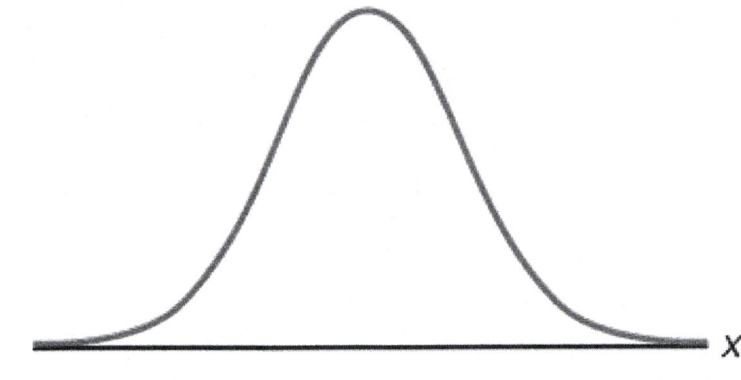

a.

Figure 7.16

b. $P($_____ $< x <$ _____$) =$ _____

4. Find the probability that the **mean** of a month's reviews will take Yoonie from 3.5 to 4.25 hrs. Sketch the graph, labeling and scaling the horizontal axis. Shade the region corresponding to the probability.

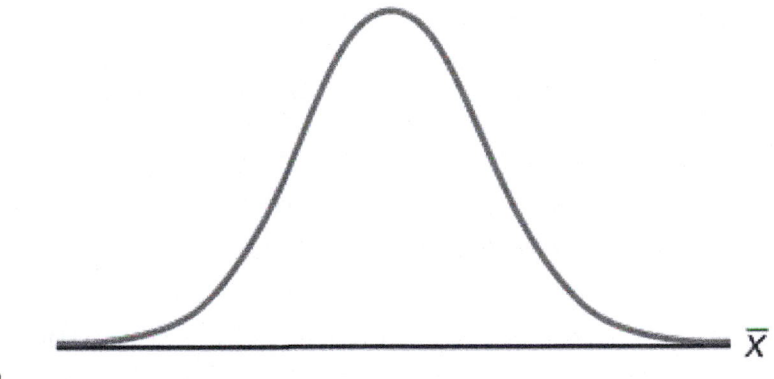

a.

Figure 7.17

b. $P($_____$) =$ _____

5. What causes the probabilities in **Exercise 7.3** and **Exercise 7.4** to be different?

6. Find the 95th percentile for the mean time to complete one month's reviews. Sketch the graph.

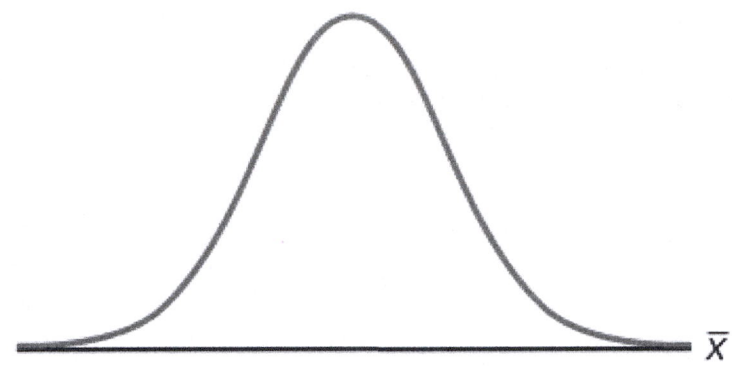

a.

Figure 7.18

b. The 95th Percentile =_____

7.2 The Central Limit Theorem for Sums

Use the following information to answer the next four exercises: An unknown distribution has a mean of 80 and a standard deviation of 12. A sample size of 95 is drawn randomly from the population.

7. Find the probability that the sum of the 95 values is greater than 7,650.

8. Find the probability that the sum of the 95 values is less than 7,400.

9. Find the sum that is two standard deviations above the mean of the sums.

10. Find the sum that is 1.5 standard deviations below the mean of the sums.

Use the following information to answer the next five exercises: The distribution of results from a cholesterol test has a mean of 180 and a standard deviation of 20. A sample size of 40 is drawn randomly.

11. Find the probability that the sum of the 40 values is greater than 7,500.

12. Find the probability that the sum of the 40 values is less than 7,000.

13. Find the sum that is one standard deviation above the mean of the sums.

14. Find the sum that is 1.5 standard deviations below the mean of the sums.

15. Find the percentage of sums between 1.5 standard deviations below the mean of the sums and one standard deviation above the mean of the sums.

Use the following information to answer the next six exercises: A researcher measures the amount of sugar in several cans of the same soda. The mean is 39.01 with a standard deviation of 0.5. The researcher randomly selects a sample of 100.

16. Find the probability that the sum of the 100 values is greater than 3,910.

17. Find the probability that the sum of the 100 values is less than 3,900.

18. Find the probability that the sum of the 100 values falls between the numbers you found in and .

19. Find the sum with a *z*–score of –2.5.

20. Find the sum with a *z*–score of 0.5.

21. Find the probability that the sums will fall between the *z*-scores –2 and 1.

Use the following information to answer the next four exercise: An unknown distribution has a mean 12 and a standard deviation of one. A sample size of 25 is taken. Let X = the object of interest.

22. What is the mean of ΣX?

23. What is the standard deviation of ΣX?

24. What is $P(\Sigma x = 290)$?

25. What is $P(\Sigma x > 290)$?

26. True or False: only the sums of normal distributions are also normal distributions.

27. In order for the sums of a distribution to approach a normal distribution, what must be true?

28. What three things must you know about a distribution to find the probability of sums?

29. An unknown distribution has a mean of 25 and a standard deviation of six. Let X = one object from this distribution. What is the sample size if the standard deviation of ΣX is 42?

30. An unknown distribution has a mean of 19 and a standard deviation of 20. Let X = the object of interest. What is the sample size if the mean of ΣX is 15,200?

Use the following information to answer the next three exercises. A market researcher analyzes how many electronics devices customers buy in a single purchase. The distribution has a mean of three with a standard deviation of 0.7. She samples 400 customers.

31. What is the z-score for $\Sigma x = 840$?

32. What is the z-score for $\Sigma x = 1,186$?

33. What is $P(\Sigma x < 1,186)$?

Use the following information to answer the next three exercises: An unkwon distribution has a mean of 100, a standard deviation of 100, and a sample size of 100. Let X = one object of interest.

34. What is the mean of ΣX?

35. What is the standard deviation of ΣX?

36. What is $P(\Sigma x > 9,000)$?

7.3 Using the Central Limit Theorem

Use the following information to answer the next ten exercises: A manufacturer produces 25-pound lifting weights. The lowest actual weight is 24 pounds, and the highest is 26 pounds. Each weight is equally likely so the distribution of weights is uniform. A sample of 100 weights is taken.

37.
   a. What is the distribution for the weights of one 25-pound lifting weight? What is the mean and standard deivation?
   b. What is the distribution for the mean weight of 100 25-pound lifting weights?
   c. Find the probability that the mean actual weight for the 100 weights is less than 24.9.

38. Draw the graph from **Exercise 7.37**

39. Find the probability that the mean actual weight for the 100 weights is greater than 25.2.

40. Draw the graph from **Exercise 7.39**

41. Find the 90th percentile for the mean weight for the 100 weights.

42. Draw the graph from **Exercise 7.41**

43.
   a. What is the distribution for the sum of the weights of 100 25-pound lifting weights?
   b. Find $P(\Sigma x < 2,450)$.

44. Draw the graph from **Exercise 7.43**

45. Find the 90th percentile for the total weight of the 100 weights.

46. Draw the graph from **Exercise 7.45**

Use the following information to answer the next five exercises: The length of time a particular smartphone's battery lasts follows an exponential distribution with a mean of ten months. A sample of 64 of these smartphones is taken.

47.

    a.   What is the standard deviation?

    b.   What is the parameter m?

48. What is the distribution for the length of time one battery lasts?

49. What is the distribution for the mean length of time 64 batteries last?

50. What is the distribution for the total length of time 64 batteries last?

51. Find the probability that the sample mean is between seven and 11.

52. Find the 80th percentile for the total length of time 64 batteries last.

53. Find the IQR for the mean amount of time 64 batteries last.

54. Find the middle 80% for the total amount of time 64 batteries last.

Use the following information to answer the next eight exercises: A uniform distribution has a minimum of six and a maximum of ten. A sample of 50 is taken.

55. Find $P(\Sigma x > 420)$.

56. Find the 90th percentile for the sums.

57. Find the 15th percentile for the sums.

58. Find the first quartile for the sums.

59. Find the third quartile for the sums.

60. Find the 80th percentile for the sums.

HOMEWORK

7.1 The Central Limit Theorem for Sample Means (Averages)

61. Previously, De Anza statistics students estimated that the amount of change daytime statistics students carry is exponentially distributed with a mean of \$0.88. Suppose that we randomly pick 25 daytime statistics students.

    a.   In words, $X = $ _____

    b. $X \sim$ _____(_____,_____)

    c.   In words, $\bar{X} = $ _____

    d. $\bar{X} \sim$ _____ (_____, _____)

    e.   Find the probability that an individual had between \$0.80 and \$1.00. Graph the situation, and shade in the area to be determined.

    f.   Find the probability that the average of the 25 students was between \$0.80 and \$1.00. Graph the situation, and shade in the area to be determined.

    g.   Explain why there is a difference in part e and part f.

62. Suppose that the distance of fly balls hit to the outfield (in baseball) is normally distributed with a mean of 250 feet and a standard deviation of 50 feet. We randomly sample 49 fly balls.

    a.   If $\bar{X}$ = average distance in feet for 49 fly balls, then $\bar{X} \sim$ _____(_____,_____)

    b.   What is the probability that the 49 balls traveled an average of less than 240 feet? Sketch the graph. Scale the horizontal axis for $\bar{X}$. Shade the region corresponding to the probability. Find the probability.

    c.   Find the 80th percentile of the distribution of the average of 49 fly balls.

63. According to the Internal Revenue Service, the average length of time for an individual to complete (keep records for, learn, prepare, copy, assemble, and send) IRS Form 1040 is 10.53 hours (without any attached schedules). The distribution is unknown. Let us assume that the standard deviation is two hours. Suppose we randomly sample 36 taxpayers.

    a.   In words, $X = $ _____

b. In words, $\bar{X}$ = _____

c. $\bar{X}$ ~ _____(_____,_____)

d. Would you be surprised if the 36 taxpayers finished their Form 1040s in an average of more than 12 hours? Explain why or why not in complete sentences.

e. Would you be surprised if one taxpayer finished his or her Form 1040 in more than 12 hours? In a complete sentence, explain why.

64. Suppose that a category of world-class runners are known to run a marathon (26 miles) in an average of 145 minutes with a standard deviation of 14 minutes. Consider 49 of the races. Let $\bar{X}$ the average of the 49 races.

a. $\bar{X}$ ~ _____(_____,_____)

b. Find the probability that the runner will average between 142 and 146 minutes in these 49 marathons.

c. Find the 80th percentile for the average of these 49 marathons.

d. Find the median of the average running times.

65. The length of songs in a collector's iTunes album collection is uniformly distributed from two to 3.5 minutes. Suppose we randomly pick five albums from the collection. There are a total of 43 songs on the five albums.

a. In words, X = _____

b. X ~ _____

c. In words, $\bar{X}$ = _____

d. $\bar{X}$ ~ _____(_____,_____)

e. Find the first quartile for the average song length.

f. The IQR(interquartile range) for the average song length is from _____–_____.

66. In 1940 the average size of a U.S. farm was 174 acres. Let's say that the standard deviation was 55 acres. Suppose we randomly survey 38 farmers from 1940.

a. In words, X = _____

b. In words, $\bar{X}$ = _____

c. $\bar{X}$ ~ _____(_____,_____)

d. The IQR for $\bar{X}$ is from _____ acres to _____ acres.

67. Determine which of the following are true and which are false. Then, in complete sentences, justify your answers.

a. When the sample size is large, the mean of $\bar{X}$ is approximately equal to the mean of X.

b. When the sample size is large, $\bar{X}$ is approximately normally distributed.

c. When the sample size is large, the standard deviation of $\bar{X}$ is approximately the same as the standard deviation of X.

68. The percent of fat calories that a person in America consumes each day is normally distributed with a mean of about 36 and a standard deviation of about ten. Suppose that 16 individuals are randomly chosen. Let $\bar{X}$ = average percent of fat calories.

a. $\bar{X}$ ~ _____(_____, _____)

b. For the group of 16, find the probability that the average percent of fat calories consumed is more than five. Graph the situation and shade in the area to be determined.

c. Find the first quartile for the average percent of fat calories.

69. The distribution of income in some Third World countries is considered wedge shaped (many very poor people, very few middle income people, and even fewer wealthy people). Suppose we pick a country with a wedge shaped distribution. Let the average salary be $2,000 per year with a standard deviation of $8,000. We randomly survey 1,000 residents of that country.

a. In words, X = _____

b. In words, $\bar{X}$ = _____

c. $\bar{X}$ ~ _____(_____,_____)

d. How is it possible for the standard deviation to be greater than the average?

e. Why is it more likely that the average of the 1,000 residents will be from $2,000 to $2,100 than from $2,100 to $2,200?

70. Which of the following is NOT TRUE about the distribution for averages?

    a. The mean, median, and mode are equal.

    b. The area under the curve is one.

    c. The curve never touches the *x*-axis.

    d. The curve is skewed to the right.

71. The cost of unleaded gasoline in the Bay Area once followed an unknown distribution with a mean of $4.59 and a standard deviation of $0.10. Sixteen gas stations from the Bay Area are randomly chosen. We are interested in the average cost of gasoline for the 16 gas stations. The distribution to use for the average cost of gasoline for the 16 gas stations is:

    a. $\bar{X}$ ~ $N(4.59, 0.10)$

    b. $\bar{X}$ ~ $N\left(4.59, \dfrac{0.10}{\sqrt{16}}\right)$

    c. $\bar{X}$ ~ $N\left(4.59, \dfrac{16}{0.10}\right)$

    d. $\bar{X}$ ~ $N\left(4.59, \dfrac{\sqrt{16}}{0.10}\right)$

7.2 The Central Limit Theorem for Sums

72. Which of the following is NOT TRUE about the theoretical distribution of sums?

    a. The mean, median and mode are equal.

    b. The area under the curve is one.

    c. The curve never touches the *x*-axis.

    d. The curve is skewed to the right.

73. Suppose that the duration of a particular type of criminal trial is known to have a mean of 21 days and a standard deviation of seven days. We randomly sample nine trials.

    a. In words, $\Sigma X = $ _____

    b. ΣX ~ _____(_____,_____)

    c. Find the probability that the total length of the nine trials is at least 225 days.

    d. Ninety percent of the total of nine of these types of trials will last at least how long?

74. Suppose that the weight of open boxes of cereal in a home with children is uniformly distributed from two to six pounds with a mean of four pounds and standard deviation of 1.1547. We randomly survey 64 homes with children.

    a. In words, $X = $ _____

    b. The distribution is _____.

    c. In words, $\Sigma X = $ _____

    d. ΣX ~ _____(_____,_____)

    e. Find the probability that the total weight of open boxes is less than 250 pounds.

    f. Find the 35th percentile for the total weight of open boxes of cereal.

75. Salaries for teachers in a particular elementary school district are normally distributed with a mean of $44,000 and a standard deviation of $6,500. We randomly survey ten teachers from that district.

    a. In words, $X = $ _____

    b. X ~ _____(_____,_____)

    c. In words, $\Sigma X = $ _____

    d. ΣX ~ _____(_____,_____)

    e. Find the probability that the teachers earn a total of over $400,000.

    f. Find the 90th percentile for an individual teacher's salary.

    g. Find the 90th percentile for the sum of ten teachers' salary.

    h. If we surveyed 70 teachers instead of ten, graphically, how would that change the distribution in part d?

    i. If each of the 70 teachers received a $3,000 raise, graphically, how would that change the distribution in part b?

7.3 Using the Central Limit Theorem

76. The attention span of a two-year-old is exponentially distributed with a mean of about eight minutes. Suppose we randomly survey 60 two-year-olds.
  a. In words, X = _____
  b. X ~ _____(_____,_____)
  c. In words, $\bar{X}$ = _____
  d. $\bar{X}$ ~ _____(_____,_____)
  e. Before doing any calculations, which do you think will be higher? Explain why.
       i. The probability that an individual attention span is less than ten minutes.
       ii. The probability that the average attention span for the 60 children is less than ten minutes?

  f. Calculate the probabilities in part e.

  g. Explain why the distribution for $\bar{X}$ is not exponential.

77. The closing stock prices of 35 U.S. semiconductor manufacturers are given as follows.

8.625; 30.25; 27.625; 46.75; 32.875; 18.25; 5; 0.125; 2.9375; 6.875; 28.25; 24.25; 21; 1.5; 30.25; 71; 43.5; 49.25; 2.5625; 31; 16.5; 9.5; 18.5; 18; 9; 10.5; 16.625; 1.25; 18; 12.87; 7; 12.875; 2.875; 60.25; 29.25

  a. In words, X = _____
  b.
       i. $\bar{x}$ = _____
       ii. s_x = _____
       iii. n = _____
  c. Construct a histogram of the distribution of the averages. Start at x = –0.0005. Use bar widths of ten.
  d. In words, describe the distribution of stock prices.
  e. Randomly average five stock prices together. (Use a random number generator.) Continue averaging five pieces together until you have ten averages. List those ten averages.
  f. Use the ten averages from part e to calculate the following.
       i. $\bar{x}$ = _____
       ii. s_x = _____
  g. Construct a histogram of the distribution of the averages. Start at x = -0.0005. Use bar widths of ten.
  h. Does this histogram look like the graph in part c?
  i. In one or two complete sentences, explain why the graphs either look the same or look different?
  j. Based upon the theory of the **central limit theorem**, $\bar{X}$ ~ _____(_____,_____)

Use the following information to answer the next three exercises: Richard's Furniture Company delivers furniture from 10 A.M. to 2 P.M. continuously and uniformly. We are interested in how long (in hours) past the 10 A.M. start time that individuals wait for their delivery.

78. X ~ _____(_____,_____)
  a. $U(0,4)$
  b. $U(10,2)$
  c. $Exp(2)$
  d. $N(2,1)$

79. The average wait time is:
  a. one hour.
  b. two hours.
  c. two and a half hours.
  d. four hours.

80. Suppose that it is now past noon on a delivery day. The probability that a person must wait at least one and a half **more** hours is:

a. $\frac{1}{4}$

b. $\frac{1}{2}$

c. $\frac{3}{4}$

d. $\frac{3}{8}$

Use the following information to answer the next two exercises: The time to wait for a particular rural bus is distributed uniformly from zero to 75 minutes. One hundred riders are randomly sampled to learn how long they waited.

81. The 90th percentile sample average wait time (in minutes) for a sample of 100 riders is:
   a. 315.0
   b. 40.3
   c. 38.5
   d. 65.2

82. Would you be surprised, based upon numerical calculations, if the sample average wait time (in minutes) for 100 riders was less than 30 minutes?
   a. yes
   b. no
   c. There is not enough information.

Use the following to answer the next two exercises: The cost of unleaded gasoline in the Bay Area once followed an unknown distribution with a mean of $4.59 and a standard deviation of $0.10. Sixteen gas stations from the Bay Area are randomly chosen. We are interested in the average cost of gasoline for the 16 gas stations.

83. What's the approximate probability that the average price for 16 gas stations is over $4.69?
   a. almost zero
   b. 0.1587
   c. 0.0943
   d. unknown

84. Find the probability that the average price for 30 gas stations is less than $4.55.
   a. 0.6554
   b. 0.3446
   c. 0.0142
   d. 0.9858
   e. 0

85. Suppose in a local Kindergarten through 12th grade (K - 12) school district, 53 percent of the population favor a charter school for grades K through five. A simple random sample of 300 is surveyed. Calculate following using the normal approximation to the binomial distribtion.
   a. Find the probability that less than 100 favor a charter school for grades K through 5.
   b. Find the probability that 170 or more favor a charter school for grades K through 5.
   c. Find the probability that no more than 140 favor a charter school for grades K through 5.
   d. Find the probability that there are fewer than 130 that favor a charter school for grades K through 5.
   e. Find the probability that exactly 150 favor a charter school for grades K through 5.

If you have access to an appropriate calculator or computer software, try calculating these probabilities using the technology.

86. Four friends, Janice, Barbara, Kathy and Roberta, decided to carpool together to get to school. Each day the driver would be chosen by randomly selecting one of the four names. They carpool to school for 96 days. Use the normal approximation to the binomial to calculate the following probabilities. Round the standard deviation to four decimal places.

   a. Find the probability that Janice is the driver at most 20 days.
   b. Find the probability that Roberta is the driver more than 16 days.
   c. Find the probability that Barbara drives exactly 24 of those 96 days.

87. $X \sim N(60, 9)$. Suppose that you form random samples of 25 from this distribution. Let $\bar{X}$ be the random variable of averages. Let ΣX be the random variable of sums. For parts c through f, sketch the graph, shade the region, label and scale the horizontal axis for $\bar{X}$, and find the probability.

    a. Sketch the distributions of X and $\bar{X}$ on the same graph.

    b. $\bar{X} \sim$ _____(_____,_____)

    c. $P(\bar{x} < 60) = $ _____

    d. Find the 30th percentile for the mean.

    e. $P(56 < \bar{x} < 62) = $ _____

    f. $P(18 < \bar{x} < 58) = $ _____

    g. $\Sigma x \sim$ _____(_____,_____)

    h. Find the minimum value for the upper quartile for the sum.

    i. $P(1{,}400 < \Sigma x < 1{,}550) = $ _____

88. Suppose that the length of research papers is uniformly distributed from ten to 25 pages. We survey a class in which 55 research papers were turned in to a professor. The 55 research papers are considered a random collection of all papers. We are interested in the average length of the research papers.

    a. In words, $X = $ _____

    b. $X \sim$ _____(_____,_____)

    c. $\mu_x = $ _____

    d. $\sigma_x = $ _____

    e. In words, $\bar{X} = $ _____

    f. $\bar{X} \sim$ _____(_____,_____)

    g. In words, $\Sigma X = $ _____

    h. $\Sigma X \sim$ _____(_____,_____)

    i. Without doing any calculations, do you think that it's likely that the professor will need to read a total of more than 1,050 pages? Why?

    j. Calculate the probability that the professor will need to read a total of more than 1,050 pages.

    k. Why is it so unlikely that the average length of the papers will be less than 12 pages?

89. Salaries for teachers in a particular elementary school district are normally distributed with a mean of $44,000 and a standard deviation of $6,500. We randomly survey ten teachers from that district.

    a. Find the 90th percentile for an individual teacher's salary.

    b. Find the 90th percentile for the average teacher's salary.

90. The average length of a maternity stay in a U.S. hospital is said to be 2.4 days with a standard deviation of 0.9 days. We randomly survey 80 women who recently bore children in a U.S. hospital.

    a. In words, $X = $ _____

    b. In words, $\bar{X} = $ _____

    c. $\bar{X} \sim$ _____(_____,_____)

    d. In words, $\Sigma X = $ _____

    e. $\Sigma X \sim$ _____(_____,_____)

    f. Is it likely that an individual stayed more than five days in the hospital? Why or why not?

    g. Is it likely that the average stay for the 80 women was more than five days? Why or why not?

    h. Which is more likely:

        i. An individual stayed more than five days.

        ii. the average stay of 80 women was more than five days.

    i. If we were to sum up the women's stays, is it likely that, collectively they spent more than a year in the hospital? Why or why not?

For each problem, wherever possible, provide graphs and use the calculator.

91. NeverReady batteries has engineered a newer, longer lasting AAA battery. The company claims this battery has an average life span of 17 hours with a standard deviation of 0.8 hours. Your statistics class questions this claim. As a class, you randomly select 30 batteries and find that the sample mean life span is 16.7 hours. If the process is working properly, what is the probability of getting a random sample of 30 batteries in which the sample mean lifetime is 16.7 hours or less? Is the company's claim reasonable?

92. Men have an average weight of 172 pounds with a standard deviation of 29 pounds.
   a. Find the probability that 20 randomly selected men will have a sum weight greater than 3600 lbs.
   b. If 20 men have a sum weight greater than 3500 lbs, then their total weight exceeds the safety limits for water taxis. Based on (a), is this a safety concern? Explain.

93. M&M candies large candy bags have a claimed net weight of 396.9 g. The standard deviation for the weight of the individual candies is 0.017 g. The following table is from a stats experiment conducted by a statistics class.

Red	Orange	Yellow	Brown	Blue	Green
0.751	0.735	0.883	0.696	0.881	0.925
0.841	0.895	0.769	0.876	0.863	0.914
0.856	0.865	0.859	0.855	0.775	0.881
0.799	0.864	0.784	0.806	0.854	0.865
0.966	0.852	0.824	0.840	0.810	0.865
0.859	0.866	0.858	0.868	0.858	1.015
0.857	0.859	0.848	0.859	0.818	0.876
0.942	0.838	0.851	0.982	0.868	0.809
0.873	0.863			0.803	0.865
0.809	0.888			0.932	0.848
0.890	0.925			0.842	0.940
0.878	0.793			0.832	0.833
0.905	0.977			0.807	0.845
	0.850			0.841	0.852
	0.830			0.932	0.778
	0.856			0.833	0.814
	0.842			0.881	0.791
	0.778			0.818	0.810
	0.786			0.864	0.881
	0.853			0.825	
	0.864			0.855	
	0.873			0.942	
	0.880			0.825	
	0.882			0.869	
	0.931			0.912	
				0.887	

Table 7.7

The bag contained 465 candies and he listed weights in the table came from randomly selected candies. Count the weights.

  a. Find the mean sample weight and the standard deviation of the sample weights of candies in the table.
  b. Find the sum of the sample weights in the table and the standard deviation of the sum the of the weights.
  c. If 465 M&Ms are randomly selected, find the probability that their weights sum to at least 396.9.
  d. Is the Mars Company's M&M labeling accurate?

94. The Screw Right Company claims their $\frac{3}{4}$ inch screws are within ±0.23 of the claimed mean diameter of 0.750 inches with a standard deviation of 0.115 inches. The following data were recorded.

0.757	0.723	0.754	0.737	0.757	0.741	0.722	0.741	0.743	0.742
0.740	0.758	0.724	0.739	0.736	0.735	0.760	0.750	0.759	0.754
0.744	0.758	0.765	0.756	0.738	0.742	0.758	0.757	0.724	0.757
0.744	0.738	0.763	0.756	0.760	0.768	0.761	0.742	0.734	0.754
0.758	0.735	0.740	0.743	0.737	0.737	0.725	0.761	0.758	0.756

Table 7.8

The screws were randomly selected from the local home repair store.

  a. Find the mean diameter and standard deviation for the sample
  b. Find the probability that 50 randomly selected screws will be within the stated tolerance levels. Is the company's diameter claim plausible?

95. Your company has a contract to perform preventive maintenance on thousands of air-conditioners in a large city. Based on service records from previous years, the time that a technician spends servicing a unit averages one hour with a standard deviation of one hour. In the coming week, your company will service a simple random sample of 70 units in the city. You plan to budget an average of 1.1 hours per technician to complete the work. Will this be enough time?

96. A typical adult has an average IQ score of 105 with a standard deviation of 20. If 20 randomly selected adults are given an IQ tesst, what is the probability that the sample mean scores will be between 85 and 125 points?

97. Certain coins have an average weight of 5.201 grams with a standard deviation of 0.065 g. If a vending machine is designed to accept coins whose weights range from 5.111 g to 5.291 g, what is the expected number of rejected coins when 280 randomly selected coins are inserted into the machine?

REFERENCES

7.1 The Central Limit Theorem for Sample Means (Averages)

Baran, Daya. "20 Percent of Americans Have Never Used Email."WebGuild, 2010. Available online at http://www.webguild.org/20080519/20-percent-of-americans-have-never-used-email (accessed May 17, 2013).

Data from The Flurry Blog, 2013. Available online at http://blog.flurry.com (accessed May 17, 2013).

Data from the United States Department of Agriculture.

7.2 The Central Limit Theorem for Sums

Farago, Peter. "The Truth About Cats and Dogs: Smartphone vs Tablet Usage Differences." The Flurry Blog, 2013. Posted October 29, 2012. Available online at http://blog.flurry.com (accessed May 17, 2013).

7.3 Using the Central Limit Theorem

Data from the Wall Street Journal.

"National Health and Nutrition Examination Survey." Center for Disease Control and Prevention. Available online at http://www.cdc.gov/nchs/nhanes.htm (accessed May 17, 2013).

SOLUTIONS

1 mean = 4 hours; standard deviation = 1.2 hours; sample size = 16

3 a. Check student's solution.
b. 3.5, 4.25, 0.2441

5 The fact that the two distributions are different accounts for the different probabilities.

7 0.3345

9 7,833.92

11 0.0089

13 7,326.49

15 77.45%

17 0.4207

19 3,888.5

21 0.8186

23 5

25 0.9772

27 The sample size, n, gets larger.

29 49

31 26.00

33 0.1587

35 1,000

37
  a. $U(24, 26)$, 25, 0.5774

  b. $N(25, 0.0577)$

  c. 0.0416

39 0.0003

41 25.07

43
  a. $N(2,500, 5.7735)$

  b. 0

45 2,507.40

47
  a. 10

  b. $\frac{1}{10}$

49 $N\left(10, \frac{10}{8}\right)$

51 0.7799

53 1.69

55 0.0072

57 391.54

59 405.51

61

  a. X = amount of change students carry

  b. $X \sim E(0.88, 0.88)$

  c. $\bar{X}$ = average amount of change carried by a sample of 25 sstudents.

  d. $\bar{X} \sim N(0.88, 0.176)$

  e. 0.0819

  f. 0.1882

  g. The distributions are different. Part a is exponential and part b is normal.

63

  a. length of time for an individual to complete IRS form 1040, in hours.

  b. mean length of time for a sample of 36 taxpayers to complete IRS form 1040, in hours.

  c. $N\left(10.53, \frac{1}{3}\right)$

  d. Yes. I would be surprised, because the probability is almost 0.

  e. No. I would not be totally surprised because the probability is 0.2312

65

  a. the length of a song, in minutes, in the collection

  b. $U(2, 3.5)$

  c. the average length, in minutes, of the songs from a sample of five albums from the collection

  d. $N(2.75, 0.0220)$

  e. 2.74 minutes

  f. 0.03 minutes

67

  a. True. The mean of a sampling distribution of the means is approximately the mean of the data distribution.

  b. True. According to the Central Limit Theorem, the larger the sample, the closer the sampling distribution of the means becomes normal.

  c. The standard deviation of the sampling distribution of the means will decrease making it approximately the same as the standard deviation of X as the sample size increases.

69

  a. X = the yearly income of someone in a third world country

  b. the average salary from samples of 1,000 residents of a third world country

  c. $\bar{X} \sim N\left(2000, \frac{8000}{\sqrt{1000}}\right)$

  d. Very wide differences in data values can have averages smaller than standard deviations.

e. The distribution of the sample mean will have higher probabilities closer to the population mean.

$P(2000 < \overline{X} < 2100) = 0.1537$

$P(2100 < \overline{X} < 2200) = 0.1317$

71 b

73

a. the total length of time for nine criminal trials

b. $N(189, 21)$

c. 0.0432

d. 162.09; ninety percent of the total nine trials of this type will last 162 days or more.

75

a. X = the salary of one elementary school teacher in the district

b. $X \sim N(44{,}000, 6{,}500)$

c. $\Sigma X \sim$ sum of the salaries of ten elementary school teachers in the sample

d. $\Sigma X \sim N(44000, 20554.80)$

e. 0.9742

f. \$52,330.09

g. 466,342.04

h. Sampling 70 teachers instead of ten would cause the distribution to be more spread out. It would be a more symmetrical normal curve.

i. If every teacher received a \$3,000 raise, the distribution of X would shift to the right by \$3,000. In other words, it would have a mean of \$47,000.

77

a. X = the closing stock prices for U.S. semiconductor manufacturers

b. i. \$20.71; ii. \$17.31; iii. 35

d. Exponential distribution, $X \sim Exp\left(\frac{1}{20.71}\right)$

e. Answers will vary.

f. i. \$20.71; ii. \$11.14

g. Answers will vary.

h. Answers will vary.

i. Answers will vary.

j. $N\left(20.71, \frac{17.31}{\sqrt{5}}\right)$

79 b

81 b

83 a

85

a. 0

b. 0.1123

c. 0.0162

d. 0.0003

e. 0.0268

87
a. Check student's solution.

b. $\bar{X} \sim N\left(60, \frac{9}{\sqrt{25}}\right)$

c. 0.5000

d. 59.06

e. 0.8536

f. 0.1333

g. $N(1500, 45)$

h. 1530.35

i. 0.6877

89
a. $52,330

b. $46,634

91
- We have $\mu = 17$, $\sigma = 0.8$, $\bar{x} = 16.7$, and $n = 30$. To calculate the probability, we use `normalcdf`(lower, upper, μ, $\frac{\sigma}{\sqrt{n}}$) = `normalcdf` $\left(E - 99, 16.7, 17, \frac{0.8}{\sqrt{30}}\right) = 0.0200$.

- If the process is working properly, then the probability that a sample of 30 batteries would have at most 16.7 lifetime hours is only 2%. Therefore, the class was justified to question the claim.

93
a. For the sample, we have $n = 100$, $\bar{x} = 0.862$, $s = 0.05$

b. $\Sigma \bar{x} = 85.65$, $\Sigma s = 5.18$

c. `normalcdf`(396.9,E99,(465)(0.8565),(0.05)($\sqrt{465}$)) ≈ 1

d. Since the probability of a sample of size 465 having at least a mean sum of 396.9 is appproximately 1, we can conclude that Mars is correctly labeling their M&M packages.

95 Use `normalcdf` $\left(E - 99, 1.1, 1, \frac{1}{\sqrt{70}}\right) = 0.7986$. This means that there is an 80% chance that the service time will be less than 1.1 hours. It could be wise to schedule more time since there is an associated 20% chance that the maintenance time will be greater than 1.1 hours.

97 Since we have `normalcdf` $\left(5.111, 5.291, 5.201, \frac{0.065}{\sqrt{280}}\right) \approx 1$, we can conclude that practically all the coins are within the limits, therefore, there should be no rejected coins out of a well selected sample of size 280.

8 | CONFIDENCE INTERVALS

Figure 8.1 Have you ever wondered what the average number of M&Ms in a bag at the grocery store is? You can use confidence intervals to answer this question. (credit: comedy_nose/flickr)

Introduction

Chapter Objectives

By the end of this chapter, the student should be able to:

- Calculate and interpret confidence intervals for estimating a population mean and a population proportion.
- Interpret the Student's t probability distribution as the sample size changes.
- Discriminate between problems applying the normal and the Student's *t* distributions.
- Calculate the sample size required to estimate a population mean and a population proportion given a desired confidence level and margin of error.

Suppose you were trying to determine the mean rent of a two-bedroom apartment in your town. You might look in the classified section of the newspaper, write down several rents listed, and average them together. You would have obtained a point estimate of the true mean. If you are trying to determine the percentage of times you make a basket when shooting a basketball, you might count the number of shots you make and divide that by the number of shots you attempted. In this case, you would have obtained a point estimate for the true proportion.

We use sample data to make generalizations about an unknown population. This part of statistics is called **inferential statistics**. **The sample data help us to make an estimate of a population parameter**. We realize that the point estimate is most likely not the exact value of the population parameter, but close to it. After calculating point estimates, we construct interval estimates, called confidence intervals.

In this chapter, you will learn to construct and interpret confidence intervals. You will also learn a new distribution, the Student's-t, and how it is used with these intervals. Throughout the chapter, it is important to keep in mind that the confidence interval is a random variable. It is the population parameter that is fixed.

If you worked in the marketing department of an entertainment company, you might be interested in the mean number of songs a consumer downloads a month from iTunes. If so, you could conduct a survey and calculate the sample mean, $\bar{x}$, and the sample standard deviation, s. You would use $\bar{x}$ to estimate the population mean and s to estimate the population standard deviation. The sample mean, $\bar{x}$, is the **point estimate** for the population mean, μ. The sample standard deviation, s, is the point estimate for the population standard deviation, σ.

Each of $\bar{x}$ and s is called a statistic.

A **confidence interval** is another type of estimate but, instead of being just one number, it is an interval of numbers. The interval of numbers is a range of values calculated from a given set of sample data. The confidence interval is likely to include an unknown population parameter.

Suppose, for the iTunes example, we do not know the population mean μ, but we do know that the population standard deviation is $\sigma = 1$ and our sample size is 100. Then, by the central limit theorem, the standard deviation for the sample mean is

$$\frac{\sigma}{\sqrt{n}} = \frac{1}{\sqrt{100}} = 0.1 \ .$$

The **empirical rule**, which applies to bell-shaped distributions, says that in approximately 95% of the samples, the sample mean, $\bar{x}$, will be within two standard deviations of the population mean μ. For our iTunes example, two standard deviations is (2)(0.1) = 0.2. The sample mean $\bar{x}$ is likely to be within 0.2 units of μ.

Because $\bar{x}$ is within 0.2 units of μ, which is unknown, then μ is likely to be within 0.2 units of $\bar{x}$ in 95% of the samples. The population mean μ is contained in an interval whose lower number is calculated by taking the sample mean and subtracting two standard deviations (2)(0.1) and whose upper number is calculated by taking the sample mean and adding two standard deviations. In other words, μ is between $\bar{x} - 0.2$ and $\bar{x} + 0.2$ in 95% of all the samples.

For the iTunes example, suppose that a sample produced a sample mean $\bar{x} = 2$. Then the unknown population mean μ is between

$$\bar{x} - 0.2 = 2 - 0.2 = 1.8 \text{ and } \bar{x} + 0.2 = 2 + 0.2 = 2.2$$

We say that we are **95% confident** that the unknown population mean number of songs downloaded from iTunes per month is between 1.8 and 2.2. **The 95% confidence interval is (1.8, 2.2).**

The 95% confidence interval implies two possibilities. Either the interval (1.8, 2.2) contains the true mean μ or our sample produced an $\bar{x}$ that is not within 0.2 units of the true mean μ. The second possibility happens for only 5% of all the samples (95–100%).

Remember that a confidence interval is created for an unknown population parameter like the population mean, μ. Confidence intervals for some parameters have the form:

(point estimate – margin of error, point estimate + margin of error)

The margin of error depends on the confidence level or percentage of confidence and the standard error of the mean.

When you read newspapers and journals, some reports will use the phrase "margin of error." Other reports will not use that phrase, but include a confidence interval as the point estimate plus or minus the margin of error. These are two ways of expressing the same concept.

NOTE

Although the text only covers symmetrical confidence intervals, there are non-symmetrical confidence intervals (for example, a confidence interval for the standard deviation).

Collaborative Exercise

Have your instructor record the number of meals each student in your class eats out in a week. Assume that the standard deviation is known to be three meals. Construct an approximate 95% confidence interval for the true mean number of meals students eat out each week.

1. Calculate the sample mean.

2. Let $\sigma = 3$ and n = the number of students surveyed.

3. Construct the interval $\left(\bar{x} - 2 \cdot \frac{\sigma}{\sqrt{n}}, \ \bar{x} + 2 \cdot \frac{\sigma}{\sqrt{n}} \right)$.

We say we are approximately 95% confident that the true mean number of meals that students eat out in a week is between _____ and _____.

8.1 | A Single Population Mean using the Normal Distribution

A confidence interval for a population mean with a known standard deviation is based on the fact that the sample means follow an approximately normal distribution. Suppose that our sample has a mean of $\bar{x} = 10$ and we have constructed the 90% confidence interval (5, 15) where $EBM = 5$.

Calculating the Confidence Interval

To construct a confidence interval for a single unknown population mean μ, **where the population standard deviation is known**, we need $\bar{x}$ as an estimate for μ and we need the margin of error. Here, the margin of error (EBM) is called the **error bound for a population mean** (abbreviated EBM). The sample mean $\bar{x}$ is the **point estimate** of the unknown population mean μ.

The confidence interval estimate will have the form:

(point estimate - error bound, point estimate + error bound) or, in symbols,($\bar{x} - EBM, \ \bar{x} + EBM$)

The margin of error (EBM) depends on the **confidence level** (abbreviated CL). The confidence level is often considered the probability that the calculated confidence interval estimate will contain the true population parameter. However, it is more accurate to state that the confidence level is the percent of confidence intervals that contain the true population parameter when repeated samples are taken. Most often, it is the choice of the person constructing the confidence interval to choose a confidence level of 90% or higher because that person wants to be reasonably certain of his or her conclusions.

There is another probability called alpha (α). α is related to the confidence level, CL. α is the probability that the interval does not contain the unknown population parameter.
Mathematically, $\alpha + CL = 1$.

Example 8.1

Suppose we have collected data from a sample. We know the sample mean but we do not know the mean for the entire population.
The sample mean is seven, and the error bound for the mean is 2.5.

$\bar{x} = 7$ and $EBM = 2.5$

The confidence interval is $(7 - 2.5, 7 + 2.5)$, and calculating the values gives $(4.5, 9.5)$.

If the confidence level (CL) is 95%, then we say that, "We estimate with 95% confidence that the true value of the population mean is between 4.5 and 9.5."

8.1 Suppose we have data from a sample. The sample mean is 15, and the error bound for the mean is 3.2.

What is the confidence interval estimate for the population mean?

A confidence interval for a population mean with a known standard deviation is based on the fact that the sample means follow an approximately normal distribution. Suppose that our sample has a mean of $\bar{x} = 10$, and we have constructed the 90% confidence interval (5, 15) where $EBM = 5$.

To get a 90% confidence interval, we must include the central 90% of the probability of the normal distribution. If we include the central 90%, we leave out a total of $\alpha = 10\%$ in both tails, or 5% in each tail, of the normal distribution.

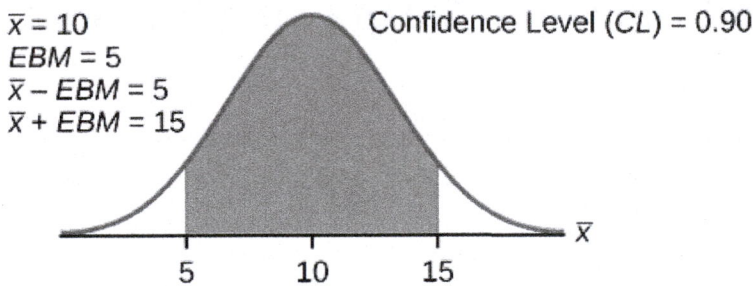

Figure 8.2

To capture the central 90%, we must go out 1.645 "standard deviations" on either side of the calculated sample mean. The value 1.645 is the z-score from a standard normal probability distribution that puts an area of 0.90 in the center, an area of 0.05 in the far left tail, and an area of 0.05 in the far right tail.

It is important that the "standard deviation" used must be appropriate for the parameter we are estimating, so in this section we need to use the standard deviation that applies to sample means, which is $\frac{\sigma}{\sqrt{n}}$. The fraction $\frac{\sigma}{\sqrt{n}}$, is commonly called the "standard error of the mean" in order to distinguish clearly the standard deviation for a mean from the population standard deviation σ.

In summary, as a result of the central limit theorem:

- $\bar{X}$ is normally distributed, that is, $\bar{X} \sim N\left(\mu_X, \frac{\sigma}{\sqrt{n}}\right)$.

- **When the population standard deviation σ is known, we use a normal distribution to calculate the error bound.**

Calculating the Confidence Interval

To construct a confidence interval estimate for an unknown population mean, we need data from a random sample. The steps to construct and interpret the confidence interval are:

- Calculate the sample mean $\bar{x}$ from the sample data. Remember, in this section we already know the population standard deviation σ.

- Find the z-score that corresponds to the confidence level.

- Calculate the error bound *EBM*.

- Construct the confidence interval.

- Write a sentence that interprets the estimate in the context of the situation in the problem. (Explain what the confidence interval means, in the words of the problem.)

We will first examine each step in more detail, and then illustrate the process with some examples.

Finding the *z*-score for the Stated Confidence Level

When we know the population standard deviation σ, we use a standard normal distribution to calculate the error bound EBM and construct the confidence interval. We need to find the value of z that puts an area equal to the confidence level (in decimal form) in the middle of the standard normal distribution $Z \sim N(0, 1)$.

The confidence level, *CL*, is the area in the middle of the standard normal distribution. $CL = 1 - \alpha$, so α is the area that is split equally between the two tails. Each of the tails contains an area equal to $\frac{\alpha}{2}$.

The z-score that has an area to the right of $\frac{\alpha}{2}$ is denoted by $z_{\frac{\alpha}{2}}$.

For example, when $CL = 0.95$, $\alpha = 0.05$ and $\frac{\alpha}{2} = 0.025$; we write $z_{\frac{\alpha}{2}} = z_{0.025}$.

The area to the right of $z_{0.025}$ is 0.025 and the area to the left of $z_{0.025}$ is $1 - 0.025 = 0.975$.

$z_{\frac{\alpha}{2}} = z_{0.025} = 1.96$, using a calculator, computer or a standard normal probability table.

  Using the TI-83, 83+, 84, 84+ Calculator
_____

$\texttt{invNorm}(0.975, 0, 1) = 1.96$

NOTE
_____

Remember to use the area to the LEFT of $z_{\frac{\alpha}{2}}$; in this chapter the last two inputs in the invNorm command are 0, 1, because you are using a standard normal distribution $Z \sim N(0, 1)$.

Calculating the Error Bound (*EBM*)

The error bound formula for an unknown population mean μ when the population standard deviation σ is known is

- $EBM = \left(z_{\frac{\alpha}{2}}\right)\left(\frac{\sigma}{\sqrt{n}}\right)$

Constructing the Confidence Interval

- The confidence interval estimate has the format $(\bar{x} - EBM,\ \bar{x} + EBM)$.

The graph gives a picture of the entire situation.

$CL + \frac{\alpha}{2} + \frac{\alpha}{2} = CL + \alpha = 1$.

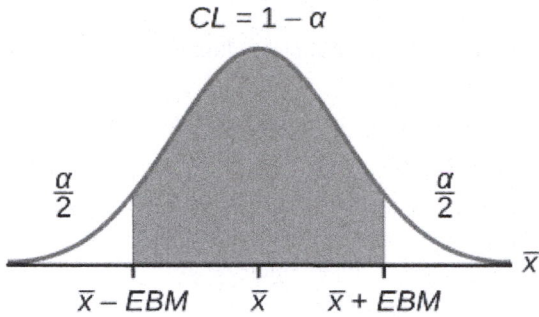

Figure 8.3

Writing the Interpretation

The interpretation should clearly state the confidence level (*CL*), explain what population parameter is being estimated (here, a **population mean**), and state the confidence interval (both endpoints). "We estimate with ___% confidence that the true population mean (include the context of the problem) is between ___ and ___ (include appropriate units)."

Example 8.2

Suppose scores on exams in statistics are normally distributed with an unknown population mean and a population standard deviation of three points. A random sample of 36 scores is taken and gives a sample mean (sample mean score) of 68. Find a confidence interval estimate for the population mean exam score (the mean score on all exams).

Find a 90% confidence interval for the true (population) mean of statistics exam scores.

Solution 8.2

- You can use technology to calculate the confidence interval directly.

- The first solution is shown step-by-step (Solution A).

- The second solution uses the TI-83, 83+, and 84+ calculators (Solution B).

Solution A

To find the confidence interval, you need the sample mean, $\bar{x}$, and the *EBM*.

$\bar{x} = 68$

$EBM = \left(z_{\frac{\alpha}{2}}\right)\left(\frac{\sigma}{\sqrt{n}}\right)$

$\sigma = 3$; $n = 36$; The confidence level is 90% (*CL* = 0.90)

$CL = 0.90$ so $\alpha = 1 - CL = 1 - 0.90 = 0.10$

$\frac{\alpha}{2} = 0.05$ $z_{\frac{\alpha}{2}} = z_{0.05}$

The area to the right of $z_{0.05}$ is 0.05 and the area to the left of $z_{0.05}$ is $1 - 0.05 = 0.95$.

$z_{\frac{\alpha}{2}} = z_{0.05} = 1.645$

using invNorm(0.95, 0, 1) on the TI-83,83+, and 84+ calculators. This can also be found using appropriate commands on other calculators, using a computer, or using a probability table for the standard normal distribution.

$EBM = (1.645)\left(\frac{3}{\sqrt{36}}\right) = 0.8225$

$\bar{x}$ - EBM = 68 - 0.8225 = 67.1775

$\bar{x}$ + EBM = 68 + 0.8225 = 68.8225

The 90% confidence interval is **(67.1775, 68.8225).**

Solution 8.2

Solution B

 Using the TI-83, 83+, 84, 84+ Calculator

Press STAT and arrow over to TESTS.
Arrow down to 7:ZInterval.
Press ENTER.
Arrow to Stats and press ENTER.

Arrow down and enter three for σ, 68 for $\bar{x}$, 36 for n, and .90 for C-level.

Arrow down to Calculate and press ENTER.
The confidence interval is (to three decimal places)(67.178, 68.822).

Interpretation

We estimate with 90% confidence that the true population mean exam score for all statistics students is between 67.18 and 68.82.

Explanation of 90% Confidence Level

Ninety percent of all confidence intervals constructed in this way contain the true mean statistics exam score. For example, if we constructed 100 of these confidence intervals, we would expect 90 of them to contain the true population mean exam score.

Try It Σ

8.2 Suppose average pizza delivery times are normally distributed with an unknown population mean and a population standard deviation of six minutes. A random sample of 28 pizza delivery restaurants is taken and has a sample mean delivery time of 36 minutes.

Find a 90% confidence interval estimate for the population mean delivery time.

Example 8.3

The Specific Absorption Rate (SAR) for a cell phone measures the amount of radio frequency (RF) energy absorbed by the user's body when using the handset. Every cell phone emits RF energy. Different phone models have different SAR measures. To receive certification from the Federal Communications Commission (FCC) for sale in the United States, the SAR level for a cell phone must be no more than 1.6 watts per kilogram. **Table 8.1** shows the highest SAR level for a random selection of cell phone models as measured by the FCC.

Phone Model	SAR	Phone Model	SAR	Phone Model	SAR
Apple iPhone 4S	1.11	LG Ally	1.36	Pantech Laser	0.74
BlackBerry Pearl 8120	1.48	LG AX275	1.34	Samsung Character	0.5
BlackBerry Tour 9630	1.43	LG Cosmos	1.18	Samsung Epic 4G Touch	0.4
Cricket TXTM8	1.3	LG CU515	1.3	Samsung M240	0.867
HP/Palm Centro	1.09	LG Trax CU575	1.26	Samsung Messager III SCH-R750	0.68
HTC One V	0.455	Motorola Q9h	1.29	Samsung Nexus S	0.51
HTC Touch Pro 2	1.41	Motorola Razr2 V8	0.36	Samsung SGH-A227	1.13
Huawei M835 Ideos	0.82	Motorola Razr2 V9	0.52	SGH-a107 GoPhone	0.3
Kyocera DuraPlus	0.78	Motorola V195s	1.6	Sony W350a	1.48
Kyocera K127 Marbl	1.25	Nokia 1680	1.39	T-Mobile Concord	1.38

Table 8.1

Find a 98% confidence interval for the true (population) mean of the Specific Absorption Rates (SARs) for cell phones. Assume that the population standard deviation is $\sigma = 0.337$.

Solution 8.3

Solution A

To find the confidence interval, start by finding the point estimate: the sample mean.

$$\bar{x} = 1.024$$

Next, find the *EBM*. Because you are creating a 98% confidence interval, $CL = 0.98$.

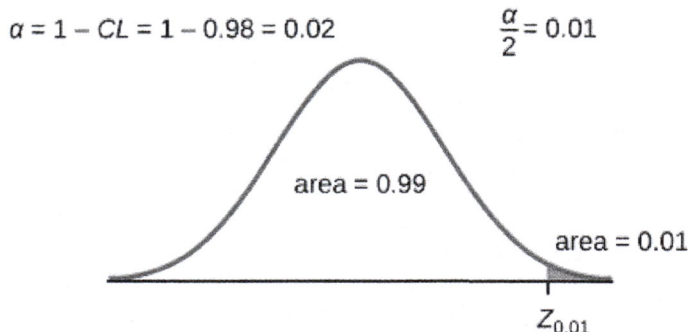

$\alpha = 1 - CL = 1 - 0.98 = 0.02 \qquad \frac{\alpha}{2} = 0.01$

area = 0.99

area = 0.01

$z_{0.01}$

Figure 8.4

You need to find $z_{0.01}$ having the property that the area under the normal density curve to the right of $z_{0.01}$ is 0.01 and the area to the left is 0.99. Use your calculator, a computer, or a probability table for the standard normal distribution to find $z_{0.01} = 2.326$.

$$EBM = (z_{0.01})\frac{\sigma}{\sqrt{n}} = (2.326)\frac{0.337}{\sqrt{30}} = 0.1431$$

To find the 98% confidence interval, find $\bar{x} \pm EBM$.

$\bar{x} - EBM = 1.024 - 0.1431 = 0.8809$

$\bar{x} - EBM = 1.024 - 0.1431 = 1.1671$

We estimate with 98% confidence that the true SAR mean for the population of cell phones in the United States is between 0.8809 and 1.1671 watts per kilogram.

Solution 8.3

Solution B

 Using the TI-83, 83+, 84, 84+ Calculator

Press STAT and arrow over to TESTS.
Arrow down to 7:ZInterval.
Press ENTER.
Arrow to Stats and press ENTER.
Arrow down and enter the following values:
σ: 0.337
$\bar{x}$: 1.024
n: 30
C-level: 0.98
Arrow down to Calculate and press ENTER.
The confidence interval is (to three decimal places) (0.881, 1.167).

Try It Σ

8.3 **Table 8.2** shows a different random sampling of 20 cell phone models. Use this data to calculate a 93% confidence interval for the true mean SAR for cell phones certified for use in the United States. As previously, assume that the population standard deviation is $\sigma = 0.337$.

Phone Model	SAR	Phone Model	SAR
Blackberry Pearl 8120	1.48	Nokia E71x	1.53
HTC Evo Design 4G	0.8	Nokia N75	0.68
HTC Freestyle	1.15	Nokia N79	1.4
LG Ally	1.36	Sagem Puma	1.24
LG Fathom	0.77	Samsung Fascinate	0.57
LG Optimus Vu	0.462	Samsung Infuse 4G	0.2
Motorola Cliq XT	1.36	Samsung Nexus S	0.51
Motorola Droid Pro	1.39	Samsung Replenish	0.3
Motorola Droid Razr M	1.3	Sony W518a Walkman	0.73
Nokia 7705 Twist	0.7	ZTE C79	0.869

Table 8.2

Notice the difference in the confidence intervals calculated in **Example 8.3** and the following **Try It** exercise. These intervals are different for several reasons: they were calculated from different samples, the samples were different sizes, and the intervals were calculated for different levels of confidence. Even though the intervals are different, they do not yield conflicting information. The effects of these kinds of changes are the subject of the next section in this chapter.

Changing the Confidence Level or Sample Size

Example 8.4

Suppose we change the original problem in **Example 8.2** by using a 95% confidence level. Find a 95% confidence interval for the true (population) mean statistics exam score.

Solution 8.4

To find the confidence interval, you need the sample mean, $\bar{x}$, and the *EBM*.

$\bar{x} = 68$

$EBM = \left(z_{\frac{\alpha}{2}}\right)\left(\frac{\sigma}{\sqrt{n}}\right)$

$\sigma = 3$; $n = 36$; The confidence level is 95% ($CL = 0.95$).

$CL = 0.95$ so $\alpha = 1 - CL = 1 - 0.95 = 0.05$

$\frac{\alpha}{2} = 0.025 \quad z_{\frac{\alpha}{2}} = z_{0.025}$

The area to the right of $z_{0.025}$ is 0.025 and the area to the left of $z_{0.025}$ is $1 - 0.025 = 0.975$.

$$z_{\frac{\alpha}{2}} = z_{0.025} = 1.96$$

when using invnorm(0.975,0,1) on the TI-83, 83+, or 84+ calculators. (This can also be found using appropriate commands on other calculators, using a computer, or using a probability table for the standard normal distribution.)

$$EBM = (1.96)\left(\frac{3}{\sqrt{36}}\right) = 0.98$$

$$\bar{x} - EBM = 68 - 0.98 = 67.02$$

$$\bar{x} + EBM = 68 + 0.98 = 68.98$$

Notice that the *EBM* is larger for a 95% confidence level in the original problem.

Interpretation

We estimate with 95% confidence that the true population mean for all statistics exam scores is between 67.02 and 68.98.

Explanation of 95% Confidence Level

Ninety-five percent of all confidence intervals constructed in this way contain the true value of the population mean statistics exam score.

Comparing the results

The 90% confidence interval is (67.18, 68.82). The 95% confidence interval is (67.02, 68.98). The 95% confidence interval is wider. If you look at the graphs, because the area 0.95 is larger than the area 0.90, it makes sense that the 95% confidence interval is wider. To be more confident that the confidence interval actually does contain the true value of the population mean for all statistics exam scores, the confidence interval necessarily needs to be wider.

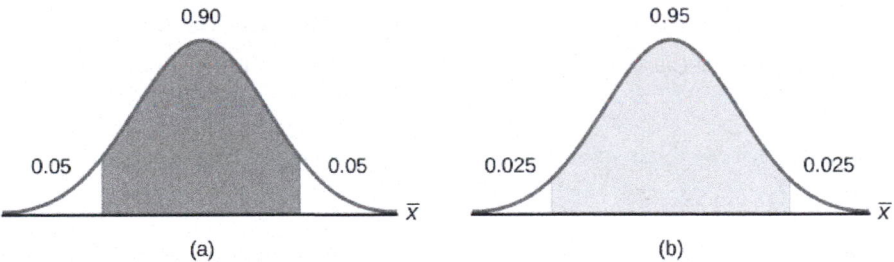

Figure 8.5

Summary: Effect of Changing the Confidence Level
- Increasing the confidence level increases the error bound, making the confidence interval wider.
- Decreasing the confidence level decreases the error bound, making the confidence interval narrower.

Try It Σ

8.4 Refer back to the pizza-delivery **Try It** exercise. The population standard deviation is six minutes and the sample mean deliver time is 36 minutes. Use a sample size of 20. Find a 95% confidence interval estimate for the true mean pizza delivery time.

Example 8.5

Suppose we change the original problem in **Example 8.2** to see what happens to the error bound if the sample size is changed.

Leave everything the same except the sample size. Use the original 90% confidence level. What happens to the error bound and the confidence interval if we increase the sample size and use $n = 100$ instead of $n = 36$? What happens if we decrease the sample size to $n = 25$ instead of $n = 36$?

- $\bar{x} = 68$

- $EBM = \left(z_{\frac{\alpha}{2}}\right)\left(\frac{\sigma}{\sqrt{n}}\right)$

- $\sigma = 3$; The confidence level is 90% ($CL=0.90$); $z_{\frac{\alpha}{2}} = z_{0.05} = 1.645$.

Solution 8.5
Solution A

If we **increase** the sample size n to 100, we **decrease** the error bound.

When $n = 100$: $EBM = \left(z_{\frac{\alpha}{2}}\right)\left(\frac{\sigma}{\sqrt{n}}\right) = (1.645)\left(\frac{3}{\sqrt{100}}\right) = 0.4935$.

Solution 8.5
Solution B

If we **decrease** the sample size n to 25, we **increase** the error bound.

When $n = 25$: $EBM = \left(z_{\frac{\alpha}{2}}\right)\left(\frac{\sigma}{\sqrt{n}}\right) = (1.645)\left(\frac{3}{\sqrt{25}}\right) = 0.987$.

Summary: Effect of Changing the Sample Size
- Increasing the sample size causes the error bound to decrease, making the confidence interval narrower.
- Decreasing the sample size causes the error bound to increase, making the confidence interval wider.

Try It

8.5 Refer back to the pizza-delivery **Try It** exercise. The mean delivery time is 36 minutes and the population standard deviation is six minutes. Assume the sample size is changed to 50 restaurants with the same sample mean. Find a 90% confidence interval estimate for the population mean delivery time.

Working Backwards to Find the Error Bound or Sample Mean

When we calculate a confidence interval, we find the sample mean, calculate the error bound, and use them to calculate the confidence interval. However, sometimes when we read statistical studies, the study may state the confidence interval only. If we know the confidence interval, we can work backwards to find both the error bound and the sample mean.

Finding the Error Bound
- From the upper value for the interval, subtract the sample mean,
- OR, from the upper value for the interval, subtract the lower value. Then divide the difference by two.

Finding the Sample Mean
- Subtract the error bound from the upper value of the confidence interval,

- OR, average the upper and lower endpoints of the confidence interval.

Notice that there are two methods to perform each calculation. You can choose the method that is easier to use with the information you know.

Example 8.6

Suppose we know that a confidence interval is **(67.18, 68.82)** and we want to find the error bound. We may know that the sample mean is 68, or perhaps our source only gave the confidence interval and did not tell us the value of the sample mean.

Calculate the Error Bound:
- If we know that the sample mean is 68: $EBM = 68.82 - 68 = 0.82$.

- If we don't know the sample mean: $EBM = \dfrac{(68.82 - 67.18)}{2} = 0.82$.

Calculate the Sample Mean:
- If we know the error bound: $\bar{x} = 68.82 - 0.82 = 68$

- If we don't know the error bound: $\bar{x} = \dfrac{(67.18 + 68.82)}{2} = 68$.

Try It Σ

8.6 Suppose we know that a confidence interval is (42.12, 47.88). Find the error bound and the sample mean.

Calculating the Sample Size n

If researchers desire a specific margin of error, then they can use the error bound formula to calculate the required sample size.

The error bound formula for a population mean when the population standard deviation is known is

$$EBM = \left(z_{\frac{\alpha}{2}} \right)\left(\frac{\sigma}{\sqrt{n}} \right).$$

The formula for sample size is $n = \dfrac{z^2 \sigma^2}{EBM^2}$, found by solving the error bound formula for n.

In this formula, z is $z_{\frac{\alpha}{2}}$, corresponding to the desired confidence level. A researcher planning a study who wants a specified confidence level and error bound can use this formula to calculate the size of the sample needed for the study.

Example 8.7

The population standard deviation for the age of Foothill College students is 15 years. If we want to be 95% confident that the sample mean age is within two years of the true population mean age of Foothill College students, how many randomly selected Foothill College students must be surveyed?

From the problem, we know that $\sigma = 15$ and $EBM = 2$.

$z = z_{0.025} = 1.96$, because the confidence level is 95%.

$n = \dfrac{z^2 \sigma^2}{EBM^2} = \dfrac{(1.96)^2 (15)^2}{2^2} = 216.09$ using the sample size equation.

Use $n = 217$: Always round the answer UP to the next higher integer to ensure that the sample size is large enough.

Therefore, 217 Foothill College students should be surveyed in order to be 95% confident that we are within two years of the true population mean age of Foothill College students.

8.7 The population standard deviation for the height of high school basketball players is three inches. If we want to be 95% confident that the sample mean height is within one inch of the true population mean height, how many randomly selected students must be surveyed?

8.2 | A Single Population Mean using the Student t Distribution

In practice, we rarely know the population **standard deviation**. In the past, when the sample size was large, this did not present a problem to statisticians. They used the sample standard deviation s as an estimate for σ and proceeded as before to calculate a **confidence interval** with close enough results. However, statisticians ran into problems when the sample size was small. A small sample size caused inaccuracies in the confidence interval.

William S. Goset (1876–1937) of the Guinness brewery in Dublin, Ireland ran into this problem. His experiments with hops and barley produced very few samples. Just replacing σ with s did not produce accurate results when he tried to calculate a confidence interval. He realized that he could not use a normal distribution for the calculation; he found that the actual distribution depends on the sample size. This problem led him to "discover" what is called the **Student's t-distribution**. The name comes from the fact that Gosset wrote under the pen name "Student."

Up until the mid-1970s, some statisticians used the **normal distribution** approximation for large sample sizes and only used the Student's t-distribution only for sample sizes of at most 30. With graphing calculators and computers, the practice now is to use the Student's t-distribution whenever s is used as an estimate for σ.

If you draw a simple random sample of size n from a population that has an approximately a normal distribution with mean μ and unknown population standard deviation σ and calculate the t-score $t = \dfrac{\overline{x} - \mu}{\left(\frac{s}{\sqrt{n}}\right)}$, then the t-scores follow a **Student's t-distribution with $n - 1$ degrees of freedom**. The t-score has the same interpretation as the **z-score**. It measures how far $\overline{x}$ is from its mean μ. For each sample size n, there is a different Student's t-distribution.

The **degrees of freedom, $n - 1$**, come from the calculation of the sample standard deviation s. In **Appendix H**, we used n deviations $(x - \overline{x}$ values$)$ to calculate s. Because the sum of the deviations is zero, we can find the last deviation once we know the other $n - 1$ deviations. The other $n - 1$ deviations can change or vary freely. **We call the number $n - 1$ the degrees of freedom (df).**

Properties of the Student's t-Distribution
- The graph for the Student's t-distribution is similar to the standard normal curve.

- The mean for the Student's t-distribution is zero and the distribution is symmetric about zero.

- The Student's t-distribution has more probability in its tails than the standard normal distribution because the spread of the t-distribution is greater than the spread of the standard normal. So the graph of the Student's t-distribution will be thicker in the tails and shorter in the center than the graph of the standard normal distribution.

- The exact shape of the Student's t-distribution depends on the degrees of freedom. As the degrees of freedom increases, the graph of Student's t-distribution becomes more like the graph of the standard normal distribution.

- The underlying population of individual observations is assumed to be normally distributed with unknown population mean μ and unknown population standard deviation σ. The size of the underlying population is generally not relevant unless it is very small. If it is bell shaped (normal) then the assumption is met and doesn't need discussion. Random sampling is assumed, but that is a completely separate assumption from normality.

Calculators and computers can easily calculate any Student's t-probabilities. The TI-83,83+, and 84+ have a tcdf function to find the probability for given values of t. The grammar for the tcdf command is tcdf(lower bound, upper bound, degrees of freedom). However for confidence intervals, we need to use **inverse** probability to find the value of t when we know the probability.

For the TI-84+ you can use the invT command on the DISTRibution menu. The invT command works similarly to the invnorm. The invT command requires two inputs: **invT(area to the left, degrees of freedom)** The output is the t-score that corresponds to the area we specified.

The TI-83 and 83+ do not have the invT command. (The TI-89 has an inverse T command.)

A probability table for the Student's t-distribution can also be used. The table gives t-scores that correspond to the confidence level (column) and degrees of freedom (row). (The TI-86 does not have an invT program or command, so if you are using that calculator, you need to use a probability table for the Student's t-Distribution.) When using a t-table, note that some tables are formatted to show the confidence level in the column headings, while the column headings in some tables may show only corresponding area in one or both tails.

A Student's t table (See **Appendix H**) gives t-scores given the degrees of freedom and the right-tailed probability. The table is very limited. **Calculators and computers can easily calculate any Student's t-probabilities.**

The notation for the Student's t-distribution (using _T_ as the random variable) is:
- $T \sim t_{df}$ where $df = n - 1$.

- For example, if we have a sample of size $n = 20$ items, then we calculate the degrees of freedom as $df = n - 1 = 20 - 1 = 19$ and we write the distribution as $T \sim t_{19}$.

If the population standard deviation is not known, the **error bound for a population mean** is:

- $EBM = \left(t_{\frac{\alpha}{2}}\right)\left(\frac{s}{\sqrt{n}}\right)$,

- $t_{\frac{\alpha}{2}}$ is the _t_-score with area to the right equal to $\frac{\alpha}{2}$,

- use $df = n - 1$ degrees of freedom, and

- s = sample standard deviation.

The format for the confidence interval is:

$(\bar{x} - EBM, \bar{x} + EBM)$.

 Using the TI-83, 83+, 84, 84+ Calculator

To calculate the confidence interval directly:
Press STAT.
Arrow over to TESTS.
Arrow down to 8:TInterval and press ENTER (or just press 8).

Example 8.8

Suppose you do a study of acupuncture to determine how effective it is in relieving pain. You measure sensory rates for 15 subjects with the results given. Use the sample data to construct a 95% confidence interval for the mean sensory rate for the population (assumed normal) from which you took the data.
The solution is shown step-by-step and by using the TI-83, 83+, or 84+ calculators.

8.6; 9.4; 7.9; 6.8; 8.3; 7.3; 9.2; 9.6; 8.7; 11.4; 10.3; 5.4; 8.1; 5.5; 6.9

Solution 8.8
- The first solution is step-by-step (Solution A).

- The second solution uses the TI-83+ and TI-84 calculators (Solution B).

Solution A

To find the confidence interval, you need the sample mean, $\bar{x}$, and the *EBM*.

$\bar{x}$ = 8.2267 s = 1.6722 n = 15

df = 15 – 1 = 14 *CL* so α = 1 – *CL* = 1 – 0.95 = 0.05

$\frac{\alpha}{2}$ = 0.025 $t_{\frac{\alpha}{2}} = t_{0.025}$

The area to the right of $t_{0.025}$ is 0.025, and the area to the left of $t_{0.025}$ is 1 – 0.025 = 0.975

$t_{\frac{\alpha}{2}} = t_{0.025} = 2.14$ using invT(.975,14) on the TI-84+ calculator.

$$EBM = \left(t_{\frac{\alpha}{2}}\right)\left(\frac{s}{\sqrt{n}}\right)$$

$$EBM = (2.14)\left(\frac{1.6722}{\sqrt{15}}\right) = 0.924$$

$\bar{x}$ – *EBM* = 8.2267 – 0.9240 = 7.3

$\bar{x}$ + *EBM* = 8.2267 + 0.9240 = 9.15

The 95% confidence interval is (7.30, 9.15).

We estimate with 95% confidence that the true population mean sensory rate is between 7.30 and 9.15.

Solution 8.8

Using the TI-83, 83+, 84, 84+ Calculator

Press STAT and arrow over to TESTS.
Arrow down to 8:TInterval and press ENTER (or you can just press 8).
Arrow to Data and press ENTER.
Arrow down to List and enter the list name where you put the data.
There should be a 1 after Freq.
Arrow down to C-level and enter 0.95
Arrow down to Calculate and press ENTER.
The 95% confidence interval is (7.3006, 9.1527)

NOTE

When calculating the error bound, a probability table for the Student's t-distribution can also be used to find the value of *t*. The table gives *t*-scores that correspond to the confidence level (column) and degrees of freedom (row); the *t*-score is found where the row and column intersect in the table.

Try It Σ

8.8 You do a study of hypnotherapy to determine how effective it is in increasing the number of hourse of sleep subjects get each night. You measure hours of sleep for 12 subjects with the following results. Construct a 95%

confidence interval for the mean number of hours slept for the population (assumed normal) from which you took the data.

8.2; 9.1; 7.7; 8.6; 6.9; 11.2; 10.1; 9.9; 8.9; 9.2; 7.5; 10.5

Example 8.9

The Human Toxome Project (HTP) is working to understand the scope of industrial pollution in the human body. Industrial chemicals may enter the body through pollution or as ingredients in consumer products. In October 2008, the scientists at HTP tested cord blood samples for 20 newborn infants in the United States. The cord blood of the "In utero/newborn" group was tested for 430 industrial compounds, pollutants, and other chemicals, including chemicals linked to brain and nervous system toxicity, immune system toxicity, and reproductive toxicity, and fertility problems. There are health concerns about the effects of some chemicals on the brain and nervous system. **Table 8.2** shows how many of the targeted chemicals were found in each infant's cord blood.

79	145	147	160	116	100	159	151	156	126
137	83	156	94	121	144	123	114	139	99

Table 8.3

Use this sample data to construct a 90% confidence interval for the mean number of targeted industrial chemicals to be found in an in infant's blood.

Solution 8.9

Solution A

From the sample, you can calculate $\bar{x} = 127.45$ and $s = 25.965$. There are 20 infants in the sample, so $n = 20$, and $df = 20 - 1 = 19$.

You are asked to calculate a 90% confidence interval: $CL = 0.90$, so $\alpha = 1 - CL = 1 - 0.90 = 0.10$ $\frac{\alpha}{2} = 0.05, t_{\frac{\alpha}{2}} = t_{0.05}$

By definition, the area to the right of $t_{0.05}$ is 0.05 and so the area to the left of $t_{0.05}$ is $1 - 0.05 = 0.95$.

Use a table, calculator, or computer to find that $t_{0.05} = 1.729$.

$$EBM = t_{\frac{\alpha}{2}}\left(\frac{s}{\sqrt{n}}\right) = 1.729\left(\frac{25.965}{\sqrt{20}}\right) \approx 10.038$$

$\bar{x} - EBM = 127.45 - 10.038 = 117.412$

$\bar{x} + EBM = 127.45 + 10.038 = 137.488$

We estimate with 90% confidence that the mean number of all targeted industrial chemicals found in cord blood in the United States is between 117.412 and 137.488.

Solution 8.9

Solution B

Using the TI-83, 83+, 84, 84+ Calculator

Enter the data as a list.
Press STAT and arrow over to TESTS.
Arrow down to 8:TInterval and press ENTER (or you can just press 8). Arrow to Data and press ENTER.
Arrow down to List and enter the list name where you put the data.
Arrow down to Freq and enter 1.
Arrow down to C-level and enter 0.90
Arrow down to Calculate and press ENTER.
The 90% confidence interval is (117.41, 137.49).

Try It Σ

8.9 A random sample of statistics students were asked to estimate the total number of hours they spend watching television in an average week. The responses are recorded in **Table 8.4**. Use this sample data to construct a 98% confidence interval for the mean number of hours statistics students will spend watching television in one week.

0	3	1	20	9
5	10	1	10	4
14	2	4	4	5

Table 8.4

8.3 | A Population Proportion

During an election year, we see articles in the newspaper that state **confidence intervals** in terms of proportions or percentages. For example, a poll for a particular candidate running for president might show that the candidate has 40% of the vote within three percentage points (if the sample is large enough). Often, election polls are calculated with 95% confidence, so, the pollsters would be 95% confident that the true proportion of voters who favored the candidate would be between 0.37 and 0.43: $(0.40 - 0.03, 0.40 + 0.03)$.

Investors in the stock market are interested in the true proportion of stocks that go up and down each week. Businesses that sell personal computers are interested in the proportion of households in the United States that own personal computers. Confidence intervals can be calculated for the true proportion of stocks that go up or down each week and for the true proportion of households in the United States that own personal computers.

The procedure to find the confidence interval, the sample size, the **error bound**, and the **confidence level** for a proportion is similar to that for the population mean, but the formulas are different.

How do you know you are dealing with a proportion problem? First, the underlying **distribution is a binomial distribution**. (There is no mention of a mean or average.) If X is a binomial random variable, then $X \sim B(n, p)$ where n is the number of trials and p is the probability of a success. To form a proportion, take X, the random variable for the number of successes and divide it by n, the number of trials (or the sample size). The random variable P' (read "P prime") is that proportion,

$$P' = \frac{X}{n}$$

(Sometimes the random variable is denoted as $\hat{P}$, read "P hat".)

When n is large and p is not close to zero or one, we can use the **normal distribution** to approximate the binomial.

$X \sim N(np, \sqrt{npq})$

If we divide the random variable, the mean, and the standard deviation by n, we get a normal distribution of proportions with P', called the estimated proportion, as the random variable. (Recall that a proportion as the number of successes divided by n.)

$$\frac{X}{n} = P' \sim N\left(\frac{np}{n}, \frac{\sqrt{npq}}{n}\right)$$

Using algebra to simplify : $\frac{\sqrt{npq}}{n} = \sqrt{\frac{pq}{n}}$

P' follows a normal distribution for proportions: $\frac{X}{n} = P' \sim N\left(\frac{np}{n}, \frac{\sqrt{npq}}{n}\right)$

The confidence interval has the form $(p' - EBP, p' + EBP)$. EBP is error bound for the proportion.

$p' = \frac{x}{n}$

p' = the **estimated proportion** of successes (p' is a **point estimate** for p, the true proportion.)

x = the **number** of successes

n = the size of the sample

The error bound for a proportion is

$EBP = \left(z_{\frac{\alpha}{2}}\right)\left(\sqrt{\frac{p'q'}{n}}\right)$ where $q' = 1 - p'$

This formula is similar to the error bound formula for a mean, except that the "appropriate standard deviation" is different. For a mean, when the population standard deviation is known, the appropriate standard deviation that we use is $\frac{\sigma}{\sqrt{n}}$. For a proportion, the appropriate standard deviation is $\sqrt{\frac{pq}{n}}$.

However, in the error bound formula, we use $\sqrt{\frac{p'q'}{n}}$ as the standard deviation, instead of $\sqrt{\frac{pq}{n}}$.

In the error bound formula, the **sample proportions p' and q' are estimates of the unknown population proportions p and q**. The estimated proportions p' and q' are used because p and q are not known. The sample proportions p' and q' are calculated from the data: p' is the estimated proportion of successes, and q' is the estimated proportion of failures.

The confidence interval can be used only if the number of successes np' and the number of failures nq' are both greater than five.

NOTE

For the normal distribution of proportions, the z-score formula is as follows.

If $P' \sim N\left(p, \sqrt{\frac{pq}{n}}\right)$ then the z-score formula is $z = \dfrac{p' - p}{\sqrt{\frac{pq}{n}}}$

Example 8.10

Suppose that a market research firm is hired to estimate the percent of adults living in a large city who have cell phones. Five hundred randomly selected adult residents in this city are surveyed to determine whether they have cell phones. Of the 500 people surveyed, 421 responded yes - they own cell phones. Using a 95% confidence level, compute a confidence interval estimate for the true proportion of adult residents of this city who have cell phones.

Solution 8.10

Solution A

- The first solution is step-by-step (Solution A).

- The second solution uses a function of the TI-83, 83+ or 84 calculators (Solution B).

Let X = the number of people in the sample who have cell phones. X is binomial. $X \sim B\left(500, \frac{421}{500}\right)$.

To calculate the confidence interval, you must find p', q', and EBP.

$n = 500$

x = the number of successes = 421

$p' = \frac{x}{n} = \frac{421}{500} = 0.842$

$p' = 0.842$ is the sample proportion; this is the point estimate of the population proportion.

$q' = 1 - p' = 1 - 0.842 = 0.158$

Since $CL = 0.95$, then $\alpha = 1 - CL = 1 - 0.95 = 0.05$ $\left(\frac{\alpha}{2}\right) = 0.025$.

Then $z_{\frac{\alpha}{2}} = z_{0.025} = 1.96$

Use the TI-83, 83+, or 84+ calculator command invNorm(0.975,0,1) to find $z_{0.025}$. Remember that the area to the right of $z_{0.025}$ is 0.025 and the area to the left of $z_{0.025}$ is 0.975. This can also be found using appropriate commands on other calculators, using a computer, or using a Standard Normal probability table.

$EBP = \left(z_{\frac{\alpha}{2}}\right)\sqrt{\frac{p'q'}{n}} = (1.96)\sqrt{\frac{(0.842)(0.158)}{500}} = 0.032$

$p' - EBP = 0.842 - 0.032 = 0.81$

$p' + EBP = 0.842 + 0.032 = 0.874$

The confidence interval for the true binomial population proportion is $(p' - EBP, p' + EBP) = (0.810, 0.874)$.

Interpretation

We estimate with 95% confidence that between 81% and 87.4% of all adult residents of this city have cell phones.

Explanation of 95% Confidence Level

Ninety-five percent of the confidence intervals constructed in this way would contain the true value for the population proportion of all adult residents of this city who have cell phones.

Solution 8.10

Solution B

 Using the TI-83, 83+, 84, 84+ Calculator

Press STAT and arrow over to TESTS.
Arrow down to A:1-PropZint. Press ENTER.
Arrow down to x and enter 421.
Arrow down to n and enter 500.
Arrow down to C-Level and enter .95.

Arrow down to `Calculate` and press `ENTER`.
The confidence interval is (0.81003, 0.87397).

Try It Σ

8.10 Suppose 250 randomly selected people are surveyed to determine if they own a tablet. Of the 250 surveyed, 98 reported owning a tablet. Using a 95% confidence level, compute a confidence interval estimate for the true proportion of people who own tablets.

Example 8.11

For a class project, a political science student at a large university wants to estimate the percent of students who are registered voters. He surveys 500 students and finds that 300 are registered voters. Compute a 90% confidence interval for the true percent of students who are registered voters, and interpret the confidence interval.

Solution 8.11

- The first solution is step-by-step (Solution A).

- The second solution uses a function of the TI-83, 83+, or 84 calculators (Solution B).

Solution A

$x = 300$ and $n = 500$

$$p' = \frac{x}{n} = \frac{300}{500} = 0.600$$

$$q' = 1 - p' = 1 - 0.600 = 0.400$$

Since $CL = 0.90$, then $\alpha = 1 - CL = 1 - 0.90 = 0.10 \left(\frac{\alpha}{2}\right) = 0.05$

$z_{\frac{\alpha}{2}} = z_{0.05} = 1.645$

Use the TI-83, 83+, or 84+ calculator command invNorm(0.95,0,1) to find $z_{0.05}$. Remember that the area to the right of $z_{0.05}$ is 0.05 and the area to the left of $z_{0.05}$ is 0.95. This can also be found using appropriate commands on other calculators, using a computer, or using a standard normal probability table.

$$EBP = \left(z_{\frac{\alpha}{2}}\right)\sqrt{\frac{p'q'}{n}} = (1.645)\sqrt{\frac{(0.60)(0.40)}{500}} = 0.036$$

$$p' - EBP = 0.60 - 0.036 = 0.564$$

$$p' + EBP = 0.60 + 0.036 = 0.636$$

The confidence interval for the true binomial population proportion is $(p' - EBP, p' + EBP) = (0.564, 0.636)$.

Interpretation

- We estimate with 90% confidence that the true percent of all students that are registered voters is between 56.4% and 63.6%.

- Alternate Wording: We estimate with 90% confidence that between 56.4% and 63.6% of ALL students are registered voters.

Explanation of 90% Confidence Level

Ninety percent of all confidence intervals constructed in this way contain the true value for the population percent of students that are registered voters.

Solution 8.11

Solution B

 Using the TI-83, 83+, 84, 84+ Calculator

Press STAT and arrow over to TESTS.
Arrow down to A:1-PropZint. Press ENTER.
Arrow down to x and enter 300.
Arrow down to n and enter 500.
Arrow down to C-Level and enter 0.90.
Arrow down to Calculate and press ENTER.
The confidence interval is (0.564, 0.636).

 Try It Σ

8.11 A student polls his school to see if students in the school district are for or against the new legislation regarding school uniforms. She surveys 600 students and finds that 480 are against the new legislation.

a. Compute a 90% confidence interval for the true percent of students who are against the new legislation, and interpret the confidence interval.

b. In a sample of 300 students, 68% said they own an iPod and a smart phone. Compute a 97% confidence interval for the true percent of students who own an iPod and a smartphone.

"Plus Four" Confidence Interval for *p*

There is a certain amount of error introduced into the process of calculating a confidence interval for a proportion. Because we do not know the true proportion for the population, we are forced to use point estimates to calculate the appropriate standard deviation of the sampling distribution. Studies have shown that the resulting estimation of the standard deviation can be flawed.

Fortunately, there is a simple adjustment that allows us to produce more accurate confidence intervals. We simply pretend that we have four additional observations. Two of these observations are successes and two are failures. The new sample size, then, is $n + 4$, and the new count of successes is $x + 2$.

Computer studies have demonstrated the effectiveness of this method. It should be used when the confidence level desired is at least 90% and the sample size is at least ten.

Example 8.12

A random sample of 25 statistics students was asked: "Have you smoked a cigarette in the past week?" Six students reported smoking within the past week. Use the "plus-four" method to find a 95% confidence interval for the true proportion of statistics students who smoke.

Solution 8.12

Solution A

Six students out of 25 reported smoking within the past week, so $x = 6$ and $n = 25$. Because we are using the "plus-four" method, we will use $x = 6 + 2 = 8$ and $n = 25 + 4 = 29$.

$$p' = \frac{x}{n} = \frac{8}{29} \approx 0.276$$

$$q' = 1 - p' = 1 - 0.276 = 0.724$$

Since $CL = 0.95$, we know $\alpha = 1 - 0.95 = 0.05$ and $\frac{\alpha}{2} = 0.025$.

$$z_{0.025} = 1.96$$

$$EPB = \left(z_{\frac{\alpha}{2}}\right)\sqrt{\frac{p'q'}{n}} = (1.96)\sqrt{\frac{0.276(0.724)}{29}} \approx 0.163$$

$$p' - EPB = 0.276 - 0.163 = 0.113$$

$$p' + EPB = 0.276 + 0.163 = 0.439$$

We are 95% confident that the true proportion of all statistics students who smoke cigarettes is between 0.113 and 0.439.

Solution 8.12

Solution B

 Using the TI-83, 83+, 84, 84+ Calculator

Press STAT and arrow over to TESTS.
Arrow down to A:1-PropZint. Press ENTER.

REMINDER

Remember that the plus-four method assume an additional four trials: two successes and two failures. You do not need to change the process for calculating the confidence interval; simply update the values of x and n to reflect these additional trials.

Arrow down to x and enter eight.
Arrow down to n and enter 29.
Arrow down to C-Level and enter 0.95.
Arrow down to Calculate and press ENTER.
The confidence interval is (0.113, 0.439).

Try It Σ

8.12 Out of a random sample of 65 freshmen at State University, 31 students have declared a major. Use the "plus-four" method to find a 96% confidence interval for the true proportion of freshmen at State University who have declared a major.

Example 8.13

The Berkman Center for Internet & Society at Harvard recently conducted a study analyzing the privacy management habits of teen internet users. In a group of 50 teens, 13 reported having more than 500 friends on Facebook. Use the "plus four" method to find a 90% confidence interval for the true proportion of teens who would report having more than 500 Facebook friends.

Solution 8.13

Solution A

Using "plus-four," we have $x = 13 + 2 = 15$ and $n = 50 + 4 = 54$.

$$p^{'} = \frac{15}{54} \approx 0.278$$

$$q^{'} = 1 - p^{'} = 1 - 0.241 = 0.722$$

Since $CL = 0.90$, we know $\alpha = 1 - 0.90 = 0.10$ and $\frac{\alpha}{2} = 0.05$.

$$z_{0.05} = 1.645$$

$$EPB = (z_{\frac{\alpha}{2}})\left(\sqrt{\frac{p^{'}q^{'}}{n}}\right) = (1.645)\left(\sqrt{\frac{(0.278)(0.722)}{54}}\right) \approx 0.100$$

$p' - EPB = 0.278 - 0.100 = 0.178$

$p' + EPB = 0.278 + 0.100 = 0.378$

We are 90% confident that between 17.8% and 37.8% of all teens would report having more than 500 friends on Facebook.

Solution 8.13

Solution B

 Using the TI-83, 83+, 84, 84+ Calculator

Press STAT and arrow over to TESTS.
Arrow down to A:1-PropZint. Press ENTER.
Arrow down to x and enter 15.
Arrow down to n and enter 54.
Arrow down to C-Level and enter 0.90.
Arrow down to Calculate and press ENTER.
The confidence interval is (0.178, 0.378).

Try It Σ

8.13 The Berkman Center Study referenced in **Example 8.13** talked to teens in smaller focus groups, but also interviewed additional teens over the phone. When the study was complete, 588 teens had answered the question about their Facebook friends with 159 saying that they have more than 500 friends. Use the "plus-four" method to find a 90% confidence interval for the true proportion of teens that would report having more than 500 Facebook friends based on this larger sample. Compare the results to those in **Example 8.13**.

Calculating the Sample Size *n*

If researchers desire a specific margin of error, then they can use the error bound formula to calculate the required sample size.

The error bound formula for a population proportion is

- $EBP = \left(z_{\frac{\alpha}{2}}\right)\left(\sqrt{\frac{p'q'}{n}}\right)$

- Solving for *n* gives you an equation for the sample size.

- $n = \dfrac{\left(z_{\frac{\alpha}{2}}\right)^2 (p'q')}{EBP^2}$

Example 8.14

Suppose a mobile phone company wants to determine the current percentage of customers aged 50+ who use text messaging on their cell phones. How many customers aged 50+ should the company survey in order to be 90% confident that the estimated (sample) proportion is within three percentage points of the true population proportion of customers aged 50+ who use text messaging on their cell phones.

Solution 8.14

From the problem, we know that **EBP = 0.03** (3%=0.03) and $z_{\frac{\alpha}{2}}$ $z_{0.05}$ = 1.645 because the confidence level is 90%.

However, in order to find *n*, we need to know the estimated (sample) proportion *p'*. Remember that *q'* = 1 − *p'*. But, we do not know *p'* yet. Since we multiply *p'* and *q'* together, we make them both equal to 0.5 because *p'q'* = (0.5)(0.5) = 0.25 results in the largest possible product. (Try other products: (0.6)(0.4) = 0.24; (0.3)(0.7) = 0.21; (0.2)(0.8) = 0.16 and so on). The largest possible product gives us the largest *n*. This gives us a large enough sample so that we can be 90% confident that we are within three percentage points of the true population proportion. To calculate the sample size *n*, use the formula and make the substitutions.

$n = \dfrac{z^2 p' q'}{EBP^2}$ gives $n = \dfrac{1.645^2 (0.5)(0.5)}{0.03^2} = 751.7$

Round the answer to the next higher value. The sample size should be 752 cell phone customers aged 50+ in order to be 90% confident that the estimated (sample) proportion is within three percentage points of the true population proportion of all customers aged 50+ who use text messaging on their cell phones.

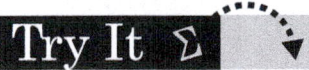

Try It Σ

8.14 Suppose an internet marketing company wants to determine the current percentage of customers who click on ads on their smartphones. How many customers should the company survey in order to be 90% confident that the estimated proportion is within five percentage points of the true population proportion of customers who click on ads on their smartphones?

8.4 | Confidence Interval (Home Costs)

Stats Lab

8.1 Confidence Interval (Home Costs)

Class Time:

Names:

Student Learning Outcomes

- The student will calculate the 90% confidence interval for the mean cost of a home in the area in which this school is located.

- The student will interpret confidence intervals.

- The student will determine the effects of changing conditions on the confidence interval.

Collect the Data

Check the Real Estate section in your local newspaper. Record the sale prices for 35 randomly selected homes recently listed in the county.

NOTE

Many newspapers list them only one day per week. Also, we will assume that homes come up for sale randomly.

1. Complete the table:

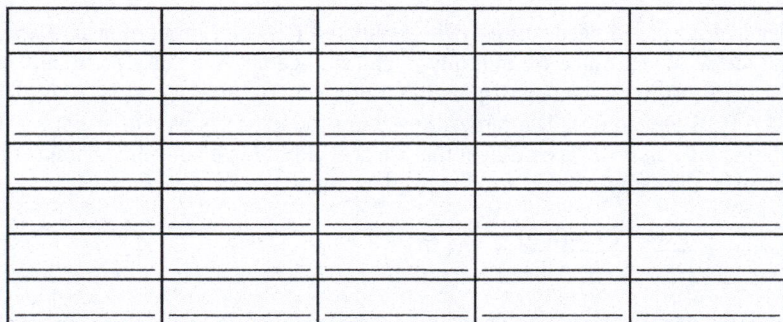

Table 8.5

Describe the Data

1. Compute the following:

   a. $\bar{x}$ = _____

   b. s_x = _____

   c. n = _____

2. In words, define the random variable $\bar{X}$.

3. State the estimated distribution to use. Use both words and symbols.

Find the Confidence Interval

1. Calculate the confidence interval and the error bound.

   a. Confidence Interval: _____

    b. Error Bound: _____

2. How much area is in both tails (combined)? α = _____

3. How much area is in each tail? $\frac{\alpha}{2}$ = _____

4. Fill in the blanks on the graph with the area in each section. Then, fill in the number line with the upper and lower limits of the confidence interval and the sample mean.

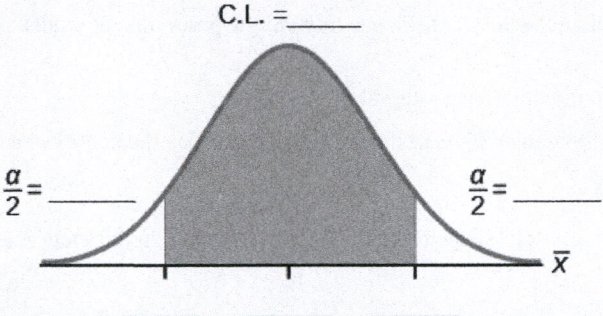

Figure 8.6

5. Some students think that a 90% confidence interval contains 90% of the data. Use the list of data on the first page and count how many of the data values lie within the confidence interval. What percent is this? Is this percent close to 90%? Explain why this percent should or should not be close to 90%.

Describe the Confidence Interval

1. In two to three complete sentences, explain what a confidence interval means (in general), as if you were talking to someone who has not taken statistics.

2. In one to two complete sentences, explain what this confidence interval means for this particular study.

Use the Data to Construct Confidence Intervals

1. Using the given information, construct a confidence interval for each confidence level given.

Confidence level	EBM/Error Bound	Confidence Interval
50%		
80%		
95%		
99%		

Table 8.6

2. What happens to the EBM as the confidence level increases? Does the width of the confidence interval increase or decrease? Explain why this happens.

8.5 | Confidence Interval (Place of Birth)

Stats Lab

8.2 Confidence Interval (Place of Birth)

Class Time:

Names:

Student Learning Outcomes

- The student will calculate the 90% confidence interval the proportion of students in this school who were born in this state.

- The student will interpret confidence intervals.

- The student will determine the effects of changing conditions on the confidence interval.

Collect the Data

1. Survey the students in your class, asking them if they were born in this state. Let X = the number that were born in this state.

   a. $n =$ _____

   b. $x =$ _____

2. In words, define the random variable P'.

3. State the estimated distribution to use.

Find the Confidence Interval and Error Bound

1. Calculate the confidence interval and the error bound.

   a. Confidence Interval: _____

   b. Error Bound: _____

2. How much area is in both tails (combined)? $\alpha =$ _____

3. How much area is in each tail? $\frac{\alpha}{2} =$ _____

4. Fill in the blanks on the graph with the area in each section. Then, fill in the number line with the upper and lower limits of the confidence interval and the sample proportion.

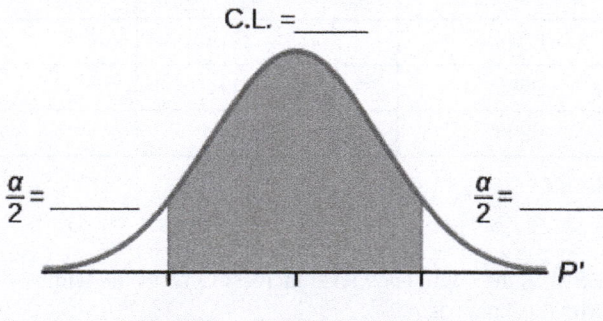

Figure 8.7

Describe the Confidence Interval

1. In two to three complete sentences, explain what a confidence interval means (in general), as though you were talking to someone who has not taken statistics.

2. In one to two complete sentences, explain what this confidence interval means for this particular study.

3. Construct a confidence interval for each confidence level given.

Confidence level	EBP/Error Bound	Confidence Interval
50%		
80%		
95%		
99%		

Table 8.7

4. What happens to the EBP as the confidence level increases? Does the width of the confidence interval increase or decrease? Explain why this happens.

8.6 | Confidence Interval (Women's Heights)

Stats Lab

8.3 Confidence Interval (Women's Heights)

Class Time:

Names:

Student Learning Outcomes

- The student will calculate a 90% confidence interval using the given data.
- The student will determine the relationship between the confidence level and the percentage of constructed intervals that contain the population mean.

Given:

59.4	71.6	69.3	65.0	62.9	66.5	61.7	55.2
67.5	67.2	63.8	62.9	63.0	63.9	68.7	65.5
61.9	69.6	58.7	63.4	61.8	60.6	69.8	60.0
64.9	66.1	66.8	60.6	65.6	63.8	61.3	59.2
64.1	59.3	64.9	62.4	63.5	60.9	63.3	66.3
61.5	64.3	62.9	60.6	63.8	58.8	64.9	65.7
62.5	70.9	62.9	63.1	62.2	58.7	64.7	66.0
60.5	64.7	65.4	60.2	65.0	64.1	61.1	65.3
64.6	59.2	61.4	62.0	63.5	61.4	65.5	62.3
65.5	64.7	58.8	66.1	64.9	66.9	57.9	69.8
58.5	63.4	69.2	65.9	62.2	60.0	58.1	62.5
62.4	59.1	66.4	61.2	60.4	58.7	66.7	67.5
63.2	56.6	67.7	62.5				

Table 8.8 Heights of 100 Women (in Inches)

1. **Table 8.8** lists the heights of 100 women. Use a random number generator to select ten data values randomly.

2. Calculate the sample mean and the sample standard deviation. Assume that the population standard deviation is known to be 3.3 inches. With these values, construct a 90% confidence interval for your sample of ten values. Write the confidence interval you obtained in the first space of **Table 8.9**.

3. Now write your confidence interval on the board. As others in the class write their confidence intervals on the board, copy them into **Table 8.9**.

Table 8.9 90% Confidence Intervals

Discussion Questions

1. The actual population mean for the 100 heights given **Table 8.8** is $\mu = 63.4$. Using the class listing of confidence intervals, count how many of them contain the population mean μ; i.e., for how many intervals does the value of μ lie between the endpoints of the confidence interval?

2. Divide this number by the total number of confidence intervals generated by the class to determine the percent of confidence intervals that contains the mean μ. Write this percent here: _____.

3. Is the percent of confidence intervals that contain the population mean μ close to 90%?

4. Suppose we had generated 100 confidence intervals. What do you think would happen to the percent of confidence intervals that contained the population mean?

5. When we construct a 90% confidence interval, we say that we are **90% confident that the true population mean lies within the confidence interval.** Using complete sentences, explain what we mean by this phrase.

6. Some students think that a 90% confidence interval contains 90% of the data. Use the list of data given (the heights of women) and count how many of the data values lie within the confidence interval that you generated based on that data. How many of the 100 data values lie within your confidence interval? What percent is this? Is this percent close to 90%?

7. Explain why it does not make sense to count data values that lie in a confidence interval. Think about the random variable that is being used in the problem.

8. Suppose you obtained the heights of ten women and calculated a confidence interval from this information. Without knowing the population mean μ, would you have any way of knowing **for certain** if your interval actually contained the value of μ? Explain.

KEY TERMS

Binomial Distribution a discrete random variable (RV) which arises from Bernoulli trials; there are a fixed number, n, of independent trials. "Independent" means that the result of any trial (for example, trial 1) does not affect the results of the following trials, and all trials are conducted under the same conditions. Under these circumstances the binomial RV X is defined as the number of successes in n trials. The notation is: $X\sim B(\mathbf{n},\mathbf{p})$. The mean is $\mu = np$ and the standard deviation is $\sigma = \sqrt{npq}$. The probability of exactly x successes in n trials is $P(X = x) = \binom{n}{x}p^x q^{n-x}$.

Confidence Interval (CI) an interval estimate for an unknown population parameter. This depends on:

- the desired confidence level,
- information that is known about the distribution (for example, known standard deviation),
- the sample and its size.

Confidence Level (CL) the percent expression for the probability that the confidence interval contains the true population parameter; for example, if the CL = 90%, then in 90 out of 100 samples the interval estimate will enclose the true population parameter.

Degrees of Freedom (*df*) the number of objects in a sample that are free to vary

Error Bound for a Population Mean (*EBM*) the margin of error; depends on the confidence level, sample size, and known or estimated population standard deviation.

Error Bound for a Population Proportion (EBP) the margin of error; depends on the confidence level, the sample size, and the estimated (from the sample) proportion of successes.

Inferential Statistics also called statistical inference or inductive statistics; this facet of statistics deals with estimating a population parameter based on a sample statistic. For example, if four out of the 100 calculators sampled are defective we might infer that four percent of the production is defective.

Normal Distribution
a continuous random variable (RV) with pdf $f(x) = \dfrac{1}{\sigma\sqrt{2\pi}}e^{-(x-\mu)^2 / 2\sigma^2}$, where μ is the mean of the distribution and σ is the standard deviation, notation: $X \sim N(\mu,\sigma)$. If $\mu = 0$ and $\sigma = 1$, the RV is called **the standard normal distribution**.

Parameter a numerical characteristic of a population

Point Estimate a single number computed from a sample and used to estimate a population parameter

Standard Deviation a number that is equal to the square root of the variance and measures how far data values are from their mean; notation: s for sample standard deviation and σ for population standard deviation

Student's t-Distribution investigated and reported by William S. Gossett in 1908 and published under the pseudonym Student; the major characteristics of the random variable (RV) are:

- It is continuous and assumes any real values.
- The pdf is symmetrical about its mean of zero. However, it is more spread out and flatter at the apex than the normal distribution.
- It approaches the standard normal distribution as n get larger.
- There is a "family of t–distributions: each representative of the family is completely defined by the number of degrees of freedom, which is one less than the number of data.

CHAPTER REVIEW

8.1 A Single Population Mean using the Normal Distribution

In this module, we learned how to calculate the confidence interval for a single population mean where the population standard deviation is known. When estimating a population mean, the margin of error is called the error bound for a population mean (*EBM*). A confidence interval has the general form:

(lower bound, upper bound) = (point estimate – *EBM*, point estimate + *EBM*)

The calculation of *EBM* depends on the size of the sample and the level of confidence desired. The confidence level is the percent of all possible samples that can be expected to include the true population parameter. As the confidence level increases, the corresponding *EBM* increases as well. As the sample size increases, the *EBM* decreases. By the central limit theorem,

$$EBM = z\frac{\sigma}{\sqrt{n}}$$

Given a confidence interval, you can work backwards to find the error bound (*EBM*) or the sample mean. To find the error bound, find the difference of the upper bound of the interval and the mean. If you do not know the sample mean, you can find the error bound by calculating half the difference of the upper and lower bounds. To find the sample mean given a confidence interval, find the difference of the upper bound and the error bound. If the error bound is unknown, then average the upper and lower bounds of the confidence interval to find the sample mean.

Sometimes researchers know in advance that they want to estimate a population mean within a specific margin of error for a given level of confidence. In that case, solve the *EBM* formula for *n* to discover the size of the sample that is needed to achieve this goal:

$$n = \frac{z^2\sigma^2}{EBM^2}$$

8.2 A Single Population Mean using the Student t Distribution

In many cases, the researcher does not know the population standard deviation, σ, of the measure being studied. In these cases, it is common to use the sample standard deviation, *s*, as an estimate of σ. The normal distribution creates accurate confidence intervals when σ is known, but it is not as accurate when *s* is used as an estimate. In this case, the Student's t-distribution is much better. Define a t-score using the following formula:

$$t = \frac{\bar{x} - \mu}{\frac{s}{\sqrt{n}}}$$

The *t*-score follows the Student's t-distribution with $n - 1$ degrees of freedom. The confidence interval under this distribution is calculated with $EBM = \left(t_{\frac{\alpha}{2}}\right)\frac{s}{\sqrt{n}}$ where $t_{\frac{\alpha}{2}}$ is the *t*-score with area to the right equal to $\frac{\alpha}{2}$, *s* is the sample standard deviation, and *n* is the sample size. Use a table, calculator, or computer to find $t_{\frac{\alpha}{2}}$ for a given α.

8.3 A Population Proportion

Some statistical measures, like many survey questions, measure qualitative rather than quantitative data. In this case, the population parameter being estimated is a proportion. It is possible to create a confidence interval for the true population proportion following procedures similar to those used in creating confidence intervals for population means. The formulas are slightly different, but they follow the same reasoning.

Let p' represent the sample proportion, x/n, where *x* represents the number of successes and *n* represents the sample size. Let $q' = 1 - p'$. Then the confidence interval for a population proportion is given by the following formula:

(lower bound, upper bound) $= (p' - EBP, p' + EBP) = \left(p' - z\sqrt{\frac{p'q'}{n}}, p' + z\sqrt{\frac{p'q'}{n}}\right)$

The "plus four" method for calculating confidence intervals is an attempt to balance the error introduced by using estimates of the population proportion when calculating the standard deviation of the sampling distribution. Simply imagine four additional trials in the study; two are successes and two are failures. Calculate $p' = \frac{x+2}{n+4}$, and proceed to find the confidence interval. When sample sizes are small, this method has been demonstrated to provide more accurate confidence intervals than the standard formula used for larger samples.

FORMULA REVIEW

8.1 A Single Population Mean using the Normal Distribution

$\bar{X} \sim N\left(\mu_X, \frac{\sigma}{\sqrt{n}}\right)$ The distribution of sample means is normally distributed with mean equal to the population mean and standard deviation given by the population standard deviation divided by the square root of the sample size.

The general form for a confidence interval for a single population mean, known standard deviation, normal distribution is given by
(lower bound, upper bound) = (point estimate – EBM, point estimate + EBM)

$= (\bar{x} - EBM, \ \bar{x} + EBM)$

$= \left(\bar{x} - z\frac{\sigma}{\sqrt{n}}, \ \bar{x} + z\frac{\sigma}{\sqrt{n}}\right)$

$EBM = z\frac{\sigma}{\sqrt{n}}$ = the error bound for the mean, or the margin of error for a single population mean; this formula is used when the population standard deviation is known.

CL = confidence level, or the proportion of confidence intervals created that are expected to contain the true population parameter

$\alpha = 1 - CL$ = the proportion of confidence intervals that will not contain the population parameter

$z_{\frac{\alpha}{2}}$ = the z-score with the property that the area to the right of the z-score is $\frac{\alpha}{2}$ this is the z-score used in the calculation of "EBM where α = 1 – CL.

$n = \dfrac{z^2 \sigma^2}{EBM^2}$ = the formula used to determine the sample size (n) needed to achieve a desired margin of error at a given level of confidence

General form of a confidence interval

(lower value, upper value) = (point estimate−error bound, point estimate + error bound)

To find the error bound when you know the confidence interval

error bound = upper value−point estimate OR error bound $= \dfrac{\text{upper value} - \text{lower value}}{2}$

Single Population Mean, Known Standard Deviation, Normal Distribution

Use the Normal Distribution for Means, Population Standard Deviation is Known $EBM = z\frac{\alpha}{2} \cdot \frac{\sigma}{\sqrt{n}}$

The confidence interval has the format ($\bar{x} - EBM$, $\bar{x} + EBM$).

8.2 A Single Population Mean using the Student t Distribution

s = the standard deviation of sample values.

$t = \dfrac{\bar{x} - \mu}{\frac{s}{\sqrt{n}}}$ is the formula for the t-score which measures how far away a measure is from the population mean in the Student's t-distribution

$df = n - 1$; the degrees of freedom for a Student's t-distribution where n represents the size of the sample

$T \sim t_{df}$ the random variable, T, has a Student's t-distribution with df degrees of freedom

$EBM = t_{\frac{\alpha}{2}} \frac{s}{\sqrt{n}}$ = the error bound for the population mean when the population standard deviation is unknown

$t_{\frac{\alpha}{2}}$ is the t-score in the Student's t-distribution with area to the right equal to $\frac{\alpha}{2}$

The general form for a confidence interval for a single mean, population standard deviation unknown, Student's t is given by (lower bound, upper bound)
= (point estimate – EBM, point estimate + EBM)
$= \left(\bar{x} - \frac{ts}{\sqrt{n}}, \ \bar{x} + \frac{ts}{\sqrt{n}}\right)$

8.3 A Population Proportion

$p' = x / n$ where x represents the number of successes and n represents the sample size. The variable p' is the sample proportion and serves as the point estimate for the true population proportion.

$q' = 1 - p'$

$p' \sim N\left(p, \sqrt{\frac{pq}{n}}\right)$ The variable p' has a binomial distribution that can be approximated with the normal distribution shown here.

EBP = the error bound for a proportion = $z_{\frac{\alpha}{2}}\sqrt{\dfrac{p'q'}{n}}$

Confidence interval for a proportion:

(lower bound, upper bound) $= (p' - EBP, p' + EBP) = \left(p' - z\sqrt{\frac{p'q'}{n}}, \quad p' + z\sqrt{\frac{p'q'}{n}}\right)$

$EBP = \left(z_{\frac{\alpha}{2}}\right)\sqrt{\frac{p'q'}{n}} \quad p' + q' = 1$

$n = \dfrac{z_{\frac{\alpha}{2}}^2 p' q'}{EBP^2}$ provides the number of participants needed to estimate the population proportion with confidence $1 - \alpha$ and margin of error EBP.

The confidence interval has the format $(p' - EBP, p' + EBP)$.

Use the normal distribution for a single population proportion $p' = \frac{x}{n}$

$\bar{x}$ is a point estimate for μ

p' is a point estimate for ρ

s is a point estimate for σ

PRACTICE

8.1 A Single Population Mean using the Normal Distribution

Use the following information to answer the next five exercises: The standard deviation of the weights of elephants is known to be approximately 15 pounds. We wish to construct a 95% confidence interval for the mean weight of newborn elephant calves. Fifty newborn elephants are weighed. The sample mean is 244 pounds. The sample standard deviation is 11 pounds.

1. Identify the following:
  a. $\bar{x}$ = _____
  b. σ = _____
  c. n = _____

2. In words, define the random variables X and $\bar{X}$.

3. Which distribution should you use for this problem?

4. Construct a 95% confidence interval for the population mean weight of newborn elephants. State the confidence interval, sketch the graph, and calculate the error bound.

5. What will happen to the confidence interval obtained, if 500 newborn elephants are weighed instead of 50? Why?

Use the following information to answer the next seven exercises: The U.S. Census Bureau conducts a study to determine the time needed to complete the short form. The Bureau surveys 200 people. The sample mean is 8.2 minutes. There is a known standard deviation of 2.2 minutes. The population distribution is assumed to be normal.

6. Identify the following:
  a. $\bar{x}$ = _____
  b. σ = _____
  c. n = _____

7. In words, define the random variables X and $\bar{X}$.

8. Which distribution should you use for this problem?

9. Construct a 90% confidence interval for the population mean time to complete the forms. State the confidence interval, sketch the graph, and calculate the error bound.

10. If the Census wants to increase its level of confidence and keep the error bound the same by taking another survey, what changes should it make?

11. If the Census did another survey, kept the error bound the same, and surveyed only 50 people instead of 200, what would happen to the level of confidence? Why?

12. Suppose the Census needed to be 98% confident of the population mean length of time. Would the Census have to survey more people? Why or why not?

Use the following information to answer the next ten exercises: A sample of 20 heads of lettuce was selected. Assume that the population distribution of head weight is normal. The weight of each head of lettuce was then recorded. The mean weight was 2.2 pounds with a standard deviation of 0.1 pounds. The population standard deviation is known to be 0.2 pounds.

13. Identify the following:

    a. $\bar{x}$ = _____

    b. σ = _____

    c. n = _____

14. In words, define the random variable X.

15. In words, define the random variable $\bar{X}$.

16. Which distribution should you use for this problem?

17. Construct a 90% confidence interval for the population mean weight of the heads of lettuce. State the confidence interval, sketch the graph, and calculate the error bound.

18. Construct a 95% confidence interval for the population mean weight of the heads of lettuce. State the confidence interval, sketch the graph, and calculate the error bound.

19. In complete sentences, explain why the confidence interval in **Exercise 8.17** is larger than in **Exercise 8.18**.

20. In complete sentences, give an interpretation of what the interval in **Exercise 8.18** means.

21. What would happen if 40 heads of lettuce were sampled instead of 20, and the error bound remained the same?

22. What would happen if 40 heads of lettuce were sampled instead of 20, and the confidence level remained the same?

Use the following information to answer the next 14 exercises: The mean age for all Foothill College students for a recent Fall term was 33.2. The population standard deviation has been pretty consistent at 15. Suppose that twenty-five Winter students were randomly selected. The mean age for the sample was 30.4. We are interested in the true mean age for Winter Foothill College students. Let X = the age of a Winter Foothill College student.

23. $\bar{x}$ = _____

24. n = _____

25. _____ = 15

26. In words, define the random variable $\bar{X}$.

27. What is $\bar{x}$ estimating?

28. Is σ_x known?

29. As a result of your answer to **Exercise 8.26**, state the exact distribution to use when calculating the confidence interval.

Construct a 95% Confidence Interval for the true mean age of Winter Foothill College students by working out then answering the next seven exercises.

30. How much area is in both tails (combined)? α = _____

31. How much area is in each tail? $\frac{\alpha}{2}$ = _____

32. Identify the following specifications:

    a. lower limit

    b. upper limit

    c. error bound

33. The 95% confidence interval is:_____.

34. Fill in the blanks on the graph with the areas, upper and lower limits of the confidence interval, and the sample mean.

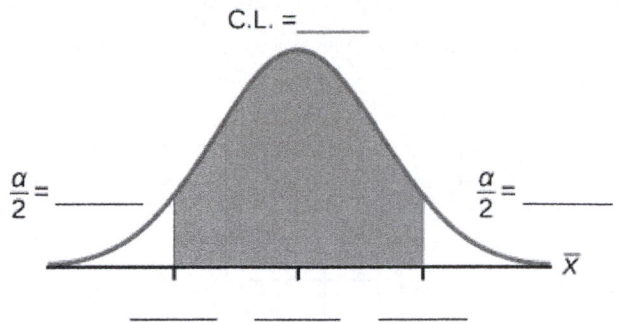

Figure 8.8

35. In one complete sentence, explain what the interval means.

36. Using the same mean, standard deviation, and level of confidence, suppose that n were 69 instead of 25. Would the error bound become larger or smaller? How do you know?

37. Using the same mean, standard deviation, and sample size, how would the error bound change if the confidence level were reduced to 90%? Why?

8.2 A Single Population Mean using the Student t Distribution

Use the following information to answer the next five exercises. A hospital is trying to cut down on emergency room wait times. It is interested in the amount of time patients must wait before being called back to be examined. An investigation committee randomly surveyed 70 patients. The sample mean was 1.5 hours with a sample standard deviation of 0.5 hours.

38. Identify the following:

    a. $\bar{x} =$_____
    b. $s_x =$_____
    c. $n =$_____
    d. $n - 1 =$_____

39. Define the random variables X and $\bar{X}$ in words.

40. Which distribution should you use for this problem?

41. Construct a 95% confidence interval for the population mean time spent waiting. State the confidence interval, sketch the graph, and calculate the error bound.

42. Explain in complete sentences what the confidence interval means.

Use the following information to answer the next six exercises: One hundred eight Americans were surveyed to determine the number of hours they spend watching television each month. It was revealed that they watched an average of 151 hours each month with a standard deviation of 32 hours. Assume that the underlying population distribution is normal.

43. Identify the following:

    a. $\bar{x} =$_____
    b. $s_x =$_____
    c. $n =$_____
    d. $n - 1 =$_____

44. Define the random variable X in words.

45. Define the random variable $\bar{X}$ in words.

46. Which distribution should you use for this problem?

47. Construct a 99% confidence interval for the population mean hours spent watching television per month. (a) State the confidence interval, (b) sketch the graph, and (c) calculate the error bound.

48. Why would the error bound change if the confidence level were lowered to 95%?

Use the following information to answer the next 13 exercises: The data in **Table 8.10** are the result of a random survey of 39 national flags (with replacement between picks) from various countries. We are interested in finding a confidence interval for the true mean number of colors on a national flag. Let X = the number of colors on a national flag.

X	Freq.
1	1
2	7
3	18
4	7
5	6

Table 8.10

49. Calculate the following:
   a. $\bar{x}$ =_____
   b. s_x =_____
   c. n =_____

50. Define the random variable $\bar{X}$ in words.

51. What is $\bar{x}$ estimating?

52. Is σ_x known?

53. As a result of your answer to **Exercise 8.52**, state the exact distribution to use when calculating the confidence interval.

Construct a 95% confidence interval for the true mean number of colors on national flags.

54. How much area is in both tails (combined)?

55. How much area is in each tail?

56. Calculate the following:
   a. lower limit
   b. upper limit
   c. error bound

57. The 95% confidence interval is_____.

58. Fill in the blanks on the graph with the areas, the upper and lower limits of the Confidence Interval and the sample mean.

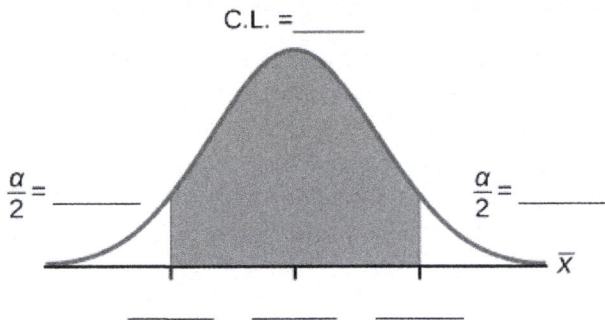

Figure 8.9

59. In one complete sentence, explain what the interval means.

60. Using the same $\bar{x}$, s_x, and level of confidence, suppose that n were 69 instead of 39. Would the error bound become larger or smaller? How do you know?

61. Using the same $\bar{x}$, s_x, and $n = 39$, how would the error bound change if the confidence level were reduced to 90%? Why?

8.3 A Population Proportion

Use the following information to answer the next two exercises: Marketing companies are interested in knowing the population percent of women who make the majority of household purchasing decisions.

62. When designing a study to determine this population proportion, what is the minimum number you would need to survey to be 90% confident that the population proportion is estimated to within 0.05?

63. If it were later determined that it was important to be more than 90% confident and a new survey were commissioned, how would it affect the minimum number you need to survey? Why?

Use the following information to answer the next five exercises: Suppose the marketing company did do a survey. They randomly surveyed 200 households and found that in 120 of them, the woman made the majority of the purchasing decisions. We are interested in the population proportion of households where women make the majority of the purchasing decisions.

64. Identify the following:
   a. $x = $ _____
   b. $n = $ _____
   c. $p' = $ _____

65. Define the random variables X and P' in words.

66. Which distribution should you use for this problem?

67. Construct a 95% confidence interval for the population proportion of households where the women make the majority of the purchasing decisions. State the confidence interval, sketch the graph, and calculate the error bound.

68. List two difficulties the company might have in obtaining random results, if this survey were done by email.

Use the following information to answer the next five exercises: Of 1,050 randomly selected adults, 360 identified themselves as manual laborers, 280 identified themselves as non-manual wage earners, 250 identified themselves as mid-level managers, and 160 identified themselves as executives. In the survey, 82% of manual laborers preferred trucks, 62% of non-manual wage earners preferred trucks, 54% of mid-level managers preferred trucks, and 26% of executives preferred trucks.

69. We are interested in finding the 95% confidence interval for the percent of executives who prefer trucks. Define random variables X and P' in words.

70. Which distribution should you use for this problem?

71. Construct a 95% confidence interval. State the confidence interval, sketch the graph, and calculate the error bound.

72. Suppose we want to lower the sampling error. What is one way to accomplish that?

73. The sampling error given in the survey is ±2%. Explain what the ±2% means.

Use the following information to answer the next five exercises: A poll of 1,200 voters asked what the most significant issue was in the upcoming election. Sixty-five percent answered the economy. We are interested in the population proportion of voters who feel the economy is the most important.

74. Define the random variable X in words.

75. Define the random variable P' in words.

76. Which distribution should you use for this problem?

77. Construct a 90% confidence interval, and state the confidence interval and the error bound.

78. What would happen to the confidence interval if the level of confidence were 95%?

Use the following information to answer the next 16 exercises: The Ice Chalet offers dozens of different beginning ice-skating classes. All of the class names are put into a bucket. The 5 P.M., Monday night, ages 8 to 12, beginning ice-skating class was picked. In that class were 64 girls and 16 boys. Suppose that we are interested in the true proportion of girls, ages 8 to 12, in all beginning ice-skating classes at the Ice Chalet. Assume that the children in the selected class are a random sample of the population.

79. What is being counted?

80. In words, define the random variable X.

81. Calculate the following:
   a. $x =$ _____
   b. $n =$ _____
   c. $p' =$ _____

82. State the estimated distribution of X. $X \sim$_____

83. Define a new random variable P'. What is p' estimating?

84. In words, define the random variable P'.

85. State the estimated distribution of P'. Construct a 92% Confidence Interval for the true proportion of girls in the ages 8 to 12 beginning ice-skating classes at the Ice Chalet.

86. How much area is in both tails (combined)?

87. How much area is in each tail?

88. Calculate the following:
   a. lower limit
   b. upper limit
   c. error bound

89. The 92% confidence interval is _____.

90. Fill in the blanks on the graph with the areas, upper and lower limits of the confidence interval, and the sample proportion.

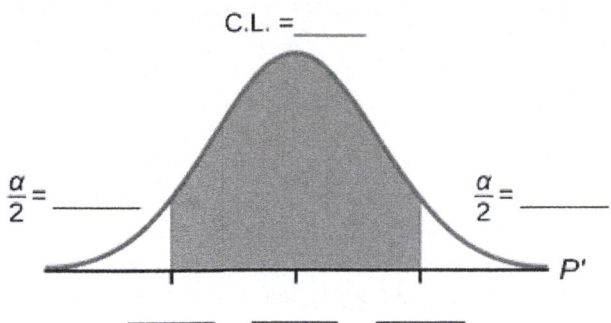

Figure 8.10

91. In one complete sentence, explain what the interval means.

92. Using the same p' and level of confidence, suppose that n were increased to 100. Would the error bound become larger or smaller? How do you know?

93. Using the same p' and $n = 80$, how would the error bound change if the confidence level were increased to 98%? Why?

94. If you decreased the allowable error bound, why would the minimum sample size increase (keeping the same level of confidence)?

HOMEWORK

8.1 A Single Population Mean using the Normal Distribution

95. Among various ethnic groups, the standard deviation of heights is known to be approximately three inches. We wish to construct a 95% confidence interval for the mean height of male Swedes. Forty-eight male Swedes are surveyed. The sample mean is 71 inches. The sample standard deviation is 2.8 inches.

    a.

        i. $\bar{x}$ =_____

        ii. σ =_____

        iii. n =_____

    b. In words, define the random variables X and $\bar{X}$.

    c. Which distribution should you use for this problem? Explain your choice.

    d. Construct a 95% confidence interval for the population mean height of male Swedes.

        i. State the confidence interval.

        ii. Sketch the graph.

        iii. Calculate the error bound.

    e. What will happen to the level of confidence obtained if 1,000 male Swedes are surveyed instead of 48? Why?

96. Announcements for 84 upcoming engineering conferences were randomly picked from a stack of IEEE Spectrum magazines. The mean length of the conferences was 3.94 days, with a standard deviation of 1.28 days. Assume the underlying population is normal.

    a. In words, define the random variables X and $\bar{X}$.

    b. Which distribution should you use for this problem? Explain your choice.

    c. Construct a 95% confidence interval for the population mean length of engineering conferences.

        i. State the confidence interval.

        ii. Sketch the graph.

        iii. Calculate the error bound.

97. Suppose that an accounting firm does a study to determine the time needed to complete one person's tax forms. It randomly surveys 100 people. The sample mean is 23.6 hours. There is a known standard deviation of 7.0 hours. The population distribution is assumed to be normal.

   a.

      i. $\bar{x}$ =_____

      ii. σ =_____

      iii. n =_____

   b. In words, define the random variables X and $\bar{X}$.

   c. Which distribution should you use for this problem? Explain your choice.

   d. Construct a 90% confidence interval for the population mean time to complete the tax forms.

      i. State the confidence interval.

      ii. Sketch the graph.

      iii. Calculate the error bound.

   e. If the firm wished to increase its level of confidence and keep the error bound the same by taking another survey, what changes should it make?

   f. If the firm did another survey, kept the error bound the same, and only surveyed 49 people, what would happen to the level of confidence? Why?

   g. Suppose that the firm decided that it needed to be at least 96% confident of the population mean length of time to within one hour. How would the number of people the firm surveys change? Why?

98. A sample of 16 small bags of the same brand of candies was selected. Assume that the population distribution of bag weights is normal. The weight of each bag was then recorded. The mean weight was two ounces with a standard deviation of 0.12 ounces. The population standard deviation is known to be 0.1 ounce.

   a.

      i. $\bar{x}$ =_____

      ii. σ =_____

      iii. s_X =_____

   b. In words, define the random variable X.

   c. In words, define the random variable $\bar{X}$.

   d. Which distribution should you use for this problem? Explain your choice.

   e. Construct a 90% confidence interval for the population mean weight of the candies.

      i. State the confidence interval.

      ii. Sketch the graph.

      iii. Calculate the error bound.

   f. Construct a 98% confidence interval for the population mean weight of the candies.

      i. State the confidence interval.

      ii. Sketch the graph.

      iii. Calculate the error bound.

   g. In complete sentences, explain why the confidence interval in part f is larger than the confidence interval in part e.

   h. In complete sentences, give an interpretation of what the interval in part f means.

99. A camp director is interested in the mean number of letters each child sends during his or her camp session. The population standard deviation is known to be 2.5. A survey of 20 campers is taken. The mean from the sample is 7.9 with a sample standard deviation of 2.8.

   a.

      i. $\bar{x}$ =_____

      ii. σ =_____

      iii. n =_____

   b. Define the random variables X and $\bar{X}$ in words.

   c. Which distribution should you use for this problem? Explain your choice.

   d. Construct a 90% confidence interval for the population mean number of letters campers send home.

      i. State the confidence interval.

      ii. Sketch the graph.

      iii. Calculate the error bound.

e. What will happen to the error bound and confidence interval if 500 campers are surveyed? Why?

100. What is meant by the term "90% confident" when constructing a confidence interval for a mean?

    a. If we took repeated samples, approximately 90% of the samples would produce the same confidence interval.

    b. If we took repeated samples, approximately 90% of the confidence intervals calculated from those samples would contain the sample mean.

    c. If we took repeated samples, approximately 90% of the confidence intervals calculated from those samples would contain the true value of the population mean.

    d. If we took repeated samples, the sample mean would equal the population mean in approximately 90% of the samples.

101. The Federal Election Commission collects information about campaign contributions and disbursements for candidates and political committees each election cycle. During the 2012 campaign season, there were 1,619 candidates for the House of Representatives across the United States who received contributions from individuals. **Table 8.11** shows the total receipts from individuals for a random selection of 40 House candidates rounded to the nearest $100. The standard deviation for this data to the nearest hundred is $\sigma = \$909,200$.

$3,600	$1,243,900	$10,900	$385,200	$581,500
$7,400	$2,900	$400	$3,714,500	$632,500
$391,000	$467,400	$56,800	$5,800	$405,200
$733,200	$8,000	$468,700	$75,200	$41,000
$13,300	$9,500	$953,800	$1,113,500	$1,109,300
$353,900	$986,100	$88,600	$378,200	$13,200
$3,800	$745,100	$5,800	$3,072,100	$1,626,700
$512,900	$2,309,200	$6,600	$202,400	$15,800

Table 8.11

a. Find the point estimate for the population mean.

b. Using 95% confidence, calculate the error bound.

c. Create a 95% confidence interval for the mean total individual contributions.

d. Interpret the confidence interval in the context of the problem.

102. The American Community Survey (ACS), part of the United States Census Bureau, conducts a yearly census similar to the one taken every ten years, but with a smaller percentage of participants. The most recent survey estimates with 90% confidence that the mean household income in the U.S. falls between $69,720 and $69,922. Find the point estimate for mean U.S. household income and the error bound for mean U.S. household income.

103. The average height of young adult males has a normal distribution with standard deviation of 2.5 inches. You want to estimate the mean height of students at your college or university to within one inch with 93% confidence. How many male students must you measure?

8.2 A Single Population Mean using the Student t Distribution

104. In six packages of "The Flintstones® Real Fruit Snacks" there were five Bam-Bam snack pieces. The total number of snack pieces in the six bags was 68. We wish to calculate a 96% confidence interval for the population proportion of Bam-Bam snack pieces.

    a. Define the random variables X and P' in words.

    b. Which distribution should you use for this problem? Explain your choice

    c. Calculate p'.

    d. Construct a 96% confidence interval for the population proportion of Bam-Bam snack pieces per bag.

      i. State the confidence interval.

     ii. Sketch the graph.

     iii. Calculate the error bound.

e. Do you think that six packages of fruit snacks yield enough data to give accurate results? Why or why not?

105. A random survey of enrollment at 35 community colleges across the United States yielded the following figures: 6,414; 1,550; 2,109; 9,350; 21,828; 4,300; 5,944; 5,722; 2,825; 2,044; 5,481; 5,200; 5,853; 2,750; 10,012; 6,357; 27,000; 9,414; 7,681; 3,200; 17,500; 9,200; 7,380; 18,314; 6,557; 13,713; 17,768; 7,493; 2,771; 2,861; 1,263; 7,285; 28,165; 5,080; 11,622. Assume the underlying population is normal.

  a.

     i. $\bar{x}$ = _____

     ii. s_x = _____

     iii. n = _____

     iv. $n - 1$ = _____

  b. Define the random variables X and $\bar{X}$ in words.

  c. Which distribution should you use for this problem? Explain your choice.

  d. Construct a 95% confidence interval for the population mean enrollment at community colleges in the United States.

     i. State the confidence interval.

     ii. Sketch the graph.

     iii. Calculate the error bound.

  e. What will happen to the error bound and confidence interval if 500 community colleges were surveyed? Why?

106. Suppose that a committee is studying whether or not there is waste of time in our judicial system. It is interested in the mean amount of time individuals waste at the courthouse waiting to be called for jury duty. The committee randomly surveyed 81 people who recently served as jurors. The sample mean wait time was eight hours with a sample standard deviation of four hours.

  a.

     i. $\bar{x}$ = _____

     ii. s_x = _____

     iii. n = _____

     iv. $n - 1$ = _____

  b. Define the random variables X and $\bar{X}$ in words.

  c. Which distribution should you use for this problem? Explain your choice.

  d. Construct a 95% confidence interval for the population mean time wasted.

     i. State the confidence interval.

     ii. Sketch the graph.

     iii. Calculate the error bound.

  e. Explain in a complete sentence what the confidence interval means.

107. A pharmaceutical company makes tranquilizers. It is assumed that the distribution for the length of time they last is approximately normal. Researchers in a hospital used the drug on a random sample of nine patients. The effective period of the tranquilizer for each patient (in hours) was as follows: 2.7; 2.8; 3.0; 2.3; 2.3; 2.2; 2.8; 2.1; and 2.4.

  a.

     i. $\bar{x}$ = _____

     ii. s_x = _____

     iii. n = _____

     iv. $n - 1$ = _____

  b. Define the random variable X in words.

  c. Define the random variable $\bar{X}$ in words.

  d. Which distribution should you use for this problem? Explain your choice.

  e. Construct a 95% confidence interval for the population mean length of time.

     i. State the confidence interval.

     ii. Sketch the graph.

        iii.   Calculate the error bound.

   f.   What does it mean to be "95% confident" in this problem?

108. Suppose that 14 children, who were learning to ride two-wheel bikes, were surveyed to determine how long they had to use training wheels. It was revealed that they used them an average of six months with a sample standard deviation of three months. Assume that the underlying population distribution is normal.

   a.

        i. $\bar{x}$ = _____

        ii. s_x = _____

        iii. n = _____

        iv. $n - 1$ = _____

   b.   Define the random variable X in words.

   c.   Define the random variable $\bar{X}$ in words.

   d.   Which distribution should you use for this problem? Explain your choice.

   e.   Construct a 99% confidence interval for the population mean length of time using training wheels.

        i.   State the confidence interval.

        ii.   Sketch the graph.

        iii.   Calculate the error bound.

   f.   Why would the error bound change if the confidence level were lowered to 90%?

109. The Federal Election Commission (FEC) collects information about campaign contributions and disbursements for candidates and political committees each election cycle. A political action committee (PAC) is a committee formed to raise money for candidates and campaigns. A Leadership PAC is a PAC formed by a federal politician (senator or representative) to raise money to help other candidates' campaigns.

The FEC has reported financial information for 556 Leadership PACs that operating during the 2011–2012 election cycle. The following table shows the total receipts during this cycle for a random selection of 20 Leadership PACs.

$46,500.00	$0	$40,966.50	$105,887.20	$5,175.00
$29,050.00	$19,500.00	$181,557.20	$31,500.00	$149,970.80
$2,555,363.20	$12,025.00	$409,000.00	$60,521.70	$18,000.00
$61,810.20	$76,530.80	$119,459.20	$0	$63,520.00
$6,500.00	$502,578.00	$705,061.10	$708,258.90	$135,810.00
$2,000.00	$2,000.00	$0	$1,287,933.80	$219,148.30

Table 8.12

$\bar{x} = \$251,854.23$

$s = \$521,130.41$

Use this sample data to construct a 96% confidence interval for the mean amount of money raised by all Leadership PACs during the 2011–2012 election cycle. Use the Student's t-distribution.

110. *Forbes* magazine published data on the best small firms in 2012. These were firms that had been publicly traded for at least a year, have a stock price of at least $5 per share, and have reported annual revenue between $5 million and $1 billion. The **Table 8.13** shows the ages of the corporate CEOs for a random sample of these firms.

48	58	51	61	56
59	74	63	53	50

59	60	60	57	46
55	63	57	47	55
57	43	61	62	49
67	67	55	55	49

Table 8.13

Use this sample data to construct a 90% confidence interval for the mean age of CEO's for these top small firms. Use the Student's t-distribution.

111. Unoccupied seats on flights cause airlines to lose revenue. Suppose a large airline wants to estimate its mean number of unoccupied seats per flight over the past year. To accomplish this, the records of 225 flights are randomly selected and the number of unoccupied seats is noted for each of the sampled flights. The sample mean is 11.6 seats and the sample standard deviation is 4.1 seats.

    a.

        i. $\bar{x} =$ _____

        ii. $s_x =$ _____

        iii. $n =$ _____

        iv. n-1 = _____

    b. Define the random variables X and $\bar{X}$ in words.

    c. Which distribution should you use for this problem? Explain your choice.

    d. Construct a 92% confidence interval for the population mean number of unoccupied seats per flight.

        i. State the confidence interval.

        ii. Sketch the graph.

        iii. Calculate the error bound.

112. In a recent sample of 84 used car sales costs, the sample mean was $6,425 with a standard deviation of $3,156. Assume the underlying distribution is approximately normal.

    a. Which distribution should you use for this problem? Explain your choice.

    b. Define the random variable $\bar{X}$ in words.

    c. Construct a 95% confidence interval for the population mean cost of a used car.

        i. State the confidence interval.

        ii. Sketch the graph.

        iii. Calculate the error bound.

    d. Explain what a "95% confidence interval" means for this study.

113. Six different national brands of chocolate chip cookies were randomly selected at the supermarket. The grams of fat per serving are as follows: 8; 8; 10; 7; 9; 9. Assume the underlying distribution is approximately normal.

    a. Construct a 90% confidence interval for the population mean grams of fat per serving of chocolate chip cookies sold in supermarkets.

        i. State the confidence interval.

        ii. Sketch the graph.

        iii. Calculate the error bound.

    b. If you wanted a smaller error bound while keeping the same level of confidence, what should have been changed in the study before it was done?

    c. Go to the store and record the grams of fat per serving of six brands of chocolate chip cookies.

    d. Calculate the mean.

    e. Is the mean within the interval you calculated in part a? Did you expect it to be? Why or why not?

114. A survey of the mean number of cents off that coupons give was conducted by randomly surveying one coupon per page from the coupon sections of a recent San Jose Mercury News. The following data were collected: 20¢; 75¢; 50¢; 65¢; 30¢; 55¢; 40¢; 40¢; 30¢; 55¢; $1.50; 40¢; 65¢; 40¢. Assume the underlying distribution is approximately normal.

    a.

   i. $\bar{x}$ = _____
   ii. s_x = _____
   iii. n = _____
   iv. n-1 = _____

b. Define the random variables X and $\bar{X}$ in words.
c. Which distribution should you use for this problem? Explain your choice.
d. Construct a 95% confidence interval for the population mean worth of coupons.
   i.   State the confidence interval.
   ii. Sketch the graph.
   iii. Calculate the error bound.

e. If many random samples were taken of size 14, what percent of the confidence intervals constructed should contain the population mean worth of coupons? Explain why.

Use the following information to answer the next two exercises: A quality control specialist for a restaurant chain takes a random sample of size 12 to check the amount of soda served in the 16 oz. serving size. The sample mean is 13.30 with a sample standard deviation of 1.55. Assume the underlying population is normally distributed.

115. Find the 95% Confidence Interval for the true population mean for the amount of soda served.
   a. (12.42, 14.18)
   b. (12.32, 14.29)
   c. (12.50, 14.10)
   d. Impossible to determine

116. What is the error bound?
   a. 0.87
   b. 1.98
   c. 0.99
   d. 1.74

8.3 A Population Proportion

117. Insurance companies are interested in knowing the population percent of drivers who always buckle up before riding in a car.
   a. When designing a study to determine this population proportion, what is the minimum number you would need to survey to be 95% confident that the population proportion is estimated to within 0.03?
   b. If it were later determined that it was important to be more than 95% confident and a new survey was commissioned, how would that affect the minimum number you would need to survey? Why?

118. Suppose that the insurance companies did do a survey. They randomly surveyed 400 drivers and found that 320 claimed they always buckle up. We are interested in the population proportion of drivers who claim they always buckle up.

a.
   i. x = _____
   ii. n = _____
   iii. p' = _____
b. Define the random variables X and P', in words.
c. Which distribution should you use for this problem? Explain your choice.
d. Construct a 95% confidence interval for the population proportion who claim they always buckle up.
   i.   State the confidence interval.
   ii. Sketch the graph.
   iii. Calculate the error bound.

e. If this survey were done by telephone, list three difficulties the companies might have in obtaining random results.

119. According to a recent survey of 1,200 people, 61% feel that the president is doing an acceptable job. We are interested in the population proportion of people who feel the president is doing an acceptable job.
   a. Define the random variables X and P' in words.
   b. Which distribution should you use for this problem? Explain your choice.

   c. Construct a 90% confidence interval for the population proportion of people who feel the president is doing an acceptable job.

      i. State the confidence interval.

      ii. Sketch the graph.

      iii. Calculate the error bound.

120. An article regarding interracial dating and marriage recently appeared in the *Washington Post*. Of the 1,709 randomly selected adults, 315 identified themselves as Latinos, 323 identified themselves as blacks, 254 identified themselves as Asians, and 779 identified themselves as whites. In this survey, 86% of blacks said that they would welcome a white person into their families. Among Asians, 77% would welcome a white person into their families, 71% would welcome a Latino, and 66% would welcome a black person.

   a. We are interested in finding the 95% confidence interval for the percent of all black adults who would welcome a white person into their families. Define the random variables X and P', in words.

   b. Which distribution should you use for this problem? Explain your choice.

   c. Construct a 95% confidence interval.

      i. State the confidence interval.

      ii. Sketch the graph.

      iii. Calculate the error bound.

121. Refer to the information in **Exercise 8.120**.

   a. Construct three 95% confidence intervals.

      i. percent of all Asians who would welcome a white person into their families.

      ii. percent of all Asians who would welcome a Latino into their families.

      iii. percent of all Asians who would welcome a black person into their families.

   b. Even though the three point estimates are different, do any of the confidence intervals overlap? Which?

   c. For any intervals that do overlap, in words, what does this imply about the significance of the differences in the true proportions?

   d. For any intervals that do not overlap, in words, what does this imply about the significance of the differences in the true proportions?

122. Stanford University conducted a study of whether running is healthy for men and women over age 50. During the first eight years of the study, 1.5% of the 451 members of the 50-Plus Fitness Association died. We are interested in the proportion of people over 50 who ran and died in the same eight-year period.

   a. Define the random variables X and P' in words.

   b. Which distribution should you use for this problem? Explain your choice.

   c. Construct a 97% confidence interval for the population proportion of people over 50 who ran and died in the same eight–year period.

      i. State the confidence interval.

      ii. Sketch the graph.

      iii. Calculate the error bound.

   d. Explain what a "97% confidence interval" means for this study.

123. A telephone poll of 1,000 adult Americans was reported in an issue of *Time Magazine*. One of the questions asked was "What is the main problem facing the country?" Twenty percent answered "crime." We are interested in the population proportion of adult Americans who feel that crime is the main problem.

   a. Define the random variables X and P' in words.

   b. Which distribution should you use for this problem? Explain your choice.

   c. Construct a 95% confidence interval for the population proportion of adult Americans who feel that crime is the main problem.

      i. State the confidence interval.

      ii. Sketch the graph.

      iii. Calculate the error bound.

   d. Suppose we want to lower the sampling error. What is one way to accomplish that?

   e. The sampling error given by Yankelovich Partners, Inc. (which conducted the poll) is ±3%. In one to three complete sentences, explain what the ±3% represents.

124. Refer to **Exercise 8.123**. Another question in the poll was "[How much are] you worried about the quality of education in our schools?" Sixty-three percent responded "a lot". We are interested in the population proportion of adult Americans who are worried a lot about the quality of education in our schools.

    a. Define the random variables X and P' in words.
    b. Which distribution should you use for this problem? Explain your choice.
    c. Construct a 95% confidence interval for the population proportion of adult Americans who are worried a lot about the quality of education in our schools.
        i. State the confidence interval.
        ii. Sketch the graph.
        iii. Calculate the error bound.

    d. The sampling error given by Yankelovich Partners, Inc. (which conducted the poll) is ±3%. In one to three complete sentences, explain what the ±3% represents.

Use the following information to answer the next three exercises: According to a Field Poll, 79% of California adults (actual results are 400 out of 506 surveyed) feel that "education and our schools" is one of the top issues facing California. We wish to construct a 90% confidence interval for the true proportion of California adults who feel that education and the schools is one of the top issues facing California.

125. A point estimate for the true population proportion is:
    a. 0.90
    b. 1.27
    c. 0.79
    d. 400

126. A 90% confidence interval for the population proportion is _____.
    a. (0.761, 0.820)
    b. (0.125, 0.188)
    c. (0.755, 0.826)
    d. (0.130, 0.183)

127. The error bound is approximately _____.
    a. 1.581
    b. 0.791
    c. 0.059
    d. 0.030

Use the following information to answer the next two exercises: Five hundred and eleven (511) homes in a certain southern California community are randomly surveyed to determine if they meet minimal earthquake preparedness recommendations. One hundred seventy-three (173) of the homes surveyed met the minimum recommendations for earthquake preparedness, and 338 did not.

128. Find the confidence interval at the 90% Confidence Level for the true population proportion of southern California community homes meeting at least the minimum recommendations for earthquake preparedness.
    a. (0.2975, 0.3796)
    b. (0.6270, 0.6959)
    c. (0.3041, 0.3730)
    d. (0.6204, 0.7025)

129. The point estimate for the population proportion of homes that do not meet the minimum recommendations for earthquake preparedness is _____.
    a. 0.6614
    b. 0.3386
    c. 173
    d. 338

130. On May 23, 2013, Gallup reported that of the 1,005 people surveyed, 76% of U.S. workers believe that they will continue working past retirement age. The confidence level for this study was reported at 95% with a ±3% margin of error.

    a. Determine the estimated proportion from the sample.
    b. Determine the sample size.
    c. Identify CL and α.
    d. Calculate the error bound based on the information provided.

e. Compare the error bound in part d to the margin of error reported by Gallup. Explain any differences between the values.
f. Create a confidence interval for the results of this study.
g. A reporter is covering the release of this study for a local news station. How should she explain the confidence interval to her audience?

131. A national survey of 1,000 adults was conducted on May 13, 2013 by Rasmussen Reports. It concluded with 95% confidence that 49% to 55% of Americans believe that big-time college sports programs corrupt the process of higher education.
a. Find the point estimate and the error bound for this confidence interval.
b. Can we (with 95% confidence) conclude that more than half of all American adults believe this?
c. Use the point estimate from part a and $n = 1,000$ to calculate a 75% confidence interval for the proportion of American adults that believe that major college sports programs corrupt higher education.
d. Can we (with 75% confidence) conclude that at least half of all American adults believe this?

132. Public Policy Polling recently conducted a survey asking adults across the U.S. about music preferences. When asked, 80 of the 571 participants admitted that they have illegally downloaded music.
a. Create a 99% confidence interval for the true proportion of American adults who have illegally downloaded music.
b. This survey was conducted through automated telephone interviews on May 6 and 7, 2013. The error bound of the survey compensates for sampling error, or natural variability among samples. List some factors that could affect the survey's outcome that are not covered by the margin of error.
c. Without performing any calculations, describe how the confidence interval would change if the confidence level changed from 99% to 90%.

133. You plan to conduct a survey on your college campus to learn about the political awareness of students. You want to estimate the true proportion of college students on your campus who voted in the 2012 presidential election with 95% confidence and a margin of error no greater than five percent. How many students must you interview?

134. In a recent Zogby International Poll, nine of 48 respondents rated the likelihood of a terrorist attack in their community as "likely" or "very likely." Use the "plus four" method to create a 97% confidence interval for the proportion of American adults who believe that a terrorist attack in their community is likely or very likely. Explain what this confidence interval means in the context of the problem.

REFERENCES

8.1 A Single Population Mean using the Normal Distribution

"American Fact Finder." U.S. Census Bureau. Available online at http://factfinder2.census.gov/faces/nav/jsf/pages/searchresults.xhtml?refresh=t (accessed July 2, 2013).

"Disclosure Data Catalog: Candidate Summary Report 2012." U.S. Federal Election Commission. Available online at http://www.fec.gov/data/index.jsp (accessed July 2, 2013).

"Headcount Enrollment Trends by Student Demographics Ten-Year Fall Trends to Most Recently Completed Fall." Foothill De Anza Community College District. Available online at http://research.fhda.edu/factbook/FH_Demo_Trends/FoothillDemographicTrends.htm (accessed September 30,2013).

Kuczmarski, Robert J., Cynthia L. Ogden, Shumei S. Guo, Laurence M. Grummer-Strawn, Katherine M. Flegal, Zuguo Mei, Rong Wei, Lester R. Curtin, Alex F. Roche, Clifford L. Johnson. "2000 CDC Growth Charts for the United States: Methods and Development." Centers for Disease Control and Prevention. Available online at http://www.cdc.gov/growthcharts/2000growthchart-us.pdf (accessed July 2, 2013).

La, Lynn, Kent German. "Cell Phone Radiation Levels." c|net part of CBX Interactive Inc. Available online at http://reviews.cnet.com/cell-phone-radiation-levels/ (accessed July 2, 2013).

"Mean Income in the Past 12 Months (in 2011 Inflaction-Adjusted Dollars): 2011 American Community Survey 1-Year Estimates." American Fact Finder, U.S. Census Bureau. Available online at http://factfinder2.census.gov/faces/tableservices/jsf/pages/productview.xhtml?pid=ACS_11_1YR_S1902&prodType=table (accessed July 2, 2013).

"Metadata Description of Candidate Summary File." U.S. Federal Election Commission. Available online at http://www.fec.gov/finance/disclosure/metadata/metadataforcandidatesummary.shtml (accessed July 2, 2013).

"National Health and Nutrition Examination Survey." Centers for Disease Control and Prevention. Available online at http://www.cdc.gov/nchs/nhanes.htm (accessed July 2, 2013).

8.2 A Single Population Mean using the Student t Distribution

"America's Best Small Companies." Forbes, 2013. Available online at http://www.forbes.com/best-small-companies/list/ (accessed July 2, 2013).

Data from *Microsoft Bookshelf*.

Data from http://www.businessweek.com/.

Data from http://www.forbes.com/.

"Disclosure Data Catalog: Leadership PAC and Sponsors Report, 2012." Federal Election Commission. Available online at http://www.fec.gov/data/index.jsp (accessed July 2,2013).

"Human Toxome Project: Mapping the Pollution in People." Environmental Working Group. Available online at http://www.ewg.org/sites/humantoxome/participants/participant-group.php?group=in+utero%2Fnewborn (accessed July 2, 2013).

"Metadata Description of Leadership PAC List." Federal Election Commission. Available online at http://www.fec.gov/finance/disclosure/metadata/metadataLeadershipPacList.shtml (accessed July 2, 2013).

8.3 A Population Proportion

Jensen, Tom. "Democrats, Republicans Divided on Opinion of Music Icons." Public Policy Polling. Available online at http://www.publicpolicypolling.com/Day2MusicPoll.pdf (accessed July 2, 2013).

Madden, Mary, Amanda Lenhart, Sandra Coresi, Urs Gasser, Maeve Duggan, Aaron Smith, and Meredith Beaton. "Teens, Social Media, and Privacy." PewInternet, 2013. Available online at http://www.pewinternet.org/Reports/2013/Teens-Social-Media-And-Privacy.aspx (accessed July 2, 2013).

Prince Survey Research Associates International. "2013 Teen and Privacy Management Survey." Pew Research Center: Internet and American Life Project. Available online at http://www.pewinternet.org/~/media//Files/Questionnaire/2013/Methods%20and%20Questions_Teens%20and%20Social%20Media.pdf (accessed July 2, 2013).

Saad, Lydia. "Three in Four U.S. Workers Plan to Work Pas Retirement Age: Slightly more say they will do this by choice rather than necessity." Gallup® Economy, 2013. Available online at http://www.gallup.com/poll/162758/three-four-workers-plan-work-past-retirement-age.aspx (accessed July 2, 2013).

The Field Poll. Available online at http://field.com/fieldpollonline/subscribers/ (accessed July 2, 2013).

Zogby. "New SUNYIT/Zogby Analytics Poll: Few Americans Worry about Emergency Situations Occurring in Their Community; Only one in three have an Emergency Plan; 70% Support Infrastructure 'Investment' for National Security." Zogby Analytics, 2013. Available online at http://www.zogbyanalytics.com/news/299-americans-neither-worried-nor-prepared-in-case-of-a-disaster-sunyit-zogby-analytics-poll (accessed July 2, 2013).

"52% Say Big-Time College Athletics Corrupt Education Process." Rasmussen Reports, 2013. Available online at http://www.rasmussenreports.com/public_content/lifestyle/sports/may_2013/52_say_big_time_college_athletics_corrupt_education_process (accessed July 2, 2013).

SOLUTIONS

1

  a. 244

  b. 15

  c. 50

3 $N\left(244, \frac{15}{\sqrt{50}}\right)$

5 As the sample size increases, there will be less variability in the mean, so the interval size decreases.

7 X is the time in minutes it takes to complete the U.S. Census short form. $\bar{X}$ is the mean time it took a sample of 200 people to complete the U.S. Census short form.

9 CI: (7.9441, 8.4559)

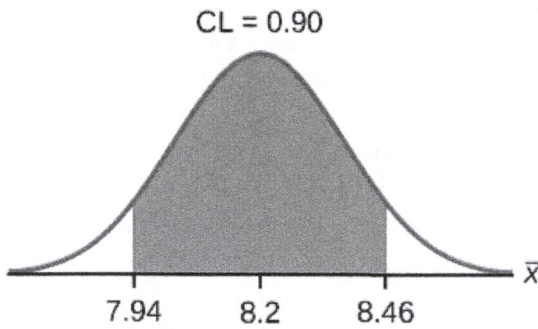

Figure 8.11

EBM = 0.26

11 The level of confidence would decrease because decreasing *n* makes the confidence interval wider, so at the same error bound, the confidence level decreases.

13

a. $\bar{x}$ = 2.2

b. $\sigma = 0.2$

c. $n = 20$

15 $\bar{X}$ is the mean weight of a sample of 20 heads of lettuce.

17 *EBM* = 0.07
CI: (2.1264, 2.2736)

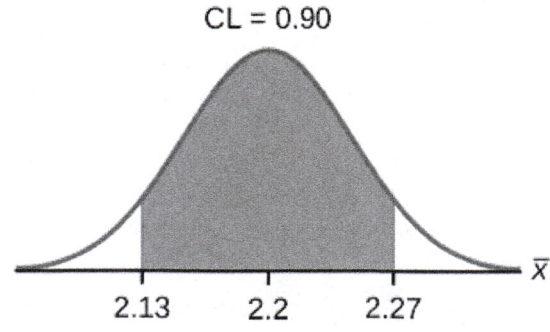

Figure 8.12

19 The interval is greater because the level of confidence increased. If the only change made in the analysis is a change in confidence level, then all we are doing is changing how much area is being calculated for the normal distribution. Therefore, a larger confidence level results in larger areas and larger intervals.

21 The confidence level would increase.

23 30.4

25 σ

27 μ

29 normal

31 0.025

33 (24.52,36.28)

35 We are 95% confident that the true mean age for Winger Foothill College students is between 24.52 and 36.28.

37 The error bound for the mean would decrease because as the CL decreases, you need less area under the normal curve (which translates into a smaller interval) to capture the true population mean.

39 X is the number of hours a patient waits in the emergency room before being called back to be examined. $\bar{X}$ is the mean wait time of 70 patients in the emergency room.

41 CI: (1.3808, 1.6192)

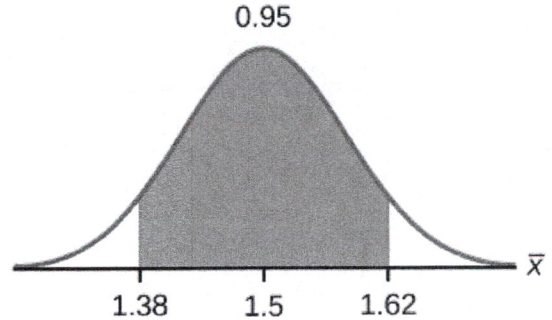

Figure 8.13

$EBM = 0.12$

43

  a. $\bar{x} = 151$

  b. $s_x = 32$

  c. $n = 108$

  d. $n - 1 = 107$

45 $\bar{X}$ is the mean number of hours spent watching television per month from a sample of 108 Americans.

47 CI: (142.92, 159.08)

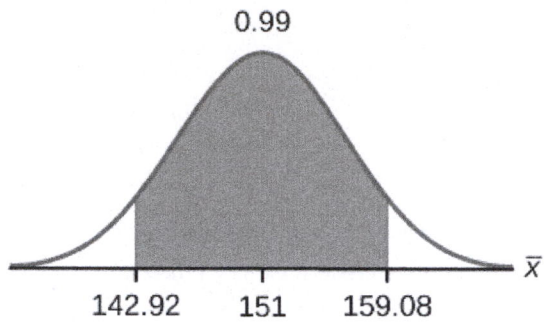

Figure 8.14

EBM = 8.08

49

  a.   3.26

  b.   1.02

  c.   39

51 μ

53 t_{38}

55 0.025

57 (2.93, 3.59)

59 We are 95% confident that the true mean number of colors for national flags is between 2.93 colors and 3.59 colors.

60 The error bound would become EBM = 0.245. This error bound decreases because as sample sizes increase, variability decreases and we need less interval length to capture the true mean.

63 It would decrease, because the z-score would decrease, which reducing the numerator and lowering the number.

65 X is the number of "successes" where the woman makes the majority of the purchasing decisions for the household. P' is the percentage of households sampled where the woman makes the majority of the purchasing decisions for the household.

67 CI: (0.5321, 0.6679)

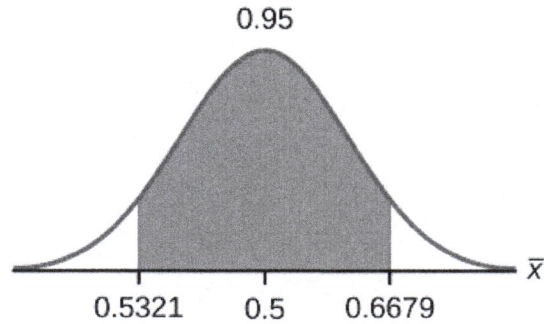

Figure 8.15

EBM: 0.0679

69 X is the number of "successes" where an executive prefers a truck. P' is the percentage of executives sampled who prefer a truck.

71 CI: (0.19432, 0.33068)

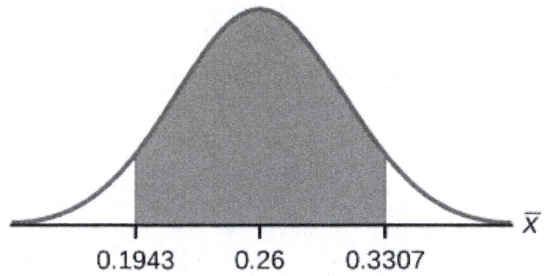

Figure 8.16

EBM: 0.0707

73 The sampling error means that the true mean can be 2% above or below the sample mean.

75 *P′* is the proportion of voters sampled who said the economy is the most important issue in the upcoming election.

77 CI: (0.62735, 0.67265) *EBM*: 0.02265

79 The number of girls, ages 8 to 12, in the 5 P.M. Monday night beginning ice-skating class.

81
   a. $x = 64$
   b. $n = 80$
   c. $p' = 0.8$

83 p

85 $P' \sim N\left(0.8, \sqrt{\frac{(0.8)(0.2)}{80}}\right)$. (0.72171, 0.87829).

87 0.04

89 (0.72; 0.88)

91 With 92% confidence, we estimate the proportion of girls, ages 8 to 12, in a beginning ice-skating class at the Ice Chalet to be between 72% and 88%.

93 The error bound would increase. Assuming all other variables are kept constant, as the confidence level increases, the area under the curve corresponding to the confidence level becomes larger, which creates a wider interval and thus a larger error.

95
   a.   i.   71
        ii. 3
        iii. 48
   b.   X is the height of a Swiss male, and is the mean height from a sample of 48 Swiss males.
   c.   Normal. We know the standard deviation for the population, and the sample size is greater than 30.
   d.   i.   CI: (70.151, 71.49)

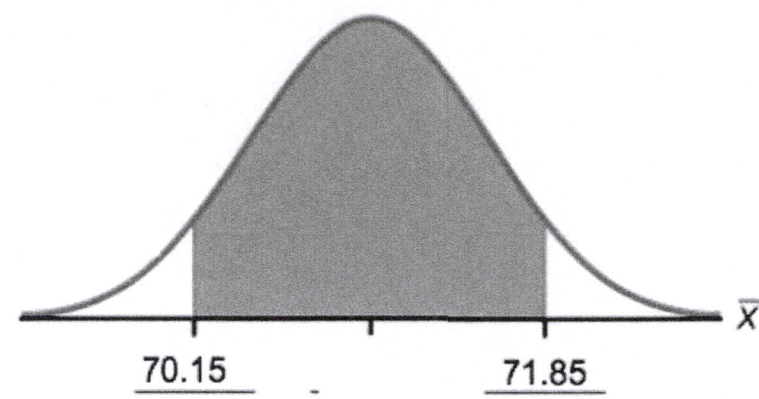

ii.

Figure 8.17

   iii. *EBM* = 0.849

e. The confidence interval will decrease in size, because the sample size increased. Recall, when all factors remain unchanged, an increase in sample size decreases variability. Thus, we do not need as large an interval to capture the true population mean.

97

a.   i. $\bar{x}$ = 23.6

    ii. σ = 7

    iii. n = 100

b. X is the time needed to complete an individual tax form. $\bar{X}$ is the mean time to complete tax forms from a sample of 100 customers.

c. $N\left(23.6, \dfrac{7}{\sqrt{100}}\right)$ because we know sigma.

d.   i. (22.228, 24.972)

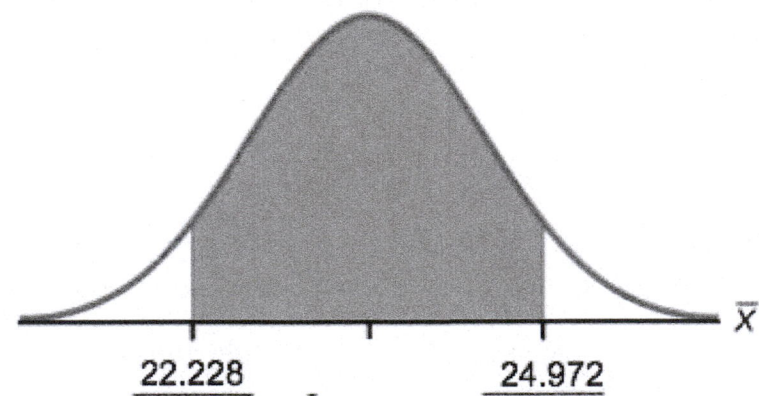

ii.

Figure 8.18

   iii. *EBM* = 1.372

e. It will need to change the sample size. The firm needs to determine what the confidence level should be, then apply the error bound formula to determine the necessary sample size.

f. The confidence level would increase as a result of a larger interval. Smaller sample sizes result in more variability. To capture the true population mean, we need to have a larger interval.

g. According to the error bound formula, the firm needs to survey 206 people. Since we increase the confidence level, we need to increase either our error bound or the sample size.

99

a.   i. 7.9

     ii. 2.5

     iii. 20

b. X is the number of letters a single camper will send home. $\bar{X}$ is the mean number of letters sent home from a sample of 20 campers.

c. $N\ 7.9\left(\dfrac{2.5}{\sqrt{20}}\right)$

d.   i. CI: (6.98, 8.82)

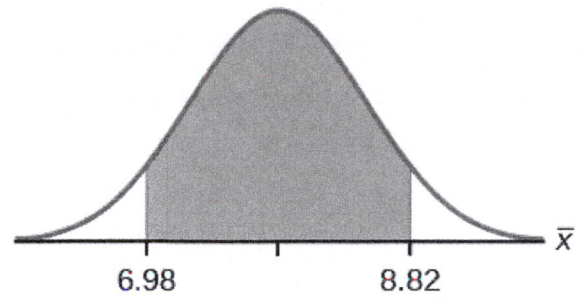

     ii.

Figure 8.19

     iii. *EBM*: 0.92

e. The error bound and confidence interval will decrease.

101

a. $\bar{x}$ = \$568,873

b. $CL = 0.95\ \alpha = 1 - 0.95 = 0.05\ z_{\frac{\alpha}{2}} = 1.96$

$$EBM = z_{0.025}\frac{\sigma}{\sqrt{n}} = 1.96\ \frac{909200}{\sqrt{40}} = \$281,764$$

c. $\bar{x} - EBM = 568,873 - 281,764 = 287,109$

     $\bar{x} + EBM = 568,873 + 281,764 = 850,637$

Alternate solution:

 Using the TI-83, 83+, 84, 84+ Calculator

1. Press **STAT** and arrow over to **TESTS**.

2. Arrow down to **7:ZInterval**.

3. Press **ENTER**.

4. Arrow to Stats and press **ENTER**.

5. Arrow down and enter the following values:

σ : 909,200

$\bar{x}$: 568,873

n: 40

CL: 0.95

6. Arrow down to Calculate and press ENTER.

7. The confidence interval is ($287,114, $850,632).

8. Notice the small difference between the two solutions–these differences are simply due to rounding error in the hand calculations.

d. We estimate with 95% confidence that the mean amount of contributions received from all individuals by House candidates is between $287,109 and $850,637.

103 Use the formula for *EBM*, solved for *n*:

$n = \dfrac{z^2\sigma^2}{EBM^2}$ From the statement of the problem, you know that σ = 2.5, and you need *EBM* = 1. $z = z_{0.035}$ = 1.812 (This is the value of *z* for which the area under the density curve to the **right** of *z* is 0.035.) $n = \dfrac{z^2\sigma^2}{EBM^2} = \dfrac{1.812^2 2.5^2}{1^2} \approx$ 20.52 You need to measure at least 21 male students to achieve your goal.

105

a.   i.   8629

ii. 6944

iii. 35

iv. 34

b. t_{34}

c.   i.   CI: (6244, 11,014)

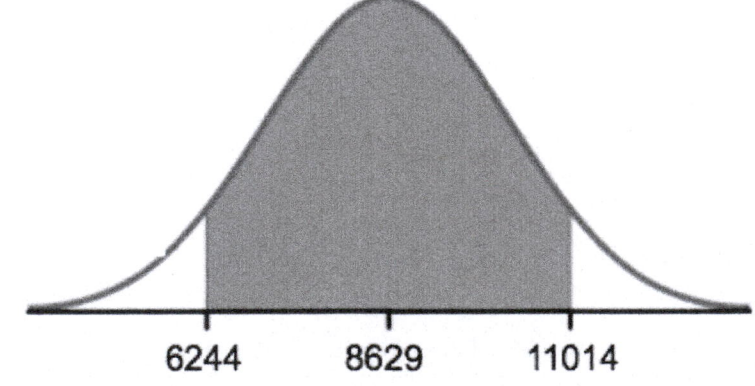

ii.

Figure 8.20

iii. EB = 2385

d. It will become smaller

107

  i. $\overline{x} = 2.51$

  ii. $s_x = 0.318$

  iii. $n = 9$

  iv. $n - 1 = 8$

b. the effective length of time for a tranquilizer

c. the mean effective length of time of tranquilizers from a sample of nine patients

d. We need to use a Student's-t distribution, because we do not know the population standard deviation.

e.   i. CI: (2.27, 2.76)

  ii. Check student's solution.

  iii. *EBM*: 0.25

f. If we were to sample many groups of nine patients, 95% of the samples would contain the true population mean length of time.

109 $\overline{x} = \$251, 854.23$ $s = \$521, 130.41$ Note that we are not given the population standard deviation, only the standard deviation of the sample. There are 30 measures in the sample, so $n = 30$, and $df = 30 - 1 = 29$ $CL = 0.96$, so $\alpha = 1 - CL = 1 - 0.96 = 0.04$ $\frac{\alpha}{2} = 0.02$ $t_{\frac{\alpha}{2}} = t_{0.02} = 2.150$ $EBM = t_{\frac{\alpha}{2}}\left(\frac{s}{\sqrt{n}}\right) = 2.150\left(\frac{521, 130.41}{\sqrt{30}}\right) \sim \$204, 561.66$ $\overline{x} -$

$EBM = \$251,854.23 - \$204,561.66 = \$47,292.57$ $\overline{x} + EBM = \$251,854.23 + \$204,561.66 = \$456,415.89$ We estimate with 96% confidence that the mean amount of money raised by all Leadership PACs during the 2011–2012 election cycle lies between $47,292.57 and $456,415.89.
Alternate Solution

Using the TI-83, 83+, 84, 84+ Calculator

Enter the data as a list.

Press STAT and arrow over to TESTS.

Arrow down to 8:TInterval.

Press ENTER.

Arrow to Data and press ENTER.

Arrow down and enter the name of the list where the data is stored.

Enter Freq: 1

Enter C-Level: 0.96

Arrow down to Calculate and press Enter.

The 96% confidence interval is ($47,262, $456,447).

The difference between solutions arises from rounding differences.

111

a.   i. $\overline{x} =$

  ii. $s_x =$

  iii. $n =$

  iv. $n - 1 =$

b. X is the number of unoccupied seats on a single flight. $\bar{X}$ is the mean number of unoccupied seats from a sample of 225 flights.

c. We will use a Student's-t distribution, because we do not know the population standard deviation.

d. i. CI: (11.12 , 12.08)

    ii. Check student's solution.

    iii. *EBM*: 0.48

113

a. i. CI: (7.64 , 9.36)

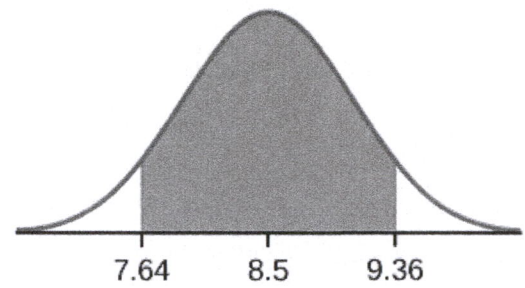

    ii.

Figure 8.21

    iii. *EBM*: 0.86

b. The sample should have been increased.

c. Answers will vary.

d. Answers will vary.

e. Answers will vary.

115 b

117

a. 1,068

b. The sample size would need to be increased since the critical value increases as the confidence level increases.

119

a. X = the number of people who feel that the president is doing an acceptable job;

    P' = the proportion of people in a sample who feel that the president is doing an acceptable job.

b. $N\left(0.61, \sqrt{\dfrac{(0.61)(0.39)}{1200}}\right)$

c. i. CI: (0.59, 0.63)

    ii. Check student's solution

    iii. *EBM*: 0.02

121

a. i. (0.72, 0.82)

    ii. (0.65, 0.76)

    iii. (0.60, 0.72)

b. Yes, the intervals (0.72, 0.82) and (0.65, 0.76) overlap, and the intervals (0.65, 0.76) and (0.60, 0.72) overlap.

Download for free at http://cnx.org/content/col11562/latest/

c. We can say that there does not appear to be a significant difference between the proportion of Asian adults who say that their families would welcome a white person into their families and the proportion of Asian adults who say that their families would welcome a Latino person into their families.

d. We can say that there is a significant difference between the proportion of Asian adults who say that their families would welcome a white person into their families and the proportion of Asian adults who say that their families would welcome a black person into their families.

123

a. X = the number of adult Americans who feel that crime is the main problem; P' = the proportion of adult Americans who feel that crime is the main problem

b. Since we are estimating a proportion, given $P' = 0.2$ and $n = 1000$, the distribution we should use is $N\left(0.2, \sqrt{\dfrac{(0.2)(0.8)}{1000}}\right)$.

c.   i. CI: (0.18, 0.22)

      ii. Check student's solution.

      iii. *EBM*: 0.02

d. One way to lower the sampling error is to increase the sample size.

e. The stated "± 3%" represents the maximum error bound. This means that those doing the study are reporting a maximum error of 3%. Thus, they estimate the percentage of adult Americans who feel that crime is the main problem to be between 18% and 22%.

125 c

127 d

129 a

131

a. $p' = \dfrac{(0.55 + 0.49)}{2} = 0.52$; $EBP = 0.55 - 0.52 = 0.03$

b. No, the confidence interval includes values less than or equal to 0.50. It is possible that less than half of the population believe this.

c. $CL = 0.75$, so $\alpha = 1 - 0.75 = 0.25$ and $\dfrac{\alpha}{2} = 0.125$ $z_{\frac{\alpha}{2}} = 1.150$. (The area to the right of this z is 0.125, so the area to the left is $1 - 0.125 = 0.875$.)

$EBP = (1.150)\sqrt{\dfrac{0.52(0.48)}{1,000}} \approx 0.018$

$(p' - EBP, p' + EBP) = (0.52 - 0.018, 0.52 + 0.018) = (0.502, 0.538)$
Alternate Solution

 Using the TI-83, 83+, 84, 84+ Calculator

STAT TESTS A: 1-PropZinterval with $x = (0.52)(1,000)$, $n = 1,000$, CL = 0.75.

Answer is (0.502, 0.538)

d. Yes – this interval does not fall less than 0.50 so we can conclude that at least half of all American adults believe that major sports programs corrupt education – but we do so with only 75% confidence.

133 $CL = 0.95$ $\alpha = 1 - 0.95 = 0.05$ $\frac{\alpha}{2} = 0.025$ $z_{\frac{\alpha}{2}} = 1.96$.   Use $p' = q' = 0.5$.

$$n = \frac{z_{\frac{\alpha}{2}}^2 p' q'}{EBP^2} = \frac{1.96^2 (0.5)(0.5)}{0.05^2} = 384.16$$ You need to interview at least 385 students to estimate the proportion to within 5% at 95% confidence.

9 | HYPOTHESIS TESTING WITH ONE SAMPLE

Figure 9.1 You can use a hypothesis test to decide if a dog breeder's claim that every Dalmatian has 35 spots is statistically sound. (Credit: Robert Neff)

Introduction

Chapter Objectives
By the end of this chapter, the student should be able to: • Differentiate between Type I and Type II Errors • Describe hypothesis testing in general and in practice • Conduct and interpret hypothesis tests for a single population mean, population standard deviation known. • Conduct and interpret hypothesis tests for a single population mean, population standard deviation unknown.

> • Conduct and interpret hypothesis tests for a single population proportion.

One job of a statistician is to make statistical inferences about populations based on samples taken from the population. **Confidence intervals** are one way to estimate a population parameter. Another way to make a statistical inference is to make a decision about a parameter. For instance, a car dealer advertises that its new small truck gets 35 miles per gallon, on average. A tutoring service claims that its method of tutoring helps 90% of its students get an A or a B. A company says that women managers in their company earn an average of $60,000 per year.

A statistician will make a decision about these claims. This process is called " **hypothesis testing**." A hypothesis test involves collecting data from a sample and evaluating the data. Then, the statistician makes a decision as to whether or not there is sufficient evidence, based upon analyses of the data, to reject the null hypothesis.

In this chapter, you will conduct hypothesis tests on single means and single proportions. You will also learn about the errors associated with these tests.

Hypothesis testing consists of two contradictory hypotheses or statements, a decision based on the data, and a conclusion. To perform a hypothesis test, a statistician will:

1. Set up two contradictory hypotheses.

2. Collect sample data (in homework problems, the data or summary statistics will be given to you).

3. Determine the correct distribution to perform the hypothesis test.

4. Analyze sample data by performing the calculations that ultimately will allow you to reject or decline to reject the null hypothesis.

5. Make a decision and write a meaningful conclusion.

NOTE

To do the hypothesis test homework problems for this chapter and later chapters, make copies of the appropriate special solution sheets. See **Appendix E**.

9.1 | Null and Alternative Hypotheses

The actual test begins by considering two **hypotheses**. They are called the **null hypothesis** and the **alternative hypothesis**. These hypotheses contain opposing viewpoints.

H_0: **The null hypothesis:** It is a statement about the population that either is believed to be true or is used to put forth an argument unless it can be shown to be incorrect beyond a reasonable doubt.

H_a: **The alternative hypothesis:** It is a claim about the population that is contradictory to H_0 and what we conclude when we reject H_0.

Since the null and alternative hypotheses are contradictory, you must examine evidence to decide if you have enough evidence to reject the null hypothesis or not. The evidence is in the form of sample data.

After you have determined which hypothesis the sample supports, you make a **decision.** There are two options for a decision. They are "reject H_0" if the sample information favors the alternative hypothesis or "do not reject H_0" or "decline to reject H_0" if the sample information is insufficient to reject the null hypothesis.

Mathematical Symbols Used in H_0 and H_a:

H_0	H_a
equal (=)	not equal (≠) **or** greater than (>) **or** less than (<)
greater than or equal to (≥)	less than (<)
less than or equal to (≤)	more than (>)

Table 9.1

NOTE

H_0 always has a symbol with an equal in it. H_a never has a symbol with an equal in it. The choice of symbol depends on the wording of the hypothesis test. However, be aware that many researchers (including one of the co-authors in research work) use = in the null hypothesis, even with > or < as the symbol in the alternative hypothesis. This practice is acceptable because we only make the decision to reject or not reject the null hypothesis.

Example 9.1

H_0: No more than 30% of the registered voters in Santa Clara County voted in the primary election. $p \leq 30$
H_a: More than 30% of the registered voters in Santa Clara County voted in the primary election. $p > 30$

 Try It Σ

9.1 A medical trial is conducted to test whether or not a new medicine reduces cholesterol by 25%. State the null and alternative hypotheses.

Example 9.2

We want to test whether the mean GPA of students in American colleges is different from 2.0 (out of 4.0). The null and alternative hypotheses are:
H_0: $\mu = 2.0$
H_a: $\mu \neq 2.0$

 Try It Σ

9.2 We want to test whether the mean height of eighth graders is 66 inches. State the null and alternative hypotheses. Fill in the correct symbol (=, ≠, ≥, <, ≤, >) for the null and alternative hypotheses.

  a. H_0: μ ___ 66

  b. H_a: μ ___ 66

Example 9.3

We want to test if college students take less than five years to graduate from college, on the average. The null and alternative hypotheses are:
H_0: $\mu \geq 5$
H_a: $\mu < 5$

Try It Σ

9.3 We want to test if it takes fewer than 45 minutes to teach a lesson plan. State the null and alternative hypotheses. Fill in the correct symbol (=, ≠, ≥, <, ≤, >) for the null and alternative hypotheses.

a. H_0: μ ___ 45

b. H_a: μ ___ 45

Example 9.4

In an issue of *U. S. News and World Report*, an article on school standards stated that about half of all students in France, Germany, and Israel take advanced placement exams and a third pass. The same article stated that 6.6% of U.S. students take advanced placement exams and 4.4% pass. Test if the percentage of U.S. students who take advanced placement exams is more than 6.6%. State the null and alternative hypotheses.

H_0: $p \leq 0.066$

H_a: $p > 0.066$

Try It Σ

9.4 On a state driver's test, about 40% pass the test on the first try. We want to test if more than 40% pass on the first try. Fill in the correct symbol ($=$, $\neq$, $\geq$, $<$, $\leq$, $>$) for the null and alternative hypotheses.

a. H_0: p ___ 0.40

b. H_a: p ___ 0.40

Collaborative Exercise

Bring to class a newspaper, some news magazines, and some Internet articles . In groups, find articles from which your group can write null and alternative hypotheses. Discuss your hypotheses with the rest of the class.

9.2 | Outcomes and the Type I and Type II Errors

When you perform a hypothesis test, there are four possible outcomes depending on the actual truth (or falseness) of the null hypothesis H_0 and the decision to reject or not. The outcomes are summarized in the following table:

ACTION	H_0 IS ACTUALLY	...
	True	False
Do not reject H_0	Correct Outcome	Type II error
Reject H_0	Type I Error	Correct Outcome

Table 9.2

The four possible outcomes in the table are:

1. The decision is **not to reject H_0** when **H_0 is true (correct decision).**

2. The decision is to **reject H_0** when **H_0 is true** (incorrect decision known as a **Type I error**).

3. The decision is **not to reject H_0** when, in fact, **H_0 is false** (incorrect decision known as a **Type II error**).

4. The decision is to **reject H_0** when **H_0 is false** (**correct decision** whose probability is called the **Power of the Test**).

Each of the errors occurs with a particular probability. The Greek letters α and β represent the probabilities.

α = probability of a Type I error = **P(Type I error)** = probability of rejecting the null hypothesis when the null hypothesis is true.

β = probability of a Type II error = **P(Type II error)** = probability of not rejecting the null hypothesis when the null hypothesis is false.

α and β should be as small as possible because they are probabilities of errors. They are rarely zero.

The Power of the Test is $1 - \beta$. Ideally, we want a high power that is as close to one as possible. Increasing the sample size can increase the Power of the Test.

The following are examples of Type I and Type II errors.

Example 9.5

Suppose the null hypothesis, H_0, is: Frank's rock climbing equipment is safe.

Type I error: Frank thinks that his rock climbing equipment may not be safe when, in fact, it really is safe. **Type II error**: Frank thinks that his rock climbing equipment may be safe when, in fact, it is not safe.

α = **probability** that Frank thinks his rock climbing equipment may not be safe when, in fact, it really is safe. β = **probability** that Frank thinks his rock climbing equipment may be safe when, in fact, it is not safe.

Notice that, in this case, the error with the greater consequence is the Type II error. (If Frank thinks his rock climbing equipment is safe, he will go ahead and use it.)

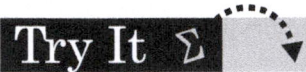 Try It Σ

9.5 Suppose the null hypothesis, H_0, is: the blood cultures contain no traces of pathogen X. State the Type I and Type II errors.

Example 9.6

Suppose the null hypothesis, H_0, is: The victim of an automobile accident is alive when he arrives at the emergency room of a hospital.

Type I error: The emergency crew thinks that the victim is dead when, in fact, the victim is alive. **Type II error**: The emergency crew does not know if the victim is alive when, in fact, the victim is dead.

α = **probability** that the emergency crew thinks the victim is dead when, in fact, he is really alive = P(Type I error). β = **probability** that the emergency crew does not know if the victim is alive when, in fact, the victim is dead = P(Type II error).

The error with the greater consequence is the Type I error. (If the emergency crew thinks the victim is dead, they will not treat him.)

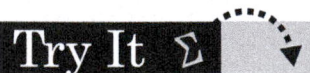 Try It Σ

9.6 Suppose the null hypothesis, H_0, is: a patient is not sick. Which type of error has the greater consequence, Type I or Type II?

Example 9.7

It's a Boy Genetic Labs claim to be able to increase the likelihood that a pregnancy will result in a boy being born. Statisticians want to test the claim. Suppose that the null hypothesis, H_0, is: It's a Boy Genetic Labs has no effect on gender outcome.

Type I error: This results when a true null hypothesis is rejected. In the context of this scenario, we would state that we believe that It's a Boy Genetic Labs influences the gender outcome, when in fact it has no effect. The probability of this error occurring is denoted by the Greek letter alpha, α.

Type II error: This results when we fail to reject a false null hypothesis. In context, we would state that It's a Boy Genetic Labs does not influence the gender outcome of a pregnancy when, in fact, it does. The probability of this error occurring is denoted by the Greek letter beta, β.

The error of greater consequence would be the Type I error since couples would use the It's a Boy Genetic Labs product in hopes of increasing the chances of having a boy.

Try It Σ

9.7 "Red tide" is a bloom of poison-producing algae–a few different species of a class of plankton called dinoflagellates. When the weather and water conditions cause these blooms, shellfish such as clams living in the area develop dangerous levels of a paralysis-inducing toxin. In Massachusetts, the Division of Marine Fisheries (DMF) monitors levels of the toxin in shellfish by regular sampling of shellfish along the coastline. If the mean level of toxin in clams exceeds 800 μg (micrograms) of toxin per kg of clam meat in any area, clam harvesting is banned there until the bloom is over and levels of toxin in clams subside. Describe both a Type I and a Type II error in this context, and state which error has the greater consequence.

Example 9.8

A certain experimental drug claims a cure rate of at least 75% for males with prostate cancer. Describe both the Type I and Type II errors in context. Which error is the more serious?

Type I: A cancer patient believes the cure rate for the drug is less than 75% when it actually is at least 75%.

Type II: A cancer patient believes the experimental drug has at least a 75% cure rate when it has a cure rate that is less than 75%.

In this scenario, the Type II error contains the more severe consequence. If a patient believes the drug works at least 75% of the time, this most likely will influence the patient's (and doctor's) choice about whether to use the drug as a treatment option.

Try It Σ

9.8 Determine both Type I and Type II errors for the following scenario:

Assume a null hypothesis, H_0, that states the percentage of adults with jobs is at least 88%.

Identify the Type I and Type II errors from these four statements.

a. Not to reject the null hypothesis that the percentage of adults who have jobs is at least 88% when that percentage is actually less than 88%

b. Not to reject the null hypothesis that the percentage of adults who have jobs is at least 88% when the percentage is actually at least 88%.

c. Reject the null hypothesis that the percentage of adults who have jobs is at least 88% when the percentage is actually at least 88%.

d. Reject the null hypothesis that the percentage of adults who have jobs is at least 88% when that percentage is actually less than 88%.

9.3 | Distribution Needed for Hypothesis Testing

Earlier in the course, we discussed sampling distributions. **Particular distributions are associated with hypothesis testing.** Perform tests of a population mean using a **normal distribution** or a **Student's *t*-distribution**. (Remember, use a Student's *t*-distribution when the population **standard deviation** is unknown and the distribution of the sample mean is approximately normal.) We perform tests of a population proportion using a normal distribution (usually *n* is large or the sample size is large).

If you are testing a **single population mean**, the distribution for the test is for **means**:

$$\bar{X} \sim N\left(\mu_X, \frac{\sigma_X}{\sqrt{n}}\right) \text{ or } t_{df}$$

The population parameter is μ. The estimated value (point estimate) for μ is $\bar{x}$, the sample mean.

If you are testing a **single population proportion**, the distribution for the test is for proportions or percentages:

$$P' \sim N\left(p, \sqrt{\frac{p \cdot q}{n}}\right)$$

The population parameter is p. The estimated value (point estimate) for p is p'. $p' = \frac{x}{n}$ where x is the number of successes and n is the sample size.

Assumptions

When you perform a **hypothesis test of a single population mean μ** using a **Student's t-distribution** (often called a t-test), there are fundamental assumptions that need to be met in order for the test to work properly. Your data should be a **simple random sample** that comes from a population that is approximately **normally distributed**. You use the sample **standard deviation** to approximate the population standard deviation. (Note that if the sample size is sufficiently large, a t-test will work even if the population is not approximately normally distributed).

When you perform a **hypothesis test of a single population mean μ** using a normal distribution (often called a z-test), you take a simple random sample from the population. The population you are testing is normally distributed or your sample size is sufficiently large. You know the value of the population standard deviation which, in reality, is rarely known.

When you perform a **hypothesis test of a single population proportion p**, you take a simple random sample from the population. You must meet the conditions for a **binomial distribution** which are: there are a certain number n of independent trials, the outcomes of any trial are success or failure, and each trial has the same probability of a success p. The shape of the binomial distribution needs to be similar to the shape of the normal distribution. To ensure this, the quantities np and nq must both be greater than five ($np > 5$ and $nq > 5$). Then the binomial distribution of a sample (estimated) proportion can be approximated by the normal distribution with $\mu = p$ and $\sigma = \sqrt{\frac{pq}{n}}$. Remember that $q = 1 - p$.

9.4 | Rare Events, the Sample, Decision and Conclusion

Establishing the type of distribution, sample size, and known or unknown standard deviation can help you figure out how to go about a hypothesis test. However, there are several other factors you should consider when working out a hypothesis test.

Rare Events

Suppose you make an assumption about a property of the population (this assumption is the **null hypothesis**). Then you gather sample data randomly. If the sample has properties that would be very **unlikely** to occur if the assumption is true, then you would conclude that your assumption about the population is probably incorrect. (Remember that your assumption is just an **assumption**—it is not a fact and it may or may not be true. But your sample data are real and the data are showing you a fact that seems to contradict your assumption.)

For example, Didi and Ali are at a birthday party of a very wealthy friend. They hurry to be first in line to grab a prize from a tall basket that they cannot see inside because they will be blindfolded. There are 200 plastic bubbles in the basket and Didi and Ali have been told that there is only one with a $100 bill. Didi is the first person to reach into the basket and pull out a bubble. Her bubble contains a $100 bill. The probability of this happening is $\frac{1}{200} = 0.005$. Because this is so unlikely, Ali is hoping that what the two of them were told is wrong and there are more $100 bills in the basket. A "rare event" has occurred (Didi getting the $100 bill) so Ali doubts the assumption about only one $100 bill being in the basket.

Using the Sample to Test the Null Hypothesis

Use the sample data to calculate the actual probability of getting the test result, called the **p-value**. The p-value is the **probability that, if the null hypothesis is true, the results from another randomly selected sample will be as extreme or more extreme as the results obtained from the given sample.**

A large p-value calculated from the data indicates that we should not reject the **null hypothesis**. The smaller the p-value, the more unlikely the outcome, and the stronger the evidence is against the null hypothesis. We would reject the null hypothesis if the evidence is strongly against it.

Draw a graph that shows the p-value. The hypothesis test is easier to perform if you use a graph because you see the problem more clearly.

Example 9.9

Suppose a baker claims that his bread height is more than 15 cm, on average. Several of his customers do not believe him. To persuade his customers that he is right, the baker decides to do a hypothesis test. He bakes 10 loaves of bread. The mean height of the sample loaves is 17 cm. The baker knows from baking hundreds of loaves of bread that the **standard deviation** for the height is 0.5 cm. and the distribution of heights is normal.

The null hypothesis could be H_0: $\mu \le 15$ The alternate hypothesis is H_a: $\mu > 15$

The words **"is more than"** translates as a ">" so "$\mu > 15$" goes into the alternate hypothesis. The null hypothesis must contradict the alternate hypothesis.

Since σ **is known** ($\sigma = 0.5$ cm.), the distribution for the population is known to be normal with mean $\mu = 15$ and standard deviation $\frac{\sigma}{\sqrt{n}} = \frac{0.5}{\sqrt{10}} = 0.16$.

Suppose the null hypothesis is true (the mean height of the loaves is no more than 15 cm). Then is the mean height (17 cm) calculated from the sample unexpectedly large? The hypothesis test works by asking the question how **unlikely** the sample mean would be if the null hypothesis were true. The graph shows how far out the sample mean is on the normal curve. The p-value is the probability that, if we were to take other samples, any other sample mean would fall at least as far out as 17 cm.

The p-value, then, is the probability that a sample mean is the same or greater than 17 cm. when the population mean is, in fact, 15 cm. We can calculate this probability using the normal distribution for means.

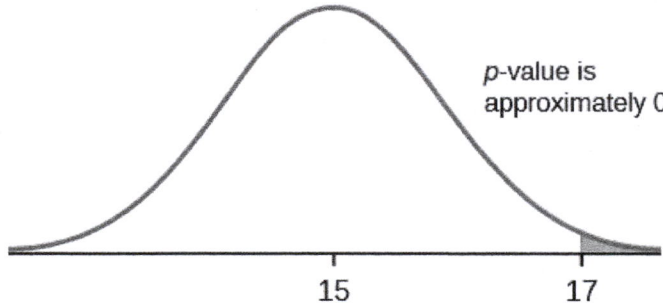

Figure 9.2

p-value= $P(\bar{x} > 17)$ which is approximately zero.

A p-value of approximately zero tells us that it is highly unlikely that a loaf of bread rises no more than 15 cm, on average. That is, almost 0% of all loaves of bread would be at least as high as 17 cm. **purely by CHANCE** had the population mean height really been 15 cm. Because the outcome of 17 cm. is so **unlikely (meaning it is happening NOT by chance alone)**, we conclude that the evidence is strongly against the null hypothesis (the mean height is at most 15 cm.). There is sufficient evidence that the true mean height for the population of the baker's loaves of bread is greater than 15 cm.

Try It Σ

9.9 A normal distribution has a standard deviation of 1. We want to verify a claim that the mean is greater than 12. A sample of 36 is taken with a sample mean of 12.5.

H_0: $\mu \le 12$
H_a: $\mu > 12$
The p-value is 0.0013
Draw a graph that shows the p-value.

Decision and Conclusion

A systematic way to make a decision of whether to reject or not reject the **null hypothesis** is to compare the p-value and a **preset or preconceived α (also called a "significance level")**. A preset α is the probability of a **Type I error** (rejecting the null hypothesis when the null hypothesis is true). It may or may not be given to you at the beginning of the problem.

When you make a **decision** to reject or not reject H_0, do as follows:

- If $\alpha > p$-value, reject H_0. The results of the sample data are significant. There is sufficient evidence to conclude that H_0 is an incorrect belief and that the **alternative hypothesis**, H_a, may be correct.

- If $\alpha \le p$-value, do not reject H_0. The results of the sample data are not significant. There is not sufficient evidence to conclude that the alternative hypothesis, H_a, may be correct.

- When you "do not reject H_0", it does not mean that you should believe that H_0 is true. It simply means that the sample data have **failed** to provide sufficient evidence to cast serious doubt about the truthfulness of H_o.

Conclusion: After you make your decision, write a thoughtful **conclusion** about the hypotheses in terms of the given problem.

Example 9.10

When using the p-value to evaluate a hypothesis test, it is sometimes useful to use the following memory device

If the p-value is low, the null must go.

If the p-value is high, the null must fly.

This memory aid relates a p-value less than the established alpha (the p is low) as rejecting the null hypothesis and, likewise, relates a p-value higher than the established alpha (the p is high) as not rejecting the null hypothesis.

Fill in the blanks.

Reject the null hypothesis when _____.

The results of the sample data _____.

Do not reject the null when hypothesis when _____.

The results of the sample data _____.

Solution 9.10

Reject the null hypothesis when **the p-value is less than the established alpha value**. The results of the sample data **support the alternative hypothesis**.

Do not reject the null hypothesis when **the p-value is greater than the established alpha value**. The results of the sample data **do not support the alternative hypothesis**.

Try It ∑

9.10 It's a Boy Genetics Labs claim their procedures improve the chances of a boy being born. The results for a test of a single population proportion are as follows:

H_0: $p = 0.50$, H_a: $p > 0.50$

$\alpha = 0.01$

p-value $= 0.025$

Interpret the results and state a conclusion in simple, non-technical terms.

9.5 | Additional Information and Full Hypothesis Test Examples

- In a **hypothesis test** problem, you may see words such as "the level of significance is 1%." The "1%" is the preconceived or preset α.
- The statistician setting up the hypothesis test selects the value of α to use **before** collecting the sample data.
- **If no level of significance is given, a common standard to use is $\alpha = 0.05$.**
- When you calculate the p-value and draw the picture, the p-value is the area in the left tail, the right tail, or split evenly between the two tails. For this reason, we call the hypothesis test left, right, or two tailed.
- The **alternative hypothesis**, H_a, tells you if the test is left, right, or two-tailed. It is the **key** to conducting the appropriate test.
- H_a **never** has a symbol that contains an equal sign.
- **Thinking about the meaning of the p-value**: A data analyst (and anyone else) should have more confidence that he made the correct decision to reject the null hypothesis with a smaller p-value (for example, 0.001 as opposed to 0.04) even if using the 0.05 level for alpha. Similarly, for a large p-value such as 0.4, as opposed to a p-value of 0.056 (alpha = 0.05 is less than either number), a data analyst should have more confidence that she made the correct decision in not rejecting the null hypothesis. This makes the data analyst use judgment rather than mindlessly applying rules.

The following examples illustrate a left-, right-, and two-tailed test.

Example 9.11

$H_0: \mu = 5$, $H_a: \mu < 5$

Test of a single population mean. H_a tells you the test is left-tailed. The picture of the p-value is as follows:

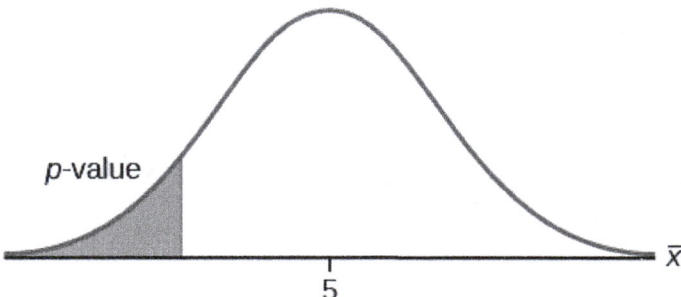

Figure 9.3

Try It

9.11 $H_0: \mu = 10$, $H_a: \mu < 10$

Assume the p-value is 0.0935. What type of test is this? Draw the picture of the p-value.

Example 9.12

$H_0: p \le 0.2$ $H_a: p > 0.2$

This is a test of a single population proportion. H_a tells you the test is **right-tailed**. The picture of the p-value is as follows:

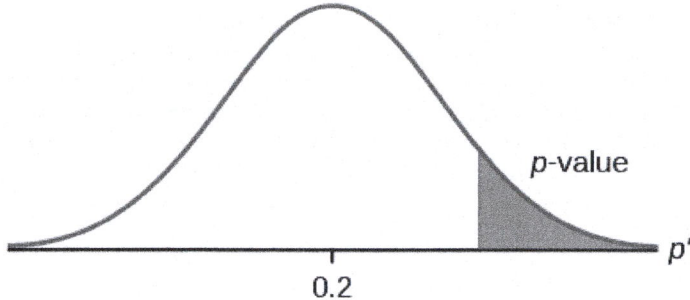

Figure 9.4

Try It Σ

9.12 H_0: $\mu \le 1$, H_a: $\mu > 1$

Assume the p-value is 0.1243. What type of test is this? Draw the picture of the p-value.

Example 9.13

H_0: $p = 50$ H_a: $p \ne 50$

This is a test of a single population mean. H_a tells you the test is **two-tailed**. The picture of the p-value is as follows.

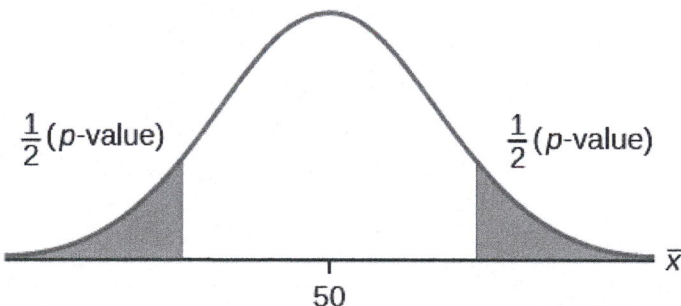

Figure 9.5

Try It Σ

9.13 H_0: $p = 0.5$, H_a: $p \ne 0.5$

Assume the p-value is 0.2564. What type of test is this? Draw the picture of the p-value.

Full Hypothesis Test Examples

Example 9.14

Jeffrey, as an eight-year old, **established a mean time of 16.43 seconds** for swimming the 25-yard freestyle, with a **standard deviation of 0.8 seconds**. His dad, Frank, thought that Jeffrey could swim the 25-yard freestyle faster using goggles. Frank bought Jeffrey a new pair of expensive goggles and timed Jeffrey for **15 25-yard freestyle swims**. For the 15 swims, **Jeffrey's mean time was 16 seconds. Frank thought that the goggles helped Jeffrey to swim faster than the 16.43 seconds.** Conduct a hypothesis test using a preset $\alpha = 0.05$. Assume that the swim times for the 25-yard freestyle are normal.

Solution 9.14

Set up the Hypothesis Test:

Since the problem is about a mean, this is a **test of a single population mean**.

$H_0: \mu = 16.43$ $H_a: \mu < 16.43$

For Jeffrey to swim faster, his time will be less than 16.43 seconds. The "<" tells you this is left-tailed.

Determine the distribution needed:

Random variable: $\bar{X}$ = the mean time to swim the 25-yard freestyle.

Distribution for the test: $\bar{X}$ is normal (population **standard deviation** is known: $\sigma = 0.8$)

$\bar{X} \sim N\left(\mu, \frac{\sigma_X}{\sqrt{n}}\right)$ Therefore, $\bar{X} \sim N\left(16.43, \frac{0.8}{\sqrt{15}}\right)$

$\mu = 16.43$ comes from H_0 and not the data. $\sigma = 0.8$, and $n = 15$.

Calculate the p-value using the normal distribution for a mean:

p-value = $P(\bar{x} < 16) = 0.0187$ where the sample mean in the problem is given as 16.

p-value = 0.0187 (This is called the **actual level of significance**.) The p-value is the area to the left of the sample mean is given as 16.

Graph:

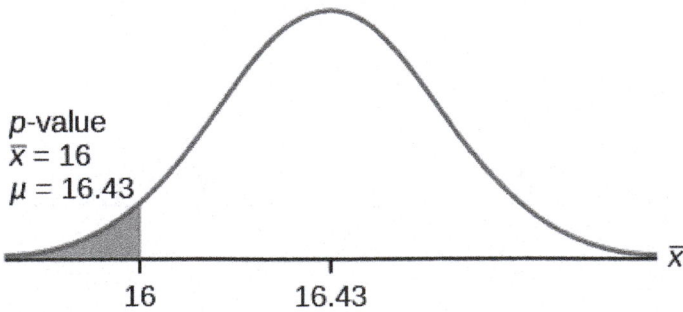

Figure 9.6

$\mu = 16.43$ comes from H_0. Our assumption is $\mu = 16.43$.

Interpretation of the p-value: If H_0 is true, there is a 0.0187 probability (1.87%)that Jeffrey's mean time to swim the 25-yard freestyle is 16 seconds or less. Because a 1.87% chance is small, the mean time of 16 seconds or less is unlikely to have happened randomly. It is a rare event.

Compare α and the p-value:

$\alpha = 0.05$ p-value $= 0.0187$ $\alpha > p$-value

Make a decision: Since $\alpha > p$-value, reject H_0.

This means that you reject $\mu = 16.43$. In other words, you do not think Jeffrey swims the 25-yard freestyle in 16.43 seconds but faster with the new goggles.

Conclusion: At the 5% significance level, we conclude that Jeffrey swims faster using the new goggles. The sample data show there is sufficient evidence that Jeffrey's mean time to swim the 25-yard freestyle is less than 16.43 seconds.

The p-value can easily be calculated.

Using the TI-83, 83+, 84, 84+ Calculator

Press **STAT** and arrow over to **TESTS**. Press **1:Z-Test**. Arrow over to **Stats** and press **ENTER**. Arrow down and enter 16.43 for μ_0 (null hypothesis), .8 for σ, 16 for the sample mean, and 15 for n. Arrow down to μ : (alternate hypothesis) and arrow over to $< \mu_0$. Press **ENTER**. Arrow down to **Calculate** and press **ENTER**. The calculator not only calculates the p-value ($p = 0.0187$) but it also calculates the test statistic (z-score) for the sample mean. $\mu < 16.43$ is the alternative hypothesis. Do this set of instructions again except arrow to **Draw**(instead of **Calculate**). Press **ENTER**. A shaded graph appears with $z = -2.08$ (test statistic) and $p = 0.0187$ (p-value). Make sure when you use **Draw** that no other equations are highlighted in $Y =$ and the plots are turned off.

When the calculator does a Z-Test, the **Z-Test** function finds the p-value by doing a normal probability calculation using the **central limit theorem**:

$$P(\overline{x} < 16) = \text{2nd DISTR normcdf}\left(-10^{\wedge}99, 16, 16.43, 0.8 / \sqrt{15}\right).$$

The Type I and Type II errors for this problem are as follows:

The Type I error is to conclude that Jeffrey swims the 25-yard freestyle, on average, in less than 16.43 seconds when, in fact, he actually swims the 25-yard freestyle, on average, in 16.43 seconds. (Reject the null hypothesis when the null hypothesis is true.)

The Type II error is that there is not evidence to conclude that Jeffrey swims the 25-yard free-style, on average, in less than 16.43 seconds when, in fact, he actually does swim the 25-yard free-style, on average, in less than 16.43 seconds. (Do not reject the null hypothesis when the null hypothesis is false.)

Try It Σ

9.14 The mean throwing distance of a football for a Marco, a high school freshman quarterback, is 40 yards, with a standard deviation of two yards. The team coach tells Marco to adjust his grip to get more distance. The coach records the distances for 20 throws. For the 20 throws, Marco's mean distance was 45 yards. The coach thought the different grip helped Marco throw farther than 40 yards. Conduct a hypothesis test using a preset $\alpha = 0.05$. Assume the throw distances for footballs are normal.

First, determine what type of test this is, set up the hypothesis test, find the p-value, sketch the graph, and state your conclusion.

Using the TI-83, 83+, 84, 84+ Calculator

Press STAT and arrow over to TESTS. Press 1:Z-Test. Arrow over to Stats and press ENTER. Arrow down and enter 40 for $\mu 0$ (null hypothesis), 2 for σ, 45 for the sample mean, and 20 for n. Arrow down to μ: (alternative hypothesis) and set it either as <, ≠, or >. Press ENTER. Arrow down to Calculate and press ENTER. The calculator not only calculates the p-value but it also calculates the test statistic (z-score) for the sample mean. Select <, ≠, or > for the alternative hypothesis. Do this set of instructions again except arrow to Draw (instead of Calculate). Press ENTER. A shaded graph appears with test statistic and p-value. Make sure when you use Draw that no other equations are highlighted in $Y =$ and the plots are turned off.

HISTORICAL NOTE (EXAMPLE 9.11)

The traditional way to compare the two probabilities, α and the p-value, is to compare the critical value (z-score from α) to the test statistic (z-score from data). The calculated test statistic for the p-value is –2.08. (From the Central Limit Theorem, the test statistic formula is $z = \dfrac{\bar{x} - \mu_X}{\left(\frac{\sigma_X}{\sqrt{n}}\right)}$. For this problem, $\bar{x}$ = 16, μ_X = 16.43 from the null hypothes is, σ_X = 0.8, and n = 15.) You can find the critical value for α = 0.05 in the normal table (see **15.Tables** in the Table of Contents). The z-score for an area to the left equal to 0.05 is midway between –1.65 and –1.64 (0.05 is midway between 0.0505 and 0.0495). The z-score is –1.645. Since –1.645 > –2.08 (which demonstrates that α > p-value), reject H_0. Traditionally, the decision to reject or not reject was done in this way. Today, comparing the two probabilities α and the p-value is very common. For this problem, the p-value, 0.0187 is considerably smaller than α, 0.05. You can be confident about your decision to reject. The graph shows α, the p-value, and the test statistics and the critical value.

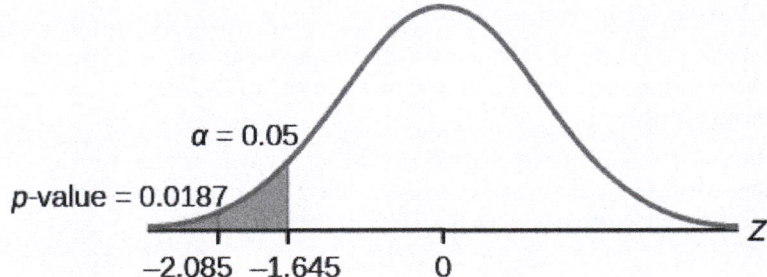

Figure 9.7

Example 9.15

A college football coach thought that his players could bench press a **mean weight of 275 pounds**. It is known that the **standard deviation is 55 pounds**. Three of his players thought that the mean weight was **more than** that amount. They asked **30** of their teammates for their estimated maximum lift on the bench press exercise. The data ranged from 205 pounds to 385 pounds. The actual different weights were (frequencies are in parentheses) 205(3); 215(3); 225(1); 241(2); 252(2); 265(2); 275(2); 313(2); 316(5); 338(2); 341(1); 345(2); 368(2); 385(1).

Conduct a hypothesis test using a 2.5% level of significance to determine if the bench press mean is **more than 275 pounds**.

Solution 9.15

Set up the Hypothesis Test:

Since the problem is about a mean weight, this is a **test of a single population mean**.

H_0: $\mu = 275$
H_a: $\mu > 275$
This is a right-tailed test.

Calculating the distribution needed:

Random variable: $\bar{X}$ = the mean weight, in pounds, lifted by the football players.

Distribution for the test: It is normal because σ is known.

$$\bar{X} \sim N\left(275, \frac{55}{\sqrt{30}}\right)$$

$\bar{x} = 286.2$ pounds (from the data).

$\sigma = 55$ pounds **(Always use σ if you know it.)** We assume $\mu = 275$ pounds unless our data shows us otherwise.

Calculate the p-value using the normal distribution for a mean and using the sample mean as input (see **Appendix G** for using the data as input):

p-value $= P(\bar{x} > 286.2) = 0.1323$.

Interpretation of the p-value: If H_0 is true, then there is a 0.1331 probability (13.23%) that the football players can lift a mean weight of 286.2 pounds or more. Because a 13.23% chance is large enough, a mean weight lift of 286.2 pounds or more is not a rare event.

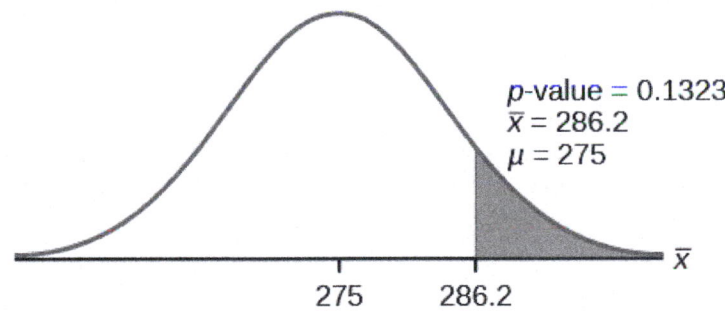

Figure 9.8

Compare α and the p-value:

$\alpha = 0.025$ p-value $= 0.1323$

Make a decision: Since $\alpha <$ p-value, do not reject H_0.

Conclusion: At the 2.5% level of significance, from the sample data, there is not sufficient evidence to conclude that the true mean weight lifted is more than 275 pounds.

The p-value can easily be calculated.

 Using the TI-83, 83+, 84, 84+ Calculator

Put the data and frequencies into lists. Press **STAT** and arrow over to **TESTS**. Press **1:Z-Test**. Arrow over to **Data** and press **ENTER**. Arrow down and enter 275 for μ_0, 55 for σ, the name of the list where you put the data, and the name of the list where you put the frequencies. Arrow down to μ: and arrow over to $> \mu_0$. Press ENTER. Arrow down to `Calculate` and press ENTER. The calculator not only calculates the p-value ($p = 0.1331$, a little different from the previous calculation - in it we used the sample mean rounded to one decimal place instead of the data) but it also calculates the test statistic (z-score) for the sample mean, the sample mean, and the sample standard deviation. $\mu > 275$ is the alternative hypothesis. Do this set of instructions again except arrow to `Draw` (instead of `Calculate`). Press ENTER. A shaded graph appears with $z = 1.112$ (test statistic) and $p = 0.1331$ (p-value). Make sure when you use `Draw` that no other equations are highlighted in $Y =$ and the plots are turned off.

Example 9.16

Statistics students believe that the mean score on the first statistics test is 65. A statistics instructor thinks the mean score is higher than 65. He samples ten statistics students and obtains the scores 65; 65; 70; 67; 66; 63; 63; 68; 72; 71. He performs a hypothesis test using a 5% level of significance. The data are assumed to be from a normal distribution.

Solution 9.16

Set up the hypothesis test:

A 5% level of significance means that $\alpha = 0.05$. This is a test of a **single population mean**.

$H_0: \mu = 65$ $H_a: \mu > 65$

Since the instructor thinks the average score is higher, use a ">". The ">" means the test is right-tailed.

Determine the distribution needed:

Random variable: $\bar{X}$ = average score on the first statistics test.

Distribution for the test: If you read the problem carefully, you will notice that there is **no population standard deviation given**. You are only given $n = 10$ sample data values. Notice also that the data come from a normal distribution. This means that the distribution for the test is a student's t.

Use t_{df}. Therefore, the distribution for the test is t_9 where $n = 10$ and $df = 10 - 1 = 9$.

Calculate the p-value using the Student's t-distribution:

p-value = $P(\bar{x} > 67) = 0.0396$ where the sample mean and sample standard deviation are calculated as 67 and 3.1972 from the data.

Interpretation of the p-value: If the null hypothesis is true, then there is a 0.0396 probability (3.96%) that the sample mean is 65 or more.

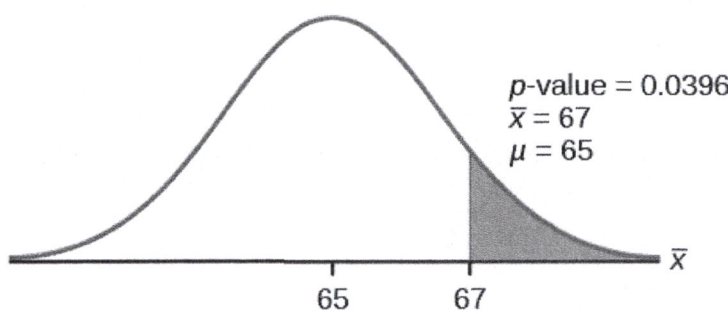

p-value = 0.0396
$\bar{x}$ = 67
μ = 65

65   67

Figure 9.9

Compare α and the p-value:

Since α = 0.05 and p-value = 0.0396. α > p-value.

Make a decision: Since α > p-value, reject H_0.

This means you reject μ = 65. In other words, you believe the average test score is more than 65.

Conclusion: At a 5% level of significance, the sample data show sufficient evidence that the mean (average) test score is more than 65, just as the math instructor thinks.

The p-value can easily be calculated.

 Using the TI-83, 83+, 84, 84+ Calculator

Put the data into a list. Press STAT and arrow over to TESTS. Press 2:T-Test. Arrow over to Data and press ENTER. Arrow down and enter 65 for μ_0, the name of the list where you put the data, and 1 for Freq:. Arrow down to μ: and arrow over to > μ_0. Press ENTER. Arrow down to Calculate and press ENTER. The calculator not only calculates the p-value (p = 0.0396) but it also calculates the test statistic (t-score) for the sample mean, the sample mean, and the sample standard deviation. μ > 65 is the alternative hypothesis. Do this set of instructions again except arrow to Draw (instead of Calculate). Press ENTER. A shaded graph appears with t = 1.9781 (test statistic) and p = 0.0396 (p-value). Make sure when you use Draw that no other equations are highlighted in Y = and the plots are turned off.

Try It Σ

9.16 It is believed that a stock price for a particular company will grow at a rate of $5 per week with a standard deviation of $1. An investor believes the stock won't grow as quickly. The changes in stock price is recorded for ten weeks and are as follows: $4, $3, $2, $3, $1, $7, $2, $1, $1, $2. Perform a hypothesis test using a 5% level of significance. State the null and alternative hypotheses, find the p-value, state your conclusion, and identify the Type I and Type II errors.

Example 9.17

Joon believes that 50% of first-time brides in the United States are younger than their grooms. She performs a hypothesis test to determine if the percentage is **the same or different from 50%**. Joon samples **100 first-time**

brides and **53** reply that they are younger than their grooms. For the hypothesis test, she uses a 1% level of significance.

Solution 9.17

Set up the hypothesis test:

The 1% level of significance means that $\alpha = 0.01$. This is a **test of a single population proportion**.

H_0: $p = 0.50$ H_a: $p \neq 0.50$

The words **"is the same or different from"** tell you this is a two-tailed test.

Calculate the distribution needed:

Random variable: P' = the percent of of first-time brides who are younger than their grooms.

Distribution for the test: The problem contains no mention of a mean. The information is given in terms of percentages. Use the distribution for P', the estimated proportion.

$$P' \sim N\left(p, \sqrt{\frac{p \cdot q}{n}}\right) \text{ Therefore, } P' \sim N\left(0.5, \sqrt{\frac{0.5 \cdot 0.5}{100}}\right)$$

where $p = 0.50$, $q = 1-p = 0.50$, and $n = 100$

Calculate the p-value using the normal distribution for proportions:

p-value = P ($p' < 0.47$ or $p' > 0.53$) = 0.5485

where $x = 53$, $p' = \frac{x}{n} = \frac{53}{100} = 0.53$.

Interpretation of the p-value: If the null hypothesis is true, there is 0.5485 probability (54.85%) that the sample (estimated) proportion p' is 0.53 or more OR 0.47 or less (see the graph in **Figure 9.9**).

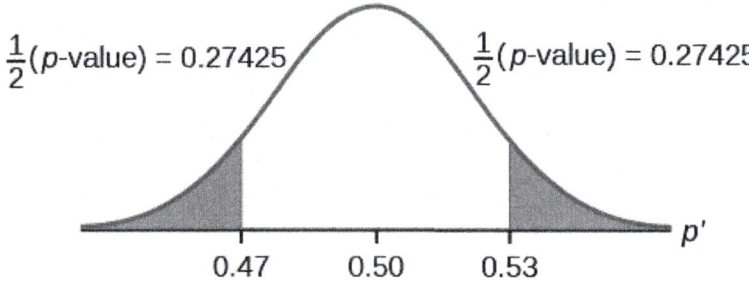

Figure 9.10

$\mu = p = 0.50$ comes from H_0, the null hypothesis.

$p' = 0.53$. Since the curve is symmetrical and the test is two-tailed, the p' for the left tail is equal to $0.50 - 0.03 = 0.47$ where $\mu = p = 0.50$. (0.03 is the difference between 0.53 and 0.50.)

Compare α and the p-value:

Since $\alpha = 0.01$ and p-value = 0.5485. $\alpha < p$-value.

Make a decision: Since $\alpha < p$-value, you cannot reject H_0.

Conclusion: At the 1% level of significance, the sample data do not show sufficient evidence that the percentage of first-time brides who are younger than their grooms is different from 50%.

The p-value can easily be calculated.

Using the TI-83, 83+, 84, 84+ Calculator

Press **STAT** and arrow over to **TESTS**. Press **5:1-PropZTest**. Enter .5 for p_0, 53 for x and 100 for n. Arrow down to **Prop** and arrow to **not equals** p_0. Press **ENTER**. Arrow down to **Calculate** and press **ENTER**. The calculator calculates the p-value ($p = 0.5485$) and the test statistic (z-score). **Prop not equals** .5 is the alternate hypothesis. Do this set of instructions again except arrow to **Draw** (instead of **Calculate**). Press **ENTER**. A shaded graph appears with $z = 0.6$ (test statistic) and $p = 0.5485$ (p-value). Make sure when you use **Draw** that no other equations are highlighted in $Y =$ and the plots are turned off.

The Type I and Type II errors are as follows:

The Type I error is to conclude that the proportion of first-time brides who are younger than their grooms is different from 50% when, in fact, the proportion is actually 50%. (Reject the null hypothesis when the null hypothesis is true).

The Type II error is there is not enough evidence to conclude that the proportion of first time brides who are younger than their grooms differs from 50% when, in fact, the proportion does differ from 50%. (Do not reject the null hypothesis when the null hypothesis is false.)

Try It Σ

9.17 A teacher believes that 85% of students in the class will want to go on a field trip to the local zoo. She performs a hypothesis test to determine if the percentage is the same or different from 85%. The teacher samples 50 students and 39 reply that they would want to go to the zoo. For the hypothesis test, use a 1% level of significance.

First, determine what type of test this is, set up the hypothesis test, find the p-value, sketch the graph, and state your conclusion.

Example 9.18

Suppose a consumer group suspects that the proportion of households that have three cell phones is 30%. A cell phone company has reason to believe that the proportion is not 30%. Before they start a big advertising campaign, they conduct a hypothesis test. Their marketing people survey 150 households with the result that 43 of the households have three cell phones.

Solution 9.18

Set up the Hypothesis Test:

H_0: $p = 0.30$ H_a: $p \neq 0.30$

Determine the distribution needed:

The **random variable** is $P' =$ proportion of households that have three cell phones.

The **distribution** for the hypothesis test is $P' \sim N\left(0.30, \sqrt{\dfrac{(0.30) \cdot (0.70)}{150}}\right)$

a. The value that helps determine the p-value is p'. Calculate p'.

Solution 9.18

a. $p' = \frac{x}{n}$ where x is the number of successes and n is the total number in the sample.

$x = 43, n = 150$

$p' = \frac{43}{150}$

b. What is a **success** for this problem?

Solution 9.18
b. A success is having three cell phones in a household.

c. What is the level of significance?

Solution 9.18
c. The level of significance is the preset α. Since α is not given, assume that $\alpha = 0.05$.

d. Draw the graph for this problem. Draw the horizontal axis. Label and shade appropriately. Calculate the p-value.

Solution 9.18
d. p-value = 0.7216

e. Make a decision. _____(Reject/Do not reject) H_0 because_____.

Solution 9.18
e. Assuming that $\alpha = 0.05$, $\alpha < p$-value. The decision is do not reject H_0 because there is not sufficient evidence to conclude that the proportion of households that have three cell phones is not 30%.

Try It Σ

9.18 Marketers believe that 92% of adults in the United States own a cell phone. A cell phone manufacturer believes that number is actually lower. 200 American adults are surveyed, of which, 174 report having cell phones. Use a 5% level of significance. State the null and alternative hypothesis, find the p-value, state your conclusion, and identify the Type I and Type II errors.

The next example is a poem written by a statistics student named Nicole Hart. The solution to the problem follows the poem. Notice that the hypothesis test is for a single population proportion. This means that the null and alternate hypotheses use the parameter p. The distribution for the test is normal. The estimated proportion p' is the proportion of fleas killed to the total fleas found on Fido. This is sample information. The problem gives a preconceived $\alpha = 0.01$, for comparison, and a 95% confidence interval computation. The poem is clever and humorous, so please enjoy it!

Example 9.19

My dog has so many fleas,
They do not come off with ease.
As for shampoo, I have tried many types
Even one called Bubble Hype,

Which only killed 25% of the fleas,
Unfortunately I was not pleased.

I've used all kinds of soap,
Until I had given up hope
Until one day I saw
An ad that put me in awe.

A shampoo used for dogs
Called GOOD ENOUGH to Clean a Hog
Guaranteed to kill more fleas.

I gave Fido a bath
And after doing the math
His number of fleas
Started dropping by 3's!

Before his shampoo
I counted 42.
At the end of his bath,
I redid the math
And the new shampoo had killed 17 fleas.
So now I was pleased.

Now it is time for you to have some fun
With the level of significance being .01,
You must help me figure out
Use the new shampoo or go without?

Solution 9.19

Set up the hypothesis test:

H_0: $p \leq 0.25$ H_a: $p > 0.25$

Determine the distribution needed:

In words, CLEARLY state what your random variable $\overline{X}$ or P' represents.

P' = The proportion of fleas that are killed by the new shampoo

State the distribution to use for the test.

Normal: $N\left(0.25, \sqrt{\dfrac{(0.25)(1 - 0.25)}{42}}\right)$

Test Statistic: $z = 2.3163$

Calculate the p-value using the normal distribution for proportions:

p-value = 0.0103

In one to two complete sentences, explain what the p-value means for this problem.

If the null hypothesis is true (the proportion is 0.25), then there is a 0.0103 probability that the sample (estimated) proportion is 0.4048 $\left(\dfrac{17}{42}\right)$ or more.

Use the previous information to sketch a picture of this situation. CLEARLY, label and scale the horizontal axis and shade the region(s) corresponding to the p-value.

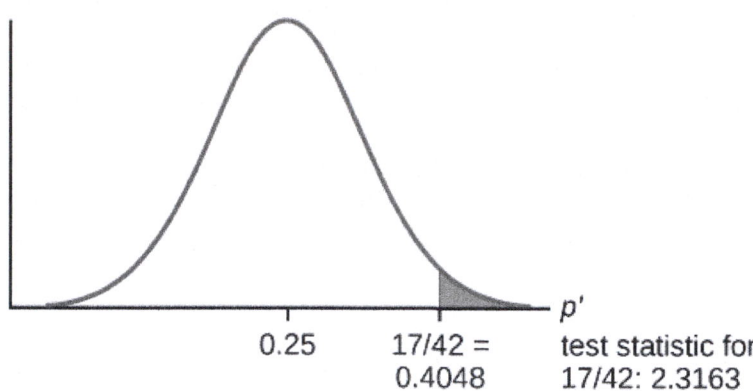

0.25   17/42 =   test statistic for
          0.4048   17/42: 2.3163

Figure 9.11

Compare α and the p-value:

Indicate the correct decision ("reject" or "do not reject" the null hypothesis), the reason for it, and write an appropriate conclusion, using complete sentences.

alpha	decision	reason for decision
0.01	Do not reject H_0	$\alpha < p$-value

Table 9.3

Conclusion: At the 1% level of significance, the sample data do not show sufficient evidence that the percentage of fleas that are killed by the new shampoo is more than 25%.

Construct a 95% confidence interval for the true mean or proportion. Include a sketch of the graph of the situation. Label the point estimate and the lower and upper bounds of the confidence interval.

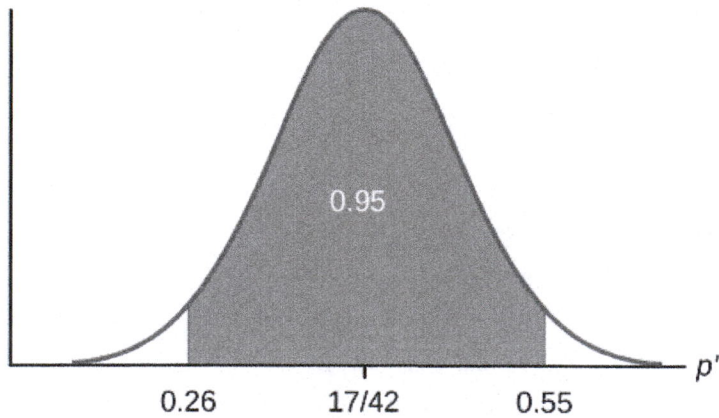

0.26   17/42   0.55

Figure 9.12

Confidence Interval: (0.26,0.55) We are 95% confident that the true population proportion p of fleas that are killed by the new shampoo is between 26% and 55%.

NOTE

This test result is not very definitive since the *p*-value is very close to alpha. In reality, one would probably do more tests by giving the dog another bath after the fleas have had a chance to return.

Example 9.20

The National Institute of Standards and Technology provides exact data on conductivity properties of materials. Following are conductivity measurements for 11 randomly selected pieces of a particular type of glass.

1.11; 1.07; 1.11; 1.07; 1.12; 1.08; .98; .98 1.02; .95; .95
Is there convincing evidence that the average conductivity of this type of glass is greater than one? Use a significance level of 0.05. Assume the population is normal.

Solution 9.20
Let's follow a four-step process to answer this statistical question.

1. **State the Question**: We need to determine if, at a 0.05 significance level, the average conductivity of the selected glass is greater than one. Our hypotheses will be

   a. $H_0: \mu \le 1$

   b. $H_a: \mu > 1$

2. **Plan**: We are testing a sample mean without a known population standard deviation. Therefore, we need to use a Student's-t distribution. Assume the underlying population is normal.

3. **Do the calculations**: We will input the sample data into the TI-83 as follows.

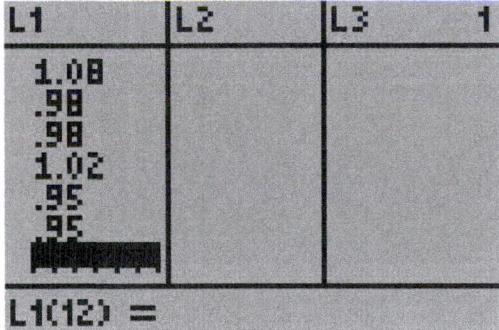

Figure 9.13

Figure 9.14

Figure 9.15

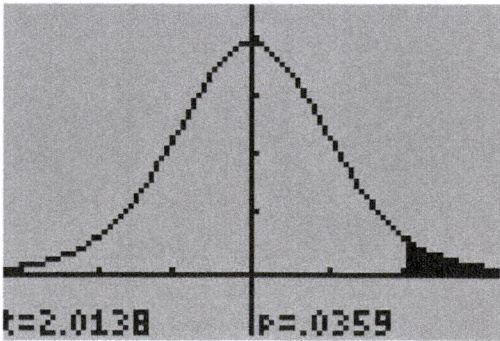

Figure 9.16

4. **State the Conclusions**: Since the *p*-value* (*p* = 0.036) is less than our alpha value, we will reject the null hypothesis. It is reasonable to state that the data supports the claim that the average conductivity level is greater than one.

Example 9.21

In a study of 420,019 cell phone users, 172 of the subjects developed brain cancer. Test the claim that cell phone users developed brain cancer at a greater rate than that for non-cell phone users (the rate of brain cancer for non-cell phone users is 0.0340%). Since this is a critical issue, use a 0.005 significance level. Explain why the significance level should be so low in terms of a Type I error.

Solution 9.21

We will follow the four-step process.

1. We need to conduct a hypothesis test on the claimed cancer rate. Our hypotheses will be
   a. H_0: $p \leq 0.00034$

   b. H_a: $p > 0.00034$

   If we commit a Type I error, we are essentially accepting a false claim. Since the claim describes cancer-causing environments, we want to minimize the chances of incorrectly identifying causes of cancer.

2. We will be testing a sample proportion with $x = 172$ and $n = 420,019$. The sample is sufficiently large because we have $np = 420,019(0.00034) = 142.8$, $nq = 420,019(0.99966) = 419,876.2$, two independent outcomes, and a fixed probability of success $p = 0.00034$. Thus we will be able to generalize our results to the population.

3. The associated TI results are

Figure 9.17

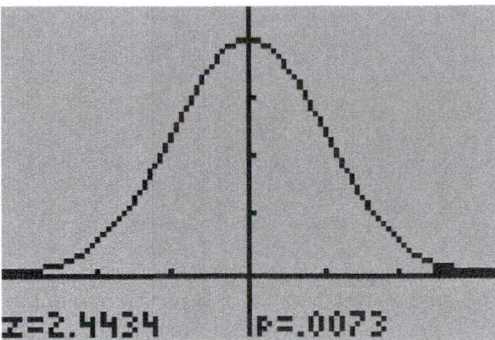

Figure 9.18

4. Since the *p*-value = 0.0073 is greater than our alpha value = 0.005, we cannot reject the null. Therefore, we conclude that there is not enough evidence to support the claim of higher brain cancer rates for the cell phone users.

Example 9.22

According to the US Census there are approximately 268,608,618 residents aged 12 and older. Statistics from the Rape, Abuse, and Incest National Network indicate that, on average, 207,754 rapes occur each year (male and female) for persons aged 12 and older. This translates into a percentage of sexual assaults of 0.078%. In Daviess

County, KY, there were reported 11 rapes for a population of 37,937. Conduct an appropriate hypothesis test to determine if there is a statistically significant difference between the local sexual assault percentage and the national sexual assault percentage. Use a significance level of 0.01.

Solution 9.22

We will follow the four-step plan.

1. We need to test whether the proportion of sexual assaults in Daviess County, KY is significantly different from the national average.

2. Since we are presented with proportions, we will use a one-proportion z-test. The hypotheses for the test will be

    a. H_0: $p = 0.00078$

    b. H_a: $p \neq 0.00078$

3. The following screen shots display the summary statistics from the hypothesis test.

Figure 9.19

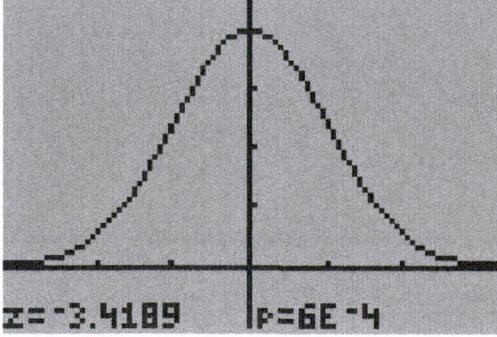

Figure 9.20

4. Since the p-value, $p = 0.00063$, is less than the alpha level of 0.01, the sample data indicates that we should reject the null hypothesis. In conclusion, the sample data support the claim that the proportion of sexual assaults in Daviess County, Kentucky is different from the national average proportion.

9.6 | Hypothesis Testing of a Single Mean and Single Proportion

Stats Lab

9.1 Hypothesis Testing of a Single Mean and Single Proportion

Class Time:

Names:

Student Learning Outcomes

- The student will select the appropriate distributions to use in each case.
- The student will conduct hypothesis tests and interpret the results.

Television Survey

In a recent survey, it was stated that Americans watch television on average four hours per day. Assume that $\sigma = 2$. Using your class as the sample, conduct a hypothesis test to determine if the average for students at your school is lower.

1. H_0: _____

2. H_a: _____

3. In words, define the random variable. _____ = _____

4. The distribution to use for the test is _____.

5. Determine the test statistic using your data.

6. Draw a graph and label it appropriately. Shade the actual level of significance.

   a. Graph:

Figure 9.21

   b. Determine the p-value.

7. Do you or do you not reject the null hypothesis? Why?

8. Write a clear conclusion using a complete sentence.

Language Survey

About 42.3% of Californians and 19.6% of all Americans over age five speak a language other than English at home. Using your class as the sample, conduct a hypothesis test to determine if the percent of the students at your school who speak a language other than English at home is different from 42.3%.

1. H_0: _____
2. H_a: _____
3. In words, define the random variable. _____ = _____
4. The distribution to use for the test is _____
5. Determine the test statistic using your data.
6. Draw a graph and label it appropriately. Shade the actual level of significance.

    a. Graph:

Figure 9.22

    b. Determine the *p*-value.
7. Do you or do you not reject the null hypothesis? Why?
8. Write a clear conclusion using a complete sentence.

Jeans Survey

Suppose that young adults own an average of three pairs of jeans. Survey eight people from your class to determine if the average is higher than three. Assume the population is normal.

1. H_0: _____
2. H_a: _____
3. In words, define the random variable. _____ = _____
4. The distribution to use for the test is _____.
5. Determine the test statistic using your data.
6. Draw a graph and label it appropriately. Shade the actual level of significance.

    a. Graph:

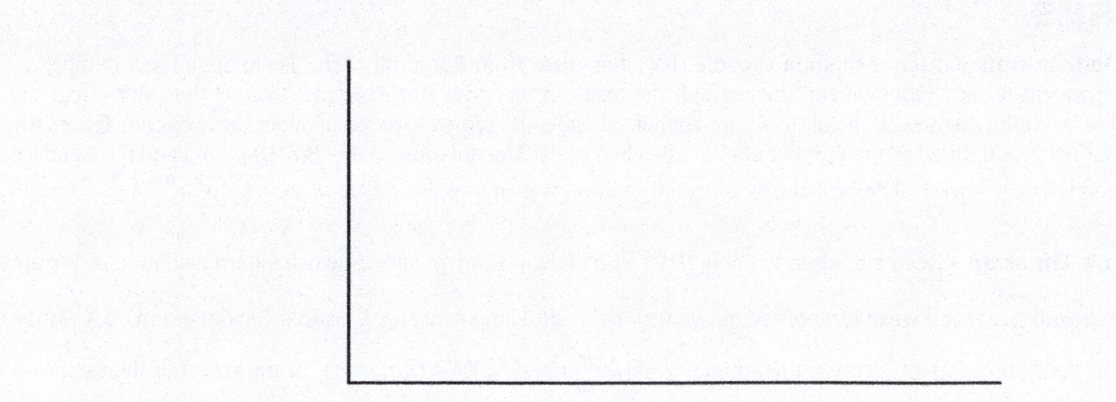

Figure 9.23

    b. Determine the *p*-value.

7. Do you or do you not reject the null hypothesis? Why?

8. Write a clear conclusion using a complete sentence.

KEY TERMS

Binomial Distribution a discrete random variable (RV) that arises from Bernoulli trials. There are a fixed number, n, of independent trials. "Independent" means that the result of any trial (for example, trial 1) does not affect the results of the following trials, and all trials are conducted under the same conditions. Under these circumstances the binomial RV X is defined as the number of successes in n trials. The notation is: $X \sim B(n, p)$ $\mu = np$ and the standard deviation is $\sigma = \sqrt{npq}$. The probability of exactly x successes in n trials is $P(X = x) = \binom{n}{x}p^x q^{n-x}$.

Central Limit Theorem Given a random variable (RV) with known mean μ and known standard deviation σ. We are sampling with size n and we are interested in two new RVs - the sample mean, $\overline{X}$, and the sample sum, ΣX. If the size n of the sample is sufficiently large, then $\overline{X} \sim N\left(\mu, \frac{\sigma}{\sqrt{n}}\right)$ and $\Sigma X \sim N(n\mu, \sqrt{n}\sigma)$. If the size n of the sample is sufficiently large, then the distribution of the sample means and the distribution of the sample sums will approximate a normal distribution regardless of the shape of the population. The mean of the sample means will equal the population mean and the mean of the sample sums will equal n times the population mean. The standard deviation of the distribution of the sample means, $\frac{\sigma}{\sqrt{n}}$, is called the standard error of the mean.

Confidence Interval (CI) an interval estimate for an unknown population parameter. This depends on:

- The desired confidence level.
- Information that is known about the distribution (for example, known standard deviation).
- The sample and its size.

Hypothesis a statement about the value of a population parameter, in case of two hypotheses, the statement assumed to be true is called the null hypothesis (notation H_0) and the contradictory statement is called the alternative hypothesis (notation H_a).

Hypothesis Testing Based on sample evidence, a procedure for determining whether the hypothesis stated is a reasonable statement and should not be rejected, or is unreasonable and should be rejected.

Level of Significance of the Test probability of a Type I error (reject the null hypothesis when it is true). Notation: α. In hypothesis testing, the Level of Significance is called the preconceived α or the preset α.

Normal Distribution a continuous random variable (RV) with pdf $f(x) = \frac{1}{\sigma\sqrt{2\pi}}e^{\frac{-(x-\mu)^2}{2\sigma^2}}$, where μ is the mean of the distribution, and σ is the standard deviation, notation: $X \sim N(\mu, \sigma)$. If $\mu = 0$ and $\sigma = 1$, the RV is called **the standard normal distribution**.

p-value the probability that an event will happen purely by chance assuming the null hypothesis is true. The smaller the p-value, the stronger the evidence is against the null hypothesis.

Standard Deviation a number that is equal to the square root of the variance and measures how far data values are from their mean; notation: s for sample standard deviation and σ for population standard deviation.

Student's t-Distribution investigated and reported by William S. Gossett in 1908 and published under the pseudonym Student. The major characteristics of the random variable (RV) are:

- It is continuous and assumes any real values.
- The pdf is symmetrical about its mean of zero. However, it is more spread out and flatter at the apex than the normal distribution.
- It approaches the standard normal distribution as n gets larger.
- There is a "family" of t distributions: every representative of the family is completely defined by the number of degrees of freedom which is one less than the number of data items.

Type 1 Error The decision is to reject the null hypothesis when, in fact, the null hypothesis is true.

Type 2 Error The decision is not to reject the null hypothesis when, in fact, the null hypothesis is false.

CHAPTER REVIEW

9.1 Null and Alternative Hypotheses

In a **hypothesis test**, sample data is evaluated in order to arrive at a decision about some type of claim. If certain conditions about the sample are satisfied, then the claim can be evaluated for a population. In a hypothesis test, we:

1. Evaluate the **null hypothesis**, typically denoted with H_0. The null is not rejected unless the hypothesis test shows otherwise. The null statement must always contain some form of equality ($=$, $\leq$ or $\geq$)

2. Always write the **alternative hypothesis**, typically denoted with H_a or H_1, using less than, greater than, or not equals symbols, i.e., ($\neq$, $>$, or $<$).

3. If we reject the null hypothesis, then we can assume there is enough evidence to support the alternative hypothesis.

4. Never state that a claim is proven true or false. Keep in mind the underlying fact that hypothesis testing is based on probability laws; therefore, we can talk only in terms of non-absolute certainties.

9.2 Outcomes and the Type I and Type II Errors

In every hypothesis test, the outcomes are dependent on a correct interpretation of the data. Incorrect calculations or misunderstood summary statistics can yield errors that affect the results. A **Type I** error occurs when a true null hypothesis is rejected. A **Type II error** occurs when a false null hypothesis is not rejected.

The probabilities of these errors are denoted by the Greek letters α and β, for a Type I and a Type II error respectively. The power of the test, $1 - \beta$, quantifies the likelihood that a test will yield the correct result of a true alternative hypothesis being accepted. A high power is desirable.

9.3 Distribution Needed for Hypothesis Testing

In order for a hypothesis test's results to be generalized to a population, certain requirements must be satisfied.

When testing for a single population mean:

1. A Student's t-test should be used if the data come from a simple, random sample and the population is approximately normally distributed, or the sample size is large, with an unknown standard deviation.

2. The normal test will work if the data come from a simple, random sample and the population is approximately normally distributed, or the sample size is large, with a known standard deviation.

When testing a single population proportion use a normal test for a single population proportion if the data comes from a simple, random sample, fill the requirements for a binomial distribution, and the mean number of success and the mean number of failures satisfy the conditions: $np > 5$ and $nq > n$ where n is the sample size, p is the probability of a success, and q is the probability of a failure.

9.4 Rare Events, the Sample, Decision and Conclusion

When the probability of an event occurring is low, and it happens, it is called a rare event. Rare events are important to consider in hypothesis testing because they can inform your willingness not to reject or to reject a null hypothesis. To test a null hypothesis, find the p-value for the sample data and graph the results. When deciding whether or not to reject the null the hypothesis, keep these two parameters in mind:

1. $\alpha > p$-value, reject the null hypothesis

2. $\alpha \leq p$-value, do not reject the null hypothesis

9.5 Additional Information and Full Hypothesis Test Examples

The **hypothesis test** itself has an established process. This can be summarized as follows:

1. Determine H_0 and H_a. Remember, they are contradictory.

2. Determine the random variable.

3. Determine the distribution for the test.

4. Draw a graph, calculate the test statistic, and use the test statistic to calculate the p-value. (A z-score and a t-score are examples of test statistics.)

5. Compare the preconceived α with the p-value, make a decision (reject or do not reject H_0), and write a clear conclusion using English sentences.

Notice that in performing the hypothesis test, you use α and not β. β is needed to help determine the sample size of the data that is used in calculating the p-value. Remember that the quantity $1 - \beta$ is called the **Power of the Test**. A high power is desirable. If the power is too low, statisticians typically increase the sample size while keeping α the same.If the power is low, the null hypothesis might not be rejected when it should be.

FORMULA REVIEW

9.1 Null and Alternative Hypotheses

H_0 and H_a are contradictory.

If H_0 has:	equal (=)	greater than or equal to ($\geq$)	less than or equal to ($\leq$)
then H_a has:	not equal ($\neq$) **or** greater than (>) **or** less than (<)	less than (<)	greater than (>)

Table 9.4

If $\alpha \leq p$-value, then do not reject H_0.

If $\alpha > p$-value, then reject H_0.

α is preconceived. Its value is set before the hypothesis test starts. The p-value is calculated from the data.

9.2 Outcomes and the Type I and Type II Errors

α = probability of a Type I error = P(Type I error) = probability of rejecting the null hypothesis when the null hypothesis is true.

β = probability of a Type II error = P(Type II error) = probability of not rejecting the null hypothesis when the null hypothesis is false.

9.3 Distribution Needed for Hypothesis Testing

If there is no given preconceived α, then use $\alpha = 0.05$.

Types of Hypothesis Tests

- Single population mean, **known** population variance (or standard deviation): **Normal test**.

- Single population mean, **unknown** population variance (or standard deviation): **Student's t-test**.

- Single population proportion: **Normal test**.

- For a **single population mean**, we may use a normal distribution with the following mean and standard deviation. Means: $\mu = \mu_{\bar{x}}$ and $\sigma_{\bar{x}} = \frac{\sigma_x}{\sqrt{n}}$

- A **single population proportion**, we may use a normal distribution with the following mean and standard deviation. Proportions: $\mu = p$ and $\sigma = \sqrt{\frac{pq}{n}}$.

PRACTICE

9.1 Null and Alternative Hypotheses

1. You are testing that the mean speed of your cable Internet connection is more than three Megabits per second. What is the random variable? Describe in words.

2. You are testing that the mean speed of your cable Internet connection is more than three Megabits per second. State the null and alternative hypotheses.

3. The American family has an average of two children. What is the random variable? Describe in words.

4. The mean entry level salary of an employee at a company is $58,000. You believe it is higher for IT professionals in the company. State the null and alternative hypotheses.

5. A sociologist claims the probability that a person picked at random in Times Square in New York City is visiting the area is 0.83. You want to test to see if the proportion is actually less. What is the random variable? Describe in words.

6. A sociologist claims the probability that a person picked at random in Times Square in New York City is visiting the area is 0.83. You want to test to see if the claim is correct. State the null and alternative hypotheses.

7. In a population of fish, approximately 42% are female. A test is conducted to see if, in fact, the proportion is less. State the null and alternative hypotheses.

8. Suppose that a recent article stated that the mean time spent in jail by a first–time convicted burglar is 2.5 years. A study was then done to see if the mean time has increased in the new century. A random sample of 26 first-time convicted burglars in a recent year was picked. The mean length of time in jail from the survey was 3 years with a standard deviation of 1.8 years. Suppose that it is somehow known that the population standard deviation is 1.5. If you were conducting a hypothesis test to determine if the mean length of jail time has increased, what would the null and alternative hypotheses be? The distribution of the population is normal.

    a. H_0: _____
    b. H_a: _____

9. A random survey of 75 death row inmates revealed that the mean length of time on death row is 17.4 years with a standard deviation of 6.3 years. If you were conducting a hypothesis test to determine if the population mean time on death row could likely be 15 years, what would the null and alternative hypotheses be?

    a. H_0: _____
    b. H_a: _____

10. The National Institute of Mental Health published an article stating that in any one-year period, approximately 9.5 percent of American adults suffer from depression or a depressive illness. Suppose that in a survey of 100 people in a certain town, seven of them suffered from depression or a depressive illness. If you were conducting a hypothesis test to determine if the true proportion of people in that town suffering from depression or a depressive illness is lower than the percent in the general adult American population, what would the null and alternative hypotheses be?

    a. H_0: _____
    b. H_a: _____

9.2 Outcomes and the Type I and Type II Errors

11. The mean price of mid-sized cars in a region is $32,000. A test is conducted to see if the claim is true. State the Type I and Type II errors in complete sentences.

12. A sleeping bag is tested to withstand temperatures of –15 °F. You think the bag cannot stand temperatures that low. State the Type I and Type II errors in complete sentences.

13. For **Exercise 9.12**, what are α and β in words?

14. In words, describe $1 - \beta$ For **Exercise 9.12**.

15. A group of doctors is deciding whether or not to perform an operation. Suppose the null hypothesis, H_0, is: the surgical procedure will go well. State the Type I and Type II errors in complete sentences.

16. A group of doctors is deciding whether or not to perform an operation. Suppose the null hypothesis, H_0, is: the surgical procedure will go well. Which is the error with the greater consequence?

17. The power of a test is 0.981. What is the probability of a Type II error?

18. A group of divers is exploring an old sunken ship. Suppose the null hypothesis, H_0, is: the sunken ship does not contain buried treasure. State the Type I and Type II errors in complete sentences.

19. A microbiologist is testing a water sample for E-coli. Suppose the null hypothesis, H_0, is: the sample does not contain E-coli. The probability that the sample does not contain E-coli, but the microbiologist thinks it does is 0.012. The probability that the sample does contain E-coli, but the microbiologist thinks it does not is 0.002. What is the power of this test?

20. A microbiologist is testing a water sample for E-coli. Suppose the null hypothesis, H_0, is: the sample contains E-coli. Which is the error with the greater consequence?

9.3 Distribution Needed for Hypothesis Testing

21. Which two distributions can you use for hypothesis testing for this chapter?

22. Which distribution do you use when you are testing a population mean and the standard deviation is known? Assume sample size is large.

23. Which distribution do you use when the standard deviation is not known and you are testing one population mean? Assume sample size is large.

24. A population mean is 13. The sample mean is 12.8, and the sample standard deviation is two. The sample size is 20. What distribution should you use to perform a hypothesis test? Assume the underlying population is normal.

25. A population has a mean is 25 and a standard deviation of five. The sample mean is 24, and the sample size is 108. What distribution should you use to perform a hypothesis test?

26. It is thought that 42% of respondents in a taste test would prefer Brand *A*. In a particular test of 100 people, 39% preferred Brand *A*. What distribution should you use to perform a hypothesis test?

27. You are performing a hypothesis test of a single population mean using a Student's *t*-distribution. What must you assume about the distribution of the data?

28. You are performing a hypothesis test of a single population mean using a Student's *t*-distribution. The data are not from a simple random sample. Can you accurately perform the hypothesis test?

29. You are performing a hypothesis test of a single population proportion. What must be true about the quantities of *np* and *nq*?

30. You are performing a hypothesis test of a single population proportion. You find out that *np* is less than five. What must you do to be able to perform a valid hypothesis test?

31. You are performing a hypothesis test of a single population proportion. The data come from which distribution?

9.4 Rare Events, the Sample, Decision and Conclusion

32. When do you reject the null hypothesis?

33. The probability of winning the grand prize at a particular carnival game is 0.005. Is the outcome of winning very likely or very unlikely?

34. The probability of winning the grand prize at a particular carnival game is 0.005. Michele wins the grand prize. Is this considered a rare or common event? Why?

35. It is believed that the mean height of high school students who play basketball on the school team is 73 inches with a standard deviation of 1.8 inches. A random sample of 40 players is chosen. The sample mean was 71 inches, and the sample standard deviation was 1.5 years. Do the data support the claim that the mean height is less than 73 inches? The *p*-value is almost zero. State the null and alternative hypotheses and interpret the *p*-value.

36. The mean age of graduate students at a University is at most 31 years with a standard deviation of two years. A random sample of 15 graduate students is taken. The sample mean is 32 years and the sample standard deviation is three years. Are the data significant at the 1% level? The *p*-value is 0.0264. State the null and alternative hypotheses and interpret the *p*-value.

37. Does the shaded region represent a low or a high *p*-value compared to a level of significance of 1%?

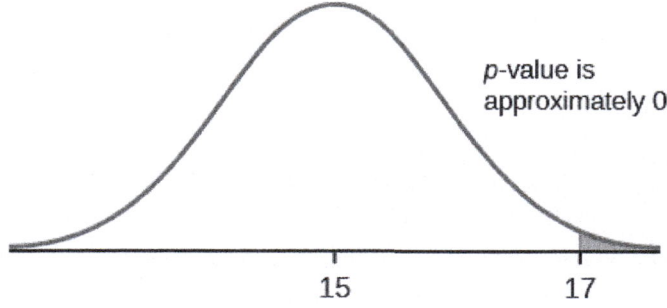

Figure 9.24

38. What should you do when $\alpha > p$-value?

39. What should you do if $\alpha = p$-value?

40. If you do not reject the null hypothesis, then it must be true. Is this statement correct? State why or why not in complete sentences.

Use the following information to answer the next seven exercises: Suppose that a recent article stated that the mean time spent in jail by a first-time convicted burglar is 2.5 years. A study was then done to see if the mean time has increased in the new century. A random sample of 26 first-time convicted burglars in a recent year was picked. The mean length of time in jail from the survey was three years with a standard deviation of 1.8 years. Suppose that it is somehow known that the population standard deviation is 1.5. Conduct a hypothesis test to determine if the mean length of jail time has increased. Assume the distribution of the jail times is approximately normal.

41. Is this a test of means or proportions?

42. What symbol represents the random variable for this test?

43. In words, define the random variable for this test.

44. Is the population standard deviation known and, if so, what is it?

45. Calculate the following:
   a. $\bar{x}$ _____
   b. σ _____
   c. s_x _____
   d. n _____

46. Since both σ and s_x are given, which should be used? In one to two complete sentences, explain why.

47. State the distribution to use for the hypothesis test.

48. A random survey of 75 death row inmates revealed that the mean length of time on death row is 17.4 years with a standard deviation of 6.3 years. Conduct a hypothesis test to determine if the population mean time on death row could likely be 15 years.
   a. Is this a test of one mean or proportion?
   b. State the null and alternative hypotheses.
       H_0: _____ H_a : _____
   c. Is this a right-tailed, left-tailed, or two-tailed test?
   d. What symbol represents the random variable for this test?
   e. In words, define the random variable for this test.
   f. Is the population standard deviation known and, if so, what is it?
   g. Calculate the following:
       i. $\bar{x}$ = _____
       ii. s = _____
       iii. n = _____
   h. Which test should be used?
   i. State the distribution to use for the hypothesis test.
   j. Find the *p*-value.
   k. At a pre-conceived $\alpha = 0.05$, what is your:
       i.   Decision:
       ii. Reason for the decision:
       iii. Conclusion (write out in a complete sentence):

9.5 Additional Information and Full Hypothesis Test Examples

49. Assume H_0: $\mu = 9$ and H_a: $\mu < 9$. Is this a left-tailed, right-tailed, or two-tailed test?

50. Assume H_0: $\mu \leq 6$ and H_a: $\mu > 6$. Is this a left-tailed, right-tailed, or two-tailed test?

51. Assume H_0: $p = 0.25$ and H_a: $p \neq 0.25$. Is this a left-tailed, right-tailed, or two-tailed test?

52. Draw the general graph of a left-tailed test.

53. Draw the graph of a two-tailed test.

54. A bottle of water is labeled as containing 16 fluid ounces of water. You believe it is less than that. What type of test would you use?

55. Your friend claims that his mean golf score is 63. You want to show that it is higher than that. What type of test would you use?

56. A bathroom scale claims to be able to identify correctly any weight within a pound. You think that it cannot be that accurate. What type of test would you use?

57. You flip a coin and record whether it shows heads or tails. You know the probability of getting heads is 50%, but you think it is less for this particular coin. What type of test would you use?

58. If the alternative hypothesis has a not equals ($\neq$) symbol, you know to use which type of test?

59. Assume the null hypothesis states that the mean is at least 18. Is this a left-tailed, right-tailed, or two-tailed test?

60. Assume the null hypothesis states that the mean is at most 12. Is this a left-tailed, right-tailed, or two-tailed test?

61. Assume the null hypothesis states that the mean is equal to 88. The alternative hypothesis states that the mean is not equal to 88. Is this a left-tailed, right-tailed, or two-tailed test?

HOMEWORK

9.1 Null and Alternative Hypotheses

62. Some of the following statements refer to the null hypothesis, some to the alternate hypothesis.

State the null hypothesis, H_0, and the alternative hypothesis. H_a, in terms of the appropriate parameter (μ or p).

    a.   The mean number of years Americans work before retiring is 34.
    b.   At most 60% of Americans vote in presidential elections.
    c.   The mean starting salary for San Jose State University graduates is at least $100,000 per year.
    d.   Twenty-nine percent of high school seniors get drunk each month.
    e.   Fewer than 5% of adults ride the bus to work in Los Angeles.
    f.   The mean number of cars a person owns in her lifetime is not more than ten.
    g.   About half of Americans prefer to live away from cities, given the choice.
    h.   Europeans have a mean paid vacation each year of six weeks.
    i.   The chance of developing breast cancer is under 11% for women.
    j.   Private universities' mean tuition cost is more than $20,000 per year.

63. Over the past few decades, public health officials have examined the link between weight concerns and teen girls' smoking. Researchers surveyed a group of 273 randomly selected teen girls living in Massachusetts (between 12 and 15 years old). After four years the girls were surveyed again. Sixty-three said they smoked to stay thin. Is there good evidence that more than thirty percent of the teen girls smoke to stay thin? The alternative hypothesis is:
    a. $p < 0.30$
    b. $p \leq 0.30$
    c. $p \geq 0.30$
    d. $p > 0.30$

64. A statistics instructor believes that fewer than 20% of Evergreen Valley College (EVC) students attended the opening night midnight showing of the latest Harry Potter movie. She surveys 84 of her students and finds that 11 attended the midnight showing. An appropriate alternative hypothesis is:
    a. $p = 0.20$
    b. $p > 0.20$
    c. $p < 0.20$
    d. $p \leq 0.20$

65. Previously, an organization reported that teenagers spent 4.5 hours per week, on average, on the phone. The organization thinks that, currently, the mean is higher. Fifteen randomly chosen teenagers were asked how many hours per week they spend on the phone. The sample mean was 4.75 hours with a sample standard deviation of 2.0. Conduct a hypothesis test. The null and alternative hypotheses are:
    a. H_0: $\bar{x} = 4.5, H_a: \bar{x} > 4.5$
    b. $H_0: \mu \geq 4.5, H_a: \mu < 4.5$

c. $H_o: \mu = 4.75$, $H_a: \mu > 4.75$
d. $H_o: \mu = 4.5$, $H_a: \mu > 4.5$

9.2 Outcomes and the Type I and Type II Errors

66. State the Type I and Type II errors in complete sentences given the following statements.
   a. The mean number of years Americans work before retiring is 34.
   b. At most 60% of Americans vote in presidential elections.
   c. The mean starting salary for San Jose State University graduates is at least $100,000 per year.
   d. Twenty-nine percent of high school seniors get drunk each month.
   e. Fewer than 5% of adults ride the bus to work in Los Angeles.
   f. The mean number of cars a person owns in his or her lifetime is not more than ten.
   g. About half of Americans prefer to live away from cities, given the choice.
   h. Europeans have a mean paid vacation each year of six weeks.
   i. The chance of developing breast cancer is under 11% for women.
   j. Private universities mean tuition cost is more than $20,000 per year.

67. For statements a-j in **Exercise 9.109**, answer the following in complete sentences.
   a. State a consequence of committing a Type I error.
   b. State a consequence of committing a Type II error.

68. When a new drug is created, the pharmaceutical company must subject it to testing before receiving the necessary permission from the Food and Drug Administration (FDA) to market the drug. Suppose the null hypothesis is "the drug is unsafe." What is the Type II Error?
   a. To conclude the drug is safe when in, fact, it is unsafe.
   b. Not to conclude the drug is safe when, in fact, it is safe.
   c. To conclude the drug is safe when, in fact, it is safe.
   d. Not to conclude the drug is unsafe when, in fact, it is unsafe.

69. A statistics instructor believes that fewer than 20% of Evergreen Valley College (EVC) students attended the opening midnight showing of the latest Harry Potter movie. She surveys 84 of her students and finds that 11 of them attended the midnight showing. The Type I error is to conclude that the percent of EVC students who attended is _____.
   a. at least 20%, when in fact, it is less than 20%.
   b. 20%, when in fact, it is 20%.
   c. less than 20%, when in fact, it is at least 20%.
   d. less than 20%, when in fact, it is less than 20%.

70. It is believed that Lake Tahoe Community College (LTCC) Intermediate Algebra students get less than seven hours of sleep per night, on average. A survey of 22 LTCC Intermediate Algebra students generated a mean of 7.24 hours with a standard deviation of 1.93 hours. At a level of significance of 5%, do LTCC Intermediate Algebra students get less than seven hours of sleep per night, on average?

The Type II error is not to reject that the mean number of hours of sleep LTCC students get per night is at least seven when, in fact, the mean number of hours

   a. is more than seven hours.
   b. is at most seven hours.
   c. is at least seven hours.
   d. is less than seven hours.

71. Previously, an organization reported that teenagers spent 4.5 hours per week, on average, on the phone. The organization thinks that, currently, the mean is higher. Fifteen randomly chosen teenagers were asked how many hours per week they spend on the phone. The sample mean was 4.75 hours with a sample standard deviation of 2.0. Conduct a hypothesis test, the Type I error is:
   a. to conclude that the current mean hours per week is higher than 4.5, when in fact, it is higher
   b. to conclude that the current mean hours per week is higher than 4.5, when in fact, it is the same
   c. to conclude that the mean hours per week currently is 4.5, when in fact, it is higher
   d. to conclude that the mean hours per week currently is no higher than 4.5, when in fact, it is not higher

9.3 Distribution Needed for Hypothesis Testing

72. It is believed that Lake Tahoe Community College (LTCC) Intermediate Algebra students get less than seven hours of sleep per night, on average. A survey of 22 LTCC Intermediate Algebra students generated a mean of 7.24 hours with a standard deviation of 1.93 hours. At a level of significance of 5%, do LTCC Intermediate Algebra students get less than seven hours of sleep per night, on average? The distribution to be used for this test is $\bar{X} \sim$ _____

a. $N(7.24, \frac{1.93}{\sqrt{22}})$

b. $N(7.24, 1.93)$

c. t_{22}

d. t_{21}

9.4 Rare Events, the Sample, Decision and Conclusion

73. The National Institute of Mental Health published an article stating that in any one-year period, approximately 9.5 percent of American adults suffer from depression or a depressive illness. Suppose that in a survey of 100 people in a certain town, seven of them suffered from depression or a depressive illness. Conduct a hypothesis test to determine if the true proportion of people in that town suffering from depression or a depressive illness is lower than the percent in the general adult American population.

a. Is this a test of one mean or proportion?
b. State the null and alternative hypotheses.
   H_0: _____ H_a: _____
c. Is this a right-tailed, left-tailed, or two-tailed test?
d. What symbol represents the random variable for this test?
e. In words, define the random variable for this test.
f. Calculate the following:
   i. $x =$ _____
   ii. $n =$ _____
   iii. $p' =$ _____

g. Calculate $\sigma_x =$ _____. Show the formula set-up.
h. State the distribution to use for the hypothesis test.
i. Find the p-value.
j. At a pre-conceived $\alpha = 0.05$, what is your:
   i. Decision:
   ii. Reason for the decision:
   iii. Conclusion (write out in a complete sentence):

9.5 Additional Information and Full Hypothesis Test Examples

For each of the word problems, use a solution sheet to do the hypothesis test. The solution sheet is found in **Appendix E**. *Please feel free to make copies of the solution sheets. For the online version of the book, it is suggested that you copy the .doc or the .pdf files.*

NOTE

If you are using a Student's-t distribution for one of the following homework problems, you may assume that the underlying population is normally distributed. (In general, you must first prove that assumption, however.)

74. A particular brand of tires claims that its deluxe tire averages at least 50,000 miles before it needs to be replaced. From past studies of this tire, the standard deviation is known to be 8,000. A survey of owners of that tire design is conducted. From the 28 tires surveyed, the mean lifespan was 46,500 miles with a standard deviation of 9,800 miles. Using alpha = 0.05, is the data highly inconsistent with the claim?

75. From generation to generation, the mean age when smokers first start to smoke varies. However, the standard deviation of that age remains constant of around 2.1 years. A survey of 40 smokers of this generation was done to see if the mean starting age is at least 19. The sample mean was 18.1 with a sample standard deviation of 1.3. Do the data support the claim at the 5% level?

76. The cost of a daily newspaper varies from city to city. However, the variation among prices remains steady with a standard deviation of 20¢. A study was done to test the claim that the mean cost of a daily newspaper is $1.00. Twelve costs yield a mean cost of 95¢ with a standard deviation of 18¢. Do the data support the claim at the 1% level?

77. An article in the *San Jose Mercury News* stated that students in the California state university system take 4.5 years, on average, to finish their undergraduate degrees. Suppose you believe that the mean time is longer. You conduct a survey of 49 students and obtain a sample mean of 5.1 with a sample standard deviation of 1.2. Do the data support your claim at the 1% level?

78. The mean number of sick days an employee takes per year is believed to be about ten. Members of a personnel department do not believe this figure. They randomly survey eight employees. The number of sick days they took for the past year are as follows: 12; 4; 15; 3; 11; 8; 6; 8. Let x = the number of sick days they took for the past year. Should the personnel team believe that the mean number is ten?

79. In 1955, *Life Magazine* reported that the 25 year-old mother of three worked, on average, an 80 hour week. Recently, many groups have been studying whether or not the women's movement has, in fact, resulted in an increase in the average work week for women (combining employment and at-home work). Suppose a study was done to determine if the mean work week has increased. 81 women were surveyed with the following results. The sample mean was 83; the sample standard deviation was ten. Does it appear that the mean work week has increased for women at the 5% level?

80. Your statistics instructor claims that 60 percent of the students who take her Elementary Statistics class go through life feeling more enriched. For some reason that she can't quite figure out, most people don't believe her. You decide to check this out on your own. You randomly survey 64 of her past Elementary Statistics students and find that 34 feel more enriched as a result of her class. Now, what do you think?

81. A Nissan Motor Corporation advertisement read, "The average man's I.Q. is 107. The average brown trout's I.Q. is 4. So why can't man catch brown trout?" Suppose you believe that the brown trout's mean I.Q. is greater than four. You catch 12 brown trout. A fish psychologist determines the I.Q.s as follows: 5; 4; 7; 3; 6; 4; 5; 3; 6; 3; 8; 5. Conduct a hypothesis test of your belief.

82. Refer to **Exercise 9.119**. Conduct a hypothesis test to see if your decision and conclusion would change if your belief were that the brown trout's mean I.Q. is **not** four.

83. According to an article in *Newsweek*, the natural ratio of girls to boys is 100:105. In China, the birth ratio is 100: 114 (46.7% girls). Suppose you don't believe the reported figures of the percent of girls born in China. You conduct a study. In this study, you count the number of girls and boys born in 150 randomly chosen recent births. There are 60 girls and 90 boys born of the 150. Based on your study, do you believe that the percent of girls born in China is 46.7?

84. A poll done for *Newsweek* found that 13% of Americans have seen or sensed the presence of an angel. A contingent doubts that the percent is really that high. It conducts its own survey. Out of 76 Americans surveyed, only two had seen or sensed the presence of an angel. As a result of the contingent's survey, would you agree with the *Newsweek* poll? In complete sentences, also give three reasons why the two polls might give different results.

85. The mean work week for engineers in a start-up company is believed to be about 60 hours. A newly hired engineer hopes that it's shorter. She asks ten engineering friends in start-ups for the lengths of their mean work weeks. Based on the results that follow, should she count on the mean work week to be shorter than 60 hours?

Data (length of mean work week): 70; 45; 55; 60; 65; 55; 55; 60; 50; 55.

86. Use the "Lap time" data for Lap 4 (see **Appendix C**) to test the claim that Terri finishes Lap 4, on average, in less than 129 seconds. Use all twenty races given.

87. Use the "Initial Public Offering" data (see **Appendix C**) to test the claim that the mean offer price was $18 per share. Do not use all the data. Use your random number generator to randomly survey 15 prices.

NOTE

The following questions were written by past students. They are excellent problems!

88. "Asian Family Reunion," by Chau Nguyen

Every two years it comes around.

We all get together from different towns.

In my honest opinion,

It's not a typical family reunion.

Not forty, or fifty, or sixty,

But how about seventy companions!

The kids would play, scream, and shout

One minute they're happy, another they'll pout.

The teenagers would look, stare, and compare

From how they look to what they wear.

The men would chat about their business

That they make more, but never less.

Money is always their subject

And there's always talk of more new projects.

The women get tired from all of the chats

They head to the kitchen to set out the mats.

Some would sit and some would stand

Eating and talking with plates in their hands.

Then come the games and the songs

And suddenly, everyone gets along!

With all that laughter, it's sad to say

That it always ends in the same old way.

They hug and kiss and say "good-bye"

And then they all begin to cry!

I say that 60 percent shed their tears

But my mom counted 35 people this year.

She said that boys and men will always have their pride,

So we won't ever see them cry.

I myself don't think she's correct,

So could you please try this problem to see if you object?

89. "The Problem with Angels," by Cyndy Dowling

Although this problem is wholly mine,

The catalyst came from the magazine, Time.

On the magazine cover I did find

The realm of angels tickling my mind.

Inside, 69% I found to be

In angels, Americans do believe.

Then, it was time to rise to the task,

Ninety-five high school and college students I did ask.

Viewing all as one group,

Random sampling to get the scoop.

So, I asked each to be true,

"Do you believe in angels?" Tell me, do!

Download for free at http://cnx.org/content/col11562/latest/

Hypothesizing at the start,

Totally believing in my heart

That the proportion who said yes

Would be equal on this test.

Lo and behold, seventy-three did arrive,

Out of the sample of ninety-five.

Now your job has just begun,

Solve this problem and have some fun.

90. "Blowing Bubbles," by Sondra Prull

Studying stats just made me tense,

I had to find some sane defense.

Some light and lifting simple play

To float my math anxiety away.

Blowing bubbles lifts me high

Takes my troubles to the sky.

POIK! They're gone, with all my stress

Bubble therapy is the best.

The label said each time I blew

The average number of bubbles would be at least 22.

I blew and blew and this I found

From 64 blows, they all are round!

But the number of bubbles in 64 blows

Varied widely, this I know.

20 per blow became the mean

They deviated by 6, and not 16.

From counting bubbles, I sure did relax

But now I give to you your task.

Was 22 a reasonable guess?

Find the answer and pass this test!

91. "Dalmatian Darnation," by Kathy Sparling

A greedy dog breeder named Spreckles

Bred puppies with numerous freckles

The Dalmatians he sought

Possessed spot upon spot

The more spots, he thought, the more shekels.

His competitors did not agree

That freckles would increase the fee.

They said, "Spots are quite nice

But they don't affect price;

One should breed for improved pedigree."

The breeders decided to prove

This strategy was a wrong move.

Breeding only for spots

Would wreak havoc, they thought.

His theory they want to disprove.

They proposed a contest to Spreckles

Comparing dog prices to freckles.

In records they looked up

One hundred one pups:

Dalmatians that fetched the most shekels.

They asked Mr. Spreckles to name

An average spot count he'd claim

To bring in big bucks.

Said Spreckles, "Well, shucks,

It's for one hundred one that I aim."

Said an amateur statistician

Who wanted to help with this mission.

"Twenty-one for the sample

Standard deviation's ample:

They examined one hundred and one

Dalmatians that fetched a good sum.

They counted each spot,

Mark, freckle and dot

And tallied up every one.

Instead of one hundred one spots

They averaged ninety six dots

Can they muzzle Spreckles'

Obsession with freckles

Based on all the dog data they've got?

92. "Macaroni and Cheese, please!!" by Nedda Misherghi and Rachelle Hall

As a poor starving student I don't have much money to spend for even the bare necessities. So my favorite and main staple food is macaroni and cheese. It's high in taste and low in cost and nutritional value.

One day, as I sat down to determine the meaning of life, I got a serious craving for this, oh, so important, food of my life. So I went down the street to Greatway to get a box of macaroni and cheese, but it was SO expensive! $2.02 !!! Can you believe it? It made me stop and think. The world is changing fast. I had thought that the mean cost of a box (the normal size, not some super-gigantic-family-value-pack) was at most $1, but now I wasn't so sure. However, I was determined to find out. I went to 53 of the closest grocery stores and surveyed the prices of macaroni and cheese. Here are the data I wrote in my notebook:

Price per box of Mac and Cheese:
- 5 stores @ $2.02
- 15 stores @ $0.25
- 3 stores @ $1.29
- 6 stores @ $0.35
- 4 stores @ $2.27
- 7 stores @ $1.50
- 5 stores @ $1.89

- 8 stores @ 0.75.

I could see that the cost varied but I had to sit down to figure out whether or not I was right. If it does turn out that this mouth-watering dish is at most $1, then I'll throw a big cheesy party in our next statistics lab, with enough macaroni and cheese for just me. (After all, as a poor starving student I can't be expected to feed our class of animals!)

93. "William Shakespeare: The Tragedy of Hamlet, Prince of Denmark," by Jacqueline Ghodsi THE CHARACTERS (in order of appearance):
- HAMLET, Prince of Denmark and student of Statistics
- POLONIUS, Hamlet's tutor
- HOROTIO, friend to Hamlet and fellow student

Scene: The great library of the castle, in which Hamlet does his lessons

Act I

(The day is fair, but the face of Hamlet is clouded. He paces the large room. His tutor, Polonius, is reprimanding Hamlet regarding the latter's recent experience. Horatio is seated at the large table at right stage.)

POLONIUS: My Lord, how cans't thou admit that thou hast seen a ghost! It is but a figment of your imagination!

HAMLET: I beg to differ; I know of a certainty that five-and-seventy in one hundred of us, condemned to the whips and scorns of time as we are, have gazed upon a spirit of health, or goblin damn'd, be their intents wicked or charitable.

POLONIUS If thou doest insist upon thy wretched vision then let me invest your time; be true to thy work and speak to me through the reason of the null and alternate hypotheses. (He turns to Horatio.) Did not Hamlet himself say, "What piece of work is man, how noble in reason, how infinite in faculties? Then let not this foolishness persist. Go, Horatio, make a survey of three-and-sixty and discover what the true proportion be. For my part, I will never succumb to this fantasy, but deem man to be devoid of all reason should thy proposal of at least five-and-seventy in one hundred hold true.

HORATIO (to Hamlet): What should we do, my Lord?

HAMLET: Go to thy purpose, Horatio.

HORATIO: To what end, my Lord?

HAMLET: That you must teach me. But let me conjure you by the rights of our fellowship, by the consonance of our youth, but the obligation of our ever-preserved love, be even and direct with me, whether I am right or no.

(Horatio exits, followed by Polonius, leaving Hamlet to ponder alone.)

Act II

(The next day, Hamlet awaits anxiously the presence of his friend, Horatio. Polonius enters and places some books upon the table just a moment before Horatio enters.)

POLONIUS: So, Horatio, what is it thou didst reveal through thy deliberations?

HORATIO: In a random survey, for which purpose thou thyself sent me forth, I did discover that one-and-forty believe fervently that the spirits of the dead walk with us. Before my God, I might not this believe, without the sensible and true avouch of mine own eyes.

POLONIUS: Give thine own thoughts no tongue, Horatio. (Polonius turns to Hamlet.) But look to't I charge you, my Lord. Come Horatio, let us go together, for this is not our test. (Horatio and Polonius leave together.)

HAMLET: To reject, or not reject, that is the question: whether 'tis nobler in the mind to suffer the slings and arrows of outrageous statistics, or to take arms against a sea of data, and, by opposing, end them. (Hamlet resignedly attends to his task.)

(Curtain falls)

94. "Untitled," by Stephen Chen

I've often wondered how software is released and sold to the public. Ironically, I work for a company that sells products with known problems. Unfortunately, most of the problems are difficult to create, which makes them difficult to fix. I usually use the test program X, which tests the product, to try to create a specific problem. When the test program is run to make an error occur, the likelihood of generating an error is 1%.

So, armed with this knowledge, I wrote a new test program Y that will generate the same error that test program X creates, but more often. To find out if my test program is better than the original, so that I can convince the management that I'm right, I ran my test program to find out how often I can generate the same error. When I ran my test program 50 times, I generated the error twice. While this may not seem much better, I think that I can convince the management to use my test program instead of the original test program. Am I right?

95. "Japanese Girls' Names"

by Kumi Furuichi

It used to be very typical for Japanese girls' names to end with "ko." (The trend might have started around my grandmothers' generation and its peak might have been around my mother's generation.) "Ko" means "child" in Chinese characters. Parents would name their daughters with "ko" attaching to other Chinese characters which have meanings that they want their daughters to become, such as Sachiko—happy child, Yoshiko—a good child, Yasuko—a healthy child, and so on.

However, I noticed recently that only two out of nine of my Japanese girlfriends at this school have names which end with "ko." More and more, parents seem to have become creative, modernized, and, sometimes, westernized in naming their children.

I have a feeling that, while 70 percent or more of my mother's generation would have names with "ko" at the end, the proportion has dropped among my peers. I wrote down all my Japanese friends', ex-classmates', co-workers', and acquaintances' names that I could remember. Following are the names. (Some are repeats.) Test to see if the proportion has dropped for this generation.

Ai, Akemi, Akiko, Ayumi, Chiaki, Chie, Eiko, Eri, Eriko, Fumiko, Harumi, Hitomi, Hiroko, Hiroko, Hidemi, Hisako, Hinako, Izumi, Izumi, Junko, Junko, Kana, Kanako, Kanayo, Kayo, Kayoko, Kazumi, Keiko, Keiko, Kei, Kumi, Kumiko, Kyoko, Kyoko, Madoka, Maho, Mai, Maiko, Maki, Miki, Miki, Mikiko, Mina, Minako, Miyako, Momoko, Nana, Naoko, Naoko, Naoko, Noriko, Rieko, Rika, Rika, Rumiko, Rei, Reiko, Reiko, Sachiko, Sachiko, Sachiyo, Saki, Sayaka, Sayoko, Sayuri, Seiko, Shiho, Shizuka, Sumiko, Takako, Takako, Tomoe, Tomoe, Tomoko, Touko, Yasuko, Yasuko, Yasuyo, Yoko, Yoko, Yoko, Yoshiko, Yoshiko, Yoshiko, Yuka, Yuki, Yuki, Yukiko, Yuko, Yuko.

96. "Phillip's Wish," by Suzanne Osorio

My nephew likes to play

Chasing the girls makes his day.

He asked his mother

If it is okay

To get his ear pierced.

She said, "No way!"

To poke a hole through your ear,

Is not what I want for you, dear.

He argued his point quite well,

Says even my macho pal, Mel,

Has gotten this done.

It's all just for fun.

C'mon please, mom, please, what the hell.

Again Phillip complained to his mother,

Saying half his friends (including their brothers)

Are piercing their ears

And they have no fears

He wants to be like the others.

She said, "I think it's much less.

We must do a hypothesis test.

And if you are right,

I won't put up a fight.

But, if not, then my case will rest."

We proceeded to call fifty guys

To see whose prediction would fly.

Nineteen of the fifty

Said piercing was nifty

And earrings they'd occasionally buy.

Then there's the other thirty-one,

Who said they'd never have this done.

So now this poem's finished.

Will his hopes be diminished,

Or will my nephew have his fun?

97. "The Craven," by Mark Salangsang

Once upon a morning dreary

In stats class I was weak and weary.

Pondering over last night's homework

Whose answers were now on the board

This I did and nothing more.

While I nodded nearly napping

Suddenly, there came a tapping.

As someone gently rapping,

Rapping my head as I snore.

Quoth the teacher, "Sleep no more."

"In every class you fall asleep,"

The teacher said, his voice was deep.

"So a tally I've begun to keep

Of every class you nap and snore.

The percentage being forty-four."

"My dear teacher I must confess,

While sleeping is what I do best.

The percentage, I think, must be less,

A percentage less than forty-four."

This I said and nothing more.

"We'll see," he said and walked away,

And fifty classes from that day

He counted till the month of May

The classes in which I napped and snored.

The number he found was twenty-four.

At a significance level of 0.05,

Please tell me am I still alive?

Or did my grade just take a dive

Plunging down beneath the floor?

Upon thee I hereby implore.

98. Toastmasters International cites a report by Gallop Poll that 40% of Americans fear public speaking. A student believes that less than 40% of students at her school fear public speaking. She randomly surveys 361 schoolmates and finds that 135 report they fear public speaking. Conduct a hypothesis test to determine if the percent at her school is less than 40%.

99. Sixty-eight percent of online courses taught at community colleges nationwide were taught by full-time faculty. To test if 68% also represents California's percent for full-time faculty teaching the online classes, Long Beach City College (LBCC) in California, was randomly selected for comparison. In the same year, 34 of the 44 online courses LBCC offered were taught by full-time faculty. Conduct a hypothesis test to determine if 68% represents California. NOTE: For more accurate results, use more California community colleges and this past year's data.

100. According to an article in *Bloomberg Businessweek*, New York City's most recent adult smoking rate is 14%. Suppose that a survey is conducted to determine this year's rate. Nine out of 70 randomly chosen N.Y. City residents reply that they smoke. Conduct a hypothesis test to determine if the rate is still 14% or if it has decreased.

101. The mean age of De Anza College students in a previous term was 26.6 years old. An instructor thinks the mean age for online students is older than 26.6. She randomly surveys 56 online students and finds that the sample mean is 29.4 with a standard deviation of 2.1. Conduct a hypothesis test.

102. Registered nurses earned an average annual salary of $69,110. For that same year, a survey was conducted of 41 California registered nurses to determine if the annual salary is higher than $69,110 for California nurses. The sample average was $71,121 with a sample standard deviation of $7,489. Conduct a hypothesis test.

103. La Leche League International reports that the mean age of weaning a child from breastfeeding is age four to five worldwide. In America, most nursing mothers wean their children much earlier. Suppose a random survey is conducted of 21 U.S. mothers who recently weaned their children. The mean weaning age was nine months (3/4 year) with a standard deviation of 4 months. Conduct a hypothesis test to determine if the mean weaning age in the U.S. is less than four years old.

104. Over the past few decades, public health officials have examined the link between weight concerns and teen girls' smoking. Researchers surveyed a group of 273 randomly selected teen girls living in Massachusetts (between 12 and 15 years old). After four years the girls were surveyed again. Sixty-three said they smoked to stay thin. Is there good evidence that more than thirty percent of the teen girls smoke to stay thin?
After conducting the test, your decision and conclusion are
   a. Reject H_0: There is sufficient evidence to conclude that more than 30% of teen girls smoke to stay thin.
   b. Do not reject H_0: There is not sufficient evidence to conclude that less than 30% of teen girls smoke to stay thin.
   c. Do not reject H_0: There is not sufficient evidence to conclude that more than 30% of teen girls smoke to stay thin.
   d. Reject H_0: There is sufficient evidence to conclude that less than 30% of teen girls smoke to stay thin.

105. A statistics instructor believes that fewer than 20% of Evergreen Valley College (EVC) students attended the opening night midnight showing of the latest Harry Potter movie. She surveys 84 of her students and finds that 11 of them attended the midnight showing.
At a 1% level of significance, an appropriate conclusion is:
   a. There is insufficient evidence to conclude that the percent of EVC students who attended the midnight showing of Harry Potter is less than 20%.
   b. There is sufficient evidence to conclude that the percent of EVC students who attended the midnight showing of Harry Potter is more than 20%.
   c. There is sufficient evidence to conclude that the percent of EVC students who attended the midnight showing of Harry Potter is less than 20%.
   d. There is insufficient evidence to conclude that the percent of EVC students who attended the midnight showing of Harry Potter is at least 20%.

106. Previously, an organization reported that teenagers spent 4.5 hours per week, on average, on the phone. The organization thinks that, currently, the mean is higher. Fifteen randomly chosen teenagers were asked how many hours per week they spend on the phone. The sample mean was 4.75 hours with a sample standard deviation of 2.0. Conduct a hypothesis test.

At a significance level of $a = 0.05$, what is the correct conclusion?
   a. There is enough evidence to conclude that the mean number of hours is more than 4.75
   b. There is enough evidence to conclude that the mean number of hours is more than 4.5
   c. There is not enough evidence to conclude that the mean number of hours is more than 4.5

d. There is not enough evidence to conclude that the mean number of hours is more than 4.75

Instructions: For the following ten exercises,
Hypothesis testing: For the following ten exercises, answer each question.

a. State the null and alternate hypothesis.

b. State the *p*-value.

c. State alpha.

d. What is your decision?

e. Write a conclusion.

f. Answer any other questions asked in the problem.

107. According to the Center for Disease Control website, in 2011 at least 18% of high school students have smoked a cigarette. An Introduction to Statistics class in Davies County, KY conducted a hypothesis test at the local high school (a medium sized–approximately 1,200 students–small city demographic) to determine if the local high school's percentage was lower. One hundred fifty students were chosen at random and surveyed. Of the 150 students surveyed, 82 have smoked. Use a significance level of 0.05 and using appropriate statistical evidence, conduct a hypothesis test and state the conclusions.

108. A recent survey in the *N.Y. Times Almanac* indicated that 48.8% of families own stock. A broker wanted to determine if this survey could be valid. He surveyed a random sample of 250 families and found that 142 owned some type of stock. At the 0.05 significance level, can the survey be considered to be accurate?

109. Driver error can be listed as the cause of approximately 54% of all fatal auto accidents, according to the American Automobile Association. Thirty randomly selected fatal accidents are examined, and it is determined that 14 were caused by driver error. Using $\alpha = 0.05$, is the AAA proportion accurate?

110. The US Department of Energy reported that 51.7% of homes were heated by natural gas. A random sample of 221 homes in Kentucky found that 115 were heated by natural gas. Does the evidence support the claim for Kentucky at the $\alpha = 0.05$ level in Kentucky? Are the results applicable across the country? Why?

111. For Americans using library services, the American Library Association claims that at most 67% of patrons borrow books. The library director in Owensboro, Kentucky feels this is not true, so she asked a local college statistic class to conduct a survey. The class randomly selected 100 patrons and found that 82 borrowed books. Did the class demonstrate that the percentage was higher in Owensboro, KY? Use $\alpha = 0.01$ level of significance. What is the possible proportion of patrons that do borrow books from the Owensboro Library?

112. The Weather Underground reported that the mean amount of summer rainfall for the northeastern US is at least 11.52 inches. Ten cities in the northeast are randomly selected and the mean rainfall amount is calculated to be 7.42 inches with a standard deviation of 1.3 inches. At the $\alpha = 0.05$ level, can it be concluded that the mean rainfall was below the reported average? What if $\alpha = 0.01$? Assume the amount of summer rainfall follows a normal distribution.

113. A survey in the *N.Y. Times Almanac* finds the mean commute time (one way) is 25.4 minutes for the 15 largest US cities. The Austin, TX chamber of commerce feels that Austin's commute time is less and wants to publicize this fact. The mean for 25 randomly selected commuters is 22.1 minutes with a standard deviation of 5.3 minutes. At the $\alpha = 0.10$ level, is the Austin, TX commute significantly less than the mean commute time for the 15 largest US cities?

114. A report by the Gallup Poll found that a woman visits her doctor, on average, at most 5.8 times each year. A random sample of 20 women results in these yearly visit totals

3; 2; 1; 3; 7; 2; 9; 4; 6; 6; 8; 0; 5; 6; 4; 2; 1; 3; 4; 1
At the $\alpha = 0.05$ level can it be concluded that the sample mean is higher than 5.8 visits per year?

115. According to the *N.Y. Times Almanac* the mean family size in the U.S. is 3.18. A sample of a college math class resulted in the following family sizes:
5; 4; 5; 4; 4; 3; 6; 4; 3; 3; 5; 5; 6; 3; 3; 2; 7; 4; 5; 2; 2; 2; 3; 2
At $\alpha = 0.05$ level, is the class' mean family size greater than the national average? Does the Almanac result remain valid? Why?

116. The student academic group on a college campus claims that freshman students study at least 2.5 hours per day, on average. One Introduction to Statistics class was skeptical. The class took a random sample of 30 freshman students and found a mean study time of 137 minutes with a standard deviation of 45 minutes. At $\alpha = 0.01$ level, is the student academic group's claim correct?

REFERENCES

9.1 Null and Alternative Hypotheses

Data from the National Institute of Mental Health. Available online at http://www.nimh.nih.gov/publicat/depression.cfm.

9.5 Additional Information and Full Hypothesis Test Examples

Data from Amit Schitai. Director of Instructional Technology and Distance Learning. LBCC.

Data from *Bloomberg Businessweek*. Available online at http://www.businessweek.com/news/2011- 09-15/nyc-smoking-rate-falls-to-record-low-of-14-bloomberg-says.html.

Data from energy.gov. Available online at http://energy.gov (accessed June 27. 2013).

Data from Gallup®. Available online at www.gallup.com (accessed June 27, 2013).

Data from *Growing by Degrees* by Allen and Seaman.

Data from La Leche League International. Available online at http://www.lalecheleague.org/Law/BAFeb01.html.

Data from the American Automobile Association. Available online at www.aaa.com (accessed June 27, 2013).

Data from the American Library Association. Available online at www.ala.org (accessed June 27, 2013).

Data from the Bureau of Labor Statistics. Available online at http://www.bls.gov/oes/current/oes291111.htm.

Data from the Centers for Disease Control and Prevention. Available online at www.cdc.gov (accessed June 27, 2013)

Data from the U.S. Census Bureau, available online at http://quickfacts.census.gov/qfd/states/00000.html (accessed June 27, 2013).

Data from the United States Census Bureau. Available online at http://www.census.gov/hhes/socdemo/language/.

Data from Toastmasters International. Available online at http://toastmasters.org/artisan/detail.asp?CategoryID=1&SubCategoryID=10&ArticleID=429&Page=1.

Data from Weather Underground. Available online at www.wunderground.com (accessed June 27, 2013).

Federal Bureau of Investigations. "Uniform Crime Reports and Index of Crime in Daviess in the State of Kentucky enforced by Daviess County from 1985 to 2005." Available online at http://www.disastercenter.com/kentucky/crime/3868.htm (accessed June 27, 2013).

"Foothill-De Anza Community College District." De Anza College, Winter 2006. Available online at http://research.fhda.edu/factbook/DAdemofs/Fact_sheet_da_2006w.pdf.

Johansen, C., J. Boice, Jr., J. McLaughlin, J. Olsen. "Cellular Telephones and Cancer—a Nationwide Cohort Study in Denmark." Institute of Cancer Epidemiology and the Danish Cancer Society, 93(3):203-7. Available online at http://www.ncbi.nlm.nih.gov/pubmed/11158188 (accessed June 27, 2013).

Rape, Abuse & Incest National Network. "How often does sexual assault occur?" RAINN, 2009. Available online at http://www.rainn.org/get-information/statistics/frequency-of-sexual-assault (accessed June 27, 2013).

SOLUTIONS

1 The random variable is the mean Internet speed in Megabits per second.

3 The random variable is the mean number of children an American family has.

5 The random variable is the proportion of people picked at random in Times Square visiting the city.

7

a. H_0: $p = 0.42$

b. H_a: $p < 0.42$

9

   a. $H_0: \mu = 15$

   b. $H_a: \mu \neq 15$

11 Type I: The mean price of mid-sized cars is \$32,000, but we conclude that it is not \$32,000. Type II: The mean price of mid-sized cars is not \$32,000, but we conclude that it is \$32,000.

13 α = the probability that you think the bag cannot withstand -15 degrees F, when in fact it can β = the probability that you think the bag can withstand -15 degrees F, when in fact it cannot

15 Type I: The procedure will go well, but the doctors think it will not. Type II: The procedure will not go well, but the doctors think it will.

17 0.019

19 0.998

21 A normal distribution or a Student's t-distribution

23 Use a Student's t-distribution

25 a normal distribution for a single population mean

27 It must be approximately normally distributed.

29 They must both be greater than five.

31 binomial distribution

33 The outcome of winning is very unlikely.

35 $H_0: \mu \geq 73$
$H_a: \mu < 73$
The p-value is almost zero, which means there is sufficient data to conclude that the mean height of high school students who play basketball on the school team is less than 73 inches at the 5% level. The data do support the claim.

37 The shaded region shows a low p-value.

39 Do not reject H_0.

41 means

43 the mean time spent in jail for 26 first time convicted burglars

45

   a.   3

   b.   1.5

   c.   1.8

   d.   26

47 $\bar{X} \sim N\left(2.5, \frac{1.5}{\sqrt{26}}\right)$

49 This is a left-tailed test.

51 This is a two-tailed test.

53

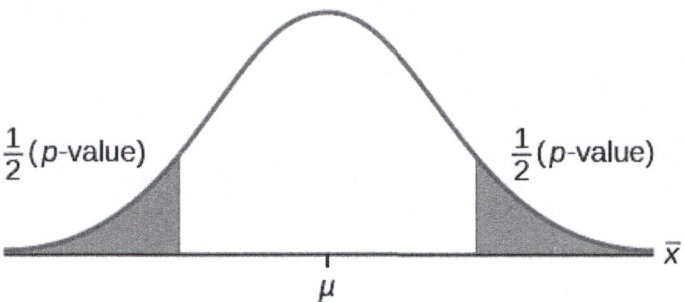

Figure 9.25

55 a right-tailed test

57 a left-tailed test

59 This is a left-tailed test.

61 This is a two-tailed test.

62

  a. $H_0: \mu = 34$; $H_a: \mu \neq 34$

  b. $H_0: p \leq 0.60$; $H_a: p > 0.60$

  c. $H_0: \mu \geq 100{,}000$; $H_a: \mu < 100{,}000$

  d. $H_0: p = 0.29$; $H_a: p \neq 0.29$

  e. $H_0: p = 0.05$; $H_a: p < 0.05$

  f. $H_0: \mu \leq 10$; $H_a: \mu > 10$

  g. $H_0: p = 0.50$; $H_a: p \neq 0.50$

  h. $H_0: \mu = 6$; $H_a: \mu \neq 6$

  i. $H_0: p \geq 0.11$; $H_a: p < 0.11$

  j. $H_0: \mu \leq 20{,}000$; $H_a: \mu > 20{,}000$

64 c

66

  a. Type I error: We conclude that the mean is not 34 years, when it really is 34 years. Type II error: We conclude that the mean is 34 years, when in fact it really is not 34 years.

  b. Type I error: We conclude that more than 60% of Americans vote in presidential elections, when the actual percentage is at most 60%.Type II error: We conclude that at most 60% of Americans vote in presidential elections when, in fact, more than 60% do.

  c. Type I error: We conclude that the mean starting salary is less than $100,000, when it really is at least $100,000. Type II error: We conclude that the mean starting salary is at least $100,000 when, in fact, it is less than $100,000.

  d. Type I error: We conclude that the proportion of high school seniors who get drunk each month is not 29%, when it really is 29%. Type II error: We conclude that the proportion of high school seniors who get drunk each month is 29% when, in fact, it is not 29%.

  e. Type I error: We conclude that fewer than 5% of adults ride the bus to work in Los Angeles, when the percentage that do is really 5% or more. Type II error: We conclude that 5% or more adults ride the bus to work in Los Angeles when, in fact, fewer that 5% do.

  f. Type I error: We conclude that the mean number of cars a person owns in his or her lifetime is more than 10, when in reality it is not more than 10. Type II error: We conclude that the mean number of cars a person owns in his or her lifetime is not more than 10 when, in fact, it is more than 10.

g. Type I error: We conclude that the proportion of Americans who prefer to live away from cities is not about half, though the actual proportion is about half. Type II error: We conclude that the proportion of Americans who prefer to live away from cities is half when, in fact, it is not half.

h. Type I error: We conclude that the duration of paid vacations each year for Europeans is not six weeks, when in fact it is six weeks. Type II error: We conclude that the duration of paid vacations each year for Europeans is six weeks when, in fact, it is not.

i. Type I error: We conclude that the proportion is less than 11%, when it is really at least 11%. Type II error: We conclude that the proportion of women who develop breast cancer is at least 11%, when in fact it is less than 11%.

j. Type I error: We conclude that the average tuition cost at private universities is more than $20,000, though in reality it is at most $20,000. Type II error: We conclude that the average tuition cost at private universities is at most $20,000 when, in fact, it is more than $20,000.

68 b

70 d

72 d

74

a. H_0: $\mu \geq 50,000$

b. H_a: $\mu < 50,000$

c. Let $\bar{X}$ = the average lifespan of a brand of tires.

d. normal distribution

e. $z = -2.315$

f. p-value = 0.0103

g. Check student's solution.

h.   i.   alpha: 0.05

  ii.   Decision: Reject the null hypothesis.

  iii.   Reason for decision: The p-value is less than 0.05.

  iv.   Conclusion: There is sufficient evidence to conclude that the mean lifespan of the tires is less than 50,000 miles.

i. (43,537, 49,463)

76

a. H_0: $\mu = \$1.00$

b. H_a: $\mu \neq \$1.00$

c. Let $\bar{X}$ = the average cost of a daily newspaper.

d. normal distribution

e. $z = -0.866$

f. p-value = 0.3865

g. Check student's solution.

h.   i.   Alpha: 0.01

  ii.   Decision: Do not reject the null hypothesis.

  iii.   Reason for decision: The p-value is greater than 0.01.

  iv.   Conclusion: There is sufficient evidence to support the claim that the mean cost of daily papers is $1. The mean cost could be $1.

i. ($0.84, $1.06)

78

a. H_0: $\mu = 10$

b. H_a: $\mu \neq 10$

c. Let $\bar{X}$ the mean number of sick days an employee takes per year.

d. Student's t-distribution

e. $t = -1.12$

f. p-value = 0.300

g. Check student's solution.

h.
   i. Alpha: 0.05

   ii. Decision: Do not reject the null hypothesis.

   iii. Reason for decision: The p-value is greater than 0.05.

   iv. Conclusion: At the 5% significance level, there is insufficient evidence to conclude that the mean number of sick days is not ten.

i. (4.9443, 11.806)

80

a. H_0: $p \geq 0.6$

b. H_a: $p < 0.6$

c. Let P' = the proportion of students who feel more enriched as a result of taking Elementary Statistics.

d. normal for a single proportion

e. 1.12

f. p-value = 0.1308

g. Check student's solution.

h.
   i. Alpha: 0.05

   ii. Decision: Do not reject the null hypothesis.

   iii. Reason for decision: The p-value is greater than 0.05.

   iv. Conclusion: There is insufficient evidence to conclude that less than 60 percent of her students feel more enriched.

i. Confidence Interval: (0.409, 0.654)
The "plus-4s" confidence interval is (0.411, 0.648)

82

a. H_0: $\mu = 4$

b. H_a: $\mu \neq 4$

c. Let $\bar{X}$ the average I.Q. of a set of brown trout.

d. two-tailed Student's t-test

e. $t = 1.95$

f. p-value = 0.076

g. Check student's solution.

h.
   i. Alpha: 0.05

   ii. Decision: Reject the null hypothesis.

   iii. Reason for decision: The p-value is greater than 0.05

    iv. Conclusion: There is insufficient evidence to conclude that the average IQ of brown trout is not four.

  i. (3.8865,5.9468)

84

a. H_0: $p \geq 0.13$

b. H_a: $p < 0.13$

c. Let P' = the proportion of Americans who have seen or sensed angels

d. normal for a single proportion

e. −2.688

f. p-value = 0.0036

g. Check student's solution.

h.   i. alpha: 0.05

    ii. Decision: Reject the null hypothesis.

    iii. Reason for decision: The p-value is less than 0.05.

    iv. Conclusion: There is sufficient evidence to conclude that the percentage of Americans who have seen or sensed an angel is less than 13%.

  i. (0, 0.0623).
The"plus-4s" confidence interval is (0.0022, 0.0978)

86

a. H_0: $\mu \geq 129$

b. H_a: $\mu < 129$

c. Let $\bar{X}$ = the average time in seconds that Terri finishes Lap 4.

d. Student's t-distribution

e. $t = 1.209$

f. 0.8792

g. Check student's solution.

h.   i. Alpha: 0.05

    ii. Decision: Do not reject the null hypothesis.

    iii. Reason for decision: The p-value is greater than 0.05.

    iv. Conclusion: There is insufficient evidence to conclude that Terri's mean lap time is less than 129 seconds.

  i. (128.63, 130.37)

88

a. H_0: $p = 0.60$

b. H_a: $p < 0.60$

c. Let P' = the proportion of family members who shed tears at a reunion.

d. normal for a single proportion

e. −1.71

f. 0.0438

g. Check student's solution.

h.   i. alpha: 0.05

    ii. Decision: Reject the null hypothesis.

iii. Reason for decision: *p*-value < alpha

iv. Conclusion: At the 5% significance level, there is sufficient evidence to conclude that the proportion of family members who shed tears at a reunion is less than 0.60. However, the test is weak because the *p*-value and alpha are quite close, so other tests should be done.

i. We are 95% confident that between 38.29% and 61.71% of family members will shed tears at a family reunion. (0.3829, 0.6171). The"plus-4s" confidence interval (see chapter 8) is (0.3861, 0.6139)

Note that here the "large-sample" 1 – PropZTest provides the approximate *p*-value of 0.0438. Whenever a *p*-value based on a normal approximation is close to the level of significance, the exact *p*-value based on binomial probabilities should be calculated whenever possible. This is beyond the scope of this course.

90

a. H_0: $\mu \geq 22$

b. H_a: $\mu < 22$

c. Let $\bar{X}$ = the mean number of bubbles per blow.

d. Student's *t*-distribution

e. −2.667

f. *p*-value = 0.00486

g. Check student's solution.

h. i. Alpha: 0.05

   ii. Decision: Reject the null hypothesis.

   iii. Reason for decision: The *p*-value is less than 0.05.

   iv. Conclusion: There is sufficient evidence to conclude that the mean number of bubbles per blow is less than 22.

i. (18.501, 21.499)

92

a. H_0: $\mu \leq 1$

b. H_a: $\mu > 1$

c. Let $\bar{X}$ = the mean cost in dollars of macaroni and cheese in a certain town.

d. Student's *t*-distribution

e. *t* = 0.340

f. *p*-value = 0.36756

g. Check student's solution.

h. i. Alpha: 0.05

   ii. Decision: Do not reject the null hypothesis.

   iii. Reason for decision: The *p*-value is greater than 0.05

   iv. Conclusion: The mean cost could be $1, or less. At the 5% significance level, there is insufficient evidence to conclude that the mean price of a box of macaroni and cheese is more than $1.

i. (0.8291, 1.241)

94

a. H_0: $p = 0.01$

b. H_a: $p > 0.01$

c. Let P' = the proportion of errors generated

d. Normal for a single proportion

e. 2.13

f. 0.0165

g. Check student's solution.

h.   i. Alpha: 0.05

    ii. Decision: Reject the null hypothesis

    iii. Reason for decision: The *p*-value is less than 0.05.

    iv. Conclusion: At the 5% significance level, there is sufficient evidence to conclude that the proportion of errors generated is more than 0.01.

i. Confidence interval: (0, 0.094).
The "plus-4s" confidence interval is (0.004, 0.144).

96

a. H_0: $p = 0.50$

b. H_a: $p < 0.50$

c. Let P' = the proportion of friends that has a pierced ear.

d. normal for a single proportion

e. −1.70

f. *p*-value = 0.0448

g. Check student's solution.

h.   i. Alpha: 0.05

    ii. Decision: Reject the null hypothesis

    iii. Reason for decision: The *p*-value is less than 0.05. (However, they are very close.)

    iv. Conclusion: There is sufficient evidence to support the claim that less than 50% of his friends have pierced ears.

i. Confidence Interval: (0.245, 0.515): The "plus-4s" confidence interval is (0.259, 0.519).

98

a. H_0: $p = 0.40$

b. H_a: $p < 0.40$

c. Let P' = the proportion of schoolmates who fear public speaking.

d. normal for a single proportion

e. −1.01

f. *p*-value = 0.1563

g. Check student's solution.

h.   i. Alpha: 0.05

    ii. Decision: Do not reject the null hypothesis.

    iii. Reason for decision: The *p*-value is greater than 0.05.

    iv. Conclusion: There is insufficient evidence to support the claim that less than 40% of students at the school fear public speaking.

i. Confidence Interval: (0.3241, 0.4240): The "plus-4s" confidence interval is (0.3257, 0.4250).

100

a. H_0: $p = 0.14$

b. H_a: $p < 0.14$

c. Let P' = the proportion of NYC residents that smoke.

    d. normal for a single proportion

    e. –0.2756

    f. *p*-value = 0.3914

    g. Check student's solution.

    h.   i. alpha: 0.05

          ii. Decision: Do not reject the null hypothesis.

          iii. Reason for decision: The *p*-value is greater than 0.05.

          iv. At the 5% significance level, there is insufficient evidence to conclude that the proportion of NYC residents who smoke is less than 0.14.

    i. Confidence Interval: (0.0502, 0.2070): The "plus-4s" confidence interval (see chapter 8) is (0.0676, 0.2297).

102

    a. H_0: $\mu = 69,110$

    b. H_a: $\mu > 69,110$

    c. Let $\bar{X}$ = the mean salary in dollars for California registered nurses.

    d. Student's *t*-distribution

    e. $t = 1.719$

    f. *p*-value: 0.0466

    g. Check student's solution.

    h.   i. Alpha: 0.05

          ii. Decision: Reject the null hypothesis.

          iii. Reason for decision: The *p*-value is less than 0.05.

          iv. Conclusion: At the 5% significance level, there is sufficient evidence to conclude that the mean salary of California registered nurses exceeds $69,110.

    i. ($68,757, $73,485)

104 c

106 c

108

    a. H_0: $p = 0.488$ H_a: $p \neq 0.488$

    b. *p*-value = 0.0114

    c. alpha = 0.05

    d. Reject the null hypothesis.

    e. At the 5% level of significance, there is enough evidence to conclude that 48.8% of families own stocks.

    f. The survey does not appear to be accurate.

110

    a. H_0: $p = 0.517$ H_a: $p \neq 0.517$

    b. *p*-value = 0.9203.

    c. alpha = 0.05.

    d. Do not reject the null hypothesis.

    e. At the 5% significance level, there is not enough evidence to conclude that the proportion of homes in Kentucky that are heated by natural gas is 0.517.

f. However, we cannot generalize this result to the entire nation. First, the sample's population is only the state of Kentucky. Second, it is reasonable to assume that homes in the extreme north and south will have extreme high usage and low usage, respectively. We would need to expand our sample base to include these possibilities if we wanted to generalize this claim to the entire nation.

112

a. H_0: $\mu \geq 11.52$ H_a: $\mu < 11.52$

b. *p*-value = 0.000002 which is almost 0.

c. alpha = 0.05.

d. Reject the null hypothesis.

e. At the 5% significance level, there is enough evidence to conclude that the mean amount of summer rain in the northeaster US is less than 11.52 inches, on average.

f. We would make the same conclusion if alpha was 1% because the *p*-value is almost 0.

114

a. H_0: $\mu \leq 5.8$ H_a: $\mu > 5.8$

b. *p*-value = 0.9987

c. alpha = 0.05

d. Do not reject the null hypothesis.

e. At the 5% level of significance, there is not enough evidence to conclude that a woman visits her doctor, on average, more than 5.8 times a year.

116

a. H_0: $\mu \geq 150$ H_a: $\mu < 150$

b. *p*-value = 0.0622

c. alpha = 0.01

d. Do not reject the null hypothesis.

e. At the 1% significance level, there is not enough evidence to conclude that freshmen students study less than 2.5 hours per day, on average.

f. The student academic group's claim appears to be correct.

10 | HYPOTHESIS TESTING WITH TWO SAMPLES

Figure 10.1 If you want to test a claim that involves two groups (the types of breakfasts eaten east and west of the Mississippi River) you can use a slightly different technique when conducting a hypothesis test. (credit: Chloe Lim)

Introduction

Chapter Objectives
By the end of this chapter, the student should be able to: • Classify hypothesis tests by type. • Conduct and interpret hypothesis tests for two population means, population standard deviations known. • Conduct and interpret hypothesis tests for two population means, population standard deviations unknown. • Conduct and interpret hypothesis tests for two population proportions. • Conduct and interpret hypothesis tests for matched or paired samples.

Studies often compare two groups. For example, researchers are interested in the effect aspirin has in preventing heart attacks. Over the last few years, newspapers and magazines have reported various aspirin studies involving two groups. Typically, one group is given aspirin and the other group is given a placebo. Then, the heart attack rate is studied over several years.

There are other situations that deal with the comparison of two groups. For example, studies compare various diet and exercise programs. Politicians compare the proportion of individuals from different income brackets who might vote for them. Students are interested in whether SAT or GRE preparatory courses really help raise their scores.

You have learned to conduct hypothesis tests on single means and single proportions. You will expand upon that in this chapter. You will compare two means or two proportions to each other. The general procedure is still the same, just expanded.

To compare two means or two proportions, you work with two groups. The groups are classified either as **independent** or **matched pairs**. **Independent groups** consist of two samples that are independent, that is, sample values selected from one population are not related in any way to sample values selected from the other population. **Matched pairs** consist of two samples that are dependent. The parameter tested using matched pairs is the population mean. The parameters tested using independent groups are either population means or population proportions.

NOTE

☞ This chapter relies on either a calculator or a computer to calculate the degrees of freedom, the test statistics, and p-values. TI-83+ and TI-84 instructions are included as well as the test statistic formulas. When using a TI-83+ or TI-84 calculator, we do not need to separate two population means, independent groups, or population variances unknown into large and small sample sizes. However, most statistical computer software has the ability to differentiate these tests.

This chapter deals with the following hypothesis tests:

Independent groups (samples are independent)
- Test of two population means.
- Test of two population proportions.

Matched or paired samples (samples are dependent)
- Test of the two population proportions by testing one population mean of differences.

10.1 | Two Population Means with Unknown Standard Deviations

1. The two independent samples are simple random samples from two distinct populations.
2. For the two distinct populations:
   ◦ if the sample sizes are small, the distributions are important (should be normal)
   ◦ if the sample sizes are large, the distributions are not important (need not be normal)

The test comparing two independent population means with unknown and possibly unequal population standard deviations is called the Aspin-Welch t-test. The degrees of freedom formula was developed by Aspin-Welch.

The comparison of two population means is very common. A difference between the two samples depends on both the means and the standard deviations. Very different means can occur by chance if there is great variation among the individual samples. In order to account for the variation, we take the difference of the sample means, $\bar{X}_1 - \bar{X}_2$, and divide by the standard error in order to standardize the difference. The result is a t-score test statistic.

Because we do not know the population standard deviations, we estimate them using the two sample standard deviations from our independent samples. For the hypothesis test, we calculate the estimated standard deviation, or **standard error**, of **the difference in sample means**, $\bar{X}_1 - \bar{X}_2$.

The standard error is:

$$\sqrt{\frac{(s_1)^2}{n_1} + \frac{(s_2)^2}{n_2}}$$

The test statistic (*t*-score) is calculated as follows:

$$\frac{(\bar{x}_1 - \bar{x}_2) - (\mu_1 - \mu_2)}{\sqrt{\frac{(s_1)^2}{n_1} + \frac{(s_2)^2}{n_2}}}$$

where:

- s_1 and s_2, the sample standard deviations, are estimates of σ_1 and σ_2, respectively.

- σ_1 and σ_1 are the unknown population standard deviations.

- $\bar{x}_1$ and $\bar{x}_2$ are the sample means. μ_1 and μ_2 are the population means.

The number of **degrees of freedom (*df*)** requires a somewhat complicated calculation. However, a computer or calculator calculates it easily. The *df* are not always a whole number. The test statistic calculated previously is approximated by the Student's *t*-distribution with *df* as follows:

Degrees of freedom

$$df = \frac{\left(\frac{(s_1)^2}{n_1} + \frac{(s_2)^2}{n_2}\right)^2}{\left(\frac{1}{n_1 - 1}\right)\left(\frac{(s_1)^2}{n_1}\right)^2 + \left(\frac{1}{n_2 - 1}\right)\left(\frac{(s_2)^2}{n_2}\right)^2}$$

When both sample sizes n_1 and n_2 are five or larger, the Student's *t* approximation is very good. Notice that the sample variances $(s_1)^2$ and $(s_2)^2$ are not pooled. (If the question comes up, do not pool the variances.)

 It is not necessary to compute this by hand. A calculator or computer easily computes it.

Example 10.1 Independent groups

The average amount of time boys and girls aged seven to 11 spend playing sports each day is believed to be the same. A study is done and data are collected, resulting in the data in **Table 10.1**. Each populations has a normal distribution.

	Sample Size	Average Number of Hours Playing Sports Per Day	Sample Standard Deviation
Girls	9	2	0.866
Boys	16	3.2	1.00

Table 10.1

Is there a difference in the mean amount of time boys and girls aged seven to 11 play sports each day? Test at the 5% level of significance.

Solution 10.1

The population standard deviations are not known. Let *g* be the subscript for girls and *b* be the subscript for boys. Then, μ_g is the population mean for girls and μ_b is the population mean for boys. This is a test of two **independent groups,** two population **means.**

Random variable: $\bar{X}_g - \bar{X}_b$ = difference in the sample mean amount of time girls and boys play sports each day.

H_0: $\mu_g = \mu_b$ H_0: $\mu_g - \mu_b = 0$

H_a: $\mu_g \neq \mu_b$ H_a: $\mu_g - \mu_b \neq 0$

The words **"the same"** tell you H_0 has an "=". Since there are no other words to indicate H_a, assume it says **"is different."** This is a two-tailed test.

Distribution for the test: Use t_{df} where df is calculated using the df formula for independent groups, two population means. Using a calculator, df is approximately 18.8462. **Do not pool the variances.**

Calculate the p-value using a Student's t-distribution: p-value = 0.0054

Graph:

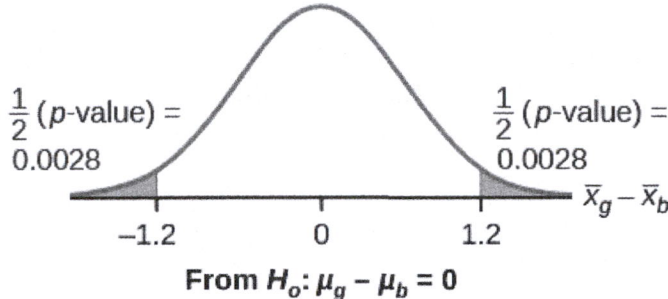

Figure 10.2

$s_g = 0.866$

$s_b = 1$

So, $\bar{x}_g - \bar{x}_b = 2 - 3.2 = -1.2$

Half the p-value is below −1.2 and half is above 1.2.

Make a decision: Since $\alpha > p$-value, reject H_0. This means you reject $\mu_g = \mu_b$. The means are different.

 Using the TI-83, 83+, 84, 84+ Calculator

Press **STAT**. Arrow over to **TESTS** and press **4:2-SampTTest**. Arrow over to Stats and press **ENTER**. Arrow down and enter **2** for the first sample mean, $\sqrt{0.866}$ for Sx1, **9** for n1, **3.2** for the second sample mean, **1** for Sx2, and **16** for n2. Arrow down to μ1: and arrow to **does not equal** μ2. Press **ENTER**. Arrow down to Pooled: and **No**. Press **ENTER**. Arrow down to **Calculate** and press **ENTER**. The p-value is $p = 0.0054$, the dfs are approximately 18.8462, and the test statistic is -3.14. Do the procedure again but instead of Calculate do Draw.

Conclusion: At the 5% level of significance, the sample data show there is sufficient evidence to conclude that the mean number of hours that girls and boys aged seven to 11 play sports per day is different (mean number of hours boys aged seven to 11 play sports per day is greater than the mean number of hours played by girls OR the mean number of hours girls aged seven to 11 play sports per day is greater than the mean number of hours played by boys).

Try It Σ

10.1 Two samples are shown in **Table 10.2**. Both have normal distributions. The means for the two populations are thought to be the same. Is there a difference in the means? Test at the 5% level of significance.

	Sample Size	Sample Mean	Sample Standard Deviation
Population A	25	5	1
Population B	16	4.7	1.2

Table 10.2

NOTE

When the sum of the sample sizes is larger than 30 ($n_1 + n_2 > 30$) you can use the normal distribution to approximate the Student's t.

Example 10.2

A study is done by a community group in two neighboring colleges to determine which one graduates students with more math classes. College A samples 11 graduates. Their average is four math classes with a standard deviation of 1.5 math classes. College B samples nine graduates. Their average is 3.5 math classes with a standard deviation of one math class. The community group believes that a student who graduates from college A **has taken more math classes,** on the average. Both populations have a normal distribution. Test at a 1% significance level. Answer the following questions.

a. Is this a test of two means or two proportions?

Solution 10.2
a. two means

b. Are the populations standard deviations known or unknown?

Solution 10.2
b. unknown

c. Which distribution do you use to perform the test?

Solution 10.2
c. Student's t

d. What is the random variable?

Solution 10.2
d. $\bar{X}_A - \bar{X}_B$

e. What are the null and alternate hypotheses? Write the null and alternate hypotheses in words and in symbols.

Solution 10.2
e.

- $H_o : \mu_A \leq \mu_B$

- $H_a : \mu_A > \mu_B$

f. Is this test right-, left-, or two-tailed?

Solution 10.2
f.

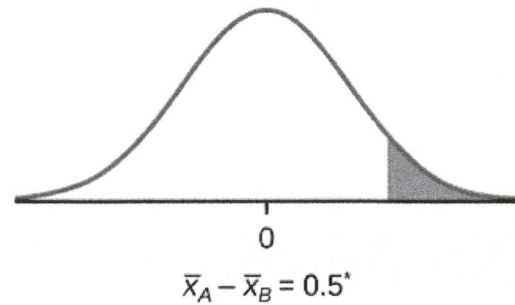

$\overline{x}_A - \overline{x}_B = 0.5^*$

Note: $\overline{x}_A - \overline{x}_B = 4 - 3.5 = 0.5$

Figure 10.3

right

g. What is the *p*-value?

Solution 10.2
g. 0.1928

h. Do you reject or not reject the null hypothesis?

Solution 10.2
h. Do not reject.

i. **Conclusion:**

Solution 10.2
i. At the 1% level of significance, from the sample data, there is not sufficient evidence to conclude that a student who graduates from college A has taken more math classes, on the average, than a student who graduates from college B.

10.2 A study is done to determine if Company A retains its workers longer than Company B. Company A samples 15 workers, and their average time with the company is five years with a standard deviation of 1.2. Company B samples

20 workers, and their average time with the company is 4.5 years with a standard deviation of 0.8. The populations are normally distributed.

   a. Are the population standard deviations known?

   b. Conduct an appropriate hypothesis test. At the 5% significance level, what is your conclusion?

Example 10.3

A professor at a large community college wanted to determine whether there is a difference in the means of final exam scores between students who took his statistics course online and the students who took his face-to-face statistics class. He believed that the mean of the final exam scores for the online class would be lower than that of the face-to-face class. Was the professor correct? The randomly selected 30 final exam scores from each group are listed in **Table 10.3** and **Table 10.4**.

67.6	41.2	85.3	55.9	82.4	91.2	73.5	94.1	64.7	64.7
70.6	38.2	61.8	88.2	70.6	58.8	91.2	73.5	82.4	35.5
94.1	88.2	64.7	55.9	88.2	97.1	85.3	61.8	79.4	79.4

Table 10.3 Online Class

77.9	95.3	81.2	74.1	98.8	88.2	85.9	92.9	87.1	88.2
69.4	57.6	69.4	67.1	97.6	85.9	88.2	91.8	78.8	71.8
98.8	61.2	92.9	90.6	97.6	100	95.3	83.5	92.9	89.4

Table 10.4 Face-to-face Class

Is the mean of the Final Exam scores of the online class lower than the mean of the Final Exam scores of the face-to-face class? Test at a 5% significance level. Answer the following questions:

   a. Is this a test of two means or two proportions?

   b. Are the population standard deviations known or unknown?

   c. Which distribution do you use to perform the test?

   d. What is the random variable?

   e. What are the null and alternative hypotheses? Write the null and alternative hypotheses in words and in symbols.

   f. Is this test right, left, or two tailed?

   g. What is the p-value?

   h. Do you reject or not reject the null hypothesis?

   i. At the ___ level of significance, from the sample data, there _____ (is/is not) sufficient evidence to conclude that _____.

(See the conclusion in **Example 10.2**, and write yours in a similar fashion)

Using the TI-83, 83+, 84, 84+ Calculator

First put the data for each group into two lists (such as L1 and L2). Press STAT. Arrow over to TESTS and press 4:2SampTTest. Make sure Data is highlighted and press ENTER. Arrow down and enter L1 for the first list and L2 for the second list. Arrow down to μ_1: and arrow to $\neq \mu_2$ (does not equal). Press ENTER. Arrow down to Pooled: No. Press ENTER. Arrow down to Calculate and press ENTER.

NOTE

Be careful not to mix up the information for Group 1 and Group 2!

Solution 10.3

a. two means

b. unknown

c. Student's t

d. $\bar{X}_1 - \bar{X}_2$

e. 1. $H_0: \mu_1 = \mu_2$ Null hypothesis: the means of the final exam scores are equal for the online and face-to-face statistics classes.

   2. $H_a: \mu_1 < \mu_2$ Alternative hypothesis: the mean of the final exam scores of the online class is less than the mean of the final exam scores of the face-to-face class.

f. left-tailed

g. p-value = 0.0011

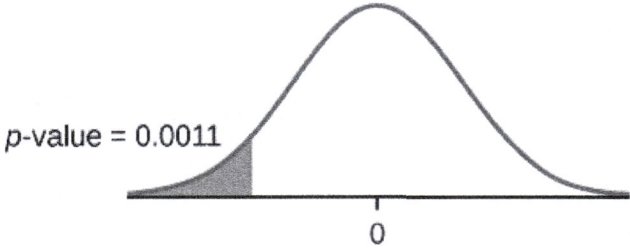

Figure 10.4

h. Reject the null hypothesis

i. The professor was correct. The evidence shows that the mean of the final exam scores for the online class is lower than that of the face-to-face class.
   At the 5% level of significance, from the sample data, there is (is/is not) sufficient evidence to conclude that the mean of the final exam scores for the online class is less than the mean of final exam scores of the face-to-face class.

Cohen's Standards for Small, Medium, and Large Effect Sizes

Cohen's d is a measure of effect size based on the differences between two means. Cohen's d, named for United States statistician Jacob Cohen, measures the relative strength of the differences between the means of two populations based on sample data. The calculated value of effect size is then compared to Cohen's standards of small, medium, and large effect sizes.

Size of effect	d
Small	0.2
medium	0.5
Large	0.8

Table 10.5 Cohen's Standard Effect Sizes

Cohen's d is the measure of the difference between two means divided by the pooled standard deviation: $d = \dfrac{\bar{x}_1 - \bar{x}_2}{s_{pooled}}$

where $s_{pooled} = \sqrt{\dfrac{(n_1-1)s_1^2 + (n_2-1)s_2^2}{n_1 + n_2 - 2}}$

Example 10.4

Calculate Cohen's d for **Example 10.2**. Is the size of the effect small, medium, or large? Explain what the size of the effect means for this problem.

Solution 10.4
$\mu_1 = 4$ $s_1 = 1.5$ $n_1 = 11$
$\mu_2 = 3.5$ $s_2 = 1$ $n_2 = 9$
$d = 0.384$
The effect is small because 0.384 is between Cohen's value of 0.2 for small effect size and 0.5 for medium effect size. The size of the differences of the means for the two colleges is small indicating that there is not a significant difference between them.

Example 10.5

Calculate Cohen's *d* for **Example 10.3**. Is the size of the effect small, medium or large? Explain what the size of the effect means for this problem.

Solution 10.5

d = 0.834; Large, because 0.834 is greater than Cohen's 0.8 for a large effect size. The size of the differences between the means of the Final Exam scores of online students and students in a face-to-face class is large indicating a significant difference.

Try It

10.5 Weighted alpha is a measure of risk-adjusted performance of stocks over a period of a year. A high positive weighted alpha signifies a stock whose price has risen while a small positive weighted alpha indicates an unchanged stock price during the time period. Weighted alpha is used to identify companies with strong upward or downward trends. The weighted alpha for the top 30 stocks of banks in the northeast and in the west as identified by Nasdaq on May 24, 2013 are listed in **Table 10.6** and **Table 10.7**, respectively.

94.2	75.2	69.6	52.0	48.0	41.9	36.4	33.4	31.5	27.6
77.3	71.9	67.5	50.6	46.2	38.4	35.2	33.0	28.7	26.5
76.3	71.7	56.3	48.7	43.2	37.6	33.7	31.8	28.5	26.0

Table 10.6 Northeast

126.0	70.6	65.2	51.4	45.5	37.0	33.0	29.6	23.7	22.6
116.1	70.6	58.2	51.2	43.2	36.0	31.4	28.7	23.5	21.6
78.2	68.2	55.6	50.3	39.0	34.1	31.0	25.3	23.4	21.5

Table 10.7 West

Is there a difference in the weighted alpha of the top 30 stocks of banks in the northeast and in the west? Test at a 5% significance level. Answer the following questions:

a. Is this a test of two means or two proportions?

b. Are the population standard deviations known or unknown?

c. Which distribution do you use to perform the test?

d. What is the random variable?

e. What are the null and alternative hypotheses? Write the null and alternative hypotheses in words and in symbols.

f. Is this test right, left, or two tailed?

g. What is the *p*-value?

h. Do you reject or not reject the null hypothesis?

i. At the ___ level of significance, from the sample data, there _____ (is/is not) sufficient evidence to conclude that _____.

j. Calculate Cohen's *d* and interpret it.

10.2 | Two Population Means with Known Standard Deviations

Even though this situation is not likely (knowing the population standard deviations is not likely), the following example illustrates hypothesis testing for independent means, known population standard deviations. The sampling distribution for the difference between the means is normal and both populations must be normal. The random variable is $\overline{X_1} - \overline{X_2}$. The normal distribution has the following format:

Normal distribution is:

$$\overline{X}_1 - \overline{X}_2 \sim N\left[\mu_1 - \mu_2, \sqrt{\frac{(\sigma_1)^2}{n_1} + \frac{(\sigma_2)^2}{n_2}}\right]$$

The standard deviation is:

$$\sqrt{\frac{(\sigma_1)^2}{n_1} + \frac{(\sigma_2)^2}{n_2}}$$

The test statistic (z-score) is:

$$z = \frac{(\overline{x}_1 - \overline{x}_2) - (\mu_1 - \mu_2)}{\sqrt{\frac{(\sigma_1)^2}{n_1} + \frac{(\sigma_2)^2}{n_2}}}$$

Example 10.6

Independent groups, population standard deviations known: The mean lasting time of two competing floor waxes is to be compared. **Twenty floors** are randomly assigned **to test each wax**. Both populations have a normal distributions. The data are recorded in **Table 10.8**.

Wax	Sample Mean Number of Months Floor Wax Lasts	Population Standard Deviation
1	3	0.33
2	2.9	0.36

Table 10.8

Does the data indicate that **wax 1 is more effective than wax 2**? Test at a 5% level of significance.

Solution 10.6

This is a test of two independent groups, two population means, population standard deviations known.

Random Variable: $\overline{X}_1 - \overline{X}_2$ = difference in the mean number of months the competing floor waxes last.

$H_0: \mu_1 \le \mu_2$

$H_a: \mu_1 > \mu_2$

The words **"is more effective"** says that **wax 1 lasts longer than wax 2**, on average. "Longer" is a ">" symbol and goes into H_a. Therefore, this is a right-tailed test.

Distribution for the test: The population standard deviations are known so the distribution is normal. Using the formula, the distribution is:

$$\overline{X}_1 - \overline{X}_2 \sim N\left(0, \sqrt{\frac{0.33^2}{20} + \frac{0.36^2}{20}}\right)$$

Since $\mu_1 \le \mu_2$ then $\mu_1 - \mu_2 \le 0$ and the mean for the normal distribution is zero.

Calculate the *p*-value using the normal distribution: *p*-value = 0.1799

Graph:

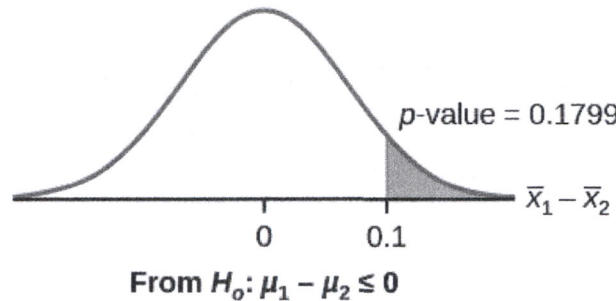

Figure 10.5

$$\overline{X}_1 - \overline{X}_2 = 3 - 2.9 = 0.1$$

Compare α and the *p*-value: $\alpha = 0.05$ and *p*-value = 0.1799. Therefore, $\alpha <$ *p*-value.

Make a decision: Since $\alpha <$ *p*-value, do not reject H_0.

Conclusion: At the 5% level of significance, from the sample data, there is not sufficient evidence to conclude that the mean time wax 1 lasts is longer (wax 1 is more effective) than the mean time wax 2 lasts.

 Using the TI-83, 83+, 84, 84+ Calculator

Press **STAT**. Arrow over to **TESTS** and press **3:2-SampZTest**. Arrow over to **Stats** and press **ENTER**. Arrow down and enter **.33** for sigma1, **.36** for sigma2, **3** for the first sample mean, **20** for n1, **2.9** for the second sample mean, and **20** for n2. Arrow down to μ1: and arrow to > μ2. Press **ENTER**. Arrow down to **Calculate** and press **ENTER**. The *p*-value is $p = 0.1799$ and the test statistic is 0.9157. Do the procedure again, but instead of **Calculate** do **Draw**.

Try It Σ

10.6 The means of the number of revolutions per minute of two competing engines are to be compared. Thirty engines are randomly assigned to be tested. Both populations have normal distributions. **Table 10.9** shows the result. Do the data indicate that Engine 2 has higher RPM than Engine 1? Test at a 5% level of significance.

Engine	Sample Mean Number of RPM	Population Standard Deviation
1	1,500	50
2	1,600	60

Table 10.9

Example 10.7

An interested citizen wanted to know if Democratic U. S. senators are older than Republican U.S. senators, on average. On May 26 2013, the mean age of 30 randomly selected Republican Senators was 61 years 247 days old (61.675 years) with a standard deviation of 10.17 years. The mean age of 30 randomly selected Democratic senators was 61 years 257 days old (61.704 years) with a standard deviation of 9.55 years.

Do the data indicate that Democratic senators are older than Republican senators, on average? Test at a 5% level of significance.

Solution 10.7

This is a test of two independent groups, two population means. The population standard deviations are unknown, but the sum of the sample sizes is 30 + 30 = 60, which is greater than 30, so we can use the normal approximation to the Student's-t distribution. Subscripts: 1: Democratic senators 2: Republican senators

Random variable: $\bar{X}_1 - \bar{X}_2$ = difference in the mean age of Democratic and Republican U.S. senators.

$H_0: \mu_1 \le \mu_2 \quad H_0: \mu_1 - \mu_2 \le 0$

$H_a: \mu_1 > \mu_2 \quad H_a: \mu_1 - \mu_2 > 0$

The words "older than" translates as a ">" symbol and goes into H_a. Therefore, this is a right-tailed test.

Distribution for the test: The distribution is the normal approximation to the Student's t for means, independent groups. Using the formula, the distribution is: $\bar{X}_1 - \bar{X}_2 \sim N[0, \sqrt{\frac{(9.55)^2}{30} + \frac{(10.17)^2}{30}}]$

Since $\mu_1 \le \mu_2$, $\mu_1 - \mu_2 \le 0$ and the mean for the normal distribution is zero.

(Calculating the p-value using the normal distribution gives p-value = 0.4040)

Graph:

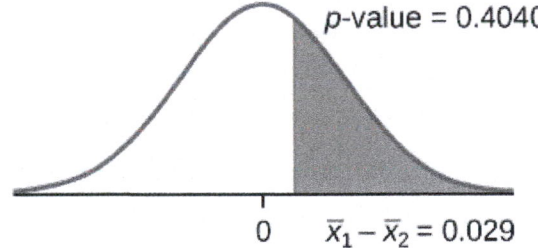

Figure 10.6

Compare α and the p-value: $\alpha = 0.05$ and p-value = 0.4040. Therefore, $\alpha < p$-value.

Make a decision: Since $\alpha < p$-value, do not reject H_0.

Conclusion: At the 5% level of significance, from the sample data, there is not sufficient evidence to conclude that the mean age of Democratic senators is greater than the mean age of the Republican senators.

10.3 | Comparing Two Independent Population Proportions

When conducting a hypothesis test that compares two independent population proportions, the following characteristics should be present:

1. The two independent samples are simple random samples that are independent.

2. The number of successes is at least five, and the number of failures is at least five, for each of the samples.

3. Growing literature states that the population must be at least ten or 20 times the size of the sample. This keeps each population from being over-sampled and causing incorrect results.

Comparing two proportions, like comparing two means, is common. If two estimated proportions are different, it may be due to a difference in the populations or it may be due to chance. A hypothesis test can help determine if a difference in the estimated proportions reflects a difference in the population proportions.

The difference of two proportions follows an approximate normal distribution. Generally, the null hypothesis states that the two proportions are the same. That is, H_0: $p_A = p_B$. To conduct the test, we use a pooled proportion, p_c.

The pooled proportion is calculated as follows:

$$p_c = \frac{x_A + x_B}{n_A + n_B}$$

The distribution for the differences is:

$$P'_A - P'_B \sim N[0, \sqrt{p_c(1 - p_c)(\frac{1}{n_A} + \frac{1}{n_B})}]$$

The test statistic (z-score) is:

$$z = \frac{(p'_A - p'_B) - (p_A - p_B)}{\sqrt{p_c(1 - p_c)(\frac{1}{n_A} + \frac{1}{n_B})}}$$

Example 10.8

Two types of medication for hives are being tested to determine if there is a **difference in the proportions of adult patient reactions. Twenty** out of a random **sample of 200** adults given medication A still had hives 30 minutes after taking the medication. **Twelve** out of another **random sample of 200 adults** given medication B still had hives 30 minutes after taking the medication. Test at a 1% level of significance.

Solution 10.8

The problem asks for a difference in proportions, making it a test of two proportions.

Let A and B be the subscripts for medication A and medication B, respectively. Then p_A and p_B are the desired population proportions.

Random Variable:

$P'_A - P'_B$ = difference in the proportions of adult patients who did not react after 30 minutes to medication A and to medication B.

H_0: $p_A = p_B$

$p_A - p_B = 0$

H_a: $p_A \neq p_B$

$p_A - p_B \neq 0$

The words **"is a difference"** tell you the test is two-tailed.

Distribution for the test: Since this is a test of two binomial population proportions, the distribution is normal:

$$p_c = \frac{x_A + x_B}{n_A + n_B} = \frac{20 + 12}{200 + 200} = 0.08 \quad 1 - p_c = 0.92$$

$$P'_A - P'_B \sim N\left[0, \sqrt{(0.08)(0.92)(\frac{1}{200} + \frac{1}{200})}\right]$$

$P'_A - P'_B$ follows an approximate normal distribution.

Calculate the *p*-value using the normal distribution: *p*-value = 0.1404.

Estimated proportion for group A: $p'_A = \frac{x_A}{n_A} = \frac{20}{200} = 0.1$

Estimated proportion for group B: $p'_B = \frac{x_B}{n_B} = \frac{12}{200} = 0.06$

Graph:

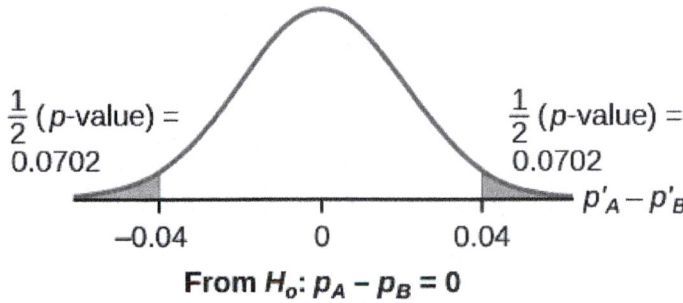

Figure 10.7

$P'_A - P'_B = 0.1 - 0.06 = 0.04$.

Half the p-value is below −0.04, and half is above 0.04.

Compare α and the p-value: $\alpha = 0.01$ and the p-value = 0.1404. $\alpha < p$-value.

Make a decision: Since $\alpha < p$-value, do not reject H_0.

Conclusion: At a 1% level of significance, from the sample data, there is not sufficient evidence to conclude that there is a difference in the proportions of adult patients who did not react after 30 minutes to medication A and medication B.

 Using the TI-83, 83+, 84, 84+ Calculator

Press **STAT**. Arrow over to **TESTS** and press **6:2-PropZTest**. Arrow down and enter **20** for x1, **200** for n1, **12** for x2, and **200** for n2. Arrow down to **p1:** and arrow to **not equal p2**. Press **ENTER**. Arrow down to **Calculate** and press **ENTER**. The p-value is $p = 0.1404$ and the test statistic is 1.47. Do the procedure again, but instead of **Calculate** do **Draw**.

Try It Σ

10.8 Two types of valves are being tested to determine if there is a difference in pressure tolerances. Fifteen out of a random sample of 100 of Valve A cracked under 4,500 psi. Six out of a random sample of 100 of Valve B cracked under 4,500 psi. Test at a 5% level of significance.

Example 10.9

A research study was conducted about gender differences in "sexting." The researcher believed that the proportion of girls involved in "sexting" is less than the proportion of boys involved. The data collected in the spring of 2010 among a random sample of middle and high school students in a large school district in the southern United States

is summarized in **Table 10.9**. Is the proportion of girls sending sexts less than the proportion of boys "sexting?"
Test at a 1% level of significance.

	Males	Females
Sent "sexts"	183	156
Total number surveyed	2231	2169

Table 10.10

Solution 10.9

This is a test of two population proportions. Let M and F be the subscripts for males and females. Then p_M and p_F are the desired population proportions.

Random variable:

$p'_F - p'_M$ = difference in the proportions of males and females who sent "sexts."

$H_0: p_F = p_M$ $H_0: p_F - p_M = 0$

$H_a: p_F < p_M$ $H_a: p_F - p_M < 0$

The words **"less than"** tell you the test is left-tailed.

Distribution for the test: Since this is a test of two population proportions, the distribution is normal:

$$p_c = \frac{x_F + x_M}{n_F + n_M} = \frac{156 + 183}{2169 + 2231} = 0.077$$

$1 - p_c = 0.923$

Therefore,

$$p'_F - p'_M \sim N\left(0, \ \sqrt{(0.077)(0.923)\left(\frac{1}{2169} + \frac{1}{2231}\right)}\right)$$

$p'_F - p'_M$ follows an approximate normal distribution.

Calculate the _p_-value using the normal distribution:
p-value = 0.1045
Estimated proportion for females: 0.0719
Estimated proportion for males: 0.082

Graph:

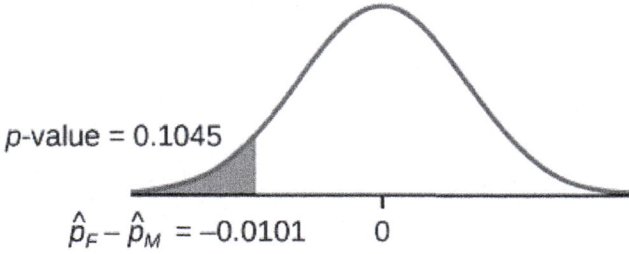

Figure 10.8

Decision: Since $\alpha < p$-value, Do not reject H_0

Conclusion: At the 1% level of significance, from the sample data, there is not sufficient evidence to conclude that the proportion of girls sending "sexts" is less than the proportion of boys sending "sexts."

Using the TI-83, 83+, 84, 84+ Calculator

Press STAT. Arrow over to TESTS and press 6:2-PropZTest. Arrow down and enter 156 for x1, 2169 for n1, 183 for x2, and 2231 for n2. Arrow down to p1: and arrow to less than p2. Press ENTER. Arrow down to Calculate and press ENTER. The *p*-value is $P = 0.1045$ and the test statistic is $z = -1.256$.

Example 10.10

Researchers conducted a study of smartphone use among adults. A cell phone company claimed that iPhone smartphones are more popular with whites (non-Hispanic) than with African Americans. The results of the survey indicate that of the 232 African American cell phone owners randomly sampled, 5% have an iPhone. Of the 1,343 white cell phone owners randomly sampled, 10% own an iPhone. Test at the 5% level of significance. Is the proportion of white iPhone owners greater than the proportion of African American iPhone owners?

Solution 10.10

This is a test of two population proportions. Let W and A be the subscripts for the whites and African Americans. Then p_W and p_A are the desired population proportions.

Random variable:

$p'_W - p'_A$ = difference in the proportions of Android and iPhone users.

$H_0: p_W = p_A$ $H_0: p_W - p_A = 0$

$H_a: p_W > p_A$ $H_a: p_W - p_A > 0$

The words "more popular" indicate that the test is right-tailed.

Distribution for the test: The distribution is approximately normal:

$$p_c = \frac{x_W + x_A}{n_W + n_A} = \frac{134 + 12}{1343 + 232} = 0.0927$$

$$1 - p_c = 0.9073$$

Therefore,

$$p'_W - p'_A \sim N\left(0, \sqrt{(0.0927)(0.9073)\left(\frac{1}{1343} + \frac{1}{232}\right)}\right)$$

$p'_W - p'_A$ follows an approximate normal distribution.

Calculate the *p*-value using the normal distribution:

p-value = 0.0077
Estimated proportion for group A: 0.10
Estimated proportion for group B: 0.05

Graph:

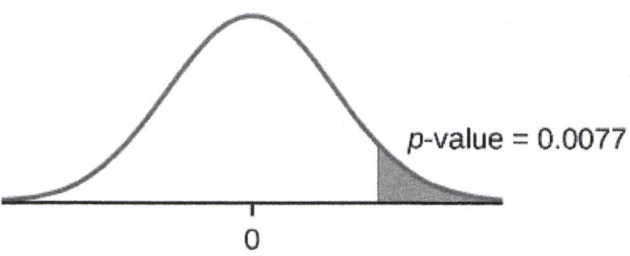

p-value = 0.0077

0

Figure 10.9

Decision: Since $\alpha > p$-value, reject the H_0.

Conclusion: At the 5% level of significance, from the sample data, there is sufficient evidence to conclude that a larger proportion of white cell phone owners use iPhones than African Americans.

 Using the TI-83, 83+, 84, 84+ Calculator

TI-83+ and TI-84: Press STAT. Arrow over to TESTS and press 6:2-PropZTest. Arrow down and enter 135 for x1, 1343 for n1, 12 for x2, and 232 for n2. Arrow down to p1: and arrow to greater than p2. Press ENTER. Arrow down to Calculate and press ENTER. The P-value is P = 0.0092 and the test statistic is Z = 2.33.

Try It Σ

10.10 A concerned group of citizens wanted to know if the proportion of forcible rapes in Texas was different in 2011 than in 2010. Their research showed that of the 113,231 violent crimes in Texas in 2010, 7,622 of them were forcible rapes. In 2011, 7,439 of the 104,873 violent crimes were in the forcible rape category. Test at a 5% significance level. Answer the following questions:

a. Is this a test of two means or two proportions?

b. Which distribution do you use to perform the test?

c. What is the random variable?

d. What are the null and alternative hypothesis? Write the null and alternative hypothesis in symbols.

e. Is this test right-, left-, or two-tailed?

f. What is the p-value?

g. Do you reject or not reject the null hypothesis?

h. At the ___ level of significance, from the sample data, there _____ (is/is not) sufficient evidence to conclude that _____.

10.4 | Matched or Paired Samples

When using a hypothesis test for matched or paired samples, the following characteristics should be present:

1.   Simple random sampling is used.

2. Sample sizes are often small.

3. Two measurements (samples) are drawn from the same pair of individuals or objects.

4. Differences are calculated from the matched or paired samples.

5. The differences form the sample that is used for the hypothesis test.

6. Either the matched pairs have differences that come from a population that is normal or the number of differences is sufficiently large so that distribution of the sample mean of differences is approximately normal.

In a hypothesis test for matched or paired samples, subjects are matched in pairs and differences are calculated. The differences are the data. The population mean for the differences, μ_d, is then tested using a Student's-t test for a single population mean with $n - 1$ degrees of freedom, where n is the number of differences.

The test statistic (*t*-score) is:

$$t = \frac{\overline{x}_d - \mu_d}{\left(\frac{s_d}{\sqrt{n}}\right)}$$

Example 10.11

A study was conducted to investigate the effectiveness of hypnotism in reducing pain. Results for randomly selected subjects are shown in **Table 10.10**. A lower score indicates less pain. The "before" value is matched to an "after" value and the differences are calculated. The differences have a normal distribution. Are the sensory measurements, on average, lower after hypnotism? Test at a 5% significance level.

Subject:	A	B	C	D	E	F	G	H
Before	6.6	6.5	9.0	10.3	11.3	8.1	6.3	11.6
After	6.8	2.4	7.4	8.5	8.1	6.1	3.4	2.0

Table 10.11

Solution 10.11

Corresponding "before" and "after" values form matched pairs. (Calculate "after" – "before.")

After Data	Before Data	Difference
6.8	6.6	0.2
2.4	6.5	-4.1
7.4	9	-1.6
8.5	10.3	-1.8
8.1	11.3	-3.2
6.1	8.1	-2
3.4	6.3	-2.9
2	11.6	-9.6

Table 10.12

The data **for the test** are the differences: {0.2, –4.1, –1.6, –1.8, –3.2, –2, –2.9, –9.6}

The sample mean and sample standard deviation of the differences are: $\overline{x_d} = -3.13$ and $s_d = 2.91$ Verify these values.

Let μ_d be the population mean for the differences. We use the subscript d to denote "differences."

Random variable: $\overline{X}_d$ = the mean difference of the sensory measurements

$H_0: \mu_d \geq 0$

The null hypothesis is zero or positive, meaning that there is the same or more pain felt after hypnotism. That means the subject shows no improvement. μ_d is the population mean of the differences.)

$H_a: \mu_d < 0$

The alternative hypothesis is negative, meaning there is less pain felt after hypnotism. That means the subject shows improvement. The score should be lower after hypnotism, so the difference ought to be negative to indicate improvement.

Distribution for the test: The distribution is a Student's t with $df = n - 1 = 8 - 1 = 7$. Use t_7. **(Notice that the test is for a single population mean.)**

Calculate the p-value using the Student's-t distribution: p-value = 0.0095

Graph:

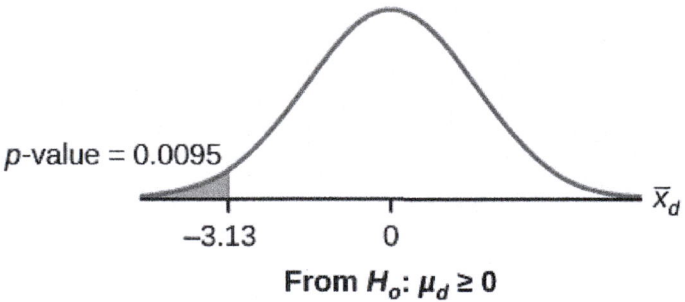

Figure 10.10

$\overline{X}_d$ is the random variable for the differences.

The sample mean and sample standard deviation of the differences are:

$\overline{x}_d = -3.13$

$\overline{s}_d = 2.91$

Compare α and the p-value: $\alpha = 0.05$ and p-value = 0.0095. $\alpha > p$-value.

Make a decision: Since $\alpha > p$-value, reject H_0. This means that $\mu_d < 0$ and there is improvement.

Conclusion: At a 5% level of significance, from the sample data, there is sufficient evidence to conclude that the sensory measurements, on average, are lower after hypnotism. Hypnotism appears to be effective in reducing pain.

NOTE

 For the TI-83+ and TI-84 calculators, you can either calculate the differences ahead of time (**after - before**) and put the differences into a list or you can put the **after** data into a first list and the **before**

data into a second list. Then go to a third list and arrow up to the name. Enter 1st list name - 2nd list name. The calculator will do the subtraction, and you will have the differences in the third list.

 Using the TI-83, 83+, 84, 84+ Calculator

Use your list of differences as the data. Press STAT and arrow over to TESTS. Press 2:T-Test. Arrow over to Data and press ENTER. Arrow down and enter 0 for μ_0, the name of the list where you put the data, and 1 for Freq:. Arrow down to μ: and arrow over to < μ_0. Press ENTER. Arrow down to Calculate and press ENTER. The *p*-value is 0.0094, and the test statistic is -3.04. Do these instructions again except, arrow to Draw (instead of Calculate). Press ENTER.

Try It Σ

10.11 A study was conducted to investigate how effective a new diet was in lowering cholesterol. Results for the randomly selected subjects are shown in the table. The differences have a normal distribution. Are the subjects' cholesterol levels lower on average after the diet? Test at the 5% level.

Subject	A	B	C	D	E	F	G	H	I
Before	209	210	205	198	216	217	238	240	222
After	199	207	189	209	217	202	211	223	201

Table 10.13

Example 10.12

A college football coach was interested in whether the college's strength development class increased his players' maximum lift (in pounds) on the bench press exercise. He asked four of his players to participate in a study. The amount of weight they could each lift was recorded before they took the strength development class. After completing the class, the amount of weight they could each lift was again measured. The data are as follows:

Weight (in pounds)	Player 1	Player 2	Player 3	Player 4
Amount of weight lifted prior to the class	205	241	338	368
Amount of weight lifted after the class	295	252	330	360

Table 10.14

The coach wants to know if the strength development class makes his players stronger, on average.
Record the **differences** data. Calculate the differences by subtracting the amount of weight lifted prior to the class from the weight lifted after completing the class. The data for the differences are: {90, 11, -8, -8}. Assume the differences have a normal distribution.

Using the differences data, calculate the sample mean and the sample standard deviation.

$\overline{x}_d = 21.3$, $s_d = 46.7$

NOTE

The data given here would indicate that the distribution is actually right-skewed. The difference 90 may be an extreme outlier? It is pulling the sample mean to be 21.3 (positive). The means of the other three data values are actually negative.

Using the difference data, this becomes a test of a single _____ (fill in the blank).

Define the random variable: $\overline{X}_d$ mean difference in the maximum lift per player.

The distribution for the hypothesis test is t_3.

H_0: $\mu_d \le 0$, H_a: $\mu_d > 0$

Graph:

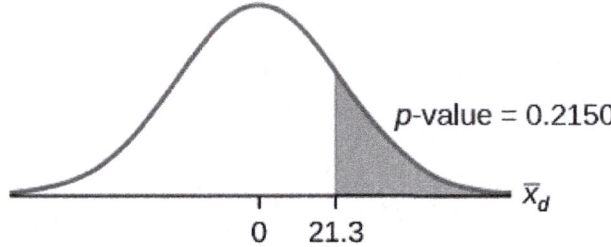

Figure 10.11

Calculate the p-value: The p-value is 0.2150

Decision: If the level of significance is 5%, the decision is not to reject the null hypothesis, because $\alpha <$ p-value.

What is the conclusion?

At a 5% level of significance, from the sample data, there is not sufficient evidence to conclude that the strength development class helped to make the players stronger, on average.

Try It Σ

10.12 A new prep class was designed to improve SAT test scores. Five students were selected at random. Their scores on two practice exams were recorded, one before the class and one after. The data recorded in **Table 10.15**. Are the scores, on average, higher after the class? Test at a 5% level.

SAT Scores	Student 1	Student 2	Student 3	Student 4
Score before class	1840	1960	1920	2150
Score after class	1920	2160	2200	2100

Table 10.15

Example 10.13

Seven eighth graders at Kennedy Middle School measured how far they could push the shot-put with their dominant (writing) hand and their weaker (non-writing) hand. They thought that they could push equal distances with either hand. The data were collected and recorded in **Table 10.16**.

Distance (in feet) using	Student 1	Student 2	Student 3	Student 4	Student 5	Student 6	Student 7
Dominant Hand	30	26	34	17	19	26	20
Weaker Hand	28	14	27	18	17	26	16

Table 10.16

Conduct a hypothesis test to determine whether the mean difference in distances between the children's dominant versus weaker hands is significant.

Record the **differences** data. Calculate the differences by subtracting the distances with the weaker hand from the distances with the dominant hand. The data for the differences are: {2, 12, 7, –1, 2, 0, 4}. The differences have a normal distribution.

Using the differences data, calculate the sample mean and the sample standard deviation. $\bar{x}_d = 3.71$, $s_d = 4.5$.

Random variable: $\bar{X}_d$ = mean difference in the distances between the hands.

Distribution for the hypothesis test: t_6

$H_0: \mu_d = 0$ $H_a: \mu_d \neq 0$

Graph:

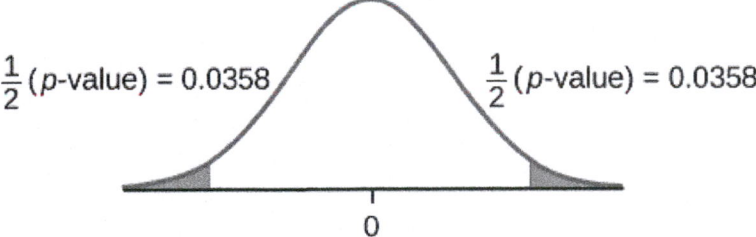

Figure 10.12

Calculate the *p*-value: The *p*-value is 0.0716 (using the data directly).

(test statistic = 2.18. *p*-value = 0.0719 using $\left(\bar{x}_d = 3.71, \ s_d = 4.5. \right)$

Decision: Assume $\alpha = 0.05$. Since $\alpha < p$-value, Do not reject H_0.

Conclusion: At the 5% level of significance, from the sample data, there is not sufficient evidence to conclude that there is a difference in the children's weaker and dominant hands to push the shot-put.

10.13 Five ball players think they can throw the same distance with their dominant hand (throwing) and off-hand (catching hand). The data were collected and recorded in **Table 10.17**. Conduct a hypothesis test to determine whether the mean difference in distances between the dominant and off-hand is significant. Test at the 5% level.

	Player 1	Player 2	Player 3	Player 4	Player 5
Dominant Hand	120	111	135	140	125
Off-hand	105	109	98	111	99

Table 10.17

10.5 | Hypothesis Testing for Two Means and Two Proportions

Stats Lab

10.1 Hypothesis Testing for Two Means and Two Proportions

Class Time:

Names:

Student Learning Outcomes

- The student will select the appropriate distributions to use in each case.
- The student will conduct hypothesis tests and interpret the results.

Supplies:

- the business section from two consecutive days' newspapers
- three small packages of M&Ms®
- five small packages of Reese's Pieces®

Increasing Stocks Survey

Look at yesterday's newspaper business section. Conduct a hypothesis test to determine if the proportion of New York Stock Exchange (NYSE) stocks that increased is greater than the proportion of NASDAQ stocks that increased. As randomly as possible, choose 40 NYSE stocks, and 32 NASDAQ stocks and complete the following statements.

1. H_0: _____
2. H_a: _____
3. In words, define the random variable.
4. The distribution to use for the test is _____.
5. Calculate the test statistic using your data.
6. Draw a graph and label it appropriately. Shade the actual level of significance.
   a. Graph:

Figure 10.13

   b. Calculate the p-value.
7. Do you reject or not reject the null hypothesis? Why?

8. Write a clear conclusion using a complete sentence.

Decreasing Stocks Survey

Randomly pick eight stocks from the newspaper. Using two consecutive days' business sections, test whether the stocks went down, on average, for the second day.

1. H_0: _____
2. H_a: _____
3. In words, define the random variable.
4. The distribution to use for the test is _____.
5. Calculate the test statistic using your data.
6. Draw a graph and label it appropriately. Shade the actual level of significance.

   a. Graph:

   Figure 10.14

   b. Calculate the *p*-value:
7. Do you reject or not reject the null hypothesis? Why?
8. Write a clear conclusion using a complete sentence.

Candy Survey

Buy three small packages of M&Ms and five small packages of Reese's Pieces (same net weight as the M&Ms). Test whether or not the mean number of candy pieces per package is the same for the two brands.

1. H_0: _____
2. H_a: _____
3. In words, define the random variable.
4. What distribution should be used for this test?
5. Calculate the test statistic using your data.
6. Draw a graph and label it appropriately. Shade the actual level of significance.

   a. Graph:

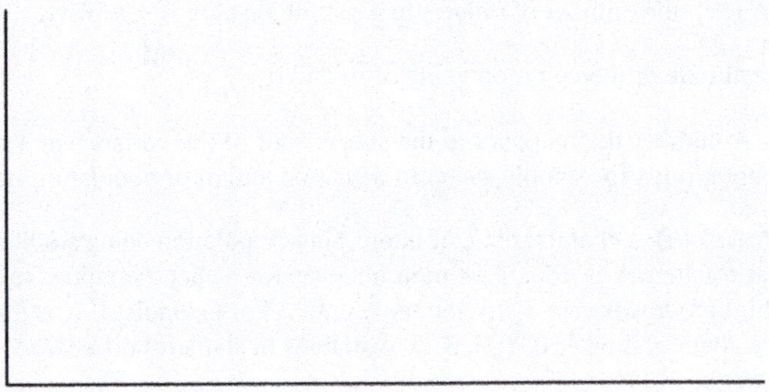

Figure 10.15

    b. Calculate the *p*-value.

7. Do you reject or not reject the null hypothesis? Why?

8. Write a clear conclusion using a complete sentence.

Shoe Survey

Test whether women have, on average, more pairs of shoes than men. Include all forms of sneakers, shoes, sandals, and boots. Use your class as the sample.

1. H_0: _____

2. H_a: _____

3. In words, define the random variable.

4. The distribution to use for the test is _____.

5. Calculate the test statistic using your data.

6. Draw a graph and label it appropriately. Shade the actual level of significance.

    a. Graph:

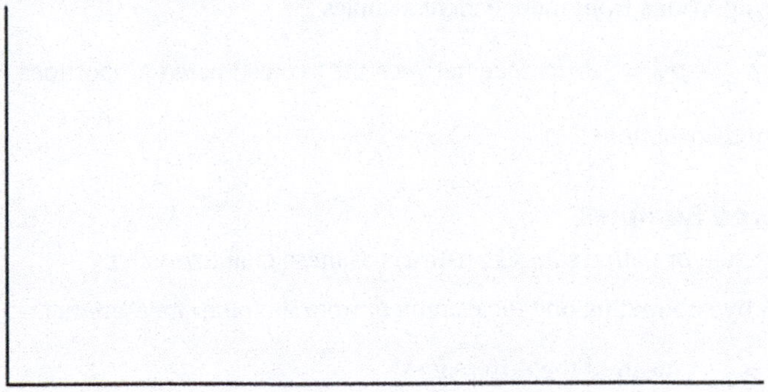

Figure 10.16

    b. Calculate the *p*-value.

7. Do you reject or not reject the null hypothesis? Why?

8. Write a clear conclusion using a complete sentence.

KEY TERMS

Degrees of Freedom (*df*) the number of objects in a sample that are free to vary.

Pooled Proportion estimate of the common value of p_1 and p_2.

Standard Deviation A number that is equal to the square root of the variance and measures how far data values are from their mean; notation: s for sample standard deviation and σ for population standard deviation.

Variable (Random Variable) a characteristic of interest in a population being studied. Common notation for variables are upper-case Latin letters X, Y, Z,... Common notation for a specific value from the domain (set of all possible values of a variable) are lower-case Latin letters x, y, z,.... For example, if X is the number of children in a family, then x represents a specific integer 0, 1, 2, 3, ... Variables in statistics differ from variables in intermediate algebra in two following ways.

- The domain of the random variable (RV) is not necessarily a numerical set; the domain may be expressed in words; for example, if X = hair color, then the domain is {black, blond, gray, green, orange}.

- We can tell what specific value x of the random variable X takes only after performing the experiment.

CHAPTER REVIEW

10.1 Two Population Means with Unknown Standard Deviations

Two population means from independent samples where the population standard deviations are not known

- Random Variable: $\bar{X}_1 - \bar{X}_2$ = the difference of the sampling means

- Distribution: Student's t-distribution with degrees of freedom (variances not pooled)

10.2 Two Population Means with Known Standard Deviations

A hypothesis test of two population means from independent samples where the population standard deviations are known (typically approximated with the sample standard deviations), will have these characteristics:

- Random variable: $\bar{X}_1 - \bar{X}_2$ = the difference of the means

- Distribution: normal distribution

10.3 Comparing Two Independent Population Proportions

Test of two population proportions from independent samples.

- Random variable: $\hat{p}_A - \hat{p}_B$ = difference between the two estimated proportions

- Distribution: normal distribution

10.4 Matched or Paired Samples

A hypothesis test for matched or paired samples (t-test) has these characteristics:

- Test the differences by subtracting one measurement from the other measurement

- Random Variable: $\bar{x}_d$ = mean of the differences

- Distribution: Student's-t distribution with $n - 1$ degrees of freedom

- If the number of differences is small (less than 30), the differences must follow a normal distribution.

- Two samples are drawn from the same set of objects.

- Samples are dependent.

FORMULA REVIEW

10.1 Two Population Means with Unknown Standard Deviations

Standard error: $SE = \sqrt{\dfrac{(s_1)^2}{n_1} + \dfrac{(s_2)^2}{n_2}}$

Test statistic (t-score): $t = \dfrac{(\bar{x}_1 - \bar{x}_2) - (\mu_1 - \mu_2)}{\sqrt{\dfrac{(s_1)^2}{n_1} + \dfrac{(s_2)^2}{n_2}}}$

Degrees of freedom:

$$df = \dfrac{\left(\dfrac{(s_1)^2}{n_1} + \dfrac{(s_2)^2}{n_2}\right)^2}{\left(\dfrac{1}{n_1 - 1}\right)\left(\dfrac{(s_1)^2}{n_1}\right)^2 + \left(\dfrac{1}{n_2 - 1}\right)\left(\dfrac{(s_2)^2}{n_2}\right)^2}$$

where:

s_1 and s_2 are the sample standard deviations, and n_1 and n_2 are the sample sizes.

$\bar{x}_1$ and $\bar{x}_2$ are the sample means.

Cohen's d is the measure of effect size:

$$d = \dfrac{\bar{x}_1 - \bar{x}_2}{s_{pooled}}$$

where $s_{pooled} = \sqrt{\dfrac{(n_1 - 1)s_1^2 + (n_2 - 1)s_2^2}{n_1 + n_2 - 2}}$

10.2 Two Population Means with Known Standard Deviations

Normal Distribution:

$$\bar{X}_1 - \bar{X}_2 \sim N\left[\mu_1 - \mu_2, \sqrt{\dfrac{(\sigma_1)^2}{n_1} + \dfrac{(\sigma_2)^2}{n_2}}\right].$$

Generally $\mu_1 - \mu_2 = 0.$

Test Statistic (z-score):

$$z = \dfrac{(\bar{x}_1 - \bar{x}_2) - (\mu_1 - \mu_2)}{\sqrt{\dfrac{(\sigma_1)^2}{n_1} + \dfrac{(\sigma_2)^2}{n_2}}}$$

Generally $\mu_1 - \mu_2 = 0.$

where:

σ_1 and σ_2 are the known population standard deviations. n_1 and n_2 are the sample sizes. $\bar{x}_1$ and $\bar{x}_2$ are the sample means. μ_1 and μ_2 are the population means.

10.3 Comparing Two Independent Population Proportions

Pooled Proportion: $p_c = \dfrac{x_F + x_M}{n_F + n_M}$

Distribution for the differences:

$$p'_A - p'_B \sim N\left[0, \sqrt{p_c(1 - p_c)\left(\dfrac{1}{n_A} + \dfrac{1}{n_B}\right)}\right]$$

where the null hypothesis is $H_0: p_A = p_B$ or $H_0: p_A - p_B = 0.$

Test Statistic (z-score): $z = \dfrac{(p'_A - p'_B)}{\sqrt{p_c(1 - p_c)\left(\dfrac{1}{n_A} + \dfrac{1}{n_B}\right)}}$

where the null hypothesis is $H_0: p_A = p_B$ or $H_0: p_A - p_B = 0.$

where

p'_A and p'_B are the sample proportions, p_A and p_B are the population proportions,

P_c is the pooled proportion, and n_A and n_B are the sample sizes.

10.4 Matched or Paired Samples

Test Statistic (t-score): $t = \dfrac{\bar{x}_d - \mu_d}{\left(\dfrac{s_d}{\sqrt{n}}\right)}$

where:

$\bar{x}_d$ is the mean of the sample differences. μ_d is the mean of the population differences. s_d is the sample standard deviation of the differences. n is the sample size.

PRACTICE

10.1 Two Population Means with Unknown Standard Deviations

Use the following information to answer the next 15 exercises: Indicate if the hypothesis test is for

a. independent group means, population standard deviations, and/or variances known

b. independent group means, population standard deviations, and/or variances unknown

c. matched or paired samples

d. single mean

e. two proportions

f. single proportion

1. It is believed that 70% of males pass their drivers test in the first attempt, while 65% of females pass the test in the first attempt. Of interest is whether the proportions are in fact equal.

2. A new laundry detergent is tested on consumers. Of interest is the proportion of consumers who prefer the new brand over the leading competitor. A study is done to test this.

3. A new windshield treatment claims to repel water more effectively. Ten windshields are tested by simulating rain without the new treatment. The same windshields are then treated, and the experiment is run again. A hypothesis test is conducted.

4. The known standard deviation in salary for all mid-level professionals in the financial industry is $11,000. Company A and Company B are in the financial industry. Suppose samples are taken of mid-level professionals from Company A and from Company B. The sample mean salary for mid-level professionals in Company A is $80,000. The sample mean salary for mid-level professionals in Company B is $96,000. Company A and Company B management want to know if their mid-level professionals are paid differently, on average.

5. The average worker in Germany gets eight weeks of paid vacation.

6. According to a television commercial, 80% of dentists agree that Ultrafresh toothpaste is the best on the market.

7. It is believed that the average grade on an English essay in a particular school system for females is higher than for males. A random sample of 31 females had a mean score of 82 with a standard deviation of three, and a random sample of 25 males had a mean score of 76 with a standard deviation of four.

8. The league mean batting average is 0.280 with a known standard deviation of 0.06. The Rattlers and the Vikings belong to the league. The mean batting average for a sample of eight Rattlers is 0.210, and the mean batting average for a sample of eight Vikings is 0.260. There are 24 players on the Rattlers and 19 players on the Vikings. Are the batting averages of the Rattlers and Vikings statistically different?

9. In a random sample of 100 forests in the United States, 56 were coniferous or contained conifers. In a random sample of 80 forests in Mexico, 40 were coniferous or contained conifers. Is the proportion of conifers in the United States statistically more than the proportion of conifers in Mexico?

10. A new medicine is said to help improve sleep. Eight subjects are picked at random and given the medicine. The means hours slept for each person were recorded before starting the medication and after.

11. It is thought that teenagers sleep more than adults on average. A study is done to verify this. A sample of 16 teenagers has a mean of 8.9 hours slept and a standard deviation of 1.2. A sample of 12 adults has a mean of 6.9 hours slept and a standard deviation of 0.6.

12. Varsity athletes practice five times a week, on average.

13. A sample of 12 in-state graduate school programs at school A has a mean tuition of $64,000 with a standard deviation of $8,000. At school B, a sample of 16 in-state graduate programs has a mean of $80,000 with a standard deviation of $6,000. On average, are the mean tuitions different?

14. A new WiFi range booster is being offered to consumers. A researcher tests the native range of 12 different routers under the same conditions. The ranges are recorded. Then the researcher uses the new WiFi range booster and records the new ranges. Does the new WiFi range booster do a better job?

15. A high school principal claims that 30% of student athletes drive themselves to school, while 4% of non-athletes drive themselves to school. In a sample of 20 student athletes, 45% drive themselves to school. In a sample of 35 non-athlete students, 6% drive themselves to school. Is the percent of student athletes who drive themselves to school more than the percent of nonathletes?

Use the following information to answer the next three exercises: A study is done to determine which of two soft drinks has more sugar. There are 13 cans of Beverage A in a sample and six cans of Beverage B. The mean amount of sugar in Beverage A is 36 grams with a standard deviation of 0.6 grams. The mean amount of sugar in Beverage B is 38 grams with a standard deviation of 0.8 grams. The researchers believe that Beverage B has more sugar than Beverage A, on average. Both populations have normal distributions.

16. Are standard deviations known or unknown?

17. What is the random variable?

18. Is this a one-tailed or two-tailed test?

Use the following information to answer the next 12 exercises: The U.S. Center for Disease Control reports that the mean life expectancy was 47.6 years for whites born in 1900 and 33.0 years for nonwhites. Suppose that you randomly survey death records for people born in 1900 in a certain county. Of the 124 whites, the mean life span was 45.3 years with a standard deviation of 12.7 years. Of the 82 nonwhites, the mean life span was 34.1 years with a standard deviation of 15.6 years. Conduct a hypothesis test to see if the mean life spans in the county were the same for whites and nonwhites.

19. Is this a test of means or proportions?

20. State the null and alternative hypotheses.
   a. H_0: _____
   b. H_a: _____

21. Is this a right-tailed, left-tailed, or two-tailed test?

22. In symbols, what is the random variable of interest for this test?

23. In words, define the random variable of interest for this test.

24. Which distribution (normal or Student's t) would you use for this hypothesis test?

25. Explain why you chose the distribution you did for **Exercise 10.24**.

26. Calculate the test statistic and p-value.

27. Sketch a graph of the situation. Label the horizontal axis. Mark the hypothesized difference and the sample difference. Shade the area corresponding to the p-value.

28. Find the p-value.

29. At a pre-conceived $\alpha = 0.05$, what is your:
   a. Decision:
   b. Reason for the decision:
   c. Conclusion (write out in a complete sentence):

30. Does it appear that the means are the same? Why or why not?

10.2 Two Population Means with Known Standard Deviations

Use the following information to answer the next five exercises. The mean speeds of fastball pitches from two different baseball pitchers are to be compared. A sample of 14 fastball pitches is measured from each pitcher. The populations have normal distributions. **Table 10.18** shows the result. Scouters believe that Rodriguez pitches a speedier fastball.

Pitcher	Sample Mean Speed of Pitches (mph)	Population Standard Deviation
Wesley	86	3
Rodriguez	91	7

Table 10.18

31. What is the random variable?

32. State the null and alternative hypotheses.

33. What is the test statistic?

34. What is the p-value?

35. At the 1% significance level, what is your conclusion?

Use the following information to answer the next five exercises. A researcher is testing the effects of plant food on plant growth. Nine plants have been given the plant food. Another nine plants have not been given the plant food. The heights of the plants are recorded after eight weeks. The populations have normal distributions. The following table is the result. The researcher thinks the food makes the plants grow taller.

Plant Group	Sample Mean Height of Plants (inches)	Population Standard Deviation
Food	16	2.5
No food	14	1.5

Table 10.19

36. Is the population standard deviation known or unknown?

37. State the null and alternative hypotheses.

38. What is the *p*-value?

39. Draw the graph of the *p*-value.

40. At the 1% significance level, what is your conclusion?

Use the following information to answer the next five exercises. Two metal alloys are being considered as material for ball bearings. The mean melting point of the two alloys is to be compared. 15 pieces of each metal are being tested. Both populations have normal distributions. The following table is the result. It is believed that Alloy Zeta has a different melting point.

	Sample Mean Melting Temperatures (°F)	Population Standard Deviation
Alloy Gamma	800	95
Alloy Zeta	900	105

Table 10.20

41. State the null and alternative hypotheses.

42. Is this a right-, left-, or two-tailed test?

43. What is the *p*-value?

44. Draw the graph of the *p*-value.

45. At the 1% significance level, what is your conclusion?

10.3 Comparing Two Independent Population Proportions

Use the following information for the next five exercises. Two types of phone operating system are being tested to determine if there is a difference in the proportions of system failures (crashes). Fifteen out of a random sample of 150 phones with OS_1 had system failures within the first eight hours of operation. Nine out of another random sample of 150 phones with OS_2 had system failures within the first eight hours of operation. OS_2 is believed to be more stable (have fewer crashes) than OS_1.

46. Is this a test of means or proportions?

47. What is the random variable?

48. State the null and alternative hypotheses.

49. What is the *p*-value?

50. What can you conclude about the two operating systems?

Use the following information to answer the next twelve exercises. In the recent Census, three percent of the U.S. population reported being of two or more races. However, the percent varies tremendously from state to state. Suppose that two random surveys are conducted. In the first random survey, out of 1,000 North Dakotans, only nine people reported being of two or more races. In the second random survey, out of 500 Nevadans, 17 people reported being of two or more races. Conduct a hypothesis test to determine if the population percents are the same for the two states or if the percent for Nevada is statistically higher than for North Dakota.

51. Is this a test of means or proportions?

52. State the null and alternative hypotheses.
   a. H_0: _____
   b. H_a: _____

53. Is this a right-tailed, left-tailed, or two-tailed test? How do you know?

54. What is the random variable of interest for this test?

55. In words, define the random variable for this test.

56. Which distribution (normal or Student's t) would you use for this hypothesis test?

57. Explain why you chose the distribution you did for the **Exercise 10.56**.

58. Calculate the test statistic.

59. Sketch a graph of the situation. Mark the hypothesized difference and the sample difference. Shade the area corresponding to the p-value.

$$p'_N - p'_{ND}$$

Figure 10.17

60. Find the p-value.

61. At a pre-conceived $\alpha = 0.05$, what is your:
   a. Decision:
   b. Reason for the decision:
   c. Conclusion (write out in a complete sentence):

62. Does it appear that the proportion of Nevadans who are two or more races is higher than the proportion of North Dakotans? Why or why not?

10.4 Matched or Paired Samples

Use the following information to answer the next five exercises. A study was conducted to test the effectiveness of a software patch in reducing system failures over a six-month period. Results for randomly selected installations are shown in **Table 10.21**. The "before" value is matched to an "after" value, and the differences are calculated. The differences have a normal distribution. Test at the 1% significance level.

Installation	A	B	C	D	E	F	G	H
Before	3	6	4	2	5	8	2	6
After	1	5	2	0	1	0	2	2

Table 10.21

63. What is the random variable?

64. State the null and alternative hypotheses.

65. What is the p-value?

66. Draw the graph of the p-value.

67. What conclusion can you draw about the software patch?

Use the following information to answer next five exercises. A study was conducted to test the effectiveness of a juggling class. Before the class started, six subjects juggled as many balls as they could at once. After the class, the same six subjects juggled as many balls as they could. The differences in the number of balls are calculated. The differences have a normal distribution. Test at the 1% significance level.

Subject	A	B	C	D	E	F
Before	3	4	3	2	4	5
After	4	5	6	4	5	7

Table 10.22

68. State the null and alternative hypotheses.

69. What is the *p*-value?

70. What is the sample mean difference?

71. Draw the graph of the *p*-value.

72. What conclusion can you draw about the juggling class?

Use the following information to answer the next five exercises. A doctor wants to know if a blood pressure medication is effective. Six subjects have their blood pressures recorded. After twelve weeks on the medication, the same six subjects have their blood pressure recorded again. For this test, only systolic pressure is of concern. Test at the 1% significance level.

Patient	A	B	C	D	E	F
Before	161	162	165	162	166	171
After	158	159	166	160	167	169

Table 10.23

73. State the null and alternative hypotheses.

74. What is the test statistic?

75. What is the *p*-value?

76. What is the sample mean difference?

77. What is the conclusion?

HOMEWORK

10.1 Two Population Means with Unknown Standard Deviations

DIRECTIONS: For each of the word problems, use a solution sheet to do the hypothesis test. The solution sheet is found in **Appendix E**. *Please feel free to make copies of the solution sheets. For the online version of the book, it is suggested that you copy the .doc or the .pdf files.*

NOTE

If you are using a Student's t-distribution for a homework problem in what follows, including for paired data, you may assume that the underlying population is normally distributed. (When using these tests in a real situation, you must first prove that assumption, however.)

78. The mean number of English courses taken in a two–year time period by male and female college students is believed to be about the same. An experiment is conducted and data are collected from 29 males and 16 females. The males took an average of three English courses with a standard deviation of 0.8. The females took an average of four English courses with a standard deviation of 1.0. Are the means statistically the same?

79. A student at a four-year college claims that mean enrollment at four–year colleges is higher than at two–year colleges in the United States. Two surveys are conducted. Of the 35 two–year colleges surveyed, the mean enrollment was 5,068 with a standard deviation of 4,777. Of the 35 four-year colleges surveyed, the mean enrollment was 5,466 with a standard deviation of 8,191.

80. At Rachel's 11th birthday party, eight girls were timed to see how long (in seconds) they could hold their breath in a relaxed position. After a two-minute rest, they timed themselves while jumping. The girls thought that the mean difference between their jumping and relaxed times would be zero. Test their hypothesis.

Relaxed time (seconds)	Jumping time (seconds)
26	21
47	40
30	28
22	21
23	25
45	43
37	35
29	32

Table 10.24

81. Mean entry-level salaries for college graduates with mechanical engineering degrees and electrical engineering degrees are believed to be approximately the same. A recruiting office thinks that the mean mechanical engineering salary is actually lower than the mean electrical engineering salary. The recruiting office randomly surveys 50 entry level mechanical engineers and 60 entry level electrical engineers. Their mean salaries were $46,100 and $46,700, respectively. Their standard deviations were $3,450 and $4,210, respectively. Conduct a hypothesis test to determine if you agree that the mean entry-level mechanical engineering salary is lower than the mean entry-level electrical engineering salary.

82. Marketing companies have collected data implying that teenage girls use more ring tones on their cellular phones than teenage boys do. In one particular study of 40 randomly chosen teenage girls and boys (20 of each) with cellular phones, the mean number of ring tones for the girls was 3.2 with a standard deviation of 1.5. The mean for the boys was 1.7 with a standard deviation of 0.8. Conduct a hypothesis test to determine if the means are approximately the same or if the girls' mean is higher than the boys' mean.

Use the information from **Appendix C** *to answer the next four exercises.*

83. Using the data from Lap 1 only, conduct a hypothesis test to determine if the mean time for completing a lap in races is the same as it is in practices.

84. Repeat the test in **Exercise 10.83**, but use Lap 5 data this time.

85. Repeat the test in **Exercise 10.83**, but this time combine the data from Laps 1 and 5.

86. In two to three complete sentences, explain in detail how you might use Terri Vogel's data to answer the following question. "Does Terri Vogel drive faster in races than she does in practices?"

Use the following information to answer the next two exercises. The Eastern and Western Major League Soccer conferences have a new Reserve Division that allows new players to develop their skills. Data for a randomly picked date showed the following annual goals.

Western	Eastern
Los Angeles 9	D.C. United 9
FC Dallas 3	Chicago 8
Chivas USA 4	Columbus 7
Real Salt Lake 3	New England 6
Colorado 4	MetroStars 5
San Jose 4	Kansas City 3

Table 10.25

Conduct a hypothesis test to answer the next two exercises.

87. The **exact** distribution for the hypothesis test is:
   a. the normal distribution
   b. the Student's *t*-distribution
   c. the uniform distribution
   d. the exponential distribution

88. If the level of significance is 0.05, the conclusion is:
   a. There is sufficient evidence to conclude that the **W** Division teams score fewer goals, on average, than the **E** teams
   b. There is insufficient evidence to conclude that the **W** Division teams score more goals, on average, than the **E** teams.
   c. There is insufficient evidence to conclude that the **W** teams score fewer goals, on average, than the **E** teams score.
   d. Unable to determine

89. Suppose a statistics instructor believes that there is no significant difference between the mean class scores of statistics day students on Exam 2 and statistics night students on Exam 2. She takes random samples from each of the populations. The mean and standard deviation for 35 statistics day students were 75.86 and 16.91. The mean and standard deviation for 37 statistics night students were 75.41 and 19.73. The "day" subscript refers to the statistics day students. The "night" subscript refers to the statistics night students. A concluding statement is:
   a. There is sufficient evidence to conclude that statistics night students' mean on Exam 2 is better than the statistics day students' mean on Exam 2.
   b. There is insufficient evidence to conclude that the statistics day students' mean on Exam 2 is better than the statistics night students' mean on Exam 2.
   c. There is insufficient evidence to conclude that there is a significant difference between the means of the statistics day students and night students on Exam 2.
   d. There is sufficient evidence to conclude that there is a significant difference between the means of the statistics day students and night students on Exam 2.

90. Researchers interviewed street prostitutes in Canada and the United States. The mean age of the 100 Canadian prostitutes upon entering prostitution was 18 with a standard deviation of six. The mean age of the 130 United States prostitutes upon entering prostitution was 20 with a standard deviation of eight. Is the mean age of entering prostitution in Canada lower than the mean age in the United States? Test at a 1% significance level.

91. A powder diet is tested on 49 people, and a liquid diet is tested on 36 different people. Of interest is whether the liquid diet yields a higher mean weight loss than the powder diet. The powder diet group had a mean weight loss of 42 pounds with a standard deviation of 12 pounds. The liquid diet group had a mean weight loss of 45 pounds with a standard deviation of 14 pounds.

92. Suppose a statistics instructor believes that there is no significant difference between the mean class scores of statistics day students on Exam 2 and statistics night students on Exam 2. She takes random samples from each of the populations. The mean and standard deviation for 35 statistics day students were 75.86 and 16.91, respectively. The mean and standard deviation for 37 statistics night students were 75.41 and 19.73. The "day" subscript refers to the statistics day students. The "night" subscript refers to the statistics night students. An appropriate alternative hypothesis for the hypothesis test is:

a. $\mu_{day} > \mu_{night}$
b. $\mu_{day} < \mu_{night}$
c. $\mu_{day} = \mu_{night}$
d. $\mu_{day} \neq \mu_{night}$

10.2 Two Population Means with Known Standard Deviations

DIRECTIONS: For each of the word problems, use a solution sheet to do the hypothesis test. The solution sheet is found in **Appendix E**. *Please feel free to make copies of the solution sheets. For the online version of the book, it is suggested that you copy the .doc or the .pdf files.*

NOTE

If you are using a Student's t-distribution for one of the following homework problems, including for paired data, you may assume that the underlying population is normally distributed. (When using these tests in a real situation, you must first prove that assumption, however.)

93. A study is done to determine if students in the California state university system take longer to graduate, on average, than students enrolled in private universities. One hundred students from both the California state university system and private universities are surveyed. Suppose that from years of research, it is known that the population standard deviations are 1.5811 years and 1 year, respectively. The following data are collected. The California state university system students took on average 4.5 years with a standard deviation of 0.8. The private university students took on average 4.1 years with a standard deviation of 0.3.

94. Parents of teenage boys often complain that auto insurance costs more, on average, for teenage boys than for teenage girls. A group of concerned parents examines a random sample of insurance bills. The mean annual cost for 36 teenage boys was $679. For 23 teenage girls, it was $559. From past years, it is known that the population standard deviation for each group is $180. Determine whether or not you believe that the mean cost for auto insurance for teenage boys is greater than that for teenage girls.

95. A group of transfer bound students wondered if they will spend the same mean amount on texts and supplies each year at their four-year university as they have at their community college. They conducted a random survey of 54 students at their community college and 66 students at their local four-year university. The sample means were $947 and $1,011, respectively. The population standard deviations are known to be $254 and $87, respectively. Conduct a hypothesis test to determine if the means are statistically the same.

96. Some manufacturers claim that non-hybrid sedan cars have a lower mean miles-per-gallon (mpg) than hybrid ones. Suppose that consumers test 21 hybrid sedans and get a mean of 31 mpg with a standard deviation of seven mpg. Thirty-one non-hybrid sedans get a mean of 22 mpg with a standard deviation of four mpg. Suppose that the population standard deviations are known to be six and three, respectively. Conduct a hypothesis test to evaluate the manufacturers claim.

97. A baseball fan wanted to know if there is a difference between the number of games played in a World Series when the American League won the series versus when the National League won the series. From 1922 to 2012, the population standard deviation of games won by the American League was 1.14, and the population standard deviation of games won by the National League was 1.11. Of 19 randomly selected World Series games won by the American League, the mean number of games won was 5.76. The mean number of 17 randomly selected games won by the National League was 5.42. Conduct a hypothesis test.

98. One of the questions in a study of marital satisfaction of dual-career couples was to rate the statement "I'm pleased with the way we divide the responsibilities for childcare." The ratings went from one (strongly agree) to five (strongly disagree). **Table 10.26** contains ten of the paired responses for husbands and wives. Conduct a hypothesis test to see if the mean difference in the husband's versus the wife's satisfaction level is negative (meaning that, within the partnership, the husband is happier than the wife).

Wife's Score	2	2	3	3	4	2	1	1	2	4
Husband's Score	2	2	1	3	2	1	1	1	2	4

Table 10.26

10.3 Comparing Two Independent Population Proportions

DIRECTIONS: For each of the word problems, use a solution sheet to do the hypothesis test. The solution sheet is found in **Appendix E**. *Please feel free to make copies of the solution sheets. For the online version of the book, it is suggested that you copy the .doc or the .pdf files.*

NOTE

If you are using a Student's t-distribution for one of the following homework problems, including for paired data, you may assume that the underlying population is normally distributed. (In general, you must first prove that assumption, however.)

99. A recent drug survey showed an increase in the use of drugs and alcohol among local high school seniors as compared to the national percent. Suppose that a survey of 100 local seniors and 100 national seniors is conducted to see if the proportion of drug and alcohol use is higher locally than nationally. Locally, 65 seniors reported using drugs or alcohol within the past month, while 60 national seniors reported using them.

100. We are interested in whether the proportions of female suicide victims for ages 15 to 24 are the same for the whites and the blacks races in the United States. We randomly pick one year, 1992, to compare the races. The number of suicides estimated in the United States in 1992 for white females is 4,930. Five hundred eighty were aged 15 to 24. The estimate for black females is 330. Forty were aged 15 to 24. We will let female suicide victims be our population.

101. Elizabeth Mjelde, an art history professor, was interested in whether the value from the Golden Ratio formula, $\left(\dfrac{\text{larger} + \text{smaller dimension}}{\text{larger dimension}}\right)$ was the same in the Whitney Exhibit for works from 1900 to 1919 as for works from 1920 to 1942. Thirty-seven early works were sampled, averaging 1.74 with a standard deviation of 0.11. Sixty-five of the later works were sampled, averaging 1.746 with a standard deviation of 0.1064. Do you think that there is a significant difference in the Golden Ratio calculation?

102. A recent year was randomly picked from 1985 to the present. In that year, there were 2,051 Hispanic students at Cabrillo College out of a total of 12,328 students. At Lake Tahoe College, there were 321 Hispanic students out of a total of 2,441 students. In general, do you think that the percent of Hispanic students at the two colleges is basically the same or different?

Use the following information to answer the next three exercises. Neuroinvasive West Nile virus is a severe disease that affects a person's nervous system . It is spread by the Culex species of mosquito. In the United States in 2010 there were 629 reported cases of neuroinvasive West Nile virus out of a total of 1,021 reported cases and there were 486 neuroinvasive reported cases out of a total of 712 cases reported in 2011. Is the 2011 proportion of neuroinvasive West Nile virus cases more than the 2010 proportion of neuroinvasive West Nile virus cases? Using a 1% level of significance, conduct an appropriate hypothesis test.

- "2011" subscript: 2011 group.

- "2010" subscript: 2010 group

103. This is:
   a. a test of two proportions
   b. a test of two independent means
   c. a test of a single mean
   d. a test of matched pairs.

104. An appropriate null hypothesis is:
   a. $p_{2011} \leq p_{2010}$

b. $p_{2011} \geq p_{2010}$

c. $\mu_{2011} \leq \mu_{2010}$

d. $p_{2011} > p_{2010}$

105. The *p*-value is 0.0022. At a 1% level of significance, the appropriate conclusion is

a. There is sufficient evidence to conclude that the proportion of people in the United States in 2011 who contracted neuroinvasive West Nile disease is less than the proportion of people in the United States in 2010 who contracted neuroinvasive West Nile disease.

b. There is insufficient evidence to conclude that the proportion of people in the United States in 2011 who contracted neuroinvasive West Nile disease is more than the proportion of people in the United States in 2010 who contracted neuroinvasive West Nile disease.

c. There is insufficient evidence to conclude that the proportion of people in the United States in 2011 who contracted neuroinvasive West Nile disease is less than the proportion of people in the United States in 2010 who contracted neuroinvasive West Nile disease.

d. There is sufficient evidence to conclude that the proportion of people in the United States in 2011 who contracted neuroinvasive West Nile disease is more than the proportion of people in the United States in 2010 who contracted neuroinvasive West Nile disease.

106. Researchers conducted a study to find out if there is a difference in the use of eReaders by different age groups. Randomly selected participants were divided into two age groups. In the 16- to 29-year-old group, 7% of the 628 surveyed use eReaders, while 11% of the 2,309 participants 30 years old and older use eReaders.

107. Adults aged 18 years old and older were randomly selected for a survey on obesity. Adults are considered obese if their body mass index (BMI) is at least 30. The researchers wanted to determine if the proportion of women who are obese in the south is less than the proportion of southern men who are obese. The results are shown in **Table 10.27**. Test at the 1% level of significance.

	Number who are obese	Sample size
Men	42,769	155,525
Women	67,169	248,775

Table 10.27

108. Two computer users were discussing tablet computers. A higher proportion of people ages 16 to 29 use tablets than the proportion of people age 30 and older. **Table 10.28** details the number of tablet owners for each age group. Test at the 1% level of significance.

	16–29 year olds	30 years old and older
Own a Tablet	69	231
Sample Size	628	2,309

Table 10.28

109. A group of friends debated whether more men use smartphones than women. They consulted a research study of smartphone use among adults. The results of the survey indicate that of the 973 men randomly sampled, 379 use smartphones. For women, 404 of the 1,304 who were randomly sampled use smartphones. Test at the 5% level of significance.

110. While her husband spent 2½ hours picking out new speakers, a statistician decided to determine whether the percent of men who enjoy shopping for electronic equipment is higher than the percent of women who enjoy shopping for electronic equipment. The population was Saturday afternoon shoppers. Out of 67 men, 24 said they enjoyed the activity. Eight of the 24 women surveyed claimed to enjoy the activity. Interpret the results of the survey.

111. We are interested in whether children's educational computer software costs less, on average, than children's entertainment software. Thirty-six educational software titles were randomly picked from a catalog. The mean cost was

$31.14 with a standard deviation of $4.69. Thirty-five entertainment software titles were randomly picked from the same catalog. The mean cost was $33.86 with a standard deviation of $10.87. Decide whether children's educational software costs less, on average, than children's entertainment software.

112. Joan Nguyen recently claimed that the proportion of college-age males with at least one pierced ear is as high as the proportion of college-age females. She conducted a survey in her classes. Out of 107 males, 20 had at least one pierced ear. Out of 92 females, 47 had at least one pierced ear. Do you believe that the proportion of males has reached the proportion of females?

113. Use the data sets found in **Appendix C** to answer this exercise. Is the proportion of race laps Terri completes slower than 130 seconds less than the proportion of practice laps she completes slower than 135 seconds?

114. "To Breakfast or Not to Breakfast?" by Richard Ayore

In the American society, birthdays are one of those days that everyone looks forward to. People of different ages and peer groups gather to mark the 18th, 20th, ..., birthdays. During this time, one looks back to see what he or she has achieved for the past year and also focuses ahead for more to come.

If, by any chance, I am invited to one of these parties, my experience is always different. Instead of dancing around with my friends while the music is booming, I get carried away by memories of my family back home in Kenya. I remember the good times I had with my brothers and sister while we did our daily routine.

Every morning, I remember we went to the shamba (garden) to weed our crops. I remember one day arguing with my brother as to why he always remained behind just to join us an hour later. In his defense, he said that he preferred waiting for breakfast before he came to weed. He said, "This is why I always work more hours than you guys!"

And so, to prove him wrong or right, we decided to give it a try. One day we went to work as usual without breakfast, and recorded the time we could work before getting tired and stopping. On the next day, we all ate breakfast before going to work. We recorded how long we worked again before getting tired and stopping. Of interest was our mean increase in work time. Though not sure, my brother insisted that it was more than two hours. Using the data in **Table 10.29**, solve our problem.

Work hours with breakfast	Work hours without breakfast
8	6
7	5
9	5
5	4
9	7
8	7
10	7
7	5
6	6
9	5

Table 10.29

10.4 Matched or Paired Samples

*DIRECTIONS: For each of the word problems, use a solution sheet to do the hypothesis test. The solution sheet is found in **Appendix E**. Please feel free to make copies of the solution sheets. For the online version of the book, it is suggested that you copy the .doc or the .pdf files.*

NOTE

If you are using a Student's t-distribution for the homework problems, including for paired data, you may assume that the underlying population is normally distributed. (When using these tests in a real situation, you must first prove that assumption, however.)

115. Ten individuals went on a low–fat diet for 12 weeks to lower their cholesterol. The data are recorded in **Table 10.30**. Do you think that their cholesterol levels were significantly lowered?

Starting cholesterol level	Ending cholesterol level
140	140
220	230
110	120
240	220
200	190
180	150
190	200
360	300
280	300
260	240

Table 10.30

Use the following information to answer the next two exercises. A new AIDS prevention drug was tried on a group of 224 HIV positive patients. Forty-five patients developed AIDS after four years. In a control group of 224 HIV positive patients, 68 developed AIDS after four years. We want to test whether the method of treatment reduces the proportion of patients that develop AIDS after four years or if the proportions of the treated group and the untreated group stay the same.

Let the subscript t = treated patient and ut = untreated patient.

116. The appropriate hypotheses are:
   a. H_0: $p_t < p_{ut}$ and H_a: $p_t \geq p_{ut}$
   b. H_0: $p_t \leq p_{ut}$ and H_a: $p_t > p_{ut}$
   c. H_0: $p_t = p_{ut}$ and H_a: $p_t \neq p_{ut}$
   d. H_0: $p_t = p_{ut}$ and H_a: $p_t < p_{ut}$

117. If the p-value is 0.0062 what is the conclusion (use $\alpha = 0.05$)?
   a.   The method has no effect.
   b.   There is sufficient evidence to conclude that the method reduces the proportion of HIV positive patients who develop AIDS after four years.
   c.   There is sufficient evidence to conclude that the method increases the proportion of HIV positive patients who develop AIDS after four years.
   d.   There is insufficient evidence to conclude that the method reduces the proportion of HIV positive patients who develop AIDS after four years.

Use the following information to answer the next two exercises. An experiment is conducted to show that blood pressure can be consciously reduced in people trained in a "biofeedback exercise program." Six subjects were randomly selected and blood pressure measurements were recorded before and after the training. The difference between blood pressures was calculated (after - before) producing the following results: $\bar{x}_d = -10.2$ $s_d = 8.4$. Using the data, test the hypothesis that the blood pressure has decreased after the training.

118. The distribution for the test is:
   a. t_5
   b. t_6
   c. $N(-10.2, 8.4)$
   d. $N(-10.2, \frac{8.4}{\sqrt{6}})$

119. If $\alpha = 0.05$, the *p*-value and the conclusion are
   a. 0.0014; There is sufficient evidence to conclude that the blood pressure decreased after the training.
   b. 0.0014; There is sufficient evidence to conclude that the blood pressure increased after the training.
   c. 0.0155; There is sufficient evidence to conclude that the blood pressure decreased after the training.
   d. 0.0155; There is sufficient evidence to conclude that the blood pressure increased after the training.

120. A golf instructor is interested in determining if her new technique for improving players' golf scores is effective. She takes four new students. She records their 18-hole scores before learning the technique and then after having taken her class. She conducts a hypothesis test. The data are as follows.

	Player 1	Player 2	Player 3	Player 4
Mean score before class	83	78	93	87
Mean score after class	80	80	86	86

Table 10.31

The correct decision is:

   a. Reject H_0.
   b. Do not reject the H_0.

121. A local cancer support group believes that the estimate for new female breast cancer cases in the south is higher in 2013 than in 2012. The group compared the estimates of new female breast cancer cases by southern state in 2012 and in 2013. The results are in **Table 10.32**.

Southern States	2012	2013
Alabama	3,450	3,720
Arkansas	2,150	2,280
Florida	15,540	15,710
Georgia	6,970	7,310
Kentucky	3,160	3,300
Louisiana	3,320	3,630
Mississippi	1,990	2,080
North Carolina	7,090	7,430
Oklahoma	2,630	2,690
South Carolina	3,570	3,580
Tennessee	4,680	5,070
Texas	15,050	14,980
Virginia	6,190	6,280

122. A traveler wanted to know if the prices of hotels are different in the ten cities that he visits the most often. The list of the cities with the corresponding hotel prices for his two favorite hotel chains is in **Table 10.33**. Test at the 1% level of significance.

Cities	Hyatt Regency prices in dollars	Hilton prices in dollars
Atlanta	107	169
Boston	358	289
Chicago	209	299
Dallas	209	198
Denver	167	169
Indianapolis	179	214
Los Angeles	179	169
New York City	625	459
Philadelphia	179	159
Washington, DC	245	239

Table 10.33

123. A politician asked his staff to determine whether the underemployment rate in the northeast decreased from 2011 to 2012. The results are in **Table 10.34**.

Northeastern States	2011	2012
Connecticut	17.3	16.4
Delaware	17.4	13.7
Maine	19.3	16.1
Maryland	16.0	15.5
Massachusetts	17.6	18.2
New Hampshire	15.4	13.5
New Jersey	19.2	18.7
New York	18.5	18.7
Ohio	18.2	18.8
Pennsylvania	16.5	16.9
Rhode Island	20.7	22.4
Vermont	14.7	12.3
West Virginia	15.5	17.3

Table 10.34

BRINGING IT TOGETHER: HOMEWORK

Use the following information to answer the next ten exercises. indicate which of the following choices best identifies the hypothesis test.

a. independent group means, population standard deviations and/or variances known

b. independent group means, population standard deviations and/or variances unknown

c. matched or paired samples

d. single mean

e. two proportions

f. single proportion

124. A powder diet is tested on 49 people, and a liquid diet is tested on 36 different people. The population standard deviations are two pounds and three pounds, respectively. Of interest is whether the liquid diet yields a higher mean weight loss than the powder diet.

125. A new chocolate bar is taste-tested on consumers. Of interest is whether the proportion of children who like the new chocolate bar is greater than the proportion of adults who like it.

126. The mean number of English courses taken in a two–year time period by male and female college students is believed to be about the same. An experiment is conducted and data are collected from nine males and 16 females.

127. A football league reported that the mean number of touchdowns per game was five. A study is done to determine if the mean number of touchdowns has decreased.

128. A study is done to determine if students in the California state university system take longer to graduate than students enrolled in private universities. One hundred students from both the California state university system and private universities are surveyed. From years of research, it is known that the population standard deviations are 1.5811 years and one year, respectively.

129. According to a YWCA Rape Crisis Center newsletter, 75% of rape victims know their attackers. A study is done to verify this.

130. According to a recent study, U.S. companies have a mean maternity-leave of six weeks.

131. A recent drug survey showed an increase in use of drugs and alcohol among local high school students as compared to the national percent. Suppose that a survey of 100 local youths and 100 national youths is conducted to see if the proportion of drug and alcohol use is higher locally than nationally.

132. A new SAT study course is tested on 12 individuals. Pre-course and post-course scores are recorded. Of interest is the mean increase in SAT scores. The following data are collected:

Pre-course score	Post-course score
1	300
960	920
1010	1100
840	880
1100	1070
1250	1320
860	860
1330	1370
790	770
990	1040

Pre-course score	Post-course score
1110	1200
740	850

Table 10.35

133. University of Michigan researchers reported in the *Journal of the National Cancer Institute* that quitting smoking is especially beneficial for those under age 49. In this American Cancer Society study, the risk (probability) of dying of lung cancer was about the same as for those who had never smoked.

134. Lesley E. Tan investigated the relationship between left-handedness vs. right-handedness and motor competence in preschool children. Random samples of 41 left-handed preschool children and 41 right-handed preschool children were given several tests of motor skills to determine if there is evidence of a difference between the children based on this experiment. The experiment produced the means and standard deviations shown **Table 10.36**. Determine the appropriate test and best distribution to use for that test.

	Left-handed	Right-handed
Sample size	41	41
Sample mean	97.5	98.1
Sample standard deviation	17.5	19.2

Table 10.36

a. Two independent means, normal distribution
b. Two independent means, Student's-t distribution
c. Matched or paired samples, Student's-t distribution
d. Two population proportions, normal distribution

135. A golf instructor is interested in determining if her new technique for improving players' golf scores is effective. She takes four (4) new students. She records their 18-hole scores before learning the technique and then after having taken her class. She conducts a hypothesis test. The data are as **Table 10.37**.

	Player 1	Player 2	Player 3	Player 4
Mean score before class	83	78	93	87
Mean score after class	80	80	86	86

Table 10.37

This is:

a. a test of two independent means.
b. a test of two proportions.
c. a test of a single mean.
d. a test of a single proportion.

REFERENCES

10.1 Two Population Means with Unknown Standard Deviations

Data from Graduating Engineer + Computer Careers. Available online at http://www.graduatingengineer.com

Data from *Microsoft Bookshelf*.

Data from the United States Senate website, available online at www.Senate.gov (accessed June 17, 2013).

"List of current United States Senators by Age." Wikipedia. Available online at http://en.wikipedia.org/wiki/List_of_current_United_States_Senators_by_age (accessed June 17, 2013).

"Sectoring by Industry Groups." Nasdaq. Available online at http://www.nasdaq.com/markets/barchart-sectors.aspx?page=sectors&base=industry (accessed June 17, 2013).

"Strip Clubs: Where Prostitution and Trafficking Happen." Prostitution Research and Education, 2013. Available online at www.prostitutionresearch.com/ProsViolPosttrauStress.html (accessed June 17, 2013).

"World Series History." Baseball-Almanac, 2013. Available online at http://www.baseball-almanac.com/ws/wsmenu.shtml (accessed June 17, 2013).

10.2 Two Population Means with Known Standard Deviations

Data from the United States Census Bureau. Available online at http://www.census.gov/prod/cen2010/briefs/c2010br-02.pdf

Hinduja, Sameer. "Sexting Research and Gender Differences." Cyberbulling Research Center, 2013. Available online at http://cyberbullying.us/blog/sexting-research-and-gender-differences/ (accessed June 17, 2013).

"Smart Phone Users, By the Numbers." Visually, 2013. Available online at http://visual.ly/smart-phone-users-numbers (accessed June 17, 2013).

Smith, Aaron. "35% of American adults own a Smartphone." Pew Internet, 2013. Available online at http://www.pewinternet.org/~/media/Files/Reports/2011/PIP_Smartphones.pdf (accessed June 17, 2013).

"State-Specific Prevalence of Obesity AmongAduls—Unites States, 2007." MMWR, CDC. Available online at http://www.cdc.gov/mmwr/preview/mmwrhtml/mm5728a1.htm (accessed June 17, 2013).

"Texas Crime Rates 1960–1012." FBI, Uniform Crime Reports, 2013. Available online at: http://www.disastercenter.com/crime/txcrime.htm (accessed June 17, 2013).

10.3 Comparing Two Independent Population Proportions

Data from *Educational Resources*, December catalog.

Data from Hilton Hotels. Available online at http://www.hilton.com (accessed June 17, 2013).

Data from Hyatt Hotels. Available online at http://hyatt.com (accessed June 17, 2013).

Data from Statistics, United States Department of Health and Human Services.

Data from Whitney Exhibit on loan to San Jose Museum of Art.

Data from the American Cancer Society. Available online at http://www.cancer.org/index (accessed June 17, 2013).

Data from the Chancellor's Office, California Community Colleges, November 1994.

"State of the States." Gallup, 2013. Available online at http://www.gallup.com/poll/125066/State-States.aspx?ref=interactive (accessed June 17, 2013).

"West Nile Virus." Centers for Disease Control and Prevention. Available online at http://www.cdc.gov/ncidod/dvbid/westnile/index.htm (accessed June 17, 2013).

SOLUTIONS

1 two proportions

3 matched or paired samples

5 single mean

7 independent group means, population standard deviations and/or variances unknown

9 two proportions

11 independent group means, population standard deviations and/or variances unknown

13 independent group means, population standard deviations and/or variances unknown

15 two proportions

17 The random variable is the difference between the mean amounts of sugar in the two soft drinks.

19 means

21 two-tailed

23 the difference between the mean life spans of whites and nonwhites

25 This is a comparison of two population means with unknown population standard deviations.

27 Check student's solution.

29
  a. Reject the null hypothesis

  b. p-value < 0.05

  c. There is not enough evidence at the 5% level of significance to support the claim that life expectancy in the 1900s is different between whites and nonwhites.

31 The difference in mean speeds of the fastball pitches of the two pitchers

33 -2.46

35 At the 1% significance level, we can reject the null hypothesis. There is sufficient data to conclude that the mean speed of Rodriguez's fastball is faster than Wesley's.

37 Subscripts: 1 = Food, 2 = No Food
H_0: $\mu_1 \leq \mu_2$
H_a: $\mu_1 > \mu_2$

39

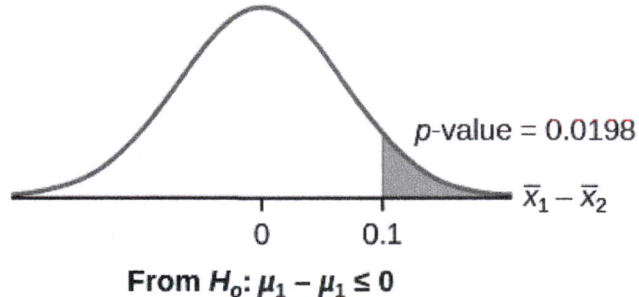

Figure 10.18

41 Subscripts: 1 = Gamma, 2 = Zeta
H_0: $\mu_1 = \mu_2$
H_a: $\mu_1 \neq \mu_2$

43 0.0062

45 There is sufficient evidence to reject the null hypothesis. The data support that the melting point for Alloy Zeta is different from the melting point of Alloy Gamma.

47 $P'_{OS1} - P'_{OS2}$ = difference in the proportions of phones that had system failures within the first eight hours of operation with OS_1 and OS_2.

49 0.1018

51 proportions

53 right-tailed

55 The random variable is the difference in proportions (percents) of the populations that are of two or more races in Nevada and North Dakota.

57 Our sample sizes are much greater than five each, so we use the normal for two proportions distribution for this hypothesis test.

59 Check student's solution.

61
a. Reject the null hypothesis.

b. *p*-value < alpha

c. At the 5% significance level, there is sufficient evidence to conclude that the proportion (percent) of the population that is of two or more races in Nevada is statistically higher than that in North Dakota.

63 the mean difference of the system failures

65 0.0067

67 With a *p*-value 0.0067, we can reject the null hypothesis. There is enough evidence to support that the software patch is effective in reducing the number of system failures.

69 0.0021

71

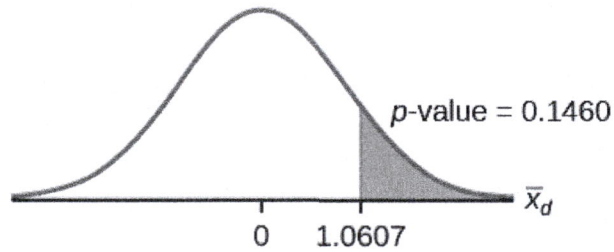

p-value = 0.1460

0   1.0607 $\bar{X}_d$

Figure 10.19

73 $H_0: \mu_d \geq 0 \ H_a: \mu_d < 0$

75 0.0699

77 We decline to reject the null hypothesis. There is not sufficient evidence to support that the medication is effective.

79 Subscripts: 1: two-year colleges; 2: four-year colleges
a. $H_0: \mu_1 \geq \mu_2$

b. $H_a: \mu_1 < \mu_2$

c. $\bar{X}_1 - \bar{X}_2$ is the difference between the mean enrollments of the two-year colleges and the four-year colleges.

d. Student's-*t*

e. test statistic: -0.2480

f. *p*-value: 0.4019

g. Check student's solution.

h.   i. Alpha: 0.05

   ii. Decision: Do not reject

   iii. Reason for Decision: *p*-value > alpha

iv. Conclusion: At the 5% significance level, there is sufficient evidence to conclude that the mean enrollment at four-year colleges is higher than at two-year colleges.

81 Subscripts: 1: mechanical engineering; 2: electrical engineering

a. H_0: $\mu_1 \geq \mu_2$

b. H_a: $\mu_1 < \mu_2$

c. $\bar{X}_1 - \bar{X}_2$ is the difference between the mean entry level salaries of mechanical engineers and electrical engineers.

d. t_{108}

e. test statistic: $t = -0.82$

f. p-value: 0.2061

g. Check student's solution.

h. i. Alpha: 0.05

ii. Decision: Do not reject the null hypothesis.

iii. Reason for Decision: p-value > alpha

iv. Conclusion: At the 5% significance level, there is insufficient evidence to conclude that the mean entry-level salaries of mechanical engineers is lower than that of electrical engineers.

83

a. H_0: $\mu_1 = \mu_2$

b. H_a: $\mu_1 \neq \mu_2$

c. $\bar{X}_1 - \bar{X}_2$ is the difference between the mean times for completing a lap in races and in practices.

d. $t_{20.32}$

e. test statistic: -4.70

f. p-value: 0.0001

g. Check student's solution.

h. i. Alpha: 0.05

ii. Decision: Reject the null hypothesis.

iii. Reason for Decision: p-value < alpha

iv. Conclusion: At the 5% significance level, there is sufficient evidence to conclude that the mean time for completing a lap in races is different from that in practices.

85

a. H_0: $\mu_1 = \mu_2$

b. H_a: $\mu_1 \neq \mu_2$

c. is the difference between the mean times for completing a lap in races and in practices.

d. $t_{40.94}$

e. test statistic: -5.08

f. p-value: zero

g. Check student's solution.

h. i. Alpha: 0.05

ii. Decision: Reject the null hypothesis.

iii. Reason for Decision: p-value < alpha

    iv.   Conclusion: At the 5% significance level, there is sufficient evidence to conclude that the mean time for completing a lap in races is different from that in practices.

88 c

90 Test: two independent sample means, population standard deviations unknown. Random variable: $\bar{X}_1 - \bar{X}_2$ Distribution: H_0: $\mu_1 = \mu_2$ H_a: $\mu_1 < \mu_2$ The mean age of entering prostitution in Canada is lower than the mean age in the United States.

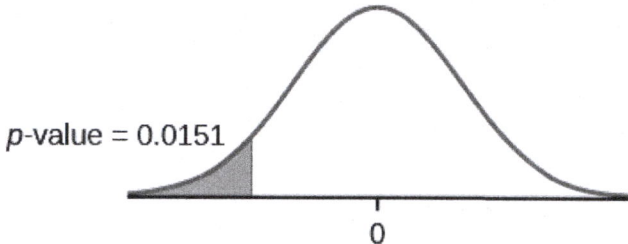

Figure 10.20

Graph: left-tailed p-value : 0.0151 Decision: Do not reject H_0. Conclusion: At the 1% level of significance, from the sample data, there is not sufficient evidence to conclude that the mean age of entering prostitution in Canada is lower than the mean age in the United States.

92 d

94 Subscripts: 1 = boys, 2 = girls
  a. H_0: $\mu_1 \le \mu_2$

  b. H_a: $\mu_1 > \mu_2$

  c. The random variable is the difference in the mean auto insurance costs for boys and girls.

  d. normal

  e. test statistic: $z = 2.50$

  f. p-value: 0.0062

  g. Check student's solution.

  h.   i.   Alpha: 0.05

      ii. Decision: Reject the null hypothesis.

      iii. Reason for Decision: p-value < alpha

      iv. Conclusion: At the 5% significance level, there is sufficient evidence to conclude that the mean cost of auto insurance for teenage boys is greater than that for girls.

96 Subscripts: 1 = non-hybrid sedans, 2 = hybrid sedans
  a. H_0: $\mu_1 \ge \mu_2$

  b. H_a: $\mu_1 < \mu_2$

  c. The random variable is the difference in the mean miles per gallon of non-hybrid sedans and hybrid sedans.

  d. normal

  e. test statistic: 6.36

  f. p-value: 0

  g. Check student's solution.

  h.   i.   Alpha: 0.05

    ii.   Decision: Reject the null hypothesis.

    iii.   Reason for decision: p-value < alpha

    iv.   Conclusion: At the 5% significance level, there is sufficient evidence to conclude that the mean miles per gallon of non-hybrid sedans is less than that of hybrid sedans.

98

a. H_0: $\mu_d = 0$

b. H_a: $\mu_d < 0$

c.   The random variable X_d is the average difference between husband's and wife's satisfaction level.

d. t_9

e.   test statistic: $t = -1.86$

f. p-value: 0.0479

g.   Check student's solution

h.   i.   Alpha: 0.05

    ii.   Decision: Reject the null hypothesis, but run another test.

    iii.   Reason for Decision: p-value < alpha

    iv.   Conclusion: This is a weak test because alpha and the p-value are close. However, there is insufficient evidence to conclude that the mean difference is negative.

100

a. H_0: $P_W = P_B$

b. H_a: $P_W \neq P_B$

c.   The random variable is the difference in the proportions of white and black suicide victims, aged 15 to 24.

d.   normal for two proportions

e.   test statistic: -0.1944

f. p-value: 0.8458

g.   Check student's solution.

h.   i.   Alpha: 0.05

    ii.   Decision: Reject the null hypothesis.

    iii.   Reason for decision: p-value > alpha

    iv.   Conclusion: At the 5% significance level, there is insufficient evidence to conclude that the proportions of white and black female suicide victims, aged 15 to 24, are different.

102 Subscripts: 1 = Cabrillo College, 2 = Lake Tahoe College

a. H_0: $p_1 = p_2$

b. H_a: $p_1 \neq p_2$

c.   The random variable is the difference between the proportions of Hispanic students at Cabrillo College and Lake Tahoe College.

d.   normal for two proportions

e.   test statistic: 4.29

f. p-value: 0.00002

g.   Check student's solution.

h.   i.   Alpha: 0.05

    ii.   Decision: Reject the null hypothesis.

    iii. Reason for decision: *p*-value < alpha

    iv. Conclusion: There is sufficient evidence to conclude that the proportions of Hispanic students at Cabrillo College and Lake Tahoe College are different.

104 a

106 Test: two independent sample proportions. Random variable: $p'_1 - p'_2$ Distribution:
H_0: $p_1 = p_2$
H_a: $p_1 \neq p_2$ The proportion of eReader users is different for the 16- to 29-year-old users from that of the 30 and older users.
Graph: two-tailed

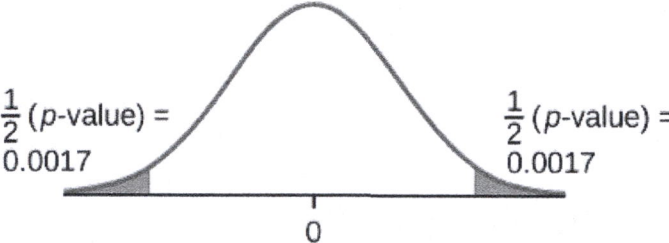

Figure 10.21

p-value : 0.0033 Decision: Reject the null hypothesis. Conclusion: At the 5% level of significance, from the sample data, there is sufficient evidence to conclude that the proportion of eReader users 16 to 29 years old is different from the proportion of eReader users 30 and older.

108 Test: two independent sample proportions Random variable: $p'_1 - p'_2$ Distribution: H_0: $p_1 = p_2$
H_a: $p_1 > p_2$ A higher proportion of tablet owners are aged 16 to 29 years old than are 30 years old and older. Graph: right-tailed

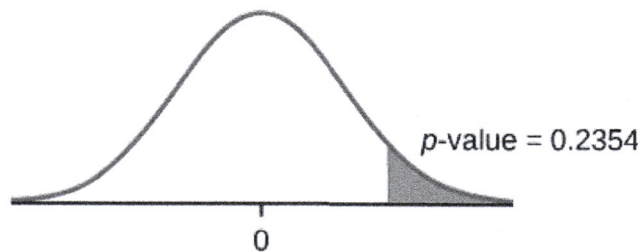

Figure 10.22

p-value: 0.2354 Decision: Do not reject the H_0. Conclusion: At the 1% level of significance, from the sample data, there is not sufficient evidence to conclude that a higher proportion of tablet owners are aged 16 to 29 years old than are 30 years old and older.

110 Subscripts: 1: men; 2: women

  a. H_0: $p_1 \leq p_2$

  b. H_a: $p_1 > p_2$

  c. $P'_1 - P'_2$ is the difference between the proportions of men and women who enjoy shopping for electronic equipment.

  d. normal for two proportions

  e. test statistic: 0.22

  f. *p*-value: 0.4133

  g. Check student's solution.

    i. Alpha: 0.05

    ii. Decision: Do not reject the null hypothesis.

    iii. Reason for Decision: p-value > alpha

    iv. Conclusion: At the 5% significance level, there is insufficient evidence to conclude that the proportion of men who enjoy shopping for electronic equipment is more than the proportion of women.

112

a. H_0: $p_1 = p_2$

b. H_a: $p_1 \neq p_2$

c. $P'_1 - P'_2$ is the difference between the proportions of men and women that have at least one pierced ear.

d. normal for two proportions

e. test statistic: –4.82

f. p-value: zero

g. Check student's solution.

h.   i. Alpha: 0.05

    ii. Decision: Reject the null hypothesis.

    iii. Reason for Decision: p-value < alpha

    iv. Conclusion: At the 5% significance level, there is sufficient evidence to conclude that the proportions of males and females with at least one pierced ear is different.

114

a. H_0: $\mu_d = 0$

b. H_a: $\mu_d > 0$

c. The random variable X_d is the mean difference in work times on days when eating breakfast and on days when not eating breakfast.

d. t_9

e. test statistic: 4.8963

f. p-value: 0.0004

g. Check student's solution.

h.   i. Alpha: 0.05

    ii. Decision: Reject the null hypothesis.

    iii. Reason for Decision: p-value < alpha

    iv. Conclusion: At the 5% level of significance, there is sufficient evidence to conclude that the mean difference in work times on days when eating breakfast and on days when not eating breakfast has increased.

115 p-value = 0.1494 At the 5% significance level, there is insufficient evidence to conclude that the medication lowered cholesterol levels after 12 weeks.

117 b

119 c

121 Test: two matched pairs or paired samples (t-test) Random variable: $\overline{X}_d$ Distribution: t_{12} H_0: $\mu_d = 0$ H_a: $\mu_d > 0$ The mean of the differences of new female breast cancer cases in the south between 2013 and 2012 is greater than zero. The estimate for new female breast cancer cases in the south is higher in 2013 than in 2012. Graph: right-tailed p-value: 0.0004

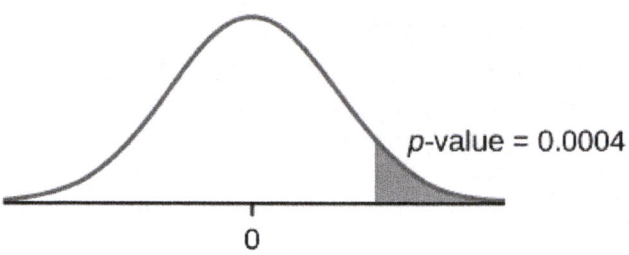

Figure 10.23

Decision: Reject H_0 Conclusion: At the 5% level of significance, from the sample data, there is sufficient evidence to conclude that there was a higher estimate of new female breast cancer cases in 2013 than in 2012.

123 Test: matched or paired samples (t-test) Difference data: {−0.9, −3.7, −3.2, −0.5, 0.6, −1.9, −0.5, 0.2, 0.6, 0.4, 1.7, −2.4, 1.8} Random Variable: $\overline{X}_d$ Distribution: H_0: $\mu_d = 0$ H_a: $\mu_d < 0$ The mean of the differences of the rate of underemployment in the northeastern states between 2012 and 2011 is less than zero. The underemployment rate went down from 2011 to 2012. Graph: left-tailed.

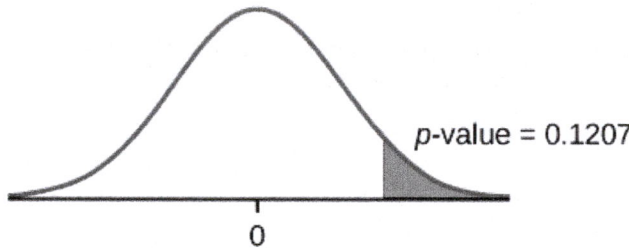

Figure 10.24

p-value: 0.1207 Decision: Do not reject H_0. Conclusion: At the 5% level of significance, from the sample data, there is not sufficient evidence to conclude that there was a decrease in the underemployment rates of the northeastern states from 2011 to 2012.

125 e

127 d

129 f

131 e

133 f

135 a

11 | THE CHI-SQUARE DISTRIBUTION

Figure 11.1 The chi-square distribution can be used to find relationships between two things, like grocery prices at different stores. (credit: Pete/flickr)

Introduction

Chapter Objectives

By the end of this chapter, the student should be able to:

- Interpret the chi-square probability distribution as the sample size changes.
- Conduct and interpret chi-square goodness-of-fit hypothesis tests.
- Conduct and interpret chi-square test of independence hypothesis tests.
- Conduct and interpret chi-square homogeneity hypothesis tests.
- Conduct and interpret chi-square single variance hypothesis tests.

Have you ever wondered if lottery numbers were evenly distributed or if some numbers occurred with a greater frequency? How about if the types of movies people preferred were different across different age groups? What about if a coffee machine was dispensing approximately the same amount of coffee each time? You could answer these questions by conducting a hypothesis test.

You will now study a new distribution, one that is used to determine the answers to such questions. This distribution is called the chi-square distribution.

In this chapter, you will learn the three major applications of the chi-square distribution:

1. the goodness-of-fit test, which determines if data fit a particular distribution, such as in the lottery example

2. the test of independence, which determines if events are independent, such as in the movie example

3. the test of a single variance, which tests variability, such as in the coffee example

NOTE

 Though the chi-square distribution depends on calculators or computers for most of the calculations, there is a table available (see **Appendix G**). TI-83+ and TI-84 calculator instructions are included in the text.

Collaborative Exercise

Look in the sports section of a newspaper or on the Internet for some sports data (baseball averages, basketball scores, golf tournament scores, football odds, swimming times, and the like). Plot a histogram and a boxplot using your data. See if you can determine a probability distribution that your data fits. Have a discussion with the class about your choice.

11.1 | Facts About the Chi-Square Distribution

The notation for the **chi-square distribution** is:

$$\chi \sim \chi^2_{df}$$

where df = degrees of freedom which depends on how chi-square is being used. (If you want to practice calculating chi-square probabilities then use $df = n - 1$. The degrees of freedom for the three major uses are each calculated differently.)

For the χ^2 distribution, the population mean is $\mu = df$ and the population standard deviation is $\sigma = \sqrt{2(df)}$.

The random variable is shown as χ^2, but may be any upper case letter.

The random variable for a chi-square distribution with k degrees of freedom is the sum of k independent, squared standard normal variables.

$\chi^2 = (Z_1)^2 + (Z_2)^2 + ... + (Z_k)^2$

1. The curve is nonsymmetrical and skewed to the right.

2. There is a different chi-square curve for each df.

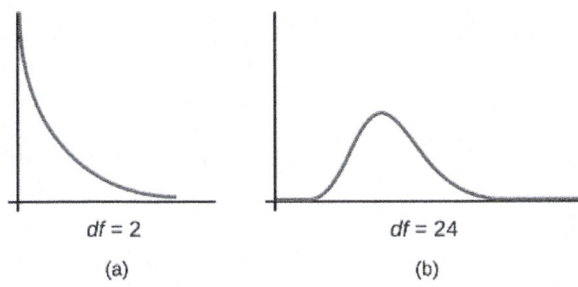

Figure 11.2

3. The test statistic for any test is always greater than or equal to zero.

4. When $df > 90$, the chi-square curve approximates the normal distribution. For $X \sim \chi^2_{1,000}$ the mean, $\mu = df = 1,000$ and the standard deviation, $\sigma = \sqrt{2(1,000)} = 44.7$. Therefore, $X \sim N(1,000, 44.7)$, approximately.

5. The mean, μ, is located just to the right of the peak.

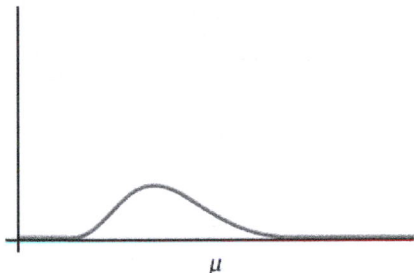

Figure 11.3

11.2 | Goodness-of-Fit Test

In this type of hypothesis test, you determine whether the data "fit" a particular distribution or not. For example, you may suspect your unknown data fit a binomial distribution. You use a chi-square test (meaning the distribution for the hypothesis test is chi-square) to determine if there is a fit or not. **The null and the alternative hypotheses for this test may be written in sentences or may be stated as equations or inequalities.**

The test statistic for a goodness-of-fit test is:

$$\sum_k \frac{(O - E)^2}{E}$$

where:

- O = **observed values** (data)
- E = **expected values** (from theory)
- k = the number of different data cells or categories

The observed values are the data values and the expected values are the values you would expect to get if the null hypothesis were true. There are n terms of the form $\frac{(O - E)^2}{E}$.

The number of degrees of freedom is $df =$ (number of categories − 1).

The goodness-of-fit test is almost always right-tailed. If the observed values and the corresponding expected values are not close to each other, then the test statistic can get very large and will be way out in the right tail of the chi-square curve.

NOTE

The expected value for each cell needs to be at least five in order for you to use this test.

Example 11.1

Absenteeism of college students from math classes is a major concern to math instructors because missing class appears to increase the drop rate. Suppose that a study was done to determine if the actual student absenteeism rate follows faculty perception. The faculty expected that a group of 100 students would miss class according to **Table 11.1**.

Number of absences per term	Expected number of students
0–2	50
3–5	30
6–8	12
9–11	6
12+	2

Table 11.1

A random survey across all mathematics courses was then done to determine the actual number **(observed)** of absences in a course. The chart in **Table 11.2** displays the results of that survey.

Number of absences per term	Actual number of students
0–2	35
3–5	40
6–8	20
9–11	1
12+	4

Table 11.2

Determine the null and alternative hypotheses needed to conduct a goodness-of-fit test.

H_0: Student absenteeism **fits** faculty perception.

The alternative hypothesis is the opposite of the null hypothesis.

H_a: Student absenteeism **does not fit** faculty perception.

a. Can you use the information as it appears in the charts to conduct the goodness-of-fit test?

Solution 11.1
a. **No.** Notice that the expected number of absences for the "12+" entry is less than five (it is two). Combine that group with the "9–11" group to create new tables where the number of students for each entry are at least five. The new results are in **Table 11.2** and **Table 11.3**.

Number of absences per term	Expected number of students
0–2	50
3–5	30
6–8	12
9+	8

Table 11.3

Number of absences per term	Actual number of students
0–2	35
3–5	40
6–8	20
9+	5

Table 11.4

b. What is the number of degrees of freedom (*df*)?

Solution 11.1

b. There are four "cells" or categories in each of the new tables.

df = number of cells − 1 = 4 − 1 = 3

Try It Σ

11.1 A factory manager needs to understand how many products are defective versus how many are produced. The number of expected defects is listed in **Table 11.5**.

Number produced	Number defective
0–100	5
101–200	6
201–300	7
301–400	8
401–500	10

Table 11.5

A random sample was taken to determine the actual number of defects. **Table 11.6** shows the results of the survey.

Number produced	Number defective
0–100	5
101–200	7
201–300	8
301–400	9
401–500	11

Table 11.6

State the null and alternative hypotheses needed to conduct a goodness-of-fit test, and state the degrees of freedom.

Example 11.2

Employers want to know which days of the week employees are absent in a five-day work week. Most employers would like to believe that employees are absent equally during the week. Suppose a random sample of 60 managers were asked on which day of the week they had the highest number of employee absences. The results were distributed as in **Table 11.6**. For the population of employees, do the days for the highest number of absences occur with equal frequencies during a five-day work week? Test at a 5% significance level.

	Monday	Tuesday	Wednesday	Thursday	Friday
Number of Absences	15	12	9	9	15

Table 11.7 Day of the Week Employees were Most Absent

Solution 11.2

The null and alternative hypotheses are:

- H_0: The absent days occur with equal frequencies, that is, they fit a uniform distribution.

- H_a: The absent days occur with unequal frequencies, that is, they do not fit a uniform distribution.

If the absent days occur with equal frequencies, then, out of 60 absent days (the total in the sample: 15 + 12 + 9 + 9 + 15 = 60), there would be 12 absences on Monday, 12 on Tuesday, 12 on Wednesday, 12 on Thursday, and 12 on Friday. These numbers are the **expected** (E) values. The values in the table are the **observed** (O) values or data.

This time, calculate the χ^2 test statistic by hand. Make a chart with the following headings and fill in the columns:

- Expected (E) values (12, 12, 12, 12, 12)

- Observed (O) values (15, 12, 9, 9, 15)

- $(O - E)$

- $(O - E)^2$

- $\dfrac{(O - E)^2}{E}$

Now add (sum) the last column. The sum is three. This is the χ^2 test statistic.

To find the *p*-value, calculate $P(\chi^2 > 3)$. This test is right-tailed. (Use a computer or calculator to find the *p*-value. You should get *p*-value = 0.5578.)

The *dfs* are the number of cells $- 1 = 5 - 1 = 4$

 Using the TI-83, 83+, 84, 84+ Calculator

Press **2nd DISTR**. Arrow down to χ^2cdf. Press **ENTER**. Enter (3,10^99,4). Rounded to four decimal places, you should see 0.5578, which is the p-value.

Next, complete a graph like the following one with the proper labeling and shading. (You should shade the right tail.)

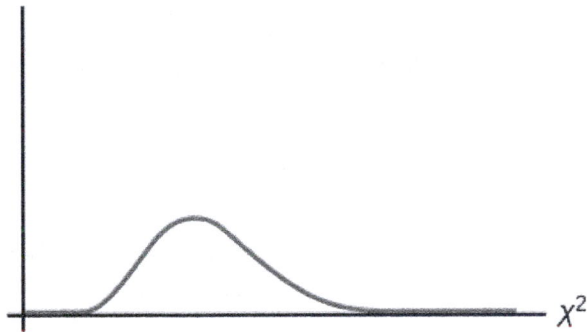

Figure 11.4

The decision is not to reject the null hypothesis.

Conclusion: At a 5% level of significance, from the sample data, there is not sufficient evidence to conclude that the absent days do not occur with equal frequencies.

 Using the TI-83, 83+, 84, 84+ Calculator

TI-83+ and some TI-84 calculators do not have a special program for the test statistic for the goodness-of-fit test. The next example **Example 11.3** has the calculator instructions. The newer TI-84 calculators have in **STAT TESTS** the test Chi2 GOF. To run the test, put the observed values (the data) into a first list and the expected values (the values you expect if the null hypothesis is true) into a second list. Press **STAT TESTS** and Chi2 GOF. Enter the list names for the Observed list and the Expected list. Enter the degrees of freedom and press calculate or draw. Make sure you clear any lists before you start. **To Clear Lists in the calculators:** Go into **STAT EDIT** and arrow up to the list name area of the particular list. Press **CLEAR** and then arrow down. The list will be cleared. Alternatively, you can press **STAT** and press 4 (for ClrList). Enter the list name and press **ENTER**.

 **Try It Σ**

11.2 Teachers want to know which night each week their students are doing most of their homework. Most teachers think that students do homework equally throughout the week. Suppose a random sample of 49 students were asked on which night of the week they did the most homework. The results were distributed as in **Table 11.8**.

	Sunday	Monday	Tuesday	Wednesday	Thursday	Friday	Saturday
Number of Students	11	8	10	7	10	5	5

Table 11.8

From the population of students, do the nights for the highest number of students doing the majority of their homework occur with equal frequencies during a week? What type of hypothesis test should you use?

Example 11.3

One study indicates that the number of televisions that American families have is distributed (this is the **given** distribution for the American population) as in **Table 11.9**.

Number of Televisions	Percent
0	10
1	16
2	55
3	11
4+	8

Table 11.9

The table contains expected (E) percents.

A random sample of 600 families in the far western United States resulted in the data in **Table 11.10**.

Number of Televisions	Frequency
0	66
1	119
2	340
3	60
4+	15
	Total = 600

Table 11.10

The table contains observed (O) frequency values.

At the 1% significance level, does it appear that the distribution "number of televisions" of far western United States families is different from the distribution for the American population as a whole?

Solution 11.3

This problem asks you to test whether the far western United States families distribution fits the distribution of the American families. This test is always right-tailed.

The first table contains expected percentages. To get expected (E) frequencies, multiply the percentage by 600. The expected frequencies are shown in **Table 11.10**.

Number of Televisions	Percent	Expected Frequency
0	10	(0.10)(600) = 60
1	16	(0.16)(600) = 96
2	55	(0.55)(600) = 330
3	11	(0.11)(600) = 66
over 3	8	(0.08)(600) = 48

Table 11.11

Therefore, the expected frequencies are 60, 96, 330, 66, and 48. In the TI calculators, you can let the calculator do the math. For example, instead of 60, enter 0.10*600.

H_0: The "number of televisions" distribution of far western United States families is the same as the "number of televisions" distribution of the American population.

H_a: The "number of televisions" distribution of far western United States families is different from the "number of televisions" distribution of the American population.

Distribution for the test: χ^2_4 where df = (the number of cells) – 1 = 5 – 1 = 4.

NOTE

$df \neq 600 - 1$

Calculate the test statistic: $\chi 2 = 29.65$

Graph:

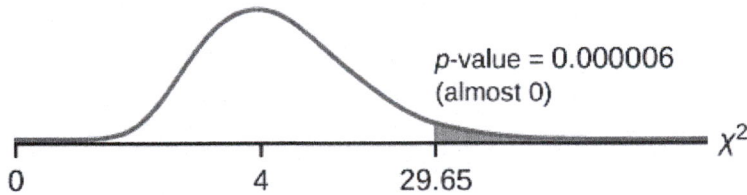

Figure 11.5

Probability statement: p-value = $P(\chi^2 > 29.65) = 0.000006$

Compare α and the p-value:

- $\alpha = 0.01$

- p-value = 0.000006

So, $\alpha > p$-value.

Make a decision: Since $\alpha > p$-value, reject H_0.

This means you reject the belief that the distribution for the far western states is the same as that of the American population as a whole.

Conclusion: At the 1% significance level, from the data, there is sufficient evidence to conclude that the "number of televisions" distribution for the far western United States is different from the "number of televisions" distribution for the American population as a whole.

 Using the TI-83, 83+, 84, 84+ Calculator

Press STAT and ENTER. Make sure to clear lists L1, L2, and L3 if they have data in them (see the note at the end of **Example 11.2**). Into L1, put the observed frequencies 66, 119, 349, 60, 15. Into L2, put the expected frequencies .10*600, .16*600, .55*600, .11*600, .08*600. Arrow over to list L3 and up to the name area "L3". Enter (L1-L2)^2/L2 and ENTER. Press 2nd QUIT. Press 2nd LIST and arrow over to MATH. Press 5. You should see "sum" (Enter L3). Rounded to 2 decimal places, you should see 29.65. Press 2nd DISTR. Press 7 or Arrow down to 7:χ2cdf and press ENTER. Enter (29.65,1E99,4). Rounded to four places, you should see 5.77E-6 = .000006 (rounded to six decimal places), which is the p-value.

The newer TI-84 calculators have in STAT TESTS the test Chi2 GOF. To run the test, put the observed values (the data) into a first list and the expected values (the values you expect if the null hypothesis is true) into a second list. Press STAT TESTS and Chi2 GOF. Enter the list names for the Observed list and the Expected list. Enter the degrees of freedom and press calculate or draw. Make sure you clear any lists before you start.

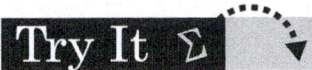

11.3 The expected percentage of the number of pets students have in their homes is distributed (this is the given distribution for the student population of the United States) as in **Table 11.12**.

Number of Pets	Percent
0	18
1	25
2	30
3	18
4+	9

Table 11.12

A random sample of 1,000 students from the Eastern United States resulted in the data in **Table 11.13**.

Number of Pets	Frequency
0	210
1	240
2	320
3	140
4+	90

Table 11.13

At the 1% significance level, does it appear that the distribution "number of pets" of students in the Eastern United States is different from the distribution for the United States student population as a whole? What is the *p*-value?

Example 11.4

Suppose you flip two coins 100 times. The results are 20 *HH*, 27 *HT*, 30 *TH*, and 23 *TT*. Are the coins fair? Test at a 5% significance level.

Solution 11.4

This problem can be set up as a goodness-of-fit problem. The sample space for flipping two fair coins is {*HH*, *HT*, *TH*, *TT*}. Out of 100 flips, you would expect 25 *HH*, 25 *HT*, 25 *TH*, and 25 *TT*. This is the expected distribution. The question, "Are the coins fair?" is the same as saying, "Does the distribution of the coins (20 *HH*, 27 *HT*, 30 *TH*, 23 *TT*) fit the expected distribution?"

Random Variable: Let X = the number of heads in one flip of the two coins. X takes on the values 0, 1, 2. (There are 0, 1, or 2 heads in the flip of two coins.) Therefore, the **number of cells is three**. Since X = the number of heads, the observed frequencies are 20 (for two heads), 57 (for one head), and 23 (for zero heads or both tails). The expected frequencies are 25 (for two heads), 50 (for one head), and 25 (for zero heads or both tails). This test is right-tailed.

H_0: The coins are fair.

H_a: The coins are not fair.

Distribution for the test: χ_2^2 where $df = 3 - 1 = 2$.

Calculate the test statistic: $\chi^2 = 2.14$

Graph:

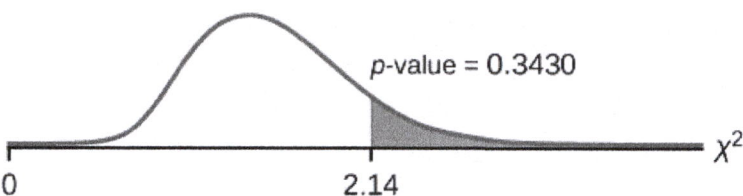

Figure 11.6

Probability statement: p-value = $P(\chi^2 > 2.14) = 0.3430$

Compare α and the p-value:

- $\alpha = 0.05$

- p-value = 0.3430

$\alpha < p$-value.

Make a decision: Since $\alpha < p$-value, do not reject H_0.

Conclusion: There is insufficient evidence to conclude that the coins are not fair.

 Using the TI-83, 83+, 84, 84+ Calculator

Press **STAT** and **ENTER**. Make sure you clear lists **L1**, **L2**, and **L3** if they have data in them. Into **L1**, put the observed frequencies **20, 57, 23**. Into **L2**, put the expected frequencies **25, 50, 25**. Arrow over to list **L3** and up to the name area "**L3**". Enter **(L1-L2)^2/L2** and **ENTER**. Press **2nd QUIT**. Press **2nd LIST** and arrow over to **MATH**. Press **5**. You should see "**sum**".**Enter L3**. Rounded to two decimal places, you should see **2.14**. Press **2nd DISTR**. Arrow down to **7:χ2cdf** (or press **7**). Press **ENTER**. Enter **2.14,1E99,2)**. Rounded to four places, you should see **.3430**, which is the p-value.

The newer TI-84 calculators have in **STAT TESTS** the test **Chi2 GOF**. To run the test, put the observed values (the data) into a first list and the expected values (the values you expect if the null hypothesis is true) into a second list. Press **STAT TESTS** and **Chi2 GOF**. Enter the list names for the Observed list and the Expected list. Enter the degrees of freedom and press **calculate** or **draw**. Make sure you clear any lists before you start.

Try It Σ

☞ **11.4** Students in a social studies class hypothesize that the literacy rates across the world for every region are 82%. **Table 11.14** shows the actual literacy rates across the world broken down by region. What are the test statistic and the degrees of freedom?

MDG Region	Adult Literacy Rate (%)
Developed Regions	99.0
Commonwealth of Independent States	99.5
Northern Africa	67.3
Sub-Saharan Africa	62.5
Latin America and the Caribbean	91.0
Eastern Asia	93.8
Southern Asia	61.9
South-Eastern Asia	91.9
Western Asia	84.5
Oceania	66.4

Table 11.14

11.3 | Test of Independence

Tests of independence involve using a **contingency table** of observed (data) values.

The test statistic for a **test of independence** is similar to that of a goodness-of-fit test:

$$\sum_{(i \cdot j)} \frac{(O - E)^2}{E}$$

where:

- O = observed values
- E = expected values
- i = the number of rows in the table
- j = the number of columns in the table

There are $i \cdot j$ terms of the form $\frac{(O - E)^2}{E}$.

A test of independence determines whether two factors are independent or not. You first encountered the term independence in **Probability Topics**. As a review, consider the following example.

NOTE

The expected value for each cell needs to be at least five in order for you to use this test.

Example 11.5

Suppose A = a speeding violation in the last year and B = a cell phone user while driving. If A and B are independent then $P(A \text{ AND } B) = P(A)P(B)$. A AND B is the event that a driver received a speeding violation last year and also used a cell phone while driving. Suppose, in a study of drivers who received speeding violations in the last year, and who used cell phone while driving, that 755 people were surveyed. Out of the 755, 70 had a speeding violation and 685 did not; 305 used cell phones while driving and 450 did not.

Let y = expected number of drivers who used a cell phone while driving and received speeding violations.

If A and B are independent, then $P(A \text{ AND } B) = P(A)P(B)$. By substitution,

$$\frac{y}{755} = \left(\frac{70}{755}\right)\left(\frac{305}{755}\right)$$

Solve for y: $y = \frac{(70)(305)}{755} = 28.3$

About 28 people from the sample are expected to use cell phones while driving and to receive speeding violations.

In a test of independence, we state the null and alternative hypotheses in words. Since the contingency table consists of **two factors**, the null hypothesis states that the factors are **independent** and the alternative hypothesis states that they are **not independent (dependent)**. If we do a test of independence using the example, then the null hypothesis is:

H_0: Being a cell phone user while driving and receiving a speeding violation are independent events.

If the null hypothesis were true, we would expect about 28 people to use cell phones while driving and to receive a speeding violation.

The test of independence is always right-tailed because of the calculation of the test statistic. If the expected and observed values are not close together, then the test statistic is very large and way out in the right tail of the chi-square curve, as it is in a goodness-of-fit.

The number of degrees of freedom for the test of independence is:

df = (number of columns - 1)(number of rows - 1)

The following formula calculates the **expected number** (E):

$$E = \frac{(\text{row total})(\text{column total})}{\text{total number surveyed}}$$

Try It Σ

11.5 A sample of 300 students is taken. Of the students surveyed, 50 were music students, while 250 were not. Ninety-seven were on the honor roll, while 203 were not. If we assume being a music student and being on the honor roll are independent events, what is the expected number of music students who are also on the honor roll?

Example 11.6

In a volunteer group, adults 21 and older volunteer from one to nine hours each week to spend time with a disabled senior citizen. The program recruits among community college students, four-year college students, and nonstudents. In **Table 11.15** is a **sample** of the adult volunteers and the number of hours they volunteer per week.

Type of Volunteer	1–3 Hours	4–6 Hours	7–9 Hours	Row Total
Community College Students	111	96	48	255
Four-Year College Students	96	133	61	290
Nonstudents	91	150	53	294
Column Total	298	379	162	839

Table 11.15 Number of Hours Worked Per Week by Volunteer Type (Observed) The table contains **observed (O)** values (data).

Is the number of hours volunteered **independent** of the type of volunteer?

Solution 11.6

The **observed table** and the question at the end of the problem, "Is the number of hours volunteered independent of the type of volunteer?" tell you this is a test of independence. The two factors are **number of hours volunteered** and **type of volunteer**. This test is always right-tailed.

H_0: The number of hours volunteered is **independent** of the type of volunteer.

H_a: The number of hours volunteered is **dependent** on the type of volunteer.

The expected result are in **Table 11.15**.

Type of Volunteer	1-3 Hours	4-6 Hours	7-9 Hours
Community College Students	90.57	115.19	49.24
Four-Year College Students	103.00	131.00	56.00
Nonstudents	104.42	132.81	56.77

Table 11.16 Number of Hours Worked Per Week by Volunteer Type (Expected) The table contains **expected (E)** values (data).

For example, the calculation for the expected frequency for the top left cell is

$$E = \frac{(\text{row total})(\text{column total})}{\text{total number surveyed}} = \frac{(255)(298)}{839} = 90.57$$

Calculate the test statistic: $\chi^2 = 12.99$ (calculator or computer)

Distribution for the test: χ^2_4

$df = (3 \text{ columns} - 1)(3 \text{ rows} - 1) = (2)(2) = 4$

Graph:

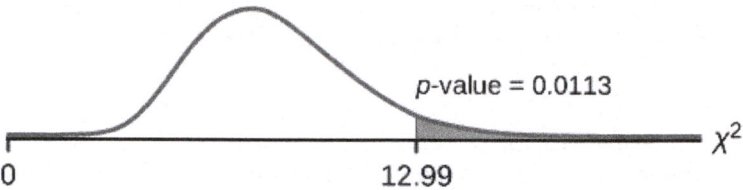

Figure 11.7

Probability statement: p-value$=P(\chi^2 > 12.99) = 0.0113$

Compare α and the p-value: Since no α is given, assume $\alpha = 0.05$. p-value = 0.0113. $\alpha > p$-value.

Make a decision: Since $\alpha > p$-value, reject H_0. This means that the factors are not independent.

Conclusion: At a 5% level of significance, from the data, there is sufficient evidence to conclude that the number of hours volunteered and the type of volunteer are dependent on one another.

For the example in **Table 11.15**, if there had been another type of volunteer, teenagers, what would the degrees of freedom be?

 Using the TI-83, 83+, 84, 84+ Calculator

Press the MATRX key and arrow over to EDIT. Press 1:[A]. Press 3 ENTER 3 ENTER. Enter the table values by row from **Table 11.15**. Press ENTER after each. Press 2nd QUIT. Press STAT and arrow over to TESTS. Arrow down to C:χ2-TEST. Press ENTER. You should see Observed:[A] and Expected:[B]. Arrow down to Calculate. Press ENTER. The test statistic is 12.9909 and the p-value = 0.0113. Do the procedure a second time, but arrow down to Draw instead of calculate.

Try It Σ

☞ **11.6** The Bureau of Labor Statistics gathers data about employment in the United States. A sample is taken to calculate the number of U.S. citizens working in one of several industry sectors over time. **Table 11.17** shows the results:

Industry Sector	2000	2010	2020	Total
Nonagriculture wage and salary	13,243	13,044	15,018	41,305
Goods-producing, excluding agriculture	2,457	1,771	1,950	6,178
Services-providing	10,786	11,273	13,068	35,127
Agriculture, forestry, fishing, and hunting	240	214	201	655
Nonagriculture self-employed and unpaid family worker	931	894	972	2,797
Secondary wage and salary jobs in agriculture and private household industries	14	11	11	36
Secondary jobs as a self-employed or unpaid family worker	196	144	152	492
Total	27,867	27,351	31,372	86,590

Table 11.17

We want to know if the change in the number of jobs is independent of the change in years. State the null and alternative hypotheses and the degrees of freedom.

Example 11.7

De Anza College is interested in the relationship between anxiety level and the need to succeed in school. A random sample of 400 students took a test that measured anxiety level and need to succeed in school. **Table 11.18** shows the results. De Anza College wants to know if anxiety level and need to succeed in school are independent events.

Need to Succeed in School	High Anxiety	Med-high Anxiety	Medium Anxiety	Med-low Anxiety	Low Anxiety	Row Total
High Need	35	42	53	15	10	155
Medium Need	18	48	63	33	31	193
Low Need	4	5	11	15	17	52
Column Total	57	95	127	63	58	400

Table 11.18 Need to Succeed in School vs. Anxiety Level

a. How many high anxiety level students are expected to have a high need to succeed in school?

Solution 11.7

a. The column total for a high anxiety level is 57. The row total for high need to succeed in school is 155. The sample size or total surveyed is 400.

$$E = \frac{(\text{row total})(\text{column total})}{\text{total surveyed}} = \frac{155 \cdot 57}{400} = 22.09$$

The expected number of students who have a high anxiety level and a high need to succeed in school is about 22.

b. If the two variables are independent, how many students do you expect to have a low need to succeed in school and a med-low level of anxiety?

Solution 11.7
b. The column total for a med-low anxiety level is 63. The row total for a low need to succeed in school is 52. The sample size or total surveyed is 400.

c. $E = \frac{(\text{row total})(\text{column total})}{\text{total surveyed}} = \underline{\hspace{1cm}}$

Solution 11.7
c. $E = \frac{(\text{row total})(\text{column total})}{\text{total surveyed}} = 8.19$

d. The expected number of students who have a med-low anxiety level and a low need to succeed in school is about _____.

Solution 11.7
d. 8

Try It Σ

11.7 Refer back to the information in **Try It**. How many service providing jobs are there expected to be in 2020? How many nonagriculture wage and salary jobs are there expected to be in 2020?

11.4 | Test for Homogeneity

The goodness–of–fit test can be used to decide whether a population fits a given distribution, but it will not suffice to decide whether two populations follow the same unknown distribution. A different test, called the **test for homogeneity**, can be used to draw a conclusion about whether two populations have the same distribution. To calculate the test statistic for a test for homogeneity, follow the same procedure as with the test of independence.

NOTE

The expected value for each cell needs to be at least five in order for you to use this test.

Hypotheses

H_0: The distributions of the two populations are the same.

H_a: The distributions of the two populations are not the same.

Test Statistic

Use a χ^2 test statistic. It is computed in the same way as the test for independence.

Degrees of Freedom (df)

df = number of columns - 1

Requirements

All values in the table must be greater than or equal to five.

Common Uses

Comparing two populations. For example: men vs. women, before vs. after, east vs. west. The variable is categorical with more than two possible response values.

Example 11.8

Do male and female college students have the same distribution of living arrangements? Use a level of significance of 0.05. Suppose that 250 randomly selected male college students and 300 randomly selected female college students were asked about their living arrangements: dormitory, apartment, with parents, other. The results are shown in **Table 11.18**. Do male and female college students have the same distribution of living arrangements?

	Dormitory	Apartment	With Parents	Other
Males	72	84	49	45
Females	91	86	88	35

Table 11.19 Distribution of Living Arragements for College Males and College Females

Solution 11.8

H_0: The distribution of living arrangements for male college students is the same as the distribution of living arrangements for female college students.

H_a: The distribution of living arrangements for male college students is not the same as the distribution of living arrangements for female college students.

Degrees of Freedom (df):
df = number of columns $- 1 = 4 - 1 = 3$

Distribution for the test: χ_3^2

Calculate the test statistic: $\chi^2 = 10.1287$ (calculator or computer)

Probability statement: p-value $= P(\chi^2 > 10.1287) = 0.0175$

🖩 Using the TI-83, 83+, 84, 84+ Calculator

Press the MATRX key and arrow over to EDIT. Press 1:[A]. Press 2 ENTER 4 ENTER. Enter the table values by row. Press ENTER after each. Press 2nd QUIT. Press STAT and arrow over to TESTS. Arrow down to C:χ2-TEST. Press ENTER. You should see Observed:[A] and Expected:[B]. Arrow down to Calculate. Press ENTER. The test statistic is 10.1287 and the p-value = 0.0175. Do the procedure a second time but arrow down to Draw instead of calculate.

Compare α and the p-value: Since no α is given, assume $\alpha = 0.05$. p-value = 0.0175. $\alpha > p$-value.

Make a decision: Since $\alpha >$ p-value, reject H_0. This means that the distributions are not the same.

Conclusion: At a 5% level of significance, from the data, there is sufficient evidence to conclude that the distributions of living arrangements for male and female college students are not the same.

Notice that the conclusion is only that the distributions are not the same. We cannot use the test for homogeneity to draw any conclusions about how they differ.

11.8 Do families and singles have the same distribution of cars? Use a level of significance of 0.05. Suppose that 100 randomly selected families and 200 randomly selected singles were asked what type of car they drove: sport, sedan, hatchback, truck, van/SUV. The results are shown in **Table 11.20**. Do families and singles have the same distribution of cars? Test at a level of significance of 0.05.

	Sport	Sedan	Hatchback	Truck	Van/SUV
Family	5	15	35	17	28
Single	45	65	37	46	7

Table 11.20

Example 11.9

Both before and after a recent earthquake, surveys were conducted asking voters which of the three candidates they planned on voting for in the upcoming city council election. Has there been a change since the earthquake? Use a level of significance of 0.05. **Table 11.20** shows the results of the survey. Has there been a change in the distribution of voter preferences since the earthquake?

	Perez	Chung	Stevens
Before	167	128	135
After	214	197	225

Table 11.21

Solution 11.9

H_0: The distribution of voter preferences was the same before and after the earthquake.

H_a: The distribution of voter preferences was not the same before and after the earthquake.

Degrees of Freedom (*df*):
df = number of columns − 1 = 3 − 1 = 2

Distribution for the test: χ_2^2

Calculate the test statistic: χ^2 = 3.2603 (calculator or computer)

Probability statement: p-value=$P(\chi^2 > 3.2603)$ = 0.1959

 Using the TI-83, 83+, 84, 84+ Calculator

Press the MATRX key and arrow over to EDIT. Press 1:[A]. Press 2 ENTER 3 ENTER. Enter the table values by row. Press ENTER after each. Press 2nd QUIT. Press STAT and arrow over to TESTS. Arrow down to C:χ2-TEST. Press ENTER. You should see Observed:[A] and Expected:[B]. Arrow down to Calculate. Press ENTER. The test statistic is 3.2603 and the p-value = 0.1959. Do the procedure a second time but arrow down to Draw instead of calculate.

Compare α and the p-value: α = 0.05 and the p-value = 0.1959. $\alpha < p$-value.

Make a decision: Since $\alpha < p$-value, do not reject H_o.

Conclusion: At a 5% level of significance, from the data, there is insufficient evidence to conclude that the distribution of voter preferences was not the same before and after the earthquake.

☞ **11.9** Ivy League schools receive many applications, but only some can be accepted. At the schools listed in **Table 11.22**, two types of applications are accepted: regular and early decision.

Application Type Accepted	Brown	Columbia	Cornell	Dartmouth	Penn	Yale
Regular	2,115	1,792	5,306	1,734	2,685	1,245
Early Decision	577	627	1,228	444	1,195	761

Table 11.22

We want to know if the number of regular applications accepted follows the same distribution as the number of early applications accepted. State the null and alternative hypotheses, the degrees of freedom and the test statistic, sketch the graph of the p-value, and draw a conclusion about the test of homogeneity.

11.5 | Comparison of the Chi-Square Tests

You have seen the χ^2 test statistic used in three different circumstances. The following bulleted list is a summary that will help you decide which χ^2 test is the appropriate one to use.

- **Goodness-of-Fit:** Use the goodness-of-fit test to decide whether a population with an unknown distribution "fits" a known distribution. In this case there will be a single qualitative survey question or a single outcome of an experiment from a single population. Goodness-of-Fit is typically used to see if the population is uniform (all outcomes occur with equal frequency), the population is normal, or the population is the same as another population with a known distribution. The null and alternative hypotheses are:

  H_0: The population fits the given distribution.

  H_a: The population does not fit the given distribution.

- **Independence:** Use the test for independence to decide whether two variables (factors) are independent or dependent. In this case there will be two qualitative survey questions or experiments and a contingency table will be constructed. The goal is to see if the two variables are unrelated (independent) or related (dependent). The null and alternative hypotheses are:

  H_0: The two variables (factors) are independent.

  H_a: The two variables (factors) are dependent.

- **Homogeneity:** Use the test for homogeneity to decide if two populations with unknown distributions have the same distribution as each other. In this case there will be a single qualitative survey question or experiment given to two different populations. The null and alternative hypotheses are:

  H_0: The two populations follow the same distribution.

  H_a: The two populations have different distributions.

11.6 | Test of a Single Variance

A **test of a single variance** assumes that the underlying distribution is **normal**. The null and alternative hypotheses are stated in terms of the **population variance** (or population standard deviation). The test statistic is:

$$\frac{(n-1)s^2}{\sigma^2}$$

where:

- n = the total number of data

- s^2 = sample variance

- σ^2 = population variance

You may think of s as the random variable in this test. The number of degrees of freedom is $df = n - 1$. **A test of a single variance may be right-tailed, left-tailed, or two-tailed. Example 11.10** will show you how to set up the null and alternative hypotheses. The null and alternative hypotheses contain statements about the population variance.

Example 11.10

Math instructors are not only interested in how their students do on exams, on average, but how the exam scores vary. To many instructors, the variance (or standard deviation) may be more important than the average.

Suppose a math instructor believes that the standard deviation for his final exam is five points. One of his best students thinks otherwise. The student claims that the standard deviation is more than five points. If the student were to conduct a hypothesis test, what would the null and alternative hypotheses be?

Solution 11.10

Even though we are given the population standard deviation, we can set up the test using the population variance as follows.

- H_0: $\sigma^2 = 5^2$
- H_a: $\sigma^2 > 5^2$

 Try It Σ

11.10 A SCUBA instructor wants to record the collective depths each of his students dives during their checkout. He is interested in how the depths vary, even though everyone should have been at the same depth. He believes the standard deviation is three feet. His assistant thinks the standard deviation is less than three feet. If the instructor were to conduct a test, what would the null and alternative hypotheses be?

Example 11.11

With individual lines at its various windows, a post office finds that the standard deviation for normally distributed waiting times for customers on Friday afternoon is 7.2 minutes. The post office experiments with a single, main waiting line and finds that for a random sample of 25 customers, the waiting times for customers have a standard deviation of 3.5 minutes.

With a significance level of 5%, test the claim that **a single line causes lower variation among waiting times (shorter waiting times) for customers**.

Solution 11.11

Since the claim is that a single line causes less variation, this is a test of a single variance. The parameter is the population variance, σ^2, or the population standard deviation, σ.

Random Variable: The sample standard deviation, s, is the random variable. Let s = standard deviation for the waiting times.

- H_0: $\sigma^2 = 7.2^2$
- H_a: $\sigma^2 < 7.2^2$

The word **"less"** tells you this is a left-tailed test.

Distribution for the test: χ^2_{24}, where:

- n = the number of customers sampled
- $df = n - 1 = 25 - 1 = 24$

Calculate the test statistic:

$$\chi^2 = \frac{(n-1)s^2}{\sigma^2} = \frac{(25-1)(3.5)^2}{7.2^2} = 5.67$$

where $n = 25$, $s = 3.5$, and $\sigma = 7.2$.

Graph:

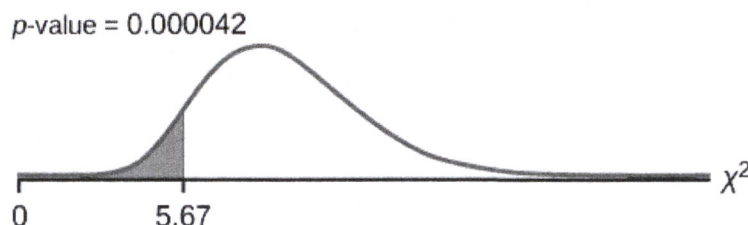

p-value = 0.000042

0   5.67

Figure 11.8

Probability statement: *p*-value = $P(\chi^2 < 5.67) = 0.000042$

Compare α and the *p*-value:
$\alpha = 0.05$; *p*-value = 0.000042; $\alpha > $ *p*-value

Make a decision: Since $\alpha > $ *p*-value, reject H_0. This means that you reject $\sigma^2 = 7.2^2$. In other words, you do not think the variation in waiting times is 7.2 minutes; you think the variation in waiting times is less.

Conclusion: At a 5% level of significance, from the data, there is sufficient evidence to conclude that a single line causes a lower variation among the waiting times **or** with a single line, the customer waiting times vary less than 7.2 minutes.

 Using the TI-83, 83+, 84, 84+ Calculator

In 2nd DISTR, use 7:χ2cdf. The syntax is (lower, upper, df) for the parameter list. For **Example 11.11**, χ2cdf(-1E99,5.67,24). The *p*-value = 0.000042.

Try It Σ

11.11 The FCC conducts broadband speed tests to measure how much data per second passes between a consumer's computer and the internet. As of August of 2012, the standard deviation of Internet speeds across Internet Service Providers (ISPs) was 12.2 percent. Suppose a sample of 15 ISPs is taken, and the standard deviation is 13.2. An analyst claims that the standard deviation of speeds is more than what was reported. State the null and alternative hypotheses, compute the degrees of freedom, the test statistic, sketch the graph of the *p*-value, and draw a conclusion. Test at the 1% significance level.

11.7 | Lab 1: Chi-Square Goodness-of-Fit

Stats Lab

11.1 Lab 1: Chi-Square Goodness-of-Fit

Class Time:

Names:

Student Learning Outcome

- The student will evaluate data collected to determine if they fit either the uniform or exponential distributions.

Collect the Data

Go to your local supermarket. Ask 30 people as they leave for the total amount on their grocery receipts. (Or, ask three cashiers for the last ten amounts. Be sure to include the express lane, if it is open.)

NOTE

You may need to combine two categories so that each cell has an expected value of at least five.

1. Record the values.

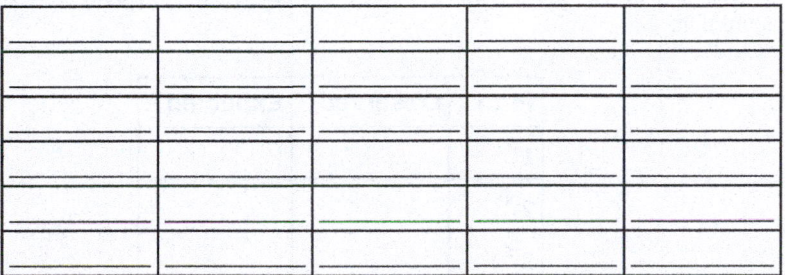

Table 11.23

2. Construct a histogram of the data. Make five to six intervals. Sketch the graph using a ruler and pencil. Scale the axes.

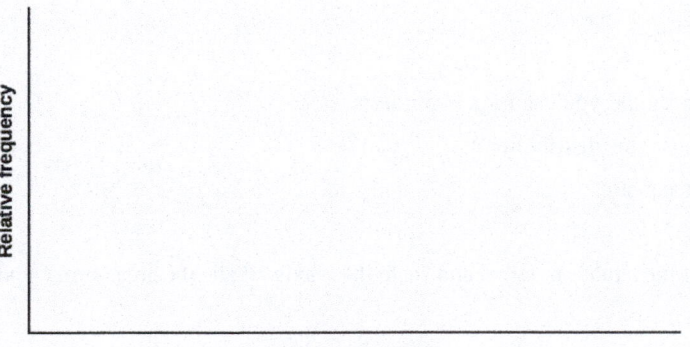

Figure 11.9

3. Calculate the following:

a. $\bar{x} =$ _____

b. $s =$ _____

c. $s^2 =$ _____

Uniform Distribution

Test to see if grocery receipts follow the uniform distribution.

1. Using your lowest and highest values, $X \sim U$ (_____, _____)

2. Divide the distribution into fifths.

3. Calculate the following:

   a. lowest value = _____

   b. 20th percentile = _____

   c. 40th percentile = _____

   d. 60th percentile = _____

   e. 80th percentile = _____

   f. highest value = _____

4. For each fifth, count the observed number of receipts and record it. Then determine the expected number of receipts and record that.

Fifth	Observed	Expected
1st		
2nd		
3rd		
4th		
5th		

Table 11.24

5. H_0: _____

6. H_a: _____

7. What distribution should you use for a hypothesis test?

8. Why did you choose this distribution?

9. Calculate the test statistic.

10. Find the p-value.

11. Sketch a graph of the situation. Label and scale the x-axis. Shade the area corresponding to the p-value.

Figure 11.10

12. State your decision.

13. State your conclusion in a complete sentence.

Exponential Distribution

Test to see if grocery receipts follow the exponential distribution with decay parameter $\frac{1}{x}$.

1. Using $\frac{1}{x}$ as the decay parameter, $X \sim Exp(\underline{\hspace{1.5cm}})$.

2. Calculate the following:

    a. lowest value = _____

    b. first quartile = _____

    c. 37th percentile = _____

    d. median = _____

    e. 63rd percentile = _____

    f. 3rd quartile = _____

    g. highest value = _____

3. For each cell, count the observed number of receipts and record it. Then determine the expected number of receipts and record that.

Cell	Observed	Expected
1st		
2nd		
3rd		
4th		
5th		
6th		

Table 11.25

4. H_0: _____

5. H_a: _____

6. What distribution should you use for a hypothesis test?

7. Why did you choose this distribution?

8. Calculate the test statistic.

9. Find the p-value.

10. Sketch a graph of the situation. Label and scale the x-axis. Shade the area corresponding to the p-value.

Figure 11.11

11. State your decision.

12. State your conclusion in a complete sentence.

Discussion Questions

1. Did your data fit either distribution? If so, which?

2. In general, do you think it's likely that data could fit more than one distribution? In complete sentences, explain why or why not.

11.8 | Lab 2: Chi-Square Test of Independence

Stats Lab

11.2 Lab 2: Chi-Square Test of Independence

Class Time:

Names:

Student Learning Outcome

- The student will evaluate if there is a significant relationship between favorite type of snack and gender.

Collect the Data

1. Using your class as a sample, complete the following chart. Ask each other what your favorite snack is, then total the results.

    ### NOTE

    You may need to combine two food categories so that each cell has an expected value of at least five.

	sweets (candy & baked goods)	ice cream	chips & pretzels	fruits & vegetables	Total
male					
female					
Total					

    Table 11.26 Favorite type of snack

2. Looking at **Table 11.26**, does it appear to you that there is a dependence between gender and favorite type of snack food? Why or why not?

Hypothesis Test

Conduct a hypothesis test to determine if the factors are independent:

1. H_0: _____
2. H_a: _____
3. What distribution should you use for a hypothesis test?
4. Why did you choose this distribution?
5. Calculate the test statistic.
6. Find the p-value.
7. Sketch a graph of the situation. Label and scale the x-axis. Shade the area corresponding to the p-value.

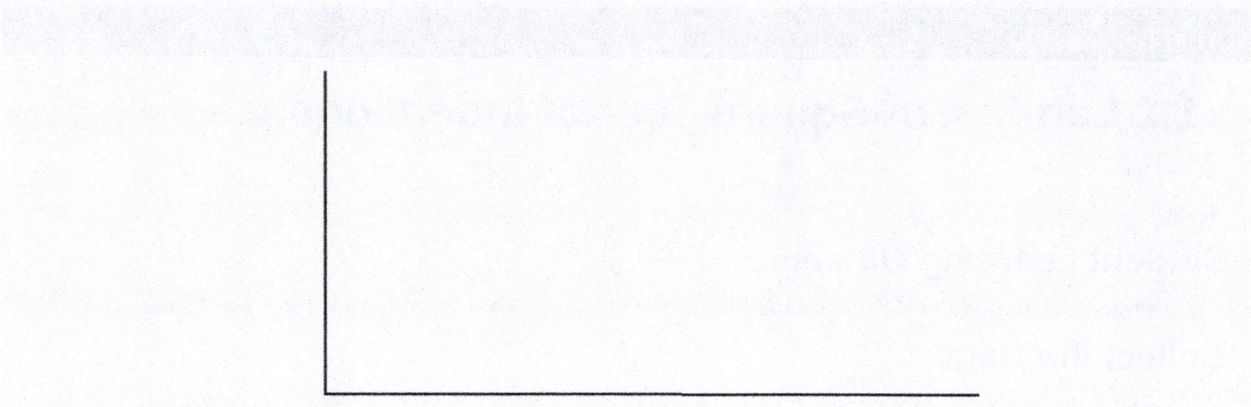

Figure 11.12

8. State your decision.
9. State your conclusion in a complete sentence.

Discussion Questions

1. Is the conclusion of your study the same as or different from your answer to answer to question two under **Collect the Data**?
2. Why do you think that occurred?

KEY TERMS

Contingency Table a table that displays sample values for two different factors that may be dependent or contingent on one another; it facilitates determining conditional probabilities.

CHAPTER REVIEW

11.1 Facts About the Chi-Square Distribution

The chi-square distribution is a useful tool for assessment in a series of problem categories. These problem categories include primarily (i) whether a data set fits a particular distribution, (ii) whether the distributions of two populations are the same, (iii) whether two events might be independent, and (iv) whether there is a different variability than expected within a population.

An important parameter in a chi-square distribution is the degrees of freedom df in a given problem. The random variable in the chi-square distribution is the sum of squares of df standard normal variables, which must be independent. The key characteristics of the chi-square distribution also depend directly on the degrees of freedom.

The chi-square distribution curve is skewed to the right, and its shape depends on the degrees of freedom df. For $df > 90$, the curve approximates the normal distribution. Test statistics based on the chi-square distribution are always greater than or equal to zero. Such application tests are almost always right-tailed tests.

11.2 Goodness-of-Fit Test

To assess whether a data set fits a specific distribution, you can apply the goodness-of-fit hypothesis test that uses the chi-square distribution. The null hypothesis for this test states that the data come from the assumed distribution. The test compares observed values against the values you would expect to have if your data followed the assumed distribution. The test is almost always right-tailed. Each observation or cell category must have an expected value of at least five.

11.3 Test of Independence

To assess whether two factors are independent or not, you can apply the test of independence that uses the chi-square distribution. The null hypothesis for this test states that the two factors are independent. The test compares observed values to expected values. The test is right-tailed. Each observation or cell category must have an expected value of at least 5.

11.4 Test for Homogeneity

To assess whether two data sets are derived from the same distribution—which need not be known, you can apply the test for homogeneity that uses the chi-square distribution. The null hypothesis for this test states that the populations of the two data sets come from the same distribution. The test compares the observed values against the expected values if the two populations followed the same distribution. The test is right-tailed. Each observation or cell category must have an expected value of at least five.

11.5 Comparison of the Chi-Square Tests

The goodness-of-fit test is typically used to determine if data fits a particular distribution. The test of independence makes use of a contingency table to determine the independence of two factors. The test for homogeneity determines whether two populations come from the same distribution, even if this distribution is unknown.

11.6 Test of a Single Variance

To test variability, use the chi-square test of a single variance. The test may be left-, right-, or two-tailed, and its hypotheses are always expressed in terms of the variance (or standard deviation).

FORMULA REVIEW

11.1 Facts About the Chi-Square Distribution

$\chi^2 = (Z_1)^2 + (Z_2)^2 + \ldots (Z_{df})^2$ chi-square distribution random variable

$\mu_{\chi^2} = df$ chi-square distribution population mean

$\sigma_{\chi^2} = \sqrt{2(df)}$ Chi-Square distribution population standard deviation

11.2 Goodness-of-Fit Test

$\sum_k \frac{(O-E)^2}{E}$ goodness-of-fit test statistic where:

O: observed values
E: expected values

k: number of different data cells or categories

$df = k - 1$ degrees of freedom

11.3 Test of Independence

Test of Independence
- The number of degrees of freedom is equal to (number of columns - 1)(number of rows - 1).

- The test statistic is $\sum_{(i \cdot j)} \frac{(O-E)^2}{E}$ where O = observed values, E = expected values, i = the number of rows in the table, and j = the number of columns in the table.

- If the null hypothesis is true, the expected number $E = \frac{(\text{row total})(\text{column total})}{\text{total surveyed}}$.

11.4 Test for Homogeneity

$\sum_{i \cdot j} \frac{(O-E)^2}{E}$ Homogeneity test statistic where: O = observed values
E = expected values
i = number of rows in data contingency table
j = number of columns in data contingency table

$df = (i-1)(j-1)$ Degrees of freedom

11.6 Test of a Single Variance

$\chi^2 = \frac{(n-1) \cdot s^2}{\sigma^2}$ Test of a single variance statistic where:
n: sample size
s: sample standard deviation
σ: population standard deviation

$df = n - 1$ Degrees of freedom

Test of a Single Variance
- Use the test to determine variation.

- The degrees of freedom is the number of samples – 1.

- The test statistic is $\frac{(n-1) \cdot s^2}{\sigma^2}$, where n = the total number of data, s^2 = sample variance, and σ^2 = population variance.

- The test may be left-, right-, or two-tailed.

PRACTICE

11.1 Facts About the Chi-Square Distribution

1. If the number of degrees of freedom for a chi-square distribution is 25, what is the population mean and standard deviation?

2. If $df > 90$, the distribution is _____. If $df = 15$, the distribution is _____.

3. When does the chi-square curve approximate a normal distribution?

4. Where is μ located on a chi-square curve?

5. Is it more likely the df is 90, 20, or two in the graph?

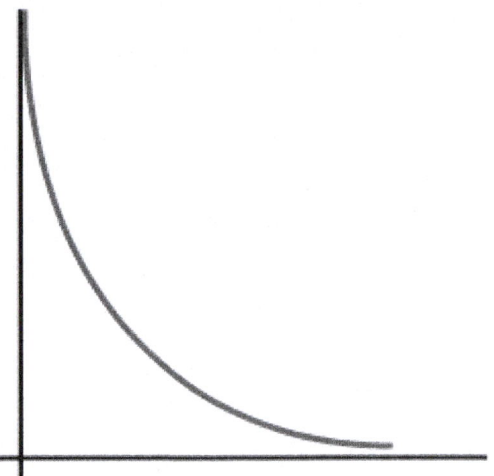

Figure 11.13

11.2 Goodness-of-Fit Test

Determine the appropriate test to be used in the next three exercises.

6. An archeologist is calculating the distribution of the frequency of the number of artifacts she finds in a dig site. Based on previous digs, the archeologist creates an expected distribution broken down by grid sections in the dig site. Once the site has been fully excavated, she compares the actual number of artifacts found in each grid section to see if her expectation was accurate.

7. An economist is deriving a model to predict outcomes on the stock market. He creates a list of expected points on the stock market index for the next two weeks. At the close of each day's trading, he records the actual points on the index. He wants to see how well his model matched what actually happened.

8. A personal trainer is putting together a weight-lifting program for her clients. For a 90-day program, she expects each client to lift a specific maximum weight each week. As she goes along, she records the actual maximum weights her clients lifted. She wants to know how well her expectations met with what was observed.

Use the following information to answer the next five exercises: A teacher predicts that the distribution of grades on the final exam will be and they are recorded in **Table 11.27**.

Grade	Proportion
A	0.25
B	0.30
C	0.35
D	0.10

Table 11.27

The actual distribution for a class of 20 is in **Table 11.28**.

Grade	Frequency
A	7

Grade	Frequency
B	7
C	5
D	1

Table 11.28

9. $df =$ _____

10. State the null and alternative hypotheses.

11. χ^2 test statistic = _____

12. p-value = _____

13. At the 5% significance level, what can you conclude?

Use the following information to answer the next nine exercises: The following data are real. The cumulative number of AIDS cases reported for Santa Clara County is broken down by ethnicity as in **Table 11.29**.

Ethnicity	Number of Cases
White	2,229
Hispanic	1,157
Black/African-American	457
Asian, Pacific Islander	232
	Total = 4,075

Table 11.29

The percentage of each ethnic group in Santa Clara County is as in **Table 11.30**.

Ethnicity	Percentage of total county population	Number expected (round to two decimal places)
White	42.9%	1748.18
Hispanic	26.7%	
Black/African-American	2.6%	
Asian, Pacific Islander	27.8%	
	Total = 100%	

Table 11.30

14. If the ethnicities of AIDS victims followed the ethnicities of the total county population, fill in the expected number of cases per ethnic group.
Perform a goodness-of-fit test to determine whether the occurrence of AIDS cases follows the ethnicities of the general population of Santa Clara County.

15. H_0: _____

16. H_a: _____

17. Is this a right-tailed, left-tailed, or two-tailed test?

18. degrees of freedom = _____

19. χ^2 test statistic = _____

20. p-value = _____

21. Graph the situation. Label and scale the horizontal axis. Mark the mean and test statistic. Shade in the region corresponding to the p-value.

Figure 11.14

Let $\alpha = 0.05$

Decision: _____

Reason for the Decision: _____

Conclusion (write out in complete sentences): _____

22. Does it appear that the pattern of AIDS cases in Santa Clara County corresponds to the distribution of ethnic groups in this county? Why or why not?

11.3 Test of Independence

Determine the appropriate test to be used in the next three exercises.

23. A pharmaceutical company is interested in the relationship between age and presentation of symptoms for a common viral infection. A random sample is taken of 500 people with the infection across different age groups.

24. The owner of a baseball team is interested in the relationship between player salaries and team winning percentage. He takes a random sample of 100 players from different organizations.

25. A marathon runner is interested in the relationship between the brand of shoes runners wear and their run times. She takes a random sample of 50 runners and records their run times as well as the brand of shoes they were wearing.

Use the following information to answer the next seven exercises: Transit Railroads is interested in the relationship between travel distance and the ticket class purchased. A random sample of 200 passengers is taken. **Table 11.31** shows the results. The railroad wants to know if a passenger's choice in ticket class is independent of the distance they must travel.

Traveling Distance	Third class	Second class	First class	Total
1–100 miles	21	14	6	41

Traveling Distance	Third class	Second class	First class	Total
101–200 miles	18	16	8	42
201–300 miles	16	17	15	48
301–400 miles	12	14	21	47
401–500 miles	6	6	10	22
Total	73	67	60	200

Table 11.31

26. State the hypotheses.

H_0: _____

H_a: _____

27. $df =$ _____

28. How many passengers are expected to travel between 201 and 300 miles and purchase second-class tickets?

29. How many passengers are expected to travel between 401 and 500 miles and purchase first-class tickets?

30. What is the test statistic?

31. What is the p-value?

32. What can you conclude at the 5% level of significance?

Use the following information to answer the next eight exercises: An article in the New England Journal of Medicine, discussed a study on smokers in California and Hawaii. In one part of the report, the self-reported ethnicity and smoking levels per day were given. Of the people smoking at most ten cigarettes per day, there were 9,886 African Americans, 2,745 Native Hawaiians, 12,831 Latinos, 8,378 Japanese Americans and 7,650 whites. Of the people smoking 11 to 20 cigarettes per day, there were 6,514 African Americans, 3,062 Native Hawaiians, 4,932 Latinos, 10,680 Japanese Americans, and 9,877 whites. Of the people smoking 21 to 30 cigarettes per day, there were 1,671 African Americans, 1,419 Native Hawaiians, 1,406 Latinos, 4,715 Japanese Americans, and 6,062 whites. Of the people smoking at least 31 cigarettes per day, there were 759 African Americans, 788 Native Hawaiians, 800 Latinos, 2,305 Japanese Americans, and 3,970 whites.

33. Complete the table.

Smoking Level Per Day	African American	Native Hawaiian	Latino	Japanese Americans	White	TOTALS
1-10						
11-20						
21-30						
31+						
TOTALS						

Table 11.32 Smoking Levels by Ethnicity (Observed)

34. State the hypotheses.

H_0: _____

H_a: _____

35. Enter expected values in **Table 11.32**. Round to two decimal places.

Calculate the following values:

36. $df =$ _____

37. χ^2 test statistic = _____

38. p-value = _____

39. Is this a right-tailed, left-tailed, or two-tailed test? Explain why.

40. Graph the situation. Label and scale the horizontal axis. Mark the mean and test statistic. Shade in the region corresponding to the p-value.

Figure 11.15

State the decision and conclusion (in a complete sentence) for the following preconceived levels of α.

41. $\alpha = 0.05$
   a. Decision: _____
   b. Reason for the decision: _____
   c. Conclusion (write out in a complete sentence): _____

42. $\alpha = 0.01$
   a. Decision: _____
   b. Reason for the decision: _____
   c. Conclusion (write out in a complete sentence): _____

11.4 Test for Homogeneity

43. A math teacher wants to see if two of her classes have the same distribution of test scores. What test should she use?

44. What are the null and alternative hypotheses for **Exercise 11.43**?

45. A market researcher wants to see if two different stores have the same distribution of sales throughout the year. What type of test should he use?

46. A meteorologist wants to know if East and West Australia have the same distribution of storms. What type of test should she use?

47. What condition must be met to use the test for homogeneity?

Use the following information to answer the next five exercises: Do private practice doctors and hospital doctors have the same distribution of working hours? Suppose that a sample of 100 private practice doctors and 150 hospital doctors are selected at random and asked about the number of hours a week they work. The results are shown in **Table 11.33**.

	20–30	30–40	40–50	50–60
Private Practice	16	40	38	6
Hospital	8	44	59	39

48. State the null and alternative hypotheses.

49. $df =$ _____

50. What is the test statistic?

51. What is the p-value?

52. What can you conclude at the 5% significance level?

11.5 Comparison of the Chi-Square Tests

53. Which test do you use to decide whether an observed distribution is the same as an expected distribution?

54. What is the null hypothesis for the type of test from **Exercise 11.53**?

55. Which test would you use to decide whether two factors have a relationship?

56. Which test would you use to decide if two populations have the same distribution?

57. How are tests of independence similar to tests for homogeneity?

58. How are tests of independence different from tests for homogeneity?

11.6 Test of a Single Variance

Use the following information to answer the next three exercises: An archer's standard deviation for his hits is six (data is measured in distance from the center of the target). An observer claims the standard deviation is less.

59. What type of test should be used?

60. State the null and alternative hypotheses.

61. Is this a right-tailed, left-tailed, or two-tailed test?

Use the following information to answer the next three exercises: The standard deviation of heights for students in a school is 0.81. A random sample of 50 students is taken, and the standard deviation of heights of the sample is 0.96. A researcher in charge of the study believes the standard deviation of heights for the school is greater than 0.81.

62. What type of test should be used?

63. State the null and alternative hypotheses.

64. $df =$ _____

Use the following information to answer the next four exercises: The average waiting time in a doctor's office varies. The standard deviation of waiting times in a doctor's office is 3.4 minutes. A random sample of 30 patients in the doctor's office has a standard deviation of waiting times of 4.1 minutes. One doctor believes the variance of waiting times is greater than originally thought.

65. What type of test should be used?

66. What is the test statistic?

67. What is the p-value?

68. What can you conclude at the 5% significance level?

HOMEWORK

11.1 Facts About the Chi-Square Distribution

Decide whether the following statements are true or false.

69. As the number of degrees of freedom increases, the graph of the chi-square distribution looks more and more symmetrical.

70. The standard deviation of the chi-square distribution is twice the mean.

71. The mean and the median of the chi-square distribution are the same if $df = 24$.

11.2 Goodness-of-Fit Test

*For each problem, use a solution sheet to solve the hypothesis test problem. Go to **Appendix E** for the chi-square solution sheet. Round expected frequency to two decimal places.*

72. A six-sided die is rolled 120 times. Fill in the expected frequency column. Then, conduct a hypothesis test to determine if the die is fair. The data in **Table 11.34** are the result of the 120 rolls.

Face Value	Frequency	Expected Frequency
1	15	
2	29	
3	16	
4	15	
5	30	
6	15	

Table 11.34

73. The marital status distribution of the U.S. male population, ages 15 and older, is as shown in **Table 11.35**.

Marital Status	Percent	Expected Frequency
never married	31.3	
married	56.1	
widowed	2.5	
divorced/separated	10.1	

Table 11.35

Suppose that a random sample of 400 U.S. young adult males, 18 to 24 years old, yielded the following frequency distribution. We are interested in whether this age group of males fits the distribution of the U.S. adult population. Calculate the frequency one would expect when surveying 400 people. Fill in **Table 11.36**, rounding to two decimal places.

Marital Status	Frequency
never married	140
married	238
widowed	2
divorced/separated	20

Table 11.36

Use the following information to answer the next two exercises: The columns in **Table 11.37** contain the Race/Ethnicity of U.S. Public Schools for a recent year, the percentages for the Advanced Placement Examinee Population for that class, and the Overall Student Population. Suppose the right column contains the result of a survey of 1,000 local students from that year who took an AP Exam.

Race/Ethnicity	AP Examinee Population	Overall Student Population	Survey Frequency
Asian, Asian American, or Pacific Islander	10.2%	5.4%	113
Black or African-American	8.2%	14.5%	94
Hispanic or Latino	15.5%	15.9%	136
American Indian or Alaska Native	0.6%	1.2%	10
White	59.4%	61.6%	604
Not reported/other	6.1%	1.4%	43

Table 11.37

74. Perform a goodness-of-fit test to determine whether the local results follow the distribution of the U.S. overall student population based on ethnicity.

75. Perform a goodness-of-fit test to determine whether the local results follow the distribution of U.S. AP examinee population, based on ethnicity.

76. The City of South Lake Tahoe, CA, has an Asian population of 1,419 people, out of a total population of 23,609. Suppose that a survey of 1,419 self-reported Asians in the Manhattan, NY, area yielded the data in **Table 11.38**. Conduct a goodness-of-fit test to determine if the self-reported sub-groups of Asians in the Manhattan area fit that of the Lake Tahoe area.

Race	Lake Tahoe Frequency	Manhattan Frequency
Asian Indian	131	174
Chinese	118	557
Filipino	1,045	518
Japanese	80	54
Korean	12	29
Vietnamese	9	21
Other	24	66

Table 11.38

Use the following information to answer the next two exercises: UCLA conducted a survey of more than 263,000 college freshmen from 385 colleges in fall 2005. The results of students' expected majors by gender were reported in *The Chronicle of Higher Education (2/2/2006)*. Suppose a survey of 5,000 graduating females and 5,000 graduating males was done as a follow-up last year to determine what their actual majors were. The results are shown in the tables for **Exercise 11.77** and **Exercise 11.78**. The second column in each table does not add to 100% because of rounding.

77. Conduct a goodness-of-fit test to determine if the actual college majors of graduating females fit the distribution of their expected majors.

Major	Women - Expected Major	Women - Actual Major
Arts & Humanities	14.0%	670
Biological Sciences	8.4%	410

Major	Women - Expected Major	Women - Actual Major
Business	13.1%	685
Education	13.0%	650
Engineering	2.6%	145
Physical Sciences	2.6%	125
Professional	18.9%	975
Social Sciences	13.0%	605
Technical	0.4%	15
Other	5.8%	300
Undecided	8.0%	420

Table 11.39

78. Conduct a goodness-of-fit test to determine if the actual college majors of graduating males fit the distribution of their expected majors.

Major	Men - Expected Major	Men - Actual Major
Arts & Humanities	11.0%	600
Biological Sciences	6.7%	330
Business	22.7%	1130
Education	5.8%	305
Engineering	15.6%	800
Physical Sciences	3.6%	175
Professional	9.3%	460
Social Sciences	7.6%	370
Technical	1.8%	90
Other	8.2%	400
Undecided	6.6%	340

Table 11.40

Read the statement and decide whether it is true or false.

79. In a goodness-of-fit test, the expected values are the values we would expect if the null hypothesis were true.

80. In general, if the observed values and expected values of a goodness-of-fit test are not close together, then the test statistic can get very large and on a graph will be way out in the right tail.

81. Use a goodness-of-fit test to determine if high school principals believe that students are absent equally during the week or not.

82. The test to use to determine if a six-sided die is fair is a goodness-of-fit test.

83. In a goodness-of fit test, if the *p*-value is 0.0113, in general, do not reject the null hypothesis.

84. A sample of 212 commercial businesses was surveyed for recycling one commodity; a commodity here means any one type of recyclable material such as plastic or aluminum. **Table 11.41** shows the business categories in the survey, the sample size of each category, and the number of businesses in each category that recycle one commodity. Based on the study, on average half of the businesses were expected to be recycling one commodity. As a result, the last column shows the expected number of businesses in each category that recycle one commodity. At the 5% significance level, perform a hypothesis test to determine if the observed number of businesses that recycle one commodity follows the uniform distribution of the expected values.

Business Type	Number in class	Observed Number that recycle one commodity	Expected number that recycle one commodity
Office	35	19	17.5
Retail/ Wholesale	48	27	24
Food/ Restaurants	53	35	26.5
Manufacturing/ Medical	52	21	26
Hotel/Mixed	24	9	12

Table 11.41

85. Table 11.42 contains information from a survey among 499 participants classified according to their age groups. The second column shows the percentage of obese people per age class among the study participants. The last column comes from a different study at the national level that shows the corresponding percentages of obese people in the same age classes in the USA. Perform a hypothesis test at the 5% significance level to determine whether the survey participants are a representative sample of the USA obese population.

Age Class (Years)	Obese (Percentage)	Expected USA average (Percentage)
20–30	75.0	32.6
31–40	26.5	32.6
41–50	13.6	36.6
51–60	21.9	36.6
61–70	21.0	39.7

Table 11.42

11.3 Test of Independence

*For each problem, use a solution sheet to solve the hypothesis test problem. Go to **Appendix E** for the chi-square solution sheet. Round expected frequency to two decimal places.*

86. A recent debate about where in the United States skiers believe the skiing is best prompted the following survey. Test to see if the best ski area is independent of the level of the skier.

U.S. Ski Area	Beginner	Intermediate	Advanced
Tahoe	20	30	40
Utah	10	30	60
Colorado	10	40	50

Table 11.43

87. Car manufacturers are interested in whether there is a relationship between the size of car an individual drives and the number of people in the driver's family (that is, whether car size and family size are independent). To test this, suppose that 800 car owners were randomly surveyed with the results in **Table 11.44**. Conduct a test of independence.

Family Size	Sub & Compact	Mid-size	Full-size	Van & Truck
1	20	35	40	35
2	20	50	70	80
3–4	20	50	100	90
5+	20	30	70	70

Table 11.44

88. College students may be interested in whether or not their majors have any effect on starting salaries after graduation. Suppose that 300 recent graduates were surveyed as to their majors in college and their starting salaries after graduation. **Table 11.45** shows the data. Conduct a test of independence.

Major	< $50,000	$50,000 – $68,999	$69,000 +
English	5	20	5
Engineering	10	30	60
Nursing	10	15	15
Business	10	20	30
Psychology	20	30	20

Table 11.45

89. Some travel agents claim that honeymoon hot spots vary according to age of the bride. Suppose that 280 recent brides were interviewed as to where they spent their honeymoons. The information is given in **Table 11.46**. Conduct a test of independence.

Location	20–29	30–39	40–49	50 and over
Niagara Falls	15	25	25	20
Poconos	15	25	25	10
Europe	10	25	15	5
Virgin Islands	20	25	15	5

Table 11.46

90. A manager of a sports club keeps information concerning the main sport in which members participate and their ages. To test whether there is a relationship between the age of a member and his or her choice of sport, 643 members of the sports club are randomly selected. Conduct a test of independence.

Sport	18 - 25	26 - 30	31 - 40	41 and over
racquetball	42	58	30	46

Sport	18 - 25	26 - 30	31 - 40	41 and over
tennis	58	76	38	65
swimming	72	60	65	33

Table 11.47

91. A major food manufacturer is concerned that the sales for its skinny french fries have been decreasing. As a part of a feasibility study, the company conducts research into the types of fries sold across the country to determine if the type of fries sold is independent of the area of the country. The results of the study are shown in **Table 11.48**. Conduct a test of independence.

Type of Fries	Northeast	South	Central	West
skinny fries	70	50	20	25
curly fries	100	60	15	30
steak fries	20	40	10	10

Table 11.48

92. According to Dan Lenard, an independent insurance agent in the Buffalo, N.Y. area, the following is a breakdown of the amount of life insurance purchased by males in the following age groups. He is interested in whether the age of the male and the amount of life insurance purchased are independent events. Conduct a test for independence.

Age of Males	None	< $200,000	$200,000–$400,000	$401,001–$1,000,000	$1,000,001+
20–29	40	15	40	0	5
30–39	35	5	20	20	10
40–49	20	0	30	0	30
50+	40	30	15	15	10

Table 11.49

93. Suppose that 600 thirty-year-olds were surveyed to determine whether or not there is a relationship between the level of education an individual has and salary. Conduct a test of independence.

Annual Salary	Not a high school graduate	High school graduate	College graduate	Masters or doctorate
< $30,000	15	25	10	5
$30,000–$40,000	20	40	70	30
$40,000–$50,000	10	20	40	55
$50,000–$60,000	5	10	20	60
$60,000+	0	5	10	150

Table 11.50

Read the statement and decide whether it is true or false.

94. The number of degrees of freedom for a test of independence is equal to the sample size minus one.

95. The test for independence uses tables of observed and expected data values.

96. The test to use when determining if the college or university a student chooses to attend is related to his or her socioeconomic status is a test for independence.

97. In a test of independence, the expected number is equal to the row total multiplied by the column total divided by the total surveyed.

98. An ice cream maker performs a nationwide survey about favorite flavors of ice cream in different geographic areas of the U.S. Based on **Table 11.51**, do the numbers suggest that geographic location is independent of favorite ice cream flavors? Test at the 5% significance level.

U.S. region/ Flavor	Strawberry	Chocolate	Vanilla	Rocky Road	Mint Chocolate Chip	Pistachio	Row total
West	12	21	22	19	15	8	97
Midwest	10	32	22	11	15	6	96
East	8	31	27	8	15	7	96
South	15	28	30	8	15	6	102
Column Total	45	112	101	46	60	27	391

Table 11.51

99. Table 11.52 provides a recent survey of the youngest online entrepreneurs whose net worth is estimated at one million dollars or more. Their ages range from 17 to 30. Each cell in the table illustrates the number of entrepreneurs who correspond to the specific age group and their net worth. Are the ages and net worth independent? Perform a test of independence at the 5% significance level.

Age Group\ Net Worth Value (in millions of US dollars)	1–5	6–24	≥25	Row Total
17–25	8	7	5	20
26–30	6	5	9	20
Column Total	14	12	14	40

Table 11.52

100. A 2013 poll in California surveyed people about taxing sugar-sweetened beverages. The results are presented in **Table 11.53**, and are classified by ethnic group and response type. Are the poll responses independent of the participants' ethnic group? Conduct a test of independence at the 5% significance level.

Opinion/ Ethnicity	Asian-American	White/Non-Hispanic	African-American	Latino	Row Total
Against tax	48	433	41	160	628
In Favor of tax	54	234	24	147	459
No opinion	16	43	16	19	84
Column Total	118	710	71	272	1171

Table 11.53

11.4 Test for Homogeneity

*For each word problem, use a solution sheet to solve the hypothesis test problem. Go to **Appendix E** for the chi-square solution sheet. Round expected frequency to two decimal places.*

101. A psychologist is interested in testing whether there is a difference in the distribution of personality types for business majors and social science majors. The results of the study are shown in **Table 11.54**. Conduct a test of homogeneity. Test at a 5% level of significance.

	Open	Conscientious	Extrovert	Agreeable	Neurotic
Business	41	52	46	61	58
Social Science	72	75	63	80	65

Table 11.54

102. Do men and women select different breakfasts? The breakfasts ordered by randomly selected men and women at a popular breakfast place is shown in **Table 11.55**. Conduct a test for homogeneity at a 5% level of significance.

	French Toast	Pancakes	Waffles	Omelettes
Men	47	35	28	53
Women	65	59	55	60

Table 11.55

103. A fisherman is interested in whether the distribution of fish caught in Green Valley Lake is the same as the distribution of fish caught in Echo Lake. Of the 191 randomly selected fish caught in Green Valley Lake, 105 were rainbow trout, 27 were other trout, 35 were bass, and 24 were catfish. Of the 293 randomly selected fish caught in Echo Lake, 115 were rainbow trout, 58 were other trout, 67 were bass, and 53 were catfish. Perform a test for homogeneity at a 5% level of significance.

104. In 2007, the United States had 1.5 million homeschooled students, according to the U.S. National Center for Education Statistics. In **Table 11.56** you can see that parents decide to homeschool their children for different reasons, and some reasons are ranked by parents as more important than others. According to the survey results shown in the table, is the distribution of applicable reasons the same as the distribution of the most important reason? Provide your assessment at the 5% significance level. Did you expect the result you obtained?

Reasons for Homeschooling	Applicable Reason (in thousands of respondents)	Most Important Reason (in thousands of respondents)	Row Total
Concern about the environment of other schools	1,321	309	1,630
Dissatisfaction with academic instruction at other schools	1,096	258	1,354
To provide religious or moral instruction	1,257	540	1,797
Child has special needs, other than physical or mental	315	55	370

Table 11.56

Reasons for Homeschooling	Applicable Reason (in thousands of respondents)	Most Important Reason (in thousands of respondents)	Row Total
Nontraditional approach to child's education	984	99	1,083
Other reasons (e.g., finances, travel, family time, etc.)	485	216	701
Column Total	5,458	1,477	6,935

Table 11.56

105. When looking at energy consumption, we are often interested in detecting trends over time and how they correlate among different countries. The information in **Table 11.57** shows the average energy use (in units of kg of oil equivalent per capita) in the USA and the joint European Union countries (EU) for the six-year period 2005 to 2010. Do the energy use values in these two areas come from the same distribution? Perform the analysis at the 5% significance level.

Year	European Union	United States	Row Total
2010	3,413	7,164	10,557
2009	3,302	7,057	10,359
2008	3,505	7,488	10,993
2007	3,537	7,758	11,295
2006	3,595	7,697	11,292
2005	3,613	7,847	11,460
Column Total	45,011	20,965	65,976

Table 11.57

106. The Insurance Institute for Highway Safety collects safety information about all types of cars every year, and publishes a report of Top Safety Picks among all cars, makes, and models. **Table 11.58** presents the number of Top Safety Picks in six car categories for the two years 2009 and 2013. Analyze the table data to conclude whether the distribution of cars that earned the Top Safety Picks safety award has remained the same between 2009 and 2013. Derive your results at the 5% significance level.

Year \ Car Type	Small	Mid-Size	Large	Small SUV	Mid-Size SUV	Large SUV	Row Total
2009	12	22	10	10	27	6	87
2013	31	30	19	11	29	4	124
Column Total	43	52	29	21	56	10	211

Table 11.58

11.5 Comparison of the Chi-Square Tests

*For each word problem, use a solution sheet to solve the hypothesis test problem. Go to **Appendix E** for the chi-square solution sheet. Round expected frequency to two decimal places.*

107. Is there a difference between the distribution of community college statistics students and the distribution of university statistics students in what technology they use on their homework? Of some randomly selected community college students, 43 used a computer, 102 used a calculator with built in statistics functions, and 65 used a table from the textbook. Of some randomly selected university students, 28 used a computer, 33 used a calculator with built in statistics functions, and 40 used a table from the textbook. Conduct an appropriate hypothesis test using a 0.05 level of significance.

Read the statement and decide whether it is true or false.

108. If $df = 2$, the chi-square distribution has a shape that reminds us of the exponential.

11.6 Test of a Single Variance

Use the following information to answer the next twelve exercises: Suppose an airline claims that its flights are consistently on time with an average delay of at most 15 minutes. It claims that the average delay is so consistent that the variance is no more than 150 minutes. Doubting the consistency part of the claim, a disgruntled traveler calculates the delays for his next 25 flights. The average delay for those 25 flights is 22 minutes with a standard deviation of 15 minutes.

109. Is the traveler disputing the claim about the average or about the variance?

110. A sample standard deviation of 15 minutes is the same as a sample variance of _____ minutes.

111. Is this a right-tailed, left-tailed, or two-tailed test?

112. H_0: _____

113. $df = $ _____

114. chi-square test statistic = _____

115. p-value = _____

116. Graph the situation. Label and scale the horizontal axis. Mark the mean and test statistic. Shade the p-value.

117. Let $\alpha = 0.05$
Decision: _____
Conclusion (write out in a complete sentence.): _____

118. How did you know to test the variance instead of the mean?

119. If an additional test were done on the claim of the average delay, which distribution would you use?

120. If an additional test were done on the claim of the average delay, but 45 flights were surveyed, which distribution would you use?

*For each word problem, use a solution sheet to solve the hypothesis test problem. Go to **Appendix E** for the chi-square solution sheet. Round expected frequency to two decimal places.*

121. A plant manager is concerned her equipment may need recalibrating. It seems that the actual weight of the 15 oz. cereal boxes it fills has been fluctuating. The standard deviation should be at most 0.5 oz. In order to determine if the machine needs to be recalibrated, 84 randomly selected boxes of cereal from the next day's production were weighed. The standard deviation of the 84 boxes was 0.54. Does the machine need to be recalibrated?

122. Consumers may be interested in whether the cost of a particular calculator varies from store to store. Based on surveying 43 stores, which yielded a sample mean of $84 and a sample standard deviation of $12, test the claim that the standard deviation is greater than $15.

123. Isabella, an accomplished **Bay to Breakers** runner, claims that the standard deviation for her time to run the 7.5 mile race is at most three minutes. To test her claim, Rupinder looks up five of her race times. They are 55 minutes, 61 minutes, 58 minutes, 63 minutes, and 57 minutes.

124. Airline companies are interested in the consistency of the number of babies on each flight, so that they have adequate safety equipment. They are also interested in the variation of the number of babies. Suppose that an airline executive believes the average number of babies on flights is six with a variance of nine at most. The airline conducts a survey. The results of the 18 flights surveyed give a sample average of 6.4 with a sample standard deviation of 3.9. Conduct a hypothesis test of the airline executive's belief.

125. The number of births per woman in China is 1.6 down from 5.91 in 1966. This fertility rate has been attributed to the law passed in 1979 restricting births to one per woman. Suppose that a group of students studied whether or not the standard deviation of births per woman was greater than 0.75. They asked 50 women across China the number of births they had had. The results are shown in **Table 11.59**. Does the students' survey indicate that the standard deviation is greater than 0.75?

# of births	Frequency
0	5
1	30
2	10
3	5

Table 11.59

126. According to an avid aquarist, the average number of fish in a 20-gallon tank is 10, with a standard deviation of two. His friend, also an aquarist, does not believe that the standard deviation is two. She counts the number of fish in 15 other 20-gallon tanks. Based on the results that follow, do you think that the standard deviation is different from two? Data: 11; 10; 9; 10; 10; 11; 11; 10; 12; 9; 7; 9; 11; 10; 11

127. The manager of "Frenchies" is concerned that patrons are not consistently receiving the same amount of French fries with each order. The chef claims that the standard deviation for a ten-ounce order of fries is at most 1.5 oz., but the manager thinks that it may be higher. He randomly weighs 49 orders of fries, which yields a mean of 11 oz. and a standard deviation of two oz.

128. You want to buy a specific computer. A sales representative of the manufacturer claims that retail stores sell this computer at an average price of $1,249 with a very narrow standard deviation of $25. You find a website that has a price comparison for the same computer at a series of stores as follows: $1,299; $1,229.99; $1,193.08; $1,279; $1,224.95; $1,229.99; $1,269.95; $1,249. Can you argue that pricing has a larger standard deviation than claimed by the manufacturer? Use the 5% significance level. As a potential buyer, what would be the practical conclusion from your analysis?

129. A company packages apples by weight. One of the weight grades is Class A apples. Class A apples have a mean weight of 150 g, and there is a maximum allowed weight tolerance of 5% above or below the mean for apples in the same consumer package. A batch of apples is selected to be included in a Class A apple package. Given the following apple weights of the batch, does the fruit comply with the Class A grade weight tolerance requirements. Conduct an appropriate hypothesis test.

(a) at the 5% significance level

(b) at the 1% significance level

Weights in selected apple batch (in grams): 158; 167; 149; 169; 164; 139; 154; 150; 157; 171; 152; 161; 141; 166; 172;

BRINGING IT TOGETHER: HOMEWORK

130.

  a. Explain why a goodness-of-fit test and a test of independence are generally right-tailed tests.
  b. If you did a left-tailed test, what would you be testing?

REFERENCES

11.1 Facts About the Chi-Square Distribution

Data from *Parade Magazine*.

"HIV/AIDS Epidemiology Santa Clara County."Santa Clara County Public Health Department, May 2011.

11.2 Goodness-of-Fit Test

Data from the U.S. Census Bureau

Data from the College Board. Available online at http://www.collegeboard.com.

Data from the U.S. Census Bureau, Current Population Reports.

Ma, Y., E.R. Bertone, E.J. Stanek III, G.W. Reed, J.R. Hebert, N.L. Cohen, P.A. Merriam, I.S. Ockene, "Association between Eating Patterns and Obesity in a Free-living US Adult Population." *American Journal of Epidemiology* volume 158, no. 1, pages 85-92.

Ogden, Cynthia L., Margaret D. Carroll, Brian K. Kit, Katherine M. Flegal, "Prevalence of Obesity in the United States, 2009–2010." NCHS Data Brief no. 82, January 2012. Available online at http://www.cdc.gov/nchs/data/databriefs/db82.pdf (accessed May 24, 2013).

Stevens, Barbara J., "Multi-family and Commercial Solid Waste and Recycling Survey." Arlington Count, VA. Available online at http://www.arlingtonva.us/departments/EnvironmentalServices/SW/file84429.pdf (accessed May 24,2013).

11.3 Test of Independence

DiCamilo, Mark, Mervin Field, "Most Californians See a Direct Linkage between Obesity and Sugary Sodas. Two in Three Voters Support Taxing Sugar-Sweetened Beverages If Proceeds are Tied to Improving School Nutrition and Physical Activity Programs." The Field Poll, released Feb. 14, 2013. Available online at http://field.com/fieldpollonline/subscribers/Rls2436.pdf (accessed May 24, 2013).

Harris Interactive, "Favorite Flavor of Ice Cream." Available online at http://www.statisticbrain.com/favorite-flavor-of-ice-cream (accessed May 24, 2013)

"Youngest Online Entrepreneurs List." Available online at http://www.statisticbrain.com/youngest-online-entrepreneur-list (accessed May 24, 2013).

11.4 Test for Homogeneity

Data from the Insurance Institute for Highway Safety, 2013. Available online at www.iihs.org/iihs/ratings (accessed May 24, 2013).

"Energy use (kg of oil equivalent per capita)." The World Bank, 2013. Available online at http://data.worldbank.org/indicator/EG.USE.PCAP.KG.OE/countries (accessed May 24, 2013).

"Parent and Family Involvement Survey of 2007 National Household Education Survey Program (NHES)," U.S. Department of Education, National Center for Education Statistics. Available online at http://nces.ed.gov/pubsearch/pubsinfo.asp?pubid=2009030 (accessed May 24, 2013).

"Parent and Family Involvement Survey of 2007 National Household Education Survey Program (NHES)," U.S. Department of Education, National Center for Education Statistics. Available online at http://nces.ed.gov/pubs2009/2009030_sup.pdf (accessed May 24, 2013).

11.6 Test of a Single Variance

"AppleInsider Price Guides." Apple Insider, 2013. Available online at http://appleinsider.com/mac_price_guide (accessed May 14, 2013).

Data from the World Bank, June 5, 2012.

SOLUTIONS

1 mean = 25 and standard deviation = 7.0711

3 when the number of degrees of freedom is greater than 90

5 $df = 2$

7 a goodness-of-fit test

9 3

11 2.04

13 We decline to reject the null hypothesis. There is not enough evidence to suggest that the observed test scores are significantly different from the expected test scores.

15 H_0: the distribution of AIDS cases follows the ethnicities of the general population of Santa Clara County.

17 right-tailed

19 88,621

21 Graph: Check student's solution. Decision: Reject the null hypothesis. Reason for the Decision: p-value < alpha Conclusion (write out in complete sentences): The make-up of AIDS cases does not fit the ethnicities of the general population of Santa Clara County.

23 a test of independence

25 a test of independence

27 8

29 6.6

31 0.0435

33

Smoking Level Per Day	African American	Native Hawaiian	Latino	Japanese Americans	White	Totals
1-10	9,886	2,745	12,831	8,378	7,650	41,490
11-20	6,514	3,062	4,932	10,680	9,877	35,065
21-30	1,671	1,419	1,406	4,715	6,062	15,273
31+	759	788	800	2,305	3,970	8,622
Totals	18,830	8,014	19,969	26,078	27,559	10,0450

Table 11.60

35

Smoking Level Per Day	African American	Native Hawaiian	Latino	Japanese Americans	White
1-10	7777.57	3310.11	8248.02	10771.29	11383.01
11-20	6573.16	2797.52	6970.76	9103.29	9620.27
21-30	2863.02	1218.49	3036.20	3965.05	4190.23
31+	1616.25	687.87	1714.01	2238.37	2365.49

Table 11.61

37 10,301.8

39 right

41
 a. Reject the null hypothesis.

 b. p-value < alpha

c.   There is sufficient evidence to conclude that smoking level is dependent on ethnic group.

43 test for homogeneity

45 test for homogeneity

47 All values in the table must be greater than or equal to five.

49 3

51 0.00005

53 a goodness-of-fit test

55 a test for independence

57 Answers will vary. Sample answer: Tests of independence and tests for homogeneity both calculate the test statistic the same way $\sum_{(ij)} \frac{(O - E)^2}{E}$. In addition, all values must be greater than or equal to five.

59 a test of a single variance

61 a left-tailed test

63 $H_0: \sigma^2 = 0.81^2; H_a: \sigma^2 > 0.81^2$

65 a test of a single variance

67 0.0542

69 true

71 false

73

Marital Status	Percent	Expected Frequency
never married	31.3	125.2
married	56.1	224.4
widowed	2.5	10
divorced/separated	10.1	40.4

Table 11.62

a.   The data fits the distribution.

b.   The data does not fit the distribution.

c.   3

d.   chi-square distribution with $df = 3$

e.   19.27

f.   0.0002

g.   Check student's solution.

h.   i.   Alpha = 0.05

   ii.   Decision: Reject null

   iii.   Reason for decision: p-value < alpha

   iv.   Conclusion: Data does not fit the distribution.

75

    a. H_0: The local results follow the distribution of the U.S. AP examinee population

    b. H_a: The local results do not follow the distribution of the U.S. AP examinee population

    c. $df = 5$

    d.   chi-square distribution with $df = 5$

    e.   chi-square test statistic = 13.4

    f. p-value = 0.0199

    g.   Check student's solution.

    h.   i.   Alpha = 0.05

          ii.   Decision: Reject null when $a = 0.05$

         iii.   Reason for Decision: p-value < alpha

         iv.   Conclusion: Local data do not fit the AP Examinee Distribution.

          v.   Decision: Do not reject null when $a = 0.01$

         vi.   Conclusion: There is insufficient evidence to conclude that local data do not follow the distribution of the U.S. AP examinee distribution.

77

    a. H_0: The actual college majors of graduating females fit the distribution of their expected majors

    b. H_a: The actual college majors of graduating females do not fit the distribution of their expected majors

    c. $df = 10$

    d.   chi-square distribution with $df = 10$

    e.   test statistic = 11.48

    f. p-value = 0.3211

    g.   Check student's solution.

    h.   i.   Alpha = 0.05

          ii.   Decision: Do not reject null when $a = 0.05$ and $a = 0.01$

         iii.   Reason for decision: p-value > alpha

         iv.   Conclusion: There is insufficient evidence to conclude that the distribution of actual college majors of graduating females fits the distribution of their expected majors.

79 true

81 true

83 false

85

    a. H_0: Surveyed obese fit the distribution of expected obese

    b. H_a: Surveyed obese do not fit the distribution of expected obese

    c. $df = 4$

    d.   chi-square distribution with $df = 4$

    e.   test statistic = 54.01

    f. p-value = 0

    g.   Check student's solution.

    h.   i.   Alpha: 0.05

          ii.   Decision: Reject the null hypothesis.

         iii.   Reason for decision: p-value < alpha

iv. Conclusion: At the 5% level of significance, from the data, there is sufficient evidence to conclude that the surveyed obese do not fit the distribution of expected obese.

87

a. H_0: Car size is independent of family size.

b. H_a: Car size is dependent on family size.

c. $df = 9$

d. chi-square distribution with $df = 9$

e. test statistic = 15.8284

f. p-value = 0.0706

g. Check student's solution.

h. i. Alpha: 0.05

ii. Decision: Do not reject the null hypothesis.

iii. Reason for decision: p-value > alpha

iv. Conclusion: At the 5% significance level, there is insufficient evidence to conclude that car size and family size are dependent.

89

a. H_0: Honeymoon locations are independent of bride's age.

b. H_a: Honeymoon locations are dependent on bride's age.

c. $df = 9$

d. chi-square distribution with $df = 9$

e. test statistic = 15.7027

f. p-value = 0.0734

g. Check student's solution.

h. i. Alpha: 0.05

ii. Decision: Do not reject the null hypothesis.

iii. Reason for decision: p-value > alpha

iv. Conclusion: At the 5% significance level, there is insufficient evidence to conclude that honeymoon location and bride age are dependent.

91

a. H_0: The types of fries sold are independent of the location.

b. H_a: The types of fries sold are dependent on the location.

c. $df = 6$

d. chi-square distribution with $df = 6$

e. test statistic =18.8369

f. p-value = 0.0044

g. Check student's solution.

h. i. Alpha: 0.05

ii. Decision: Reject the null hypothesis.

iii. Reason for decision: p-value < alpha

iv. Conclusion: At the 5% significance level, There is sufficient evidence that types of fries and location are dependent.

93

a. H_0: Salary is independent of level of education.

b. H_a: Salary is dependent on level of education.

c. $df = 12$

d. chi-square distribution with $df = 12$

e. test statistic = 255.7704

f. p-value = 0

g. Check student's solution.

h. Alpha: 0.05

Decision: Reject the null hypothesis.

Reason for decision: p-value < alpha

Conclusion: At the 5% significance level, there is sufficient evidence to conclude that salary and level of education are dependent.

95 true

97 true

99

a. H_0: Age is independent of the youngest online entrepreneurs' net worth.

b. H_a: Age is dependent on the net worth of the youngest online entrepreneurs.

c. $df = 2$

d. chi-square distribution with $df = 2$

e. test statistic = 1.76

f. p-value 0.4144

g. Check student's solution.

h.   i. Alpha: 0.05

    ii. Decision: Do not reject the null hypothesis.

    iii. Reason for decision: p-value > alpha

    iv. Conclusion: At the 5% significance level, there is insufficient evidence to conclude that age and net worth for the youngest online entrepreneurs are dependent.

101

a. H_0: The distribution for personality types is the same for both majors

b. H_a: The distribution for personality types is not the same for both majors

c. $df = 4$

d. chi-square with $df = 4$

e. test statistic = 3.01

f. p-value = 0.5568

g. Check student's solution.

h.   i. Alpha: 0.05

    ii. Decision: Do not reject the null hypothesis.

    iii. Reason for decision: p-value > alpha

    iv. Conclusion: There is insufficient evidence to conclude that the distribution of personality types is different for business and social science majors.

103
a. H_0: The distribution for fish caught is the same in Green Valley Lake and in Echo Lake.

b. H_a: The distribution for fish caught is not the same in Green Valley Lake and in Echo Lake.

c. 3

d. chi-square with $df = 3$

e. 11.75

f. p-value = 0.0083

g. Check student's solution.

h. i. Alpha: 0.05

   ii. Decision: Reject the null hypothesis.

   iii. Reason for decision: p-value < alpha

   iv. Conclusion: There is evidence to conclude that the distribution of fish caught is different in Green Valley Lake and in Echo Lake

105
a. H_0: The distribution of average energy use in the USA is the same as in Europe between 2005 and 2010.

b. H_a: The distribution of average energy use in the USA is not the same as in Europe between 2005 and 2010.

c. $df = 4$

d. chi-square with $df = 4$

e. test statistic = 2.7434

f. p-value = 0.7395

g. Check student's solution.

h. i. Alpha: 0.05

   ii. Decision: Do not reject the null hypothesis.

   iii. Reason for decision: p-value > alpha

   iv. Conclusion: At the 5% significance level, there is insufficient evidence to conclude that the average energy use values in the US and EU are not derived from different distributions for the period from 2005 to 2010.

107
a. H_0: The distribution for technology use is the same for community college students and university students.

b. H_a: The distribution for technology use is not the same for community college students and university students.

c. 2

d. chi-square with $df = 2$

e. 7.05

f. p-value = 0.0294

g. Check student's solution.

h. i. Alpha: 0.05

   ii. Decision: Reject the null hypothesis.

   iii. Reason for decision: p-value < alpha

   iv. Conclusion: There is sufficient evidence to conclude that the distribution of technology use for statistics homework is not the same for statistics students at community colleges and at universities.

110 225

112 H_0: $\sigma^2 \le 150$

114 36

116 Check student's solution.

118 The claim is that the variance is no more than 150 minutes.

120 a Student's *t*- or normal distribution

122

a. H_0: $\sigma = 15$

b. H_a: $\sigma > 15$

c. $df = 42$

d. chi-square with $df = 42$

e. test statistic = 26.88

f. *p*-value = 0.9663

g. Check student's solution.

h.
   i. Alpha = 0.05
   ii. Decision: Do not reject null hypothesis.
   iii. Reason for decision: *p*-value > alpha
   iv. Conclusion: There is insufficient evidence to conclude that the standard deviation is greater than 15.

124

a. H_0: $\sigma \leq 3$

b. H_a: $\sigma > 3$

c. $df = 17$

d. chi-square distribution with $df = 17$

e. test statistic = 28.73

f. *p*-value = 0.0371

g. Check student's solution.

h.
   i. Alpha: 0.05
   ii. Decision: Reject the null hypothesis.
   iii. Reason for decision: *p*-value < alpha
   iv. Conclusion: There is sufficient evidence to conclude that the standard deviation is greater than three.

126

a. H_0: $\sigma = 2$

b. H_a: $\sigma \neq 2$

c. $df = 14$

d. chi-square distiribution with $df = 14$

e. chi-square test statistic = 5.2094

f. *p*-value = 0.0346

g. Check student's solution.

h.
   i. Alpha = 0.05
   ii. Decision: Reject the null hypothesis
   iii. Reason for decision: *p*-value < alpha
   iv. Conclusion: There is sufficient evidence to conclude that the standard deviation is different than 2.

128 The sample standard deviation is \$34.29. $H_0 : \sigma^2 = 25^2$

$H_a : \sigma^2 > 25^2$

$df = n - 1 = 7.$

test statistic: $x^2 = x_7^2 = \dfrac{(n-1)s^2}{25^2} = \dfrac{(8-1)(34.29)^2}{25^2} = 13.169$;

p-value: $P\left(x_7^2 > 13.169\right) = 1 - P\left(x_7^2 \leq 13.169\right) = 0.0681$

Alpha: 0.05

Decision: Do not reject the null hypothesis.

Reason for decision: p-value > alpha

Conclusion: At the 5% level, there is insufficient evidence to conclude that the variance is more than 625.

130

a. The test statistic is always positive and if the expected and observed values are not close together, the test statistic is large and the null hypothesis will be rejected.

b. Testing to see if the data fits the distribution "too well" or is too perfect.

12 | LINEAR REGRESSION AND CORRELATION

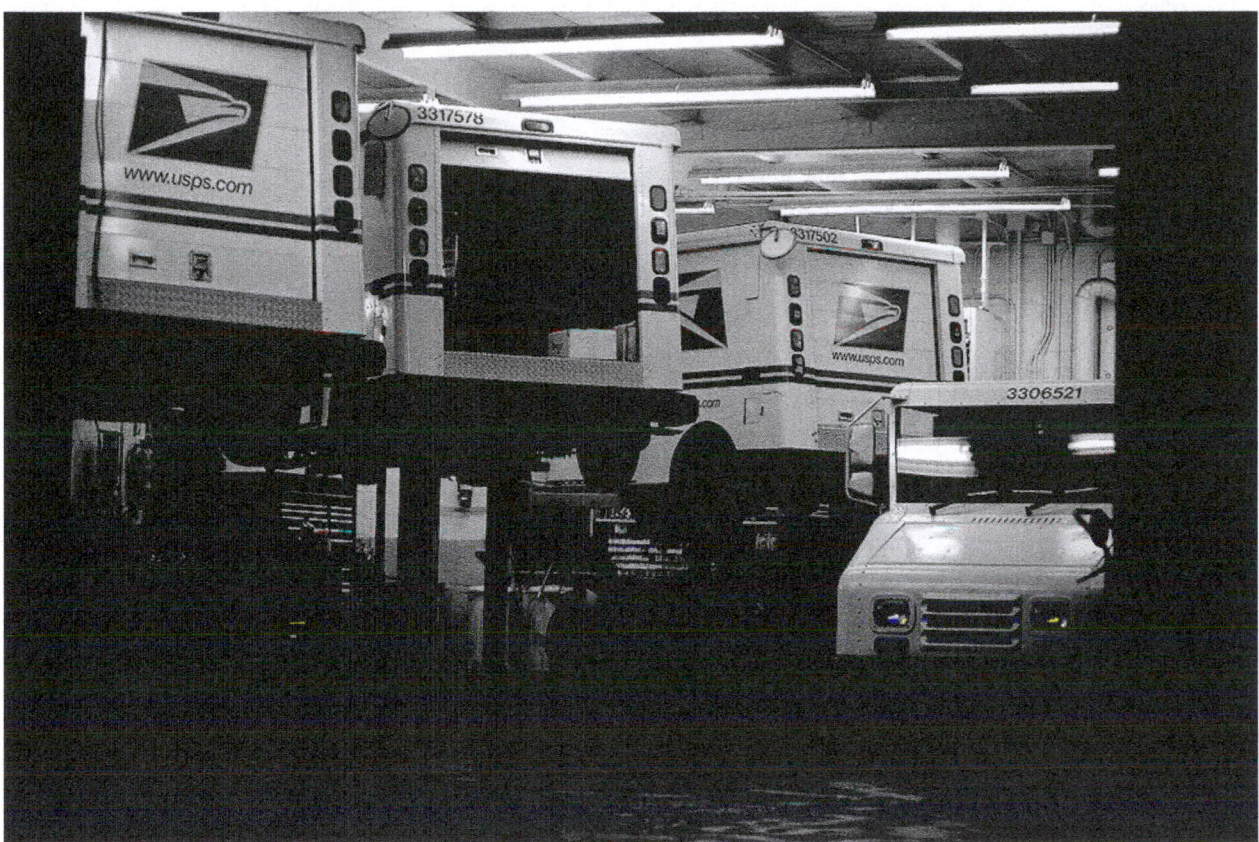

Figure 12.1 Linear regression and correlation can help you determine if an auto mechanic's salary is related to his work experience. (credit: Joshua Rothhaas)

Introduction

Professionals often want to know how two or more numeric variables are related. For example, is there a relationship between the grade on the second math exam a student takes and the grade on the final exam? If there is a relationship, what is the relationship and how strong is it?

In another example, your income may be determined by your education, your profession, your years of experience, and your ability. The amount you pay a repair person for labor is often determined by an initial amount plus an hourly fee.

The type of data described in the examples is **bivariate** data — "bi" for two variables. In reality, statisticians use **multivariate** data, meaning many variables.

In this chapter, you will be studying the simplest form of regression, "linear regression" with one independent variable (x). This involves data that fits a line in two dimensions. You will also study correlation which measures how strong the relationship is.

12.1 | Linear Equations

Linear regression for two variables is based on a linear equation with one independent variable. The equation has the form:

$$y = a + \text{b}x$$

where a and b are constant numbers.

The variable x **is the independent variable, and** y **is the dependent variable.** Typically, you choose a value to substitute for the independent variable and then solve for the dependent variable.

Example 12.1

The following examples are linear equations.

$$y = 3 + 2\text{x}$$
$$y = -0.01 + 1.2\text{x}$$

Try It Σ

12.1 Is the following an example of a linear equation?

$y = -0.125 - 3.5x$

The graph of a linear equation of the form $y = a + bx$ is a **straight line**. Any line that is not vertical can be described by this equation.

Example 12.2

Graph the equation $y = -1 + 2x$.

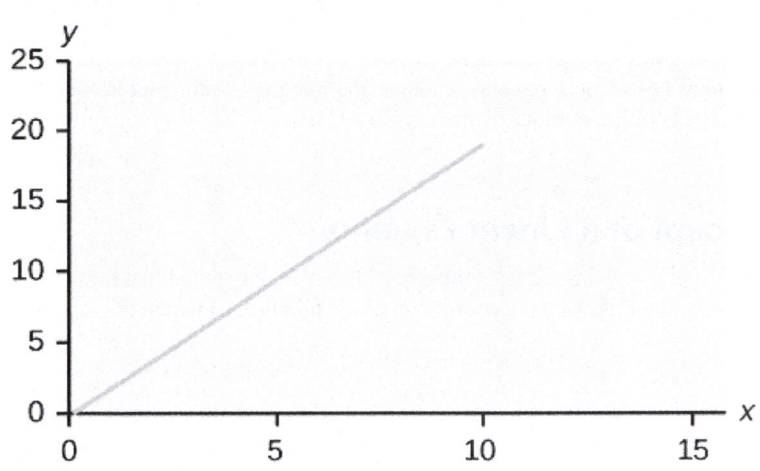

Figure 12.2

12.2 Is the following an example of a linear equation? Why or why not?

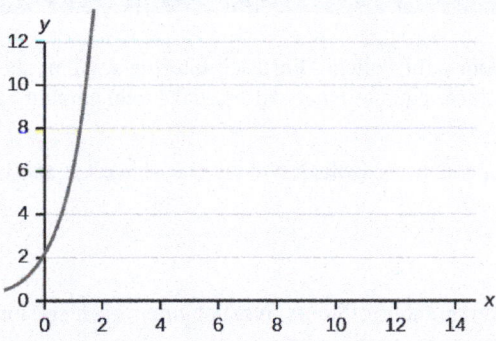

Figure 12.3

Example 12.3

Aaron's Word Processing Service (AWPS) does word processing. The rate for services is $32 per hour plus a $31.50 one-time charge. The total cost to a customer depends on the number of hours it takes to complete the job.

Find the equation that expresses the **total cost** in terms of the **number of hours** required to complete the job.

Solution 12.3

Let x = the number of hours it takes to get the job done.
Let y = the total cost to the customer.

The $31.50 is a fixed cost. If it takes x hours to complete the job, then $(32)(x)$ is the cost of the word processing only. The total cost is: $y = 31.50 + 32x$

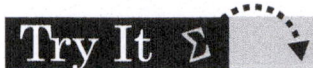

12.3 Emma's Extreme Sports hires hang-gliding instructors and pays them a fee of $50 per class as well as $20 per student in the class. The total cost Emma pays depends on the number of students in a class. Find the equation that expresses the total cost in terms of the number of students in a class.

Slope and *Y*-Intercept of a Linear Equation

For the linear equation $y = a + bx$, b = slope and a = y-intercept. From algebra recall that the slope is a number that describes the steepness of a line, and the y-intercept is the y coordinate of the point $(0, a)$ where the line crosses the y-axis.

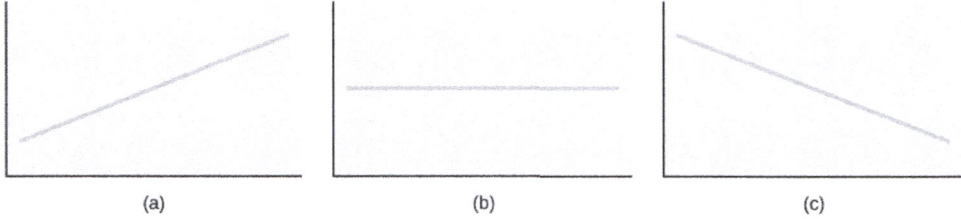

Figure 12.4 Three possible graphs of $y = a + bx$. (a) If $b > 0$, the line slopes upward to the right. (b) If $b = 0$, the line is horizontal. (c) If $b < 0$, the line slopes downward to the right.

Example 12.4

Svetlana tutors to make extra money for college. For each tutoring session, she charges a one-time fee of $25 plus $15 per hour of tutoring. A linear equation that expresses the total amount of money Svetlana earns for each session she tutors is $y = 25 + 15x$.

What are the independent and dependent variables? What is the y-intercept and what is the slope? Interpret them using complete sentences.

Solution 12.4

The independent variable (x) is the number of hours Svetlana tutors each session. The dependent variable (y) is the amount, in dollars, Svetlana earns for each session.

The y-intercept is 25 ($a = 25$). At the start of the tutoring session, Svetlana charges a one-time fee of $25 (this is when $x = 0$). The slope is 15 ($b = 15$). For each session, Svetlana earns $15 for each hour she tutors.

12.4 Ethan repairs household appliances like dishwashers and refrigerators. For each visit, he charges $25 plus $20 per hour of work. A linear equation that expresses the total amount of money Ethan earns per visit is $y = 25 + 20x$.

What are the independent and dependent variables? What is the y-intercept and what is the slope? Interpret them using complete sentences.

12.2 | Scatter Plots

Before we take up the discussion of linear regression and correlation, we need to examine a way to display the relation between two variables x and y. The most common and easiest way is a **scatter plot**. The following example illustrates a scatter plot.

Example 12.5

In Europe and Asia, m-commerce is popular. M-commerce users have special mobile phones that work like electronic wallets as well as provide phone and Internet services. Users can do everything from paying for parking to buying a TV set or soda from a machine to banking to checking sports scores on the Internet. For the years 2000 through 2004, was there a relationship between the year and the number of m-commerce users? Construct a scatter plot. Let x = the year and let y = the number of m-commerce users, in millions.

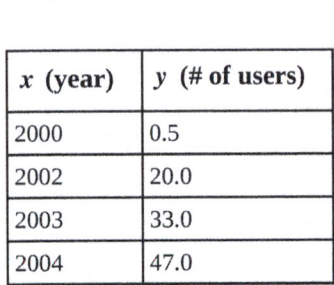

x (year)	y (# of users)
2000	0.5
2002	20.0
2003	33.0
2004	47.0

Table 12.1

(a) Table showing the number of m-commerce users (in millions) by year.

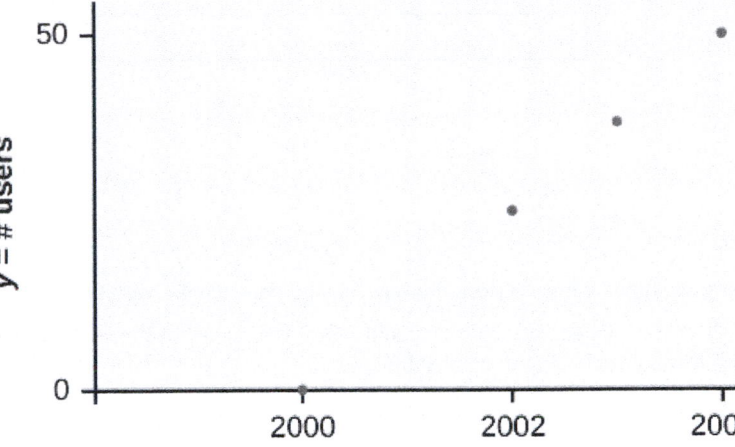

(b) Scatter plot showing the number of m-commerce users (in millions) by year.

Figure 12.5

 Using the TI-83, 83+, 84, 84+ Calculator

To create a scatter plot:

1. Enter your X data into list L1 and your Y data into list L2.

2. Press 2nd STATPLOT ENTER to use Plot 1. On the input screen for PLOT 1, highlight On and press ENTER. (Make sure the other plots are OFF.)

3. For TYPE: highlight the very first icon, which is the scatter plot, and press ENTER.

4. For Xlist:, enter L1 ENTER and for Ylist: L2 ENTER.

5. For Mark: it does not matter which symbol you highlight, but the square is the easiest to see. Press ENTER.

6. Make sure there are no other equations that could be plotted. Press Y = and clear any equations out.

7. Press the ZOOM key and then the number 9 (for menu item "ZoomStat") ; the calculator will fit the window to the data. You can press WINDOW to see the scaling of the axes.

Try It

12.5 Amelia plays basketball for her high school. She wants to improve to play at the college level. She notices that the number of points she scores in a game goes up in response to the number of hours she practices her jump shot each week. She records the following data:

X (hours practicing jump shot)	Y (points scored in a game)
5	15
7	22
9	28
10	31
11	33
12	36

Table 12.2

Construct a scatter plot and state if what Amelia thinks appears to be true.

A scatter plot shows the **direction** of a relationship between the variables. A clear direction happens when there is either:

- High values of one variable occurring with high values of the other variable or low values of one variable occurring with low values of the other variable.

- High values of one variable occurring with low values of the other variable.

You can determine the **strength** of the relationship by looking at the scatter plot and seeing how close the points are to a line, a power function, an exponential function, or to some other type of function. For a linear relationship there is an exception. Consider a scatter plot where all the points fall on a horizontal line providing a "perfect fit." The horizontal line would in fact show no relationship.

When you look at a scatterplot, you want to notice the **overall pattern** and any **deviations** from the pattern. The following scatterplot examples illustrate these concepts.

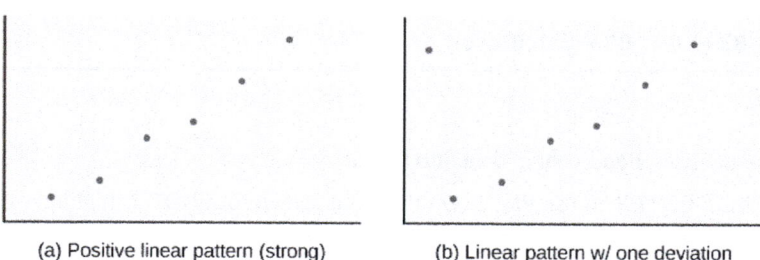

(a) Positive linear pattern (strong)   (b) Linear pattern w/ one deviation

Figure 12.6

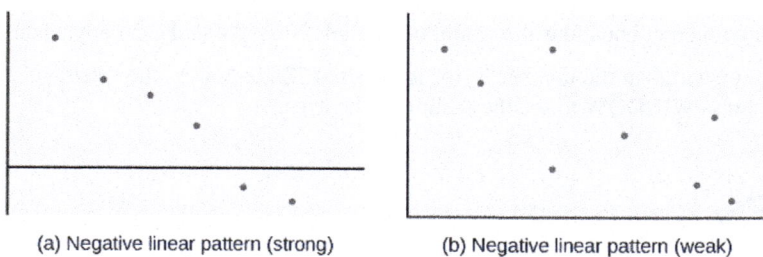

(a) Negative linear pattern (strong)   (b) Negative linear pattern (weak)

Figure 12.7

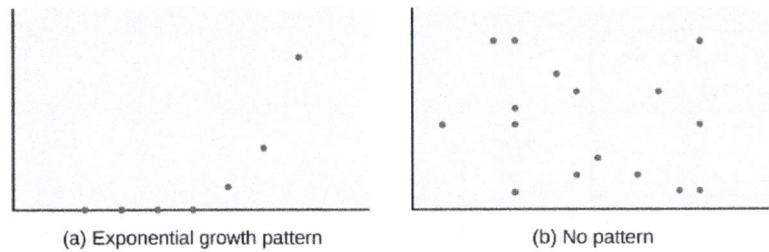

(a) Exponential growth pattern   (b) No pattern

Figure 12.8

In this chapter, we are interested in scatter plots that show a linear pattern. Linear patterns are quite common. The linear relationship is strong if the points are close to a straight line, except in the case of a horizontal line where there is no relationship. If we think that the points show a linear relationship, we would like to draw a line on the scatter plot. This line can be calculated through a process called **linear regression**. However, we only calculate a regression line if one of the variables helps to explain or predict the other variable. If x is the independent variable and y the dependent variable, then we can use a regression line to predict y for a given value of x

12.3 | The Regression Equation

Data rarely fit a straight line exactly. Usually, you must be satisfied with rough predictions. Typically, you have a set of data whose scatter plot appears to **"fit"** a straight line. This is called a **Line of Best Fit or Least-Squares Line**.

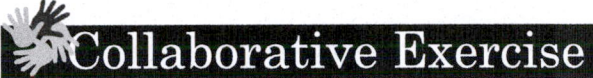
Collaborative Exercise

If you know a person's pinky (smallest) finger length, do you think you could predict that person's height? Collect data from your class (pinky finger length, in inches). The independent variable, x, is pinky finger length and the dependent variable, y, is height. For each set of data, plot the points on graph paper. Make your graph big enough and **use a ruler**. Then "by eye" draw a line that appears to "fit" the data. For your line, pick two convenient points and use them to find the slope of the line. Find the y-intercept of the line by extending your line so it crosses the y-axis. Using the slopes and the y-intercepts, write your equation of "best fit." Do you think everyone will have the same equation? Why or why not? According to your equation, what is the predicted height for a pinky length of 2.5 inches?

Example 12.6

A random sample of 11 statistics students produced the following data, where x is the third exam score out of 80, and y is the final exam score out of 200. Can you predict the final exam score of a random student if you know the third exam score?

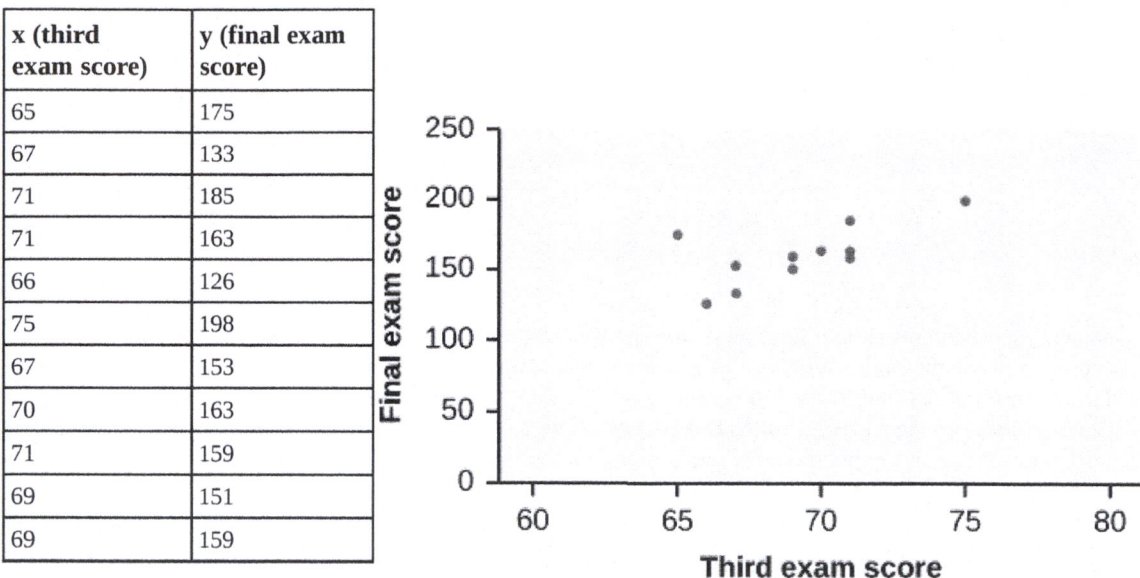

x (third exam score)	y (final exam score)
65	175
67	133
71	185
71	163
66	126
75	198
67	153
70	163
71	159
69	151
69	159

Table 12.3

(b) Scatter plot showing the scores on the final exam based on scores from the third exam.

(a) Table showing the scores on the final exam based on scores from the third exam.

Figure 12.9

Try It Σ

☞ **12.6** SCUBA divers have maximum dive times they cannot exceed when going to different depths. The data in **Table 12.4** show different depths with the maximum dive times in minutes. Use your calculator to find the least squares regression line and predict the maximum dive time for 110 feet.

X (depth in feet)	Y (maximum dive time)
50	80
60	55
70	45
80	35
90	25
100	22

Table 12.4

The third exam score, *x*, is the independent variable and the final exam score, *y*, is the dependent variable. We will plot a regression line that best "fits" the data. If each of you were to fit a line "by eye," you would draw different lines. We can use what is called a **least-squares regression line** to obtain the best fit line.

Consider the following diagram. Each point of data is of the the form (x, y) and each point ofthe line of best fit using least-squares linear regression has the form $(x, \hat{y})$.

The $\hat{y}$ is read **"y hat"** and is the **estimated value of y**. It is the value of y obtained using the regression line. It is not generally equal to y from data.

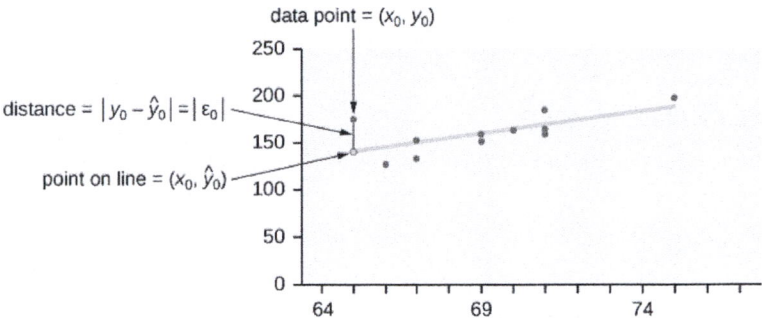

Figure 12.10

The term $y_0 - \hat{y}_0 = \varepsilon_0$ is called the **"error" or residual**. It is not an error in the sense of a mistake. The **absolute value of a residual** measures the vertical distance between the actual value of y and the estimated value of y. In other words, it measures the vertical distance between the actual data point and the predicted point on the line.

If the observed data point lies above the line, the residual is positive, and the line underestimates the actual data value for y. If the observed data point lies below the line, the residual is negative, and the line overestimates that actual data value for y.

In the diagram in **Figure 12.10**, $y_0 - \hat{y}_0 = \varepsilon_0$ is the residual for the point shown. Here the point lies above the line and the residual is positive.

ε = the Greek letter **epsilon**

For each data point, you can calculate the residuals or errors, $y_i - \hat{y}_i = \varepsilon_i$ for i = 1, 2, 3, ..., 11.

Each $|\varepsilon|$ is a vertical distance.

For the example about the third exam scores and the final exam scores for the 11 statistics students, there are 11 data points. Therefore, there are 11 ε values. If yousquare each ε and add, you get

$$(\varepsilon_1)^2 + (\varepsilon_2)^2 + ... + (\varepsilon_{11})^2 = \sum_{i=1}^{11} \varepsilon^2$$

This is called the **Sum of Squared Errors (SSE)**.

Using calculus, you can determine the values of a and b that make the **SSE** a minimum. When you make the **SSE** a minimum, you have determined the points that are on the line of best fit. It turns out that the line of best fit has the equation:

$$\hat{y} = a + bx$$

where $a = \bar{y} - b\bar{x}$ and $b = \dfrac{\Sigma(x - \bar{x})(y - \bar{y})}{\Sigma(x - \bar{x})^2}$.

The sample means of the x values and the y values are $\bar{x}$ and $\bar{y}$, respectively. The best fit line always passes through the point $(\bar{x}, \bar{y})$.

The slope b can be written as $b = r\left(\dfrac{s_y}{s_x}\right)$ where s_y = the standard deviation of the y values and s_x = the standard deviation of the x values. r is the correlation coefficient, which is discussed in the next section.

Least Squares Criteria for Best Fit

The process of fitting the best-fit line is called **linear regression**. The idea behind finding the best-fit line is based on the assumption that the data are scattered about a straight line. The criteria for the best fit line is that the sum of the squared errors (SSE) is minimized, that is, made as small as possible. Any other line you might choose would have a higher SSE than the best fit line. This best fit line is called the **least-squares regression line** .

NOTE

👉 Computer spreadsheets, statistical software, and many calculators can quickly calculate the best-fit line and create the graphs. The calculations tend to be tedious if done by hand. Instructions to use the TI-83, TI-83+, and TI-84+ calculators to find the best-fit line and create a scatterplot are shown at the end of this section.

THIRD EXAM vs FINAL EXAM EXAMPLE:

The graph of the line of best fit for the third-exam/final-exam example is as follows:

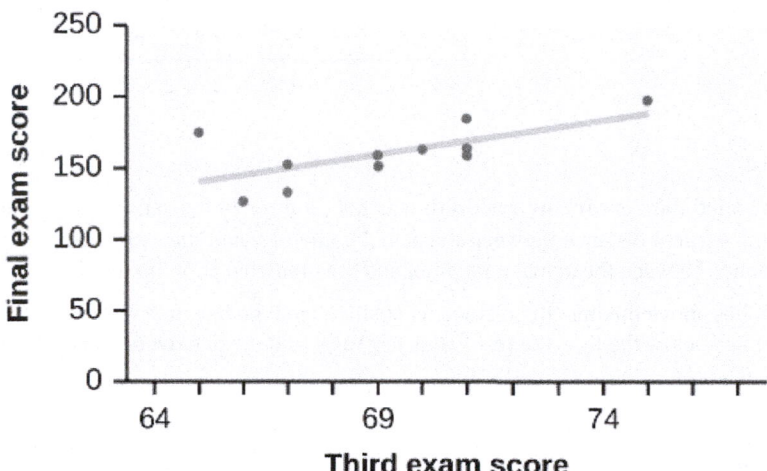

Figure 12.11

The least squares regression line (best-fit line) for the third-exam/final-exam example has the equation:

$$\hat{y} = -173.51 + 4.83x$$

REMINDER

Remember, it is always important to plot a scatter diagram first. If the scatter plot indicates that there is a linear relationship between the variables, then it is reasonable to use a best fit line to make predictions for y given x within the domain of x-values in the sample data, **but not necessarily for x-values outside that domain.** You could use the line to predict the final exam score for a student who earned a grade of 73 on the third exam. You should NOT use the line to predict the final exam score for a student who earned a grade of 50 on the third exam, because 50 is not within the domain of the x-values in the sample data, which are between 65 and 75.

UNDERSTANDING SLOPE

The slope of the line, b, describes how changes in the variables are related. It is important to interpret the slope of the line in the context of the situation represented by the data. You should be able to write a sentence interpreting the slope in plain English.

INTERPRETATION OF THE SLOPE: The slope of the best-fit line tells us how the dependent variable (y) changes for every one unit increase in the independent (x) variable, on average.

THIRD EXAM vs FINAL EXAM EXAMPLE

Slope: The slope of the line is $b = 4.83$.
Interpretation: For a one-point increase in the score on the third exam, the final exam score increases by 4.83 points, on average.

Using the TI-83, 83+, 84, 84+ Calculator

Using the Linear Regression T Test: LinRegTTest

1. In the STAT list editor, enter the X data in list L1 and the Y data in list L2, paired so that the corresponding (x,y) values are next to each other in the lists. (If a particular pair of values is repeated, enter it as many times as it appears in the data.)

2. On the STAT TESTS menu, scroll down with the cursor to select the LinRegTTest. (Be careful to select LinRegTTest, as some calculators may also have a different item called LinRegTInt.)

3. On the LinRegTTest input screen enter: Xlist: L1 ; Ylist: L2 ; Freq: 1

4. On the next line, at the prompt β or ρ, highlight "≠ 0" and press ENTER

5. Leave the line for "RegEq:" blank

6. Highlight Calculate and press ENTER.

LinRegTTest Input Screen and Output Screen

LinRegTTest
Xlist: L1
Ylist: L2
Freq: 1
β or ρ: ≠0 <0 >0
RegEQ:
Calculate

TI-83+ and TI-84+
calculators

LinRegTTest
y = a + bx
$\beta \neq 0$ and $\rho \neq 0$
t = 2.657560155
p = .0261501512
df = 9
↓a = −173.513363
b = 4.827394209
s = 16.41237711
r^2 = .4396931104
r = .663093591

Figure 12.12

The output screen contains a lot of information. For now we will focus on a few items from the output, and will return later to the other items.

The second line says $y = a + bx$. Scroll down to find the values $a = -173.513$, and $b = 4.8273$; the equation of the best fit line is $\hat{y} = -173.51 + 4.83x$

The two items at the bottom are $r_2 = 0.43969$ and $r = 0.663$. For now, just note where to find these values; we will discuss them in the next two sections.

Graphing the Scatterplot and Regression Line

1. We are assuming your X data is already entered in list L1 and your Y data is in list L2

2. Press 2nd STATPLOT ENTER to use Plot 1

3. On the input screen for PLOT 1, highlight **On**, and press ENTER

4. For TYPE: highlight the very first icon which is the scatterplot and press ENTER

5. Indicate Xlist: L1 and Ylist: L2

6. For Mark: it does not matter which symbol you highlight.

7. Press the ZOOM key and then the number 9 (for menu item "ZoomStat") ; the calculator will fit the window to the data

8. To graph the best-fit line, press the "Y=" key and type the equation –173.5 + 4.83X into equation Y1. (The X key is immediately left of the STAT key). Press ZOOM 9 again to graph it.

9. Optional: If you want to change the viewing window, press the WINDOW key. Enter your desired window using Xmin, Xmax, Ymin, Ymax

---

NOTE

Another way to graph the line after you create a scatter plot is to use LinRegTTest.

1. Make sure you have done the scatter plot. Check it on your screen.

2. Go to LinRegTTest and enter the lists.

3. At RegEq: press VARS and arrow over to Y-VARS. Press 1 for 1:Function. Press 1 for 1:Y1. Then arrow down to Calculate and do the calculation for the line of best fit.

4. Press Y = (you will see the regression equation).

5. Press GRAPH. The line will be drawn."

The Correlation Coefficient *r*

Besides looking at the scatter plot and seeing that a line seems reasonable, how can you tell if the line is a good predictor? Use the correlation coefficient as another indicator (besides the scatterplot) of the strength of the relationship between x and y.

The **correlation coefficient, *r*,** developed by Karl Pearson in the early 1900s, is numerical and provides a measure of strength and direction of the linear association between the independent variable x and the dependent variable y.

The correlation coefficient is calculated as

$$r = \frac{n\Sigma(xy) - (\Sigma x)(\Sigma y)}{\sqrt{\left[n\Sigma x^2 - (\Sigma x)^2\right]\left[n\Sigma y^2 - (\Sigma y)^2\right]}}$$

where n = the number of data points.

If you suspect a linear relationship between x and y, then r can measure how strong the linear relationship is.

What the VALUE of *r* tells us:

- The value of r is always between –1 and +1: $-1 \le r \le 1$.

- The size of the correlation r indicates the strength of the linear relationship between x and y. Values of r close to –1 or to +1 indicate a stronger linear relationship between x and y.

- If $r = 0$ there is absolutely no linear relationship between x and y **(no linear correlation)**.

- If $r = 1$, there is perfect positive correlation. If $r = -1$, there is perfect negativecorrelation. In both these cases, all of the original data points lie on a straight line. Of course,in the real world, this will not generally happen.

What the SIGN of *r* tells us

- A positive value of r means that when x increases, y tends to increase and when x decreases, y tends to decrease **(positive correlation)**.

- A negative value of r means that when x increases, y tends to decrease and when x decreases, y tends to increase **(negative correlation)**.

- The sign of r is the same as the sign of the slope, b, of the best-fit line.

NOTE

Strong correlation does not suggest that x causes y or y causes x. We say **"correlation does not imply causation."**

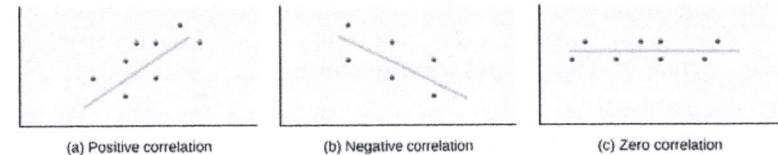

(a) Positive correlation   (b) Negative correlation   (c) Zero correlation

Figure 12.13 (a) A scatter plot showing data with a positive correlation. $0 < r < 1$ (b) A scatter plot showing data with a negative correlation. $-1 < r < 0$ (c) A scatter plot showing data with zero correlation. $r = 0$

The formula for r looks formidable. However, computer spreadsheets, statistical software, and many calculators can quickly calculate r. The correlation coefficient r is the bottom item in the output screens for the LinRegTTest on the TI-83, TI-83+, or TI-84+ calculator (see previous section for instructions).

The Coefficient of Determination

The variable r^2 is called the coefficient of determination and is the square of the correlation coefficient, but is usually stated as a percent, rather than in decimal form. It has an interpretation in the context of the data:

- r^2, when expressed as a percent, represents the percent of variation in the dependent (predicted) variable y that can be explained by variation in the independent (explanatory) variable x using the regression (best-fit) line.

- $1 - r^2$, when expressed as a percentage, represents the percent of variation in y that is NOT explained by variation in x using the regression line. This can be seen as the scattering of the observed data points about the regression line.

Consider the **third exam/final exam example** introduced in the previous section

- The line of best fit is: $\hat{y} = -173.51 + 4.83x$

- The correlation coefficient is $r = 0.6631$

- The coefficient of determination is $r^2 = 0.6631^2 = 0.4397$

- **Interpretation of r^2 in the context of this example:**

- Approximately 44% of the variation (0.4397 is approximately 0.44) in the final-exam grades can be explained by the variation in the grades on the third exam, using the best-fit regression line.

- Therefore, approximately 56% of the variation $(1 - 0.44 = 0.56)$ in the final exam grades can NOT be explained by the variation in the grades on the third exam, using the best-fit regression line. (This is seen as the scattering of the points about the line.)

12.4 | Testing the Significance of the Correlation Coefficient

The correlation coefficient, r, tells us about the strength and direction of the linear relationship between x and y. However, the reliability of the linear model also depends on how many observed data points are in the sample. We need to look at both the value of the correlation coefficient r and the sample size n, together.

We perform a hypothesis test of the **"significance of the correlation coefficient"** to decide whether the linear relationship in the sample data is strong enough to use to model the relationship in the population.

The sample data are used to compute r, the correlation coefficient for the sample. If we had data for the entire population, we could find the population correlation coefficient. But because we have only have sample data, we cannot calculate the population correlation coefficient. The sample correlation coefficient, r, is our estimate of the unknown population correlation coefficient.

The symbol for the population correlation coefficient is ρ, the Greek letter "rho."

ρ = population correlation coefficient (unknown)

r = sample correlation coefficient (known; calculated from sample data)

The hypothesis test lets us decide whether the value of the population correlation coefficient ρ is "close to zero" or "significantly different from zero". We decide this based on the sample correlation coefficient r and the sample size n.

If the test concludes that the correlation coefficient is significantly different from zero, we say that the correlation coefficient is "significant."

- Conclusion: There is sufficient evidence to conclude that there is a significant linear relationship between x and y because the correlation coefficient is significantly different from zero.

- What the conclusion means: There is a significant linear relationship between x and y. We can use the regression line to model the linear relationship between x and y in the population.

If the test concludes that the correlation coefficient is not significantly different from zero (it is close to zero), we say that correlation coefficient is "not significant".

- Conclusion: "There is insufficient evidence to conclude that there is a significant linear relationship between x and y because the correlation coefficient is not significantly different from zero."

- What the conclusion means: There is not a significant linear relationship between x and y. Therefore, we CANNOT use the regression line to model a linear relationship between x and y in the population.

NOTE

- If r is significant and the scatter plot shows a linear trend, the line can be used to predict the value of y for values of x that are within the domain of observed x values.

- If r is not significant OR if the scatter plot does not show a linear trend, the line should not be used for prediction.

- If r is significant and if the scatter plot shows a linear trend, the line may NOT be appropriate or reliable for prediction OUTSIDE the domain of observed x values in the data.

PERFORMING THE HYPOTHESIS TEST

- **Null Hypothesis:** H_0: $\rho = 0$

- **Alternate Hypothesis:** H_a: $\rho \neq 0$

WHAT THE HYPOTHESES MEAN IN WORDS:

- **Null Hypothesis H_0:** The population correlation coefficient IS NOT significantly different from zero. There IS NOT a significant linear relationship(correlation) between x and y in the population.

- **Alternate Hypothesis H_a:** The population correlation coefficient IS significantly DIFFERENT FROM zero. There IS A SIGNIFICANT LINEAR RELATIONSHIP (correlation) between x and y in the population.

DRAWING A CONCLUSION:

There are two methods of making the decision. The two methods are equivalent and give the same result.

- **Method 1: Using the p-value**

- **Method 2: Using a table of critical values**

In this chapter of this textbook, we will always use a significance level of 5%, $\alpha = 0.05$

NOTE

Using the p-value method, you could choose any appropriate significance level you want; you are not limited to using $\alpha = 0.05$. But the table of critical values provided in this textbook assumes that we are using a significance level of 5%, $\alpha = 0.05$. (If we wanted to use a different significance level than 5% with the critical value method, we would need different tables of critical values that are not provided in this textbook.)

METHOD 1: Using a p-value to make a decision

Using the TI-83, 83+, 84, 84+ Calculator

To calculate the p-value using LinRegTTEST:
On the LinRegTTEST input screen, on the line prompt for β or ρ, highlight "$\neq$ **0**"
The output screen shows the p-value on the line that reads "p =".
(Most computer statistical software can calculate the p-value.)

If the *p*-value is less than the significance level (α = 0.05):
- Decision: Reject the null hypothesis.
- Conclusion: "There is sufficient evidence to conclude that there is a significant linear relationship between *x* and *y* because the correlation coefficient is significantly different from zero."

If the *p*-value is NOT less than the significance level (α = 0.05)
- Decision: DO NOT REJECT the null hypothesis.
- Conclusion: "There is insufficient evidence to conclude that there is a significant linear relationship between *x* and *y* because the correlation coefficient is NOT significantly different from zero."

You will use technology to calculate the *p*-value. The following describes the calculations to compute the test statistics and the *p*-value:

The *p*-value is calculated using a *t*-distribution with *n* - 2 degrees of freedom.

The formula for the test statistic is $t = \frac{r\sqrt{n-2}}{\sqrt{1-r^2}}$. The value of the test statistic, *t*, is shown in the computer or calculator output along with the *p*-value. The test statistic *t* has the same sign as the correlation coefficient *r*. The *p*-value is the combined area in both tails.

An alternative way to calculate the *p*-value (**p**) given by LinRegTTest is the command 2*tcdf(abs(t),10^99, n-2) in 2nd DISTR.

THIRD-EXAM vs FINAL-EXAM EXAMPLE: *p*-value method
- Consider the **third exam/final exam example**.
- The line of best fit is: $\hat{y}$ = -173.51 + 4.83*x* with *r* = 0.6631 and there are *n* = 11 data points.
- Can the regression line be used for prediction? **Given a third exam score (*x* value), can we use the line to predict the final exam score (predicted *y* value)?**

H_0: ρ = 0

H_a: $\rho \neq 0$

α = 0.05

- The *p*-value is 0.026 (from LinRegTTest on your calculator or from computer software).
- The *p*-value, 0.026, is less than the significance level of α = 0.05.
- Decision: Reject the Null Hypothesis H_0
- Conclusion: There is sufficient evidence to conclude that there is a significant linear relationship between the third exam score (*x*) and the final exam score (*y*) because the correlation coefficient is significantly different from zero.

Because *r* is significant and the scatter plot shows a linear trend, the regression line can be used to predict final exam scores.

METHOD 2: Using a table of Critical Values to make a decision

The **95% Critical Values of the Sample Correlation Coefficient Table** can be used to give you a good idea of whether the computed value of *r* **is significant or not**. Compare *r* to the appropriate critical value in the table. If *r* is not between the positive and negative critical values, then the correlation coefficient is significant. If *r* is significant, then you may want to use the line for prediction.

Example 12.7

Suppose you computed *r* = 0.801 using *n* = 10 data points.*df* = *n* - 2 = 10 - 2 = 8. The critical values associated with *df* = 8 are -0.632 and + 0.632. If *r* < negative critical value or *r* > positive critical value, then *r* issignificant. Since *r* = 0.801 and 0.801 > 0.632, *r* is significant and the line may be usedfor prediction. If you view this example on a number line, it will help you.

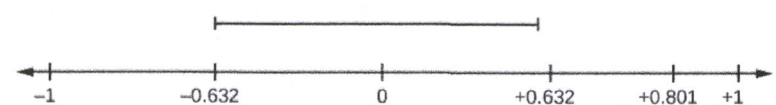

Figure 12.14 *r* is not significant between -0.632 and +0.632. *r* = 0.801 > +0.632. Therefore, *r* is significant.

Try It Σ

12.7 For a given line of best fit, you computed that *r* = 0.6501 using *n* = 12 data points and the critical value is 0.576. Can the line be used for prediction? Why or why not?

Example 12.8

Suppose you computed *r* = –0.624 with 14 data points. *df* = 14 – 2 = 12. The critical values are –0.532 and 0.532. Since –0.624 < –0.532, *r* is significant and the line can be used for prediction

Figure 12.15 r = –0.624-0.532. Therefore, *r* is significant.

Try It Σ

12.8 For a given line of best fit, you compute that *r* = 0.5204 using *n* = 9 data points, and the critical value is 0.666. Can the line be used for prediction? Why or why not?

Example 12.9

Suppose you computed *r* = 0.776 and *n* = 6. *df* = 6 – 2 = 4. The critical values are –0.811 and 0.811. Since –0.811 < 0.776 < 0.811, *r* is not significant, and the line should not be used for prediction.

Figure 12.16 -0.811 < *r* = 0.776 < 0.811. Therefore, *r* is not significant.

Try It Σ

12.9 For a given line of best fit, you compute that *r* = –0.7204 using *n* = 8 data points, and the critical value is = 0.707. Can the line be used for prediction? Why or why not?

THIRD-EXAM vs FINAL-EXAM EXAMPLE: critical value method

Consider the **third exam/final exam example**. The line of best fit is: $\hat{y} = -173.51 + 4.83x$ with $r = 0.6631$ and there are $n = 11$ data points. Can the regression line be used for prediction? **Given a third-exam score (x value), can we use the line to predict the final exam score (predicted y value)?**

H_0: $\rho = 0$

H_a: $\rho \neq 0$

$\alpha = 0.05$

- Use the "95% Critical Value" table for r with $df = n - 2 = 11 - 2 = 9$.

- The critical values are -0.602 and $+0.602$

- Since $0.6631 > 0.602$, r is significant.

- Decision: Reject the null hypothesis.

- Conclusion:There is sufficient evidence to conclude that there is a significant linear relationship between the third exam score (x) and the final exam score (y) because the correlation coefficient is significantly different from zero.

Because r is significant and the scatter plot shows a linear trend, the regression line can be used to predict final exam scores.

Example 12.10

Suppose you computed the following correlation coefficients. Using the table at the end of the chapter, determine if r is significant and the line of best fit associated with each r can be used to predict a y value. If it helps, draw a number line.

a. $r = -0.567$ and the sample size, n, is 19. The $df = n - 2 = 17$. The critical value is -0.456. $-0.567 < -0.456$ so r is significant.

b. $r = 0.708$ and the sample size, n, is nine. The $df = n - 2 = 7$. The critical value is 0.666. $0.708 > 0.666$ so r is significant.

c. $r = 0.134$ and the sample size, n, is 14. The $df = 14 - 2 = 12$. The critical value is 0.532. 0.134 is between -0.532 and 0.532 so r is not significant.

d. $r = 0$ and the sample size, n, is five. No matter what the dfs are, $r = 0$ is between the two critical values so r is not significant.

Try It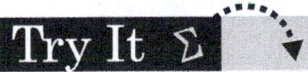

12.10 For a given line of best fit, you compute that $r = 0$ using $n = 100$ data points. Can the line be used for prediction? Why or why not?

Assumptions in Testing the Significance of the Correlation Coefficient

Testing the significance of the correlation coefficient requires that certain assumptions about the data are satisfied. The premise of this test is that the data are a sample of observed points taken from a larger population. We have not examined the entire population because it is not possible or feasible to do so. We are examining the sample to draw a conclusion about whether the linear relationship that we see between x and y in the sample data provides strong enough evidence so that we can conclude that there is a linear relationship between x and y in the population.

The regression line equation that we calculate from the sample data gives the best-fit line for our particular sample. We want to use this best-fit line for the sample as an estimate of the best-fit line for the population. Examining the scatterplot and testing the significance of the correlation coefficient helps us determine if it is appropriate to do this.

The assumptions underlying the test of significance are:
- There is a linear relationship in the population that models the average value of y for varying values of x. In other words, the expected value of y for each particular value lies on a straight line in the population. (We do not know the

equation for the line for the population. Our regression line from the sample is our best estimate of this line in the population.)

- The *y* values for any particular *x* value are normally distributed about the line. This implies that there are more *y* values scattered closer to the line than are scattered farther away. Assumption (1) implies that these normal distributions are centered on the line: the means of these normal distributions of *y* values lie on the line.

- The standard deviations of the population *y* values about the line are equal for each value of *x*. In other words, each of these normal distributions of *y* values has the same shape and spread about the line.

- The residual errors are mutually independent (no pattern).

- The data are produced from a well-designed, random sample or randomized experiment.

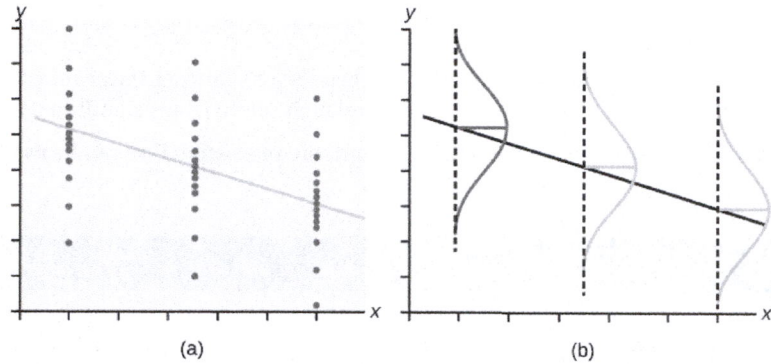

(a)   (b)

Figure 12.17 The *y* values for each *x* value are normally distributed about the line with the same standard deviation. For each *x* value, the mean of the *y* values lies on the regression line. More *y* values lie near the line than are scattered further away.

12.5 | Prediction

Recall the **third exam/final exam example**.

We examined the scatterplot and showed that the correlation coefficient is significant. We found the equation of the best-fit line for the final exam grade as a function of the grade on the third-exam. We can now use the least-squares regression line for prediction.

Suppose you want to estimate, or predict, the mean final exam score of statistics students who received 73 on the third exam. The exam scores **(*x*-values)** range from 65 to 75. **Since 73 is between the *x*-values 65 and 75**, substitute *x* = 73 into the equation. Then:

$$\hat{y} = -173.51 + 4.83(73) = 179.08$$

We predict that statistics students who earn a grade of 73 on the third exam will earn a grade of 179.08 on the final exam, on average.

Example 12.11

Recall the **third exam/final exam example**.

a. What would you predict the final exam score to be for a student who scored a 66 on the third exam?

Solution 12.11
a. 145.27

b. What would you predict the final exam score to be for a student who scored a 90 on the third exam?

Solution 12.11

b. The *x* values in the data are between 65 and 75. Ninety is outside of the domain of the observed *x* values in the data (independent variable), so you cannot reliably predict the final exam score for this student. (Even though it is possible to enter 90 into the equation for *x* and calculate a corresponding *y* value, the *y* value that you get will not be reliable.)

To understand really how unreliable the prediction can be outside of the observed *x* values observed in the data, make the substitution *x* = 90 into the equation.

$$\hat{y} = -173.51 + 4.83\left(90\right) = 261.19$$

The final-exam score is predicted to be 261.19. The largest the final-exam score can be is 200.

> **NOTE**
>
> The process of predicting inside of the observed *x* values observed in the data is called **interpolation**. The process of predicting outside of the observed *x* values observed in the data is called **extrapolation**.

12.11 Data are collected on the relationship between the number of hours per week practicing a musical instrument and scores on a math test. The line of best fit is as follows:

$\hat{y} = 72.5 + 2.8x$

What would you predict the score on a math test would be for a student who practices a musical instrument for five hours a week?

12.6 | Outliers

In some data sets, there are values **(observed data points)** called **outliers**. **Outliers are observed data points that are far from the least squares line.** They have large "errors", where the "error" or residual is the vertical distance from the line to the point.

Outliers need to be examined closely. Sometimes, for some reason or another, they should not be included in the analysis of the data. It is possible that an outlier is a result of erroneous data. Other times, an outlier may hold valuable information about the population under study and should remain included in the data. The key is to examine carefully what causes a data point to be an outlier.

Besides outliers, a sample may contain one or a few points that are called **influential points**. Influential points are observed data points that are far from the other observed data points in the horizontal direction. These points may have a big effect on the slope of the regression line. To begin to identify an influential point, you can remove it from the data set and see if the slope of the regression line is changed significantly.

Computers and many calculators can be used to identify outliers from the data. Computer output for regression analysis will often identify both outliers and influential points so that you can examine them.

Identifying Outliers

We could guess at outliers by looking at a graph of the scatterplot and best fit-line. However, we would like some guideline as to how far away a point needs to be in order to be considered an outlier. **As a rough rule of thumb, we can flag any point that is located further than two standard deviations above or below the best-fit line as an outlier.** The standard deviation used is the standard deviation of the residuals or errors.

We can do this visually in the scatter plot by drawing an extra pair of lines that are two standard deviations above and below the best-fit line. Any data points that are outside this extra pair of lines are flagged as potential outliers. Or we can do this numerically by calculating each residual and comparing it to twice the standard deviation. On the TI-83, 83+, or 84+, the graphical approach is easier. The graphical procedure is shown first, followed by the numerical calculations. You would generally need to use only one of these methods.

Example 12.12

In the **third exam/final exam example**, you can determine if there is an outlier or not. If there is an outlier, as an exercise, delete it and fit the remaining data to a new line. For this example, the new line ought to fit the remaining data better. This means the **SSE** should be smaller and the correlation coefficient ought to be closer to 1 or –1.

Solution 12.12

Graphical Identification of Outliers

With the TI-83, 83+, 84+ graphing calculators, it is easy to identify the outliers graphically and visually. If we were to measure the vertical distance from any data point to the corresponding point on the line of best fit and that distance were equal to $2s$ or more, then we would consider the data point to be "too far" from the line of best fit. We need to find and graph the lines that are two standard deviations below and above the regression line. Any points that are outside these two lines are outliers. We will call these lines Y2 and Y3:

As we did with the equation of the regression line and the correlation coefficient, we will use technology to calculate this standard deviation for us. Using the **LinRegTTest** with this data, scroll down through the output screens to find $s = \mathbf{16.412}$.

Line Y2 = $-173.5 + 4.83x - 2(16.4)$ and line Y3 = $-173.5 + 4.83x + 2(16.4)$

where $\hat{y} = -173.5 + 4.83x$ is the line of best fit. Y2 and Y3 have the same slope as the line of best fit.

Graph the scatterplot with the best fit line in equation Y1, then enter the two extra lines as Y2 and Y3 in the "Y="equation editor and press ZOOM 9. You will find that the only data point that is not between lines Y2 and Y3 is the point $x = 65$, $y = 175$. On the calculator screen it is just barely outside these lines. The outlier is the student who had a grade of 65 on the third exam and 175 on the final exam; this point is further than two standard deviations away from the best-fit line.

Sometimes a point is so close to the lines used to flag outliers on the graph that it is difficult to tell if the point is between or outside the lines. On a computer, enlarging the graph may help; on a small calculator screen, zooming in may make the graph clearer. Note that when the graph does not give a clear enough picture, you can use the numerical comparisons to identify outliers.

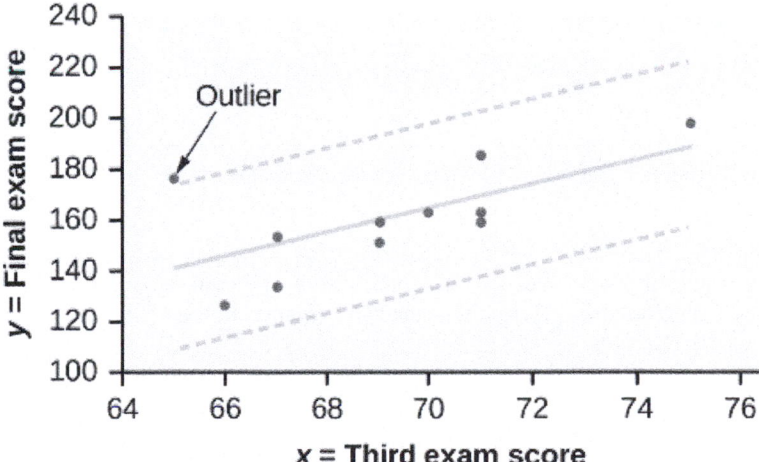

Figure 12.18

Try It Σ

12.12 Identify the potential outlier in the scatter plot. The standard deviation of the residuals or errors is approximately 8.6.

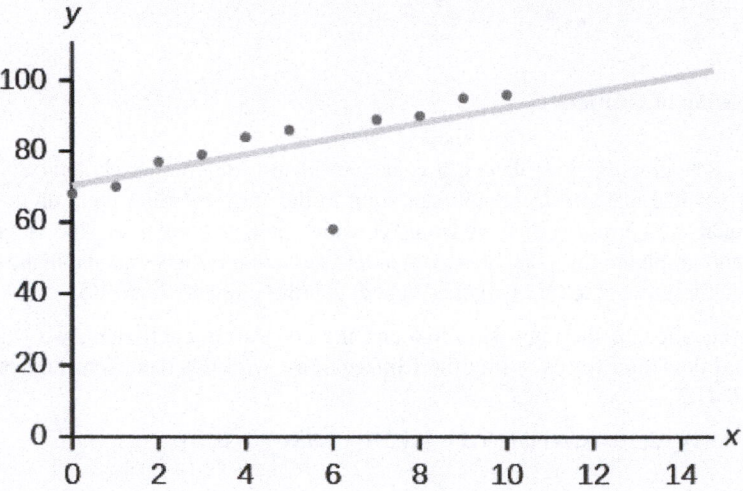

Figure 12.19

Numerical Identification of Outliers

In **Table 12.5**, the first two columns are the third-exam and final-exam data. The third column shows the predicted $\hat{y}$ values calculated from the line of best fit: $\hat{y} = -173.5 + 4.83x$. The residuals, or errors, have been calculated in the fourth column of the table: observed y value–predicted y value $= y - \hat{y}$.

s is the standard deviation of all the $y - \hat{y} = \varepsilon$ values where n = the total number of data points. If each residual is calculated and squared, and the results are added, we get the SSE. The standard deviation of the residuals is calculated from the SSE as:

$$s = \sqrt{\frac{SSE}{n-2}}$$

NOTE

We divide by $(n - 2)$ because the regression model involves two estimates.

Rather than calculate the value of s ourselves, we can find s using the computer or calculator. For this example, the calculator function LinRegTTest found $s = 16.4$ as the standard deviation of the residuals 35; –17; 16; –6; –19; 9; 3; –1; –10; –9; –1.

x	y	$\hat{y}$	$y - \hat{y}$
65	175	140	175 – 140 = 35
67	133	150	133 – 150= –17
71	185	169	185 – 169 = 16
71	163	169	163 – 169 = –6

Table 12.5

x	y	$\hat{y}$	$y - \hat{y}$
66	126	145	$126 - 145 = -19$
75	198	189	$198 - 189 = 9$
67	153	150	$153 - 150 = 3$
70	163	164	$163 - 164 = -1$
71	159	169	$159 - 169 = -10$
69	151	160	$151 - 160 = -9$
69	159	160	$159 - 160 = -1$

Table 12.5

We are looking for all data points for which the residual is greater than $2s = 2(16.4) = 32.8$ or less than -32.8. Compare these values to the residuals in column four of the table. The only such data point is the student who had a grade of 65 on the third exam and 175 on the final exam; the residual for this student is 35.

How does the outlier affect the best fit line?

Numerically and graphically, we have identified the point (65, 175) as an outlier. We should re-examine the data for this point to see if there are any problems with the data. If there is an error, we should fix the error if possible, or delete the data. If the data is correct, we would leave it in the data set. For this problem, we will suppose that we examined the data and found that this outlier data was an error. Therefore we will continue on and delete the outlier, so that we can explore how it affects the results, as a learning experience.

Compute a new best-fit line and correlation coefficient using the ten remaining points:

On the TI-83, TI-83+, TI-84+ calculators, delete the outlier from L1 and L2. Using the LinRegTTest, the new line of best fit and the correlation coefficient are:

$\hat{y} = -355.19 + 7.39x$ and $r = 0.9121$

The new line with $r = 0.9121$ is a stronger correlation than the original ($r = 0.6631$) because $r = 0.9121$ is closer to one. This means that the new line is a better fit to the ten remaining data values. The line can better predict the final exam score given the third exam score.

Numerical Identification of Outliers: Calculating s and Finding Outliers Manually

If you do not have the function LinRegTTest, then you can calculate the outlier in the first example by doing the following.

First, **square each $|y - \hat{y}|$**

The squares are 35^2; 17^2; 16^2; 6^2; 19^2; 9^2; 3^2; 1^2; 10^2; 9^2; 1^2

Then, add (sum) all the $|y - \hat{y}|$ squared terms using the formula

$$\sum_{i=1}^{11} \left(|y_i - \hat{y}_i|\right)^2 = \sum_{i=1}^{11} \varepsilon_i^2 \text{ (Recall that } y_i - \hat{y}_i = \varepsilon_i.)$$

$= 35^2 + 17^2 + 16^2 + 6^2 + 19^2 + 9^2 + 3^2 + 1^2 + 10^2 + 9^2 + 1^2$

$= 2440 = $ **SSE**. The result, **SSE** is the Sum of Squared Errors.

Next, calculate s, the standard deviation of all the $y - \hat{y} = \varepsilon$ values where n = the total number of data points.

The calculation is $s = \sqrt{\dfrac{SSE}{n-2}}$.

For the third exam/final exam problem, $s = \sqrt{\dfrac{2440}{11-2}} = 16.47$.

Next, multiply *s* by 2:

(2)(16.47) = 32.94

32.94 is 2 standard deviations away from the mean of the $y - \hat{y}$ values.

If we were to measure the vertical distance from any data point to the corresponding point on the line of best fit and that distance is at least 2*s*, then we would consider the data point to be "too far" from the line of best fit. We call that point a **potential outlier**.

For the example, if any of the $|y - \hat{y}|$ values are **at least** 32.94, the corresponding (x, y) data point is a potential outlier.

For the third exam/final exam problem, all the $|y - \hat{y}|$'s are less than 31.29 except for the first one which is 35.

35 > 31.29 That is, $|y - \hat{y}| \geq (2)(s)$

The point which corresponds to $|y - \hat{y}| = 35$ is (65, 175). **Therefore, the data point (65,175) is a potential outlier.** For this example, we will delete it. (Remember, we do not always delete an outlier.)

NOTE

When outliers are deleted, the researcher should either record that data was deleted, and why, or the researcher should provide results both with and without the deleted data. If data is erroneous and the correct values are known (e.g., student one actually scored a 70 instead of a 65), then this correction can be made to the data.

The next step is to compute a new best-fit line using the ten remaining points. The new line of best fit and the correlation coefficient are:

$\hat{y} = -355.19 + 7.39x$ and $r = 0.9121$

Example 12.13

Using this new line of best fit (based on the remaining ten data points in the **third exam/final exam example**), what would a student who receives a 73 on the third exam expect to receive on the final exam? Is this the same as the prediction made using the original line?

Solution 12.13

Using the new line of best fit, $\hat{y} = -355.19 + 7.39(73) = 184.28$. A student who scored 73 points on the third exam would expect to earn 184 points on the final exam.

The original line predicted $\hat{y} = -173.51 + 4.83(73) = 179.08$ so the prediction using the new line with the outlier eliminated differs from the original prediction.

 Try It

12.13 The data points for the graph from the **third exam/final exam example** are as follows: (1, 5), (2, 7), (2, 6), (3, 9), (4, 12), (4, 13), (5, 18), (6, 19), (7, 12), and (7, 21). Remove the outlier and recalculate the line of best fit. Find the value of $\hat{y}$ when *x* = 10.

Example 12.14

The Consumer Price Index (CPI) measures the average change over time in the prices paid by urban consumers for consumer goods and services. The CPI affects nearly all Americans because of the many ways it is used. One of its biggest uses is as a measure of inflation. By providing information about price changes in the Nation's economy to government, business, and labor, the CPI helps them to make economic decisions. The President,

Congress, and the Federal Reserve Board use the CPI's trends to formulate monetary and fiscal policies. In the following table, x is the year and y is the CPI.

x	y	x	y
1915	10.1	1969	36.7
1926	17.7	1975	49.3
1935	13.7	1979	72.6
1940	14.7	1980	82.4
1947	24.1	1986	109.6
1952	26.5	1991	130.7
1964	31.0	1999	166.6

Table 12.6 Data

a. Draw a scatterplot of the data.

b. Calculate the least squares line. Write the equation in the form $\hat{y} = a + bx$.

c. Draw the line on the scatterplot.

d. Find the correlation coefficient. Is it significant?

e. What is the average CPI for the year 1990?

Solution 12.14
a. See **Figure 12.19**.

b. $\hat{y} = -3204 + 1.662x$ is the equation of the line of best fit.

c. $r = 0.8694$

d. The number of data points is $n = 14$. Use the 95% Critical Values of the Sample Correlation Coefficient table at the end of Chapter 12. $n - 2 = 12$. The corresponding critical value is 0.532. Since $0.8694 > 0.532$, r is significant.
$\hat{y} = -3204 + 1.662(1990) = 103.4$ CPI

e. Using the calculator LinRegTTest, we find that $s = 25.4$; graphing the lines Y2 $= -3204 + 1.662$X $- 2(25.4)$ and Y3 $= -3204 + 1.662$X $+ 2(25.4)$ shows that no data values are outside those lines, identifying no outliers. (Note that the year 1999 was very close to the upper line, but still inside it.)

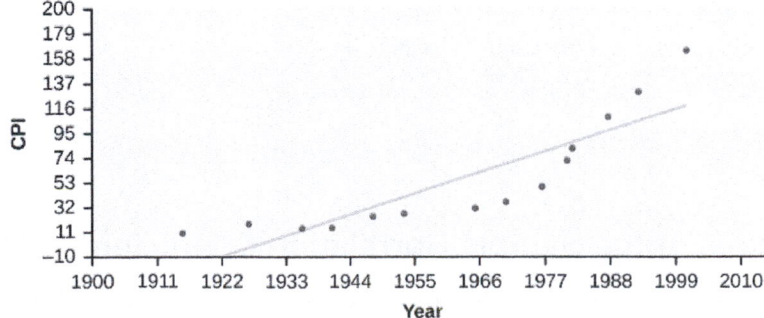

Figure 12.20

NOTE

In the example, notice the pattern of the points compared to the line. Although the correlation coefficient is significant, the pattern in the scatterplot indicates that a curve would be a more appropriate model to use than a line. In this example, a statistician should prefer to use other methods to fit a curve to this data, rather than model the data with the line we found. In addition to doing the calculations, it is always important to look at the scatterplot when deciding whether a linear model is appropriate.

If you are interested in seeing more years of data, visit the Bureau of Labor Statistics CPI website ftp://ftp.bls.gov/pub/special.requests/cpi/cpiai.txt; our data is taken from the column entitled "Annual Avg." (third column from the right). For example you could add more current years of data. Try adding the more recent years: 2004: CPI = 188.9; 2008: CPI = 215.3; 2011: CPI = 224.9. See how it affects the model. (Check: $\hat{y}$ = –4436 + 2.295x; r = 0.9018. Is r significant? Is the fit better with the addition of the new points?)

12.14 The following table shows economic development measured in per capita income PCINC.

Year	PCINC	Year	PCINC
1870	340	1920	1050
1880	499	1930	1170
1890	592	1940	1364
1900	757	1950	1836
1910	927	1960	2132

Table 12.7

a. What are the independent and dependent variables?

b. Draw a scatter plot.

c. Use regression to find the line of best fit and the correlation coefficient.

d. Interpret the significance of the correlation coefficient.

e. Is there a linear relationship between the variables?

f. Find the coefficient of determination and interpret it.

g. What is the slope of the regression equation? What does it mean?

h. Use the line of best fit to estimate PCINC for 1900, for 2000.

i. Determine if there are any outliers.

95% Critical Values of the Sample Correlation Coefficient Table

Degrees of Freedom: $n - 2$	Critical Values: (+ and –)
1	0.997

Table 12.8

Degrees of Freedom: $n - 2$	Critical Values: (+ and –)
2	0.950
3	0.878
4	0.811
5	0.754
6	0.707
7	0.666
8	0.632
9	0.602
10	0.576
11	0.555
12	0.532
13	0.514
14	0.497
15	0.482
16	0.468
17	0.456
18	0.444
19	0.433
20	0.423
21	0.413
22	0.404
23	0.396
24	0.388
25	0.381
26	0.374
27	0.367
28	0.361
29	0.355
30	0.349
40	0.304
50	0.273
60	0.250
70	0.232
80	0.217
90	0.205
100	0.195

Table 12.8

12.7 | Regression (Distance from School)

Stats Lab

12.1 Regression (Distance from School)

Class Time:

Names:

Student Learning Outcomes

- The student will calculate and construct the line of best fit between two variables.

- The student will evaluate the relationship between two variables to determine if that relationship is significant.

Collect the Data

Use eight members of your class for the sample. Collect bivariate data (distance an individual lives from school, the cost of supplies for the current term).

1. Complete the table.

Distance from school	Cost of supplies this term

Table 12.9

2. Which variable should be the dependent variable and which should be the independent variable? Why?

3. Graph "distance" vs. "cost." Plot the points on the graph. Label both axes with words. Scale both axes.

Figure 12.21

Analyze the Data

Enter your data into your calculator or computer. Write the linear equation, rounding to four decimal places.

1. Calculate the following:

   a. $a = $ _____

   b. $b = $ _____

   c. correlation = _____

   d. $n = $ _____

   e. equation: $\hat{y} = $ _____

   f. Is the correlation significant? Why or why not? (Answer in one to three complete sentences.)

2. Supply an answer for the following senarios:

   a. For a person who lives eight miles from campus, predict the total cost of supplies this term:

   b. For a person who lives eighty miles from campus, predict the total cost of supplies this term:

3. Obtain the graph on your calculator or computer. Sketch the regression line.

Figure 12.22

Discussion Questions

1. Answer each question in complete sentences.

   a. Does the line seem to fit the data? Why?

   b. What does the correlation imply about the relationship between the distance and the cost?

2. Are there any outliers? If so, which point is an outlier?

3. Should the outlier, if it exists, be removed? Why or why not?

12.8 | Regression (Textbook Cost)

Stats l.ab

12.2 Regression (Textbook Cost)

Class Time:

Names:

Student Learning Outcomes

- The student will calculate and construct the line of best fit between two variables.
- The student will evaluate the relationship between two variables to determine if that relationship is significant.

Collect the Data

Survey ten textbooks. Collect bivariate data (number of pages in a textbook, the cost of the textbook).

1. Complete the table.

Number of pages	Cost of textbook

Table 12.10

2. Which variable should be the dependent variable and which should be the independent variable? Why?
3. Graph "pages" vs. "cost." Plot the points on the graph in **Analyze the Data**. Label both axes with words. Scale both axes.

Analyze the Data

Enter your data into your calculator or computer. Write the linear equation, rounding to four decimal places.

1. Calculate the following:
   a. $a = $ _____
   b. $b = $ _____
   c. correlation = _____
   d. $n = $ _____
   e. equation: $y = $ _____
   f. Is the correlation significant? Why or why not? (Answer in complete sentences.)
2. Supply an answer for the following senarios:
   a. For a textbook with 400 pages, predict the cost.
   b. For a textbook with 600 pages, predict the cost.

3. Obtain the graph on your calculator or computer. Sketch the regression line.

Figure 12.23

Discussion Questions

1. Answer each question in complete sentences.

    a. Does the line seem to fit the data? Why?

    b. What does the correlation imply about the relationship between the number of pages and the cost?

2. Are there any outliers? If so, which point(s) is an outlier?

3. Should the outlier, if it exists, be removed? Why or why not?

12.9 | Regression (Fuel Efficiency)

Stats Lab

12.3 Regression (Fuel Efficiency)

Class Time:

Names:

Student Learning Outcomes

- The student will calculate and construct the line of best fit between two variables.
- The student will evaluate the relationship between two variables to determine if that relationship is significant.

Collect the Data

Use the most recent April issue of Consumer Reports. It will give the total fuel efficiency (in miles per gallon) and weight (in pounds) of new model cars with automatic transmissions. We will use this data to determine the relationship, if any, between the fuel efficiency of a car and its weight.

1. Using your random number generator, randomly select 20 cars from the list and record their weights and fuel efficiency into **Table 12.11**.

Weight	Fuel Efficiency

Table 12.11

2. Which variable should be the dependent variable and which should be the independent variable? Why?

3. By hand, do a scatterplot of "weight" vs. "fuel efficiency". Plot the points on graph paper. Label both axes with words. Scale both axes accurately.

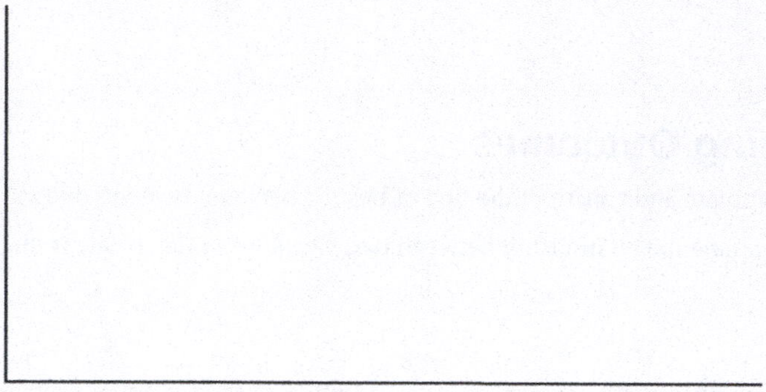

Figure 12.24

Analyze the Data

Enter your data into your calculator or computer. Write the linear equation, rounding to 4 decimal places.

1. Calculate the following:

    a. $a = $ _____

    b. $b = $ _____

    c. correlation = _____

    d. $n = $ _____

    e. equation: $\hat{y} = $ _____

2. Obtain the graph of the regression line on your calculator. Sketch the regression line on the same axes as your scatter plot.

Discussion Questions

1. Is the correlation significant? Explain how you determined this in complete sentences.

2. Is the relationship a positive one or a negative one? Explain how you can tell and what this means in terms of weight and fuel efficiency.

3. In one or two complete sentences, what is the practical interpretation of the slope of the least squares line in terms of fuel efficiency and weight?

4. For a car that weighs 4,000 pounds, predict its fuel efficiency. Include units.

5. Can we predict the fuel efficiency of a car that weighs 10,000 pounds using the least squares line? Explain why or why not.

6. Answer each question in complete sentences.

    a. Does the line seem to fit the data? Why or why not?

    b. What does the correlation imply about the relationship between fuel efficiency and weight of a car? Is this what you expected?

7. Are there any outliers? If so, which point is an outlier?

KEY TERMS

Coefficient of Correlation a measure developed by Karl Pearson (early 1900s) that gives the strength of association between the independent variable and the dependent variable; the formula is:

$$r = \frac{n\sum xy - (\sum x)(\sum y)}{\sqrt{[n\sum x^2 - (\sum x)^2][n\sum y^2 - (\sum y)^2]}}$$

where n is the number of data points. The coefficient cannot be more then 1 and less then −1. The closer the coefficient is to ±1, the stronger the evidence of a significant linear relationship between x and y.

Outlier an observation that does not fit the rest of the data

CHAPTER REVIEW

12.1 Linear Equations

The most basic type of association is a linear association. This type of relationship can be defined algebraically by the equations used, numerically with actual or predicted data values, or graphically from a plotted curve. (Lines are classified as straight curves.) Algebraically, a linear equation typically takes the form $y = mx + b$, where m and b are constants, x is the independent variable, y is the dependent variable. In a statistical context, a linear equation is written in the form $y = a + bx$, where a and b are the constants. This form is used to help readers distinguish the statistical context from the algebraic context. In the equation $y = a + bx$, the constant b that multiplies the x variable (b is called a coefficient) is called as the **slope**. The slope describes the rate of change between the independent and dependent variables; in other words, the rate of change describes the change that occurs in the dependent variable as the independent variable is changed. In the equation $y = a + bx$, the constant a is called as the y-intercept. Graphically, the y-intercept is the y coordinate of the point where the graph of the line crosses the y axis. At this point $x = 0$.

The **slope of a line** is a value that describes the rate of change between the independent and dependent variables. The **slope** tells us how the dependent variable (y) changes for every one unit increase in the independent (x) variable, on average. The **y-intercept** is used to describe the dependent variable when the independent variable equals zero. Graphically, the slope is represented by three line types in elementary statistics.

12.2 Scatter Plots

Scatter plots are particularly helpful graphs when we want to see if there is a linear relationship among data points. They indicate both the direction of the relationship between the x variables and the y variables, and the strength of the relationship. We calculate the strength of the relationship between an independent variable and a dependent variable using linear regression.

12.3 The Regression Equation

A regression line, or a line of best fit, can be drawn on a scatter plot and used to predict outcomes for the x and y variables in a given data set or sample data. There are several ways to find a regression line, but usually the least-squares regression line is used because it creates a uniform line. Residuals, also called "errors," measure the distance from the actual value of y and the estimated value of y. The Sum of Squared Errors, when set to its minimum, calculates the points on the line of best fit. Regression lines can be used to predict values within the given set of data, but should not be used to make predictions for values outside the set of data.

The correlation coefficient r measures the strength of the linear association between x and y. The variable r has to be between −1 and +1. When r is positive, the x and y will tend to increase and decrease together. When r is negative, x will increase and y will decrease, or the opposite, x will decrease and y will increase. The coefficient of determination r^2, is equal to the square of the correlation coefficient. When expressed as a percent, r^2 represents the percent of variation in the dependent variable y that can be explained by variation in the independent variable x using the regression line.

12.4 Testing the Significance of the Correlation Coefficient

Linear regression is a procedure for fitting a straight line of the form $\hat{y} = a + bx$ to data. The conditions for regression are:

- **Linear** In the population, there is a linear relationship that models the average value of y for different values of x.

- **Independent** The residuals are assumed to be independent.
- **Normal** The y values are distributed normally for any value of x.
- **Equal variance** The standard deviation of the y values is equal for each x value.
- **Random** The data are produced from a well-designed random sample or randomized experiment.

The slope b and intercept a of the least-squares line estimate the slope β and intercept α of the population (true) regression line. To estimate the population standard deviation of y, σ, use the standard deviation of the residuals, s. $s = \sqrt{\frac{SEE}{n-2}}$. The variable ρ (rho) is the population correlation coefficient. To test the null hypothesis H_0: $\rho = hypothesized\ value$, use a linear regression t-test. The most common null hypothesis is H_0: $\rho = 0$ which indicates there is no linear relationship between x and y in the population. The TI-83, 83+, 84, 84+ calculator function LinRegTTest can perform this test (STATS TESTS LinRegTTest).

12.5 Prediction

After determining the presence of a strong correlation coefficient and calculating the line of best fit, you can use the least squares regression line to make predictions about your data.

12.6 Outliers

To determine if a point is an outlier, do one of the following:

1. Input the following equations into the TI 83, 83+,84, 84+:

   $y_1 = a + bx$

   $y_2 = (2s)a + bx$ where s is the standard deviation of the residuals

   $y_3 = -(2s)a + bx$

   If any point is above y_2 or below y_3 then the point is considered to be an outlier.

2. Use the residuals and compare their absolute values to $2s$ where s is the standard deviation of the residuals. If the absolute value of any residual is greater than or equal to $2s$, then the corresponding point is an outlier.

3. Note: The calculator function LinRegTTest (STATS TESTS LinRegTTest) calculates s.

FORMULA REVIEW

12.1 Linear Equations

$y = a + bx$ where a is the y-intercept and b is the slope. The variable x is the independent variable and y is the dependent variable.

12.4 Testing the Significance of the Correlation Coefficient

Least Squares Line or Line of Best Fit:

$\hat{y} = a + bx$

where

a = y-intercept

b = slope

Standard deviation of the residuals:

$s = \sqrt{\frac{SEE}{n-2}}.$

where

SEE = sum of squared errors

n = the number of data points

PRACTICE

12.1 Linear Equations

Use the following information to answer the next three exercises. A vacation resort rents SCUBA equipment to certified divers. The resort charges an up-front fee of $25 and another fee of $12.50 an hour.

1. What are the dependent and independent variables?

2. Find the equation that expresses the total fee in terms of the number of hours the equipment is rented.

3. Graph the equation from **Exercise 12.2**.

Use the following information to answer the next two exercises. A credit card company charges $10 when a payment is late, and $5 a day each day the payment remains unpaid.

4. Find the equation that expresses the total fee in terms of the number of days the payment is late.

5. Graph the equation from **Exercise 12.4**.

6. Is the equation $y = 10 + 5x - 3x^2$ linear? Why or why not?

7. Which of the following equations are linear?

a. $y = 6x + 8$

b. $y + 7 = 3x$

c. $y - x = 8x^2$

d. $4y = 8$

8. Does the graph show a linear equation? Why or why not?

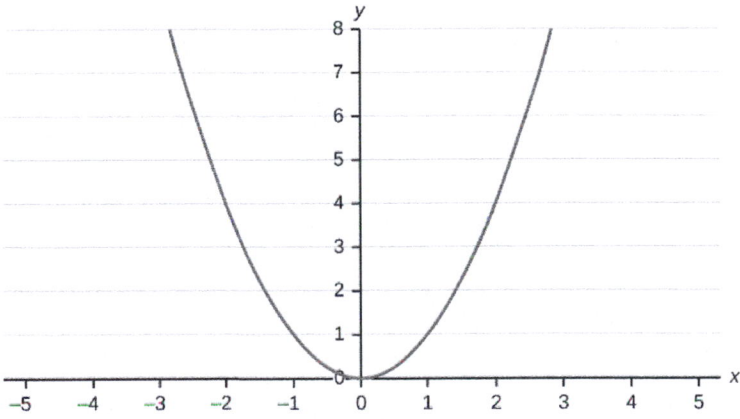

Figure 12.25

Table 12.12 contains real data for the first two decades of AIDS reporting.

Year	# AIDS cases diagnosed	# AIDS deaths
Pre-1981	91	29
1981	319	121
1982	1,170	453
1983	3,076	1,482
1984	6,240	3,466
1985	11,776	6,878

1986	19,032	11,987
1987	28,564	16,162
1988	35,447	20,868
1989	42,674	27,591
1990	48,634	31,335
1991	59,660	36,560
1992	78,530	41,055
1993	78,834	44,730
1994	71,874	49,095
1995	68,505	49,456
1996	59,347	38,510
1997	47,149	20,736
1998	38,393	19,005
1999	25,174	18,454
2000	25,522	17,347
2001	25,643	17,402
2002	26,464	16,371
Total	**802,118**	**489,093**

Table 12.12 Adults and Adolescents only, United States

9. Use the columns "year" and "# AIDS cases diagnosed. Why is "year" the independent variable and "# AIDS cases diagnosed." the dependent variable (instead of the reverse)?

Use the following information to answer the next two exercises. A specialty cleaning company charges an equipment fee and an hourly labor fee. A linear equation that expresses the total amount of the fee the company charges for each session is $y = 50 + 100x$.

10. What are the independent and dependent variables?

11. What is the y-intercept and what is the slope? Interpret them using complete sentences.

Use the following information to answer the next three questions. Due to erosion, a river shoreline is losing several thousand pounds of soil each year. A linear equation that expresses the total amount of soil lost per year is $y = 12,000x$.

12. What are the independent and dependent variables?

13. How many pounds of soil does the shoreline lose in a year?

14. What is the y-intercept? Interpret its meaning.

Use the following information to answer the next two exercises. The price of a single issue of stock can fluctuate throughout the day. A linear equation that represents the price of stock for Shipment Express is $y = 15 - 1.5x$ where x is the number of hours passed in an eight-hour day of trading.

15. What are the slope and y-intercept? Interpret their meaning.

16. If you owned this stock, would you want a positive or negative slope? Why?

12.2 Scatter Plots

17. Does the scatter plot appear linear? Strong or weak? Positive or negative?

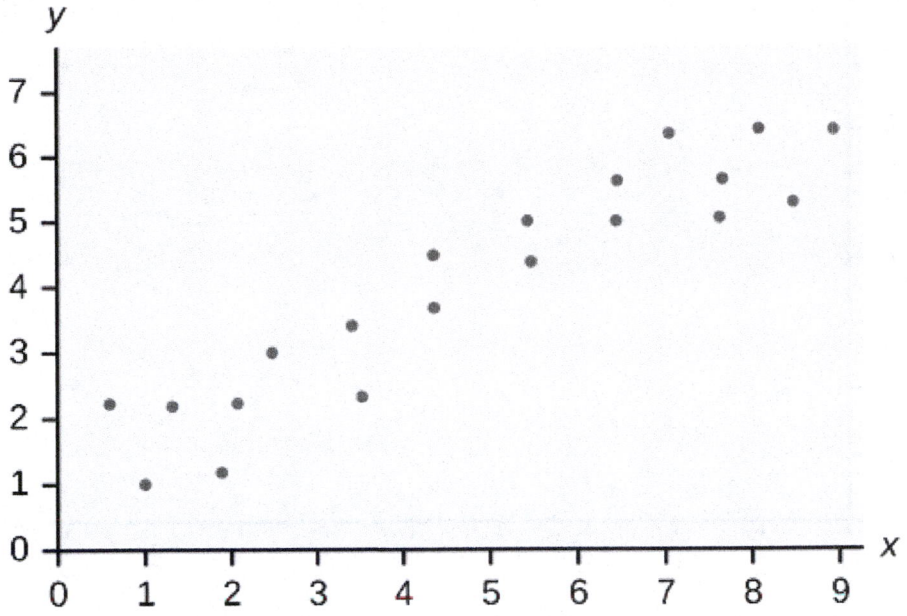

Figure 12.26

18. Does the scatter plot appear linear? Strong or weak? Positive or negative?

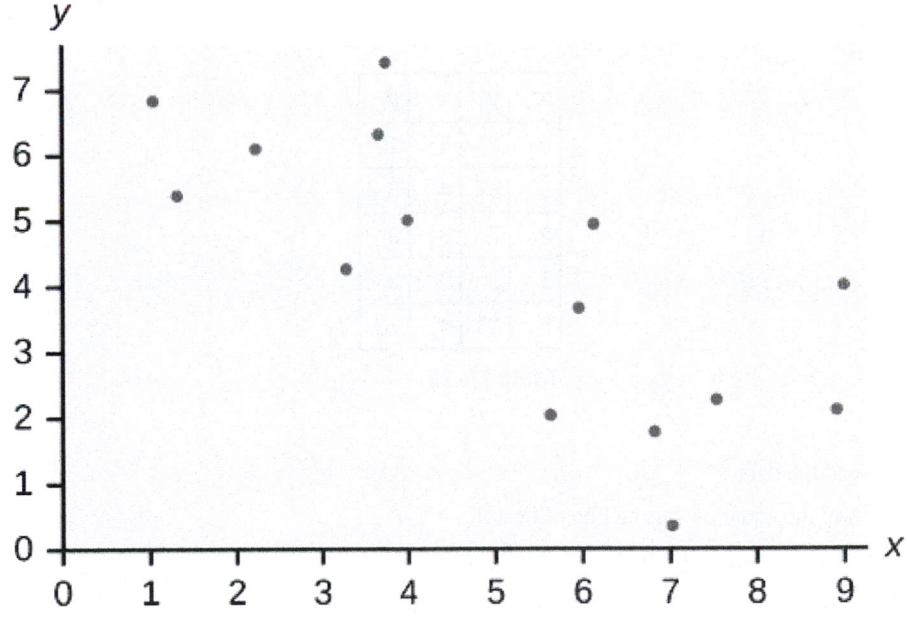

Figure 12.27

19. Does the scatter plot appear linear? Strong or weak? Positive or negative?

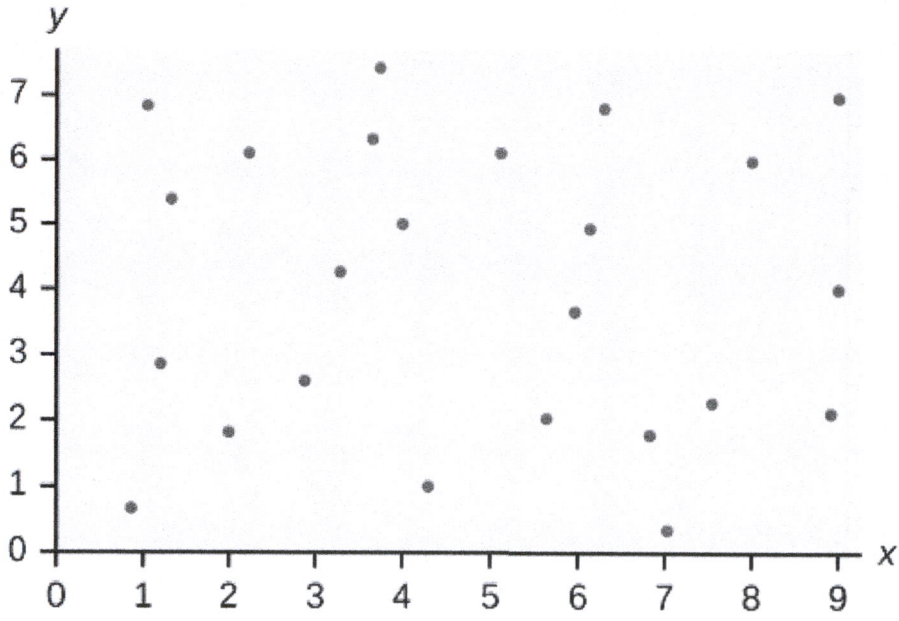

Figure 12.28

12.3 The Regression Equation

Use the following information to answer the next five exercises. A random sample of ten professional athletes produced the following data where x is the number of endorsements the player has and y is the amount of money made (in millions of dollars).

x	y	x	y
0	2	5	12
3	8	4	9
2	7	3	9
1	3	0	3
5	13	4	10

Table 12.13

20. Draw a scatter plot of the data.

21. Use regression to find the equation for the line of best fit.

22. Draw the line of best fit on the scatter plot.

23. What is the slope of the line of best fit? What does it represent?

24. What is the y-intercept of the line of best fit? What does it represent?

25. What does an r value of zero mean?

26. When $n = 2$ and $r = 1$, are the data significant? Explain.

27. When $n = 100$ and $r = -0.89$, is there a significant correlation? Explain.

12.4 Testing the Significance of the Correlation Coefficient

28. When testing the significance of the correlation coefficient, what is the null hypothesis?

29. When testing the significance of the correlation coefficient, what is the alternative hypothesis?

30. If the level of significance is 0.05 and the *p*-value is 0.04, what conclusion can you draw?

12.5 Prediction

Use the following information to answer the next two exercises. An electronics retailer used regression to find a simple model to predict sales growth in the first quarter of the new year (January through March). The model is good for 90 days, where *x* is the day. The model can be written as follows:

$\hat{y} = 101.32 + 2.48x$ where $\hat{y}$ is in thousands of dollars.

31. What would you predict the sales to be on day 60?

32. What would you predict the sales to be on day 90?

Use the following information to answer the next three exercises. A landscaping company is hired to mow the grass for several large properties. The total area of the properties combined is 1,345 acres. The rate at which one person can mow is as follows:

$\hat{y} = 1350 - 1.2x$ where *x* is the number of hours and $\hat{y}$ represents the number of acres left to mow.

33. How many acres will be left to mow after 20 hours of work?

34. How many acres will be left to mow after 100 hours of work?

35. How many hours will it take to mow all of the lawns? (When is $\hat{y} = 0$?)

Table 12.14 contains real data for the first two decades of AIDS reporting.

Year	# AIDS cases diagnosed	# AIDS deaths
Pre-1981	91	29
1981	319	121
1982	1,170	453
1983	3,076	1,482
1984	6,240	3,466
1985	11,776	6,878
1986	19,032	11,987
1987	28,564	16,162
1988	35,447	20,868
1989	42,674	27,591
1990	48,634	31,335
1991	59,660	36,560
1992	78,530	41,055
1993	78,834	44,730
1994	71,874	49,095
1995	68,505	49,456
1996	59,347	38,510
1997	47,149	20,736
1998	38,393	19,005

1999	25,174	18,454
2000	25,522	17,347
2001	25,643	17,402
2002	26,464	16,371
Total	**802,118**	**489,093**

Table 12.14 Adults and Adolescents only, United States

36. Graph "year" versus "# AIDS cases diagnosed" (plot the scatter plot). Do not include pre-1981 data.

37. Perform linear regression. What is the linear equation? Round to the nearest whole number.

38. Write the equations:

    a. Linear equation: _____
    b. $a =$ _____
    c. $b =$ _____
    d. $r =$ _____
    e. $n =$ _____

39. Solve.
    a. When $x = 1985$, $\hat{y} =$ _____
    b. When $x = 1990$, $\hat{y} =$ _____
    c. When $x = 1970$, $\hat{y} =$ _____ Why doesn't this answer make sense?

40. Does the line seem to fit the data? Why or why not?

41. What does the correlation imply about the relationship between time (years) and the number of diagnosed AIDS cases reported in the U.S.?

42. Plot the two given points on the following graph. Then, connect the two points to form the regression line.

Figure 12.29

Obtain the graph on your calculator or computer.

43. Write the equation: $\hat{y} =$ _____

44. Hand draw a smooth curve on the graph that shows the flow of the data.

45. Does the line seem to fit the data? Why or why not?

46. Do you think a linear fit is best? Why or why not?

47. What does the correlation imply about the relationship between time (years) and the number of diagnosed AIDS cases reported in the U.S.?

48. Graph "year" vs. "# AIDS cases diagnosed." Do not include pre-1981. Label both axes with words. Scale both axes.

49. Enter your data into your calculator or computer. The pre-1981 data should not be included. Why is that so?

Write the linear equation, rounding to four decimal places:

50. Calculate the following:
  a. $a =$ _____
  b. $b =$ _____
  c. correlation = _____
  d. $n =$ _____

12.6 Outliers

Use the following information to answer the next four exercises. The scatter plot shows the relationship between hours spent studying and exam scores. The line shown is the calculated line of best fit. The correlation coefficient is 0.69.

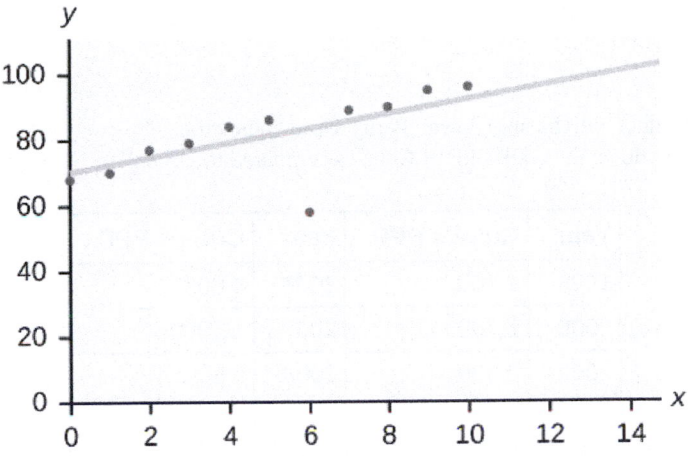

Figure 12.30

51. Do there appear to be any outliers?

52. A point is removed, and the line of best fit is recalculated. The new correlation coefficient is 0.98. Does the point appear to have been an outlier? Why?

53. What effect did the potential outlier have on the line of best fit?

54. Are you more or less confident in the predictive ability of the new line of best fit?

55. The Sum of Squared Errors for a data set of 18 numbers is 49. What is the standard deviation?

56. The Standard Deviation for the Sum of Squared Errors for a data set is 9.8. What is the cutoff for the vertical distance that a point can be from the line of best fit to be considered an outlier?

HOMEWORK

12.1 Linear Equations

57. For each of the following situations, state the independent variable and the dependent variable.
  a. A study is done to determine if elderly drivers are involved in more motor vehicle fatalities than other drivers. The number of fatalities per 100,000 drivers is compared to the age of drivers.
  b. A study is done to determine if the weekly grocery bill changes based on the number of family members.
  c. Insurance companies base life insurance premiums partially on the age of the applicant.
  d. Utility bills vary according to power consumption.
  e. A study is done to determine if a higher education reduces the crime rate in a population.

58. Piece-rate systems are widely debated incentive payment plans. In a recent study of loan officer effectiveness, the following piece-rate system was examined:

% of goal reached	< 80	80	100	120
Incentive	n/a	$4,000 with an additional $125 added per percentage point from 81–99%	$6,500 with an additional $125 added per percentage point from 101–119%	$9,500 with an additional $125 added per percentage point starting at 121%

Table 12.15

If a loan officer makes 95% of his or her goal, write the linear function that applies based on the incentive plan table. In context, explain the *y*-intercept and slope.

12.2 Scatter Plots

59. The Gross Domestic Product Purchasing Power Parity is an indication of a country's currency value compared to another country. **Table 12.16** shows the GDP PPP of Cuba as compared to US dollars. Construct a scatter plot of the data.

Year	Cuba's PPP	Year	Cuba's PPP
1999	1,700	2006	4,000
2000	1,700	2007	11,000
2002	2,300	2008	9,500
2003	2,900	2009	9,700
2004	3,000	2010	9,900
2005	3,500		

Table 12.16

60. The following table shows the poverty rates and cell phone usage in the United States. Construct a scatter plot of the data

Year	Poverty Rate	Cellular Usage per Capita
2003	12.7	54.67
2005	12.6	74.19
2007	12	84.86
2009	12	90.82

Table 12.17

61. Does the higher cost of tuition translate into higher-paying jobs? The table lists the top ten colleges based on mid-career salary and the associated yearly tuition costs. Construct a scatter plot of the data.

School	Mid-Career Salary (in thousands)	Yearly Tuition
Princeton	137	28,540
Harvey Mudd	135	40,133
CalTech	127	39,900
US Naval Academy	122	0
West Point	120	0
MIT	118	42,050
Lehigh University	118	43,220
NYU-Poly	117	39,565
Babson College	117	40,400
Stanford	114	54,506

Table 12.18

62. If the level of significance is 0.05 and the p-value is 0.06, what conclusion can you draw?

63. If there are 15 data points in a set of data, what is the number of degree of freedom?

12.3 The Regression Equation

64. What is the process through which we can calculate a line that goes through a scatter plot with a linear pattern?

65. Explain what it means when a correlation has an r^2 of 0.72.

66. Can a coefficient of determination be negative? Why or why not?

12.4 Testing the Significance of the Correlation Coefficient

67. If the level of significance is 0.05 and the p-value is 0.06, what conclusion can you draw?

68. If there are 15 data points in a set of data, what is the number of degree of freedom?

12.5 Prediction

69. Recently, the annual number of driver deaths per 100,000 for the selected age groups was as follows:

Age	Number of Driver Deaths per 100,000
17.5	38
22	36
29.5	24
44.5	20
64.5	18
80	28

Table 12.19

    a. For each age group, pick the midpoint of the interval for the x value. (For the 75+ group, use 80.)

    b. Using "ages" as the independent variable and "Number of driver deaths per 100,000" as the dependent variable, make a scatter plot of the data.

    c. Calculate the least squares (best–fit) line. Put the equation in the form of: $\hat{y} = a + bx$

d.   Find the correlation coefficient. Is it significant?
e.   Predict the number of deaths for ages 40 and 60.
f.   Based on the given data, is there a linear relationship between age of a driver and driver fatality rate?
g.   What is the slope of the least squares (best-fit) line? Interpret the slope.

70. Table 12.20 shows the life expectancy for an individual born in the United States in certain years.

Year of Birth	Life Expectancy
1930	59.7
1940	62.9
1950	70.2
1965	69.7
1973	71.4
1982	74.5
1987	75
1992	75.7
2010	78.7

Table 12.20

a.   Decide which variable should be the independent variable and which should be the dependent variable.
b.   Draw a scatter plot of the ordered pairs.
c.   Calculate the least squares line. Put the equation in the form of: $\hat{y} = a + bx$
d.   Find the correlation coefficient. Is it significant?
e.   Find the estimated life expectancy for an individual born in 1950 and for one born in 1982.
f.   Why aren't the answers to part e the same as the values in **Table 12.20** that correspond to those years?
g.   Use the two points in part e to plot the least squares line on your graph from part b.
h.   Based on the data, is there a linear relationship between the year of birth and life expectancy?
i.   Are there any outliers in the data?
j.   Using the least squares line, find the estimated life expectancy for an individual born in 1850. Does the least squares line give an accurate estimate for that year? Explain why or why not.
k.   What is the slope of the least-squares (best-fit) line? Interpret the slope.

71. The maximum discount value of the Entertainment® card for the "Fine Dining" section, Edition ten, for various pages is given in **Table 12.21**

Page number	Maximum value ($)
4	16
14	19
25	15
32	17
43	19
57	15
72	16
85	15

Page number	Maximum value ($)
90	17

Table 12.21

a. Decide which variable should be the independent variable and which should be the dependent variable.
b. Draw a scatter plot of the ordered pairs.
c. Calculate the least-squares line. Put the equation in the form of: $\hat{y} = a + bx$
d. Find the correlation coefficient. Is it significant?
e. Find the estimated maximum values for the restaurants on page ten and on page 70.
f. Does it appear that the restaurants giving the maximum value are placed in the beginning of the "Fine Dining" section? How did you arrive at your answer?
g. Suppose that there were 200 pages of restaurants. What do you estimate to be the maximum value for a restaurant listed on page 200?
h. Is the least squares line valid for page 200? Why or why not?
i. What is the slope of the least-squares (best-fit) line? Interpret the slope.

72. Table 12.22 gives the gold medal times for every other Summer Olympics for the women's 100-meter freestyle (swimming).

Year	Time (seconds)
1912	82.2
1924	72.4
1932	66.8
1952	66.8
1960	61.2
1968	60.0
1976	55.65
1984	55.92
1992	54.64
2000	53.8
2008	53.1

Table 12.22

a. Decide which variable should be the independent variable and which should be the dependent variable.
b. Draw a scatter plot of the data.
c. Does it appear from inspection that there is a relationship between the variables? Why or why not?
d. Calculate the least squares line. Put the equation in the form of: $\hat{y} = a + bx$.
e. Find the correlation coefficient. Is the decrease in times significant?
f. Find the estimated gold medal time for 1932. Find the estimated time for 1984.
g. Why are the answers from part f different from the chart values?
h. Does it appear that a line is the best way to fit the data? Why or why not?
i. Use the least-squares line to estimate the gold medal time for the next Summer Olympics. Do you think that your answer is reasonable? Why or why not?

73.

State	# letters in name	Year entered the Union	Rank for entering the Union	Area (square miles)
Alabama	7	1819	22	52,423
Colorado	8	1876	38	104,100
Hawaii	6	1959	50	10,932
Iowa	4	1846	29	56,276
Maryland	8	1788	7	12,407
Missouri	8	1821	24	69,709
New Jersey	9	1787	3	8,722
Ohio	4	1803	17	44,828
South Carolina	13	1788	8	32,008
Utah	4	1896	45	84,904
Wisconsin	9	1848	30	65,499

Table 12.23

We are interested in whether or not the number of letters in a state name depends upon the year the state entered the Union.

   a. Decide which variable should be the independent variable and which should be the dependent variable.
   b. Draw a scatter plot of the data.
   c. Does it appear from inspection that there is a relationship between the variables? Why or why not?
   d. Calculate the least-squares line. Put the equation in the form of: $\hat{y} = a + bx$.
   e. Find the correlation coefficient. What does it imply about the significance of the relationship?
   f. Find the estimated number of letters (to the nearest integer) a state would have if it entered the Union in 1900. Find the estimated number of letters a state would have if it entered the Union in 1940.
   g. Does it appear that a line is the best way to fit the data? Why or why not?
   h. Use the least-squares line to estimate the number of letters a new state that enters the Union this year would have. Can the least squares line be used to predict it? Why or why not?

12.6 Outliers

74. The height (sidewalk to roof) of notable tall buildings in America is compared to the number of stories of the building (beginning at street level).

Height (in feet)	Stories
1,050	57
428	28
362	26
529	40
790	60
401	22
380	38
1,454	110

Height (in feet)	Stories
1,127	100
700	46

Table 12.24

a. Using "stories" as the independent variable and "height" as the dependent variable, make a scatter plot of the data.
b. Does it appear from inspection that there is a relationship between the variables?
c. Calculate the least squares line. Put the equation in the form of: $\hat{y} = a + bx$
d. Find the correlation coefficient. Is it significant?
e. Find the estimated heights for 32 stories and for 94 stories.
f. Based on the data in **Table 12.24**, is there a linear relationship between the number of stories in tall buildings and the height of the buildings?
g. Are there any outliers in the data? If so, which point(s)?
h. What is the estimated height of a building with six stories? Does the least squares line give an accurate estimate of height? Explain why or why not.
i. Based on the least squares line, adding an extra story is predicted to add about how many feet to a building?
j. What is the slope of the least squares (best-fit) line? Interpret the slope.

75. Ornithologists, scientists who study birds, tag sparrow hawks in 13 different colonies to study their population. They gather data for the percent of new sparrow hawks in each colony and the percent of those that have returned from migration.

Percent return: 74; 66; 81; 52; 73; 62; 52; 45; 62; 46; 60; 46; 38
Percent new: 5; 6; 8; 11; 12; 15; 16; 17; 18; 18; 19; 20; 20

a. Enter the data into your calculator and make a scatter plot.
b. Use your calculator's regression function to find the equation of the least-squares regression line. Add this to your scatter plot from part a.
c. Explain in words what the slope and y-intercept of the regression line tell us.
d. How well does the regression line fit the data? Explain your response.
e. Which point has the largest residual? Explain what the residual means in context. Is this point an outlier? An influential point? Explain.
f. An ecologist wants to predict how many birds will join another colony of sparrow hawks to which 70% of the adults from the previous year have returned. What is the prediction?

76. The following table shows data on average per capita wine consumption and heart disease rate in a random sample of 10 countries.

Yearly wine consumption in liters	2.5	3.9	2.9	2.4	2.9	0.8	9.1	2.7	0.8	0.7
Death from heart diseases	221	167	131	191	220	297	71	172	211	300

Table 12.25

a. Enter the data into your calculator and make a scatter plot.
b. Use your calculator's regression function to find the equation of the least-squares regression line. Add this to your scatter plot from part a.
c. Explain in words what the slope and y-intercept of the regression line tell us.
d. How well does the regression line fit the data? Explain your response.
e. Which point has the largest residual? Explain what the residual means in context. Is this point an outlier? An influential point? Explain.
f. Do the data provide convincing evidence that there is a linear relationship between the amount of alcohol consumed and the heart disease death rate? Carry out an appropriate test at a significance level of 0.05 to help answer this question.

77. The following table consists of one student athlete's time (in minutes) to swim 2000 yards and the student's heart rate (beats per minute) after swimming on a random sample of 10 days:

Swim Time	Heart Rate
34.12	144
35.72	152
34.72	124
34.05	140
34.13	152
35.73	146
36.17	128
35.57	136
35.37	144
35.57	148

Table 12.26

a. Enter the data into your calculator and make a scatter plot.
b. Use your calculator's regression function to find the equation of the least-squares regression line. Add this to your scatter plot from part a.
c. Explain in words what the slope and y-intercept of the regression line tell us.
d. How well does the regression line fit the data? Explain your response.
e. Which point has the largest residual? Explain what the residual means in context. Is this point an outlier? An influential point? Explain.

78. A researcher is investigating whether non-white minorities commit a disproportionate number of homicides. He uses demographic data from Detroit, MI to compare homicide rates and the number of the population that are white males.

White Males	Homicide rate per 100,000 people
558,724	8.6
538,584	8.9
519,171	8.52
500,457	8.89
482,418	13.07
465,029	14.57
448,267	21.36
432,109	28.03
416,533	31.49
401,518	37.39
387,046	46.26
373,095	47.24

White Males	Homicide rate per 100,000 people
359,647	52.33

Table 12.27

a. Use your calculator to construct a scatter plot of the data. What should the independent variable be? Why?
b. Use your calculator's regression function to find the equation of the least-squares regression line. Add this to your scatter plot.
c. Discuss what the following mean in context.
    i. The slope of the regression equation
    ii. The y-intercept of the regression equation
    iii. The correlation r
    iv. The coefficient of determination r2.
d. Do the data provide convincing evidence that there is a linear relationship between the number of white males in the population and the homicide rate? Carry out an appropriate test at a significance level of 0.05 to help answer this question.

79.

School	Mid-Career Salary (in thousands)	Yearly Tuition
Princeton	137	28,540
Harvey Mudd	135	40,133
CalTech	127	39,900
US Naval Academy	122	0
West Point	120	0
MIT	118	42,050
Lehigh University	118	43,220
NYU-Poly	117	39,565
Babson College	117	40,400
Stanford	114	54,506

Table 12.28

Using the data to determine the linear-regression line equation with the outliers removed. Is there a linear correlation for the data set with outliers removed? Justify your answer.

REFERENCES

12.1 Linear Equations

Data from the Centers for Disease Control and Prevention.

Data from the National Center for HIV, STD, and TB Prevention.

12.5 Prediction

Data from the Centers for Disease Control and Prevention.

Data from the National Center for HIV, STD, and TB Prevention.

Data from the United States Census Bureau. Available online at http://www.census.gov/compendia/statab/cats/transportation/motor_vehicle_accidents_and_fatalities.html

Data from the National Center for Health Statistics.

12.6 Outliers

Data from the House Ways and Means Committee, the Health and Human Services Department.

Data from Microsoft Bookshelf.

Data from the United States Department of Labor, the Bureau of Labor Statistics.

Data from the Physician's Handbook, 1990.

Data from the United States Department of Labor, the Bureau of Labor Statistics.

BRINGING IT TOGETHER: HOMEWORK

80. The average number of people in a family that received welfare for various years is given in **Table 12.29**.

Year	Welfare family size
1969	4.0
1973	3.6
1975	3.2
1979	3.0
1983	3.0
1988	3.0
1991	2.9

Table 12.29

a. Using "year" as the independent variable and "welfare family size" as the dependent variable, draw a scatter plot of the data.
b. Calculate the least-squares line. Put the equation in the form of: $\hat{y} = a + bx$
c. Find the correlation coefficient. Is it significant?
d. Pick two years between 1969 and 1991 and find the estimated welfare family sizes.
e. Based on the data in **Table 12.29**, is there a linear relationship between the year and the average number of people in a welfare family?
f. Using the least-squares line, estimate the welfare family sizes for 1960 and 1995. Does the least-squares line give an accurate estimate for those years? Explain why or why not.
g. Are there any outliers in the data?
h. What is the estimated average welfare family size for 1986? Does the least squares line give an accurate estimate for that year? Explain why or why not.
i. What is the slope of the least squares (best-fit) line? Interpret the slope.

81. The percent of female wage and salary workers who are paid hourly rates is given in **Table 12.30** for the years 1979 to 1992.

Year	Percent of workers paid hourly rates
1979	61.2
1980	60.7
1981	61.3
1982	61.3
1983	61.8
1984	61.7
1985	61.8
1986	62.0
1987	62.7
1990	62.8
1992	62.9

Table 12.30

a. Using "year" as the independent variable and "percent" as the dependent variable, draw a scatter plot of the data.
b. Does it appear from inspection that there is a relationship between the variables? Why or why not?
c. Calculate the least-squares line. Put the equation in the form of: $\hat{y} = a + bx$
d. Find the correlation coefficient. Is it significant?
e. Find the estimated percents for 1991 and 1988.
f. Based on the data, is there a linear relationship between the year and the percent of female wage and salary earners who are paid hourly rates?
g. Are there any outliers in the data?
h. What is the estimated percent for the year 2050? Does the least-squares line give an accurate estimate for that year? Explain why or why not.
i. What is the slope of the least-squares (best-fit) line? Interpret the slope.

Use the following information to answer the next two exercises. The cost of a leading liquid laundry detergent in different sizes is given in **Table 12.31**.

Size (ounces)	Cost ($)	Cost per ounce
16	3.99	
32	4.99	
64	5.99	
200	10.99	

Table 12.31

82.
a. Using "size" as the independent variable and "cost" as the dependent variable, draw a scatter plot.
b. Does it appear from inspection that there is a relationship between the variables? Why or why not?
c. Calculate the least-squares line. Put the equation in the form of: $\hat{y} = a + bx$
d. Find the correlation coefficient. Is it significant?
e. If the laundry detergent were sold in a 40-ounce size, find the estimated cost.
f. If the laundry detergent were sold in a 90-ounce size, find the estimated cost.
g. Does it appear that a line is the best way to fit the data? Why or why not?

     h. Are there any outliers in the given data?

     i. Is the least-squares line valid for predicting what a 300-ounce size of the laundry detergent would you cost? Why or why not?

     j. What is the slope of the least-squares (best-fit) line? Interpret the slope.

83.

     a. Complete **Table 12.31** for the cost per ounce of the different sizes.

     b. Using "size" as the independent variable and "cost per ounce" as the dependent variable, draw a scatter plot of the data.

     c. Does it appear from inspection that there is a relationship between the variables? Why or why not?

     d. Calculate the least-squares line. Put the equation in the form of: $\hat{y} = a + bx$

     e. Find the correlation coefficient. Is it significant?

     f. If the laundry detergent were sold in a 40-ounce size, find the estimated cost per ounce.

     g. If the laundry detergent were sold in a 90-ounce size, find the estimated cost per ounce.

     h. Does it appear that a line is the best way to fit the data? Why or why not?

     i. Are there any outliers in the the data?

     j. Is the least-squares line valid for predicting what a 300-ounce size of the laundry detergent would cost per ounce? Why or why not?

     k. What is the slope of the least-squares (best-fit) line? Interpret the slope.

84. According to a flyer by a Prudential Insurance Company representative, the costs of approximate probate fees and taxes for selected net taxable estates are as follows:

Net Taxable Estate ($)	Approximate Probate Fees and Taxes ($)
600,000	30,000
750,000	92,500
1,000,000	203,000
1,500,000	438,000
2,000,000	688,000
2,500,000	1,037,000
3,000,000	1,350,000

Table 12.32

     a. Decide which variable should be the independent variable and which should be the dependent variable.

     b. Draw a scatter plot of the data.

     c. Does it appear from inspection that there is a relationship between the variables? Why or why not?

     d. Calculate the least-squares line. Put the equation in the form of: $\hat{y} = a + bx$.

     e. Find the correlation coefficient. Is it significant?

     f. Find the estimated total cost for a next taxable estate of $1,000,000. Find the cost for $2,500,000.

     g. Does it appear that a line is the best way to fit the data? Why or why not?

     h. Are there any outliers in the data?

     i. Based on these results, what would be the probate fees and taxes for an estate that does not have any assets?

     j. What is the slope of the least-squares (best-fit) line? Interpret the slope.

85. The following are advertised sale prices of color televisions at Anderson's.

Size (inches)	Sale Price ($)
9	147

Size (inches)	Sale Price ($)
20	197
27	297
31	447
35	1177
40	2177
60	2497

Table 12.33

a. Decide which variable should be the independent variable and which should be the dependent variable.
b. Draw a scatter plot of the data.
c. Does it appear from inspection that there is a relationship between the variables? Why or why not?
d. Calculate the least-squares line. Put the equation in the form of: $\hat{y} = a + bx$
e. Find the correlation coefficient. Is it significant?
f. Find the estimated sale price for a 32 inch television. Find the cost for a 50 inch television.
g. Does it appear that a line is the best way to fit the data? Why or why not?
h. Are there any outliers in the data?
i. What is the slope of the least-squares (best-fit) line? Interpret the slope.

86. Table 12.34 shows the average heights for American boy s in 1990.

Age (years)	Height (cm)
birth	50.8
2	83.8
3	91.4
5	106.6
7	119.3
10	137.1
14	157.5

Table 12.34

a. Decide which variable should be the independent variable and which should be the dependent variable.
b. Draw a scatter plot of the data.
c. Does it appear from inspection that there is a relationship between the variables? Why or why not?
d. Calculate the least-squares line. Put the equation in the form of: $\hat{y} = a + bx$
e. Find the correlation coefficient. Is it significant?
f. Find the estimated average height for a one-year-old. Find the estimated average height for an eleven-year-old.
g. Does it appear that a line is the best way to fit the data? Why or why not?
h. Are there any outliers in the data?
i. Use the least squares line to estimate the average height for a sixty-two-year-old man. Do you think that your answer is reasonable? Why or why not?
j. What is the slope of the least-squares (best-fit) line? Interpret the slope.

87.

State	# letters in name	Year entered the Union	Ranks for entering the Union	Area (square miles)
Alabama	7	1819	22	52,423
Colorado	8	1876	38	104,100
Hawaii	6	1959	50	10,932
Iowa	4	1846	29	56,276
Maryland	8	1788	7	12,407
Missouri	8	1821	24	69,709
New Jersey	9	1787	3	8,722
Ohio	4	1803	17	44,828
South Carolina	13	1788	8	32,008
Utah	4	1896	45	84,904
Wisconsin	9	1848	30	65,499

Table 12.35

We are interested in whether there is a relationship between the ranking of a state and the area of the state.

a. What are the independent and dependent variables?
b. What do you think the scatter plot will look like? Make a scatter plot of the data.
c. Does it appear from inspection that there is a relationship between the variables? Why or why not?
d. Calculate the least-squares line. Put the equation in the form of: $\hat{y} = a + bx$
e. Find the correlation coefficient. What does it imply about the significance of the relationship?
f. Find the estimated areas for Alabama and for Colorado. Are they close to the actual areas?
g. Use the two points in part f to plot the least-squares line on your graph from part b.
h. Does it appear that a line is the best way to fit the data? Why or why not?
i. Are there any outliers?
j. Use the least squares line to estimate the area of a new state that enters the Union. Can the least-squares line be used to predict it? Why or why not?
k. Delete "Hawaii" and substitute "Alaska" for it. Alaska is the forty-ninth, state with an area of 656,424 square miles.
l. Calculate the new least-squares line.
m. Find the estimated area for Alabama. Is it closer to the actual area with this new least-squares line or with the previous one that included Hawaii? Why do you think that's the case?
n. Do you think that, in general, newer states are larger than the original states?

SOLUTIONS

1 dependent variable: fee amount; independent variable: time

3

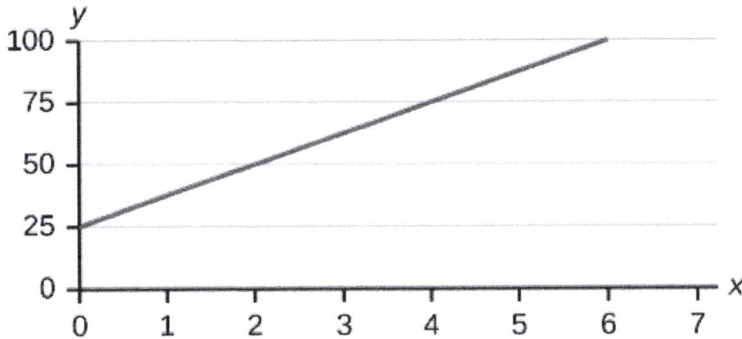

Figure 12.31

5

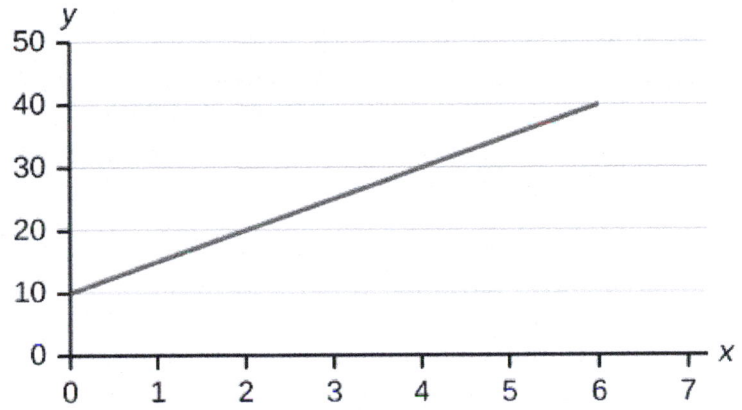

Figure 12.32

7 $y = 6x + 8$, $4y = 8$, and $y + 7 = 3x$ are all linear equations.

9 The number of AIDS cases depends on the year. Therefore, year becomes the independent variable and the number of AIDS cases is the dependent variable.

11 The y-intercept is 50 ($a = 50$). At the start of the cleaning, the company charges a one-time fee of $50 (this is when $x = 0$). The slope is 100 ($b = 100$). For each session, the company charges $100 for each hour they clean.

13 12,000 pounds of soil

15 The slope is –1.5 ($b = -1.5$). This means the stock is losing value at a rate of $1.50 per hour. The y-intercept is $15 ($a = 15$). This means the price of stock before the trading day was $15.

17 The data appear to be linear with a strong, positive correlation.

19 The data appear to have no correlation.

21 $\hat{y} = 2.23 + 1.99x$

23 The slope is 1.99 ($b = 1.99$). It means that for every endorsement deal a professional player gets, he gets an average of another $1.99 million in pay each year.

25 It means that there is no correlation between the data sets.

27 Yes, there are enough data points and the value of r is strong enough to show that there is a strong negative correlation between the data sets.

29 $H_a: \rho \neq 0$

31 $250,120

33 1,326 acres

35 1,125 hours, or when $x = 1,125$

37 Check student's solution.

39
  a. When $x = 1985$, $\hat{y} = 25,52$

  b. When $x = 1990$, $\hat{y} = 34,275$

  c. When $x = 1970$, $\hat{y} = -725$ Why doesn't this answer make sense? The range of x values was 1981 to 2002; the year 1970 is not in this range. The regression equation does not apply, because predicting for the year 1970 is extrapolation, which requires a different process. Also, a negative number does not make sense in this context, where we are predicting AIDS cases diagnosed.

41 Also, the correlation $r = 0.4526$. If r is compared to the value in the 95% Critical Values of the Sample Correlation Coefficient Table, because $r > 0.423$, r is significant, and you would think that the line could be used for prediction. But the scatter plot indicates otherwise.

43 $\hat{y} = 3,448,225 + 1750x$

45 There was an increase in AIDS cases diagnosed until 1993. From 1993 through 2002, the number of AIDS cases diagnosed declined each year. It is not appropriate to use a linear regression line to fit to the data.

47 Since there is no linear association between year and # of AIDS cases diagnosed, it is not appropriate to calculate a linear correlation coefficient. When there is a linear association and it is appropriate to calculate a correlation, we cannot say that one variable "causes" the other variable.

49 We don't know if the pre-1981 data was collected from a single year. So we don't have an accurate x value for this figure. Regression equation: $\hat{y}$ (#AIDS Cases) $= -3,448,225 + 1749.777$ (year)

	Coefficients
Intercept	$-3,448,225$
X Variable 1	1,749.777

Table 12.36

51 Yes, there appears to be an outlier at (6, 58).

53 The potential outlier flattened the slope of the line of best fit because it was below the data set. It made the line of best fit less accurate is a predictor for the data.

55 $s = 1.75$

57
  a. independent variable: age; dependent variable: fatalities

  b. independent variable: # of family members; dependent variable: grocery bill

  c. independent variable: age of applicant; dependent variable: insurance premium

  d. independent variable: power consumption; dependent variable: utility

  e. independent variable: higher education (years); dependent variable: crime rates

59 Check student's solution.

61 For graph: check student's solution. Note that tuition is the independent variable and salary is the dependent variable.

63 13

65 It means that 72% of the variation in the dependent variable (y) can be explained by the variation in the independent variable (x).

67 We do not reject the null hypothesis. There is not sufficient evidence to conclude that there is a significant linear relationship between x and y because the correlation coefficient is not significantly different from zero.

69

a.

Age	Number of Driver Deaths per 100,000
16–19	38
20–24	36
25–34	24
35–54	20
55–74	18
75+	28

Table 12.37

b. Check student's solution.

c. $\hat{y} = 35.5818045 - 0.19182491x$

d. $r = -0.57874$
For four *df* and alpha = 0.05, the LinRegTTest gives *p*-value = 0.2288 so we do not reject the null hypothesis; there is not a significant linear relationship between deaths and age.
Using the table of critical values for the correlation coefficient, with four *df*, the critical value is 0.811. The correlation coefficient $r = -0.57874$ is not less than −0.811, so we do not reject the null hypothesis.

e. if age = 40, $\hat{y}$ (deaths) = 35.5818045 − 0.19182491(40) = 27.9
if age = 60, $\hat{y}$ (deaths) = 35.5818045 − 0.19182491(60) = 24.1

f. For entire dataset, there is a linear relationship for the ages up to age 74. The oldest age group shows an increase in deaths from the prior group, which is not consistent with the younger ages.

g. slope = −0.19182491

71

a. We wonder if the better discounts appear earlier in the book so we select page as *X* and discount as *Y*.

b. Check student's solution.

c. $\hat{y} = 17.21757 - 0.01412x$

d. $r = -0.2752$
For seven *df* and alpha = 0.05, using LinRegTTest *p*-value = 0.4736 so we do not reject; there is a not a significant linear relationship between page and discount.
Using the table of critical values for the correlation coefficient, with seven *df*, the critical value is 0.666. The correlation coefficient $xi = -0.2752$ is not less than 0.666 so we do not reject.

e. page 10: 17.08 page 70: 16.23

f. There is not a significant linear correlation so it appears there is no relationship between the page and the amount of the discount.

g. page 200: 14.39

h. No, using the regression equation to predict for page 200 is extrapolation.

i. slope = −0.01412

As the page number increases by one page, the discount decreases by $0.01412

73

a. Year is the independent or *x* variable; the number of letters is the dependent or *y* variable.

b. Check student's solution.

c. no

d. $\hat{y} = 47.03 - 0.0216x$

e. −0.4280

f. 6; 5

g. No, the relationship does not appear to be linear; the correlation is not significant.

h. current year: 2013: 3.55 or four letters; this is not an appropriate use of the least squares line. It is extrapolation.

75 a. and b. Check student's solution. c. The slope of the regression line is -0.3179 with a y-intercept of 32.966. In context, the y-intercept indicates that when there are no returning sparrow hawks, there will be almost 31% new sparrow hawks, which doesn't make sense since if there are no returning birds, then the new percentage would have to be 100% (this is an example of why we do not extrapolate). The slope tells us that for each percentage increase in returning birds, the percentage of new birds in the colony decreases by 0.3179%. d. If we examine r2, we see that only 50.238% of the variation in the percent of new birds is explained by the model and the correlation coefficient, r = 0.71 only indicates a somewhat strong correlation between returning and new percentages. e. The ordered pair (66, 6) generates the largest residual of 6.0. This means that when the observed return percentage is 66%, our observed new percentage, 6%, is almost 6% less than the predicted new value of 11.98%. If we remove this data pair, we see only an adjusted slope of -0.2723 and an adjusted intercept of 30.606. In other words, even though this data generates the largest residual, it is not an outlier, nor is the data pair an influential point. f. If there are 70% returning birds, we would expect to see y = -0.2723(70) + 30.606 = 0.115 or 11.5% new birds in the colony.

77

a. Check student's solution.

b. Check student's solution.

c. We have a slope of −1.4946 with a y-intercept of 193.88. The slope, in context, indicates that for each additional minute added to the swim time, the heart rate will decrease by 1.5 beats per minute. If the student is not swimming at all, the y-intercept indicates that his heart rate will be 193.88 beats per minute. While the slope has meaning (the longer it takes to swim 2,000 meters, the less effort the heart puts out), the y-intercept does not make sense. If the athlete is not swimming (resting), then his heart rate should be very low.

d. Since only 1.5% of the heart rate variation is explained by this regression equation, we must conclude that this association is not explained with a linear relationship.

e. The point (34.72, 124) generates the largest residual of −11.82. This means that our observed heart rate is almost 12 beats less than our predicted rate of 136 beats per minute. When this point is removed, the slope becomes 1.6914 with the y-intercept changing to 83.694. While the linear association is still very weak, we see that the removed data pair can be considered an influential point in the sense that the y-intercept becomes more meaningful.

79 If we remove the two service academies (the tuition is $0.00), we construct a new regression equation of y = −0.0009x + 160 with a correlation coefficient of 0.71397 and a coefficient of determination of 0.50976. This allows us to say there is a fairly strong linear association between tuition costs and salaries if the service academies are removed from the data set.

81

a. Check student's solution.

b. yes

c. $\hat{y} = -266.8863 + 0.1656x$

d. 0.9448; Yes

e. 62.8233; 62.3265

f. yes

g. yes; (1987, 62.7)

h. 72.5937; no

i. slope = 0.1656.
    As the year increases by one, the percent of workers paid hourly rates tends to increase by 0.1656.

83

a.

Size (ounces)	Cost ($)	cents/oz
16	3.99	24.94
32	4.99	15.59
64	5.99	9.36
200	10.99	5.50

Table 12.38

b. Check student's solution.

c. There is a linear relationship for the sizes 16 through 64, but that linear trend does not continue to the 200-oz size.

d. $\hat{y} = 20.2368 - 0.0819x$

e. $r = -0.8086$

f. 40-oz: 16.96 cents/oz

g. 90-oz: 12.87 cents/oz

h. The relationship is not linear; the least squares line is not appropriate.

i. no outliers

j. No, you would be extrapolating. The 300-oz size is outside the range of x.

k. slope = –0.08194; for each additional ounce in size, the cost per ounce decreases by 0.082 cents.

85

a. Size is x, the independent variable, price is y, the dependent variable.

b. Check student's solution.

c. The relationship does not appear to be linear.

d. $\hat{y} = -745.252 + 54.75569x$

e. $r = 0.8944$, yes it is significant

f. 32-inch: $1006.93, 50-inch: $1992.53

g. No, the relationship does not appear to be linear. However, r is significant.

h. yes, the 60-inch TV

i. For each additional inch, the price increases by $54.76

87

a. Let rank be the independent variable and area be the dependent variable.

b. Check student's solution.

c. There appears to be a linear relationship, with one outlier.

d. $\hat{y}$ (area) = 24177.06 + 1010.478x

e. $r = 0.50047$, r is not significant so there is no relationship between the variables.

f. Alabama: 46407.576 Colorado: 62575.224

g. Alabama estimate is closer than Colorado estimate.

h. If the outlier is removed, there is a linear relationship.

i. There is one outlier (Hawaii).

j. rank 51: 75711.4; no

k.

Alabama	7	1819	22	52,423
Colorado	8	1876	38	104,100
Alaska	6	1959	51	656,424
Iowa	4	1846	29	56,276
Maryland	8	1788	7	12,407
Missouri	8	1821	24	69,709
New Jersey	9	1787	3	8,722
Ohio	4	1803	17	44,828
South Carolina	13	1788	8	32,008
Utah	4	1896	45	84,904
Wisconsin	9	1848	30	65,499

Table 12.39

l. $\hat{y} = -87065.3 + 7828.532x$

m. Alabama: 85,162.404; the prior estimate was closer. Alaska is an outlier.

n. yes, with the exception of Hawaii

81

a. Check student's solution.

b. yes

c. $\hat{y} = -266.8863 + 0.1656x$

d. 0.9448; Yes

e. 62.8233; 62.3265

f. yes

g. yes; (1987, 62.7)

h. 72.5937; no

i. slope = 0.1656.
    As the year increases by one, the percent of workers paid hourly rates tends to increase by 0.1656.

83

a.

Size (ounces)	Cost ($)	cents/oz
16	3.99	24.94
32	4.99	15.59
64	5.99	9.36
200	10.99	5.50

Table 12.40

b. Check student's solution.

c. There is a linear relationship for the sizes 16 through 64, but that linear trend does not continue to the 200-oz size.

d. $\hat{y} = 20.2368 - 0.0819x$

e. $r = -0.8086$

f. 40-oz: 16.96 cents/oz

g. 90-oz: 12.87 cents/oz

h. The relationship is not linear; the least squares line is not appropriate.

i. no outliers

j. No, you would be extrapolating. The 300-oz size is outside the range of x.

k. slope = –0.08194; for each additional ounce in size, the cost per ounce decreases by 0.082 cents.

85

a. Size is x, the independent variable, price is y, the dependent variable.

b. Check student's solution.

c. The relationship does not appear to be linear.

d. $\hat{y} = -745.252 + 54.75569x$

e. $r = 0.8944$, yes it is significant

f. 32-inch: $1006.93, 50-inch: $1992.53

g. No, the relationship does not appear to be linear. However, r is significant.

h. yes, the 60-inch TV

i. For each additional inch, the price increases by $54.76

87

a. Let rank be the independent variable and area be the dependent variable.

b. Check student's solution.

c. There appears to be a linear relationship, with one outlier.

d. $\hat{y}$ (area) = 24177.06 + 1010.478x

e. r = 0.50047, r is not significant so there is no relationship between the variables.

f. Alabama: 46407.576 Colorado: 62575.224

g. Alabama estimate is closer than Colorado estimate.

h. If the outlier is removed, there is a linear relationship.

i. There is one outlier (Hawaii).

j. rank 51: 75711.4; no

k.

Alabama	7	1819	22	52,423
Colorado	8	1876	38	104,100
Alaska	6	1959	51	656,424
Iowa	4	1846	29	56,276
Maryland	8	1788	7	12,407
Missouri	8	1821	24	69,709
New Jersey	9	1787	3	8,722
Ohio	4	1803	17	44,828
South Carolina	13	1788	8	32,008
Utah	4	1896	45	84,904
Wisconsin	9	1848	30	65,499

Table 12.41

l. $\hat{y}$ = –87065.3 + 7828.532x

m. Alabama: 85,162.404; the prior estimate was closer. Alaska is an outlier.

n. yes, with the exception of Hawaii

13 | F DISTRIBUTION AND ONE-WAY ANOVA

Figure 13.1 One-way ANOVA is used to measure information from several groups.

Introduction

Chapter Objectives

By the end of this chapter, the student should be able to:

- Interpret the F probability distribution as the number of groups and the sample size change.
- Discuss two uses for the F distribution: one-way ANOVA and the test of two variances.
- Conduct and interpret one-way ANOVA.
- Conduct and interpret hypothesis tests of two variances.

Many statistical applications in psychology, social science, business administration, and the natural sciences involve several groups. For example, an environmentalist is interested in knowing if the average amount of pollution varies in several bodies of water. A sociologist is interested in knowing if the amount of income a person earns varies according to his or her upbringing. A consumer looking for a new car might compare the average gas mileage of several models.

For hypothesis tests comparing averages between more than two groups, statisticians have developed a method called "Analysis of Variance" (abbreviated ANOVA). In this chapter, you will study the simplest form of ANOVA called single factor or one-way ANOVA. You will also study the F distribution, used for one-way ANOVA, and the test of two variances. This is just a very brief overview of one-way ANOVA. You will study this topic in much greater detail in future statistics courses. One-Way ANOVA, as it is presented here, relies heavily on a calculator or computer.

13.1 | One-Way ANOVA

The purpose of a one-way ANOVA test is to determine the existence of a statistically significant difference among several group means. The test actually uses **variances** to help determine if the means are equal or not. In order to perform a one-way ANOVA test, there are five basic **assumptions** to be fulfilled:

1. Each population from which a sample is taken is assumed to be normal.

2. All samples are randomly selected and independent.

3. The populations are assumed to have **equal standard deviations (or variances)**.

4. The factor is a categorical variable.

5. The response is a numerical variable.

The Null and Alternative Hypotheses

The null hypothesis is simply that all the group population means are the same. The alternative hypothesis is that at least one pair of means is different. For example, if there are k groups:

$H_0: \mu_1 = \mu_2 = \mu_3 = ... = \mu_k$

H_a: At least two of the group means $\mu_1, \mu_2, \mu_3, ..., \mu_k$ are not equal.

The graphs, a set of box plots representing the distribution of values with the group means indicated by a horizontal line through the box, help in the understanding of the hypothesis test. In the first graph (red box plots), $H_0: \mu_1 = \mu_2 = \mu_3$ and the three populations have the same distribution if the null hypothesis is true. The variance of the combined data is approximately the same as the variance of each of the populations.

If the null hypothesis is false, then the variance of the combined data is larger which is caused by the different means as shown in the second graph (green box plots).

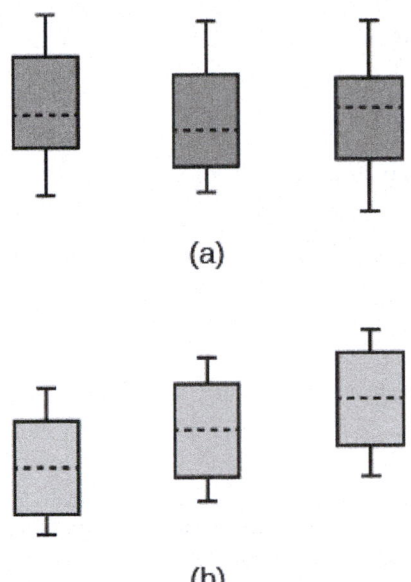

(a)

(b)

Figure 13.2 (a) H_0 is true. All means are the same; the differences are due to random variation. (b) H_0 is not true. All means are not the same; the differences are too large to be due to random variation.

13.2 | The F Distribution and the F-Ratio

The distribution used for the hypothesis test is a new one. It is called the **F distribution**, named after Sir Ronald Fisher, an English statistician. The F statistic is a ratio (a fraction). There are two sets of degrees of freedom; one for the numerator and one for the denominator.

For example, if F follows an F distribution and the number of degrees of freedom for the numerator is four, and the number of degrees of freedom for the denominator is ten, then $F \sim F_{4,10}$.

NOTE

The F distribution is derived from the Student's t-distribution. The values of the F distribution are squares of the corresponding values of the t-distribution. One-Way ANOVA expands the t-test for comparing more than two groups. The scope of that derivation is beyond the level of this course.

To calculate the **F ratio**, two estimates of the variance are made.

1. **Variance between samples:** An estimate of σ^2 that is the variance of the sample means multiplied by n (when the sample sizes are the same.). If the samples are different sizes, the variance between samples is weighted to account for the different sample sizes. The variance is also called **variation due to treatment or explained variation.**

2. **Variance within samples:** An estimate of σ^2 that is the average of the sample variances (also known as a pooled variance). When the sample sizes are different, the variance within samples is weighted. The variance is also called the **variation due to error or unexplained variation.**

- $SS_{between}$ = the **sum of squares** that represents the variation among the different samples
- SS_{within} = the sum of squares that represents the variation within samples that is due to chance.

To find a "sum of squares" means to add together squared quantities that, in some cases, may be weighted. We used sum of squares to calculate the sample variance and the sample standard deviation in **Descriptive Statistics**.

MS means " **mean square**." $MS_{between}$ is the variance between groups, and MS_{within} is the variance within groups.

Calculation of Sum of Squares and Mean Square

- k = the number of different groups

- n_j = the size of the jth group

- s_j = the sum of the values in the jth group

- n = total number of all the values combined (total sample size: $\sum n_j$)

- x = one value: $\sum x = \sum s_j$

- Sum of squares of all values from every group combined: $\sum x^2$

- Between group variability: $SS_{total} = \sum x^2 - \dfrac{\left(\sum x^2\right)}{n}$

- Total sum of squares: $\sum x^2 - \dfrac{\left(\sum x\right)^2}{n}$

- Explained variation: sum of squares representing variation among the different samples: $SS_{between} = \sum \left[\dfrac{(sj)^2}{n_j}\right] - \dfrac{\left(\sum s_j\right)^2}{n}$

- Unexplained variation: sum of squares representing variation within samples due to chance: $SS_{within} = SS_{total} - SS_{between}$

- df's for different groups (df's for the numerator): $df = k - 1$

- Equation for errors within samples (df's for the denominator): $df_{within} = n - k$

- Mean square (variance estimate) explained by the different groups: $MS_{between} = \dfrac{SS_{between}}{df_{between}}$

- Mean square (variance estimate) that is due to chance (unexplained): $MS_{within} = \dfrac{SS_{within}}{df_{within}}$

$MS_{between}$ and MS_{within} can be written as follows:

- $MS_{between} = \dfrac{SS_{between}}{df_{between}} = \dfrac{SS_{between}}{k - 1}$

- $MS_{within} = \dfrac{SS_{within}}{df_{within}} = \dfrac{SS_{within}}{n - k}$

The one-way ANOVA test depends on the fact that $MS_{between}$ can be influenced by population differences among means of the several groups. Since MS_{within} compares values of each group to its own group mean, the fact that group means might be different does not affect MS_{within}.

The null hypothesis says that all groups are samples from populations having the same normal distribution. The alternate hypothesis says that at least two of the sample groups come from populations with different normal distributions. If the null hypothesis is true, $MS_{between}$ and MS_{within} should both estimate the same value.

NOTE

The null hypothesis says that all the group population means are equal. The hypothesis of equal means implies that the populations have the same normal distribution, because it is assumed that the populations are normal and that they have equal variances.

F-Ratio or *F* Statistic

$$F = \dfrac{MS_{between}}{MS_{within}}$$

If $MS_{between}$ and MS_{within} estimate the same value (following the belief that H_0 is true), then the *F*-ratio should be approximately equal to one. Mostly, just sampling errors would contribute to variations away from one. As it turns out, $MS_{between}$ consists of the population variance plus a variance produced from the differences between the samples. MS_{within}

is an estimate of the population variance. Since variances are always positive, if the null hypothesis is false, MS_{between} will generally be larger than MS_{within}. Then the F-ratio will be larger than one. However, if the population effect is small, it is not unlikely that MS_{within} will be larger in a given sample.

The foregoing calculations were done with groups of different sizes. If the groups are the same size, the calculations simplify somewhat and the F-ratio can be written as:

_F_-Ratio Formula when the groups are the same size

$$F = \frac{n \cdot s_{\bar{x}}^{2}}{s^{2}_{\text{pooled}}}$$

where ...

- n = the sample size
- $df_{\text{numerator}} = k - 1$
- $df_{\text{denominator}} = n - k$
- s^{2} pooled = the mean of the sample variances (pooled variance)
- $s_{\bar{x}}^{2}$ = the variance of the sample means

Data are typically put into a table for easy viewing. One-Way ANOVA results are often displayed in this manner by computer software.

Source of Variation	Sum of Squares (_SS_)	Degrees of Freedom (_df_)	Mean Square (_MS_)	_F_
Factor (Between)	SS(Factor)	$k - 1$	MS(Factor) = SS(Factor)/($k - 1$)	F = MS(Factor)/MS(Error)
Error (Within)	SS(Error)	$n - k$	MS(Error) = SS(Error)/($n - k$)	
Total	SS(Total)	$n - 1$		

Table 13.1

Example 13.1

Three different diet plans are to be tested for mean weight loss. The entries in the table are the weight losses for the different plans. The one-way ANOVA results are shown in **Table 13.2**.

Plan 1: $n_1 = 4$	Plan 2: $n_2 = 3$	Plan 3: $n_3 = 3$
5	3.5	8
4.5	7	4
4		3.5
3	4.5	

Table 13.2

$s_1 = 16.5$, $s_2 = 15$, $s_3 = 15.7$

Following are the calculations needed to fill in the one-way ANOVA table. The table is used to conduct a hypothesis test.

$$SS(between) = \sum \left[\frac{(s_j)^2}{n_j} \right] - \frac{\left(\sum s_j \right)^2}{n}$$

$$= \frac{s_1^2}{4} + \frac{s_2^2}{3} + \frac{s_3^2}{3} - \frac{(s_1 + s_2 + s_3)^2}{10}$$

where $n_1 = 4$, $n_2 = 3$, $n_3 = 3$ and $n = n_1 + n_2 + n_3 = 10$

$$= \frac{(16.5)^2}{4} + \frac{(15)^2}{3} + \frac{(5.5)^2}{3} - \frac{(16.5 + 15 + 15.5)^2}{10}$$

$$SS(between) = 2.2458$$

$$S(total) = \sum x^2 - \frac{\left(\sum x \right)^2}{n}$$

$$= \left(5^2 + 4.5^2 + 4^2 + 3^2 + 3.5^2 + 7^2 + 4.5^2 + 8^2 + 4^2 + 3.5^2 \right)$$

$$- \frac{(5 + 4.5 + 4 + 3 + 3.5 + 7 + 4.5 + 8 + 4 + 3.5)^2}{10}$$

$$= 244 - \frac{47^2}{10} = 244 - 220.9$$

$$SS(total) = 23.1$$

$$SS(within) = SS(total) - SS(between)$$

$$= 23.1 - 2.2458$$

$$SS(within) = 20.8542$$

Using the TI-83, 83+, 84, 84+ Calculator

One-Way ANOVA Table: The formulas for SS(Total), SS(Factor) = SS(Between) and SS(Error) = SS(Within) as shown previously. The same information is provided by the TI calculator hypothesis test function ANOVA in STAT TESTS (syntax is ANOVA(L1, L2, L3) where L1, L2, L3 have the data from Plan 1, Plan 2, Plan 3 respectively).

Source of Variation	Sum of Squares (SS)	Degrees of Freedom (df)	Mean Square (MS)	F
Factor (Between)	SS(Factor) = SS(Between) = 2.2458	$k - 1$ = 3 groups − 1 = 2	MS(Factor) = SS(Factor)/(k − 1) = 2.2458/2 = 1.1229	$F = MS$(Factor)/MS(Error) = 1.1229/2.9792 = 0.3769
Error (Within)	SS(Error) = SS(Within) = 20.8542	$n - k$ = 10 total data − 3 groups = 7	MS(Error) = SS(Error)/($n - k$) = 20.8542/7 = 2.9792	
Total	SS(Total) = 2.2458 + 20.8542 = 23.1	$n - 1$ = 10 total data − 1 = 9		

Table 13.3

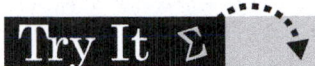

☞ **13.1** As part of an experiment to see how different types of soil cover would affect slicing tomato production, Marist College students grew tomato plants under different soil cover conditions. Groups of three plants each had one of the following treatments

- bare soil

- a commercial ground cover

- black plastic

- straw

- compost

All plants grew under the same conditions and were the same variety. Students recorded the weight (in grams) of tomatoes produced by each of the $n = 15$ plants:

Bare: $n_1 = 3$	Ground Cover: $n_2 = 3$	Plastic: $n_3 = 3$	Straw: $n_4 = 3$	Compost: $n_5 = 3$
2,625	5,348	6,583	7,285	6,277
2,997	5,682	8,560	6,897	7,818
4,915	5,482	3,830	9,230	8,677

Table 13.4

Create the one-way ANOVA table.

The one-way ANOVA hypothesis test is always right-tailed because larger F-values are way out in the right tail of the F-distribution curve and tend to make us reject H_0.

Notation

The notation for the F distribution is $F \sim F_{df(num),df(denom)}$

where $df(num) = df_{between}$ and $df(denom) = df_{within}$

The mean for the F distribution is $\mu = \dfrac{df(num)}{df(denom) - 1}$

13.3 | Facts About the F Distribution

Here are some facts about the F distribution.

1. The curve is not symmetrical but skewed to the right.

2. There is a different curve for each set of dfs.

3. The F statistic is greater than or equal to zero.

4. As the degrees of freedom for the numerator and for the denominator get larger, the curve approximates the normal.

5. Other uses for the F distribution include comparing two variances and two-way Analysis of Variance. Two-Way Analysis is beyond the scope of this chapter.

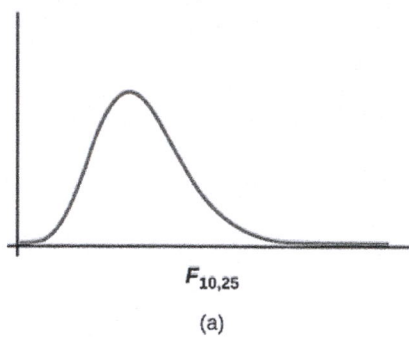

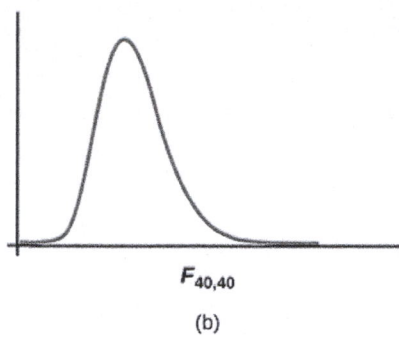

$F_{10,25}$

(a)

$F_{40,40}$

(b)

Figure 13.3

Example 13.2

Let's return to the slicing tomato exercise in **Try It**. The means of the tomato yields under the five mulching conditions are represented by μ_1, μ_2, μ_3, μ_4, μ_5. We will conduct a hypothesis test to determine if all means are the same or at least one is different. Using a significance level of 5%, test the null hypothesis that there is no difference in mean yields among the five groups against the alternative hypothesis that at least one mean is different from the rest.

Solution 13.2

The null and alternative hypotheses are:

H_0: $\mu_1 = \mu_2 = \mu_3 = \mu_4 = \mu_5$

H_a: $\mu_i \neq \mu_j$ some $i \neq j$

The one-way ANOVA results are shown in **Table 13.4**

Source of Variation	Sum of Squares (SS)	Degrees of Freedom (df)	Mean Square (MS)	F
Factor (Between)	36,648,561	$5 - 1 = 4$	$\dfrac{36{,}648{,}561}{4} = 9{,}162{,}140$	$\dfrac{9{,}162{,}140}{2{,}044{,}672.6} = 4.4810$
Error (Within)	20,446,726	$15 - 5 = 10$	$\dfrac{20{,}446{,}726}{10} = 2{,}044{,}672.6$	
Total	57,095,287	$15 - 1 = 14$		

Table 13.5

Distribution for the test: $F_{4,10}$

$df(num) = 5 - 1 = 4$

$df(denom) = 15 - 5 = 10$

Test statistic: $F = 4.4810$

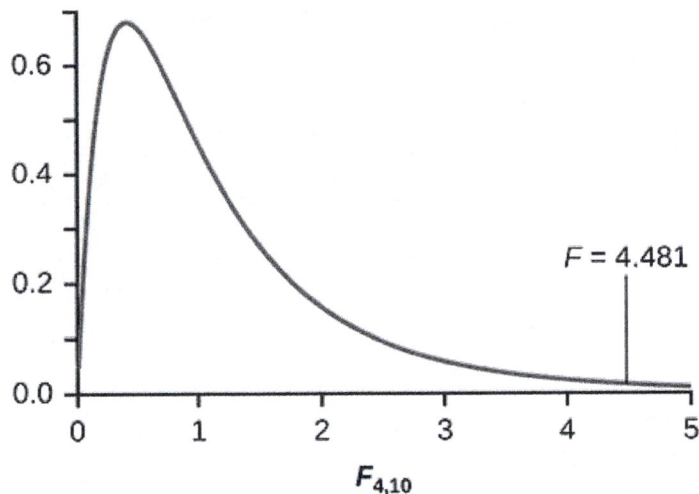

Figure 13.4

Probability Statement: p-value = $P(F > 4.481) = 0.0248$.

Compare α and the p-value: $\alpha = 0.05$, p-value = 0.0248

Make a decision: Since $\alpha > p$-value, we reject H_0.

Conclusion: At the 5% significance level, we have reasonably strong evidence that differences in mean yields for slicing tomato plants grown under different mulching conditions are unlikely to be due to chance alone. We may conclude that at least some of mulches led to different mean yields.

 Using the TI-83, 83+, 84, 84+ Calculator

To find these results on the calculator:

Press STAT. Press 1:EDIT. Put the data into the lists L_1, L_2, L_3, L_4, L_5.

Press STAT, and arrow over to TESTS, and arrow down to ANOVA. Press ENTER, and then enter L_1, L_2, L_3, L_4, L_5). Press ENTER. You will see that the values in the foregoing ANOVA table are easily produced by the calculator, including the test statistic and the p-value of the test.

The calculator displays:
$F = 4.4810$
$p = 0.0248$ (p-value)
Factor
$df = 4$
$SS = 36648560.9$
$MS = 9162140.23$
Error
$df = 10$
$SS = 20446726$
$MS = 2044672.6$

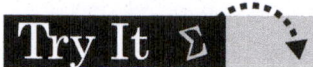

13.2 MRSA, or *Staphylococcus aureus*, can cause a serious bacterial infections in hospital patients. **Table 13.6** shows various colony counts from different patients who may or may not have MRSA.

Conc = 0.6	Conc = 0.8	Conc = 1.0	Conc = 1.2	Conc = 1.4
9	16	22	30	27
66	93	147	199	168
98	82	120	148	132

Table 13.6

Plot of the data for the different concentrations:

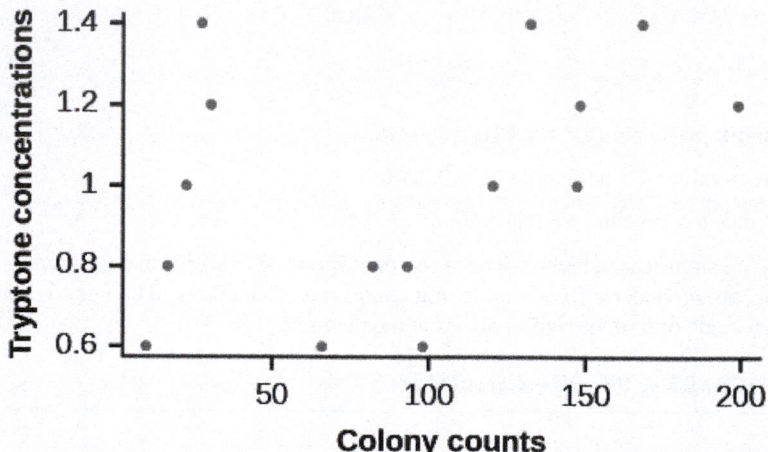

Figure 13.5

Test whether the mean number of colonies are the same or are different. Construct the ANOVA table (by hand or by using a TI-83, 83+, or 84+ calculator), find the *p*-value, and state your conclusion. Use a 5% significance level.

Example 13.3

Four sororities took a random sample of sisters regarding their grade means for the past term. The results are shown in **Table 13.7**.

Sorority 1	Sorority 2	Sorority 3	Sorority 4
2.17	2.63	2.63	3.79
1.85	1.77	3.78	3.45
2.83	3.25	4.00	3.08
1.69	1.86	2.55	2.26
3.33	2.21	2.45	3.18

Table 13.7 MEAN GRADES FOR FOUR SORORITIES

Using a significance level of 1%, is there a difference in mean grades among the sororities?

Solution 13.3

Let μ_1, μ_2, μ_3, μ_4 be the population means of the sororities. Remember that the null hypothesis claims that the sorority groups are from the same normal distribution. The alternate hypothesis says that at least two of the sorority groups come from populations with different normal distributions. Notice that the four sample sizes are each five.

NOTE

This is an example of a **balanced design**, because each factor (i.e., sorority) has the same number of observations.

H_0: $\mu_1 = \mu_2 = \mu_3 = \mu_4$

H_a: Not all of the means μ_1, μ_2, μ_3, μ_4 are equal.

Distribution for the test: $F_{3,16}$

where $k = 4$ groups and $n = 20$ samples in total

$df(num) = k - 1 = 4 - 1 = 3$

$df(denom) = n - k = 20 - 4 = 16$

Calculate the test statistic: $F = 2.23$

Graph:

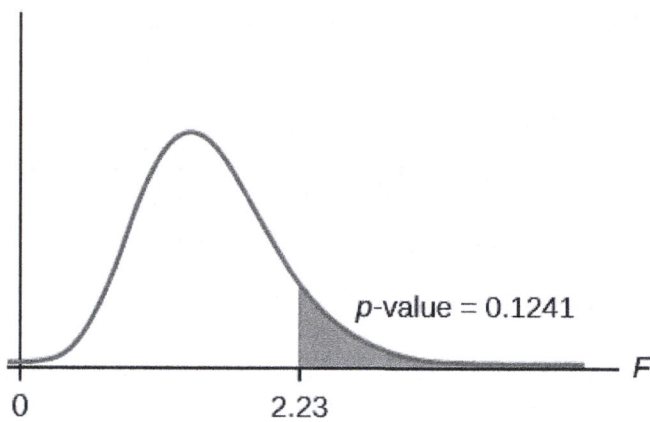

Figure 13.6

Probability statement: p-value = $P(F > 2.23) = 0.1241$

Compare α and the p-value: $\alpha = 0.01$
p-value = 0.1241
$\alpha < p$-value

Make a decision: Since $\alpha < p$-value, you cannot reject H_0.

Conclusion: There is not sufficient evidence to conclude that there is a difference among the mean grades for the sororities.

 Using the TI-83, 83+, 84, 84+ Calculator

Put the data into lists L_1, L_2, L_3, and L_4. Press **STAT** and arrow over to **TESTS**. Arrow down to **F:ANOVA**. Press **ENTER** and Enter (L1,L2,L3,L4).

The calculator displays the F statistic, the p-value and the values for the one-way ANOVA table:
$F = 2.2303$
$p = 0.1241$ (p-value)
Factor
$df = 3$
$SS = 2.88732$
$MS = 0.96244$
Error
$df = 16$
$SS = 6.9044$
$MS = 0.431525$

 **Try It** Σ

13.3 Four sports teams took a random sample of players regarding their GPAs for the last year. The results are shown in **Table 13.8**.

Basketball	Baseball	Hockey	Lacrosse
3.6	2.1	4.0	2.0
2.9	2.6	2.0	3.6
2.5	3.9	2.6	3.9
3.3	3.1	3.2	2.7
3.8	3.4	3.2	2.5

Table 13.8 GPAs FOR FOUR SPORTS TEAMS

Use a significance level of 5%, and determine if there is a difference in GPA among the teams.

Example 13.4

A fourth grade class is studying the environment. One of the assignments is to grow bean plants in different soils. Tommy chose to grow his bean plants in soil found outside his classroom mixed with dryer lint. Tara chose to grow her bean plants in potting soil bought at the local nursery. Nick chose to grow his bean plants in soil from his mother's garden. No chemicals were used on the plants, only water. They were grown inside the classroom next to a large window. Each child grew five plants. At the end of the growing period, each plant was measured, producing the data (in inches) in **Table 13.9**.

Tommy's Plants	Tara's Plants	Nick's Plants
24	25	23
21	31	27
23	23	22
30	20	30
23	28	20

Table 13.9

Does it appear that the three media in which the bean plants were grown produce the same mean height? Test at a 3% level of significance.

Solution 13.4

This time, we will perform the calculations that lead to the F' statistic. Notice that each group has the same number of plants, so we will use the formula $F' = \dfrac{n \cdot s_{\bar{x}}^2}{s^2_{pooled}}$.

First, calculate the sample mean and sample variance of each group.

	Tommy's Plants	Tara's Plants	Nick's Plants
Sample Mean	24.2	25.4	24.4
Sample Variance	11.7	18.3	16.3

Table 13.10

Next, calculate the variance of the three group means (Calculate the variance of 24.2, 25.4, and 24.4). **Variance of the group means = 0.413 =** $s_{\bar{x}}^2$

Then $MS_{between} = ns_{\bar{x}}^2 = (5)(0.413)$ where $n = 5$ is the sample size (number of plants each child grew).

Calculate the mean of the three sample variances (Calculate the mean of 11.7, 18.3, and 16.3). **Mean of the sample variances = 15.433 =** s^2 **pooled**

Then $MS_{within} = s^2_{pooled} = 15.433$.

The F statistic (or F ratio) is $F = \dfrac{MS_{between}}{MS_{within}} = \dfrac{ns_{\bar{x}}^2}{s^2_{pooled}} = \dfrac{(5)(0.413)}{15.433} = 0.134$

The dfs for the numerator = the number of groups − 1 = 3 − 1 = 2.

The *dfs* for the denominator = the total number of samples – the number of groups = $15 - 3 = 12$

The distribution for the test is $F_{2,12}$ and the F statistic is $F = 0.134$

The p-value is $P(F > 0.134) = 0.8759$.

Decision: Since $\alpha = 0.03$ and the p-value = 0.8759, do not reject H_0. (Why?)

Conclusion: With a 3% level of significance, from the sample data, the evidence is not sufficient to conclude that the mean heights of the bean plants are different.

 Using the TI-83, 83+, 84, 84+ Calculator

To calculate the p-value:

*Press **2nd DISTR**

*Arrow down to **Fcdf**(and press **ENTER**.

*Enter 0.134, **E99**, 2, 12)

*Press **ENTER**

The p-value is 0.8759.

Try It Σ

13.4 Another fourth grader also grew bean plants, but this time in a jelly-like mass. The heights were (in inches) 24, 28, 25, 30, and 32. Do a one-way ANOVA test on the four groups. Are the heights of the bean plants different? Use the same method as shown in **Example 13.4**.

Collaborative Exercise

From the class, create four groups of the same size as follows: men under 22, men at least 22, women under 22, women at least 22. Have each member of each group record the number of states in the United States he or she has visited. Run an ANOVA test to determine if the average number of states visited in the four groups are the same. Test at a 1% level of significance. Use one of the solution sheets in **Appendix E**.

13.4 | Test of Two Variances

Another of the uses of the F distribution is testing two variances. It is often desirable to compare two variances rather than two averages. For instance, college administrators would like two college professors grading exams to have the same variation in their grading. In order for a lid to fit a container, the variation in the lid and the container should be the same. A supermarket might be interested in the variability of check-out times for two checkers.

In order to perform a F test of two variances, it is important that the following are true:

1. The populations from which the two samples are drawn are normally distributed.

2. The two populations are independent of each other.

Unlike most other tests in this book, the F test for equality of two variances is very sensitive to deviations from normality. If the two distributions are not normal, the test can give higher p-values than it should, or lower ones, in ways that are unpredictable. Many texts suggest that students not use this test at all, but in the interest of completeness we include it here.

Suppose we sample randomly from two independent normal populations. Let σ_1^2 and σ_2^2 be the population variances and s_1^2 and s_2^2 be the sample variances. Let the sample sizes be n_1 and n_2. Since we are interested in comparing the two sample variances, we use the F ratio:

$$F = \frac{\left[\frac{(s_1)^2}{(\sigma_1)^2}\right]}{\left[\frac{(s_2)^2}{(\sigma_2)^2}\right]}$$

F has the distribution $F \sim F(n_1 - 1, n_2 - 1)$

where $n_1 - 1$ are the degrees of freedom for the numerator and $n_2 - 1$ are the degrees of freedom for the denominator.

If the null hypothesis is $\sigma_1^2 = \sigma_2^2$, then the F Ratio becomes $F = \dfrac{\left[\frac{(s_1)^2}{(\sigma_1)^2}\right]}{\left[\frac{(s_2)^2}{(\sigma_2)^2}\right]} = \dfrac{(s_1)^2}{(s_2)^2}$.

NOTE

The F ratio could also be $\dfrac{(s_2)^2}{(s_1)^2}$. It depends on H_a and on which sample variance is larger.

If the two populations have equal variances, then s_1^2 and s_2^2 are close in value and $F = \dfrac{(s_1)^2}{(s_2)^2}$ is close to one. But if the two population variances are very different, s_1^2 and s_2^2 tend to be very different, too. Choosing s_1^2 as the larger sample variance causes the ratio $\dfrac{(s_1)^2}{(s_2)^2}$ to be greater than one. If s_1^2 and s_2^2 are far apart, then $F = \dfrac{(s_1)^2}{(s_2)^2}$ is a large number.

Therefore, if F is close to one, the evidence favors the null hypothesis (the two population variances are equal). But if F is much larger than one, then the evidence is against the null hypothesis. **A test of two variances may be left, right, or two-tailed.**

Example 13.5

Two college instructors are interested in whether or not there is any variation in the way they grade math exams. They each grade the same set of 30 exams. The first instructor's grades have a variance of 52.3. The second instructor's grades have a variance of 89.9. Test the claim that the first instructor's variance is smaller. (In most colleges, it is desirable for the variances of exam grades to be nearly the same among instructors.) The level of significance is 10%.

Solution 13.5

Let 1 and 2 be the subscripts that indicate the first and second instructor, respectively.

$n_1 = n_2 = 30$.

H_0: $\sigma_1^2 = \sigma_2^2$ and H_a: $\sigma_1^2 < \sigma_2^2$

Calculate the test statistic: By the null hypothesis $(\sigma_1^2 = \sigma_2^2)$, the F statistic is:

$$F = \frac{\left[\frac{(s_1)^2}{(\sigma_1)^2}\right]}{\left[\frac{(s_2)^2}{(\sigma_2)^2}\right]} = \frac{(s_1)^2}{(s_2)^2} = \frac{52.3}{89.9} = 0.5818$$

Distribution for the test: $F_{29,29}$ where $n_1 - 1 = 29$ and $n_2 - 1 = 29$.

Graph: This test is left tailed.

Draw the graph labeling and shading appropriately.

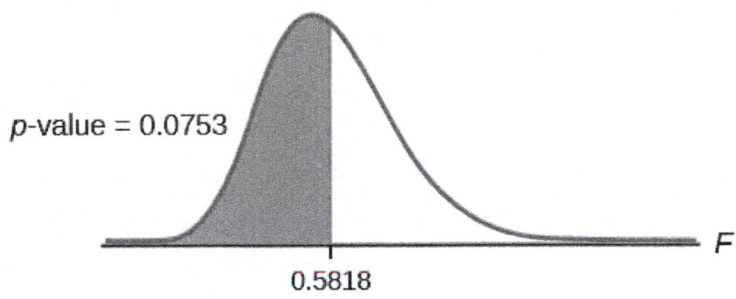

p-value = 0.0753

0.5818

F

Figure 13.7

Probability statement: p-value = $P(F < 0.5818) = 0.0753$

Compare α and the p-value: $\alpha = 0.10$ $\alpha > p$-value.

Make a decision: Since $\alpha > p$-value, reject H_0.

Conclusion: With a 10% level of significance, from the data, there is sufficient evidence to conclude that the variance in grades for the first instructor is smaller.

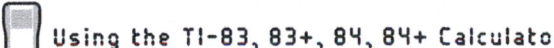 **Using the TI-83, 83+, 84, 84+ Calculator**

Press **STAT** and arrow over to **TESTS**. Arrow down to **D:2-SampFTest**. Press **ENTER**. Arrow to **Stats** and press **ENTER**. For **Sx1, n1, Sx2,** and **n2**, enter $\sqrt{(52.3)}$, **30**, $\sqrt{(89.9)}$, and **30**. Press **ENTER** after each. Arrow to **σ1:** and **<σ2**. Press **ENTER**. Arrow down to **Calculate** and press **ENTER**. $F = 0.5818$ and p-value = 0.0753. Do the procedure again and try **Draw** instead of **Calculate**.

13.5 The New York Choral Society divides male singers up into four categories from highest voices to lowest: Tenor1, Tenor2, Bass1, Bass2. In the table are heights of the men in the Tenor1 and Bass2 groups. One suspects that taller men will have lower voices, and that the variance of height may go up with the lower voices as well. Do we have good evidence that the variance of the heights of singers in each of these two groups (Tenor1 and Bass2) are different?

Tenor1	Bass2	Tenor 1	Bass 2	Tenor 1	Bass 2
69	72	67	72	68	67
72	75	70	74	67	70
71	67	65	70	64	70
66	75	72	66		69
76	74	70	68		72
74	72	68	75		71
71	72	64	68		74
66	74	73	70		75
68	72	66	72		

Table 13.11

13.5 | Lab: One-Way ANOVA

Stats Lab

13.1 One-Way ANOVA

Class Time:

Names:

Student Learning Outcome

- The student will conduct a simple one-way ANOVA test involving three variables.

Collect the Data

1. Record the price per pound of eight fruits, eight vegetables, and eight breads in your local supermarket.

Fruits	Vegetables	Breads

Table 13.12

2. Explain how you could try to collect the data randomly.

Analyze the Data and Conduct a Hypothesis Test

1. Compute the following:

    a. Fruit:

      i. $\bar{x}$ = _____

      ii. s_x = _____

      iii. n = _____

    b. Vegetables:

      i. $\bar{x}$ = _____

      ii. s_x = _____

      iii. n = _____

    c. Bread:

      i. $\bar{x}$ = _____

      ii. s_x = _____

      iii. n = _____

2. Find the following:

      a. *df(num)* = _____

      b. *df(denom)* = _____

3. State the approximate distribution for the test.

4. Test statistic: F = _____

5. Sketch a graph of this situation. CLEARLY, label and scale the horizontal axis and shade the region(s) corresponding to the *p*-value.

6. *p*-value = _____

7. Test at $\alpha = 0.05$. State your decision and conclusion.

8.   a. Decision: Why did you make this decision?

      b. Conclusion (write a complete sentence).

      c. Based on the results of your study, is there a need to investigate any of the food groups' prices? Why or why not?

KEY TERMS

Analysis of Variance also referred to as ANOVA, is a method of testing whether or not the means of three or more populations are equal. The method is applicable if:

- all populations of interest are normally distributed.
- the populations have equal standard deviations.
- samples (not necessarily of the same size) are randomly and independently selected from each population.

The test statistic for analysis of variance is the F-ratio.

One-Way ANOVA a method of testing whether or not the means of three or more populations are equal; the method is applicable if:

- all populations of interest are normally distributed.
- the populations have equal standard deviations.
- samples (not necessarily of the same size) are randomly and independently selected from each population.

The test statistic for analysis of variance is the F-ratio.

Variance mean of the squared deviations from the mean; the square of the standard deviation. For a set of data, a deviation can be represented as $x - \bar{x}$ where x is a value of the data and $\bar{x}$ is the sample mean. The sample variance is equal to the sum of the squares of the deviations divided by the difference of the sample size and one.

CHAPTER REVIEW

13.1 One-Way ANOVA

Analysis of variance extends the comparison of two groups to several, each a level of a categorical variable (factor). Samples from each group are independent, and must be randomly selected from normal populations with equal variances. We test the null hypothesis of equal means of the response in every group versus the alternative hypothesis of one or more group means being different from the others. A one-way ANOVA hypothesis test determines if several population means are equal. The distribution for the test is the F distribution with two different degrees of freedom.

Assumptions:

1. Each population from which a sample is taken is assumed to be normal.
2. All samples are randomly selected and independent.
3. The populations are assumed to have equal standard deviations (or variances).

13.2 The F Distribution and the F-Ratio

Analysis of variance compares the means of a response variable for several groups. ANOVA compares the variation within each group to the variation of the mean of each group. The ratio of these two is the F statistic from an F distribution with (number of groups $-$ 1) as the numerator degrees of freedom and (number of observations $-$ number of groups) as the denominator degrees of freedom. These statistics are summarized in the ANOVA table.

13.3 Facts About the F Distribution

The graph of the F distribution is always positive and skewed right, though the shape can be mounded or exponential depending on the combination of numerator and denominator degrees of freedom. The F statistic is the ratio of a measure of the variation in the group means to a similar measure of the variation within the groups. If the null hypothesis is correct, then the numerator should be small compared to the denominator. A small F statistic will result, and the area under the F curve to the right will be large, representing a large p-value. When the null hypothesis of equal group means is incorrect, then the numerator should be large compared to the denominator, giving a large F statistic and a small area (small p-value) to the right of the statistic under the F curve.

When the data have unequal group sizes (unbalanced data), then techniques from **Section 13.2** need to be used for hand calculations. In the case of balanced data (the groups are the same size) however, simplified calculations based on group means and variances may be used. In practice, of course, software is usually employed in the analysis. As in any analysis, graphs of various sorts should be used in conjunction with numerical techniques. Always look of your data!

13.4 Test of Two Variances

The F test for the equality of two variances rests heavily on the assumption of normal distributions. The test is unreliable if this assumption is not met. If both distributions are normal, then the ratio of the two sample variances is distributed as an F statistic, with numerator and denominator degrees of freedom that are one less than the samples sizes of the corresponding two groups. A **test of two variances** hypothesis test determines if two variances are the same. The distribution for the hypothesis test is the F distribution with two different degrees of freedom.

Assumptions:

1. The populations from which the two samples are drawn are normally distributed.
2. The two populations are independent of each other.

FORMULA REVIEW

13.2 The F Distribution and the F-Ratio

$$SS_{between} = \sum \left[\frac{(s_j)^2}{n_j} \right] - \frac{\left(\sum s_j \right)^2}{n}$$

$$SS_{total} = \sum x^2 - \frac{\left(\sum x \right)^2}{n}$$

$$SS_{within} = SS_{total} - SS_{between}$$

$$df_{between} = df(num) = k - 1$$

$$df_{within} = df(denom) = n - k$$

$$MS_{between} = \frac{SS_{between}}{df_{between}}$$

$$MS_{within} = \frac{SS_{within}}{df_{within}}$$

$$F = \frac{MS_{between}}{MS_{within}}$$

F ratio when the groups are the same size: $F = \dfrac{ns_{\bar{x}}^2}{s^2_{pooled}}$

Mean of the F distribution: $\mu = \dfrac{df(num)}{df(denom) - 1}$

where:

- k = the number of groups
- n_j = the size of the jth group
- s_j = the sum of the values in the jth group
- n = the total number of all values (observations) combined
- x = one value (one observation) from the data
- $s_{\bar{x}}^2$ = the variance of the sample means
- s^2_{pooled} = the mean of the sample variances (pooled variance)

13.4 Test of Two Variances

F has the distribution $F \sim F(n_1 - 1, n_2 - 1)$

$$F = \frac{\frac{s_1^2}{\sigma_1^2}}{\frac{s_2^2}{\sigma_2^2}}$$

If $\sigma_1 = \sigma_2$, then $F = \dfrac{s_1^2}{s_2^2}$

PRACTICE

13.1 One-Way ANOVA

Use the following information to answer the next five exercises. There are five basic assumptions that must be fulfilled in order to perform a one-way ANOVA test. What are they?

1. Write one assumption.

2. Write another assumption.

3. Write a third assumption.

4. Write a fourth assumption.

5. Write the final assumption.

6. State the null hypothesis for a one-way ANOVA test if there are four groups.

7. State the alternative hypothesis for a one-way ANOVA test if there are three groups.

8. When do you use an ANOVA test?

13.2 The F Distribution and the F-Ratio

Use the following information to answer the next eight exercises. Groups of men from three different areas of the country are to be tested for mean weight. The entries in the table are the weights for the different groups. The one-way ANOVA results are shown in **Table 13.13**.

Group 1	Group 2	Group 3
216	202	170
198	213	165
240	284	182
187	228	197
176	210	201

Table 13.13

9. What is the Sum of Squares Factor?

10. What is the Sum of Squares Error?

11. What is the *df* for the numerator?

12. What is the *df* for the denominator?

13. What is the Mean Square Factor?

14. What is the Mean Square Error?

15. What is the *F* statistic?

Use the following information to answer the next eight exercises. Girls from four different soccer teams are to be tested for mean goals scored per game. The entries in the table are the goals per game for the different teams. The one-way ANOVA results are shown in **Table 13.14**.

Team 1	Team 2	Team 3	Team 4
1	2	0	3
2	3	1	4
0	2	1	4
3	4	0	3
2	4	0	2

Table 13.14

16. What is $SS_{between}$?

17. What is the df for the numerator?

18. What is $MS_{between}$?

19. What is SS_{within}?

20. What is the df for the denominator?

21. What is MS_{within}?

22. What is the F statistic?

23. Judging by the F statistic, do you think it is likely or unlikely that you will reject the null hypothesis?

13.3 Facts About the F Distribution

24. An F statistic can have what values?

25. What happens to the curves as the degrees of freedom for the numerator and the denominator get larger?

Use the following information to answer the next seven exercise. Four basketball teams took a random sample of players regarding how high each player can jump (in inches). The results are shown in **Table 13.15**.

Team 1	Team 2	Team 3	Team 4	Team 5
36	32	48	38	41
42	35	50	44	39
51	38	39	46	40

Table 13.15

26. What is the $df(num)$?

27. What is the $df(denom)$?

28. What are the Sum of Squares and Mean Squares Factors?

29. What are the Sum of Squares and Mean Squares Errors?

30. What is the F statistic?

31. What is the p-value?

32. At the 5% significance level, is there a difference in the mean jump heights among the teams?

Use the following information to answer the next seven exercises. A video game developer is testing a new game on three different groups. Each group represents a different target market for the game. The developer collects scores from a random sample from each group. The results are shown in **Table 13.16**

Group A	Group B	Group C
101	151	101
108	149	109
98	160	198
107	112	186
111	126	160

Table 13.16

33. What is the $df(num)$?

34. What is the *df(denom)*?

35. What are the $SS_{between}$ and $MS_{between}$?

36. What are the SS_{within} and MS_{within}?

37. What is the F Statistic?

38. What is the *p*-value?

39. At the 10% significance level, are the scores among the different groups different?

Use the following information to answer the next three exercises. Suppose a group is interested in determining whether teenagers obtain their drivers licenses at approximately the same average age across the country. Suppose that the following data are randomly collected from five teenagers in each region of the country. The numbers represent the age at which teenagers obtained their drivers licenses.

	Northeast	South	West	Central	East
	16.3	16.9	16.4	16.2	17.1
	16.1	16.5	16.5	16.6	17.2
	16.4	16.4	16.6	16.5	16.6
	16.5	16.2	16.1	16.4	16.8
$\bar{x} =$	_____	_____	_____	_____	_____
$s^2 =$	_____	_____	_____	_____	_____

Table 13.17

Enter the data into your calculator or computer.

40. *p*-value = _____

State the decisions and conclusions (in complete sentences) for the following preconceived levels of α.

41. $\alpha = 0.05$

a. Decision: _____

b. Conclusion: _____

42. $\alpha = 0.01$

a. Decision: _____

b. Conclusion: _____

13.4 Test of Two Variances

Use the following information to answer the next two exercises. There are two assumptions that must be true in order to perform an F test of two variances.

43. Name one assumption that must be true.

44. What is the other assumption that must be true?

Use the following information to answer the next five exercises. Two coworkers commute from the same building. They are interested in whether or not there is any variation in the time it takes them to drive to work. They each record their times for 20 commutes. The first worker's times have a variance of 12.1. The second worker's times have a variance of 16.9. The first worker thinks that he is more consistent with his commute times and that his commute time is shorter. Test the claim at the 10% level.

45. State the null and alternative hypotheses.

46. What is s_1 in this problem?

47. What is s_2 in this problem?

48. What is n?

49. What is the F statistic?

50. What is the p-value?

51. Is the claim accurate?

Use the following information to answer the next four exercises. Two students are interested in whether or not there is variation in their test scores for math class. There are 15 total math tests they have taken so far. The first student's grades have a standard deviation of 38.1. The second student's grades have a standard deviation of 22.5. The second student thinks his scores are lower.

52. State the null and alternative hypotheses.

53. What is the F Statistic?

54. What is the p-value?

55. At the 5% significance level, do we reject the null hypothesis?

Use the following information to answer the next three exercises. Two cyclists are comparing the variances of their overall paces going uphill. Each cyclist records his or her speeds going up 35 hills. The first cyclist has a variance of 23.8 and the second cyclist has a variance of 32.1. The cyclists want to see if their variances are the same or different.

56. State the null and alternative hypotheses.

57. What is the F Statistic?

58. At the 5% significance level, what can we say about the cyclists' variances?

HOMEWORK

13.1 One-Way ANOVA

59. Three different traffic routes are tested for mean driving time. The entries in the table are the driving times in minutes on the three different routes. The one-way ANOVA results are shown in **Table 13.18**.

Route 1	Route 2	Route 3
30	27	16
32	29	41
27	28	22
35	36	31

Table 13.18

State $SS_{between}$, SS_{within}, and the F statistic.

60. Suppose a group is interested in determining whether teenagers obtain their drivers licenses at approximately the same average age across the country. Suppose that the following data are randomly collected from five teenagers in each region of the country. The numbers represent the age at which teenagers obtained their drivers licenses.

	Northeast	South	West	Central	East
	16.3	16.9	16.4	16.2	17.1
	16.1	16.5	16.5	16.6	17.2
	16.4	16.4	16.6	16.5	16.6
	16.5	16.2	16.1	16.4	16.8
$\bar{x} =$	_____	_____	_____	_____	_____
$s^2 =$	_____	_____	_____	_____	_____

Table 13.19

State the hypotheses.

H_0: _____

H_a: _____

13.2 The F Distribution and the F-Ratio

Use the following information to answer the next three exercises. Suppose a group is interested in determining whether teenagers obtain their drivers licenses at approximately the same average age across the country. Suppose that the following data are randomly collected from five teenagers in each region of the country. The numbers represent the age at which teenagers obtained their drivers licenses.

	Northeast	South	West	Central	East
	16.3	16.9	16.4	16.2	17.1
	16.1	16.5	16.5	16.6	17.2
	16.4	16.4	16.6	16.5	16.6
	16.5	16.2	16.1	16.4	16.8
$\bar{x} =$	_____	_____	_____	_____	_____
$s^2 =$	_____	_____	_____	_____	_____

Table 13.20

H_0: $\mu_1 = \mu_2 = \mu_3 = \mu_4 = \mu_5$

H_a: At least any two of the group means μ_1, μ_2, …, μ_5 are not equal.

61. degrees of freedom – numerator: $df(num) =$ _____

62. degrees of freedom – denominator: $df(denom) =$ _____

63. F statistic = _____

13.3 Facts About the F Distribution

DIRECTIONS

Use a solution sheet to conduct the following hypothesis tests. The solution sheet can be found in **Appendix E**.

64. Three students, Linda, Tuan, and Javier, are given five laboratory rats each for a nutritional experiment. Each rat's weight is recorded in grams. Linda feeds her rats Formula A, Tuan feeds his rats Formula B, and Javier feeds his rats Formula C. At the end of a specified time period, each rat is weighed again, and the net gain in grams is recorded. Using a significance level of 10%, test the hypothesis that the three formulas produce the same mean weight gain.

Linda's rats	Tuan's rats	Javier's rats
43.5	47.0	51.2
39.4	40.5	40.9
41.3	38.9	37.9
46.0	46.3	45.0
38.2	44.2	48.6

Table 13.21 Weights of Student Lab Rats

65. A grassroots group opposed to a proposed increase in the gas tax claimed that the increase would hurt working-class people the most, since they commute the farthest to work. Suppose that the group randomly surveyed 24 individuals and asked them their daily one-way commuting mileage. The results are in **Table 13.22**. Using a 5% significance level, test the hypothesis that the three mean commuting mileages are the same.

working-class	professional (middle incomes)	professional (wealthy)
17.8	16.5	8.5
26.7	17.4	6.3
49.4	22.0	4.6
9.4	7.4	12.6
65.4	9.4	11.0
47.1	2.1	28.6
19.5	6.4	15.4
51.2	13.9	9.3

Table 13.22

66. Examine the seven practice laps from **Table 13.1**. Determine whether the mean lap time is statistically the same for the seven practice laps, or if there is at least one lap that has a different mean time from the others.

Use the following information to answer the next two exercises. **Table 13.23** lists the number of pages in four different types of magazines.

home decorating	news	health	computer
172	87	82	104
286	94	153	136
163	123	87	98
205	106	103	207
197	101	96	146

67. Using a significance level of 5%, test the hypothesis that the four magazine types have the same mean length.

68. Eliminate one magazine type that you now feel has a mean length different from the others. Redo the hypothesis test, testing that the remaining three means are statistically the same. Use a new solution sheet. Based on this test, are the mean lengths for the remaining three magazines statistically the same?

69. A researcher wants to know if the mean times (in minutes) that people watch their favorite news station are the same. Suppose that **Table 13.24** shows the results of a study.

CNN	FOX	Local
45	15	72
12	43	37
18	68	56
38	50	60
23	31	51
35	22	

Table 13.24

Assume that all distributions are normal, the four population standard deviations are approximately the same, and the data were collected independently and randomly. Use a level of significance of 0.05.

70. Are the means for the final exams the same for all statistics class delivery types? **Table 13.25** shows the scores on final exams from several randomly selected classes that used the different delivery types.

Online	Hybrid	Face-to-Face
72	83	80
84	73	78
77	84	84
80	81	81
81		86
		79
		82

Table 13.25

Assume that all distributions are normal, the four population standard deviations are approximately the same, and the data were collected independently and randomly. Use a level of significance of 0.05.

71. Are the mean number of times a month a person eats out the same for whites, blacks, Hispanics and Asians? Suppose that **Table 13.26** shows the results of a study.

White	Black	Hispanic	Asian
6	4	7	8
8	1	3	3
2	5	5	5

White	Black	Hispanic	Asian
4	2	4	1
6		6	7

Table 13.26

Assume that all distributions are normal, the four population standard deviations are approximately the same, and the data were collected independently and randomly. Use a level of significance of 0.05.

72. Are the mean numbers of daily visitors to a ski resort the same for the three types of snow conditions? Suppose that **Table 13.27** shows the results of a study.

Powder	Machine Made	Hard Packed
1,210	2,107	2,846
1,080	1,149	1,638
1,537	862	2,019
941	1,870	1,178
	1,528	2,233
	1,382	

Table 13.27

Assume that all distributions are normal, the four population standard deviations are approximately the same, and the data were collected independently and randomly. Use a level of significance of 0.05.

73. Sanjay made identical paper airplanes out of three different weights of paper, light, medium and heavy. He made four airplanes from each of the weights, and launched them himself across the room. Here are the distances (in meters) that his planes flew.

Paper Type/Trial	Trial 1	Trial 2	Trial 3	Trial 4
Heavy	5.1 meters	3.1 meters	4.7 meters	5.3 meters
Medium	4 meters	3.5 meters	4.5 meters	6.1 meters
Light	3.1 meters	3.3 meters	2.1 meters	1.9 meters

Table 13.28

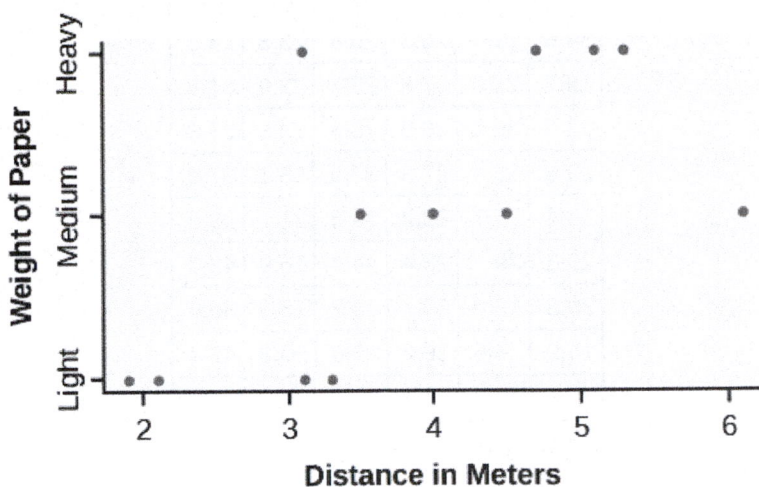

Figure 13.8

    a. Take a look at the data in the graph. Look at the spread of data for each group (light, medium, heavy). Does it seem reasonable to assume a normal distribution with the same variance for each group? Yes or No.

    b. Why is this a balanced design?

    c. Calculate the sample mean and sample standard deviation for each group.

    d. Does the weight of the paper have an effect on how far the plane will travel? Use a 1% level of significance. Complete the test using the method shown in the bean plant example in **Example 13.4**.

- variance of the group means _____
- $MS_{between}$= _____
- mean of the three sample variances _____
- MS_{within} = _____
- F statistic = _____
- $df(num)$ = _____, $df(denom)$ = _____
- number of groups _____
- number of observations _____
- p-value = _____ $(P(F >$ _____$) =$ _____$)$
- Graph the p-value.
- decision: _____
- conclusion: _____

74. DDT is a pesticide that has been banned from use in the United States and most other areas of the world. It is quite effective, but persisted in the environment and over time became seen as harmful to higher-level organisms. Famously, egg shells of eagles and other raptors were believed to be thinner and prone to breakage in the nest because of ingestion of DDT in the food chain of the birds.

An experiment was conducted on the number of eggs (fecundity) laid by female fruit flies. There are three groups of flies. One group was bred to be resistant to DDT (the RS group). Another was bred to be especially susceptible to DDT (SS). Finally there was a control line of non-selected or typical fruitflies (NS). Here are the data:

RS	SS	NS	RS	SS	NS
12.8	38.4	35.4	22.4	23.1	22.6
21.6	32.9	27.4	27.5	29.4	40.4
14.8	48.5	19.3	20.3	16	34.4
23.1	20.9	41.8	38.7	20.1	30.4

RS	SS	NS	RS	SS	NS
34.6	11.6	20.3	26.4	23.3	14.9
19.7	22.3	37.6	23.7	22.9	51.8
22.6	30.2	36.9	26.1	22.5	33.8
29.6	33.4	37.3	29.5	15.1	37.9
16.4	26.7	28.2	38.6	31	29.5
20.3	39	23.4	44.4	16.9	42.4
29.3	12.8	33.7	23.2	16.1	36.6
14.9	14.6	29.2	23.6	10.8	47.4
27.3	12.2	41.7			

Table 13.29

The values are the average number of eggs laid daily for each of 75 flies (25 in each group) over the first 14 days of their lives. Using a 1% level of significance, are the mean rates of egg selection for the three strains of fruitfly different? If so, in what way? Specifically, the researchers were interested in whether or not the selectively bred strains were different from the nonselected line, and whether the two selected lines were different from each other.

Here is a chart of the three groups:

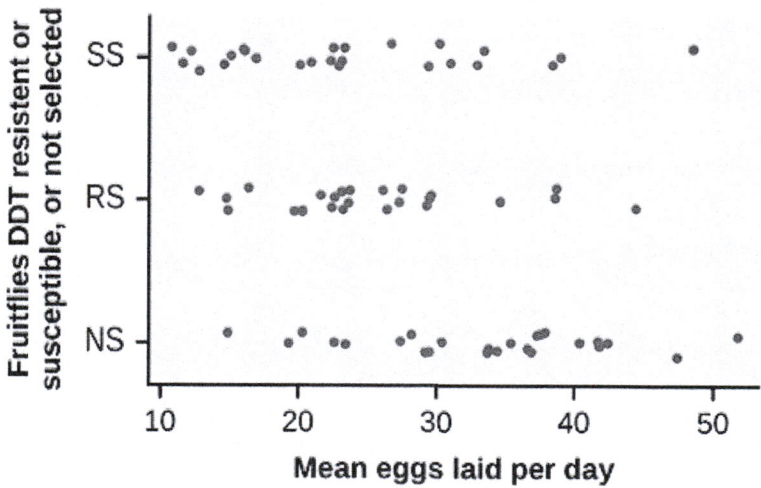

Figure 13.9

75. The data shown is the recorded body temperatures of 130 subjects as estimated from available histograms.

Traditionally we are taught that the normal human body temperature is 98.6 F. This is not quite correct for everyone. Are the mean temperatures among the four groups different?

Calculate 95% confidence intervals for the mean body temperature in each group and comment about the confidence intervals.

FL	FH	ML	MH	FL	FH	ML	MH
96.4	96.8	96.3	96.9	98.4	98.6	98.1	98.6

FL	FH	ML	MH	FL	FH	ML	MH
96.7	97.7	96.7	97	98.7	98.6	98.1	98.6
97.2	97.8	97.1	97.1	98.7	98.6	98.2	98.7
97.2	97.9	97.2	97.1	98.7	98.7	98.2	98.8
97.4	98	97.3	97.4	98.7	98.7	98.2	98.8
97.6	98	97.4	97.5	98.8	98.8	98.2	98.8
97.7	98	97.4	97.6	98.8	98.8	98.3	98.9
97.8	98	97.4	97.7	98.8	98.8	98.4	99
97.8	98.1	97.5	97.8	98.8	98.9	98.4	99
97.9	98.3	97.6	97.9	99.2	99	98.5	99
97.9	98.3	97.6	98	99.3	99	98.5	99.2
98	98.3	97.8	98		99.1	98.6	99.5
98.2	98.4	97.8	98		99.1	98.6	
98.2	98.4	97.8	98.3		99.2	98.7	
98.2	98.4	97.9	98.4		99.4	99.1	
98.2	98.4	98	98.4		99.9	99.3	
98.2	98.5	98	98.6		100	99.4	
98.2	98.6	98	98.6		100.8		

Table 13.30

13.4 Test of Two Variances

76. Three students, Linda, Tuan, and Javier, are given five laboratory rats each for a nutritional experiment. Each rat's weight is recorded in grams. Linda feeds her rats Formula A, Tuan feeds his rats Formula B, and Javier feeds his rats Formula C. At the end of a specified time period, each rat is weighed again and the net gain in grams is recorded.

Linda's rats	Tuan's rats	Javier's rats
43.5	47.0	51.2
39.4	40.5	40.9
41.3	38.9	37.9
46.0	46.3	45.0
38.2	44.2	48.6

Table 13.31

Determine whether or not the variance in weight gain is statistically the same among Javier's and Linda's rats. Test at a significance level of 10%.

77. A grassroots group opposed to a proposed increase in the gas tax claimed that the increase would hurt working-class people the most, since they commute the farthest to work. Suppose that the group randomly surveyed 24 individuals and asked them their daily one-way commuting mileage. The results are as follows.

working-class	professional (middle incomes)	professional (wealthy)
17.8	16.5	8.5
26.7	17.4	6.3
49.4	22.0	4.6
9.4	7.4	12.6
65.4	9.4	11.0
47.1	2.1	28.6
19.5	6.4	15.4
51.2	13.9	9.3

Table 13.32

Determine whether or not the variance in mileage driven is statistically the same among the working class and professional (middle income) groups. Use a 5% significance level.

78. Refer to the data from Table 13.1.

Examine practice laps 3 and 4. Determine whether or not the variance in lap time is statistically the same for those practice laps.

Use the following information to answer the next two exercises. The following table lists the number of pages in four different types of magazines.

home decorating	news	health	computer
172	87	82	104
286	94	153	136
163	123	87	98
205	106	103	207
197	101	96	146

Table 13.33

79. Which two magazine types do you think have the same variance in length?

80. Which two magazine types do you think have different variances in length?

81. Is the variance for the amount of money, in dollars, that shoppers spend on Saturdays at the mall the same as the variance for the amount of money that shoppers spend on Sundays at the mall? Suppose that the **Table 13.34** shows the results of a study.

Saturday	Sunday	Saturday	Sunday
75	44	62	137
18	58	0	82
150	61	124	39
94	19	50	127
62	99	31	141

Saturday	Sunday	Saturday	Sunday
73	60	118	73
	89		

Table 13.34

82. Are the variances for incomes on the East Coast and the West Coast the same? Suppose that **Table 13.35** shows the results of a study. Income is shown in thousands of dollars. Assume that both distributions are normal. Use a level of significance of 0.05.

East	West
38	71
47	126
30	42
82	51
75	44
52	90
115	88
67	

Table 13.35

83. Thirty men in college were taught a method of finger tapping. They were randomly assigned to three groups of ten, with each receiving one of three doses of caffeine: 0 mg, 100 mg, 200 mg. This is approximately the amount in no, one, or two cups of coffee. Two hours after ingesting the caffeine, the men had the rate of finger tapping per minute recorded. The experiment was double blind, so neither the recorders nor the students knew which group they were in. Does caffeine affect the rate of tapping, and if so how?

Here are the data:

0 mg	100 mg	200 mg	0 mg	100 mg	200 mg
242	248	246	245	246	248
244	245	250	248	247	252
247	248	248	248	250	250
242	247	246	244	246	248
246	243	245	242	244	250

Table 13.36

84. King Manuel I, Komnenus ruled the Byzantine Empire from Constantinople (Istanbul) during the years 1145 to 1180 A.D. The empire was very powerful during his reign, but declined significantly afterwards. Coins minted during his era were found in Cyprus, an island in the eastern Mediterranean Sea. Nine coins were from his first coinage, seven from the second, four from the third, and seven from a fourth. These spanned most of his reign. We have data on the silver content of the coins:

First Coinage	Second Coinage	Third Coinage	Fourth Coinage
5.9	6.9	4.9	5.3
6.8	9.0	5.5	5.6
6.4	6.6	4.6	5.5
7.0	8.1	4.5	5.1
6.6	9.3		6.2
7.7	9.2		5.8
7.2	8.6		5.8
6.9			
6.2			

Table 13.37

Did the silver content of the coins change over the course of Manuel's reign?

Here are the means and variances of each coinage. The data are unbalanced.

	First	Second	Third	Fourth
Mean	6.7444	8.2429	4.875	5.6143
Variance	0.2953	1.2095	0.2025	0.1314

Table 13.38

85. The American League and the National League of Major League Baseball are each divided into three divisions: East, Central, and West. Many years, fans talk about some divisions being stronger (having better teams) than other divisions. This may have consequences for the postseason. For instance, in 2012 Tampa Bay won 90 games and did not play in the postseason, while Detroit won only 88 and did play in the postseason. This may have been an oddity, but is there good evidence that in the 2012 season, the American League divisions were significantly different in overall records? Use the following data to test whether the mean number of wins per team in the three American League divisions were the same or not. Note that the data are not balanced, as two divisions had five teams, while one had only four.

Division	Team	Wins
East	NY Yankees	95
East	Baltimore	93
East	Tampa Bay	90
East	Toronto	73
East	Boston	69

Table 13.39

Division	Team	Wins
Central	Detroit	88

Division	Team	Wins
Central	Chicago Sox	85
Central	Kansas City	72
Central	Cleveland	68
Central	Minnesota	66

Table 13.40

Division	Team	Wins
West	Oakland	94
West	Texas	93
West	LA Angels	89
West	Seattle	75

Table 13.41

REFERENCES

13.2 The F Distribution and the F-Ratio

Tomato Data, Marist College School of Science (unpublished student research)

13.3 Facts About the F Distribution

Data from a fourth grade classroom in 1994 in a private K – 12 school in San Jose, CA.

Hand, D.J., F. Daly, A.D. Lunn, K.J. McConway, and E. Ostrowski. *A Handbook of Small Datasets: Data for Fruitfly Fecundity.* London: Chapman & Hall, 1994.

Hand, D.J., F. Daly, A.D. Lunn, K.J. McConway, and E. Ostrowski. *A Handbook of Small Datasets.* London: Chapman & Hall, 1994, pg. 50.

Hand, D.J., F. Daly, A.D. Lunn, K.J. McConway, and E. Ostrowski. A Handbook of Small Datasets. London: Chapman & Hall, 1994, pg. 118.

"MLB Standings – 2012." Available online at http://espn.go.com/mlb/standings/_/year/2012.

Mackowiak, P. A., Wasserman, S. S., and Levine, M. M. (1992), "A Critical Appraisal of 98.6 Degrees F, the Upper Limit of the Normal Body Temperature, and Other Legacies of Carl Reinhold August Wunderlich," *Journal of the American Medical Association*, 268, 1578-1580.

13.4 Test of Two Variances

"MLB Vs. Division Standings – 2012." Available online at http://espn.go.com/mlb/standings/_/year/2012/type/vs-division/order/true.

SOLUTIONS

1 Each population from which a sample is taken is assumed to be normal.

3 The populations are assumed to have equal standard deviations (or variances).

5 The response is a numerical value.

7 H_a: At least two of the group means μ_1, μ_2, μ_3 are not equal.

9 4,939.2

11 2

13 2,469.6

15 3.7416

17 3

19 13.2

21 0.825

23 Because a one-way ANOVA test is always right-tailed, a high F statistic corresponds to a low p-value, so it is likely that we will reject the null hypothesis.

25 The curves approximate the normal distribution.

27 ten

29 $SS = 237.33$; $MS = 23.73$

31 0.1614

33 two

35 $SS = 5,700.4$; $MS = 2,850.2$

37 3.6101

39 Yes, there is enough evidence to show that the scores among the groups are statistically significant at the 10% level.

43 The populations from which the two samples are drawn are normally distributed.

45 H_0: $\sigma_1 = \sigma_2$ H_a: $\sigma_1 < \sigma_2$ or H_0: $\sigma_1^2 = \sigma_2^2$ H_a: $\sigma_1^2 < \sigma_2^2$

47 4.11

49 0.7159

51 No, at the 10% level of significance, we do not reject the null hypothesis and state that the data do not show that the variation in drive times for the first worker is less than the variation in drive times for the second worker.

53 2.8674

55 Reject the null hypothesis. There is enough evidence to say that the variance of the grades for the first student is higher than the variance in the grades for the second student.

57 0.7414

59 $SS_{between} = 26$
$SS_{within} = 441$
$F = 0.2653$

62 $df(denom) = 15$

64

  a. H_0: $\mu_L = \mu_T = \mu_J$

  b. at least any two of the means are different

  c. $df(num) = 2$; $df(denom) = 12$

  d. F distribution

  e. 0.67

  f. 0.5305

  g. Check student's solution.

   h.   Decision: Do not reject null hypothesis; Conclusion: There is insufficient evidence to conclude that the means are different.

66

   a. H_0: $\mu_1 = \mu_2 = \mu_3 = \mu_4 = \mu_5 = \mu_6 = \mu_7$

   b.   At least two mean lap times are different.

   c.   *df(num)* = 6; *df(denom)* = 98

   d.   *F* distribution

   e.   1.69

   f.   0.1319

   g.   Check student's solution.

   h.   Decision: Do not reject null hypothesis; Conclusion: There is insufficient evidence to conclude that the mean lap times are different.

68

   a. H_a: $\mu_d = \mu_n = \mu_h$

   b.   At least any two of the magazines have different mean lengths.

   c.   *df(num)* = 2, *df(denom)* = 12

   d.   *F* distribtuion

   e.   *F* = 15.28

   f.   *p*-value = 0.001

   g.   Check student's solution.

   h.   i.   Alpha: 0.05

         ii.   Decision: Reject the Null Hypothesis.

         iii.   Reason for decision: *p*-value < alpha

         iv.   Conclusion: There is sufficient evidence to conclude that the mean lengths of the magazines are different.

70

   a. H_0: $\mu_o = \mu_h = \mu_f$

   b.   At least two of the means are different.

   c.   *df(n)* = 2, *df(d)* = 13

   d. $F_{2,13}$

   e.   0.64

   f.   0.5437

   g.   Check student's solution.

   h.   i.   Alpha: 0.05

         ii.   Decision: Do not reject the null hypothesis.

         iii.   Reason for decision: *p*-value > alpha

         iv.   Conclusion: The mean scores of different class delivery are not different.

72

   a. H_0: $\mu_p = \mu_m = \mu_h$

   b.   At least any two of the means are different.

   c.   *df(n)* = 2, *df(d)* = 12

   d. $F_{2,12}$

e. 3.13

f. 0.0807

g. Check student's solution.

h.   i.   Alpha: 0.05

      ii.   Decision: Do not reject the null hypothesis.

     iii.   Reason for decision: p-value > alpha

      iv.   Conclusion: There is not sufficient evidence to conclude that the mean numbers of daily visitors are different.

74   The data appear normally distributed from the chart and of similar spread. There do not appear to be any serious outliers, so we may proceed with our ANOVA calculations, to see if we have good evidence of a difference between the three groups. H_0: $\mu_1 = \mu_2 = \mu_3$; H_a: $\mu_i \neq \mu_j$ some $i \neq j$. Define μ_1, μ_2, μ_3, as the population mean number of eggs laid by the three groups of fruit flies. F statistic = 8.6657; p-value = 0.0004

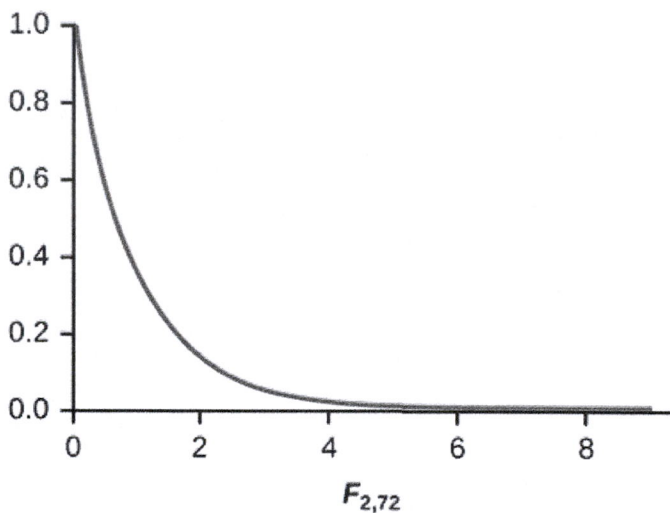

Figure 13.10

Decision: Since the p-value is less than the level of significance of 0.01, we reject the null hypothesis. **Conclusion:** We have good evidence that the average number of eggs laid during the first 14 days of life for these three strains of fruitflies are different. Interestingly, if you perform a two sample t-test to compare the RS and NS groups they are significantly different (p = 0.0013). Similarly, SS and NS are significantly different (p = 0.0006). However, the two selected groups, RS and SS are *not* significantly different (p = 0.5176). Thus we appear to have good evidence that selection either for resistance or for susceptibility involves a reduced rate of egg production (for these specific strains) as compared to flies that were not selected for resistance or susceptibility to DDT. Here, genetic selection has apparently involved a loss of fecundity.

76

a. $H_0: \sigma_1^2 = \sigma_2^2$

b. $H_a: \ \sigma_1^2 \neq \sigma_1^2$

c. $df(num) = 4$; $df(denom) = 4$

d. $F_{4,\,4}$

e. 3.00

f. 2(0.1563) = 0.3126. Using the TI-83+/84+ function 2-SampFtest, you get the test statistic as 2.9986 and p-value directly as 0.3127. If you input the lists in a different order, you get a test statistic of 0.3335 but the p-value is the same because this is a two-tailed test.

g. Check student't solution.

h.   Decision: Do not reject the null hypothesis; Conclusion: There is insufficient evidence to conclude that the variances are different.

78

a. H_0: $\sigma_1^2 = \sigma_2^2$

b. H_a: $\sigma_1^2 \neq \sigma_1^2$

c. $df(n) = 19$, $df(d) = 19$

d. $F_{19,19}$

e.   1.13

f.   0.786

g.   Check student's solution.

h.   i.   Alpha:0.05

   ii.   Decision: Do not reject the null hypothesis.

   iii.   Reason for decision: p-value > alpha

   iv.   Conclusion: There is not sufficient evidence to conclude that the variances are different.

80 The answers may vary. Sample answer: Home decorating magazines and news magazines have different variances.

82

a. H_0: = $\sigma_1^2 = \sigma_2^2$

b. H_a: $\sigma_1^2 \neq \sigma_1^2$

c. $df(n) = 7$, $df(d) = 6$

d. $F_{7,6}$

e.   0.8117

f.   0.7825

g.   Check student's solution.

h.   i.   Alpha: 0.05

   ii.   Decision: Do not reject the null hypothesis.

   iii.   Reason for decision: p-value > alpha

   iv.   Conclusion: There is not sufficient evidence to conclude that the variances are different.

84 Here is a strip chart of the silver content of the coins:

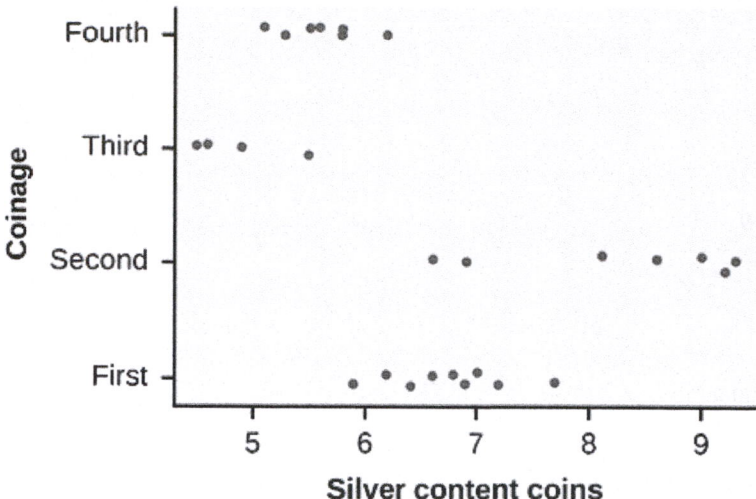

Figure 13.11

While there are differences in spread, it is not unreasonable to use ANOVA techniques. Here is the completed ANOVA table:

Source of Variation	Sum of Squares (*SS*)	Degrees of Freedom (*df*)	Mean Square (*MS*)	*F*
Factor (Between)	37.748	4 − 1 = 3	12.5825	26.272
Error (Within)	11.015	27 − 4 = 23	0.4789	
Total	48.763	27 − 1 = 26		

Table 13.42

$P(F > 26.272) = 0$; Reject the null hypothesis for any alpha. There is sufficient evidence to conclude that the mean silver content among the four coinages are different. From the strip chart, it appears that the first and second coinages had higher silver contents than the third and fourth.

85 Here is a stripchart of the number of wins for the 14 teams in the AL for the 2012 season.

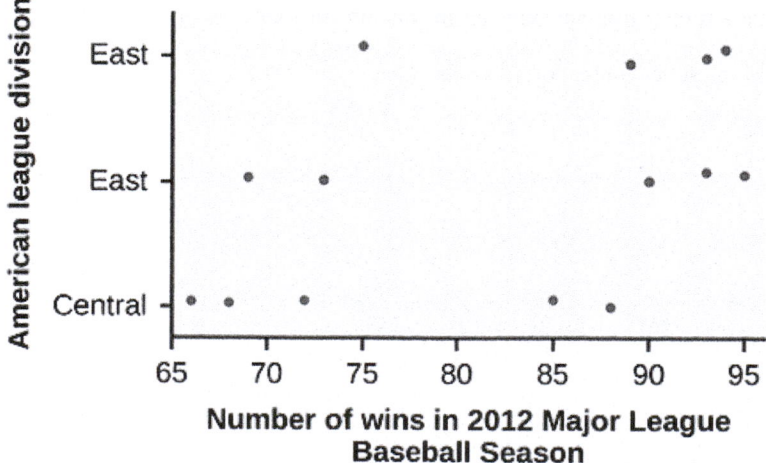

Figure 13.12

While the spread seems similar, there may be some question about the normality of the data, given the wide gaps in the middle near the 0.500 mark of 82 games (teams play 162 games each season in MLB). However, one-way ANOVA is robust. Here is the ANOVA table for the data:

Source of Variation	Sum of Squares (*SS*)	Degrees of Freedom (*df*)	Mean Square (*MS*)	*F*
Factor (Between)	344.16	$3 - 1 = 2$	172.08	26.272
Error (Within)	1,219.55	$14 - 3 = 11$	110.87	1.5521
Total	1,563.71	$14 - 1 = 13$		

Table 13.43

$P(F > 1.5521) = 0.2548$

Since the *p*-value is so large, there is not good evidence against the null hypothesis of equal means. We decline to reject the null hypothesis. Thus, for 2012, there is not any have any good evidence of a significant difference in mean number of wins between the divisions of the American League.

APPENDIX A: REVIEW EXERCISES (CH 3-13)

These review exercises are designed to provide extra practice on concepts learned before a particular chapter. For example, the review exercises for Chapter 3, cover material learned in chapters 1 and 2.

Chapter 3

Use the following information to answer the next six exercises: In a survey of 100 stocks on NASDAQ, the average percent increase for the past year was 9% for NASDAQ stocks.

1. The "average increase" for all NASDAQ stocks is the:

a. population

b. statistic

c. parameter

d. sample

e. variable

2. All of the NASDAQ stocks are the:

a. population

b. statistics

c. parameter

d. sample

e. variable

3. Nine percent is the:

a. population

b. statistics

c. parameter

d. sample

e. variable

4. The 100 NASDAQ stocks in the survey are the:

a. population

b. statistic

c. parameter

d. sample

e. variable

5. The percent increase for one stock in the survey is the:

a. population

b. statistic

c. parameter

d. sample

e. variable

6. Would the data collected by qualitative, quantitative discrete, or quantitative continuous?

Use the following information to answer the next two exercises: Thirty people spent two weeks around Mardi Gras in New Orleans. Their two-week weight gain is below. (Note: a loss is shown by a negative weight gain.)

Weight Gain	Frequency
−2	3
−1	5
0	2
1	4
4	13
6	2
11	1

Table A1

7. Calculate the following values:

a. the average weight gain for the two weeks

b. the standard deviation

c. the first, second, and third quartiles

8. Construct a histogram and box plot of the data.

Chapter 4

Use the following information to answer the next two exercises: A recent poll concerning credit cards found that 35 percent of respondents use a credit card that gives them a mile of air travel for every dollar they charge. Thirty percent of the respondents charge more than $2,000 per month. Of those respondents who charge more than $2,000, 80 percent use a credit card that gives them a mile of air travel for every dollar they charge.

9. What is the probability that a randomly selected respondent will spend more than $2,000 AND use a credit card that gives them a mile of air travel for every dollar they charge?

a. (0.30)(0.35)

b. (0.80)(0.35)

c. (0.80)(0.30)

d. (0.80)

10. Are using a credit card that gives a mile of air travel for each dollar spent AND charging more than $2,000 per month independent events?

a. Yes

b. No, and they are not mutually exclusive either.

c. No, but they are mutually exclusive.

d. Not enough information given to determine the answer

11. A sociologist wants to know the opinions of employed adult women about government funding for day care. She obtains a list of 520 members of a local business and professional women's club and mails a questionnaire to 100 of these women selected at random. Sixty-eight questionnaires are returned. What is the population in this study?

a. all employed adult women

b. all the members of a local business and professional women's club

c. the 100 women who received the questionnaire

d. all employed women with children

Use the following information to answer the next two exercises: The next two questions refer to the following: An article from The San Jose Mercury News was concerned with the racial mix of the 1500 students at Prospect High School in Saratoga, CA. The table summarizes the results. (Male and female values are approximate.) Suppose one Prospect High School student is randomly selected.

Gender/Ethnic group	White	Asian	Hispanic	Black	American Indian
Male	400	468	115	35	16
Female	440	132	140	40	14

Table A2

12. Find the probability that a student is Asian or Male.

13. Find the probability that a student is Black given that the student is female.

14. A sample of pounds lost, in a certain month, by individual members of a weight reducing clinic produced the following statistics:

- Mean = 5 lbs.
- Median = 4.5 lbs.
- Mode = 4 lbs.
- Standard deviation = 3.8 lbs.
- First quartile = 2 lbs.
- Third quartile = 8.5 lbs.

The correct statement is:

a. One fourth of the members lost exactly two pounds.

b. The middle fifty percent of the members lost from two to 8.5 lbs.

c. Most people lost 3.5 to 4.5 lbs.

d. All of the choices above are correct.

15. What does it mean when a data set has a standard deviation equal to zero?

a. All values of the data appear with the same frequency.

b. The mean of the data is also zero.

c. All of the data have the same value.

d. There are no data to begin with.

16. The statement that describe the illustration is:

Figure A1

a.   the mean is equal to the median.

b.   There is no first quartile.

c.   The lowest data value is the median.

d.   The median equals $\dfrac{Q_1 + Q_3}{2}$.

17. According to a recent article in the *San Jose Mercury News* the average number of babies born with significant hearing loss (deafness) is approximately 2 per 1000 babies in a healthy baby nursery. The number climbs to an average of 30 per 1000 babies in an intensive care nursery. Suppose that 1,000 babies from healthy baby nurseries were randomly surveyed. Find the probability that exactly two babies were born deaf.

18. A "friend" offers you the following "deal." For a $10 fee, you may pick an envelope from a box containing 100 seemingly identical envelopes. However, each envelope contains a coupon for a free gift.

- Ten of the coupons are for a free gift worth $6.

- Eighty of the coupons are for a free gift worth $8.

- Six of the coupons are for a free gift worth $12.

- Four of the coupons are for a free gift worth $40.

Based upon the financial gain or loss over the long run, should you play the game?

a.   Yes, I expect to come out ahead in money.

b.   No, I expect to come out behind in money.

c.   It doesn't matter. I expect to break even.

Use the following information to answer the next four exercises: Recently, a nurse commented that when a patient calls the medical advice line claiming to have the flu, the chance that he/she truly has the flu (and not just a nasty cold) is only about 4%. Of the next 25 patients calling in claiming to have the flu, we are interested in how many actually have the flu.

19. Define the random variable and list its possible values.

20. State the distribution of X.

21. Find the probability that at least four of the 25 patients actually have the flu.

22. On average, for every 25 patients calling in, how many do you expect to have the flu?

Use the following information to answer the next two exercises: Different types of writing can sometimes be distinguished by the number of letters in the words used. A student interested in this fact wants to study the number of letters of words used by Tom Clancy in his novels. She opens a Clancy novel at random and records the number of letters of the first 250 words on the page.

23. What kind of data was collected?

a.   qualitative

b.   quantitative continuous

c.   quantitative discrete

24. What is the population under study?

Chapter 5

Use the following information to answer the next seven exercises: A recent study of mothers of junior high school children in Santa Clara County reported that 76% of the mothers are employed in paid positions. Of those mothers who are employed, 64% work full-time (over 35 hours per week), and 36% work part-time. However, out of all of the mothers in the population, 49% work full-time. The population under study is made up of mothers of junior high school children in Santa Clara County. Let E = employed and F = full-time employment.

25.

a. Find the percent of all mothers in the population that are NOT employed.

b. Find the percent of mothers in the population that are employed part-time.

26. The "type of employment" is considered to be what type of data?

27. Find the probability that a randomly selected mother works part-time given that she is employed.

28. Find the probability that a randomly selected person from the population will be employed or work full-time.

29. Being employed and working part-time:

a. mutually exclusive events? Why or why not?

b. independent events? Why or why not?

Use the following additional information to answer the next two exercises: We randomly pick ten mothers from the above population. We are interested in the number of the mothers that are employed. Let X = number of mothers that are employed.

30. State the distribution for X.

31. Find the probability that at least six are employed.

32. We expect the statistics discussion board to have, on average, 14 questions posted to it per week. We are interested in the number of questions posted to it per day.

a. Define X.

b. What are the values that the random variable may take on?

c. State the distribution for X.

d. Find the probability that from ten to 14 (inclusive) questions are posted to the listserv on a randomly picked day.

33. A person invests $1,000 into stock of a company that hopes to go public in one year. The probability that the person will lose all his money after one year (i.e. his stock will be worthless) is 35%. The probability that the person's stock will still have a value of $1,000 after one year (i.e. no profit and no loss) is 60%. The probability that the person's stock will increase in value by $10,000 after one year (i.e. will be worth $11,000) is 5%. Find the expected profit after one year.

34. Rachel's piano cost $3,000. The average cost for a piano is $4,000 with a standard deviation of $2,500. Becca's guitar cost $550. The average cost for a guitar is $500 with a standard deviation of $200. Matt's drums cost $600. The average cost for drums is $700 with a standard deviation of $100. Whose cost was lowest when compared to his or her own instrument?

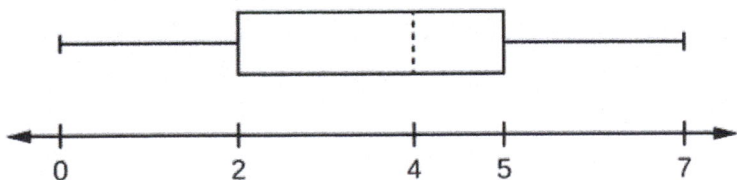

Figure A2

35. Explain why each statement is either true or false given the box plot in **Figure A2**.

a. Twenty-five percent of the data re at most five.

b. There is the same amount of data from 4–5 as there is from 5–7.

c. There are no data values of three.

d. Fifty percent of the data are four.

Using the following information to answer the next two exercises: 64 faculty members were asked the number of cars they owned (including spouse and children's cars). The results are given in the following graph:

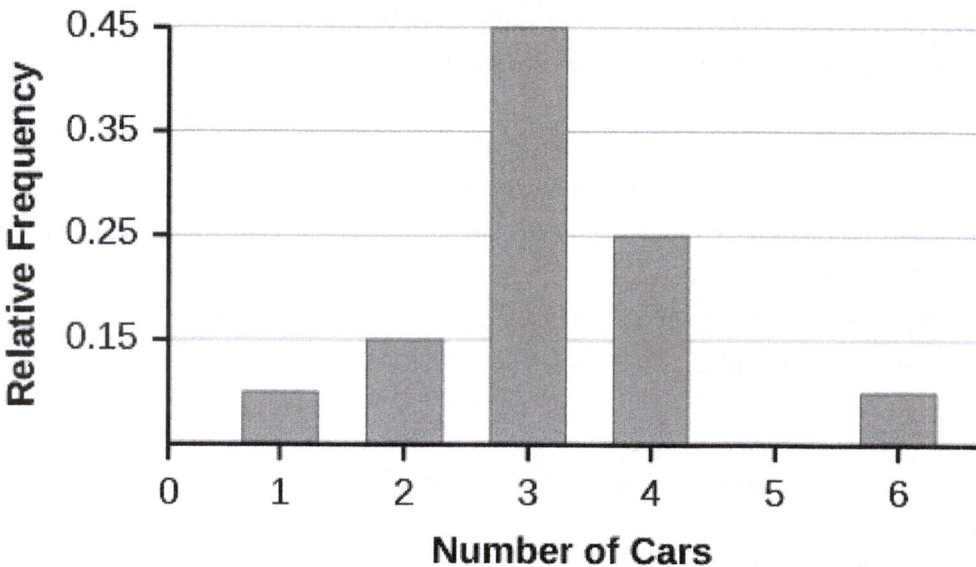

Figure A3

36. Find the approximate number of responses that were three.

37. Find the first, second and third quartiles. Use them to construct a box plot of the data.

Use the following information to answer the next three exercises: **Table A3** shows data gathered from 15 girls on the Snow Leopard soccer team when they were asked how they liked to wear their hair. Supposed one girl from the team is randomly selected.

Hair Style/Hair Color	Blond	Brown	Black
Ponytail	3	2	5
Plain	2	2	1

Table A3

38. Find the probability that the girl has black hair GIVEN that she wears a ponytail.

39. Find the probability that the girl wears her hair plain OR has brown hair.

40. Find the probability that the girl has blond hair AND that she wears her hair plain.

Chapter 6

Use the following information to answer the next two exercises: $X \sim U(3, 13)$

41. Explain which of the following are false and which are true.

a. $f(x) = \frac{1}{10}, 3 \le x \le 13$

b. There is no mode

   c.   The median is less than the mean.

   d. $P(x > 10) = P(x \le 6)$

42. Calculate:

   a.   the mean.

   b.   the median.

   c.   the 65th percentile.

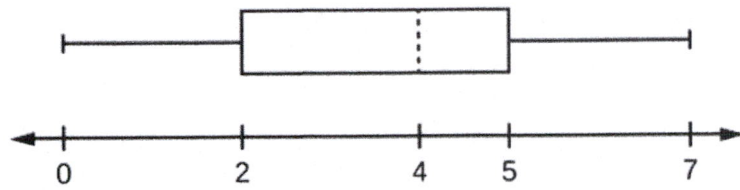

Figure A4

43. Which of the following is true for the box plot in **Figure A4**?

   a.   Twenty-five percent of the data are at most five.

   b.   There is about the same amount of data from 4–5 as there is from 5–7.

   c.   There are no data values of three.

   d.   Fifty percent of the data are four.

44. If $P(G|H) = P(G)$, then which of the following is correct?

   a. G and H are mutually exclusive events.

   b. $P(G) = P(H)$

   c.   Knowing that H has occurred will affect the chance that G will happen.

   d. G and H are independent events.

45. If $P(J) = 0.3$, $P(K) = 0.63$, and J and K are independent events, then explain which are correct and which are incorrect.

   a. $P(J \text{ AND } K) = 0$

   b. $P(J \text{ OR } K) = 0.9$

   c. $P(J \text{ OR } K) = 0.72$

   d. $P(J) \ne P(J|K)$

46. On average, five students from each high school class get full scholarships to four-year colleges. Assume that most high school classes have about 500 students. X = the number of students from a high school class that get full scholarships to four-year schools. Which of the following is the distribution of X?

   a. $P(5)$

   b. $B(500, 5)$

   c. $Exp\left(\frac{1}{5}\right)$

   d. $N\left(5, \frac{(0.01)(0.99)}{500}\right)$

Chapter 7

Use the following information to answer the next three exercises: Richard's Furniture Company delivers furniture from 10 A.M. to 2 P.M. continuously and uniformly. We are interested in how long (in hours) past the 10 A.M. start time that individuals wait for their delivery.

47. $X \sim$ _____

a. $U(0, 4)$

b. $U(10, 20)$

c. $Exp(2)$

d. $N(2, 1)$

48. The average wait time is:

a. 1 hour.

b. 2 hours.

c. 2.5 hours.

d. 4 hours.

49. Suppose that it is now past noon on a delivery day. The probability that a person must wait at least 1.5 more hours is:

a. $\frac{1}{4}$

b. $\frac{1}{2}$

c. $\frac{3}{4}$

d. $\frac{3}{8}$

50. Given: $X \sim Exp\left(\frac{1}{3}\right)$

a. Find $P(x > 1)$.

b. Calculate the minimum value for the upper quartile.

c. Find $P\left(x = \frac{1}{3}\right)$

51.

- 40% of full-time students took 4 years to graduate
- 30% of full-time students took 5 years to graduate
- 20% of full-time students took 6 years to graduate
- 10% of full-time students took 7 years to graduate

The expected time for full-time students to graduate is:

a. 4 years

b. 4.5 years

c. 5 years

d. 5.5 years

52. Which of the following distributions is described by the following example?
Many people can run a short distance of under two miles, but as the distance increases, fewer people can run that far.

a. binomial

b. uniform

c. exponential

d. normal

53. The length of time to brush one's teeth is generally thought to be exponentially distributed with a mean of $\frac{3}{4}$ minutes.

Find the probability that a randomly selected person brushes his or her teeth less than $\frac{3}{4}$ minutes.

a. 0.5

b. $\frac{3}{4}$

c. 0.43

d. 0.63

54. Which distribution accurately describes the following situation?
The chance that a teenage boy regularly gives his mother a kiss goodnight is about 20%. Fourteen teenage boys are randomly surveyed. Let X = the number of teenage boys that regularly give their mother a kiss goodnight.

a. $B(14,0.20)$

b. $P(2.8)$

c. $N(2.8,2.24)$

d. $Exp\left(\frac{1}{0.20}\right)$

55. A 2008 report on technology use states that approximately 20% of U.S. households have never sent an e-mail. Suppose that we select a random sample of fourteen U.S. households. Let X = the number of households in a 2008 sample of 14 households that have never sent an email

a. $B(14,0.20)$

b. $P(2.8)$

c. $N(2.8,2.24)$

d. $Exp\left(\frac{1}{0.20}\right)$

Chapter 8

Use the following information to answer the next three exercises: Suppose that a sample of 15 randomly chosen people were put on a special weight loss diet. The amount of weight lost, in pounds, follows an unknown distribution with mean equal to 12 pounds and standard deviation equal to three pounds. Assume that the distribution for the weight loss is normal.

56. To find the probability that the mean amount of weight lost by 15 people is no more than 14 pounds, the random variable should be:

a. number of people who lost weight on the special weight loss diet.

b. the number of people who were on the diet.

c. the mean amount of weight lost by 15 people on the special weight loss diet.

d. the total amount of weight lost by 15 people on the special weight loss diet.

57. Find the probability asked for in **Question 56**.

58. Find the 90th percentile for the mean amount of weight lost by 15 people.

Using the following information to answer the next three exercises: The time of occurrence of the first accident during rush-hour traffic at a major intersection is uniformly distributed between the three hour interval 4 p.m. to 7 p.m. Let X = the amount of time (hours) it takes for the first accident to occur.

59. What is the probability that the time of occurrence is within the first half-hour or the last hour of the period from 4 to 7 p.m.?

a. cannot be determined from the information given

b. $\frac{1}{6}$

c. $\frac{1}{2}$

d. $\frac{1}{3}$

60. The 20th percentile occurs after how many hours?

a. 0.20

b. 0.60

c. 0.50

d. 1

61. Assume Ramon has kept track of the times for the first accidents to occur for 40 different days. Let C = the total cumulative time. Then C follows which distribution?

a. $U(0,3)$

b. $Exp(13)$

c. $N(60, 5.477)$

d. $N(1.5, 0.01875)$

62. Using the information in **Question 61**, find the probability that the total time for all first accidents to occur is more than 43 hours.

Use the following information to answer the next two exercises: The length of time a parent must wait for his children to clean their rooms is uniformly distributed in the time interval from one to 15 days.

63. How long must a parent expect to wait for his children to clean their rooms?

a. eight days

b. three days

c. 14 days

d. six days

64. What is the probability that a parent will wait more than six days given that the parent has already waited more than three days?

a. 0.5174

b. 0.0174

c. 0.7500

d. 0.2143

Use the following information to answer the next five exercises: Twenty percent of the students at a local community college live in within five miles of the campus. Thirty percent of the students at the same community college receive some kind of financial aid. Of those who live within five miles of the campus, 75% receive some kind of financial aid.

65. Find the probability that a randomly chosen student at the local community college does not live within five miles of the campus.

a. 80%

b. 20%

c. 30%

d. cannot be determined

66. Find the probability that a randomly chosen student at the local community college lives within five miles of the campus or receives some kind of financial aid.

a. 50%

b. 35%

c. 27.5%

d. 75%

67. Are living in student housing within five miles of the campus and receiving some kind of financial aid mutually exclusive?

a. yes

b. no

c. cannot be determined

68. The interest rate charged on the financial aid is _____ data.

a. quantitative discrete

b. quantitative continuous

c. qualitative discrete

d. qualitative

69. The following information is about the students who receive financial aid at the local community college.

- 1st quartile = $250

- 2nd quartile = $700

- 3rd quartile = $1200

These amounts are for the school year. If a sample of 200 students is taken, how many are expected to receive $250 or more?

a. 50

b. 250

c. 150

d. cannot be determined

Use the following information to answer the next two exercises: $P(A) = 0.2$, $P(B) = 0.3$; A and B are independent events.

70. $P(A$ AND $B) =$ _____

  a. 0.5

  b. 0.6

  c. 0

  d. 0.06

71. $P(A$ OR $B) =$ _____

  a. 0.56

  b. 0.5

  c. 0.44

  d. 1

72. If H and D are mutually exclusive events, $P(H) = 0.25$, $P(D) = 0.15$, then $P(H|D)$.

  a. 1

  b. 0

  c. 0.40

  d. 0.0375

Chapter 9

73. Rebecca and Matt are 14 year old twins. Matt's height is two standard deviations below the mean for 14 year old boys' height. Rebecca's height is 0.10 standard deviations above the mean for 14 year old girls' height. Interpret this.

  a. Matt is 2.1 inches shorter than Rebecca.

  b. Rebecca is very tall compared to other 14 year old girls.

  c. Rebecca is taller than Matt.

  d. Matt is shorter than the average 14 year old boy.

74. Construct a histogram of the IPO data (see **Appendix C**).

Use the following information to answer the next three exercises: Ninety homeowners were asked the number of estimates they obtained before having their homes fumigated. Let $X =$ the number of estimates.

x	Relative Frequency	Cumulative Relative Frequency
1	0.3	
2	0.2	
4	0.4	
5	0.1	

Table A4

75. Complete the cumulative frequency column.

76. Calculate the sample mean (a), the sample standard deviation (b) and the percent of the estimates that fall at or below four (c).

77. Calculate the median, M, the first quartile, Q_1, the third quartile, Q_3. Then construct a box plot of the data.

78. The middle 50% of the data are between _____ and _____.

Use the following information to answer the next three exercises: Seventy 5th and 6th graders were asked their favorite dinner.

	Pizza	Hamburgers	Spaghetti	Fried shrimp
5th grader	15	6	9	0
6th grader	15	7	10	8

Table A5

79. Find the probability that one randomly chosen child is in the 6th grade and prefers fried shrimp.

a. $\frac{32}{70}$

b. $\frac{8}{32}$

c. $\frac{8}{8}$

d. $\frac{8}{70}$

80. Find the probability that a child does not prefer pizza.

a. $\frac{30}{70}$

b. $\frac{30}{40}$

c. $\frac{40}{70}$

d. 1

81. Find the probability a child is in the 5th grade given that the child prefers spaghetti.

a. $\frac{9}{19}$

b. $\frac{9}{70}$

c. $\frac{9}{30}$

d. $\frac{19}{70}$

82. A sample of convenience is a random sample.

a. true

b. false

83. A statistic is a number that is a property of the population.

a. true

b. false

84. You should always throw out any data that are outliers.

a. true

b. false

85. Lee bakes pies for a small restaurant in Felton, CA. She generally bakes 20 pies in a day, on average. Of interest is the number of pies she bakes each day.

a. Define the random variable X.

b. State the distribution for X.

c. Find the probability that Lee bakes more than 25 pies in any given day.

86. Six different brands of Italian salad dressing were randomly selected at a supermarket. The grams of fat per serving are 7, 7, 9, 6, 8, 5. Assume that the underlying distribution is normal. Calculate a 95% confidence interval for the population mean grams of fat per serving of Italian salad dressing sold in supermarkets.

87. Given: uniform, exponential, normal distributions. Match each to a statement below.

a. mean = median ≠ mode

b. mean > median > mode

c. mean = median = mode

Chapter 10

Use the following information to answer the next three exercises: In a survey at Kirkwood Ski Resort the following information was recorded:

	0–10	11–20	21–40	40+
Ski	10	12	30	8
Snowboard	6	17	12	5

Table A6

Suppose that one person from **Table A6** was randomly selected.

88. Find the probability that the person was a skier or was age 11–20.

89. Find the probability that the person was a snowboarder given he or she was age 21–40.

90. Explain which of the following are true and which are false.

a. Sport and age are independent events.

b. Ski and age 11–20 are mutually exclusive events.

c. P(Ski AND age 21–40) < P(Ski|age 21–40)

d. P(Snowboard OR age 0–10) < P(Snowboard|age 0–10)

91. The average length of time a person with a broken leg wears a cast is approximately six weeks. The standard deviation is about three weeks. Thirty people who had recently healed from broken legs were interviewed. State the distribution that most accurately reflects total time to heal for the thirty people.

92. The distribution for X is uniform. What can we say for certain about the distribution for $\bar{X}$ when $n = 1$?

a. The distribution for $\bar{X}$ is still uniform with the same mean and standard deviation as the distribution for X.

b. The distribution for $\bar{X}$ is normal with the different mean and a different standard deviation as the distribution for X.

c. The distribution for $\bar{X}$ is normal with the same mean but a larger standard deviation than the distribution for X.

d. The distribution for $\bar{X}$ is normal with the same mean but a smaller standard deviation than the distribution for X.

93. The distribution for X is uniform. What can we say for certain about the distribution for $\sum X$ when $n = 50$?

a. distribution for $\sum X$ is still uniform with the same mean and standard deviation as the distribution for X.

b. The distribution for $\sum X$ is normal with the same mean but a larger standard deviation as the distribution for X.

c. The distribution for $\sum X$ is normal with a larger mean and a larger standard deviation than the distribution for X.

d. The distribution for $\sum X$ is normal with the same mean but a smaller standard deviation than the distribution for X.

Use the following information to answer the next three exercises: A group of students measured the lengths of all the carrots in a five-pound bag of baby carrots. They calculated the average length of baby carrots to be 2.0 inches with a standard deviation of 0.25 inches. Suppose we randomly survey 16 five-pound bags of baby carrots.

94. State the approximate distribution for $\bar{X}$, the distribution for the average lengths of baby carrots in 16 five-pound bags.

$\bar{X} \sim$ _____

95. Explain why we cannot find the probability that one individual randomly chosen carrot is greater than 2.25 inches.

96. Find the probability that $\bar{x}$ is between two and 2.25 inches.

Use the following information to answer the next three exercises: At the beginning of the term, the amount of time a student waits in line at the campus store is normally distributed with a mean of five minutes and a standard deviation of two minutes.

97. Find the 90th percentile of waiting time in minutes.

98. Find the median waiting time for one student.

99. Find the probability that the average waiting time for 40 students is at least 4.5 minutes.

Chapter 11

Use the following information to answer the next four exercises: Suppose that the time that owners keep their cars (purchased new) is normally distributed with a mean of seven years and a standard deviation of two years. We are interested in how long an individual keeps his car (purchased new). Our population is people who buy their cars new.

100. Sixty percent of individuals keep their cars **at most** how many years?

101. Suppose that we randomly survey one person. Find the probability that person keeps his or her car **less than** 2.5 years.

102. If we are to pick individuals ten at a time, find the distribution for the **mean** car length ownership.

103. If we are to pick ten individuals, find the probability that the **sum** of their ownership time is more than 55 years.

104. For which distribution is the median not equal to the mean?

a. Uniform

b. Exponential

c. Normal

d. Student *t*

105. Compare the standard normal distribution to the Student's *t*-distribution, centered at zero. Explain which of the following are true and which are false.

a. As the number surveyed increases, the area to the left of –1 for the Student's *t*-distribution approaches the area for the standard normal distribution.

b. As the degrees of freedom decrease, the graph of the Student's *t*-distribution looks more like the graph of the standard normal distribution.

c. If the number surveyed is 15, the normal distribution should never be used.

Use the following information to answer the next five exercises: We are interested in the checking account balance of twenty-year-old college students. We randomly survey 16 twenty-year-old college students. We obtain a sample mean of $640 and a sample standard deviation of $150. Let X = checking account balance of an individual twenty year old college student.

106. Explain why we cannot determine the distribution of X.

107. If you were to create a confidence interval or perform a hypothesis test for the population mean checking account balance of twenty-year-old college students, what distribution would you use?

108. Find the 95% confidence interval for the true mean checking account balance of a twenty-year-old college student.

109. What type of data is the balance of the checking account considered to be?

110. What type of data is the number of twenty-year-olds considered to be?

111. On average, a busy emergency room gets a patient with a shotgun wound about once per week. We are interested in the number of patients with a shotgun wound the emergency room gets per 28 days.

a. Define the random variable X.

b. State the distribution for X.

c. Find the probability that the emergency room gets no patients with shotgun wounds in the next 28 days.

Use the following information to answer the next two exercises: The probability that a certain slot machine will pay back money when a quarter is inserted is 0.30. Assume that each play of the slot machine is independent from each other. A person puts in 15 quarters for 15 plays.

112. Is the expected number of plays of the slot machine that will pay back money greater than, less than or the same as the median? Explain your answer.

113. Is it likely that exactly eight of the 15 plays would pay back money? Justify your answer numerically.

114. A game is played with the following rules:

*   it costs $10 to enter.
*   a fair coin is tossed four times.
*   if you do not get four heads or four tails, you lose your $10.
*   if you get four heads or four tails, you get back your $10, plus $30 more.

Over the long run of playing this game, what are your expected earnings?

115.

*   The mean grade on a math exam in Rachel's class was 74, with a standard deviation of five. Rachel earned an 80.
*   The mean grade on a math exam in Becca's class was 47, with a standard deviation of two. Becca earned a 51.
*   The mean grade on a math exam in Matt's class was 70, with a standard deviation of eight. Matt earned an 83.

Find whose score was the best, compared to his or her own class. Justify your answer numerically.

Use the following information to answer the next two exercises: A random sample of 70 compulsive gamblers were asked the number of days they go to casinos per week. The results are given in the following graph:

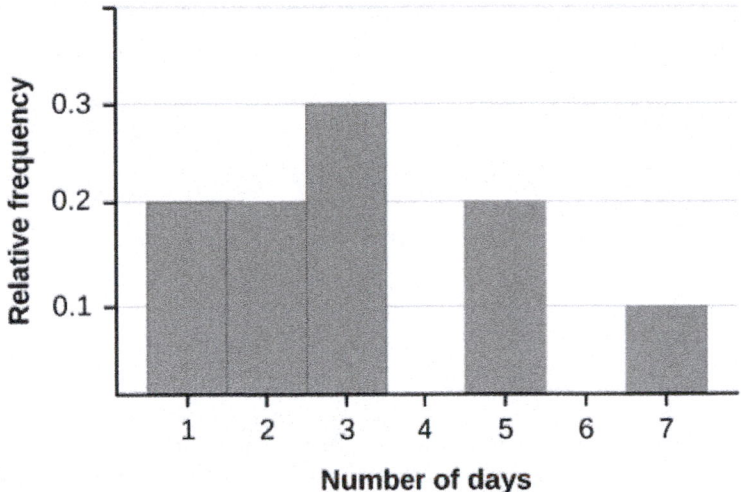

Figure A5

116. Find the number of responses that were five.

117. Find the mean, standard deviation, the median, the first quartile, the third quartile and the *IQR*.

118. Based upon research at De Anza College, it is believed that about 19% of the student population speaks a language other than English at home. Suppose that a study was done this year to see if that percent has decreased. Ninety-eight students were randomly surveyed with the following results. Fourteen said that they speak a language other than English at home.

a.   State an appropriate null hypothesis.

b.   State an appropriate alternative hypothesis.

c.   Define the random variable, P'.

d.   Calculate the test statistic.

e.   Calculate the *p*-value.

f.   At the 5% level of decision, what is your decision about the null hypothesis?

g.   What is the Type I error?

h.   What is the Type II error?

119. Assume that you are an emergency paramedic called in to rescue victims of an accident. You need to help a patient who is bleeding profusely. The patient is also considered to be a high risk for contracting AIDS. Assume that the null hypothesis is that the patient does **not** have the HIV virus. What is a Type I error?

120. It is often said that Californians are more casual than the rest of Americans. Suppose that a survey was done to see if the proportion of Californian professionals that wear jeans to work is greater than the proportion of non-Californian professionals. Fifty of each was surveyed with the following results. Fifteen Californians wear jeans to work and six non-Californians wear jeans to work.
Let C = Californian professional; NC = non-Californian professional

a.   State appropriate null and alternate hypotheses.

b.   Define the random variable.

c.   Calculate the test statistic and *p*-value.

d.   At the 5% significance level, what is your decision?

e.   What is the Type I error?

   f.   What is the Type II error?

Use the following information to answer the next two exercises: A group of Statistics students have developed a technique that they feel will lower their anxiety level on statistics exams. They measured their anxiety level at the start of the quarter and again at the end of the quarter. Recorded is the paired data in that order: (1000, 900); (1200, 1050); (600, 700); (1300, 1100); (1000, 900); (900, 900).

121. This is a test of (pick the best answer):

  a. large samples, independent means

  b. small samples, independent means

  c. dependent means

122. State the distribution to use for the test.

Chapter 12

Use the following information to answer the next two exercises: A recent survey of U.S. teenage pregnancy was answered by 720 girls, age 12–19. Six percent of the girls surveyed said they have been pregnant. We are interested in the true proportion of U.S. girls, age 12–19, who have been pregnant.

123. Find the 95% confidence interval for the true proportion of U.S. girls, age 12–19, who have been pregnant.

124. The report also stated that the results of the survey are accurate to within ±3.7% at the 95% confidence level. Suppose that a new study is to be done. It is desired to be accurate to within 2% of the 95% confidence level. What is the minimum number that should be surveyed?

125. Given: $X \sim Exp\left(\frac{1}{3}\right)$. Sketch the graph that depicts: $P(x > 1)$.

Use the following information to answer the next three exercises: The amount of money a customer spends in one trip to the supermarket is known to have an exponential distribution. Suppose the mean amount of money a customer spends in one trip to the supermarket is $72.

126. Find the probability that one customer spends less than $72 in one trip to the supermarket?

127. Suppose five customers pool their money. How much money altogether would you expect the five customers to spend in one trip to the supermarket (in dollars)?

128. State the distribution to use if you want to find the probability that the **mean** amount spent by five customers in one trip to the supermarket is less than $60.

Chapter 13

Use the following information to answer the next two exercises: Suppose that the probability of a drought in any independent year is 20%. Out of those years in which a drought occurs, the probability of water rationing is 10%. However, in any year, the probability of water rationing is 5%.

129. What is the probability of both a drought **and** water rationing occurring?

130. Out of the years with water rationing, find the probability that there is a drought.

Use the following information to answer the next three exercises:

	Apple	Pumpkin	Pecan
Female	40	10	30
Male	20	30	10

Table A7

131. Suppose that one individual is randomly chosen. Find the probability that the person's favorite pie is apple **or** the person is male.

132. Suppose that one male is randomly chosen. Find the probability his favorite pie is pecan.

133. Conduct a hypothesis test to determine if favorite pie type and gender are independent.

Use the following information to answer the next two exercises: Let's say that the probability that an adult watches the news at least once per week is 0.60.

134. We randomly survey 14 people. On average, how many people do we expect to watch the news at least once per week?

135. We randomly survey 14 people. Of interest is the number that watch the news at least once per week. State the distribution of X. $X \sim$ _____

136. The following histogram is most likely to be a result of sampling from which distribution?

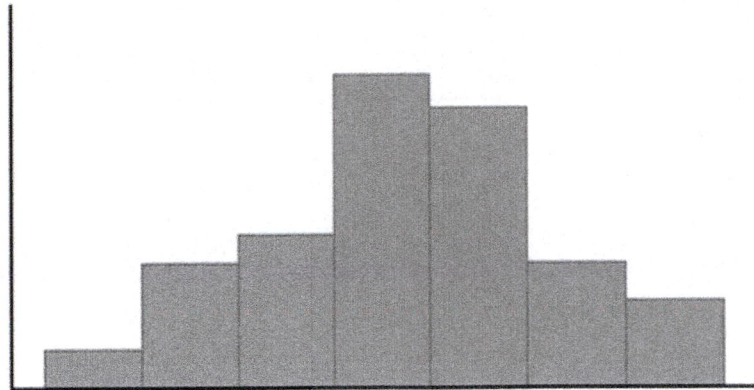

Figure A6

   a. Chi-Square

   b. Geometric

   c. Uniform

   d. Binomial

137. The ages of De Anza evening students is known to be normally distributed with a population mean of 40 and a population standard deviation of six. A sample of six De Anza evening students reported their ages (in years) as: 28; 35; 47; 45; 30; 50. Find the probability that the mean of six ages of randomly chosen students is less than 35 years. Hint: Find the sample mean.

138. A math exam was given to all the fifth grade children attending Country School. Two random samples of scores were taken. The null hypothesis is that the mean math scores for boys and girls in fifth grade are the same. Conduct a hypothesis test.

	n	$\bar{x}$	s^2
Boys	55	82	29
Girls	60	86	46

Table A8

139. In a survey of 80 males, 55 had played an organized sport growing up. Of the 70 females surveyed, 25 had played an organized sport growing up. We are interested in whether the proportion for males is higher than the proportion for females. Conduct a hypothesis test.

140. Which of the following is preferable when designing a hypothesis test?

   a. Maximize α and minimize β

   b. Minimize α and maximize β

c. Maximize α and β

d. Minimize α and β

Use the following information to answer the next three exercises: 120 people were surveyed as to their favorite beverage (non-alcoholic). The results are below.

Beverage/Age	0–9	10–19	20–29	30+	Totals
Milk	14	10	6	0	30
Soda	3	8	26	15	52
Juice	7	12	12	7	38
Totals	24	330	44	22	120

Table A9

141. Are the events of milk and 30+:

a. independent events? Justify your answer.

b. mutually exclusive events? Justify your answer.

142. Suppose that one person is randomly chosen. Find the probability that person is 10–19 given that he or she prefers juice.

143. Are "Preferred Beverage" and "Age" independent events? Conduct a hypothesis test.

144. Given the following histogram, which distribution is the data most likely to come from?

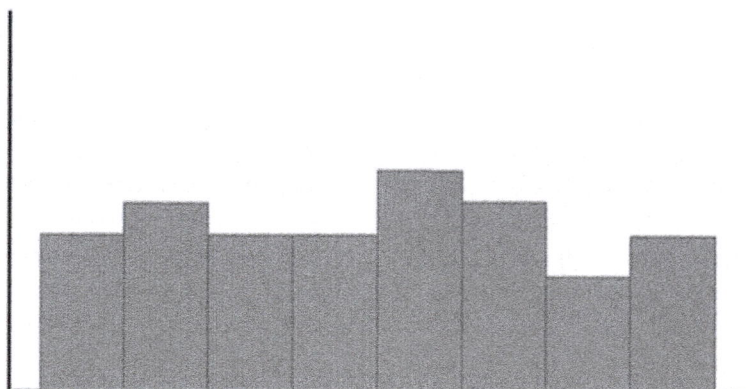

Figure A7

a. uniform

b. exponential

c. normal

d. chi-square

Solutions
Chapter 3

1. c. parameter

2. a. population

3. b. statistic

4. d. sample

5. e. variable

6. quantitative continuous

7.

  a. 2.27

  b. 3.04

  c. −1, 4, 4

8. Answers will vary.

Chapter 4

9. c. (0.80)(0.30)

10. b. No, and they are not mutually exclusive either.

11. a. all employed adult women

12. 0.5773

13. 0.0522

14. b. The middle fifty percent of the members lost from 2 to 8.5 lbs.

15. c. All of the data have the same value.

16. c. The lowest data value is the median.

17. 0.279

18. b. No, I expect to come out behind in money.

19. X = the number of patients calling in claiming to have the flu, who actually have the flu. X = 0, 1, 2, …25

20. $B(25, 0.04)$

21. 0.0165

22. 1

23. c. quantitative discrete

24. all words used by Tom Clancy in his novels

Chapter 5

25.

  a. 24%

  b. 27%

26. qualitative

27. 0.36

28. 0.7636

29.

  a. No

  b. No

30. $B(10, 0.76)$

31. 0.9330

32.

   a. X = the number of questions posted to the statistics listserv per day.

   b. $X = 0, 1, 2,…$

   c. $X \sim P(2)$

   d. 0

33. $150

34. Matt

35.

   a. false

   b. true

   c. false

   d. false

36. 16

37. first quartile: 2
second quartile: 2
third quartile: 3

38. 0.5

39. $\frac{7}{15}$

40. $\frac{2}{15}$

Chapter 6

41.

   a. true

   b. true

   c. False – the median and the mean are the same for this symmetric distribution.

   d. true

42.

   a. 8

   b. 8

   c. $P(x < k) = 0.65 = (k - 3)\left(\frac{1}{10}\right)$. $k = 9.5$

43.

   a. False – $\frac{3}{4}$ of the data are at most five.

   b. True – each quartile has 25% of the data.

   c. False – that is unknown.

803

d. False – 50% of the data are four or less.

44. d. *G* and *H* are independent events.

45.

a. False – *J* and *K* are independent so they are not mutually exclusive which would imply dependency (meaning $P(J \text{ AND } K)$ is not 0).

b. False – see answer c.

c. True – $P(J \text{ OR } K) = P(J) + P(K) - P(J \text{ AND } K) = P(J) + P(K) - P(J)P(K) = 0.3 + 0.6 - (0.3)(0.6) = 0.72$. Note the $P(J \text{ AND } K) = P(J)P(K)$ because *J* and *K* are independent.

d. False – *J* and *K* are independent so $P(J) = P(J|K)$

46. a. $P(5)$

Chapter 7

47. a. $U(0, 4)$

48. b. 2 hour

49. a. $\frac{1}{4}$

50.

a. 0.7165

b. 4.16

c. 0

51. c. 5 years

52. c. exponential

53. 0.63

54. $B(14, 0.20)$

55. $B(14, 0.20)$

Chapter 8

56. c. the mean amount of weight lost by 15 people on the special weight loss diet.

57. 0.9951

58. 12.99

59. c. $\frac{1}{2}$

60. b. 0.60

61. c. $N(60, 5.477)$

62. 0.9990

63. a. eight days

64. c. 0.7500

65. a. 80%

66. b. 35%

67. b. no

68. b. quantitative continuous

69. c. 150

70. d. 0.06

71. c. 0.44

72. b. 0

Chapter 9

73. d. Matt is shorter than the average 14 year old boy.

74. Answers will vary.

75.

x	Relative Frequency	Cumulative Relative Frequency
1	0.3	0.3
2	0.2	0.2
4	0.4	0.4
5	0.1	0.1

Table A10

76.

a. 2.8

b. 1.48

c. 90%

77. $M = 3$; $Q_1 = 1$; $Q_3 = 4$

78. 1 and 4

79. d. $\frac{8}{70}$

80. c. $\frac{40}{70}$

81. a. $\frac{9}{19}$

82. b. false

83. b. false

84. b. false

85.

a. X = the number of pies Lee bakes every day.

b. $P(20)$

c. 0.1122

86. CI: (5.25, 8.48)

87.

a. uniform

b. exponential

c. normal

Chapter 10

88. $\frac{77}{100}$

89. $\frac{12}{42}$

90.

 a. false

 b. false

 c. true

 d. false

91. $N(180, 16.43)$

92. a. The distribution for $\bar{X}$ is still uniform with the same mean and standard deviation as the distribution for X.

93. c. The distribution for $\sum X$ is normal with a larger mean and a larger standard deviation than the distribution for X.

94. $N\left(2, \frac{0.25}{\sqrt{16}}\right)$

95. Answers will vary.

96. 0.5000

97. 7.6

98. 5

99. 0.9431

Chapter 11

100. 7.5

101. 0.0122

102. $N(7, 0.63)$

103. 0.9911

104. b. Exponential

105.

 a. true

 b. false

 c. false

106. Answers will vary.

107. Student's t with $df = 15$

108. (560.07, 719.93)

109. quantitative continuous data

110. quantitative discrete data

111.

 a. $X =$ the number of patients with a shotgun wound the emergency room gets per 28 days

 b. $P(4)$

c. 0.0183

112. greater than

113. No; $P(x = 8) = 0.0348$

114. You will lose $5.

115. Becca

116. 14

117. Sample mean = 3.2
Sample standard deviation = 1.85
Median = 3
$Q_1 = 2$
$Q_3 = 5$
$IQR = 3$

118. d. $z = -1.19$
e. 0.1171
f. Do not reject the null hypothesis.

119. We conclude that the patient does have the HIV virus when, in fact, the patient does not.

120. c. $z = 2.21$; $p = 0.0136$
d. Reject the null hypothesis.
e. We conclude that the proportion of Californian professionals that wear jeans to work is greater than the proportion of non-Californian professionals when, in fact, it is not greater.
f. We cannot conclude that the proportion of Californian professionals that wear jeans to work is greater than the proportion of non-Californian professionals when, in fact, it is greater.

121. c. dependent means

122. t_5

Chapter 12

123. (0.0424, 0.0770)

124. 2,401

125. Check student's solution.

126. 0.6321

127. $360

128. $N\left(72, \ \frac{72}{\sqrt{5}}\right)$

Chapter 13

129. 0.02

130. 0.40

131. $\frac{100}{140}$

132. $\frac{10}{60}$

133. p-value = 0; Reject the null hypothesis; conclude that they are dependent events

134. 8.4

135. $B(14, 0.60)$

136. d. Binomial

137. 0.3669

138. *p*-value = 0.0006; reject the null hypothesis; conclude that the averages are not equal

139. *p*-value = 0; reject the null hypothesis; conclude that the proportion of males is higher

140. Minimize α and β

141.

a. No

b. Yes, $P(M \text{ AND } 30+) = 0$

142. $\frac{12}{38}$

143. No; *p*-value = 0

144. a. uniform

References

Data from the *San Jose Mercury News*.

Baran, Daya. "20 Percent of Americans Have Never Used Email." Webguild.org, 2010. Available online at: http://www.webguild.org/20080519/20-percent-of-americans-have-never-used-email (accessed October 17, 2013).

Data from *Parade Magazine*.

APPENDIX B: PRACTICE TESTS (1-4) AND FINAL EXAMS

Practice Test 1
1.1: Definitions of Statistics, Probability, and Key Terms

Use the following information to answer the next three exercises. A grocery store is interested in how much money, on average, their customers spend each visit in the produce department. Using their store records, they draw a sample of 1,000 visits and calculate each customer's average spending on produce.

1. Identify the population, sample, parameter, statistic, variable, and data for this example.

  a. population

  b. sample

  c. parameter

  d. statistic

  e. variable

  f. data

2. What kind of data is "amount of money spent on produce per visit"?

  a. qualitative

  b. quantitative-continuous

  c. quantitative-discrete

3. The study finds that the mean amount spent on produce per visit by the customers in the sample is $12.84. This is an example of a:

  a. population

  b. sample

  c. parameter

  d. statistic

  e. variable

1.2: Data, Sampling, and Variation in Data and Sampling

Use the following information to answer the next two exercises. A health club is interested in knowing how many times a typical member uses the club in a week. They decide to ask every tenth customer on a specified day to complete a short survey including information about how many times they have visited the club in the past week.

4. What kind of a sampling design is this?

  a. cluster

  b. stratified

c. simple random

d. systematic

5. "Number of visits per week" is what kind of data?

a. qualitative

b. quantitative-continuous

c. quantitative-discrete

6. Describe a situation in which you would calculate a parameter, rather than a statistic.

7. The U.S. federal government conducts a survey of high school seniors concerning their plans for future education and employment. One question asks whether they are planning to attend a four-year college or university in the following year. Fifty percent answer yes to this question; that fifty percent is a:

a. parameter

b. statistic

c. variable

d. data

8. Imagine that the U.S. federal government had the means to survey all high school seniors in the U.S. concerning their plans for future education and employment, and found that 50 percent were planning to attend a 4-year college or university in the following year. This 50 percent is an example of a:

a. parameter

b. statistic

c. variable

d. data

Use the following information to answer the next three exercises. A survey of a random sample of 100 nurses working at a large hospital asked how many years they had been working in the profession. Their answers are summarized in the following (incomplete) table.

9. Fill in the blanks in the table and round your answers to two decimal places for the Relative Frequency and Cumulative Relative Frequency cells.

# of years	Frequency	Relative Frequency	Cumulative Relative Frequency
< 5	25		
5–10	30		
> 10	empty		

Table B1

10. What proportion of nurses have five or more years of experience?

11. What proportion of nurses have ten or fewer years of experience?

12. Describe how you might draw a random sample of 30 students from a lecture class of 200 students.

13. Describe how you might draw a stratified sample of students from a college, where the strata are the students' class standing (freshman, sophomore, junior, or senior).

14. A manager wants to draw a sample, without replacement, of 30 employees from a workforce of 150. Describe how the chance of being selected will change over the course of drawing the sample.

15. The manager of a department store decides to measure employee satisfaction by selecting four departments at random, and conducting interviews with all the employees in those four departments. What type of survey design is this?

   a. cluster

   b. stratified

   c. simple random

   d. systematic

16. A popular American television sports program conducts a poll of viewers to see which team they believe will win the NFL (National Football League) championship this year. Viewers vote by calling a number displayed on the television screen and telling the operator which team they think will win. Do you think that those who participate in this poll are representative of all football fans in America?

17. Two researchers studying vaccination rates independently draw samples of 50 children, ages 3–18 months, from a large urban area, and determine if they are up to date on their vaccinations. One researcher finds that 84 percent of the children in her sample are up to date, and the other finds that 86 percent in his sample are up to date. Assuming both followed proper sampling procedures and did their calculations correctly, what is a likely explanation for this discrepancy?

18. A high school increased the length of the school day from 6.5 to 7.5 hours. Students who wished to attend this high school were required to sign contracts pledging to put forth their best effort on their school work and to obey the school rules; if they did not wish to do so, they could attend another high school in the district. At the end of one year, student performance on statewide tests had increased by ten percentage points over the previous year. Does this improvement prove that a longer school day improves student achievement?

19. You read a newspaper article reporting that eating almonds leads to increased life satisfaction. The study was conducted by the Almond Growers Association, and was based on a randomized survey asking people about their consumption of various foods, including almonds, and also about their satisfaction with different aspects of their life. Does anything about this poll lead you to question its conclusion?

20. Why is non-response a problem in surveys?

1.3: Frequency, Frequency Tables, and Levels of Measurement

21. Compute the mean of the following numbers, and report your answer using one more decimal place than is present in the original data:
14, 5, 18, 23, 6

1.4: Experimental Design and Ethics

22. A psychologist is interested in whether the size of tableware (bowls, plates, etc.) influences how much college students eat. He randomly assigns 100 college students to one of two groups: the first is served a meal using normal-sized tableware, while the second is served the same meal, but using tableware that it 20 percent smaller than normal. He records how much food is consumed by each group. Identify the following components of this study.

   a. population

   b. sample

   c. experimental units

   d. explanatory variable

   e. treatment

   f. response variable

23. A researcher analyzes the results of the SAT (Scholastic Aptitude Test) over a five-year period and finds that male students on average score higher on the math section, and female students on average score higher on the verbal section. She concludes that these observed differences in test performance are due to genetic factors. Explain how lurking variables could offer an alternative explanation for the observed differences in test scores.

24. Explain why it would not be possible to use random assignment to study the health effects of smoking.

25. A professor conducts a telephone survey of a city's population by drawing a sample of numbers from the phone book and having her student assistants call each of the selected numbers once to administer the survey. What are some sources of bias with this survey?

26. A professor offers extra credit to students who take part in her research studies. What is an ethical problem with this method of recruiting subjects?

2.1: Stem-and Leaf Graphs (Stemplots), Line Graphs, and Bar Graphs

Use the following information to answer the next four exercises. The midterm grades on a chemistry exam, graded on a scale of 0 to 100, were:
62, 64, 65, 65, 68, 70, 72, 72, 74, 75, 75, 75, 76,78, 78, 81, 83, 83, 84, 85, 87, 88, 92, 95, 98, 98, 100, 100, 740

27. Do you see any outliers in this data? If so, how would you address the situation?

28. Construct a stem plot for this data, using only the values in the range 0–100.

29. Describe the distribution of exam scores.

2.2: Histograms, Frequency Polygons, and Time Series Graphs

30. In a class of 35 students, seven students received scores in the 70–79 range. What is the relative frequency of scores in this range?

Use the following information to answer the next three exercises. You conduct a poll of 30 students to see how many classes they are taking this term. Your results are:
1; 1; 1; 1
2; 2; 2; 2; 2
3; 3; 3; 3; 3; 3; 3; 3
4; 4; 4; 4; 4; 4; 4; 4; 4
5; 5; 5; 5

31. You decide to construct a histogram of this data. What will be the range of your first bar, and what will be the central point?

32. What will be the widths and central points of the other bars?

33. Which bar in this histogram will be the tallest, and what will be its height?

34. You get data from the U.S. Census Bureau on the median household income for your city, and decide to display it graphically. Which is the better choice for this data, a bar graph or a histogram?

35. You collect data on the color of cars driven by students in your statistics class, and want to display this information graphically. Which is the better choice for this data, a bar graph or a histogram?

2.3: Measures of the Location of the Data

36. Your daughter brings home test scores showing that she scored in the 80th percentile in math and the 76th percentile in reading for her grade. Interpret these scores.

37. You have to wait 90 minutes in the emergency room of a hospital before you can see a doctor. You learn that your wait time was in the 82nd percentile of all wait times. Explain what this means, and whether you think it is good or bad.

2.4: Box Plots

Use the following information to answer the next three exercises. 1; 1; 2; 3; 4; 4; 5; 5; 6; 7; 7; 8; 9

38. What is the median for this data?

39. What is the first quartile for this data?

40. What is the third quartile for this data?

Use the following information to answer the next four exercises. This box plot represents scores on the final exam for a physics class.

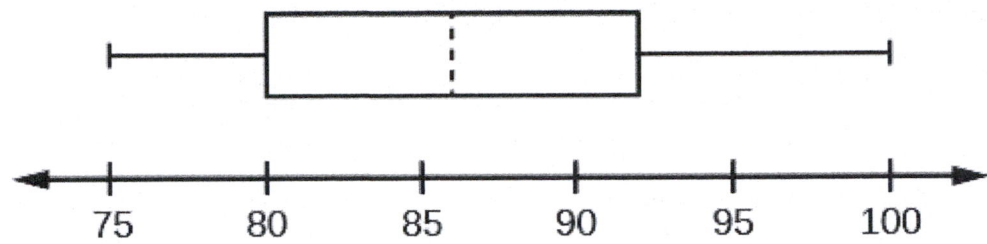

Figure B1

41. What is the median for this data, and how do you know?

42. What are the first and third quartiles for this data, and how do you know?

43. What is the interquartile range for this data?

44. What is the range for this data?

2.5: Measures of the Center of the Data

45. In a marathon, the median finishing time was 3:35:04 (three hours, 35 minutes, and four seconds). You finished in 3:34:10. Interpret the meaning of the median time, and discuss your time in relation to it.

Use the following information to answer the next three exercises. The value, in thousands of dollars, for houses on a block, are: 45; 47; 47.5; 51; 53.5; 125.

46. Calculate the mean for this data.

47. Calculate the median for this data.

48. Which do you think better reflects the average value of the homes on this block?

2.6: Skewness and the Mean, Median, and Mode

49. In a left-skewed distribution, which is greater?

a. the mean

b. the media

c. the mode

50. In a right-skewed distribution, which is greater?

a. the mean

b. the median

c. the mode

51. In a symmetrical distribution what will be the relationship among the mean, median, and mode?

2.7: Measures of the Spread of the Data

Use the following information to answer the next four exercises. 10; 11; 15; 15; 17; 22

52. Compute the mean and standard deviation for this data; use the sample formula for the standard deviation.

53. What number is two standard deviations above the mean of this data?

54. Express the number 13.7 in terms of the mean and standard deviation of this data.

55. In a biology class, the scores on the final exam were normally distributed, with a mean of 85, and a standard deviation of five. Susan got a final exam score of 95. Express her exam result as a z-score, and interpret its meaning.

3.1: Terminology

Use the following information to answer the next two exercises. You have a jar full of marbles: 50 are red, 25 are blue, and 15 are yellow. Assume you draw one marble at random for each trial, and replace it before the next trial. Let $P(R)$ = the probability of drawing a red marble.

Let $P(B)$ = the probability of drawing a blue marble.
Let $P(Y)$ = the probability of drawing a yellow marble.

56. Find $P(B)$.

57. Which is more likely, drawing a red marble or a yellow marble? Justify your answer numerically.

Use the following information to answer the next two exercises. The following are probabilities describing a group of college students.
Let $P(M)$ = the probability that the student is male
Let $P(F)$ = the probability that the student is female
Let $P(E)$ = the probability the student is majoring in education
Let $P(S)$ = the probability the student is majoring in science

58. Write the symbols for the probability that a student, selected at random, is both female and a science major.

59. Write the symbols for the probability that the student is an education major, given that the student is male.

3.2: Independent and Mutually Exclusive Events

60. Events A and B are independent.
If $P(A) = 0.3$ and $P(B) = 0.5$, find $P(A$ AND $B)$.

61. C and D are mutually exclusive events.
If $P(C) = 0.18$ and $P(D) = 0.03$, find $P(C$ OR $D)$.

3.3: Two Basic Rules of Probability

62. In a high school graduating class of 300, 200 students are going to college, 40 are planning to work full-time, and 80 are taking a gap year. Are these events mutually exclusive?

Use the following information to answer the next two exercises. An archer hits the center of the target (the bullseye) 70 percent of the time. However, she is a streak shooter, and if she hits the center on one shot, her probability of hitting it on the shot immediately following is 0.85. Written in probability notation:
$P(A) = P(B) = P$(hitting the center on one shot) $= 0.70$
$P(B|A) = P$(hitting the center on a second shot, given that she hit it on the first) $= 0.85$

63. Calculate the probability that she will hit the center of the target on two consecutive shots.

64. Are $P(A)$ and $P(B)$ independent in this example?

3.4: Contingency Tables

Use the following information to answer the next three exercises. The following contingency table displays the number of students who report studying at least 15 hours per week, and how many made the honor roll in the past semester.

	Honor roll	No honor roll	Total
Study at least 15 hours/week		200	
Study less than 15 hours/week	125	193	
Total			1,000

Table B2

65. Complete the table.

66. Find P(honor roll|study at least 15 hours per week).

67. What is the probability a student studies less than 15 hours per week?

68. Are the events "study at least 15 hours per week" and "makes the honor roll" independent? Justify your answer numerically.

3.5: Tree and Venn Diagrams

69. At a high school, some students play on the tennis team, some play on the soccer team, but neither plays both tennis and soccer. Draw a Venn diagram illustrating this.

70. At a high school, some students play tennis, some play soccer, and some play both. Draw a Venn diagram illustrating this.

Practice Test 1 Solutions
1.1: Definitions of Statistics, Probability, and Key Terms

1.

a. population: all the shopping visits by all the store's customers

b. sample: the 1,000 visits drawn for the study

c. parameter: the average expenditure on produce per visit by all the store's customers

d. statistic: the average expenditure on produce per visit by the sample of 1,000

e. variable: the expenditure on produce for each visit

f. data: the dollar amounts spent on produce; for instance, $15.40, $11.53, etc

2. c

3. d

1.2: Data, Sampling, and Variation in Data and Sampling

4. d

5. c

6. Answers will vary.
Sample Answer: Any solution in which you use data from the entire population is acceptable. For instance, a professor might calculate the average exam score for her class: because the scores of all members of the class were used in the calculation, the average is a parameter.

7. b

8. a

9.

# of years	Frequency	Relative Frequency	Cumulative Relative Frequency
< 5	25	0.25	0.25
5–10	30	0.30	0.55
> 10	45	0.45	1.00

Table B3

10. 0.75

11. 0.55

12. Answers will vary.
Sample Answer: One possibility is to obtain the class roster and assign each student a number from 1 to 200. Then use a random number generator or table of random number to generate 30 numbers between 1 and 200, and select the students matching the random numbers. It would also be acceptable to write each student's name on a card, shuffle them in a box, and draw 30 names at random.

13. One possibility would be to obtain a roster of students enrolled in the college, including the class standing for each student. Then you would draw a proportionate random sample from within each class (for instance, if 30 percent of the students in the college are freshman, then 30 percent of your sample would be drawn from the freshman class).

14. For the first person picked, the chance of any individual being selected is one in 150. For the second person, it is one in 149, for the third it is one in 148, and so on. For the 30th person selected, the chance of selection is one in 121.

15. a

16. No. There are at least two chances for bias. First, the viewers of this particular program may not be representative of American football fans as a whole. Second, the sample will be self-selected, because people have to make a phone call in order to take part, and those people are probably not representative of the American football fan population as a whole.

17. These results (84 percent in one sample, 86 percent in the other) are probably due to sampling variability. Each researcher drew a different sample of children, and you would not expect them to get exactly the same result, although you would expect the results to be similar, as they are in this case.

18. No. The improvement could also be due to self-selection: only motivated students were willing to sign the contract, and they would have done well even in a school with 6.5 hour days. Because both changes were implemented at the same time, it is not possible to separate out their influence.

19. At least two aspects of this poll are troublesome. The first is that it was conducted by a group who would benefit by the result—almond sales are likely to increase if people believe that eating almonds will make them happier. The second is that this poll found that almond consumption and life satisfaction are correlated, but does not establish that eating almonds causes satisfaction. It is equally possible, for instance, that people with higher incomes are more likely to eat almonds, and are also more satisfied with their lives.

20. You want the sample of people who take part in a survey to be representative of the population from which they are drawn. People who refuse to take part in a survey often have different views than those who do participate, and so even a random sample may produce biased results if a large percentage of those selected refuse to participate in a survey.

1.3: Frequency, Frequency Tables, and Levels of Measurement

21. 13.2

1.4: Experimental Design and Ethics

22.

a. population: all college students

b. sample: the 100 college students in the study

c. experimental units: each individual college student who participated

d. explanatory variable: the size of the tableware

e. treatment: tableware that is 20 percent smaller than normal

f. response variable: the amount of food eaten

23. There are many lurking variables that could influence the observed differences in test scores. Perhaps the boys, on average, have taken more math courses than the girls, and the girls have taken more English classes than the boys. Perhaps the boys have been encouraged by their families and teachers to prepare for a career in math and science, and thus have put more effort into studying math, while the girls have been encouraged to prepare for fields like communication and psychology that are more focused on language use. A study design would have to control for these and other potential lurking variables (anything that could explain the observed difference in test scores, other than the genetic explanation) in order to draw a scientifically sound conclusion about genetic differences.

24. To use random assignment, you would have to be able to assign people to either smoke or not smoke. Because smoking has many harmful effects, this would not be an ethical experiment. Instead, we study people who have chosen to smoke, and compare them to others who have chosen not to smoke, and try to control for the other ways those two groups may differ (lurking variables).

25. Sources of bias include the fact that not everyone has a telephone, that cell phone numbers are often not listed in published directories, and that an individual might not be at home at the time of the phone call; all these factors make it likely that the respondents to the survey will not be representative of the population as a whole.

26. Research subjects should not be coerced into participation, and offering extra credit in exchange for participation could be construed as coercion. In addition, this method will result in a volunteer sample, which cannot be assumed to be representative of the population as a whole.

2.1: Stem-and Leaf Graphs (Stemplots), Line Graphs, and Bar Graphs

27. The value 740 is an outlier, because the exams were graded on a scale of 0 to 100, and 740 is far outside that range. It may be a data entry error, with the actual score being 74, so the professor should check that exam again to see what the actual score was.

28.

Stem	Leaf
6	2 4 5 5 8
7	0 2 2 4 5 5 5 6 8 8
8	1 3 3 4 5 7 8
9	2 5 8 8
10	0 0

Table B4

29. Most scores on this exam were in the range of 70–89, with a few scoring in the 60–69 range, and a few in the 90–100 range.

2.2: Histograms, Frequency Polygons, and Time Series Graphs

30. $RF = \frac{7}{35} = 0.2$

31. The range will be 0.5–1.5, and the central point will be 1.

32. Range 1.5–2.5, central point 2; range 2.5–3.5, central point 3; range 3.5–4.5, central point 4; range 4.5–5.5., central point 5.

33. The bar from 3.5 to 4.5, with a central point of 4, will be tallest; its height will be nine, because there are nine students taking four courses.

34. The histogram is a better choice, because income is a continuous variable.

35. A bar graph is the better choice, because this data is categorical rather than continuous.

2.3: Measures of the Location of the Data

36. Your daughter scored better than 80 percent of the students in her grade on math and better than 76 percent of the students in reading. Both scores are very good, and place her in the upper quartile, but her math score is slightly better in relation to her peers than her reading score.

37. You had an unusually long wait time, which is bad: 82 percent of patients had a shorter wait time than you, and only 18 percent had a longer wait time.

2.4: Box Plots

38. 5

39. 3

40. 7

41. The median is 86, as represented by the vertical line in the box.

42. The first quartile is 80, and the third quartile is 92, as represented by the left and right boundaries of the box.

43. $IQR = 92 - 80 = 12$

44. Range = $100 - 75 = 25$

2.5: Measures of the Center of the Data

45. Half the runners who finished the marathon ran a time faster than 3:35:04, and half ran a time slower than 3:35:04. Your time is faster than the median time, so you did better than more than half of the runners in this race.

46. 61.5, or $61,500

47. 49.25 or $49,250

48. The median, because the mean is distorted by the high value of one house.

2.6: Skewness and the Mean, Median, and Mode

49. c

50. a

51. They will all be fairly close to each other.

2.7: Measures of the Spread of the Data

52. Mean: 15
Standard deviation: 4.3

$$\mu = \frac{10 + 11 + 15 + 15 + 17 + 22}{6} = 15$$

$$s = \sqrt{\frac{\sum(x - \bar{x})^2}{n - 1}} = \sqrt{\frac{94}{5}} = 4.3$$

53. 15 + (2)(4.3) = 23.6

54. 13.7 is one standard deviation below the mean of this data, because 15 − 4.3 = 10.7

55. $z = \frac{95 - 85}{5} = 2.0$

Susan's z-score was 2.0, meaning she scored two standard deviations above the class mean for the final exam.

3.1: Terminology

56. $P(B) = \frac{25}{90} = 0.28$

57. Drawing a red marble is more likely.
$P(R) = \frac{50}{80} = 0.62$

$P(Y) = \frac{15}{80} = 0.19$

58. $P(F \text{ AND } S)$

59. $P(E|M)$

3.2: Independent and Mutually Exclusive Events

60. $P(A \text{ AND } B) = (0.3)(0.5) = 0.15$

61. $P(C \text{ OR } D) = 0.18 + 0.03 = 0.21$

3.3: Two Basic Rules of Probability

62. No, they cannot be mutually exclusive, because they add up to more than 300. Therefore, some students must fit into two or more categories (e.g., both going to college and working full time).

63. $P(A \text{ and } B) = (P(B|A))(P(A)) = (0.85)(0.70) = 0.595$

64. No. If they were independent, $P(B)$ would be the same as $P(B|A)$. We know this is not the case, because $P(B) = 0.70$ and $P(B|A) = 0.85$.

3.4: Contingency Tables

65.

	Honor roll	No honor roll	Total
Study at least 15 hours/week	482	200	682
Study less than 15 hours/week	125	193	318
Total	607	393	1,000

Table B5

66. P(honor roll|study at least 15 hours word per week) $= \dfrac{482}{1000} = 0.482$

67. P(studies less than 15 hours word per week) $= \dfrac{125 + 193}{1000} = 0.318$

68. Let $P(S)$ = study at least 15 hours per week
Let $P(H)$ = makes the honor roll
From the table, $P(S) = 0.682$, $P(H) = 0.607$, and $P(S \text{ AND } H) = 0.482$.
If $P(S)$ and $P(H)$ were independent, then $P(S \text{ AND } H)$ would equal $(P(S))(P(H))$.
However, $(P(S))(P(H)) = (0.682)(0.607) = 0.414$, while $P(S \text{ AND } H) = 0.482$.
Therefore, $P(S)$ and $P(H)$ are not independent.

3.5: Tree and Venn Diagrams

69.

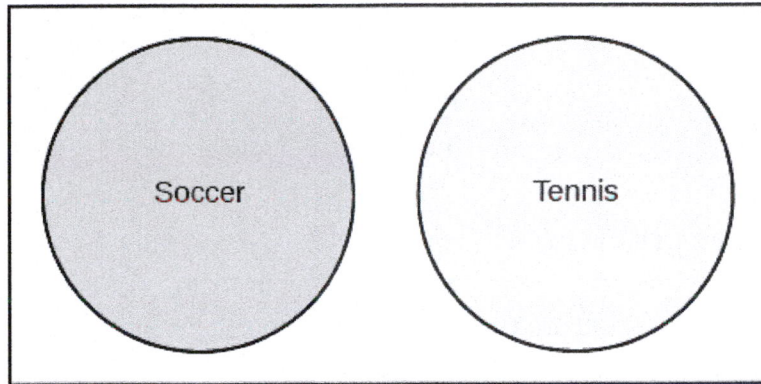

Figure B2

70.

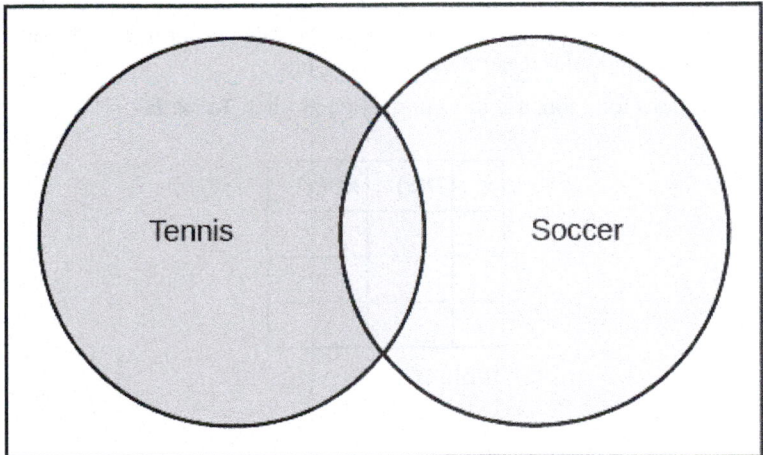

Figure B3

Practice Test 2
4.1: Probability Distribution Function (PDF) for a Discrete Random Variable

Use the following information to answer the next five exercises. You conduct a survey among a random sample of students at a particular university. The data collected includes their major, the number of classes they took the previous semester, and amount of money they spent on books purchased for classes in the previous semester.

1. If X = student's major, then what is the domain of X?

2. If Y = the number of classes taken in the previous semester, what is the domain of Y?

3. If Z = the amount of money spent on books in the previous semester, what is the domain of Z?

4. Why are X, Y, and Z in the previous example random variables?

5. After collecting data, you find that for one case, $z = -7$. Is this a possible value for Z?

6. What are the two essential characteristics of a discrete probability distribution?

Use this discrete probability distribution represented in this table to answer the following six questions. The university library records the number of books checked out by each patron over the course of one day, with the following result:

x	P(x)
0	0.20
1	0.45
2	0.20
3	0.10
4	0.05

Table B6

7. Define the random variable X for this example.

8. What is $P(x > 2)$?

9. What is the probability that a patron will check out at least one book?

10. What is the probability a patron will take out no more than three books?

11. If the table listed $P(x)$ as 0.15, how would you know that there was a mistake?

12. What is the average number of books taken out by a patron?

4.2: Mean or Expected Value and Standard Deviation

Use the following information to answer the next four exercises. Three jobs are open in a company: one in the accounting department, one in the human resources department, and one in the sales department. The accounting job receives 30 applicants, and the human resources and sales department 60 applicants.

13. If X = the number of applications for a job, use this information to fill in **Table B7**.

x	P(x)	xP(x)

Table B7

14. What is the mean number of applicants?

15. What is the PDF for X?

16. Add a fourth column to the table, for $(x - \mu)^2 P(x)$.

17. What is the standard deviation of X?

4.3: Binomial Distribution

18. In a binomial experiment, if $p = 0.65$, what does q equal?

19. What are the required characteristics of a binomial experiment?

20. Joe conducts an experiment to see how many times he has to flip a coin before he gets four heads in a row. Does this qualify as a binomial experiment?

Use the following information to answer the next three exercises. In a particularly community, 65 percent of households include at least one person who has graduated from college. You randomly sample 100 households in this community. Let X = the number of households including at least one college graduate.

21. Describe the probability distribution of X.

22. What is the mean of X?

23. What is the standard deviation of X?

Use the following information to answer the next four exercises. Joe is the star of his school's baseball team. His batting average is 0.400, meaning that for every ten times he comes to bat (an at-bat), four of those times he gets a hit. You decide to track his batting performance his next 20 at-bats.

24. Define the random variable X in this experiment.

25. Assuming Joe's probability of getting a hit is independent and identical across all 20 at-bats, describe the distribution of X.

26. Given this information, what number of hits do you predict Joe will get?

27. What is the standard deviation of X?

4.4: Geometric Distribution

28. What are the three major characteristics of a geometric experiment?

29. You decide to conduct a geometric experiment by flipping a coin until it comes up heads. This takes five trials. Represent the outcomes of this trial, using H for heads and T for tails.

30. You are conducting a geometric experiment by drawing cards from a normal 52-card pack, with replacement, until you draw the Queen of Hearts. What is the domain of X for this experiment?

31. You are conducting a geometric experiment by drawing cards from a normal 52-card deck, without replacement, until you draw a red card. What is the domain of X for this experiment?

Use the following information to answer the next three exercises. In a particular university, 27 percent of students are engineering majors. You decide to select students at random until you choose one that is an engineering major. Let X = the number of students you select until you find one that is an engineering major.

32. What is the probability distribution of X?

33. What is the mean of X?

34. What is the standard deviation of X?

4.5: Hypergeometric Distribution

35. You draw a random sample of ten students to participate in a survey, from a group of 30, consisting of 16 boys and 14 girls. You are interested in the probability that seven of the students chosen will be boys. Does this qualify as a hypergeometric experiment? List the conditions and whether or not they are met.

36. You draw five cards, without replacement, from a normal 52-card deck of playing cards, and are interested in the probability that two of the cards are spades. What are the group of interest, size of the group of interest, and sample size for this example?

4.6: Poisson Distribution

37. What are the key characteristics of the Poisson distribution?

Use the following information to answer the next three exercises. The number of drivers to arrive at a toll booth in an hour can be modeled by the Poisson distribution.

38. If X = the number of drivers, and the average numbers of drivers per hour is four, how would you express this distribution?

39. What is the domain of X?

40. What are the mean and standard deviation of X?

5.1: Continuous Probability Functions

41. You conduct a survey of students to see how many books they purchased the previous semester, the total amount they paid for those books, the number they sold after the semester was over, and the amount of money they received for the books they sold. Which variables in this survey are discrete, and which are continuous?

42. With continuous random variables, we never calculate the probability that X has a particular value, but always speak in terms of the probability that X has a value within a particular range. Why is this?

43. For a continuous random variable, why are $P(x < c)$ and $P(x \leq c)$ equivalent statements?

44. For a continuous probability function, $P(x < 5) = 0.35$. What is $P(x > 5)$, and how do you know?

45. Describe how you would draw the continuous probability distribution described by the function $f(x) = \frac{1}{10}$ for $0 \leq x \leq 10$. What type of a distribution is this?

46. For the continuous probability distribution described by the function $f(x) = \frac{1}{10}$ for $0 \leq x \leq 10$, what is the $P(0 < x < 4)$?

5.2: The Uniform Distribution

47. For the continuous probability distribution described by the function $f(x) = \frac{1}{10}$ for $0 \leq x \leq 10$, what is the $P(2 < x < 5)$?

Use the following information to answer the next four exercises. The number of minutes that a patient waits at a medical clinic to see a doctor is represented by a uniform distribution between zero and 30 minutes, inclusive.

48. If X equals the number of minutes a person waits, what is the distribution of X?

49. Write the probability density function for this distribution.

50. What is the mean and standard deviation for waiting time?

51. What is the probability that a patient waits less than ten minutes?

5.3: The Exponential Distribution

52. The distribution of the variable X, representing the average time to failure for an automobile battery, can be written as: $X \sim Exp(m)$. Describe this distribution in words.

53. If the value of m for an exponential distribution is ten, what are the mean and standard deviation for the distribution?

54. Write the probability density function for a variable distributed as: $X \sim Exp(0.2)$.

6.1: The Standard Normal Distribution

55. Translate this statement about the distribution of a random variable X into words: $X \sim (100, 15)$.

56. If the variable X has the standard normal distribution, express this symbolically.

Use the following information for the next six exercises. According to the World Health Organization, distribution of height in centimeters for girls aged five years and no months has the distribution: $X \sim N(109, 4.5)$.

57. What is the z-score for a height of 112 inches?

58. What is the z-score for a height of 100 centimeters?

59. Find the z-score for a height of 105 centimeters and explain what that means In the context of the population.

60. What height corresponds to a z-score of 1.5 in this population?

61. Using the empirical rule, we expect about 68 percent of the values in a normal distribution to lie within one standard deviation above or below the mean. What does this mean, in terms of a specific range of values, for this distribution?

62. Using the empirical rule, about what percent of heights in this distribution do you expect to be between 95.5 cm and 122.5 cm?

6.2: Using the Normal Distribution

Use the following information to answer the next four exercises. The distributor of lotto tickets claims that 20 percent of the tickets are winners. You draw a sample of 500 tickets to test this proposition.

63. Can you use the normal approximation to the binomial for your calculations? Why or why not.

64. What are the expected mean and standard deviation for your sample, assuming the distributor's claim is true?

65. What is the probability that your sample will have a mean greater than 100?

66. If the z-score for your sample result is -2.00, explain what this means, using the empirical rule.

7.1: The Central Limit Theorem for Sample Means (Averages)

67. What does the central limit theorem state with regard to the distribution of sample means?

68. The distribution of results from flipping a fair coin is uniform: heads and tails are equally likely on any flip, and over a large number of trials, you expect about the same number of heads and tails. Yet if you conduct a study by flipping 30 coins and recording the number of heads, and repeat this 100 times, the distribution of the mean number of heads will be approximately normal. How is this possible?

69. The mean of a normally-distributed population is 50, and the standard deviation is four. If you draw 100 samples of size 40 from this population, describe what you would expect to see in terms of the sampling distribution of the sample mean.

70. X is a random variable with a mean of 25 and a standard deviation of two. Write the distribution for the sample mean of samples of size 100 drawn from this population.

71. Your friend is doing an experiment drawing samples of size 50 from a population with a mean of 117 and a standard deviation of 16. This sample size is large enough to allow use of the central limit theorem, so he says the standard deviation of the sampling distribution of sample means will also be 16. Explain why this is wrong, and calculate the correct value.

72. You are reading a research article that refers to "the standard error of the mean." What does this mean, and how is it calculated?

Use the following information to answer the next six exercises. You repeatedly draw samples of $n = 100$ from a population with a mean of 75 and a standard deviation of 4.5.

73. What is the expected distribution of the sample means?

74. One of your friends tries to convince you that the standard error of the mean should be 4.5. Explain what error your friend made.

75. What is the z-score for a sample mean of 76?

76. What is the z-score for a sample mean of 74.7?

77. What sample mean corresponds to a z-score of 1.5?

78. If you decrease the sample size to 50, will the standard error of the mean be smaller or larger? What would be its value?

Use the following information to answer the next two questions. We use the empirical rule to analyze data for samples of size 60 drawn from a population with a mean of 70 and a standard deviation of 9.

79. What range of values would you expect to include 68 percent of the sample means?

80. If you increased the sample size to 100, what range would you expect to contain 68 percent of the sample means, applying the empirical rule?

7.2: The Central Limit Theorem for Sums

81. How does the central limit theorem apply to sums of random variables?

82. Explain how the rules applying the central limit theorem to sample means, and to sums of a random variable, are similar.

83. If you repeatedly draw samples of size 50 from a population with a mean of 80 and a standard deviation of four, and calculate the sum of each sample, what is the expected distribution of these sums?

Use the following information to answer the next four exercises. You draw one sample of size 40 from a population with a mean of 125 and a standard deviation of seven.

84. Compute the sum. What is the probability that the sum for your sample will be less than 5,000?

85. If you drew samples of this size repeatedly, computing the sum each time, what range of values would you expect to contain 95 percent of the sample sums?

86. What value is one standard deviation below the mean?

87. What value corresponds to a z-score of 2.2?

7.3: Using the Central Limit Theorem

88. What does the law of large numbers say about the relationship between the sample mean and the population mean?

89. Applying the law of large numbers, which sample mean would expect to be closer to the population mean, a sample of size ten or a sample of size 100?

Use this information for the next three questions. A manufacturer makes screws with a mean diameter of 0.15 cm (centimeters) and a range of 0.10 cm to 0.20 cm; within that range, the distribution is uniform.

90. If X = the diameter of one screw, what is the distribution of X?

91. Suppose you repeatedly draw samples of size 100 and calculate their mean. Applying the central limit theorem, what is the distribution of these sample means?

92. Suppose you repeatedly draw samples of 60 and calculate their sum. Applying the central limit theorem, what is the distribution of these sample sums?

Practice Test 2 Solutions
Probability Distribution Function (PDF) for a Discrete Random Variable

1. The domain of X = {English, Mathematics,....], i.e., a list of all the majors offered at the university, plus "undeclared."

2. The domain of Y = {0, 1, 2, ...}, i.e., the integers from 0 to the upper limit of classes allowed by the university.

3. The domain of Z = any amount of money from 0 upwards.

4. Because they can take any value within their domain, and their value for any particular case is not known until the survey is completed.

5. No, because the domain of Z includes only positive numbers (you can't spend a negative amount of money). Possibly the value –7 is a data entry error, or a special code to indicated that the student did not answer the question.

6. The probabilities must sum to 1.0, and the probabilities of each event must be between 0 and 1, inclusive.

7. Let X = the number of books checked out by a patron.

8. $P(x > 2) = 0.10 + 0.05 = 0.15$

9. $P(x \geq 0) = 1 - 0.20 = 0.80$

10. $P(x \leq 3) = 1 - 0.05 = 0.95$

11. The probabilities would sum to 1.10, and the total probability in a distribution must always equal 1.0.

12. $\bar{x} = 0(0.20) + 1(0.45) + 2(0.20) + 3(0.10) + 4(0.05) = 1.35$

Mean or Expected Value and Standard Deviation

13.

x	P(x)	xP(x)
30	0.33	9.90
40	0.33	13.20
60	0.33	19.80

Table B8

14. $\bar{x} = 9.90 + 13.20 + 19.80 = 42.90$

15. $P(x = 30) = 0.33$
$P(x = 40) = 0.33$
$P(x = 60) = 0.33$

16.

x	P(x)	xP(x)	$(x - \mu)^2 P(x)$
30	0.33	9.90	$(30 - 42.90)^2(0.33) = 54.91$
40	0.33	13.20	$(40 - 42.90)^2(0.33) = 2.78$
60	0.33	19.90	$(60 - 42.90)^2(0.33) = 96.49$

Table B9

17. $\sigma_x = \sqrt{54.91 + 2.78 + 96.49} = 12.42$

Binomial Distribution

18. $q = 1 - 0.65 = 0.35$

19.

1. There are a fixed number of trials.

2. There are only two possible outcomes, and they add up to 1.

3. The trials are independent and conducted under identical conditions.

20. No, because there are not a fixed number of trials

21. $X \sim B(100, 0.65)$

22. $\mu = np = 100(0.65) = 65$

23. $\sigma_x = \sqrt{npq} = \sqrt{100(0.65)(0.35)} = 4.77$

24. X = Joe gets a hit in one at-bat (in one occasion of his coming to bat)

25. $X \sim B(20, 0.4)$

26. $\mu = np = 20(0.4) = 8$

27. $\sigma_x = \sqrt{npq} = \sqrt{20(0.40)(0.60)} = 2.19$

4.4: Geometric Distribution

28.

1. A series of Bernoulli trials are conducted until one is a success, and then the experiment stops.

2. At least one trial is conducted, but there is no upper limit to the number of trials.

3. The probability of success or failure is the same for each trial.

29. $TTTTH$

30. The domain of X = {1, 2, 3, 4, 5, ...n}. Because you are drawing with replacement, there is no upper bound to the number of draws that may be necessary.

31. The domain of X = {1, 2, 3, 4, 5, 6, 7, 8., 9, 10, 11, 12...27}. Because you are drawing without replacement, and 26 of the 52 cards are red, you have to draw a red card within the first 17 draws.

32. $X \sim G(0.24)$

33. $\mu = \frac{1}{p} = \frac{1}{0.27} = 3.70$

34. $\sigma = \sqrt{\frac{1-p}{p^2}} = \sqrt{\frac{1-0.27}{0.27^2}} = 3.16$

4.5: Hypergeometric Distribution

35. Yes, because you are sampling from a population composed of two groups (boys and girls), have a group of interest (boys), and are sampling without replacement (hence, the probabilities change with each pick, and you are not performing Bernoulli trials).

36. The group of interest is the cards that are spades, the size of the group of interest is 13, and the sample size is five.

4.6: Poisson Distribution

37. A Poisson distribution models the number of events occurring in a fixed interval of time or space, when the events are independent and the average rate of the events is known.

38. $X \sim P(4)$

39. The domain of $X = \{0, 1, 2, 3, \ldots.\}$ i.e., any integer from 0 upwards.

40. $\mu = 4$

$\sigma = \sqrt{4} = 2$

5.1: Continuous Probability Functions

41. The discrete variables are the number of books purchased, and the number of books sold after the end of the semester. The continuous variables are the amount of money spent for the books, and the amount of money received when they were sold.

42. Because for a continuous random variable, $P(x = c) = 0$, where c is any single value. Instead, we calculate $P(c < x < d)$, i.e., the probability that the value of x is between the values c and d.

43. Because $P(x = c) = 0$ for any continuous random variable.

44. $P(x > 5) = 1 - 0.35 = 0.65$, because the total probability of a continuous probability function is always 1.

45. This is a uniform probability distribution. You would draw it as a rectangle with the vertical sides at 0 and 20, and the horizontal sides at $\frac{1}{10}$ and 0.

46. $P(0 < x < 4) = (4 - 0)\left(\frac{1}{10}\right) = 0.4$

5.2: The Uniform Distribution

47. $P(2 < x < 5) = (5 - 2)\left(\frac{1}{10}\right) = 0.3$

48. $X \sim U(0, 15)$

49. $f(x) = \frac{1}{b - a}$ for $(a \leq x \leq b)$ so $f(x) = \frac{1}{30}$ for $(0 \leq x \leq 30)$

50. $\mu = \frac{a + b}{2} = \frac{0 + 30}{5} = 15.0$

$\sigma = \sqrt{\frac{(b - a)^2}{12}} = \sqrt{\frac{(30 - 0)^2}{12}} = 8.66$

51. $P(x < 10) = (10)\left(\frac{1}{30}\right) = 0.33$

5.3: The Exponential Distribution

52. X has an exponential distribution with decay parameter m and mean and standard deviation $\frac{1}{m}$. In this distribution, there will be a relatively large numbers of small values, with values becoming less common as they become larger.

53. $\mu = \sigma = \frac{1}{m} = \frac{1}{10} = 0.1$

54. $f(x) = 0.2e^{-0.2x}$ where $x \geq 0$.

6.1: The Standard Normal Distribution

55. The random variable X has a normal distribution with a mean of 100 and a standard deviation of 15.

56. $X \sim N(0,1)$

57. $z = \frac{x - \mu}{\sigma}$ so $z = \frac{112 - 109}{4.5} = 0.67$

58. $z = \frac{x - \mu}{\sigma}$ so $z = \frac{100 - 109}{4.5} = -2.00$

59. $z = \frac{105 - 109}{4.5} = -0.89$

This girl is shorter than average for her age, by 0.89 standard deviations.

60. $109 + (1.5)(4.5) = 115.75$ cm

61. We expect about 68 percent of the heights of girls of age five years and zero months to be between 104.5 cm and 113.5 cm.

62. We expect 99.7 percent of the heights in this distribution to be between 95.5 cm and 122.5 cm, because that range represents the values three standard deviations above and below the mean.

6.2: Using the Normal Distribution

63. Yes, because both np and nq are greater than five.
$np = (500)(0.20) = 100$ and $nq = 500(0.80) = 400$

64. $\mu = np = (500)(0.20) = 100$

$\sigma = \sqrt{npq} = \sqrt{500(0.20)(0.80)} = 8.94$

65. Fifty percent, because in a normal distribution, half the values lie above the mean.

66. The results of our sample were two standard deviations below the mean, suggesting it is unlikely that 20 percent of the lotto tickets are winners, as claimed by the distributor, and that the true percent of winners is lower. Applying the Empirical Rule, If that claim were true, we would expect to see a result this far below the mean only about 2.5 percent of the time.

7.1: The Central Limit Theorem for Sample Means (Averages)

67. The central limit theorem states that if samples of sufficient size drawn from a population, the distribution of sample means will be normal, even if the distribution of the population is not normal.

68. The sample size of 30 is sufficiently large in this example to apply the central limit theorem. This theorem] states that for samples of sufficient size drawn from a population, the sampling distribution of the sample mean will approach normality, regardless of the distribution of the population from which the samples were drawn.

69. You would not expect each sample to have a mean of 50, because of sampling variability. However, you would expect the sampling distribution of the sample means to cluster around 50, with an approximately normal distribution, so that values close to 50 are more common than values further removed from 50.

70. $\bar{X} \sim N(25, 0.2)$ because $\bar{X} \sim N\left(\mu_x, \frac{\sigma_x}{\sqrt{n}}\right)$

71. The standard deviation of the sampling distribution of the sample means can be calculated using the formula $\left(\frac{\sigma_x}{\sqrt{n}}\right)$, which in this case is $\left(\frac{16}{\sqrt{50}}\right)$. The correct value for the standard deviation of the sampling distribution of the sample means is therefore 2.26.

72. The standard error of the mean is another name for the standard deviation of the sampling distribution of the sample mean. Given samples of size n drawn from a population with standard deviation σ_x, the standard error of the mean is $\left(\frac{\sigma_x}{\sqrt{n}}\right)$.

73. $X \sim N(75, 0.45)$

74. Your friend forgot to divide the standard deviation by the square root of n.

75. $z = \frac{\bar{x} - \mu_x}{\sigma_x} = \frac{76 - 75}{4.5} = 2.2$

76. $z = \frac{\bar{x} - \mu_x}{\sigma_x} = \frac{74.7 - 75}{4.5} = -0.67$

77. $75 + (1.5)(0.45) = 75.675$

78. The standard error of the mean will be larger, because you will be dividing by a smaller number. The standard error of the mean for samples of size $n = 50$ is:

$$\left(\frac{\sigma_x}{\sqrt{n}}\right) = \frac{4.5}{\sqrt{50}} = 0.64$$

79. You would expect this range to include values up to one standard deviation above or below the mean of the sample means. In this case:

$70 + \frac{9}{\sqrt{60}} = 71.16$ and $70 - \frac{9}{\sqrt{60}} = 68.84$ so you would expect 68 percent of the sample means to be between 68.84 and 71.16.

80. $70 + \frac{9}{\sqrt{100}} = 70.9$ and $70 - \frac{9}{\sqrt{100}} = 69.1$ so you would expect 68 percent of the sample means to be between 69.1 and 70.9. Note that this is a narrower interval due to the increased sample size.

7.2: The Central Limit Theorem for Sums

81. For a random variable X, the random variable ΣX will tend to become normally distributed as the size n of the samples used to compute the sum increases.

82. Both rules state that the distribution of a quantity (the mean or the sum) calculated on samples drawn from a population will tend to have a normal distribution, as the sample size increases, regardless of the distribution of population from which the samples are drawn.

83. $\Sigma X \sim N(n\mu_x, (\sqrt{n})(\sigma_x))$ so $\Sigma X \sim N(4000, 28.3)$

84. The probability is 0.50, because 5,000 is the mean of the sampling distribution of sums of size 40 from this population. Sums of random variables computed from a sample of sufficient size are normally distributed, and in a normal distribution, half the values lie below the mean.

85. Using the empirical rule, you would expect 95 percent of the values to be within two standard deviations of the mean. Using the formula for the standard deviation is for a sample sum: $(\sqrt{n})(\sigma_x) = (\sqrt{40})(7) = 44.3$ so you would expect 95 percent of the values to be between $5,000 + (2)(44.3)$ and $5,000 - (2)(44.3)$, or between 4,911.4 and 588.6.

86. $\mu - (\sqrt{n})(\sigma_x) = 5000 - (\sqrt{40})(7) = 4955.7$

87. $5000 + (2.2)(\sqrt{40})(7) = 5097.4$

7.3: Using the Central Limit Theorem

88. The law of large numbers says that as sample size increases, the sample mean tends to get nearer and nearer to the population mean.

89. You would expect the mean from a sample of size 100 to be nearer to the population mean, because the law of large numbers says that as sample size increases, the sample mean tends to approach the population mea.

90. $X \sim N(0.10, 0.20)$

91. $\bar{X} \sim N\left(\mu_x, \frac{\sigma_x}{\sqrt{n}}\right)$ and the standard deviation of a uniform distribution is $\frac{b-a}{\sqrt{12}}$. In this example, the standard deviation of the distribution is $\frac{b-a}{\sqrt{12}} = \frac{0.10}{\sqrt{12}} = 0.03$

so $\bar{X} \sim N(0.15, 0.003)$

92. $\Sigma X \sim N((n)(\mu_x), (\sqrt{n})(\sigma_x))$ so $\Sigma X \sim N(9.0, 0.23)$

Practice Test 3

8.1: Confidence Interval, Single Population Mean, Population Standard Deviation Known, Normal

Use the following information to answer the next seven exercises. You draw a sample of size 30 from a normally distributed population with a standard deviation of four.

1. What is the standard error of the sample mean in this scenario, rounded to two decimal places?

2. What is the distribution of the sample mean?

3. If you want to construct a two-sided 95% confidence interval, how much probability will be in each tail of the distribution?

4. What is the appropriate *z*-score and error bound or margin of error (*EBM*) for a 95% confidence interval for this data?

5. Rounding to two decimal places, what is the 95% confidence interval if the sample mean is 41?

6. What is the 90% confidence interval if the sample mean is 41? Round to two decimal places

7. Suppose the sample size in this study had been 50, rather than 30. What would the 95% confidence interval be if the sample mean is 41? Round your answer to two decimal places.

8. For any given data set and sampling situation, which would you expect to be wider: a 95% confidence interval or a 99% confidence interval?

8.2: Confidence Interval, Single Population Mean, Standard Deviation Unknown, Student's *t*

9. Comparing graphs of the standard normal distribution (*z*-distribution) and a *t*-distribution with 15 degrees of freedom (*df*), how do they differ?

10. Comparing graphs of the standard normal distribution (*z*-distribution) and a *t*-distribution with 15 degrees of freedom (*df*), how are they similar?

Use the following information to answer the next five exercises. Body temperature is known to be distributed normally among healthy adults. Because you do not know the population standard deviation, you use the t-distribution to study body temperature. You collect data from a random sample of 20 healthy adults and find that your sample temperatures have a mean of 98.4 and a sample standard deviation of 0.3 (both in degrees Fahrenheit).

11. What is the degrees of freedom (*df*) for this study?

12. For a two-tailed 95% confidence interval, what is the appropriate *t*-value to use in the formula?

13. What is the 95% confidence interval?

14. What is the 99% confidence interval? Round to two decimal places.

15. Suppose your sample size had been 30 rather than 20. What would the 95% confidence interval be then? Round to two decimal places

8.3: Confidence Interval for a Population Proportion

Use this information to answer the next four exercises. You conduct a poll of 500 randomly selected city residents, asking them if they own an automobile. 280 say they do own an automobile, and 220 say they do not.

16. Find the sample proportion and sample standard deviation for this data.

17. What is the 95% two-sided confidence interval? Round to four decimal places.

18. Calculate the 90% confidence interval. Round to four decimal places.

19. Calculate the 99% confidence interval. Round to four decimal places.

Use the following information to answer the next three exercises. You are planning to conduct a poll of community members age 65 and older, to determine how many own mobile phones. You want to produce an estimate whose 95% confidence interval will be within four percentage points (plus or minus) the true population proportion. Use an estimated population proportion of 0.5.

20. What sample size do you need?

21. Suppose you knew from prior research that the population proportion was 0.6. What sample size would you need?

22. Suppose you wanted a 95% confidence interval within three percentage points of the population. Assume the population proportion is 0.5. What sample size do you need?

9.1: Null and Alternate Hypotheses

23. In your state, 58 percent of registered voters in a community are registered as Republicans. You want to conduct a study to see if this also holds up in your community. State the null and alternative hypotheses to test this.

24. You believe that at least 58 percent of registered voters in a community are registered as Republicans. State the null and alternative hypotheses to test this.

25. The mean household value in a city is $268,000. You believe that the mean household value in a particular neighborhood is lower than the city average. Write the null and alternative hypotheses to test this.

26. State the appropriate alternative hypothesis to this null hypothesis: $H_0: \mu = 107$

27. State the appropriate alternative hypothesis to this null hypothesis: $H_0: p < 0.25$

9.2: Outcomes and the Type I and Type II Errors

28. If you reject H_0 when H_0 is correct, what type of error is this?

29. If you fail to reject H_0 when H_0 is false, what type of error is this?

30. What is the relationship between the Type II error and the power of a test?

31. A new blood test is being developed to screen patients for cancer. Positive results are followed up by a more accurate (and expensive) test. It is assumed that the patient does not have cancer. Describe the null hypothesis, the Type I and Type II errors for this situation, and explain which type of error is more serious.

32. Explain in words what it means that a screening test for TB has an α level of 0.10. The null hypothesis is that the patient does not have TB.

33. Explain in words what it means that a screening test for TB has a β level of 0.20. The null hypothesis is that the patient does not have TB.

34. Explain in words what it means that a screening test for TB has a power of 0.80.

9.3: Distribution Needed for Hypothesis Testing

35. If you are conducting a hypothesis test of a single population mean, and you do not know the population variance, what test will you use if the sample size is 10 and the population is normal?

36. If you are conducting a hypothesis test of a single population mean, and you know the population variance, what test will you use?

37. If you are conducting a hypothesis test of a single population proportion, with np and nq greater than or equal to five, what test will you use, and with what parameters?

38. Published information indicates that, on average, college students spend less than 20 hours studying per week. You draw a sample of 25 students from your college, and find the sample mean to be 18.5 hours, with a standard deviation of 1.5 hours. What distribution will you use to test whether study habits at your college are the same as the national average, and why?

39. A published study says that 95 percent of American children are vaccinated against measles, with a standard deviation of 1.5 percent. You draw a sample of 100 children from your community and check their vaccination records, to see if the vaccination rate in your community is the same as the national average. What distribution will you use for this test, and why?

9.4: Rare Events, the Sample, Decision, and Conclusion

40. You are conducting a study with an α level of 0.05. If you get a result with a p-value of 0.07, what will be your decision?

41. You are conducting a study with $\alpha = 0.01$. If you get a result with a p-value of 0.006, what will be your decision?

Use the following information to answer the next five exercises. According to the World Health Organization, the average height of a one-year-old child is 29". You believe children with a particular disease are smaller than average, so you draw a sample of 20 children with this disease and find a mean height of 27.5" and a sample standard deviation of 1.5".

42. What are the null and alternative hypotheses for this study?

43. What distribution will you use to test your hypothesis, and why?

44. What is the test statistic and the *p*-value?

45. Based on your sample results, what is your decision?

46. Suppose the mean for your sample was 25.0. Redo the calculations and describe what your decision would be.

9.5: Additional Information and Full Hypothesis Test Examples

47. You conduct a study using $\alpha = 0.05$. What is the level of significance for this study?

48. You conduct a study, based on a sample drawn from a normally distributed population with a known variance, with the following hypotheses:
$H_0: \mu = 35.5$
$H_a: \mu \neq 35.5$
Will you conduct a one-tailed or two-tailed test?

49. You conduct a study, based on a sample drawn from a normally distributed population with a known variance, with the following hypotheses:
$H_0: \mu \geq 35.5$
$H_a: \mu < 35.5$
Will you conduct a one-tailed or two-tailed test?

Use the following information to answer the next three exercises. Nationally, 80 percent of adults own an automobile. You are interested in whether the same proportion in your community own cars. You draw a sample of 100 and find that 75 percent own cars.

50. What are the null and alternative hypotheses for this study?

51. What test will you use, and why?

10.1: Comparing Two Independent Population Means with Unknown Population Standard Deviations

52. You conduct a poll of political opinions, interviewing both members of 50 married couples. Are the groups in this study independent or matched?

53. You are testing a new drug to treat insomnia. You randomly assign 80 volunteer subjects to either the experimental (new drug) or control (standard treatment) conditions. Are the groups in this study independent or matched?

54. You are investigating the effectiveness of a new math textbook for high school students. You administer a pretest to a group of students at the beginning of the semester, and a posttest at the end of a year's instruction using this textbook, and compare the results. Are the groups in this study independent or matched?

Use the following information to answer the next two exercises. You are conducting a study of the difference in time at two colleges for undergraduate degree completion. At College A, students take an average of 4.8 years to complete an undergraduate degree, while at College B, they take an average of 4.2 years. The pooled standard deviation for this data is 1.6 years

55. Calculate Cohen's *d* and interpret it.

56. Suppose the mean time to earn an undergraduate degree at College A was 5.2 years. Calculate the effect size and interpret it.

57. You conduct an independent-samples t-test with sample size ten in each of two groups. If you are conducting a two-tailed hypothesis test with $\alpha = 0.01$, what p-values will cause you to reject the null hypothesis?

58. You conduct an independent samples *t*-test with sample size 15 in each group, with the following hypotheses:
$H_0: \mu \geq 110$
$H_a: \mu < 110$
If $\alpha = 0.05$, what *t*-values will cause you to reject the null hypothesis?

10.2: Comparing Two Independent Population Means with Known Population Standard Deviations

Use the following information to answer the next six exercises. College students in the sciences often complain that they must spend more on textbooks each semester than students in the humanities. To test this, you draw random samples of 50 science and 50 humanities students from your college, and record how much each spent last semester on textbooks. Consider the science students to be group one, and the humanities students to be group two.

59. What is the random variable for this study?

60. What are the null and alternative hypotheses for this study?

61. If the 50 science students spent an average of $530 with a sample standard deviation of $20 and the 50 humanities students spent an average of $380 with a sample standard deviation of $15, would you not reject or reject the null hypothesis? Use an alpha level of 0.05. What is your conclusion?

62. What would be your decision, if you were using $\alpha = 0.01$?

10.3: Comparing Two Independent Population Proportions

Use the information to answer the next six exercises. You want to know if proportion of homes with cable television service differs between Community A and Community B. To test this, you draw a random sample of 100 for each and record whether they have cable service.

63. What are the null and alternative hypotheses for this study

64. If 65 households in Community A have cable service, and 78 households in community B, what is the pooled proportion?

65. At $\alpha = 0.03$, will you reject the null hypothesis? What is your conclusion? 65 households in Community A have cable service, and 78 households in community B. 100 households in each community were surveyed.

66. Using an alpha value of 0.01, would you reject the null hypothesis? What is your conclusion? 65 households in Community A have cable service, and 78 households in community B. 100 households in each community were surveyed.

10.4: Matched or Paired Samples

Use the following information to answer the next five exercises. You are interested in whether a particular exercise program helps people lose weight. You conduct a study in which you weigh the participants at the start of the study, and again at the conclusion, after they have participated in the exercise program for six months. You compare the results using a matched-pairs t-test, in which the data is {weight at conclusion – weight at start}. You believe that, on average, the participants will have lost weight after six months on the exercise program.

67. What are the null and alternative hypotheses for this study?

68. Calculate the test statistic, assuming that $\bar{x}_d = -5$, $s_d = 6$, and $n = 30$ (pairs).

69. What are the degrees of freedom for this statistic?

70. Using $\alpha = 0.05$, what is your decision regarding the effectiveness of this program in causing weight loss? What is the conclusion?

71. What would it mean if the *t*-statistic had been 4.56, and what would have been your decision in that case?

11.1: Facts About the Chi-Square Distribution

72. What is the mean and standard deviation for a chi-square distribution with 20 degrees of freedom?

11.2: Goodness-of-Fit Test

Use the following information to answer the next four exercises. Nationally, about 66 percent of high school graduates enroll in higher education. You perform a chi-square goodness of fit test to see if this same proportion applies to your high school's most recent graduating class of 200. Your null hypothesis is that the national distribution also applies to your high school.

73. What are the expected numbers of students from your high school graduating class enrolled and not enrolled in higher education?

74. Fill out the rest of this table.

	Observed (O)	Expected (E)	O – E	(O – E)2	$\frac{(O-E)^2}{z}$
Enrolled	145				
Not enrolled	55				

Table B10

75. What are the degrees of freedom for this chi-square test?

76. What is the chi-square test statistic and the p-value. At the 5% significance level, what do you conclude?

77. For a chi-square distribution with 92 degrees of freedom, the curve _____.

78. For a chi-square distribution with five degrees of freedom, the curve is _____.

11.3: Test of Independence

Use the following information to answer the next four exercises. You are considering conducting a chi-square test of independence for the data in this table, which displays data about cell phone ownership for freshman and seniors at a high school. Your null hypothesis is that cell phone ownership is independent of class standing.

79. Compute the expected values for the cells.

	Cell = Yes	Cell = No
Freshman	100	150
Senior	200	50

Table B11

80. Compute $\dfrac{(O - E)^2}{z}$ for each cell, where O = observed and E = expected.

81. What is the chi-square statistic and degrees of freedom for this study?

82. At the $\alpha = 0.5$ significance level, what is your decision regarding the null hypothesis?

11.4: Test of Homogeneity

83. You conduct a chi-square test of homogeneity for data in a five by two table. What is the degrees of freedom for this test?

11.5: Comparison Summary of the Chi-Square Tests: Goodness-of-Fit, Independence and Homogeneity

84. A 2013 poll in the State of California surveyed people about taxing sugar-sweetened beverages. The results are presented in the following table, and are classified by ethnic group and response type. Are the poll responses independent of the participants' ethnic group? Conduct a hypothesis test at the 5% significance level.

Ethnic Group \ Response Type	Favor	Oppose	No Opinion	Row Total
White / Non-Hispanic	234	433	43	710
Latino	147	106	19	272
African American	24	41	6	71
Asian American	54	48	16	118
Column Total	459	628	84	1171

Table B12

85. In a test of homogeneity, what must be true about the expected value of each cell?

86. Stated in general terms, what are the null and alternative hypotheses for the chi-square test of independence?

87. Stated in general terms, what are the null and alternative hypotheses for the chi-square test of homogeneity?

11.6: Test of a Single Variance

88. A lab test claims to have a variance of no more than five. You believe the variance is greater. What are the null and alternative hypothesis to test this?

Practice Test 3 Solutions
8.1: Confidence Interval, Single Population Mean, Population Standard Deviation Known, Normal

1. $\frac{\sigma}{\sqrt{n}} = \frac{4}{\sqrt{30}} = 0.73$

2. normal

3. 0.025 or 2.5%; A 95% confidence interval contains 95% of the probability, and excludes five percent, and the five percent excluded is split evenly between the upper and lower tails of the distribution.

4. z-score = 1.96; $EBM = z_{\frac{\alpha}{2}}\left(\frac{\sigma}{\sqrt{n}}\right) = (1.96)(0.73) = 1.4308$

5. $41 \pm 1.43 = (39.57, 42.43)$; Using the calculator function Zinterval, answer is (40.74, 41.26. Answers differ due to rounding.

6. The z-value for a 90% confidence interval is 1.645, so $EBM = 1.645(0.73) = 1.20085$.
The 90% confidence interval is $41 \pm 1.20 = (39.80, 42.20)$.
The calculator function Zinterval answer is (40.78, 41.23). Answers differ due to rounding.

7. The standard error of measurement is: $\frac{\sigma}{\sqrt{n}} = \frac{4}{\sqrt{50}} = 0.57$

$EBM = z_{\frac{\alpha}{2}}\left(\frac{\sigma}{\sqrt{n}}\right) = (1.96)(0.57) = 1.12$

The 95% confidence interval is $41 \pm 1.12 = (39.88, 42.12)$.
The calculator function Zinterval answer is (40.84, 41.16). Answers differ due to rounding.

8. The 99% confidence interval, because it includes all but one percent of the distribution. The 95% confidence interval will be narrower, because it excludes five percent of the distribution.

8.2: Confidence Interval, Single Population Mean, Standard Deviation Unknown, Student's t

9. The t-distribution will have more probability in its tails ("thicker tails") and less probability near the mean of the distribution ("shorter in the center").

10. Both distributions are symmetrical and centered at zero.

11. $df = n - 1 = 20 - 1 = 19$

12. You can get the t-value from a probability table or a calculator. In this case, for a t-distribution with 19 degrees of freedom, and a 95% two-sided confidence interval, the value is 2.093, i.e.,
$t_{\frac{\alpha}{2}} = 2.093$. The calculator function is invT(0.975, 19).

13. $EBM = t_{\frac{\alpha}{2}}\left(\frac{s}{\sqrt{n}}\right) = (2.093)\left(\frac{0.3}{\sqrt{20}}\right) = 0.140$

$98.4 \pm 0.14 = (98.26, 98.54)$.
The calculator function Tinterval answer is (98.26, 98.54).

14. $t_{\frac{\alpha}{2}} = 2.861$. The calculator function is invT(0.995, 19).

$EBM = t_{\frac{\alpha}{2}}\left(\frac{s}{\sqrt{n}}\right) = (2.861)\left(\frac{0.3}{\sqrt{20}}\right) = 0.192$

$98.4 \pm 0.19 = (98.21, 98.59)$. The calculator function Tinterval answer is (98.21, 98.59).

15. $df = n - 1 = 30 - 1 = 29$. $t_{\frac{\alpha}{2}} = 2.045$

$EBM = z_t\left(\frac{s}{\sqrt{n}}\right) = (2.045)\left(\frac{0.3}{\sqrt{30}}\right) = 0.112$

$98.4 \pm 0.11 = (98.29, 98.51)$. The calculator function Tinterval answer is (98.29, 98.51).

8.3: Confidence Interval for a Population Proportion

16. $p' = \frac{280}{500} = 0.56$

$q' = 1 - p' = 1 - 0.56 = 0.44$

$s = \sqrt{\frac{pq}{n}} = \sqrt{\frac{0.56(0.44)}{500}} = 0.0222$

17. Because you are using the normal approximation to the binomial, $z_{\frac{\alpha}{2}} = 1.96$.

Calculate the error bound for the population (*EBP*):

$EBP = z_{\frac{\alpha}{2}}\sqrt{\frac{pq}{n}} = 1.96(0.222) = 0.0435$

Calculate the 95% confidence interval:
$0.56 \pm 0.0435 = (0.5165, 0.6035)$.
The calculator function 1-PropZint answer is (0.5165, 0.6035).

18. $z_{\frac{\alpha}{2}} = 1.64$

$EBP = z_{\frac{\alpha}{2}}\sqrt{\frac{pq}{n}} = 1.64(0.0222) = 0.0364$

$0.56 \pm 0.03 = (0.5236, 0.5964)$. The calculator function 1-PropZint answer is (0.5235, 0.5965)

19. $z_{\frac{\alpha}{2}} = 2.58$

$EBP = z_{\frac{\alpha}{2}}\sqrt{\frac{pq}{n}} = 2.58(0.0222) = 0.0573$

$0.56 \pm 0.05 = (0.5127, 0.6173)$.
The calculator function 1-PropZint answer is (0.5028, 0.6172).

20. $EBP = 0.04$ (because 4% = 0.04)
$z_{\frac{\alpha}{2}} = 1.96$ for a 95% confidence interval

$n = \frac{z^2 pq}{EBP^2} = \frac{1.96^2(0.5)(0.5)}{0.04^2} = \frac{0.9604}{0.0016} = 600.25$

You need 601 subjects (rounding upward from 600.25).

21. $n = \frac{n^2 pq}{EBP^2} = \frac{1.96^2(0.6)(0.4)}{0.04^2} = \frac{0.9220}{0.0016} = 576.24$

You need 577 subjects (rounding upward from 576.24).

22. $n = \frac{n^2 pq}{EBP^2} = \frac{1.96^2(0.5)(0.5)}{0.03^2} = \frac{0.9604}{0.0009} = 1067.11$

You need 1,068 subjects (rounding upward from 1,067.11).

9.1: Null and Alternate Hypotheses

23. H_0: $p = 0.58$
H_a: $p \neq 0.58$

24. H_0: $p \geq 0.58$
H_a: $p < 0.58$

25. H_0: $\mu \geq \$268,000$
H_a: $\mu < \$268,000$

26. H_a: $\mu \neq 107$

27. H_a: $p \geq 0.25$

9.2: Outcomes and the Type I and Type II Errors

28. a Type I error

29. a Type II error

30. Power = $1 - \beta = 1 - P$(Type II error).

31. The null hypothesis is that the patient does not have cancer. A Type I error would be detecting cancer when it is not present. A Type II error would be not detecting cancer when it is present. A Type II error is more serious, because failure to detect cancer could keep a patient from receiving appropriate treatment.

32. The screening test has a ten percent probability of a Type I error, meaning that ten percent of the time, it will detect TB when it is not present.

33. The screening test has a 20 percent probability of a Type II error, meaning that 20 percent of the time, it will fail to detect TB when it is in fact present.

34. Eighty percent of the time, the screening test will detect TB when it is actually present.

9.3: Distribution Needed for Hypothesis Testing

35. The Student's t-test.

36. The normal distribution or z-test.

37. The normal distribution with $\mu = p$ and $\sigma = \sqrt{\frac{pq}{n}}$

38. t_{24}. You use the t-distribution because you don't know the population standard deviation, and the degrees of freedom are 24 because $df = n - 1$.

39. $\bar{X} \sim N\left(0.95, \frac{0.051}{\sqrt{100}}\right)$

Because you know the population standard deviation, and have a large sample, you can use the normal distribution.

9.4: Rare Events, the Sample, Decision, and Conclusion

40. Fail to reject the null hypothesis, because $\alpha \leq p$

41. Reject the null hypothesis, because $\alpha \geq p$.

42. H_0: $\mu \geq 29.0$"
H_a: $\mu < 29.0$"

43. t_{19}. Because you do not know the population standard deviation, use the t-distribution. The degrees of freedom are 19, because $df = n - 1$.

44. The test statistic is –4.4721 and the p-value is 0.00013 using the calculator function TTEST.

45. With $\alpha = 0.05$, reject the null hypothesis.

46. With $\alpha = 0.05$, the p-value is almost zero using the calculator function TTEST so reject the null hypothesis.

9.5: Additional Information and Full Hypothesis Test Examples

47. The level of significance is five percent.

48. two-tailed

49. one-tailed

50. H_0: $p = 0.8$
H_a: $p \neq 0.8$

51. You will use the normal test for a single population proportion because np and nq are both greater than five.

10.1: Comparing Two Independent Population Means with Unknown Population Standard Deviations

52. They are matched (paired), because you interviewed married couples.

53. They are independent, because participants were assigned at random to the groups.

54. They are matched (paired), because you collected data twice from each individual.

55. $d = \dfrac{\bar{x}_1 - \bar{x}_2}{s_{pooled}} = \dfrac{4.8 - 4.2}{1.6} = 0.375$

This is a small effect size, because 0.375 falls between Cohen's small (0.2) and medium (0.5) effect sizes.

56. $d = \dfrac{\bar{x}_1 - \bar{x}_2}{s_{pooled}} = \dfrac{5.2 - 4.2}{1.6} = 0.625$

The effect size is 0.625. By Cohen's standard, this is a medium effect size, because it falls between the medium (0.5) and large (0.8) effect sizes.

57. p-value < 0.01.

58. You will only reject the null hypothesis if you get a value significantly below the hypothesized mean of 110.

10.2: Comparing Two Independent Population Means with Known Population Standard Deviations

59. $\bar{X}_1 - \bar{X}_2$, i.e., the mean difference in amount spent on textbooks for the two groups.

60. $H_0:$ $\bar{X}_1 - \bar{X}_2 \leq 0$

$H_a:$ $\bar{X}_1 - \bar{X}_2 > 0$

This could also be written as:

$H_0:$ $\bar{X}_1 \leq \bar{X}_2$

$H_a:$ $\bar{X}_1 > \bar{X}_2$

61. Using the calculator function 2-SampTtest, reject the null hypothesis. At the 5% significance level, there is sufficient evidence to conclude that the science students spend more on textbooks than the humanities students.

62. Using the calculator function 2-SampTtest, reject the null hypothesis. At the 1% significance level, there is sufficient evidence to conclude that the science students spend more on textbooks than the humanities students.

10.3: Comparing Two Independent Population Proportions

63. $H_0: p_A = p_B$
$H_a: p_A \neq p_B$

64. $p_c = \dfrac{x_A + x_A}{n_A + n_A} = \dfrac{65 + 78}{100 + 100} = 0.715$

65. Using the calculator function 2-PropZTest, the p-value = 0.0417. Reject the null hypothesis. At the 3% significance level, here is sufficient evidence to conclude that there is a difference between the proportions of households in the two communities that have cable service.

66. Using the calculator function 2-PropZTest, the p-value = 0.0417. Do not reject the null hypothesis. At the 1% significance level, there is insufficient evidence to conclude that there is a difference between the proportions of households in the two communities that have cable service.

10.4: Matched or Paired Samples

67. $H_0:$ $\bar{x}_d \geq 0$

$H_a:$ $\bar{x}_d < 0$

68. $t = -4.5644$

69. $df = 30 - 1 = 29$.

70. Using the calculator function TTEST, the p-value = 0.00004 so reject the null hypothesis. At the 5% level, there is sufficient evidence to conclude that the participants lost weight, on average.

71. A positive t-statistic would mean that participants, on average, gained weight over the six months.

11.1: Facts About the Chi-Square Distribution

72. $\mu = df = 20$

$\sigma = \sqrt{2(df)} = \sqrt{40} = 6.32$

11.2: Goodness-of-Fit Test

73. Enrolled = 200(0.66) = 132. Not enrolled = 200(0.34) = 68

74.

	Observed (O)	Expected (E)	O – E	(O – E)2	$\frac{(O-E)^2}{z}$
Enrolled	145	132	145 – 132 = 13	169	$\frac{169}{132} = 1.280$
Not enrolled	55	68	55 – 68 = –13	169	$\frac{169}{68} = 2.485$

Table B13

75. $df = n - 1 = 2 - 1 = 1$.

76. Using the calculator function Chi-square GOF – Test (in STAT TESTS), the test statistic is 3.7656 and the p-value is 0.0523. Do not reject the null hypothesis. At the 5% significance level, there is insufficient evidence to conclude that high school most recent graduating class distribution of enrolled and not enrolled does not fit that of the national distribution.

77. approximates the normal

78. skewed right

11.3: Test of Independence

79.

	Cell = Yes	Cell = No	Total
Freshman	$\frac{250(300)}{500} = 150$	$\frac{250(200)}{500} = 100$	250
Senior	$\frac{250(300)}{500} = 150$	$\frac{250(200)}{500} = 100$	250
Total	300	200	500

Table B14

80. $\frac{(100 - 150)^2}{150} = 16.67$

$\frac{(150 - 100)^2}{100} = 25$

$\frac{(200 - 100)^2}{150} = 16.67$

$\frac{(50 - 100)^2}{100} = 25$

81. Chi-square = 16.67 + 25 + 16.67 + 25 = 83.34.

$df = (r - 1)(c - 1) = 1$

82. p-value $= P$(Chi-square, 83.34) $= 0$
Reject the null hypothesis.
You could also use the calculator function STAT TESTS Chi-Square – Test.

11.4: Test of Homogeneity

83. The table has five rows and two columns. $df = (r - 1)(c - 1) = (4)(1) = 4$.

11.5: Comparison Summary of the Chi-Square Tests: Goodness-of-Fit, Independence and Homogeneity

84. Using the calculator function (STAT TESTS) Chi-square Test, the p-value $= 0$. Reject the null hypothesis. At the 5% significance level, there is sufficient evidence to conclude that the poll responses independent of the participants' ethnic group.

85. The expected value of each cell must be at least five.

86. H_0: The variables are independent.
H_a: The variables are not independent.

87. H_0: The populations have the same distribution.
H_a: The populations do not have the same distribution.

11.6: Test of a Single Variance

88. H_0: $\sigma^2 \leq 5$
H_a: $\sigma^2 > 5$

Practice Test 4
12.1 Linear Equations

1. Which of the following equations is/are linear?

  a. $y = -3x$

  b. $y = 0.2 + 0.74x$

  c. $y = -9.4 - 2x$

  d. A and B

  e. A, B, and C

2. To complete a painting job requires four hours setup time plus one hour per 1,000 square feet. How would you express this information in a linear equation?

3. A statistics instructor is paid a per-class fee of $2,000 plus $100 for each student in the class. How would you express this information in a linear equation?

4. A tutoring school requires students to pay a one-time enrollment fee of $500 plus tuition of $3,000 per year. Express this information in an equation.

12.2: Slope and Y-intercept of a Linear Equation

Use the following information to answer the next four exercises. For the labor costs of doing repairs, an auto mechanic charges a flat fee of $75 per car, plus an hourly rate of $55.

5. What are the independent and dependent variables for this situation?

6. Write the equation and identify the slope and intercept.

7. What is the labor charge for a job that takes 3.5 hours to complete?

8. One job takes 2.4 hours to complete, while another takes 6.3 hours. What is the difference in labor costs for these two jobs?

12.3: Scatter Plots

9. Describe the pattern in this scatter plot, and decide whether the X and Y variables would be good candidates for linear regression.

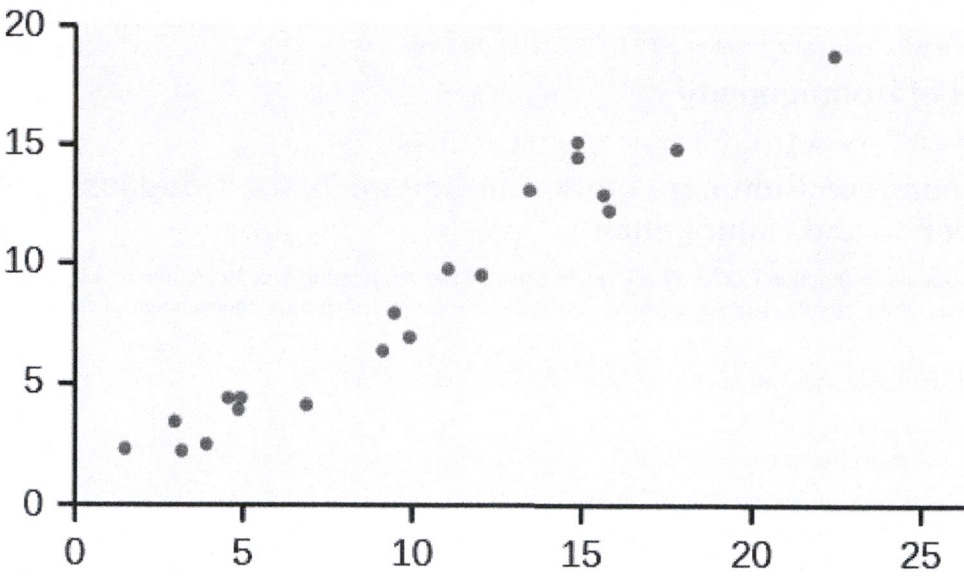

Figure B4

10. Describe the pattern in this scatter plot, and decide whether the X and Y variables would be good candidates for linear regression.

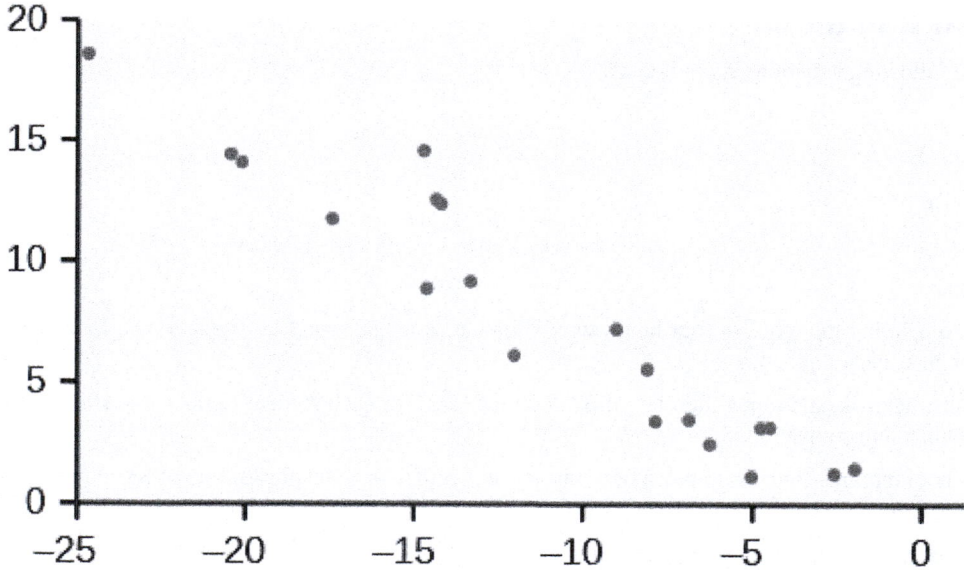

Figure B5

11. Describe the pattern in this scatter plot, and decide whether the X and Y variables would be good candidates for linear regression.

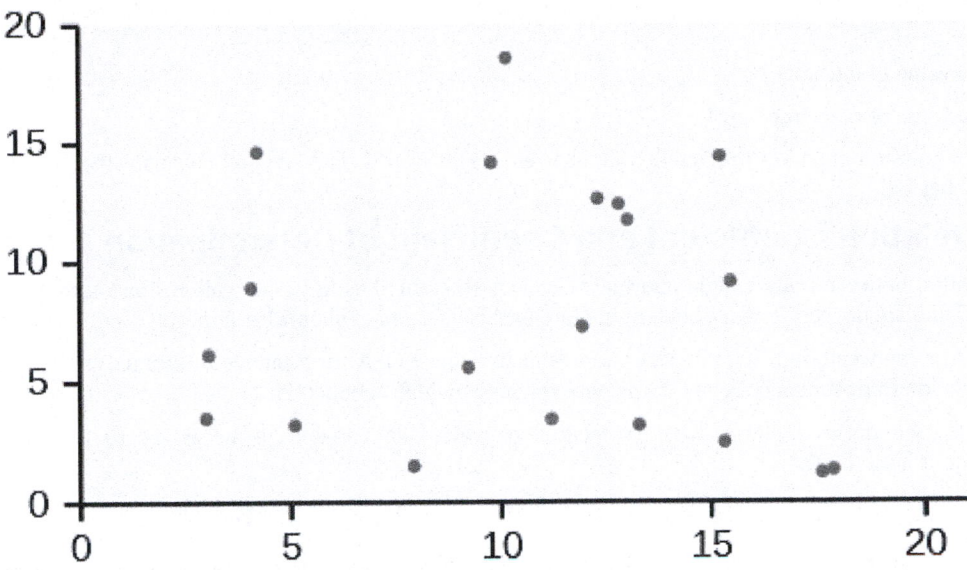

Figure B6

12. Describe the pattern in this scatter plot, and decide whether the X and Y variables would be good candidates for linear regression.

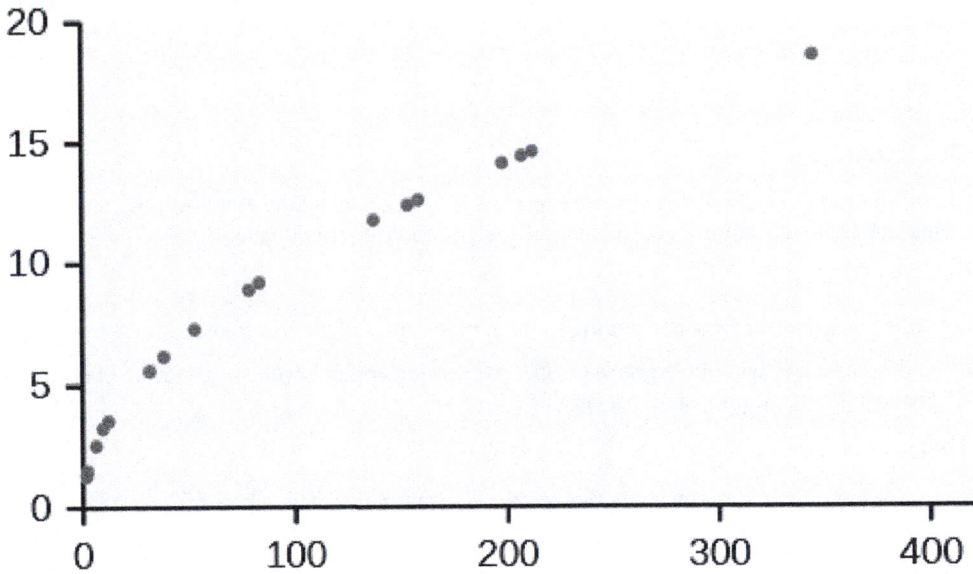

Figure B7

12.4: The Regression Equation

Use the following information to answer the next four exercises. Height (in inches) and weight (In pounds) in a sample of college freshman men have a linear relationship with the following summary statistics:

$\bar{x}$ = 68.4

$\bar{y}$ =141.6

$s_x = 4.0$
$s_y = 9.6$
$r = 0.73$

Let Y = weight and X = height, and write the regression equation in the form:
$$\hat{y} = a + bx$$

13. What is the value of the slope?

14. What is the value of the y intercept?

15. Write the regression equation predicting weight from height in this data set, and calculate the predicted weight for someone 68 inches tall.

12.5: Correlation Coefficient and Coefficient of Determination

16. The correlation between body weight and fuel efficiency (measured as miles per gallon) for a sample of 2,012 model cars is –0.56. Calculate the coefficient of determination for this data and explain what it means.

17. The correlation between high school GPA and freshman college GPA for a sample of 200 university students is 0.32. How much variation in freshman college GPA is not explained by high school GPA?

18. Rounded to two decimal places what correlation between two variables is necessary to have a coefficient of determination of at least 0.50?

12.6: Testing the Significance of the Correlation Coefficient

19. Write the null and alternative hypotheses for a study to determine if two variables are significantly correlated.

20. In a sample of 30 cases, two variables have a correlation of 0.33. Do a t-test to see if this result is significant at the $\alpha = 0.05$ level. Use the formula:
$$t = \frac{r\sqrt{n-2}}{\sqrt{1-r^2}}$$

21. In a sample of 25 cases, two variables have a correlation of 0.45. Do a t-test to see if this result is significant at the $\alpha = 0.05$ level. Use the formula:
$$t = \frac{r\sqrt{n-2}}{\sqrt{1-r^2}}$$

12.7: Prediction

Use the following information to answer the next two exercises. A study relating the grams of potassium (Y) to the grams of fiber (X) per serving in enriched flour products (bread, rolls, etc.) produced the equation:
$$\hat{y} = 25 + 16x$$

22. For a product with five grams of fiber per serving, what are the expected grams of potassium per serving?

23. Comparing two products, one with three grams of fiber per serving and one with six grams of fiber per serving, what is the expected difference in grams of potassium per serving?

12.8: Outliers

24. In the context of regression analysis, what is the definition of an outlier, and what is a rule of thumb to evaluate if a given value in a data set is an outlier?

25. In the context of regression analysis, what is the definition of an influential point, and how does an influential point differ from an outlier?

26. The least squares regression line for a data set is $\hat{y} = 5 + 0.3x$ and the standard deviation of the residuals is 0.4. Does a case with the values $x = 2, y = 6.2$ qualify as an outlier?

27. The least squares regression line for a data set is $\hat{y} = 2.3 - 0.1x$ and the standard deviation of the residuals is 0.13. Does a case with the values $x = 4.1, y = 2.34$ qualify as an outlier?

13.1: One-Way ANOVA

28. What are the five basic assumptions to be met if you want to do a one-way ANOVA?

29. You are conducting a one-way ANOVA comparing the effectiveness of four drugs in lowering blood pressure in hypertensive patients. What are the null and alternative hypotheses for this study?

30. What is the primary difference between the independent samples t-test and one-way ANOVA?

31. You are comparing the results of three methods of teaching geometry to high school students. The final exam scores X_1, X_2, X_3, for the samples taught by the different methods have the following distributions:

$X_1 \sim N(85, 3.6)$

$X_1 \sim N(82, 4.8)$

$X_1 \sim N(79, 2.9)$

Each sample includes 100 students, and the final exam scores have a range of 0–100. Assuming the samples are independent and randomly selected, have the requirements for conducting a one-way ANOVA been met? Explain why or why not for each assumption.

32. You conduct a study comparing the effectiveness of four types of fertilizer to increase crop yield on wheat farms. When examining the sample results, you find that two of the samples have an approximately normal distribution, and two have an approximately uniform distribution. Is this a violation of the assumptions for conducting a one-way ANOVA?

13.2: The F Distribution

Use the following information to answer the next seven exercises. You are conducting a study of three types of feed supplements for cattle to test their effectiveness in producing weight gain among calves whose feed includes one of the supplements. You have four groups of 30 calves (one is a control group receiving the usual feed, but no supplement). You will conduct a one-way ANOVA after one year to see if there are difference in the mean weight for the four groups.

33. What is SS_{within} in this experiment, and what does it mean?

34. What is $SS_{between}$ in this experiment, and what does it mean?

35. What are k and i for this experiment?

36. If $SS_{within} = 374.5$ and $SS_{total} = 621.4$ for this data, what is $SS_{between}$?

37. What are $MS_{between}$, and MS_{within}, for this experiment?

38. What is the F Statistic for this data?

39. If there had been 35 calves in each group, instead of 30, with the sums of squares remaining the same, would the F Statistic be larger or smaller?

13.3: Facts About the F Distribution

40. Which of the following numbers are possible F Statistics?

a. 2.47

b. 5.95

c. –3.61

d. 7.28

e. 0.97

41. Histograms $F1$ and $F2$ below display the distribution of cases from samples from two populations, one distributed $F_{3,15}$ and one distributed $F_{5,500}$. Which sample came from which population?

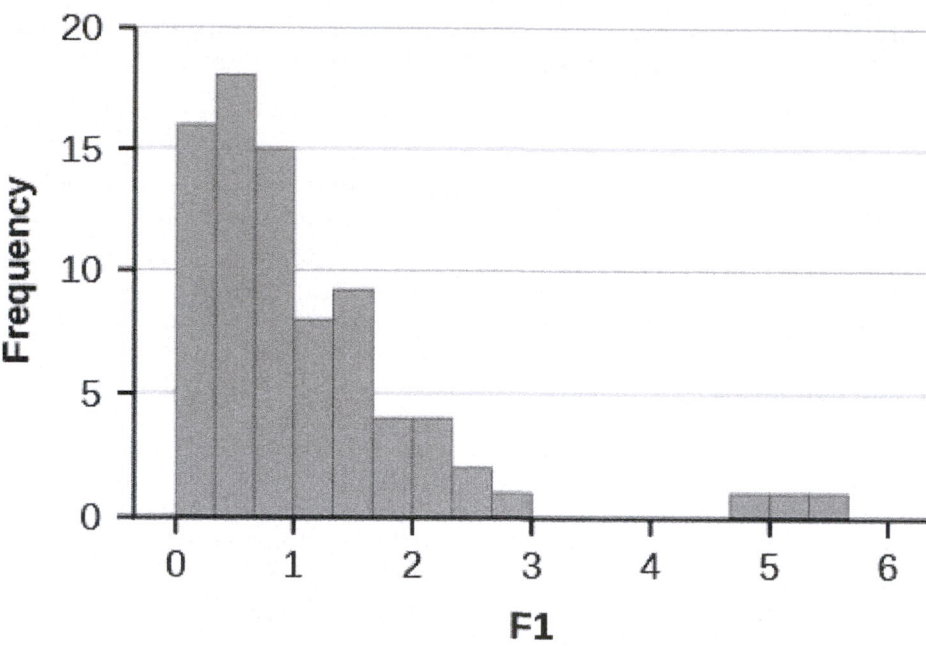

Figure B8

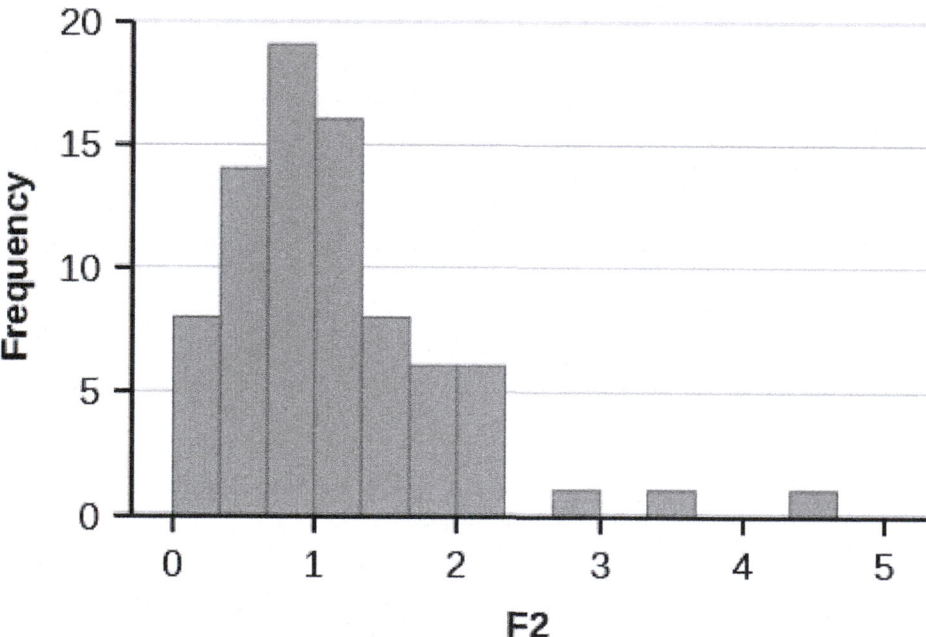

Figure B9

42. The F Statistic from an experiment with $k = 3$ and $n = 50$ is 3.67. At $\alpha = 0.05$, will you reject the null hypothesis?

43. The F Statistic from an experiment with $k = 4$ and $n = 100$ is 4.72. At $\alpha = 0.01$, will you reject the null hypothesis?

13.4: Test of Two Variances

44. What assumptions must be met to perform the F test of two variances?

45. You believe there is greater variance in grades given by the math department at your university than in the English department. You collect all the grades for undergraduate classes in the two departments for a semester, and compute the variance of each, and conduct an F test of two variances. What are the null and alternative hypotheses for this study?

Practice Test 4 Solutions
12.1 Linear Equations

1. e. A, B, and C.
All three are linear equations of the form $y = mx + b$.

2. Let y = the total number of hours required, and x the square footage, measured in units of 1,000. The equation is: $y = x + 4$

3. Let y = the total payment, and x the number of students in a class. The equation is: $y = 100(x) + 2,000$

4. Let y = the total cost of attendance, and x the number of years enrolled. The equation is: $y = 3,000(x) + 500$

12.2: Slope and Y-intercept of a Linear Equation

5. The independent variable is the hours worked on a car. The dependent variable is the total labor charges to fix a car.

6. Let y = the total charge, and x the number of hours required. The equation is: $y = 55x + 75$
The slope is 55 and the intercept is 75.

7. $y = 55(3.5) + 75 = 267.50$

8. Because the intercept is included in both equations, while you are only interested in the difference in costs, you do not need to include the intercept in the solution. The difference in number of hours required is: $6.3 - 2.4 = 3.9$.
Multiply this difference by the cost per hour: $55(3.9) = 214.5$.
The difference in cost between the two jobs is $214.50.

12.3: Scatter Plots

9. The X and Y variables have a strong linear relationship. These variables would be good candidates for analysis with linear regression.

10. The X and Y variables have a strong negative linear relationship. These variables would be good candidates for analysis with linear regression.

11. There is no clear linear relationship between the X and Y variables, so they are not good candidates for linear regression.

12. The X and Y variables have a strong positive relationship, but it is curvilinear rather than linear. These variables are not good candidates for linear regression.

12.4: The Regression Equation

13. $r\left(\frac{s_y}{s_x}\right) = 0.73\left(\frac{9.6}{4.0}\right) = 1.752 \approx 1.75$

14. $a = \bar{y} - b\bar{x} = 141.6 - 1.752(68.4) = 21.7632 \approx 21.76$

15. $\hat{y} = 21.76 + 1.75(68) = 140.76$

12.5: Correlation Coefficient and Coefficient of Determination

16. The coefficient of determination is the square of the correlation, or r^2.
For this data, $r^2 = (-0.56)2 = 0.3136 \approx 0.31$ or 31%. This means that 31 percent of the variation in fuel efficiency can be explained by the bodyweight of the automobile.

17. The coefficient of determination $= 0.32^2 = 0.1024$. This is the amount of variation in freshman college GPA that can be explained by high school GPA. The amount that cannot be explained is $1 - 0.1024 = 0.8976 \approx 0.90$. So about 90 percent of variance in freshman college GPA in this data is not explained by high school GPA.

18. $r = \sqrt{r^2}$
$\sqrt{0.5} = 0.707106781 \approx 0.71$
You need a correlation of 0.71 or higher to have a coefficient of determination of at least 0.5.

12.6: Testing the Significance of the Correlation Coefficient

19. $H_0: \rho = 0$
$H_a: \rho \neq 0$

20. $t = \dfrac{r\sqrt{n-2}}{\sqrt{1-r^2}} = \dfrac{0.33\sqrt{30-2}}{\sqrt{1-0.33^2}} = 1.85$

The critical value for $\alpha = 0.05$ for a two-tailed test using the t_{29} distribution is 2.045. Your value is less than this, so you fail to reject the null hypothesis and conclude that the study produced no evidence that the variables are significantly correlated. Using the calculator function tcdf, the p-value is 2tcdf(1.85, 10^99, 29) = 0.0373. Do not reject the null hypothesis and conclude that the study produced no evidence that the variables are significantly correlated.

21. $t = \dfrac{r\sqrt{n-2}}{\sqrt{1-r^2}} = \dfrac{0.45\sqrt{25-2}}{\sqrt{1-0.45^2}} = 2.417$

The critical value for $\alpha = 0.05$ for a two-tailed test using the t_{24} distribution is 2.064. Your value is greater than this, so you reject the null hypothesis and conclude that the study produced evidence that the variables are significantly correlated. Using the calculator function tcdf, the p-value is 2tcdf(2.417, 10^99, 24) = 0.0118. Reject the null hypothesis and conclude that the study produced evidence that the variables are significantly correlated.

12.7: Prediction

22. $\hat{y} = 25 + 16(5) = 105$

23. Because the intercept appears in both predicted values, you can ignore it in calculating a predicted difference score. The difference in grams of fiber per serving is $6 - 3 = 3$ and the predicted difference in grams of potassium per serving is $(16)(3) = 48$.

12.8: Outliers

24. An outlier is an observed value that is far from the least squares regression line. A rule of thumb is that a point more than two standard deviations of the residuals from its predicted value on the least squares regression line is an outlier.

25. An influential point is an observed value in a data set that is far from other points in the data set, in a horizontal direction. Unlike an outlier, an influential point is determined by its relationship with other values in the data set, not by its relationship to the regression line.

26. The predicted value for y is: $\hat{y} = 5 + 0.3x = 5.6$. The value of 6.2 is less than two standard deviations from the predicted value, so it does not qualify as an outlier.
Residual for (2, 6.2): $6.2 - 5.6 = 0.6 \ (0.6 < 2(0.4))$

27. The predicted value for y is: $\hat{y} = 2.3 - 0.1(4.1) = 1.89$. The value of 2.32 is more than two standard deviations from the predicted value, so it qualifies as an outlier.
Residual for (4.1, 2.34): $2.32 - 1.89 = 0.43 \ (0.43 > 2(0.13))$

13.1: One-Way ANOVA

28.

1. Each sample is drawn from a normally distributed population
2. All samples are independent and randomly selected.
3. The populations from which the samples are draw have equal standard deviations.
4. The factor is a categorical variable.
5. The response is a numerical variable.

29. H_0: $\mu 1 = \mu 2 = \mu 3 = \mu 4$
H_a: At least two of the group means $\mu 1, \mu 2, \mu 3, \mu 4$ are not equal.

30. The independent samples t-test can only compare means from two groups, while one-way ANOVA can compare means of more than two groups.

31. Each sample appears to have been drawn from a normally distributed populations, the factor is a categorical variable (method), the outcome is a numerical variable (test score), and you were told the samples were independent and randomly selected, so those requirements are met. However, each sample has a different standard deviation, and this suggests that the populations from which they were drawn also have different standard deviations, which is a violation of an assumption for one-way ANOVA. Further statistical testing will be necessary to test the assumption of equal variance before proceeding with the analysis.

32. One of the assumptions for a one-way ANOVA is that the samples are drawn from normally distributed populations. Since two of your samples have an approximately uniform distribution, this casts doubt on whether this assumption has been met. Further statistical testing will be necessary to determine if you can proceed with the analysis.

13.2: The *F* Distribution

33. SS_{within} is the sum of squares within groups, representing the variation in outcome that cannot be attributed to the different feed supplements, but due to individual or chance factors among the calves in each group.

34. $SS_{between}$ is the sum of squares between groups, representing the variation in outcome that can be attributed to the different feed supplements.

35. k = the number of groups = 4
n_1 = the number of cases in group 1 = 30
n = the total number of cases = 4(30) = 120

36. $SS_{total} = SS_{within} + SS_{between}$ so $SS_{between} = SS_{total} - SS_{within}$
621.4 − 374.5 = 246.9

37. The mean squares in an ANOVA are found by dividing each sum of squares by its respective degrees of freedom (*df*).
For SS_{total}, $df = n - 1 = 120 - 1 = 119$.
For $SS_{between}$, $df = k - 1 = 4 - 1 = 3$.
For SS_{within}, $df = 120 - 4 = 116$.

$MS_{between} = \dfrac{246.9}{3} = 82.3$

$MS_{within} = \dfrac{374.5}{116} = 3.23$

38. $F = \dfrac{MS_{between}}{MS_{within}} = \dfrac{82.3}{3.23} = 25.48$

39. It would be larger, because you would be dividing by a smaller number. The value of $MS_{between}$ would not change with a change of sample size, but the value of MS_{within} would be smaller, because you would be dividing by a larger number (df_{within} would be 136, not 116). Dividing a constant by a smaller number produces a larger result.

13.3: Facts About the *F* Distribution

40. All but choice c, −3.61. *F* Statistics are always greater than or equal to 0.

41. As the degrees of freedom increase in an *F* distribution, the distribution becomes more nearly normal. Histogram *F2* is closer to a normal distribution than histogram *F1*, so the sample displayed in histogram *F1* was drawn from the $F_{3,15}$ population, and the sample displayed in histogram *F2* was drawn from the $F_{5,500}$ population.

42. Using the calculator function Fcdf, *p*-value = Fcdf(3.67, 1E, 3,50) = 0.0182. Reject the null hypothesis.

43. Using the calculator function Fcdf, *p*-value = Fcdf(4.72, 1E, 4, 100) = 0.0016 Reject the null hypothesis.

13.4: Test of Two Variances

44. The samples must be drawn from populations that are normally distributed, and must be drawn from independent populations.

45. Let σ_M^2 = variance in math grades, and σ_E^2 = variance in English grades.

H_0: $\sigma_M^2 \leq \sigma_E^2$

H_a: $\sigma_M^2 > \sigma_E^2$

Practice Final Exam 1

Use the following information to answer the next two exercises: An experiment consists of tossing two, 12-sided dice (the numbers 1–12 are printed on the sides of each die).

- Let Event A = both dice show an even number.
- Let Event B = both dice show a number more than eight

1. Events A and B are:

a. mutually exclusive.

b. independent.

c. mutually exclusive and independent.

d. neither mutually exclusive nor independent.

2. Find $P(A|B)$.

a. $\frac{2}{4}$

b. $\frac{16}{144}$

c. $\frac{4}{16}$

d. $\frac{2}{144}$

3. Which of the following are TRUE when we perform a hypothesis test on matched or paired samples?

a. Sample sizes are almost never small.

b. Two measurements are drawn from the same pair of individuals or objects.

c. Two sample means are compared to each other.

d. Answer choices b and c are both true.

Use the following information to answer the next two exercises: One hundred eighteen students were asked what type of color their bedrooms were painted: light colors, dark colors, or vibrant colors. The results were tabulated according to gender.

	Light colors	Dark colors	Vibrant colors
Female	20	22	28
Male	10	30	8

Table B15

4. Find the probability that a randomly chosen student is male or has a bedroom painted with light colors.

a. $\frac{10}{118}$

b. $\frac{68}{118}$

c. $\frac{48}{118}$

d. $\frac{10}{48}$

5. Find the probability that a randomly chosen student is male given the student's bedroom is painted with dark colors.

a. $\frac{30}{118}$

b. $\frac{30}{48}$

c. $\frac{22}{118}$

d. $\frac{30}{52}$

Use the following information to answer the next two exercises: We are interested in the number of times a teenager must be reminded to do his or her chores each week. A survey of 40 mothers was conducted. **Table B16** shows the results of the survey.

x	P (x)
0	$\frac{2}{40}$
1	$\frac{5}{40}$
2	
3	$\frac{14}{40}$
4	$\frac{7}{40}$
5	$\frac{4}{40}$

Table B16

6. Find the probability that a teenager is reminded two times.

a. 8

b. $\frac{8}{40}$

c. $\frac{6}{40}$

d. 2

7. Find the expected number of times a teenager is reminded to do his or her chores.

a. 15

b. 2.78

c. 1.0

d. 3.13

Use the following information to answer the next two exercises: On any given day, approximately 37.5% of the cars parked in the De Anza parking garage are parked crookedly. We randomly survey 22 cars. We are interested in the number of cars that are parked crookedly.

8. For every 22 cars, how many would you expect to be parked crookedly, on average?

a. 8.25

b. 11

c. 18

d. 7.5

9. What is the probability that at least ten of the 22 cars are parked crookedly.

a. 0.1263

b. 0.1607

c. 0.2870

d. 0.8393

10. Using a sample of 15 Stanford-Binet IQ scores, we wish to conduct a hypothesis test. Our claim is that the mean IQ score on the Stanford-Binet IQ test is more than 100. It is known that the standard deviation of all Stanford-Binet IQ scores is 15 points. The correct distribution to use for the hypothesis test is:

a. Binomial

b. Student's t

c. Normal

d. Uniform

Use the following information to answer the next three exercises: De Anza College keeps statistics on the pass rate of students who enroll in math classes. In a sample of 1,795 students enrolled in Math 1A (1st quarter calculus), 1,428 passed the course. In a sample of 856 students enrolled in Math 1B (2nd quarter calculus), 662 passed. In general, are the pass rates of Math 1A and Math 1B statistically the same? Let A = the subscript for Math 1A and B = the subscript for Math 1B.

11. If you were to conduct an appropriate hypothesis test, the alternate hypothesis would be:

a. H_a: $p_A = p_B$

b. H_a: $p_A > p_B$

c. H_o: $p_A = p_B$

d. H_a: $p_A \neq p_B$

12. The Type I error is to:

a. conclude that the pass rate for Math 1A is the same as the pass rate for Math 1B when, in fact, the pass rates are different.

b. conclude that the pass rate for Math 1A is different than the pass rate for Math 1B when, in fact, the pass rates are the same.

c. conclude that the pass rate for Math 1A is greater than the pass rate for Math 1B when, in fact, the pass rate for Math 1A is less than the pass rate for Math 1B.

d. conclude that the pass rate for Math 1A is the same as the pass rate for Math 1B when, in fact, they are the same.

13. The correct decision is to:

a. reject H_0

b. not reject H_0

c. There is not enough information given to conduct the hypothesis test

Kia, Alejandra, and Iris are runners on the track teams at three different schools. Their running times, in minutes, and the statistics for the track teams at their respective schools, for a one mile run, are given in the table below:

	Running Time	School Average Running Time	School Standard Deviation
Kia	4.9	5.2	0.15
Alejandra	4.2	4.6	0.25
Iris	4.5	4.9	0.12

Table B17

14. Which student is the BEST when compared to the other runners at her school?

a. Kia

b. Alejandra

c. Iris

d. Impossible to determine

Use the following information to answer the next two exercises: The following adult ski sweater prices are from the Gorsuch Ltd. Winter catalog: $212, $292, $278, $199, $280, $236

Assume the underlying sweater price population is approximately normal. The null hypothesis is that the mean price of adult ski sweaters from Gorsuch Ltd. is at least $275.

15. The correct distribution to use for the hypothesis test is:

a. Normal

b. Binomial

c. Student's t

d. Exponential

16. The hypothesis test:

a. is two-tailed.

b. is left-tailed.

c. is right-tailed.

d. has no tails.

17. Sara, a statistics student, wanted to determine the mean number of books that college professors have in their office. She randomly selected two buildings on campus and asked each professor in the selected buildings how many books are in his or her office. Sara surveyed 25 professors. The type of sampling selected is

a. simple random sampling.

b. systematic sampling.

c. cluster sampling.

d. stratified sampling.

18. A clothing store would use which measure of the center of data when placing orders for the typical "middle" customer?

a. mean

b. median

c. mode

d. IQR

19. In a hypothesis test, the *p*-value is

a. the probability that an outcome of the data will happen purely by chance when the null hypothesis is true.

b. called the preconceived alpha.

c. compared to beta to decide whether to reject or not reject the null hypothesis.

d. Answer choices A and B are both true.

Use the following information to answer the next three exercises: A community college offers classes 6 days a week: Monday through Saturday. Maria conducted a study of the students in her classes to determine how many days per week the students who are in her classes come to campus for classes. In each of her 5 classes she randomly selected 10 students and asked them how many days they come to campus for classes. Each of her classes are the same size. The results of her survey are summarized in **Table B18**.

Number of Days on Campus	Frequency	Relative Frequency	Cumulative Relative Frequency
1	2		

Table B18

Number of Days on Campus	Frequency	Relative Frequency	Cumulative Relative Frequency
2	12	.24	
3	10	.20	
4			.98
5	0		
6	1	.02	1.00

Table B18

20. Combined with convenience sampling, what other sampling technique did Maria use?

a. simple random

b. systematic

c. cluster

d. stratified

21. How many students come to campus for classes four days a week?

a. 49

b. 25

c. 30

d. 13

22. What is the 60th percentile for the this data?

a. 2

b. 3

c. 4

d. 5

Use the following information to answer the next two exercises: The following data are the results of a random survey of 110 Reservists called to active duty to increase security at California airports.

Number of Dependents	Frequency
0	11
1	27
2	33
3	20
4	19

Table B19

23. Construct a 95% confidence interval for the true population mean number of dependents of Reservists called to active duty to increase security at California airports.

a. (1.85, 2.32)

b. (1.80, 2.36)

c. (1.97, 2.46)

d. (1.92, 2.50)

24. The 95% confidence interval above means:

a. Five percent of confidence intervals constructed this way will not contain the true population aveage number of dependents.

b. We are 95% confident the true population mean number of dependents falls in the interval.

c. Both of the above answer choices are correct.

d. None of the above.

25. $X \sim U(4, 10)$. Find the 30th percentile.

a. 0.3000

b. 3

c. 5.8

d. 6.1

26. If $X \sim Exp(0.8)$, then $P(x < \mu) = $ _____

a. 0.3679

b. 0.4727

c. 0.6321

d. cannot be determined

27. The lifetime of a computer circuit board is normally distributed with a mean of 2,500 hours and a standard deviation of 60 hours. What is the probability that a randomly chosen board will last at most 2,560 hours?

a. 0.8413

b. 0.1587

c. 0.3461

d. 0.6539

28. A survey of 123 reservists called to active duty as a result of the September 11, 2001, attacks was conducted to determine the proportion that were married. Eighty-six reported being married. Construct a 98% confidence interval for the true population proportion of reservists called to active duty that are married.

a. (0.6030, 0.7954)

b. (0.6181, 0.7802)

c. (0.5927, 0.8057)

d. (0.6312, 0.7672)

29. Winning times in 26 mile marathons run by world class runners average 145 minutes with a standard deviation of 14 minutes. A sample of the last ten marathon winning times is collected. Let x = mean winning times for ten marathons. The distribution for x is:

a. $N\left(145, \frac{14}{\sqrt{10}}\right)$

b. $N(145, 14)$

c. t_9

d. t_{10}

30. Suppose that Phi Beta Kappa honors the top one percent of college and university seniors. Assume that grade point means (GPA) at a certain college are normally distributed with a 2.5 mean and a standard deviation of 0.5. What would be the minimum GPA needed to become a member of Phi Beta Kappa at that college?

a. 3.99

b. 1.34

c. 3.00

d. 3.66

The number of people living on American farms has declined steadily during the 20th century. Here are data on the farm population (in millions of persons) from 1935 to 1980.

Year	1935	1940	1945	1950	1955	1960	1965	1970	1975	1980
Population	32.1	30.5	24.4	23.0	19.1	15.6	12.4	9.7	8.9	7.2

Table B20

31. The linear regression equation is $\hat{y} = 1166.93 - 0.5868x$. What was the expected farm population (in millions of persons) for 1980?

a. 7.2

b. 5.1

c. 6.0

d. 8.0

32. In linear regression, which is the best possible *SSE*?

a. 13.46

b. 18.22

c. 24.05

d. 16.33

33. In regression analysis, if the correlation coefficient is close to one what can be said about the best fit line?

a. It is a horizontal line. Therefore, we can not use it.

b. There is a strong linear pattern. Therefore, it is most likely a good model to be used.

c. The coefficient correlation is close to the limit. Therefore, it is hard to make a decision.

d. We do not have the equation. Therefore, we cannot say anything about it.

Use the following information to answer the next three exercises: A study of the career plans of young women and men sent questionnaires to all 722 members of the senior class in the College of Business Administration at the University of Illinois. One question asked which major within the business program the student had chosen. Here are the data from the students who responded.

	Female	Male
Accounting	68	56
Administration	91	40
Economics	5	6
Finance	61	59

Table B21 Does the data suggest that there is a relationship between the gender of students and their choice of major?

34. The distribution for the test is:

a. Chi^2_8.

b. Chi^2_3.

c. t_{721}.

d. $N(0, 1)$.

35. The expected number of female who choose finance is:

a. 37.

b. 61.

c. 60.

d. 70.

36. The p-value is 0.0127 and the level of significance is 0.05. The conclusion to the test is:

a. there is insufficient evidence to conclude that the choice of major and the gender of the student are not independent of each other.

b. there is sufficient evidence to conclude that the choice of major and the gender of the student are not independent of each other.

c. there is sufficient evidence to conclude that students find economics very hard.

d. there is in sufficient evidence to conclude that more females prefer administration than males.

37. An agency reported that the work force nationwide is composed of 10% professional, 10% clerical, 30% skilled, 15% service, and 35% semiskilled laborers. A random sample of 100 San Jose residents indicated 15 professional, 15 clerical, 40 skilled, 10 service, and 20 semiskilled laborers. At $\alpha = 0.10$ does the work force in San Jose appear to be consistent with the agency report for the nation? Which kind of test is it?

a. Chi^2 goodness of fit

b. Chi^2 test of independence

c. Independent groups proportions

d. Unable to determine

Practice Final Exam 1 Solutions
Solutions

1. b. independent

2. c. $\frac{4}{16}$

3. b. Two measurements are drawn from the same pair of individuals or objects.

4. b. $\frac{68}{118}$

5. d. $\frac{30}{52}$

6. b. $\frac{8}{40}$

7. b. 2.78

8. a. 8.25

9. c. 0.2870

10. c. Normal

11. d. H_a: $p_A \neq p_B$

12. b. conclude that the pass rate for Math 1A is different than the pass rate for Math 1B when, in fact, the pass rates are the same.

13. b. not reject H_0

14. c. Iris

15. c. Student's t

16. b. is left-tailed.

17. c. cluster sampling

18. b. median

19. a. the probability that an outcome of the data will happen purely by chance when the null hypothesis is true.

20. d. stratified

21. b. 25

22. c. 4

23. a. (1.85, 2.32)

24. c. Both above are correct.

25. c. 5.8

26. c. 0.6321

27. a. 0.8413

28. a. (0.6030, 0.7954)

29. a. $N\left(145, \frac{14}{\sqrt{10}}\right)$

30. d. 3.66

31. b. 5.1

32. a. 13.46

33. b. There is a strong linear pattern. Therefore, it is most likely a good model to be used.

34. b. $\text{Chi}^2{}_3$.

35. d. 70

36. b. There is sufficient evidence to conclude that the choice of major and the gender of the student are not independent of each other.

37. a. Chi^2 goodness-of-fit

Practice Final Exam 2

1. A study was done to determine the proportion of teenagers that own a car. The population proportion of teenagers that own a car is the:

a.   statistic.

b.   parameter.

c.   population.

d.   variable.

Use the following information to answer the next two exercises:

value	frequency
0	1
1	4

Table B22

value	frequency
2	7
3	9
6	4

Table B22

2. The box plot for the data is:

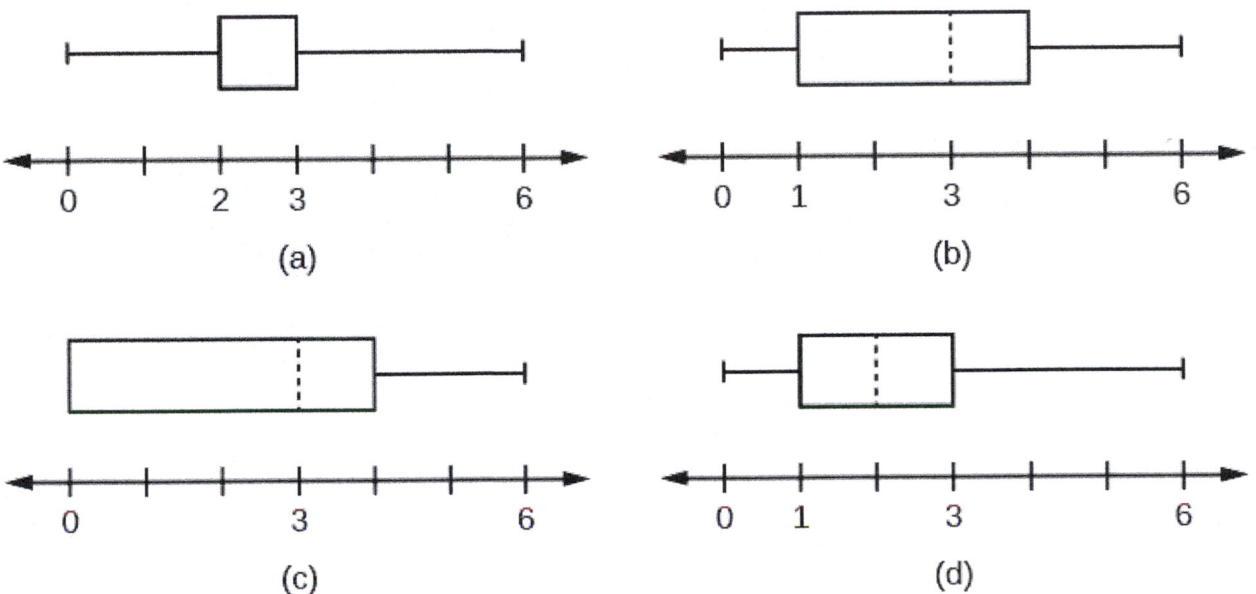

(a)   (b)

(c)   (d)

Figure B10

3. If six were added to each value of the data in the table, the 15th percentile of the new list of values is:

a. six

b. one

c. seven

d. eight

Use the following information to answer the next two exercises: Suppose that the probability of a drought in any independent year is 20%. Out of those years in which a drought occurs, the probability of water rationing is ten percent. However, in any year, the probability of water rationing is five percent.

4. What is the probability of both a drought and water rationing occurring?

a. 0.05

b. 0.01

c. 0.02

d. 0.30

5. Which of the following is true?

a. Drought and water rationing are independent events.

b. Drought and water rationing are mutually exclusive events.

c. None of the above

Use the following information to answer the next two exercises: Suppose that a survey yielded the following data:

gender	apple	pumpkin	pecan
female	40	10	30
male	20	30	10

Table B23 Favorite Pie

6. Suppose that one individual is randomly chosen. The probability that the person's favorite pie is apple or the person is male is _____.

a. $\frac{40}{60}$

b. $\frac{60}{140}$

c. $\frac{120}{140}$

d. $\frac{100}{140}$

7. Suppose H_0 is: Favorite pie and gender are independent. The *p*-value is _____.

a. ≈ 0

b. 1

c. 0.05

d. cannot be determined

Use the following information to answer the next two exercises: Let's say that the probability that an adult watches the news at least once per week is 0.60. We randomly survey 14 people. Of interest is the number of people who watch the news at least once per week.

8. Which of the following statements is FALSE?

a. $X \sim B(14\ 0.60)$

b. The values for *x* are: {1, 2, 3, ... , 14}.

c. $\mu = 8.4$

d. $P(X = 5) = 0.0408$

9. Find the probability that at least six adults watch the news at least once per week.

a. $\frac{6}{14}$

b. 0.8499

c. 0.9417

d. 0.6429

10. The following histogram is most likely to be a result of sampling from which distribution?

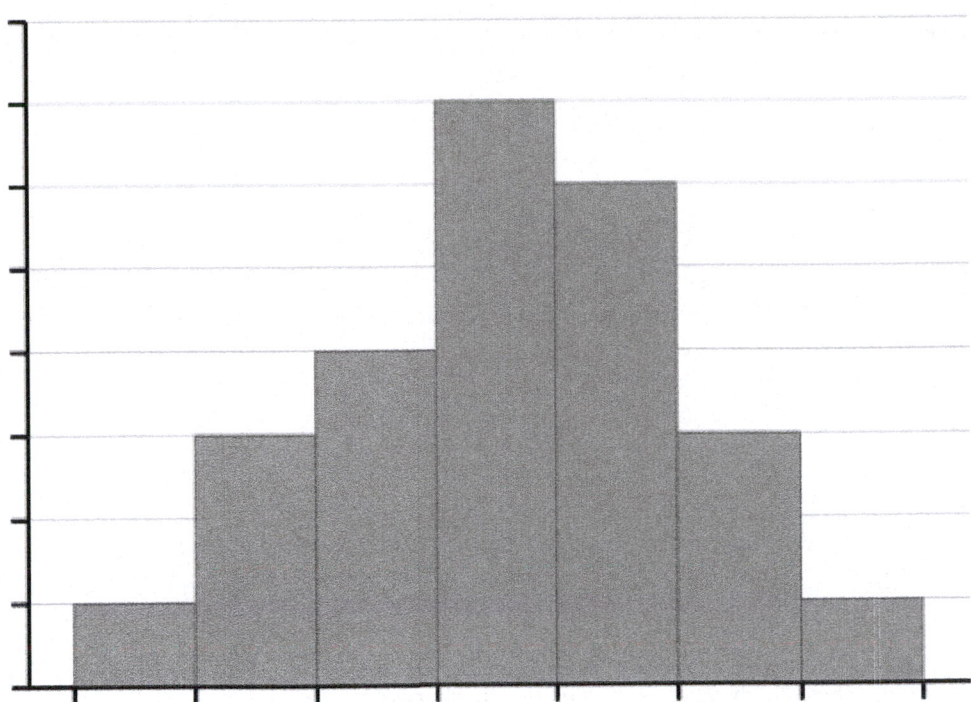

Figure B11

a. chi-square with $df = 6$

b. exponential

c. uniform

d. binomial

11. The ages of campus day and evening students is known to be normally distributed. A sample of six campus day and evening students reported their ages (in years) as: {18, 35, 27, 45, 20, 20}. What is the error bound for the 90% confidence interval of the true average age?

a. 11.2

b. 22.3

c. 17.5

d. 8.7

12. If a normally distributed random variable has $\mu = 0$ and $\sigma = 1$, then 97.5% of the population values lie above:

a. −1.96.

b. 1.96.

c. 1.

d. −1.

Use the following information to answer the next three exercises. The amount of money a customer spends in one trip to the supermarket is known to have an exponential distribution. Suppose the average amount of money a customer spends in one trip to the supermarket is $72.

13. What is the probability that one customer spends less than $72 in one trip to the supermarket?

a. 0.6321

b. 0.5000

c. 0.3714

d. 1

14. How much money altogether would you expect the next five customers to spend in one trip to the supermarket (in dollars)?

a. 72

b. $\dfrac{72^2}{5}$

c. 5184

d. 360

15. If you want to find the probability that the mean amount of money 50 customers spend in one trip to the supermarket is less than $60, the distribution to use is:

a. $N(72, 72)$

b. $N\left(72, \dfrac{72}{\sqrt{50}}\right)$

c. $Exp(72)$

d. $Exp\left(\dfrac{1}{72}\right)$

Use the following information to answer the next three exercises: The amount of time it takes a fourth grader to carry out the trash is uniformly distributed in the interval from one to ten minutes.

16. What is the probability that a randomly chosen fourth grader takes more than seven minutes to take out the trash?

a. $\dfrac{3}{9}$

b. $\dfrac{7}{9}$

c. $\dfrac{3}{10}$

d. $\dfrac{7}{10}$

17. Which graph best shows the probability that a randomly chosen fourth grader takes more than six minutes to take out the trash given that he or she has already taken more than three minutes?

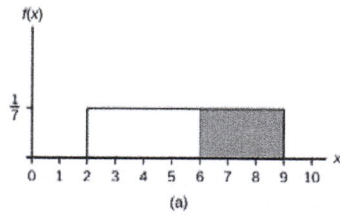
(a)

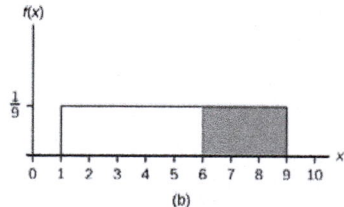
(b)

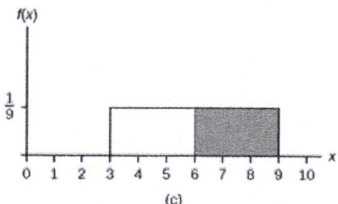
(c)

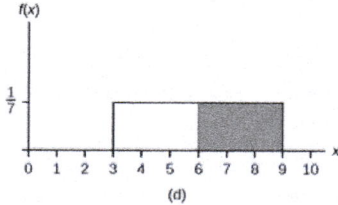
(d)

Figure B12

18. We should expect a fourth grader to take how many minutes to take out the trash?

a. 4.5

b. 5.5

c. 5

d. 10

Use the following information to answer the next three exercises: At the beginning of the quarter, the amount of time a student waits in line at the campus cafeteria is normally distributed with a mean of five minutes and a standard deviation of 1.5 minutes.

19. What is the 90th percentile of waiting times (in minutes)?

a. 1.28

b. 90

c. 7.47

d. 6.92

20. The median waiting time (in minutes) for one student is:

a. 5.

b. 50.

c. 2.5.

d. 1.5.

21. Find the probability that the average wait time for ten students is at most 5.5 minutes.

a. 0.6301

b. 0.8541

c. 0.3694

d. 0.1459

22. A sample of 80 software engineers in Silicon Valley is taken and it is found that 20% of them earn approximately $50,000 per year. A point estimate for the true proportion of engineers in Silicon Valley who earn $50,000 per year is:

a. 16.

b. 0.2.

c. 1.

d. 0.95.

23. If $P(Z < z_\alpha) = 0.1587$ where $Z \sim N(0, 1)$, then α is equal to:

a. −1.

b. 0.1587.

c. 0.8413.

d. 1.

24. A professor tested 35 students to determine their entering skills. At the end of the term, after completing the course, the same test was administered to the same 35 students to study their improvement. This would be a test of:

a. independent groups.

b. two proportions.

c. matched pairs, dependent groups.

d. exclusive groups.

A math exam was given to all the third grade children attending ABC School. Two random samples of scores were taken.

	n	$\overline{x}$	s
Boys	55	82	5
Girls	60	86	7

Table B24

25. Which of the following correctly describes the results of a hypothesis test of the claim, "There is a difference between the mean scores obtained by third grade girls and boys at the 5% level of significance"?

  a.   Do not reject H_0. There is insufficient evidence to conclude that there is a difference in the mean scores.

  b.   Do not reject H_0. There is sufficient evidence to conclude that there is a difference in the mean scores.

  c.   Reject H_0. There is insufficient evidence to conclude that there is no difference in the mean scores.

  d.   Reject H_0. There is sufficient evidence to conclude that there is a difference in the mean scores.

26. In a survey of 80 males, 45 had played an organized sport growing up. Of the 70 females surveyed, 25 had played an organized sport growing up. We are interested in whether the proportion for males is higher than the proportion for females. The correct conclusion is that:

  a.   there is insufficient information to conclude that the proportion for males is the same as the proportion for females.

  b.   there is insufficient information to conclude that the proportion for males is not the same as the proportion for females.

  c.   there is sufficient evidence to conclude that the proportion for males is higher than the proportion for females.

  d.   not enough information to make a conclusion.

27. From past experience, a statistics teacher has found that the average score on a midterm is 81 with a standard deviation of 5.2. This term, a class of 49 students had a standard deviation of 5 on the midterm. Do the data indicate that we should reject the teacher's claim that the standard deviation is 5.2? Use $\alpha = 0.05$.

  a.   Yes

  b.   No

  c.   Not enough information given to solve the problem

28. Three loading machines are being compared. Ten samples were taken for each machine. Machine I took an average of 31 minutes to load packages with a standard deviation of two minutes. Machine II took an average of 28 minutes to load packages with a standard deviation of 1.5 minutes. Machine III took an average of 29 minutes to load packages with a standard deviation of one minute. Find the *p*-value when testing that the average loading times are the same.

  a.   *p*-value is close to zero

  b.   *p*-value is close to one

  c.   not enough information given to solve the problem

Use the following information to answer the next three exercises: A corporation has offices in different parts of the country. It has gathered the following information concerning the number of bathrooms and the number of employees at seven sites:

Number of employees x	650	730	810	900	102	107	1150
Number of bathrooms y	40	50	54	61	82	110	121

Table B25

29. Is the correlation between the number of employees and the number of bathrooms significant?

  a.   Yes

  b.   No

  c.   Not enough information to answer question

30. The linear regression equation is:

a. $\hat{y} = 0.0094 - 79.96x$

b. $\hat{y} = 79.96 + 0.0094x$

c. $\hat{y} = 79.96 - 0.0094x$

d. $\hat{y} = -0.0094 + 79.96x$

31. If a site has 1,150 employees, approximately how many bathrooms should it have?

a. 69

b. 91

c. 91,954

d. We should not be estimating here.

32. Suppose that a sample of size ten was collected, with $\bar{x}$ = 4.4 and s = 1.4. H_0: σ^2 = 1.6 vs. H_a: $\sigma^2 \neq$ 1.6. Which graph best describes the results of the test?

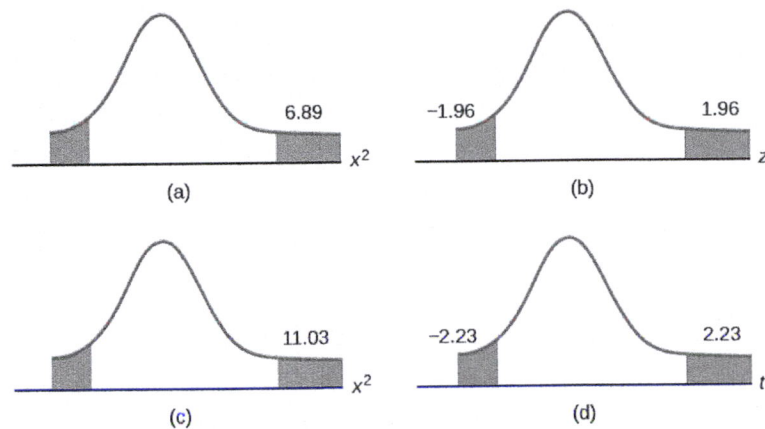

Figure B13

Sixty-four backpackers were asked the number of days since their latest backpacking trip. The number of days is given in **Table B26**:

# of days	1	2	3	4	5	6	7	8
Frequency	5	9	6	12	7	10	5	10

Table B26

33. Conduct an appropriate test to determine if the distribution is uniform.

a. The p-value is > 0.10. There is insufficient information to conclude that the distribution is not uniform.

b. The p-value is < 0.01. There is sufficient information to conclude the distribution is not uniform.

c. The p-value is between 0.01 and 0.10, but without alpha (α) there is not enough information

d. There is no such test that can be conducted.

34. Which of the following statements is true when using one-way ANOVA?

a. The populations from which the samples are selected have different distributions.

b. The sample sizes are large.

c. The test is to determine if the different groups have the same means.

d. There is a correlation between the factors of the experiment.

Practice Final Exam 2 Solutions
Solutions

1. b. parameter.

2. a.

3. c. seven

4. c. 0.02

5. c. none of the above

6. d. $\frac{100}{140}$

7. a. ≈ 0

8. b. The values for x are: {1, 2, 3,..., 14}

9. c. 0.9417.

10. d. binomial

11. d. 8.7

12. a. −1.96

13. a. 0.6321

14. d. 360

15. b. $N\left(72, \frac{72}{\sqrt{50}}\right)$

16. a. $\frac{3}{9}$

17. d.

18. b. 5.5

19. d. 6.92

20. a. 5

21. b. 0.8541

22. b. 0.2

23. a. −1.

24. c. matched pairs, dependent groups.

25. d. Reject H_0. There is sufficient evidence to conclude that there is a difference in the mean scores.

26. c. there is sufficient evidence to conclude that the proportion for males is higher than the proportion for females.

27. b. no

28. b. p-value is close to 1.

29. b. No

30. c. $\hat{y}$ = 79.96x − 0.0094

31. d. We should not be estimating here.

32. a.

33. a. The p-value is > 0.10. There is insufficient information to conclude that the distribution is not uniform.

34. c. The test is to determine if the different groups have the same means.

APPENDIX C: DATA SETS

Lap Times

The following tables provide lap times from Terri Vogel's log book. Times are recorded in seconds for 2.5-mile laps completed in a series of races and practice runs.

	Lap 1	Lap 2	Lap 3	Lap 4	Lap 5	Lap 6	Lap 7
Race 1	135	130	131	132	130	131	133
Race 2	134	131	131	129	128	128	129
Race 3	129	128	127	127	130	127	129
Race 4	125	125	126	125	124	125	125
Race 5	133	132	132	132	131	130	132
Race 6	130	130	130	129	129	130	129
Race 7	132	131	133	131	134	134	131
Race 8	127	128	127	130	128	126	128
Race 9	132	130	127	128	126	127	124
Race 10	135	131	131	132	130	131	130
Race 11	132	131	132	131	130	129	129
Race 12	134	130	130	130	131	130	130
Race 13	128	127	128	128	128	129	128
Race 14	132	131	131	131	132	130	130
Race 15	136	129	129	129	129	129	129
Race 16	129	129	129	128	128	129	129
Race 17	134	131	132	131	132	132	132
Race 18	129	129	130	130	133	133	127
Race 19	130	129	129	129	129	129	128
Race 20	131	128	130	128	129	130	130

Table C1 Race Lap Times (in seconds)

	Lap 1	Lap 2	Lap 3	Lap 4	Lap 5	Lap 6	Lap 7
Practice 1	142	143	180	137	134	134	172
Practice 2	140	135	134	133	128	128	131
Practice 3	130	133	130	128	135	133	133

Table C2 Practice Lap Times (in seconds)

	Lap 1	Lap 2	Lap 3	Lap 4	Lap 5	Lap 6	Lap 7
Practice 4	141	136	137	136	136	136	145
Practice 5	140	138	136	137	135	134	134
Practice 6	142	142	139	138	129	129	127
Practice 7	139	137	135	135	137	134	135
Practice 8	143	136	134	133	134	133	132
Practice 9	135	134	133	133	132	132	133
Practice 10	131	130	128	129	127	128	127
Practice 11	143	139	139	138	138	137	138
Practice 12	132	133	131	129	128	127	126
Practice 13	149	144	144	139	138	138	137
Practice 14	133	132	137	133	134	130	131
Practice 15	138	136	133	133	132	131	131

Table C2 Practice Lap Times (in seconds)

Stock Prices

The following table lists initial public offering (IPO) stock prices for all 1999 stocks that at least doubled in value during the first day of trading.

$17.00	$23.00	$14.00	$16.00	$12.00	$26.00
$20.00	$22.00	$14.00	$15.00	$22.00	$18.00
$18.00	$21.00	$21.00	$19.00	$15.00	$21.00
$18.00	$17.00	$15.00	$25.00	$14.00	$30.00
$16.00	$10.00	$20.00	$12.00	$16.00	$17.44
$16.00	$14.00	$15.00	$20.00	$20.00	$16.00
$17.00	$16.00	$15.00	$15.00	$19.00	$48.00
$16.00	$18.00	$9.00	$18.00	$18.00	$20.00
$8.00	$20.00	$17.00	$14.00	$11.00	$16.00
$19.00	$15.00	$21.00	$12.00	$8.00	$16.00
$13.00	$14.00	$15.00	$14.00	$13.41	$28.00
$21.00	$17.00	$28.00	$17.00	$19.00	$16.00
$17.00	$19.00	$18.00	$17.00	$15.00	
$14.00	$21.00	$12.00	$18.00	$24.00	
$15.00	$23.00	$14.00	$16.00	$12.00	
$24.00	$20.00	$14.00	$14.00	$15.00	
$14.00	$19.00	$16.00	$38.00	$20.00	

Table C3 IPO Offer Prices

$24.00	$16.00	$8.00	$18.00	$17.00	
$16.00	$15.00	$7.00	$19.00	$12.00	
$8.00	$23.00	$12.00	$18.00	$20.00	
$21.00	$34.00	$16.00	$26.00	$14.00	

Table C3 IPO Offer Prices

References

Data compiled by Jay R. Ritter of University of Florida using data from *Securities Data Co.* and *Bloomberg*.

APPENDIX D: GROUP AND PARTNER PROJECTS

Univariate Data

Student Learning Objectives

- The student will design and carry out a survey.
- The student will analyze and graphically display the results of the survey.

Instructions

As you complete each task below, check it off. Answer all questions in your summary.

____ Decide what data you are going to study.

Here are two examples, but you may **NOT** use them: number of M&M's per bag, number of pencils students have in their backpacks.

____ Are your data discrete or continuous? How do you know?

____ Decide how you are going to collect the data (for instance, buy 30 bags of M&M's; collect data from the World Wide Web).

____ Describe your sampling technique in detail. Use cluster, stratified, systematic, or simple random (using a random number generator) sampling. Do not use convenience sampling. Which method did you use? Why did you pick that method?

____ Conduct your survey. **Your data size must be at least 30.**

____ Summarize your data in a chart with columns showing **data value, frequency, relative frequency and cumulative relative frequency.**

Answer the following (rounded to two decimal places):

a. $\bar{x}$ = _____

b. s = _____

c. First quartile = _____

d. Median = _____

e. 70th percentile = _____

____ What value is two standard deviations above the mean?

____ What value is 1.5 standard deviations below the mean?

____ Construct a histogram displaying your data.

____ In complete sentences, describe the shape of your graph.

____ Do you notice any potential outliers? If so, what values are they? Show your work in how you used the potential outlier formula to determine whether or not the values might be outliers.

____ Construct a box plot displaying your data.

____ Does the middle 50% of the data appear to be concentrated together or spread apart? Explain how you determined this.

____ Looking at both the histogram and the box plot, discuss the distribution of your data.

Assignment Checklist

You need to turn in the following typed and stapled packet, with pages in the following order:

____ **Cover sheet**: name, class time, and name of your study

_____ **Summary page**: This should contain paragraphs written with complete sentences. It should include answers to all the questions above. It should also include statements describing the population under study, the sample, a parameter or parameters being studied, and the statistic or statistics produced.

_____ **URL** for data, if your data are from the World Wide Web

_____ **Chart of data, frequency, relative frequency, and cumulative relative frequency**

_____ **Page(s) of graphs:** histogram and box plot

Continuous Distributions and Central Limit Theorem

Student Learning Objectives

- The student will collect a sample of continuous data.
- The student will attempt to fit the data sample to various distribution models.
- The student will validate the central limit theorem.

Instructions

As you complete each task below, check it off. Answer all questions in your summary.

Part I: Sampling

_____ Decide what **continuous** data you are going to study. (Here are two examples, but you may NOT use them: the amount of money a student spent on college supplies this term, or the length of time distance telephone call lasts.)

_____ Describe your sampling technique in detail. Use cluster, stratified, systematic, or simple random (using a random number generator) sampling. Do not use convenience sampling. What method did you use? Why did you pick that method?

_____ Conduct your survey. Gather **at least 150 pieces of continuous, quantitative data**.

_____ Define (in words) the random variable for your data. $X =$ _____

_____ Create two lists of your data: (1) unordered data, (2) in order of smallest to largest.

_____ Find the sample mean and the sample standard deviation (rounded to two decimal places).

a. $\bar{x} =$ _____

b. $s =$ _____

_____ Construct a histogram of your data containing five to ten intervals of equal width. The histogram should be a representative display of your data. Label and scale it.

Part II: Possible Distributions

_____ Suppose that X followed the following theoretical distributions. Set up each distribution using the appropriate information from your data.

_____ Uniform: $X \sim U$ _____ Use the lowest and highest values as a and b.

_____ Normal: $X \sim N$ _____ Use $\bar{x}$ to estimate for μ and s to estimate for σ.

_____ **Must** your data fit one of the above distributions? Explain why or why not.

_____ **Could** the data fit two or three of the previous distributions (at the same time)? Explain.

_____ Calculate the value k(an X value) that is 1.75 standard deviations above the sample mean. $k =$ _____ (rounded to two decimal places) Note: $k = \bar{x} + (1.75)s$

_____ Determine the relative frequencies (*RF*) rounded to four decimal places.

NOTE

$$RF = \frac{frequency}{total\ number\ surveyed}$$

a. $RF(X < k) =$ _____

b. $RF(X > k) =$ _____

c. $RF(X = k) =$ _____

NOTE

You should have one page for the uniform distribution, one page for the exponential distribution, and one page for the normal distribution.

_____ State the distribution: $X \sim$ _____
_____ Draw a graph for each of the three theoretical distributions. Label the axes and mark them appropriately.
_____ Find the following theoretical probabilities (rounded to four decimal places).

a. $P(X < k) =$ _____

b. $P(X > k) =$ _____

c. $P(X = k) =$ _____

_____ Compare the relative frequencies to the corresponding probabilities. Are the values close?
_____ Does it appear that the data fit the distribution well? Justify your answer by comparing the probabilities to the relative frequencies, and the histograms to the theoretical graphs.

Part III: CLT Experiments

_____ From your original data (before ordering), use a random number generator to pick 40 samples of size five. For each sample, calculate the average.
_____ On a separate page, attached to the summary, include the 40 samples of size five, along with the 40 sample averages.
_____ List the 40 averages in order from smallest to largest.
_____ Define the random variable, $\bar{X}$, in words. $\bar{X} =$ _____

_____ State the approximate theoretical distribution of $\bar{X}$. $\bar{X} \sim$ _____
_____ Base this on the mean and standard deviation from your original data.
_____ Construct a histogram displaying your data. Use five to six intervals of equal width. Label and scale it.

Calculate the value $\bar{k}$ (an $\bar{X}$ value) that is 1.75 standard deviations above the sample mean. $\bar{k} =$ _____ (rounded to two decimal places)
Determine the relative frequencies (*RF*) rounded to four decimal places.

a. $RF(\bar{X} < \bar{k}) =$ _____

b. $RF(\bar{X} > \bar{k}) =$ _____

c. $RF(\bar{X} = \bar{k}) =$ _____

Find the following theoretical probabilities (rounded to four decimal places).

a. $P(\bar{X} < \bar{k}) =$ _____

b. $P(\bar{X} > \bar{k}) =$ _____

c. $P(\bar{X} = \bar{k}) =$ _____

_____ Draw the graph of the theoretical distribution of $\bar{X}$.
_____ Compare the relative frequencies to the probabilities. Are the values close?

_____ Does it appear that the data of averages fit the distribution of $\bar{X}$ well? Justify your answer by comparing the probabilities to the relative frequencies, and the histogram to the theoretical graph.
In three to five complete sentences for each, answer the following questions. Give thoughtful explanations.
_____ In summary, do your original data seem to fit the uniform, exponential, or normal distributions? Answer why or why not for each distribution. If the data do not fit any of those distributions, explain why.
_____ What happened to the shape and distribution when you averaged your data? **In theory,** what should have happened? In theory, would "it" always happen? Why or why not?

_____ Were the relative frequencies compared to the theoretical probabilities closer when comparing the X or $\bar{X}$ distributions? Explain your answer.

Assignment Checklist

You need to turn in the following typed and stapled packet, with pages in the following order:

_____ **Cover sheet**: name, class time, and name of your study

_____ **Summary pages**: These should contain several paragraphs written with complete sentences that describe the experiment, including what you studied and your sampling technique, as well as answers to all of the questions previously asked questions

_____ **URL** for data, if your data are from the World Wide Web

_____ **Pages, one for each theoretical distribution**, with the distribution stated, the graph, and the probability questions answered

_____ **Pages of the data requested**

_____ **All graphs required**

Hypothesis Testing-Article
Student Learning Objectives

- The student will identify a hypothesis testing problem in print.

- The student will conduct a survey to verify or dispute the results of the hypothesis test.

- The student will summarize the article, analysis, and conclusions in a report.

Instructions

As you complete each task, check it off. Answer all questions in your summary.

_____**Find an article** in a newspaper, magazine, or on the internet which makes a claim about **ONE** population mean or **ONE** population proportion. The claim may be based upon a survey that the article was reporting on. Decide whether this claim is the null or alternate hypothesis.

_____**Copy or print out the article** and include a copy in your project, along with the source.

_____**State how you will collect your data.** (Convenience sampling is not acceptable.)

_____**Conduct your survey. You must have more than 50 responses in your sample.** When you hand in your final project, attach the tally sheet or the packet of questionnaires that you used to collect data. Your data must be real.

_____**State the statistics** that are a result of your data collection: sample size, sample mean, and sample standard deviation, OR sample size and number of successes.

_____**Make two copies of the appropriate solution sheet.**

_____**Record the hypothesis test** on the solution sheet, based on your experiment. **Do a DRAFT solution** first on one of the solution sheets and check it over carefully. Have a classmate check your solution to see if it is done correctly. Make your decision using a 5% level of significance. Include the 95% confidence interval on the solution sheet.

_____**Create a graph that illustrates your data.** This may be a pie or bar graph or may be a histogram or box plot, depending on the nature of your data. Produce a graph that makes sense for your data and gives useful visual information about your data. You may need to look at several types of graphs before you decide which is the most appropriate for the type of data in your project.

_____**Write your summary** (in complete sentences and paragraphs, with proper grammar and correct spelling) that describes the project. The summary **MUST** include:

a. Brief discussion of the article, including the source

b. Statement of the claim made in the article (one of the hypotheses).

c. Detailed description of how, where, and when you collected the data, including the sampling technique; did you use cluster, stratified, systematic, or simple random sampling (using a random number generator)? As previously mentioned, convenience sampling is not acceptable.

d. Conclusion about the article claim in light of your hypothesis test; this is the conclusion of your hypothesis test, stated in words, in the context of the situation in your project in sentence form, as if you were writing this conclusion for a non-statistician.

e. Sentence interpreting your confidence interval in the context of the situation in your project

Assignment Checklist

Turn in the following typed (12 point) and stapled packet for your final project:

_____**Cover sheet** containing your name(s), class time, and the name of your study

_____**Summary**, which includes all items listed on summary checklist

_____**Solution sheet** neatly and completely filled out. The solution sheet does not need to be typed.

____**Graphic representation of your data**, created following the guidelines previously discussed; include only graphs which are appropriate and useful.

____**Raw data collected AND a table summarizing the sample data** (n, $\bar{x}$ and s; or x, n, and p', as appropriate for your hypotheses); the raw data does not need to be typed, but the summary does. Hand in the data as you collected it. (Either attach your tally sheet or an envelope containing your questionnaires.)

Bivariate Data, Linear Regression, and Univariate Data
Student Learning Objectives

- The students will collect a bivariate data sample through the use of appropriate sampling techniques.

- The student will attempt to fit the data to a linear model.

- The student will determine the appropriateness of linear fit of the model.

- The student will analyze and graph univariate data.

Instructions

1. As you complete each task below, check it off. Answer all questions in your introduction or summary.

2. Check your course calendar for intermediate and final due dates.

3. Graphs may be constructed by hand or by computer, unless your instructor informs you otherwise. All graphs must be neat and accurate.

4. All other responses must be done on the computer.

5. Neatness and quality of explanations are used to determine your final grade.

Part I: Bivariate Data

Introduction

____State the bivariate data your group is going to study.

_____

Here are two examples, but you may **NOT** use them: height vs. weight and age vs. running distance.

____Describe your sampling technique in detail. Use cluster, stratified, systematic, or simple random sampling (using a random number generator) sampling. Convenience sampling is **NOT** acceptable.
____Conduct your survey. Your number of pairs must be at least 30.
____Print out a copy of your data.

Analysis

____On a separate sheet of paper construct a scatter plot of the data. Label and scale both axes.
____State the least squares line and the correlation coefficient.
____On your scatter plot, in a different color, construct the least squares line.
____Is the correlation coefficient significant? Explain and show how you determined this.
____Interpret the slope of the linear regression line in the context of the data in your project. Relate the explanation to your data, and quantify what the slope tells you.
____Does the regression line seem to fit the data? Why or why not? If the data does not seem to be linear, explain if any other model seems to fit the data better.
____Are there any outliers? If so, what are they? Show your work in how you used the potential outlier formula in the Linear Regression and Correlation chapter (since you have bivariate data) to determine whether or not any pairs might be outliers.

Part II: Univariate Data

In this section, you will use the data for **ONE** variable only. Pick the variable that is more interesting to analyze. For example: if your independent variable is sequential data such as year with 30 years and one piece of data per year, your x-values might be 1971, 1972, 1973, 1974, ..., 2000. This would not be interesting to analyze. In that case, choose to use the dependent variable to analyze for this part of the project.
____Summarize your data in a chart with columns showing data value, frequency, relative frequency, and cumulative

relative frequency.

_____Answer the following question, rounded to two decimal places:

a. Sample mean = _____

b. Sample standard deviation = _____

c. First quartile = _____

d. Third quartile = _____

e. Median = _____

f. 70th percentile = _____

g. Value that is 2 standard deviations above the mean = _____

h. Value that is 1.5 standard deviations below the mean = _____

_____Construct a histogram displaying your data. Group your data into six to ten intervals of equal width. Pick regularly spaced intervals that make sense in relation to your data. For example, do NOT group data by age as 20-26,27-33,34-40,41-47,48-54,55-61 . . . Instead, maybe use age groups 19.5-24.5, 24.5-29.5, . . . or 19.5-29.5, 29.5-39.5, 39.5-49.5, . . .

_____In complete sentences, describe the shape of your histogram.

_____Are there any potential outliers? Which values are they? Show your work and calculations as to how you used the potential outlier formula in **Descriptive Statistics** (since you are now using univariate data) to determine which values might be outliers.

_____Construct a box plot of your data.

_____Does the middle 50% of your data appear to be concentrated together or spread out? Explain how you determined this.

_____Looking at both the histogram AND the box plot, discuss the distribution of your data. For example: how does the spread of the middle 50% of your data compare to the spread of the rest of the data represented in the box plot; how does this correspond to your description of the shape of the histogram; how does the graphical display show any outliers you may have found; does the histogram show any gaps in the data that are not visible in the box plot; are there any interesting features of your data that you should point out.

Due Dates

- Part I, Intro: _____ (keep a copy for your records)

- Part I, Analysis: _____ (keep a copy for your records)

- Entire Project, typed and stapled: _____

    _____ Cover sheet: names, class time, and name of your study

    _____ Part I: label the sections "Intro" and "Analysis."
    _____ Part II:
    _____ Summary page containing several paragraphs written in complete sentences describing the experiment, including what you studied and how you collected your data. The summary page should also include answers to ALL the questions asked above.
    _____ All graphs requested in the project
    _____ All calculations requested to support questions in data
    _____ Description: what you learned by doing this project, what challenges you had, how you overcame the challenges

NOTE

Include answers to ALL questions asked, even if not explicitly repeated in the items above.

APPENDIX E: SOLUTION SHEETS

Hypothesis Testing with One Sample

Class Time: _____

Name: _____

a. H_0: _____

b. H_a: _____

c. In words, **CLEARLY** state what your random variable $\overline{X}$ or P' represents.

d. State the distribution to use for the test.

e. What is the test statistic?

f. What is the *p*-value? In one or two complete sentences, explain what the *p*-value means for this problem.

g. Use the previous information to sketch a picture of this situation. CLEARLY, label and scale the horizontal axis and shade the region(s) corresponding to the *p*-value.

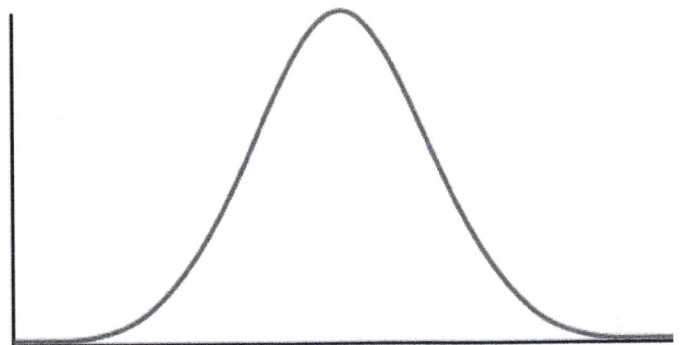

Figure E1

h. Indicate the correct decision ("reject" or "do not reject" the null hypothesis), the reason for it, and write an appropriate conclusion, using **complete sentences**.

    i. Alpha: _____

    ii. Decision: _____

    iii. Reason for decision: _____

    iv. Conclusion: _____

i. Construct a 95% confidence interval for the true mean or proportion. Include a sketch of the graph of the situation. Label the point estimate and the lower and upper bounds of the confidence interval.

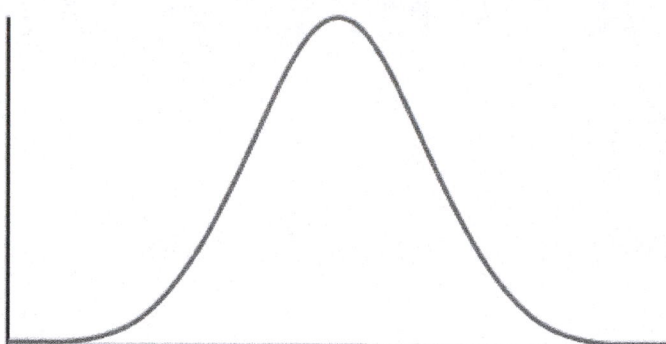

Figure E2

Hypothesis Testing with Two Samples

Class Time: _____

Name: _____

a. H_0: _____

b. H_a: _____

c. In words, **clearly** state what your random variable $\bar{X}_1 - \bar{X}_2$, $P'_1 - P'_2$ or $\bar{X}_d$ represents.

d. State the distribution to use for the test.

e. What is the test statistic?

f. What is the *p*-value? In one to two complete sentences, explain what the p-value means for this problem.

g. Use the previous information to sketch a picture of this situation. **CLEARLY** label and scale the horizontal axis and shade the region(s) corresponding to the *p*-value.

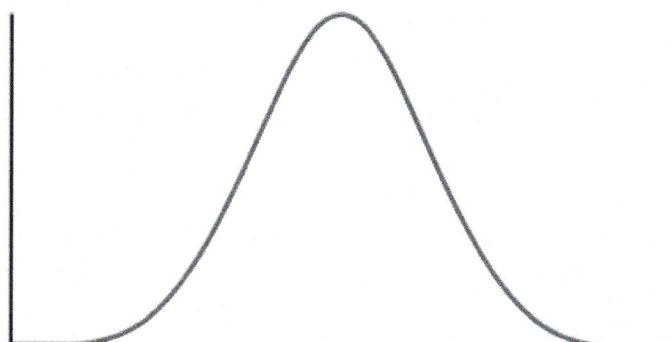

Figure E3

h. Indicate the correct decision ("reject" or "do not reject" the null hypothesis), the reason for it, and write an appropriate conclusion, using **complete sentences**.

   a. Alpha: _____

   b. Decision: _____

   c. Reason for decision: _____

   d. Conclusion: _____

i. In complete sentences, explain how you determined which distribution to use.

The Chi-Square Distribution

Class Time: _____

Name: _____

a. H_0: _____

b. H_a: _____

c. What are the degrees of freedom?

d. State the distribution to use for the test.

e. What is the test statistic?

f. What is the *p*-value? In one to two complete sentences, explain what the *p*-value means for this problem.

g. Use the previous information to sketch a picture of this situation. **Clearly** label and scale the horizontal axis and shade the region(s) corresponding to the *p*-value.

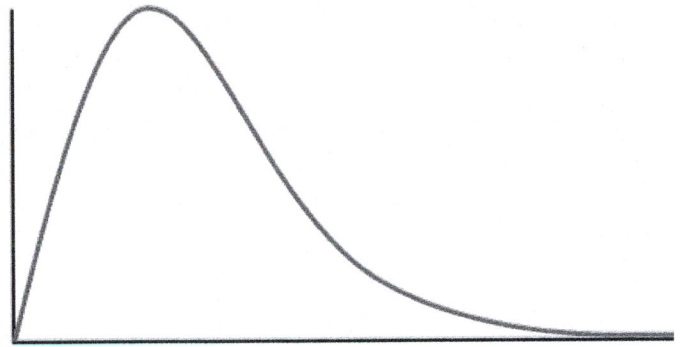

Figure E4

h. Indicate the correct decision ("reject" or "do not reject" the null hypothesis) and write appropriate conclusions, using **complete sentences.**

   i.   Alpha: _____

   ii.   Decision: _____

   iii. Reason for decision: _____

   iv.   Conclusion: _____

F Distribution and One-Way ANOVA

Class Time: _____

Name: _____

a. H_0: _____

b. H_a: _____

c. $df(n)$ = _____ $df(d)$ = _____

d. State the distribution to use for the test.

e. What is the test statistic?

f. What is the *p*-value?

g. Use the previous information to sketch a picture of this situation. **Clearly** label and scale the horizontal axis and shade the region(s) corresponding to the *p*-value.

Figure E5

h. Indicate the correct decision ("reject" or "do not reject" the null hypothesis) and write appropriate conclusions, using **complete sentences**.

   a. Alpha: _____

   b. Decision: _____

   c. Reason for decision: _____

   d. Conclusion: _____

APPENDIX F: MATHEMATICAL PHRASES, SYMBOLS, AND FORMULAS

English Phrases Written Mathematically

When the English says:	Interpret this as:
X is at least 4.	$X \geq 4$
The minimum of X is 4.	$X \geq 4$
X is no less than 4.	$X \geq 4$
X is greater than or equal to 4.	$X \geq 4$
X is at most 4.	$X \leq 4$
The maximum of X is 4.	$X \leq 4$
X is no more than 4.	$X \leq 4$
X is less than or equal to 4.	$X \leq 4$
X does not exceed 4.	$X \leq 4$
X is greater than 4.	$X > 4$
X is more than 4.	$X > 4$
X exceeds 4.	$X > 4$
X is less than 4.	$X < 4$
There are fewer X than 4.	$X < 4$
X is 4.	$X = 4$
X is equal to 4.	$X = 4$
X is the same as 4.	$X = 4$
X is not 4.	$X \neq 4$
X is not equal to 4.	$X \neq 4$
X is not the same as 4.	$X \neq 4$
X is different than 4.	$X \neq 4$

Table F1

Formulas

Formula 1: Factorial

$$n! = n(n-1)(n-2)...(1)$$

$$0! = 1$$

Formula 2: Combinations

$$\binom{n}{r} = \frac{n!}{(n-r)!r!}$$

Formula 3: Binomial Distribution

$$X \sim B(n, p)$$

$$P(X = x) = \binom{n}{x}p^x q^{n-x}, \text{ for } x = 0, 1, 2, ..., n$$

Formula 4: Geometric Distribution

$$X \sim G(p)$$

$$P(X = x) = q^{x-1}p, \text{ for } x = 1, 2, 3, ...$$

Formula 5: Hypergeometric Distribution

$$X \sim H(r, b, n)$$

$$P(X = x) = \left(\frac{\binom{r}{x}\binom{b}{n-x}}{\binom{r+b}{n}}\right)$$

Formula 6: Poisson Distribution

$$X \sim P(\mu)$$

$$P(X = x) = \frac{\mu^x e^{-\mu}}{x!}$$

Formula 7: Uniform Distribution

$$X \sim U(a, b)$$

$$f(X) = \frac{1}{b-a}, \ a < x < b$$

Formula 8: Exponential Distribution

$$X \sim Exp(m)$$

$$f(x) = me^{-mx} m > 0, \ x \geq 0$$

Formula 9: Normal Distribution

$$X \sim N(\mu, \sigma^2)$$

$$f(x) = \frac{1}{\sigma\sqrt{2\pi}}e^{\frac{-(x-\mu)^2}{2\sigma^2}} \ , \ -\infty < x < \infty$$

Formula 10: Gamma Function

$$\Gamma(z) = \int_{\infty}^{0} x^{z-1} e^{-x} dx \ \ z > 0$$

$$\Gamma\left(\tfrac{1}{2}\right) = \sqrt{\pi}$$

$\Gamma(m+1) = m!$ for m, a nonnegative integer

otherwise: $\Gamma(a+1) = a\Gamma(a)$

Formula 11: Student's *t*-distribution

$X \sim t_{df}$

$$f(x) = \frac{\left(1+\frac{x^2}{n}\right)^{\frac{-(n+1)}{2}} \Gamma\left(\frac{n+1}{2}\right)}{\sqrt{n\pi}\,\Gamma\left(\frac{n}{2}\right)}$$

$$X = \frac{Z}{\sqrt{\frac{Y}{n}}}$$

$Z \sim N(0, 1), Y \sim X_{df}^2$, n = degrees of freedom

Formula 12: Chi-Square Distribution

$X \sim X_{df}^2$

$$f(x) = \frac{x^{\frac{n-2}{2}} e^{\frac{-x}{2}}}{2^{\frac{n}{2}} \Gamma\left(\frac{n}{2}\right)}, \ x > 0 \ , \ n = \text{positive integer and degrees of freedom}$$

Formula 13: F Distribution

$X \sim F_{df(n),\, df(d)}$

$df(n) = $ degrees of freedom for the numerator

$df(d) = $ degrees of freedom for the denominator

$$f(x) = \frac{\Gamma\left(\frac{u+v}{2}\right)}{\Gamma\left(\frac{u}{2}\right)\Gamma\left(\frac{v}{2}\right)} \left(\frac{u}{v}\right)^{\frac{u}{2}} x^{\left(\frac{u}{2}-1\right)} \left[1 + \left(\frac{u}{v}\right)x\right]^{-0.5(u+v)}$$

$X = \frac{Y_u}{W_v}$, Y, W are chi-square

Symbols and Their Meanings

Chapter (1st used)	Symbol	Spoken	Meaning
Sampling and Data	$\sqrt{\ }$	The square root of	same
Sampling and Data	π	Pi	3.14159… (a specific number)
Descriptive Statistics	Q_1	Quartile one	the first quartile

Table F2 Symbols and their Meanings

Chapter (1st used)	Symbol	Spoken	Meaning
Descriptive Statistics	Q_2	Quartile two	the second quartile
Descriptive Statistics	Q_3	Quartile three	the third quartile
Descriptive Statistics	IQR	interquartile range	$Q_3 - Q_1 = IQR$
Descriptive Statistics	$\bar{x}$	x-bar	sample mean
Descriptive Statistics	μ	mu	population mean
Descriptive Statistics	$s\ s_x\ sx$	s	sample standard deviation
Descriptive Statistics	$s^2\ s_x^2$	s squared	sample variance
Descriptive Statistics	$\sigma\ \sigma_x\ \sigma x$	sigma	population standard deviation
Descriptive Statistics	$\sigma^2\ \sigma_x^2$	sigma squared	population variance
Descriptive Statistics	Σ	capital sigma	sum
Probability Topics	$\{\ \}$	brackets	set notation
Probability Topics	S	S	sample space
Probability Topics	A	Event A	event A
Probability Topics	$P(A)$	probability of A	probability of A occurring
Probability Topics	$P(A|B)$	probability of A given B	prob. of A occurring given B has occurred
Probability Topics	$P(A\ OR\ B)$	prob. of A or B	prob. of A or B or both occurring
Probability Topics	$P(A\ AND\ B)$	prob. of A and B	prob. of both A and B occurring (same time)
Probability Topics	A'	A-prime, complement of A	complement of A, not A
Probability Topics	$P(A')$	prob. of complement of A	same
Probability Topics	G_1	green on first pick	same
Probability Topics	$P(G_1)$	prob. of green on first pick	same
Discrete Random Variables	PDF	prob. distribution function	same
Discrete Random Variables	X	X	the random variable X
Discrete Random Variables	$X\sim$	the distribution of X	same
Discrete Random Variables	B	binomial distribution	same
Discrete Random Variables	G	geometric distribution	same
Discrete Random Variables	H	hypergeometric dist.	same
Discrete Random Variables	P	Poisson dist.	same
Discrete Random Variables	λ	Lambda	average of Poisson distribution
Discrete Random Variables	$\geq$	greater than or equal to	same
Discrete Random Variables	$\leq$	less than or equal to	same
Discrete Random Variables	$=$	equal to	same

Table F2 Symbols and their Meanings

Chapter (1st used)	Symbol	Spoken	Meaning
Discrete Random Variables	$\neq$	not equal to	same
Continuous Random Variables	$f(x)$	f of x	function of x
Continuous Random Variables	pdf	prob. density function	same
Continuous Random Variables	U	uniform distribution	same
Continuous Random Variables	Exp	exponential distribution	same
Continuous Random Variables	k	k	critical value
Continuous Random Variables	$f(x) =$	f of x equals	same
Continuous Random Variables	m	m	decay rate (for exp. dist.)
The Normal Distribution	N	normal distribution	same
The Normal Distribution	z	z-score	same
The Normal Distribution	Z	standard normal dist.	same
The Central Limit Theorem	CLT	Central Limit Theorem	same
The Central Limit Theorem	$\bar{X}$	X-bar	the random variable X-bar
The Central Limit Theorem	μ_x	mean of X	the average of X
The Central Limit Theorem	$\mu_{\bar{x}}$	mean of X-bar	the average of X-bar
The Central Limit Theorem	σ_x	standard deviation of X	same
The Central Limit Theorem	$\sigma_{\bar{x}}$	standard deviation of X-bar	same
The Central Limit Theorem	ΣX	sum of X	same
The Central Limit Theorem	Σx	sum of x	same
Confidence Intervals	CL	confidence level	same
Confidence Intervals	CI	confidence interval	same
Confidence Intervals	EBM	error bound for a mean	same
Confidence Intervals	EBP	error bound for a proportion	same
Confidence Intervals	t	Student's t-distribution	same
Confidence Intervals	df	degrees of freedom	same
Confidence Intervals	$t_{\frac{\alpha}{2}}$	student t with $a/2$ area in right tail	same
Confidence Intervals	p' ; $\hat{p}$	p-prime; p-hat	sample proportion of success
Confidence Intervals	q' ; $\hat{q}$	q-prime; q-hat	sample proportion of failure

Table F2 Symbols and their Meanings

Chapter (1st used)	Symbol	Spoken	Meaning
Hypothesis Testing	H_0	H-naught, H-sub 0	null hypothesis
Hypothesis Testing	H_a	H-a, H-sub a	alternate hypothesis
Hypothesis Testing	H_1	H-1, H-sub 1	alternate hypothesis
Hypothesis Testing	α	alpha	probability of Type I error
Hypothesis Testing	β	beta	probability of Type II error
Hypothesis Testing	$\overline{X1} - \overline{X2}$	X1-bar minus X2-bar	difference in sample means
Hypothesis Testing	$\mu_1 - \mu_2$	mu-1 minus mu-2	difference in population means
Hypothesis Testing	$P'_1 - P'_2$	P1-prime minus P2-prime	difference in sample proportions
Hypothesis Testing	$p_1 - p_2$	p1 minus p2	difference in population proportions
Chi-Square Distribution	X^2	Ky-square	Chi-square
Chi-Square Distribution	O	Observed	Observed frequency
Chi-Square Distribution	E	Expected	Expected frequency
Linear Regression and Correlation	$y = a + bx$	y equals a plus b-x	equation of a line
Linear Regression and Correlation	$\hat{y}$	y-hat	estimated value of y
Linear Regression and Correlation	r	correlation coefficient	same
Linear Regression and Correlation	ε	error	same
Linear Regression and Correlation	SSE	Sum of Squared Errors	same
Linear Regression and Correlation	$1.9s$	1.9 times s	cut-off value for outliers
F-Distribution and ANOVA	F	F-ratio	F-ratio

Table F2 Symbols and their Meanings

APPENDIX G: NOTES FOR THE TI-83, 83+, 84, 84+ CALCULATORS

Quick Tips

Legend

- represents a button press
- [ ] represents yellow command or green letter behind a key
- < > represents items on the screen

To adjust the contrast

Press , then hold to increase the contrast or 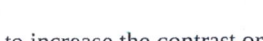 to decrease the contrast.

To capitalize letters and words

Press  to get one capital letter, or press , then to set all button presses to capital letters. You can return to the top-level button values by pressing 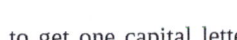 again.

To correct a mistake

If you hit a wrong button, just hit and start again.

To write in scientific notation

Numbers in scientific notation are expressed on the TI-83, 83+, 84, and 84+ using E notation, such that...

- $4.321 \text{ E } 4 = 4.321 \times 10^4$
- $4.321 \text{ E } -4 = 4.321 \times 10^{-4}$

To transfer programs or equations from one calculator to another:

Both calculators: Insert your respective end of the link cable cable and press , then [LINK].

Calculator receiving information:
1. Use the arrows to navigate to and select <RECEIVE>
2. Press .

Calculator sending information:
1. Press appropriate number or letter.

2. Use up and down arrows to access the appropriate item.

3. Press to select item to transfer.

4. Press right arrow to navigate to and select <TRANSMIT>.

5. Press .

NOTE

ERROR 35 LINK generally means that the cables have not been inserted far enough.

Both calculators: Insert your respective end of the link cable cable Both calculators: press 2nd , then [QUIT] to exit when done.

Manipulating One-Variable Statistics

NOTE

These directions are for entering data with the built-in statistical program.

Data	Frequency
−2	10
−1	3
0	4
1	5
3	8

Table G1 Sample Data We are manipulating one-variable statistics.

To begin:

1. Turn on the calculator.

2. Access statistics mode.

3. Select <4:ClrList> to clear data from lists, if desired.

,

4. Enter list [L1] to be cleared.

[2nd] , [L1] , [ENTER]

5. Display last instruction.

[2nd] , [ENTRY]

6. Continue clearing remaining lists in the same fashion, if desired.

[◀] , [2nd] , [L2] , [ENTER]

7. Access statistics mode.

[STAT]

8. Select <1:Edit . . .>

[ENTER]

9. Enter data. Data values go into [L1]. (You may need to arrow over to [L1]).

   ◦ Type in a data value and enter it. (For negative numbers, use the negate (-) key at the bottom of the keypad).

   [(−)] , [9] , [ENTER]

   ◦ Continue in the same manner until all data values are entered.

10. In [L2], enter the frequencies for each data value in [L1].

    ◦ Type in a frequency and enter it. (If a data value appears only once, the frequency is "1").

    [4] , [ENTER]

    ◦ Continue in the same manner until all data values are entered.

11. Access statistics mode.

[STAT]

12. Navigate to <CALC>.

13. Access <1:1-var Stats>.

[ENTER]

14. Indicate that the data is in [L1]...

[2nd] , [L1] , [,]

15. ...and indicate that the frequencies are in [L2].

[2nd] , [L2] , [ENTER]

16. The statistics should be displayed. You may arrow down to get remaining statistics. Repeat as necessary.

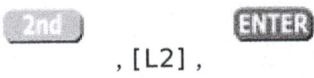

Drawing Histograms

NOTE

We will assume that the data is already entered.

We will construct two histograms with the built-in STATPLOT application. The first way will use the default ZOOM. The second way will involve customizing a new graph.

1. Access graphing mode.

    , [STAT PLOT]

2. Select `<1:plot 1>` to access plotting - first graph.

3. Use the arrows navigate go to `<ON>` to turn on Plot 1.

    `<ON>` ,

4. Use the arrows to go to the histogram picture and select the histogram.

5. Use the arrows to navigate to `<Xlist>`.
6. If "L1" is not selected, select it.

    , [L1] ,

7. Use the arrows to navigate to `<Freq>`.
8. Assign the frequencies to `[L2]`.

    , [L2] ,

9. Go back to access other graphs.

    , [STAT PLOT]

10. Use the arrows to turn off the remaining plots.
11. **Be sure to deselect or clear all equations before graphing.**

To deselect equations:

1. Access the list of equations.

2. Select each equal sign (=).

3. Continue, until all equations are deselected.

To clear equations:

1. Access the list of equations.

2. Use the arrow keys to navigate to the right of each equal sign (=) and clear them.

3. Repeat until all equations are deleted.

To draw default histogram:
1. Access the ZOOM menu.

2. Select `<9:ZoomStat>`.

3. The histogram will show with a window automatically set.

To draw custom histogram:
1. Access window mode to set the graph parameters.

2. ◦ $X_{min} = -2.5$

    ◦ $X_{max} = 3.5$

    ◦ $X_{scl} = 1$ (width of bars)

    ◦ $Y_{min} = 0$

    ◦ $Y_{max} = 10$

    ◦ $Y_{scl} = 1$ (spacing of tick marks on y-axis)

    ◦ $X_{res} = 1$

3. Access graphing mode to see the histogram.

To draw box plots:
1. Access graphing mode.

,[STAT PLOT]

2. Select `<1:Plot 1>` to access the first graph.

3. Use the arrows to select `<ON>` and turn on Plot 1.

4. Use the arrows to select the box plot picture and enable it.

5. Use the arrows to navigate to `<Xlist>`.
6. If "L1" is not selected, select it.

   , [L1] ,

7. Use the arrows to navigate to `<Freq>`.
8. Indicate that the frequencies are in [L2].

   , [L2] ,

9. Go back to access other graphs.

   , [STAT PLOT]

10. **Be sure to deselect or clear all equations before graphing** using the method mentioned above.
11. View the box plot.

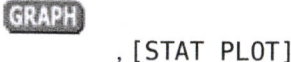

   , [STAT PLOT]

Linear Regression
Sample Data

The following data is real. The percent of declared ethnic minority students at De Anza College for selected years from 1970–1995 was:

Year	Student Ethnic Minority Percentage
1970	14.13
1973	12.27
1976	14.08
1979	18.16
1982	27.64
1983	28.72
1986	31.86
1989	33.14
1992	45.37
1995	53.1

Table G2 The independent variable is "Year," while the independent variable is "Student Ethnic Minority Percent."

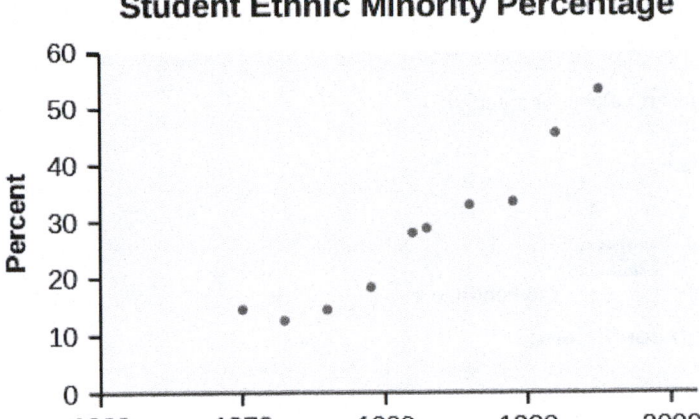

Figure G1 Student Ethnic Minority Percentage By hand, verify the scatterplot above.

NOTE
_____

The TI-83 has a built-in linear regression feature, which allows the data to be edited. The x-values will be in [L1]; the y-values in [L2].

To enter data and do linear regression:

1. ON Turns calculator on.

2. Before accessing this program, be sure to turn off all plots.

   ◦ Access graphing mode.

, [STAT PLOT]

   ◦ Turn off all plots.

,

3. Round to three decimal places. To do so:

   ◦ Access the mode menu.

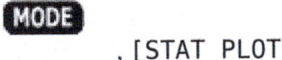, [STAT PLOT]

   ◦ Navigate to <Float> and then to the right to <3>.

   ◦ All numbers will be rounded to three decimal places until changed.

4. Enter statistics mode and clear lists [L1] and [L2], as describe previously.

,

5. Enter editing mode to insert values for *x* and *y*.

,

6. Enter each value. Press _____ to continue.

To display the correlation coefficient:

1. Access the catalog.

, [CATALOG]

2. Arrow down and select <DiagnosticOn>

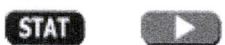

... , ,

3. *r* and r^2 will be displayed during regression calculations.

4. Access linear regression.

5. Select the form of *y* = *a* + *bx*.

,

The display will show:

LinReg

- *y* = *a* + *bx*

- *a* = −3176.909

- *b* = 1.617

- *r* = 2 0.924

- *r* = 0.961

This means the Line of Best Fit (Least Squares Line) is:

- *y* = −3176.909 + 1.617*x*

- Percent = −3176.909 + 1.617 (year #)

The correlation coefficient *r* = 0.961

To see the scatter plot:

1. Access graphing mode.

, [STAT PLOT]

2. Select `<1:plot 1>` To access plotting - first graph.

ENTER

3. Navigate and select `<ON>` to turn on Plot 1.

ENTER

`<ON>`

4. Navigate to the first picture.

5. Select the scatter plot.

ENTER

6. Navigate to `<Xlist>`.

2nd

7. If `[L1]` is not selected, press   , `[L1]` to select it.

8. Confirm that the data values are in `[L1]`.

ENTER

`<ON>`

9. Navigate to `<Ylist>`.

10. Select that the frequencies are in `[L2]`.

2nd   ENTER

     , `[L2]` ,

11. Go back to access other graphs.

2nd

     , `[STAT PLOT]`

12. Use the arrows to turn off the remaining plots.

13. Access window mode to set the graph parameters.

WINDOW

- ◦ $X_{min} = 1970$

- ◦ $X_{max} = 2000$

- ◦ $X_{scl} = 10$ (spacing of tick marks on x-axis)

- ◦ $Y_{min} = -0.05$

- ◦ $Y_{max} = 60$

- ◦ $Y_{scl} = 10$ (spacing of tick marks on y-axis)

- ◦ $X_{res} = 1$

14. Be sure to deselect or clear all equations before graphing, using the instructions above.

GRAPH

15. Press the graph button to see the scatter plot.

To see the regression graph:

1. Access the equation menu. The regression equation will be put into Y1.

2. Access the vars menu and navigate to <5: Statistics>.

   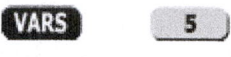

3. Navigate to <EQ>.

4. <1: RegEQ> contains the regression equation which will be entered in Y1.

5. Press the graphing mode button. The regression line will be superimposed over the scatter plot.

To see the residuals and use them to calculate the critical point for an outlier:

1. Access the list. RESID will be an item on the menu. Navigate to it.

   , [LIST], <RESID>

2. Confirm twice to view the list of residuals. Use the arrows to select them.

3. The critical point for an outlier is: $1.9V\frac{\text{SSE}}{n-2}$ where:

   ◦ n = number of pairs of data

   ◦ SSE = sum of the squared errors

   ◦ $\sum \left(\text{residual}^2\right)$

4. Store the residuals in [L3].

   , [L3] ,

5. Calculate the $\frac{(\text{residual})^2}{n-2}$. Note that $n-2=8$

   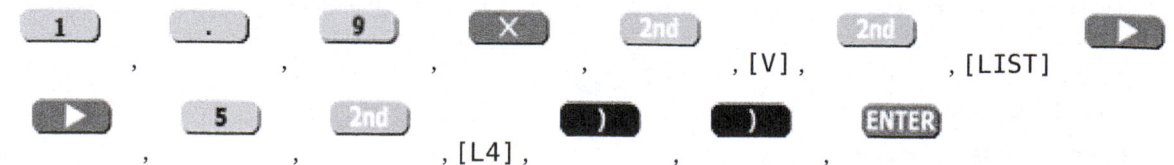
   , [L3] ,

6. Store this value in [L4].

   , [L4] ,

7. Calculate the critical value using the equation above.

   , [V] ,   , [LIST]

   , [L4] ,

8. Verify that the calculator displays: 7.642669563. This is the critical value.

9. Compare the absolute value of each residual value in [L3] to 7.64. If the absolute value is greater than 7.64, then the (x, y) corresponding point is an outlier. In this case, none of the points is an outlier.

To obtain estimates of *y* for various *x*-values:

There are various ways to determine estimates for "*y*." One way is to substitute values for "*x*" in the equation. Another way

is to use the on the graph of the regression line.

TI-83, 83+, 84, 84+ instructions for distributions and tests

Distributions

Access DISTR (for "Distributions").

For technical assistance, visit the Texas Instruments website at **http://www.ti.com (http://www.ti.com)** and enter your calculator model into the "search" box.

Binomial Distribution

- binompdf(*n*,*p*,*x*) corresponds to $P(X = x)$
- binomcdf(*n*,*p*,*x*) corresponds to $P(X \leq x)$
- To see a list of all probabilities for *x*: 0, 1, . . . , *n*, leave off the "*x*" parameter.

Poisson Distribution

- poissonpdf(λ,*x*) corresponds to $P(X = x)$
- poissoncdf(λ,*x*) corresponds to $P(X \leq x)$

Continuous Distributions (general)

- $-\infty$ uses the value –1EE99 for left bound
- $+\infty$ uses the value 1EE99 for right bound

Normal Distribution

- normalpdf(*x*,*μ*,*σ*) yields a probability density function value (only useful to plot the normal curve, in which case "*x*" is the variable)
- normalcdf(left bound, right bound, *μ*, *σ*) corresponds to $P(\text{left bound} < X < \text{right bound})$
- normalcdf(left bound, right bound) corresponds to $P(\text{left bound} < Z < \text{right bound})$ – standard normal
- invNorm(*p*,*μ*,*σ*) yields the critical value, *k*: $P(X < k) = p$
- invNorm(*p*) yields the critical value, *k*: $P(Z < k) = p$ for the standard normal

Student's *t*-Distribution

- tpdf(*x*,*df*) yields the probability density function value (only useful to plot the student-*t* curve, in which case "*x*" is the variable)
- tcdf(left bound, right bound, *df*) corresponds to $P(\text{left bound} < t < \text{right bound})$

Chi-square Distribution

- X^2pdf(*x*,*df*) yields the probability density function value (only useful to plot the chi^2 curve, in which case "*x*" is the variable)
- X^2cdf(left bound, right bound, *df*) corresponds to $P(\text{left bound} < X^2 < \text{right bound})$

F Distribution

- Fpdf(*x*,*dfnum*,*dfdenom*) yields the probability density function value (only useful to plot the *F* curve, in which case "*x*" is the variable)
- Fcdf(left bound,right bound,*dfnum*,*dfdenom*) corresponds to $P(\text{left bound} < F < \text{right bound})$

Tests and Confidence Intervals

Access STAT and TESTS.

For the confidence intervals and hypothesis tests, you may enter the data into the appropriate lists and press DATA to have the calculator find the sample means and standard deviations. Or, you may enter the sample means and sample standard deviations directly by pressing STAT once in the appropriate tests.

Confidence Intervals

- ZInterval is the confidence interval for mean when σ is known.
- TInterval is the confidence interval for mean when σ is unknown; s estimates σ.
- 1-PropZInt is the confidence interval for proportion.

NOTE

The confidence levels should be given as percents (ex. enter "95" or ".95" for a 95% confidence level).

Hypothesis Tests

- Z-Test is the hypothesis test for single mean when σ is known.
- T-Test is the hypothesis test for single mean when σ is unknown; s estimates σ.
- 2-SampZTest is the hypothesis test for two independent means when both σ's are known.
- 2-SampTTest is the hypothesis test for two independent means when both σ's are unknown.
- 1-PropZTest is the hypothesis test for single proportion.
- 2-PropZTest is the hypothesis test for two proportions.
- X^2-Test is the hypothesis test for independence.
- X^2GOF-Test is the hypothesis test for goodness-of-fit (TI-84+ only).
- LinRegTTEST is the hypothesis test for Linear Regression (TI-84+ only).

NOTE

Input the null hypothesis value in the row below "Inpt." For a test of a single mean, "$\mu\varnothing$" represents the null hypothesis. For a test of a single proportion, "$p\varnothing$" represents the null hypothesis. Enter the alternate hypothesis on the bottom row.

APPENDIX H: TABLES

The module contains links to government site tables used in statistics.

NOTE

When you are finished with the table link, use the back button on your browser to return here.

Tables (NIST/SEMATECH e-Handbook of Statistical Methods, http://www.itl.nist.gov/div898/handbook/, January 3, 2009)

- **Student t table (http://www.itl.nist.gov/div898/handbook/eda/section3/eda3672.htm)**

- **Normal table (http://www.itl.nist.gov/div898/handbook/eda/section3/eda3671.htm)**

- **Chi-Square table (http://www.itl.nist.gov/div898/handbook/eda/section3/eda3674.htm)**

- **F-table (http://www.itl.nist.gov/div898/handbook/eda/section3/eda3673.htm)**

- All **four tables (http://www.itl.nist.gov/div898/handbook/eda/section3/eda367.htm)** can be accessed by going to

95% Critical Values of the Sample Correlation Coefficient Table
- **95% Critical Values of the Sample Correlation Coefficient**

INDEX